Environmental GEOLOGY

Third Edition

James S. Reichard

Georgia Southern University

McGraw Hill Education

ENVIRONMENTAL GEOLOGY, THIRD EDITION

1 2 3 4 5 6 7 8 9 LMN 21 20 19 18 17

ISBN 978-0-07-802296-8
MHID 0-07-802296-7

Chief Product Officer, SVP Products & Markets: *G. Scott Virkler*
Vice President, General Manager, Products & Markets: *Marty Lange*
Vice President, Content Design & Delivery: *Betsy Whalen*
Managing Director: *Thomas Timp*
Director of Marketing: *Tamara Hodge*
Brand Manager: *Michael Ivanov Ph.D.*
Director, Product Development: *Rose Koos*
Product Developer: *Jodi Rhomberg*
Marketing Manager: *Noah Evans*
Digital Product Analyst: *Patrick Dillard*
Digital Product Developer: *Joan Weber*
Director, Content Design & Delivery: *Linda Avenarius*
Program Manager: *Lora Neyens*
Content Project Managers: *Sherry Kane, Rachael Hillebrand*
Buyer: *Susan K. Culbertson*
Design: *Matt Backhaus*
Content Licensing Specialists: *Lori Hancock, Jacob Sullivan*
Cover Image: *© Alexandros Maragos/Getty Images*
Compositor: *SPi Global*
Printer: *LSC Communications*

Library of Congress Cataloging-in-Publication Data
Names: Reichard, James S., author.
Title: Environmental geology / James S. Reichard, Georgia Southern University.
Description: Third edition. | New York, NY : McGraw-Hill, [2017]
Identifiers: LCCN 2016030710 | ISBN 9780078022968 (alk. paper)
Subjects: LCSH: Environmental geology—Textbooks.
Classification: LCC QE38 .R45 2017 | DDC 550—dc23 LC record available at https://lccn.loc.gov/2016030710

mheducation.com/highered

This book is dedicated to my wife Linda and children, Brett and Kristen. Their love and support has carried me through the many hours spent away from home. Words cannot express my love and gratitude.

—James Reichard

© Jim Reichard

Jim Reichard and his family on the Highline Trail in Glacier National Park, Montana. From left to right: Jim, son Brett, daughter Kristen, and wife Linda.

Brief Contents

Contents

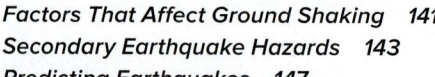

Chapter 8

© Doug Sherman/Geofile

Chapter 9

Program for the Study of Developed Shorelines, Western Carolina University.

PART THREE

Earth Resources

Chapter 10

© Glow Images

Tim McCabe, USDA, NRCS

© Neil Beer/Getty Images

U.S. Coast Guard

© George Hammerstein/Corbis/Glow Images

PART FOUR

The Health of Our Environment

Chapter 15

Pollution and Waste Disposal   481

© Jason Hawkes/The Image Bank/Getty Images

Chapter 16

Global Climate Change   517

© James Jordan Photography/ Getty Images

Preface

© Steve Cole/Getty Images

Environmental Geology, 3e focuses on the fascinating interaction between humans and the geologic processes that shape Earth's environment. Because this text emphasizes how human survival is highly dependent on the natural environment, students should find the topics to be quite relevant to their own lives and, therefore, more interesting. One of the key themes of this textbook centers on a serious challenge facing modern society: the need to continue obtaining large quantities of energy, and at the same time, make the transition from fossil fuels to clean sources of energy that do not impact the climate system. *Environmental Geology* provides extensive coverage of the problems associated with our conventional fossil fuel supplies (Chapter 13), and an equally in-depth discussion of alternative energy sources (Chapter 14). The two chapters on energy are intimately linked to a comprehensive overview of global climate change (Chapter 16), which is arguably civilization's most critical environmental challenge.

Another major theme in the third edition of *Environmental Geology* is that humans are an integral part of a complex and interactive system scientists call the Earth system. Throughout the text the author explains how the Earth system responds to human activity, and how that response then affects the very environment in which we live. A key point is that our activity often produces unintended and undesirable consequences. An example from the text is how engineers have built dams and artificial levees to control flooding on the Mississippi River. This has caused unintended changes in the geologic environment. For thousands of years, the rate at which the river deposited sediment in the Mississippi Delta was approximately equal to the rate that the sediment compacted under its own weight. The land surface remained above sea level because the two rates had been similar. However, by using dams and artificial levees to confine the Mississippi River to its channel, humans disrupted the delicate balance between sediment deposition and compaction. Today large sections of the Louisiana coast, including New Orleans, are sinking below sea level, and at the same time sea level is rising due to global warming. This has not only caused severe coastal erosion, but greatly increased the chance that New Orleans will be inundated during a major hurricane.

> *"My overall impression after reading Chapter 16 "Global Climate Change" was that of an excellent coverage of a still very controversial topic. Reichard has managed to cover the most fundamental societal and scientific issues related to global climate change in a format accessible to undergraduate students with or without strong science background. Reichard provides an unbiased representation of facts and does not shy away from a critical discussion of opposing arguments resulting from the interpretation of the facts."*
>
> —Thomas Boving, *University of Rhode Island*

Environmental Geology also includes a sufficient amount of background material on physical geology for students who have never taken a geology course. The author believes this additional coverage is critical. Without a basic understanding of physical geology, students would not be able to fully appreciate the interrelationships between humans and the geologic environment. To meet the needs of courses with a physical geology prerequisite, the book was organized so that instructors could easily omit the few chapters that contain mostly background material. In addition, *Environmental Geology* does more than provide a physical description of water, mineral, and energy resources; it explores the difficult problems associated with extracting the enormous quantities of resources needed to sustain modern societies. With respect to geologic hazards (e.g., earthquakes, volcanic eruptions, and floods), the textbook goes beyond the physical science and examines the societal impacts as well as the ways humans can minimize the risks. The author also highlights the fact that as population continues to grow, the problems related to resource depletion and hazards will become more severe.

Finally, this textbook includes learning tools designed to make it easier for students to utilize information found in the text. For example, it is unreasonable to expect students to remember everything they read. For this reason, the text often cross references topics between chapters as a reminder that additional information can be found in other parts of the book. It is hoped that cross-referencing will encourage students to make better use of the index for locating additional information.

> "...I give the author credit for excelling in a very up-to-date assessment of alternative technologies, with some delightful examples of innovative systems that should interest the student reader. The author recognizes the importance of portraying the subject within the modern world that the student lives in."
>
> —Lee Slater, *Rutgers University–Newark*

New for the Third Edition

Readers familiar with *Environmental Geology* should find that the changes to the third edition have significantly improved the already outstanding pedagogy and photo and art program of the previous editions. Perhaps the most significant improvement is the addition of six new case studies, bringing the total to nineteen. Increasing the number of case studies was a priority for the third edition because instructors commonly have students use case studies to explore chapter concepts in more detail. In addition to the new case studies, the chapter narratives have been thoroughly revised to include recent geologic events and scientific advances. Likewise, care was taken to ensure that all of the graphs and tables include the most recently available data. Many new photos and several new graphics were added to enhance the pedagogy and increase student interest. Finally, some of the existing graphics were modified to improve student comprehension.

Although changes in the third edition are too numerous to be listed individually, some of the more significant improvements are described below. Note that the chapters with the most revisions are those on energy resources (Chapters 13 and 14) and climate change (Chapter 16).

Chapter 1—In addition to updating the chapter content for recent events and scientific advances, four of the existing photos (Figures 1.1, 1.2, 1.8, and 1.18) have been replaced to help improve visual comprehension.

Chapter 2—The opening photo has been replaced with a dramatic NASA image of Earth's western hemisphere, which helps reinforce the theme that Earth is part of a much larger system. Perhaps the most significant change is that the case study on the search for life on Mars has been completely rewritten and now includes five new high-resolution photos. In addition to the case study, the discussion in the text on possible extraterrestrial life has been updated, and new graphics of Saturn's moon Enceladus (Figure 2.21) and the Kepler-62 planetary system (2.23) have been added. Also, an improved graphic of Antarctic ozone concentrations (2.25) illustrates how the ozone hole has changed over time. Finally, new photos provide students with a close-up view of the iridium-rich layer at the K/T boundary (2.31) and of comet 67P/Churyumov-Gerasimenko (2.34) as it approached the Sun in 2014.

Chapter 3—The opening photo was replaced with a classic image of a sandstone butte in Monument Valley that coincides with the chapter theme of earth materials (rocks and minerals). With respect to the chapter content, a new case study with photos and graphics (Figures B3.1 and B3.2) describes how ancient zircon crystals are providing geologists with important clues as to Earth's early history, in a period just 160 million years after the planet formed. Three photos have also been replaced (3.11, 3.17, and 3.24) to help improve student comprehension.

Chapter 4—A key graphic (Figure 4.16) showing the different types of plate boundaries has been re-labeled to improve student comprehension. Similarly, the discussion on how the movement of tectonic plates generates forces that cause buckling at convergent boundaries and rifting at divergent boundaries has been rewritten to improve clarity.

Chapter 5—Discussion of the recent earthquake in Nepal is accompanied by a new photo (Figure 5.20) taken after the quake that illustrates the hazards associated with masonry buildings. The chapter also includes the recently updated USGS seismic hazard map of the United States (5.34) and the newly released USGS hazard probability map for the San Francisco area (5.35). Finally, a new graphic (5.39) has been added to the section on earthquake early warning systems to complement the revised discussion of Japan's nationwide system and California's new *ShakeAlert* system.

Chapter 6—The opening photo has been replaced with a new, dramatic image of Mount Fuji that helps illustrate the relationship between humans and the geologic environment. Also, the discussion on the early warning system for mudflow hazards near Mount Rainier has been completely rewritten and updated.

Chapter 7—A new case study, including photos and graphics (Figures B7.3 and B7.4), describes the history behind the tragic 2014 mass wasting event that killed 43 people in Oso, Washington. In addition, the graphic that illustrates the different forces on a slope (7.4) has been relabeled to improve student comprehension, and the graphic depicting how reducing slope materials increases slope stability (7.29) has been revised for accuracy.

Chapter 8—The opening photo has been replaced with an impressive image of the Yellowstone River to help highlight the important role that streams play in the Earth system. A discussion of the thousand-year flood event in 2013 near Boulder, Colorado, has been included in the section on flash floods to help illustrate the hazards associated with these rare events.

Chapter 9—In addition to minor text changes to convey new information, the histograms showing coastal population growth (Figure 9.1) and Atlantic hurricane history (9.23) and the map of buoys and sensors in the worldwide tsunami early warning system (9.26) have all been redrawn based on the most currently available data.

Chapter 10—The opening photo has been replaced with a new image to reinforce the chapter theme—how our food supply is inextricably linked to soils. Most significant though is the addition of a new case study, including graphics (Figures B10.1 and B10.2), that describes the lessons learned regarding soil conservation practices in the aftermath of the 1930s Dust Bowl in the United States.

Chapter 11—A new case study on the Aral Sea disaster, with photos (Figure B11.1), highlights the potential problems associated with off-stream water usage. In addition, the graph showing total U.S. water withdrawals (11.4) has been updated, and the graphic illustrating saltwater intrusion of coastal aquifers (11.21) has been revised for accuracy.

Chapter 12—The opening photo has been replaced with an impressive image of a spinning bucket excavator being used to remove overburden material from a surface mine. Three data tables (Tables 12.1, 12.4, and 12.5) have been updated based on recently released USGS mineral reports. Similarly, new data from the USGS were used to update the graphs showing U.S. mineral imports (12.24) and yearly mineral consumption (12.25).

Chapter 13—Much of this chapter has undergone extensive revision that reflects recent developments with respect to conventional oil and gas resources. For example, the section on exploration and production wells now includes a discussion on deep-water oil and gas deposits along with a graphic (Figure 13.17) that includes a map of the world's deep-water fields and a diagram of a deep-water drilling operation. This section also has a new, in-depth discussion on tight oil and gas wells and a new graph (13.18) comparing depletion rates of conventional and tight gas wells. The case study on hydraulic fracturing has been updated and now includes a discussion on the significant increase in earthquake activity related to the injection of greater volumes of wastewater into deep aquifers. The map showing the location of tight oil and gas deposits (B13.1) has been updated to include deposits in Canada and Mexico. In addition, major revisions to the section on the energy crisis reflect the current volatility in oil and gas markets related to the interaction between changes in supply and demand and economic activity. Likewise, the sections on peak oil and avoiding the energy crisis have been updated based on the latest projections concerning the increase of tight oil production and the depletion of conventional oil fields. Finally, many of the graphs and charts (13.3, 13.22–13.27, 13.31, and 13.34–13.36) were updated using data from newly released reports.

Chapter 14—To help illustrate how the world is transitioning from fossil fuels to clean and renewable energy sources, the opening photo has been replaced with one showing a new housing development with rooftop solar panels. More importantly, the chapter has been thoroughly revised to reflect both minor and major developments in a wide array of alternative energy sources. Significant revisions include updates on China's increased use of coal liquefaction for transportation fuels, changes in the EPA's renewable fuel standard (RFS) and its effect on U.S. ethanol production, new battery technology for storing electricity, and home energy conservation. Perhaps the most important change is a new case study on the next generation of nuclear power plants, which can produce carbon-free electricity and eliminate the need for long-term storage of nuclear waste. This case study complements a revised discussion on how major reductions in greenhouse gas emissions could be accomplished through a combination of increased use of nuclear power and scaled-up production of electricity from wind and solar resources. Other additions include an updated wind power map of the United States (Figure 14.21) and a graphic showing the breakdown of energy usage in the typical U.S. household (14.32). Finally, the most recently available data were used to update the table listing the cost of producing electricity from difference sources (14.1), the graph of U.S. ethanol production (14.7), and the graph of world wind power generating capacity (14.36).

Chapter 15—A new graphic (Figure 15.3) in the section on U.S. environmental laws shows the total number and status of EPA Superfund sites over time. In addition, six graphs (15.9, 15.15, 15.16, 15.17, and 15.39) were updated based on recently released EPA reports, and a new version of the acid rain deposition map (15.37) has been included. Four photos (15.6, 15.17, 15.25, and 15.32) have also been replaced to help improve student comprehension. Finally, the section on reducing anthropogenic mercury emissions underwent significant revisions based on recent EPA data, and the potential health risk from recycled tire material in athletic surfaces is described in the discussion of scrap tires.

Chapter 16—The opening photo has been replaced with an impressive image of combustion gases being released from a coal-burning power plant, highlighting the connection between society's use of fossil fuels and global climate change. In addition, numerous small changes have been made throughout the chapter to reflect the results and conclusions from the most recent United Nations report (IPCC) on climate change. Perhaps the single most significant change is a new case study describing how Miami and South Florida are at severe risk from accelerated sea level rise. Also new is a NASA image (Figure 16.1) showing the dramatic loss of ice from Greenland's Jakobshavn Glacier, a graphic (16.11) illustrating the albedo effect, a photo (16.27) showing the devastating effects of pine-beetle infestations on conifer forests, a pair of photos (16.28) illustrating the effect of a long-term drought on California's water supply, and a graph (16.39) of worldwide carbon dioxide emissions over time. In addition, eight graphs (16.2, 16.19, 16.25, 16.26, 16.32, 16.37, 16.39 and 16.40) were updated based on recently released data. Finally, a detailed discussion on the recent Paris Climate Agreement has been included in the section on mitigation of climate change.

Key Features

As with all college textbooks, there are differences among the various environmental geology books currently being offered. These are some of the more significant and noticeable differences you will find in *Environmental Geology*:

- **Learning Outcomes.** Each chapter is introduced with a list that provides valuable student guidance by stating key chapter concepts. This encourages students to be "active" learners as they complete the tasks and activities that require them to use critical thinking skills.

- **Chapter 2 Is Unique.** "Earth from a Larger Perspective" describes Earth's relationship to the solar system and universe, which helps give students the broadest possible perspective on our environment. Here students learn how the Earth system is part of even larger systems before moving on to the remaining chapters that focus on our planet. Chapter 2 also gives instructors the opportunity to discuss some of the external forces that influence Earth's environment, such as solar radiation, asteroid impacts, and the effect of the Moon on our tides and climate. In addition, this chapter helps explain why Earth supports a diverse array of complex life, and why humans are so dependent on its unique and fragile environment. This sets the stage for a theme that is woven throughout the entire text—that human survival is intimately linked to the environment. Students can then see how being better stewards of the Earth is in our own best interest.

- **Case Studies.** Nearly every chapter includes a case study that is designed to give students a more in-depth look at an environmental issue. A good example is Chapter 7, where the case study examines the recurring mass wasting problems at La Conchita, California. Here students are asked to consider why some people willingly live in a hazardous area, even when the risk is well understood. In Chapter 13, the case study explores the controversy over hydraulic fracturing and the development of tight oil and gas. Students are given an objective overview of both the science and policy sides of the issue, and are then expected to draw their own conclusion as to which side of the policy debate they would support.

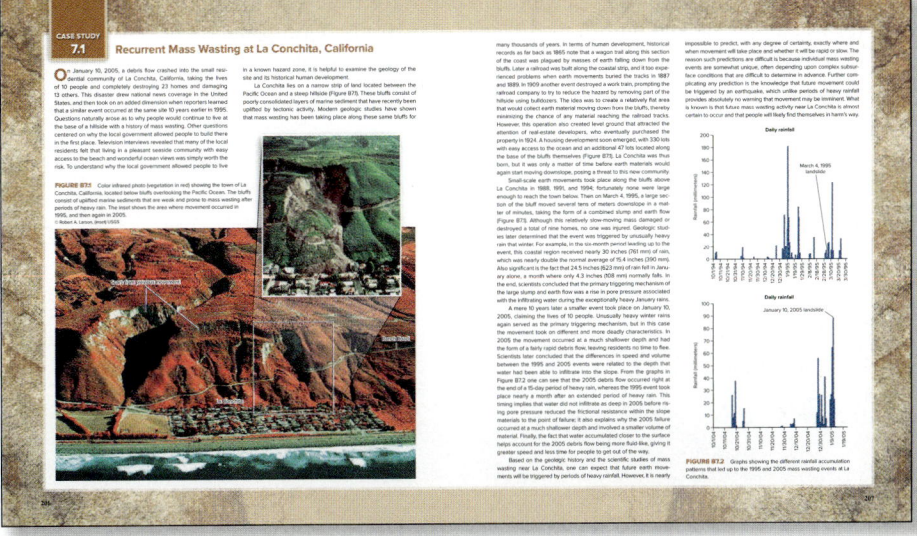

FIGURE 9.27 Rip currents (A) form when backwash from the surf zone funnels through a break in underwater sand bars. Photo (B) showing a rip current flowing back out to sea through the surf zone in the Monterey Bay area of California. Note that the rip current can be recognized by how it disrupts breaking waves within the surf zone.
(B) Wendy Carey, Delaware Sea Grant

- **Photos and Illustrations.** It is well established in the field of education that most people are predominantly visual learners. Therefore, the author integrated very relevant photos and illustrations within the narrative so that abstract and complex concepts are easier to understand. The integrated use of visual examples within a narrative writing style should not only help increase student comprehension, but it should also encourage students to read more of the text.

FIGURE 14.3 Synthetic crude oil is currently being produced from oil sand deposits (A) in Alberta, Canada. The hydrocarbons are in the form of bitumen, a highly viscous substance that is separated from the sand using steam. Photo (B) shows a bitumen sample whose viscosity has been lowered by heating. Approximately 60% of Canadian production involves strip mining (C); the remainder is produced by steam injection and pumping wells.
(B) © Syncrude Canada Ltd; (C) © Photographic Services, Shell International Ltd.

FIGURE 9.17 Storm surge (A) not only inundates areas normally above high tide, but also brings breaking waves that demolish structures. Photo (B) of Mantoloking, New Jersey, showing the effects of storm surge and waves associated with Hurricane Sandy in 2012. Arrows mark the same house that appears in both images. Notice the destroyed houses and roads and extensive beach erosion. Also note the large volume of sand that was deposited on the back side of the island.
(b) (both) USGS

- **Summary Points.** Each chapter concludes with a list of Summary Points to provide students with a list of important concepts that should be reviewed in preparation for exams.

- **Key Words.** The study of geologic processes can be daunting due to the proliferation of unfamiliar terms. Each chapter includes a list of important terms with page references, so that terms can be viewed within the context of their use. Complete definitions are also provided in the Glossary at the back of the text.

SUMMARY POINTS

1. Shorelines are unique in that they are where Earth's two most fundamental environments meet, the terrestrial (land) and marine (ocean), forming a desirable habitat for humans. Because population growth is much higher in coastal zones compared to inland areas, more people are exposed to coastal erosion and hazards such as hurricanes and tsunamis.
2. Ocean tides are caused by the gravitational pull of the Moon and Sun on Earth's oceans, whereas ocean currents move in response to winds, differences in water density, and wave action along coastlines.
3. Wave energy travels horizontally and causes water molecules to move vertically in circular paths. As a wave enters shallow water, the moving water molecules begin to drag on the bottom, causing the wave to decelerate. This causes the wave to increase in height and to become more asymmetric, eventually falling over to form a breaking wave.
4. When waves crash onto shore at an angle, water is pushed parallel to shore in a longshore current. Grains of sand in the surf zone move in a zigzagging pattern in the direction of the longshore current in the process of beach drift.
5. Irregular shorelines are where more isolated beaches are commonly found in tectonically active areas and in places where sea level is rising relative to the land surface. Erosion and deposition from wave action slowly causes shorelines to evolve into ones with longer and wider beaches. At the same time, weathering and erosion of the landscape tend to produce more low-lying terrain.
6. Hurricanes are a serious coastal hazard as they generate powerful winds, storm surge, and heavy rains. Satellite early warning systems have greatly

reduced the number of fatalities, but increased coastal development has caused property losses to escalate.
7. Tsunamis most commonly form during subduction zone earthquakes as water is displaced by movement of the seafloor. When a tsunami reaches shallow water, the tremendous wave energy translates into tall waves that break far beyond the normal surf zone, causing death and destruction along developed coastlines.
8. Rip currents pose a serious risk to swimmers as water from the surf funnels back out to sea through breaks in shallow sand bars. People are unable to swim back to shore against the strong currents, but can get out of the current by swimming parallel to shore.
9. The interaction between waves and a landmass can cause a shoreline to naturally retreat landward. The slow migration of a shoreline can also occur when there is a global rise in sea level or when the land itself becomes lower due to subsidence. Accelerated shoreline retreat is occurring in many areas due to human activity, increasing the hazards associated with ocean storms.
10. Humans attempt to protect their property and reduce shoreline retreat through engineering techniques such as seawalls, groins, and beach nourishment. Jetties are used to keep navigational channels free of sediment, and breakwaters provide quiet areas by keeping waves from impacting on the shoreline. Some of these techniques result in beach starvation and accelerated retreat in down-drift areas.

4. Oceanic crust is relatively thin and composed of basalt, whereas continental crust is thicker and composed of granitic rock. The crust and upper mantle act as a single layer and together are called the lithosphere. The lithosphere is broken up into rigid plates that move over the weak, semimolten layer called the asthenosphere.
5. Decay of radioactive elements within the Earth generates heat and helps create a large temperature difference between the core and the crust. This sets in motion large convection cells that transport both heat and plastic mantle material toward the surface.

9. Major surface features develop along plate boundaries and include ocean ridges, ocean trenches, rift valleys, island arcs, volcanic arcs, and complex mountain belts. The majority of earthquakes and volcanic eruptions also occur along plate boundaries.
10. In addition to geologic processes, plate tectonics plays a central role in the Earth system by affecting the atmosphere, hydrosphere, and biosphere. For humans, plate tectonics is important because it creates hazards (earthquakes, volcanic eruptions, landslides), regulates our climate, distributes natural resources, and was important in the development of life.

KEY WORDS

asthenosphere 97	elastic limit 95	mid-oceanic ridges 101	tectonic plates 98
compression 95	fault 96	ocean trenches 101	tension 95
continental arc 112	geothermal gradient 99	outer core 97	theory of plate tectonics 94
convection cells 99	inner core 97	rift valley 109	transform boundary 106
convergent boundary 106	island arc 111	seafloor spreading 102	
crust 98	lithosphere 98	shear 95	
divergent boundary 106	mantle 97	subduction 103	

APPLICATIONS

Student Activity Go to the website of a major news organization, such as the *New York Times*, the *Wall Street Journal*, or the BBC, and make a list of all the stories you can find that are in some way related to plate tectonics. Then, briefly state how the topic relates to plate tectonics. For example, earthquakes and volcanoes are easy to relate, but if you think about it, mineral, energy, and water resources and even climate change can be linked to plate tectonics.

Critical Thinking Questions
1. Rocks that are brittle will deform by fracturing when they exceed their elastic limit, but if they become ductile, they will deform by flowing plastically. Can you think of an everyday material that can go from being brittle to being ductile, and thus go from fracturing to flowing plastically when deformed?
2. Large-scale convection cells develop in the mantle due to differences in density caused by Earth's internal heat. Can you think of an example of density-driven convection circulation that takes place in Earth's surface environment?
3. Some disaster movies have shown lava erupting in downtown Los Angeles, California. Given the fact that Los Angeles is situated along a transform plate boundary, is a volcanic eruption there very likely? Explain.

Your Environment: YOU Decide In this Chapter you learned how the theory of plate tectonics was finally developed in the 1960s, based on data gathered from numerous geologic studies over many years. These data include maps of the seafloor, magnetic studies, earthquake locations, and similar rock and fossil sequences on the continents.

Describe the piece of evidence you found to be most convincing in showing that tectonic plates are indeed moving on top of the asthenosphere. Be sure to explain why.

- **Applications.** At the end of each chapter, sections called **Student Activity** and **Critical Thinking Questions** and **Your Environment: YOU Decide** encourage students to think about how their own lifestyles may be playing a role in environmental issues. For example, in Chapter 12 ("Mineral and Rock Resources") they are asked to think about the social implications of buying a diamond that comes from a part of the world where illegal proceeds support violent uprisings and civil war. In Chapter 15 ("Pollution and Waste Disposal") students are asked to contact their local government to determine the location of the landfill where their trash is being sent. They are then asked to investigate whether the landfill has any reported pollution problems, and if so, to describe what impacts the landfill might be having on local residents.

- **Laboratory Manual.** Twelve comprehensive laboratory exercises are available on the text website. These include a list of materials needed, questions for students to complete, and corresponding answer keys on the instructor resource website.

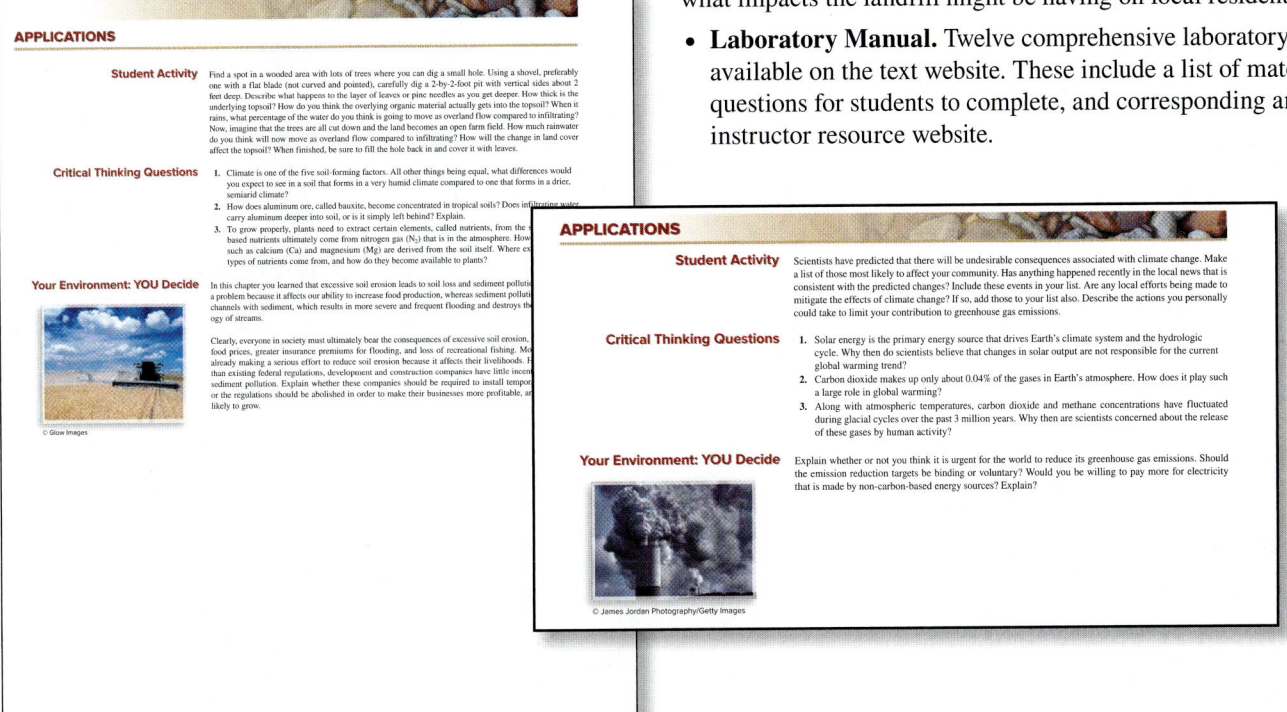

Organization

In most environmental geology courses the list of topics includes some combination of geologic hazards and resources along with waste disposal and pollution. Consequently, this book is conveniently organized so instructors can pick and choose the chapters that coincide with their particular course objectives. The chapters are organized as follows:

Part One   Fundamentals of Environmental Geology
Chapter 1 Humans and the Geologic Environment
Chapter 2 Earth from a Larger Perspective
Chapter 3 Earth Materials
Chapter 4 Earth's Structure and Plate Tectonics

Part Two   Hazardous Earth Processes
Chapter 5 Earthquakes and Related Hazards
Chapter 6 Volcanoes and Related Hazards
Chapter 7 Mass Wasting and Related Hazards
Chapter 8 Streams and Flooding
Chapter 9 Coastal Hazards

Part Three   Earth Resources
Chapter 10 Soil Resources
Chapter 11 Water Resources
Chapter 12 Mineral and Rock Resources
Chapter 13 Conventional Fossil Fuel Resources
Chapter 14 Alternative Energy Resources

Part Four The Health of Our Environment
Chapter 15 Pollution and Waste Disposal
Chapter 16 Global Climate Change

"I found the chapter [16] to overall be very well written, very interesting, and logically organized. I am especially impressed by the thorough summary the author provides on the Earth's climate system."
—John C. White, *Eastern Kentucky University*

McGraw-Hill Connect®
Learn Without Limits

Connect is a teaching and learning platform that is proven to deliver better results for students and instructors.

Connect empowers students by continually adapting to deliver precisely what they need, when they need it, and how they need it, so your class time is more engaging and effective.

73% of instructors who use Connect require it; instructor satisfaction increases by 28% when Connect is required.

Analytics

Connect Insight®

Connect Insight is Connect's new one-of-a-kind visual analytics dashboard—now available for both instructors and students—that provides at-a-glance information regarding student performance, which is immediately actionable. By presenting assignment, assessment, and topical performance results together with a time metric that is easily visible for aggregate or individual results, Connect Insight gives the user the ability to take a just-in-time approach to teaching and learning, which was never before available. Connect Insight presents data that empowers students and helps instructors improve class performance in a way that is efficient and effective.

Mobile

Connect's new, intuitive mobile interface gives students and instructors flexible and convenient, anytime–anywhere access to all components of the Connect platform.

Connect's Impact on Retention Rates, Pass Rates, and Average Exam Scores

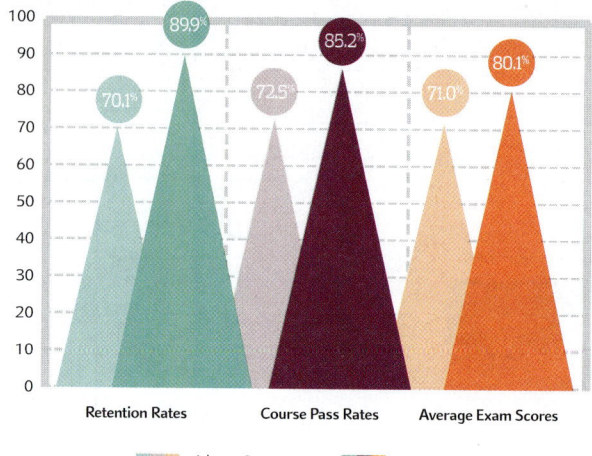

Using Connect improves retention rates by 19.8%, passing rates by 12.7%, and exam scores by 9.1%.

Impact on Final Course Grade Distribution

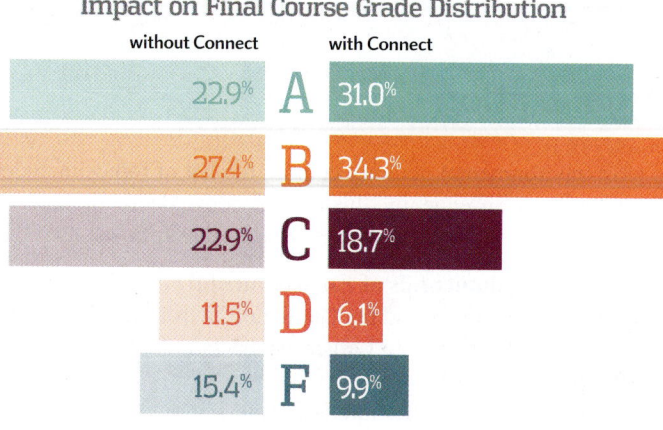

without Connect		with Connect
22.9%	A	31.0%
27.4%	B	34.3%
22.9%	C	18.7%
11.5%	D	6.1%
15.4%	F	9.9%

Students can view their results for any Connect course.

Adaptive

THE ADAPTIVE READING EXPERIENCE DESIGNED TO TRANSFORM THE WAY STUDENTS READ

More students earn **A's** and **B's** when they use McGraw-Hill Education **Adaptive** products.

SmartBook®

Proven to help students improve grades and study more efficiently, SmartBook contains the same content within the print book, but actively tailors that content to the needs of the individual. SmartBook's adaptive technology provides precise, personalized instruction on what the student should do next, guiding the student to master and remember key concepts, targeting gaps in knowledge and offering customized feedback, and driving the student toward comprehension and retention of the subject matter. Available on tablets, SmartBook puts learning at the student's fingertips—anywhere, anytime.

Over **8 billion questions** have been answered, making McGraw-Hill Education products more intelligent, reliable, and precise.

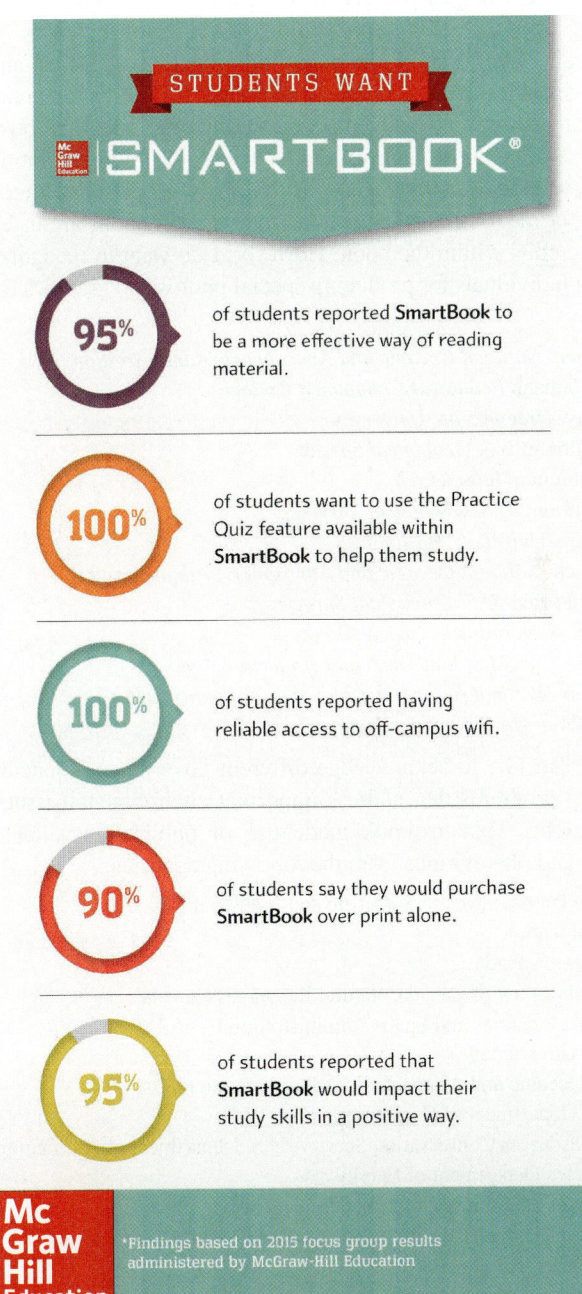

STUDENTS WANT SMARTBOOK®

95% of students reported **SmartBook** to be a more effective way of reading material.

100% of students want to use the Practice Quiz feature available within **SmartBook** to help them study.

100% of students reported having reliable access to off-campus wifi.

90% of students say they would purchase **SmartBook** over print alone.

95% of students reported that **SmartBook** would impact their study skills in a positive way.

*Findings based on 2015 focus group results administered by McGraw-Hill Education

www.mheducation.com

Acknowledgments

The third edition of *Environmental Geology* allowed me to improve upon the outstanding features of the original text. This required the help of many different people. In particular, I would like to thank the McGraw-Hill team that worked on this project, including Jodi Rhomberg (Senior Product Developer), Sherry Kane (Content Project Manager), Matt Backhaus (designer), Lori Hancock (Lead Content Licensing Specialist), Noah Evans (Associate Marketing Manager), Michael Ivanov Ph.D. (Brand Manager) and Thomas Timp (Managing Director). In addition to the publishing team, a special thanks goes to my wife, Linda. The demands placed on me by publishing deadlines, teaching schedules, and research commitments were at times overwhelming. Linda not only took on nearly all of the family responsibilities, giving me the time I needed, but her unwavering support and encouragement helped me get through it all. I can never thank her enough.

I would also like to thank the many individuals, companies, and government agencies who generously supplied the noncommercial photos. In many cases, this involved people taking time from their busy schedules to search photo archives and retrieve the high-resolution photos that I wanted to use. Because far too many people contributed to this effort for me to acknowledge here, their contributions are listed in the photo credits within the book. However, I do want to recognize the following individuals for producing special photos and graphics for this textbook:

James Bunn *National Oceanic and Atmospheric Administration*
Eleanor Camann *Red Rocks Community College*
Chris Daly *Oregon State University*
Carolyn Donlin *U.S. Geological Survey*
Lundy Gammon *IntraSearch*
Robert Gilliom *U.S. Geological Survey*
Bob Larson *University of Illinois*
Jake Crouch *National Oceanic and Atmospheric Administration*
Naomi Nakagaki *U.S. Geological Survey*
Matt Sares *Colorado Geological Survey*
Christie St. Clair *U.S. Environmental Protection Agency*
Cindy Starr *National Aeronautics and Space Administration*
Jeremy Weiss *University of Arizona*

I would also like to acknowledge different government agencies for supporting programs that address important environmental issues around the globe. This textbook made use of publically available reports, data, and photographs from the following agencies:

Australian Commonwealth Scientific and Industrial Research Organization
Environment Canada
Geological Survey of Canada, Natural Resources Canada
National Aeronautics and Space Administration (NASA), U.S. Government
National Oceanic and Atmospheric Administration (NOAA), U.S. Department of Commerce
Natural Resources Conservation Service, U.S. Department of Agriculture
United States Department of Energy
United States Environmental Protection Agency
United States Fish and Wildlife Service, Department of the Interior
United States Geological Survey, Department of the Interior

I would also like to thank Trent McDowell, who wrote and reviewed learning goal oriented content for LearnSmart.

Finally, I would like to thank all those who reviewed various parts of the manuscript during the course of this project. Their insightful comments, suggestions, and criticisms were of immense value.

Lewis Abrams *University of North Carolina–Wilmington*
Christine N. Aide *Southeast Missouri State University*
Michael T. Aide *Southeast Missouri State University*
Erin P. Argyilan *Indiana University Northwest*
Richard W. Aurisano *Wharton County Junior College*
Dirk Baron *California State University—Bakersfield*
Jessica Barone *Monroe Community College*
Mark Baskaran *Wayne State University*
Robert E. Behling *West Virginia University*
Prajukti Bhattacharyya *University of Wisconsin–Whitewater*
Thomas Boving *University of Rhode Island*
David A. Braaten *University of Kansas*
Eric C. Brevik *Dickinson State University*
Charles Brown *George Washington University*
Patrick Burkhart *Slippery Rock University of Pennsylvania*
Ernest H. Carlson *Kent State University*
James R. Carr *University of Nevada–Reno*
Patricia H. Cashman *University of Nevada–Reno*
Elizabeth Catlos *Oklahoma State University*
Robert Cicerone *Bridgewater State College*
Gary Cwick *Southeast Missouri State University*
Katherine Folk Clancy *University of Maryland*
Jim Constantopoulos *Eastern New Mexico University*
Geoffrey W. Cook *University of Rhode Island*
Heather M. Cook *University of Rhode Island*
Raymond Coveney *University of Missouri–Kansas City*
Ellen A. Cowan *Appalachian State University*
Anna M. Cruse *Oklahoma State University*
Gary J. Cwick *Southeast Missouri State University*
George E. Davis *California State University–Northridge*
Hailang Dong *Miami University of Ohio*
Yoram Eckstein *Kent State University*
Dori J. Farthing *SUNY–Geneseo*
Larry A. Fegel *Grand Valley State University*
James R. Fleming *Colby College*
Christine A. M. France *University of Maryland–College Park*
Tony Foyle *Penn State Erie, The Behrend College*
Alan Fryar *University of Kentucky*
Heather L. Gallacher *Cleveland State University*
Alexander E. Gates *Rutgers University–Newark*
David H. Griffing *Hartwick College*
John R. Griffin *University of Nebraska–Lincoln*
Syed E. Hasan *University of Missouri–Kansas City*
Chad Heinzel *Minot State University*
Donald L. Hoff *Valley City State University*
Brad Johnson *Davidson College*
Neil E. Johnson *Appalachian State University*
Steven Kadel *Glendale Community College*
Chris R. Kelson *University of Georgia*
John Keyantash *California State University–Dominguez Hills*

Karin Kirk *Carleton College*
Christopher Kofp *Mansfield University*
Gerald H. Krockover *Purdue University*
Glenn C. Kroeger *Trinity University*
Michael A. Krol *Bridgewater State College*
Jennifer Latimer *Indiana State University*
Liliana Lefticariu *Southern Illinois University*
Adrianne A. Leinbach *Wake Technical Community College*
Gene W. Lené *St. Mary's University of San Antonio*
Nathaniel Lorentz *California State University–Northridge*
James B. Maynard *University of Cincinnati*
Richard V. McGehee *Austin Community College*
Gretchen L. Miller *Wake Technical Community College*
Barry E. Muller *ERG Consult, LLC*
Klaus Neumann *Ball State University*
Barry E. Muller *ERG Consult, LLC*
Suzanne O'Brien *Stonehill College*
Duke Ophori *Montclair State University*
David L. Ozsvath *University of Wisconsin–Stevens Point*
Evangelos K. Paleologos *University of South Carolina–Columbia*
Alyson Ponomarenko *San Diego City College*
Libby Prueher *University of Northern Colorado*

Fredrick J. Rich *Georgia Southern University*
Paul Robbins *University of Arizona*
Michael Roden *University of Georgia*
Lee D. Slater *Rutgers University–Newark*
Edgar W. Spencer *Washington and Lee University*
Michelle Stoklosa *Boise State University*
Eric C. Straffin *Edinboro University of Pennsylvania*
Christiane Stidham *Stony Brook University*
Benjamin Surpless *Trinity University*
Sam Swanson *University of Georgia*
Gina Seegers Szablewski *University of Wisconsin–Milwaukee*
James V. Taranik *University of Nevada–Reno*
J. Robert Thompson *Glendale Community College*
Jody Tinsley *Clemson University*
Daniel L. Vaughn *Southern Illinois University*
Adil M. Wadia *University of Akron, Wayne College*
Miriam Weber *California State University–Monterey Bay*
David B. Wenner *University of Georgia*
John C. White *Eastern Kentucky University*
David Wilkins *Boise State University*
Ken Windom *Iowa State University*

Meet the Author

James Reichard   James Reichard is a Professor in the Department of Geology and Geography at Georgia Southern University. He obtained his Ph.D. in Geology (1995) from Purdue University, specializing in hydrogeology, and his M.S. (1984) and B.S. (1981) degrees from the University of Toledo, where he focused on structural and petroleum geology. Prior to earning his Ph.D., he worked as an environmental consultant in Cleveland, Ohio, and as a photogeologist in Denver, Colorado.

James (Jim) grew up in the flat glacial terrain of northwestern Ohio. Each summer, he went on an extended road trip with his family and traveled the American West. It was during this time that Jim was exposed to a variety of scenic landscapes. Although he had no idea how the landscapes formed, he was fascinated nonetheless. It was not until college, when Jim had to satisfy a science requirement, that he finally came across the field of geology. Here, he discovered a science that could explain how different landscapes actually form. From that moment on, he was hooked on geology. This eventually led Jim to a graduate degree in geology, after which he was able to fulfill his dream of living and working in Colorado. Then, due to one of life's many unexpected opportunities, he accepted a position with an environmental firm back in Ohio. This ultimately led to a Ph.D. from Purdue and a faculty position at Georgia Southern University, where he currently enjoys teaching and doing research in environmental geology and hydrogeology. His personal interests include hiking, camping, and sightseeing.

It is through this textbook that Professor Reichard hopes to excite students about how geology shapes the environment in which we live, similar to the way he became excited about geology in his youth. To help meet this goal, he has tried to write this book with the student's perspective in mind in order keep it more interesting and relevant. Hopefully, students who read the text will begin to share some of Professor Reichard's fascination with how geology plays an integral role in our everyday lives.

Crater Lake National Park, Oregon.

© Jim Reichard

Environmental
GEOLOGY

Chapter 1

Aerial photo showing extensive urban sprawl in the desert environment of Las Vegas, Nevada. Modern humans have been able to thrive in such harsh environments because of our ability to generate electrical power to run air conditioners and to bring in water from reservoirs and underground aquifers. However, population growth is threatening to outstrip Earth's ability to provide the resources needed to sustain our population. Humans therefore must find a way to stabilize population growth and limit our consumption of resources.

Humans and the Geologic Environment

LEARNING OUTCOMES

After reading this chapter, you should be able to:

▶ Describe the major focus of the discipline called environmental geology.
▶ Characterize how scientists develop hypotheses and theories as a means of understanding the natural world.
▶ Describe the concept of geologic time and how the geologic time scale was constructed.
▶ Explain how geologic time and the rate at which natural processes operate affect how humans respond to environmental issues.
▶ Describe how Earth operates as a system and why humans are an integral part of the system.
▶ Explain the concept of exponential population growth and how it relates to geologic hazards and resource depletion.
▶ Define the concept of sustainability in terms of the living standard of developed nations and also in terms of the human impact on the biosphere.

Introduction

© StockTrek/Getty Images

Earth is unique among the other planets in the solar system in that it has an environment where life has been able to thrive, evolving over billions of years from single-cell bacteria to complex plants and animals. There have been three critical factors that have led to the diversity of life we see today. One is that Earth's surface temperatures are in the range where water can exist in both the liquid and vapor states. The second is that our planet was able to retain its atmosphere, which in turn allows the water to move between the liquid and vapor states in a cyclic manner. Last, Earth has a natural mechanism for removing carbon dioxide from the atmosphere, namely, the formation of carbonate rocks (e.g., limestone). This has prevented a buildup of carbon dioxide and a runaway greenhouse effect, similar to what happened on Venus, where surface temperatures today exceed 800°F (425°C). With respect to humans (*Homo sapiens*), our most direct ancestors have been part of Earth's biosphere for only the past 200,000 years, whereas other hominid species go back as far as 6 to 7 million years. Compared to Earth's 4.6-billion-year history, humans have existed for a very brief period of time. However, rapid population growth combined with the Industrial Revolution has resulted in profound changes in Earth's surface environment. The focus of this textbook will be on the interaction between humans and Earth's geologic environment. We will pay particular attention to how people use resources such as soils, minerals, and fossil fuels and how we interact with natural processes, including floods, earthquakes, landslides, and so forth.

One of the key reasons humans have been able to thrive is our ability to understand and modify the environment in which we live. For example, consider that for most of history people lived directly off the land. To survive they had to be keenly aware of the environment in order to find food, water, and shelter. This forced some people to travel with migrating herds of wild animals, who in turn were following seasonal changes in their own food and water supplies. Eventually we learned to clear the land and grow crops in organized settlements. As they practiced agriculture, humans became skilled at recognizing those parts of the landscape with the most productive soils. The best soils, however, were commonly found in low-lying areas along rivers and periodically inundated by floodwaters. To reduce the risk of floods, people learned to seek out farmland on higher ground and place their homes even higher, thereby avoiding all but the most extreme floods. In addition to reducing the risk of floods and other natural hazards, we learned how to take advantage of Earth's mineral and energy resources. This led directly to the Industrial Revolution and the modern consumer societies of today.

Although humans have benefited greatly by modifying the environment and using Earth's resources, this activity has also resulted in unintended and undesirable consequences. For example, in order to grow crops and build cities it was necessary to remove forests and grasslands that once covered the natural landscape. This reduced the land's ability to absorb water, thereby increasing the frequency and severity of floods. Also, the use of mineral and energy resources by modern societies creates waste by-products that can poison our streams and foul the air we breathe. The prolific use of fossil fuels is even altering the planet's climate system and contributing to the problem of global warming. It has become abundantly clear that the human race is an integral part of the Earth system and that our actions affect the very environment upon which we depend.

While the link between environmental degradation and human activity may be clear to scientists, it is not always so obvious to large segments of the population. A well-established concept with respect to environmental degradation is known as the **tragedy of the commons**, which is where the self-interest of individuals results in the destruction of a common or shared resource. A common resource includes such things as a river used for water supply, wood in a forest, grassland for grazing animals, and fish in the sea. Consider a coastal village whose primary source of food is the local fishing grounds offshore. This resource is renewable as long as the fish are not harvested at a faster rate than they can reproduce; hence, everyone in the village benefits. However, if the village grows too large, the increased demand can make the fishing unsustainable. As the fish become scarcer, the competition for the remaining fish gets more intense as the individual fisherman try to feed their families. The fisherman's self-interest creates a downward spiral where all members of society ultimately suffer as the fishery becomes so depleted that it collapses and is unable to recover.

Another phenomenon that can contribute to environmental degradation is when citizens in consumer societies become disconnected from the natural environment. An example is the United States, where many people now live and work in climate-controlled buildings and get their food from grocery stores as opposed to growing their own (Figure 1.1). People then tend to lose their sense of being connected with the natural world, despite the fact they remain dependent upon the environment as were our ancient ancestors. As with the tragedy of the commons, a lack of environmental awareness can lead to serious problems and hardships for society.

FIGURE 1.1 In modern consumer societies few people live directly off the land, but instead buy most of their food in stores. This trend has led to a greater disconnection between people and the natural environment upon which they still depend.
(Left) © Glowimages/Punchstock; (Right) © The McGraw-Hill Companies, Inc./Andrew Resek, photographer

A   B   C

FIGURE 1.2   Rock and mineral deposits (A) provide the raw materials used for building (B) and operating our modern societies. The geologic resources known as fossil fuels provide the bulk of the energy used for powering (C) the industrial, transportation, and residential sectors of society.
(a) © Dr. Parvinder Sethi; (b) © Skip Nall/Getty Images; (c) © PhotoLink/Getty Images

A

B

FIGURE 1.3   In addition to locating resources, geologists study hazardous earth processes and use this knowledge to help society avoid or minimize the loss of life and property damage. Photo (A) shows a building that was destroyed during the 1995 earthquake in Kobe, Japan, and (B) shows the results of an earthquake-induced landslide in Las Colinas, El Salvador, in 2001.
(a) Roger Hutchinson/NOAA; (b) USGS

What Is Geology?

The science of **geology** is the study of the solid earth, which includes the materials that make up the planet and the various processes that shape it. Many students who are unfamiliar with geology tend to think it is just a study of rocks, and therefore must not be very interesting. However, this perception commonly changes once students realize how intertwined their own lives are with the geologic environment. For example, the success of our high-tech society is directly tied to certain minerals whose physical properties are used to perform vital tasks. Perhaps the most important are minerals containing the element copper, a metal whose ability to conduct electricity is absolutely essential to our modern way of life. Imagine doing without electric lights, refrigerators, televisions, cell phones, and the like. Because geologists study how minerals form, mining companies hire geologists to look for places where valuable minerals have become concentrated (Figure 1.2). Equally important is the ability of geologists to locate deposits of oil, gas, and coal, as these serve as society's primary source of energy. Geologists also provide valuable information as to how society can minimize the risk from hazardous Earth processes such as floods, landslides, earthquakes, and volcanic eruptions (Figure 1.3).

Geology has traditionally been divided into two main subdisciplines: physical geology and historical geology. **Physical geology** involves the study of the solid earth and the processes that shape and modify the planet, whereas **historical geology** interprets Earth's past by unraveling the information held in rocks. The most important geologic tool in both disciplines is Earth's 4-billion-year-old collection of rocks known as the *geologic rock record*. This vast record contains a wealth of information on topics ranging from the evolution of life-forms to the rise and fall of mountain ranges to changes in climate and sea level. Over the past 30 years or so a new subdiscipline has emerged called **environmental geology**, whereby geologic information is used to address problems arising from the interaction between humans and the geologic environment. Environmental geology is becoming increasingly important as population continues to expand, which in turn is leading to widespread pollution and shortages of certain resources, particularly water and energy. Population growth has also resulted in greater numbers of people living in areas where floods, earthquakes, volcanic eruptions, and landslides pose a serious risk to life and property.

The first step in solving our environmental problems is to understand the way in which various Earth processes operate and how humans interact with these processes. Once this interaction is understood, appropriate action can be taken to reduce or minimize the problems. The most effective way of accomplishing this is through *science,* which is the methodical approach developed by humans for learning about the natural world. Because science is critical to addressing our environmental problems, we will begin by taking a brief look at how science operates.

Scientific Inquiry

By our very nature, humans are curious about our surroundings. This natural curiosity has ultimately led to the development of a systematic and logical process that tries to explain how the physical world operates. We call this process *science,* which comes from the Latin word *scientia,* meaning "knowledge." Over the past several thousand years the human race has accumulated a staggering amount of scientific knowledge. Although we now understand certain aspects of the natural world in great detail, there is still a lot we do not understand. Throughout this period of discovery the public has generally remained fascinated with what scientists have learned about the physical world. Evidence for this fascination is the continued popularity of science programs currently available on television. It seems rather odd then that one of the common complaints in science courses is that nonscience majors find the subject boring. This raises the question of what is it about science courses that tends to cause students to lose their natural interest in science?

One reason, perhaps, for the loss of interest is that students are often required to memorize trivial facts and terminology. The problem is compounded when it is not made clear how this information is relevant to our own lives. Focusing on just the facts is unfortunate because it is the *explanation* of the facts that makes science interesting, not necessarily the facts themselves. Take, for example, the fact that coal is found on the continent of Antarctica, which sits directly over the South Pole (Figure 1.4). Because coal forms only in swamps where vegetation and liquid water are abundant, we can logically conclude that Antarctica at one time must have been relatively ice-free. This means that either the climate was much warmer in the past, or Antarctica was once located much closer to the equator. This leads us to ask the obvious: What could have caused the global climate to change so dramatically? Conversely, how could this giant landmass actually move to its present position? To answer these questions scientists must gather additional data (i.e., facts). This data will likely result in even more questions that need to be answered.

Science therefore can be thought of as a method by which people use data to discover how the natural world operates. Unlocking the secrets of nature is truly exciting, which is why most scientists love what they do. Anyone who has found a fossil or an old coin, for example, can relate to the thrill of discovery. A key point here is that nearly everyone practices science each and every day. When we observe dark clouds moving toward us we process this information (i.e., data) along with past observations, and logically conclude that a storm is approaching and that it is wise to seek shelter. A fisherman who keeps changing lures until he or she finds one that attracts a certain type of fish is also practicing science. Because science is fundamental to the topics discussed in this textbook, we will explore the actual process in more detail in the next section.

How Science Operates

Modern scientific studies of the physical world are based on the premise that the entire universe, not just planet Earth, behaves in a consistent and often predictable

FIGURE 1.4   The basic goal of science is to use facts or data to explain different aspects of our natural world. For example, the coal beds shown here in Antarctica are a scientific fact. It's also a fact that coal forms only in lush swamps, which means Antarctica must have been ice-free at some point in the geologic past. The best explanation for this is that Earth's climate was much warmer in the past, or that Antarctica was once much closer to the equator.
Dr. Barrie McKelvey and N.C.N. Stephenson, Dept of Geophysics, University of New England

manner. When an event or phenomenon is observed repeatedly and consistently, it can be described as having a pattern. Discovering patterns in nature is important because it allows us to predict future events. For example, people long ago observed that the ocean rises and falls along coastlines on a regular basis. This pattern, known as the tides, is so regular that we can accurately predict when the sea will reach its maximum and minimum heights each day. In contrast, events such as floods and volcanic eruptions occur repeatedly, but on a more irregular or random basis. The random nature of certain types of events means that scientists can only predict their future occurrences in terms of statistical probabilities. Although recognizing natural patterns is a key component of science, the goal is to explain *how* and *why* things happen in the first place.

The process by which the physical world is examined in a logical manner is commonly referred to as the **scientific method**. The basic approach is to first gather factual data about the world through observations or by conducting experiments. Examples of data include such things as temperature readings, frequency of floods, and fossils preserved within rocks. Note that all scientific data can be observed and/or physically measured. Also, data are considered to be facts provided that scientists working independently of each other are able to repeat the work and obtain similar results. Once data are collected, scientists then seek to develop an explanation for the data itself and any patterns it may contain. For example, suppose a researcher collects marine fossils from rock layers that are 10,000 feet above sea level. The next step would be to develop a scientific explanation for the fossils that is consistent with other known data. In this case, any explanation would have to be consistent with the fact that the planet does not contain enough water for sea level to ever have been 10,000 feet higher than it is today. Logic dictates then that any plausible explanation must include a mechanism for uplifting the fossil-bearing rocks from sea level to their present position.

The term **hypothesis** refers to a scientific explanation of data that can be tested in such a way that it can be shown to be false or incorrect; something scientists refer to as being *falsifiable.* Supernatural explanations are not considered scientific simply because they are not testable and cannot be shown to be false. This concept of a hypothesis being falsifiable may seem odd since people generally think about trying to prove ideas to be true rather than false. Nevertheless, this is an important concept in science because a hypothesis is considered valid so long as additional testing does not show it to be false. Take, for example, how fossil evidence shows that dinosaurs went extinct 65 million years ago, whereas the first fossils of primitive humans (hominids) do not show up in the rock record until around 7 million years ago. Scientists have logically concluded, or hypothesized, that humans never coexisted with dinosaurs. This hypothesis would be proven to be false if hominid fossils are ever found in rocks of the same age as those that contain dinosaur fossils. Because extensive searches have never yielded such hominid fossils, the hypothesis that people and dinosaurs did not coexist remains valid.

Another key aspect of the methodology we call science is that during the early stages of an investigation researchers commonly come up with more than one plausible hypothesis for a given set of data. As shown in Figure 1.5, scientists refer to these different explanations as **multiple working hypotheses**, which are all considered valid so long as they are consistent with existing data. Because the goal of science is to seek out the best possible explanation, researchers continue to collect new data as they try to disprove one or more of the hypotheses. If an individual hypothesis is shown to be false, then it must either be modified or removed from consideration. Over time, this process of eliminating and refining hypotheses by gathering new data gives scientists

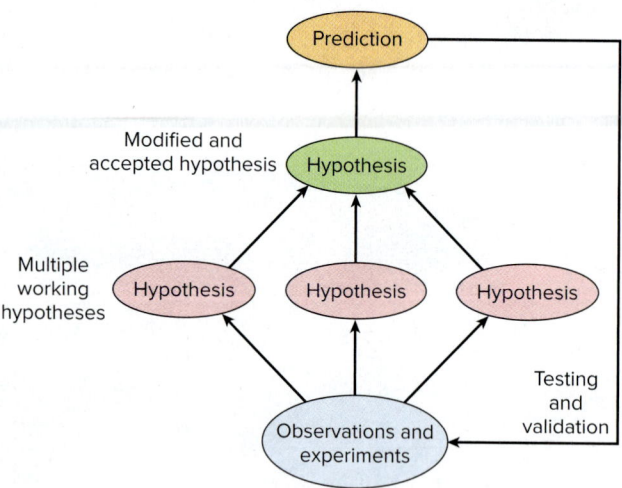

FIGURE 1.5   A scientific hypothesis is an explanation of known observations and experimental data. Multiple hypotheses are commonly developed, with most being discarded or modified as new data are gathered during testing and validation. Over time, a refined hypothesis normally emerges from the process and becomes generally accepted by the scientific community. Validation involves the ability of a hypothesis to predict future events.

greater and greater confidence in the validity of the remaining hypotheses. Note in Figure 1.5 that hypotheses are validated by their ability to predict future observations or experimental data. It should be emphasized that geology is more of an observational science than an experimental one, such as chemistry. This means that geologic hypotheses are typically tested or validated by making predictions that are confirmed through additional observations as opposed to controlled lab experiments. A good example is the hypothesis that humans and dinosaurs did not coexist, something that cannot be tested in a lab, but rather by observing more of the fossil record.

The terms *theory* and *hypothesis* are sometimes used interchangeably, but actually have different meanings. As indicated in Figure 1.6, a **theory** describes the relationship between several different and well-accepted hypotheses, providing a more comprehensive or unified explanation of how the world operates. In other words, a theory ties together seemingly unrelated hypotheses and allows us to see the "big picture." For example, the theories of atomic matter, relativity, and evolution unify various hypotheses within their respective disciplines of chemistry, physics, and biology. In geology the central unifying theory is known as the theory of plate tectonics (Chapter 4). This important theory explains how Earth's rigid crust is broken up into separate plates, which are in constant motion due to forces associated with the planet's internal heat. The movement of tectonic plates influences the location of continents and circulation of ocean currents, and consequently has a strong effect on the global climate system and biosphere on which we humans depend. As with all scientific theories, the theory of plate tectonics provides scientists with a larger context for understanding an array of different hypotheses. It should be emphasized that when scientists use the term *theory,* it has an entirely different meaning compared to its use by the general public. In common everyday language, the word "theory" is used to describe some educated guess or speculation. In science, however, a theory is a widely accepted and logical explanation of natural phenomena that has survived rigorous testing. Later we will examine how these different meanings of *theory* can impact public debate and policy considerations of environmental issues.

There are some phenomena in nature where the relationship between different data occurs so regularly and with so little deviation that scientists refer to the relationship as a **law**. In some cases a law can be expressed mathematically, as in Newton's three laws of motion and gravitational law. An example of a law in geology is the *law of superposition,* which states that in a sequence of layered rocks derived from weathering and erosion (i.e., sedimentary rocks, Chapter 3), the layer on top is the youngest and the one on the bottom is the oldest. This simple and intuitive idea that sedimentary layers become progressively older with depth has been invaluable in using the geologic rock record to unravel Earth's history. Scientific laws, therefore, are quite useful despite the fact they do not necessarily unify different hypotheses and provide grand explanations as do theories.

A good example of how knowledge is advanced through the use of science is the discovery of the planet Neptune. Early astronomers noted strange wobbles in the elliptical orbits of the planets around the Sun, but could not explain the wobbling with the existing knowledge. It was not until after Isaac Newton published his theory of gravitation in 1687 that astronomers could explain that the wobbling was caused by the gravitational effects of planets in adjoining orbits. In the 1800s, scientists remained puzzled by the wobble in the orbit of Uranus since it was the outermost known planet at the time. This led some astronomers to predict that an unknown planet existed beyond Uranus's orbit and was causing the wobble. The planet Neptune was then discovered in 1846 when astronomers pointed a telescope at the exact position in

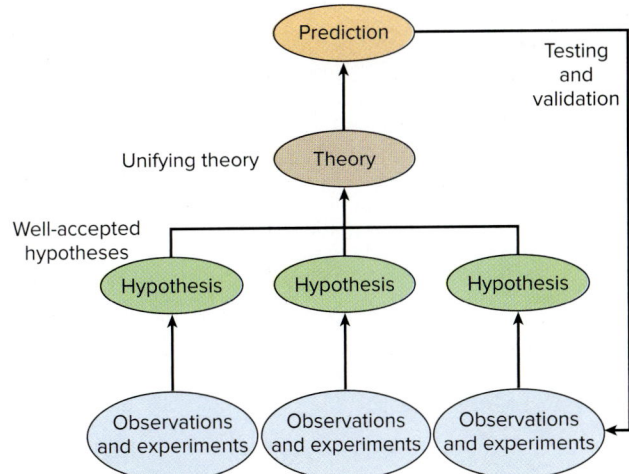

FIGURE 1.6 Scientific theories describe the relationship among different hypotheses and provide a more comprehensive or unified explanation of how the natural world operates. As with all scientific explanations, theories undergo repeated testing and validation.

the sky where Newton's laws predicted a planet would be. Because of this and other successful predictions, scientists soon accepted the validity of Newton's theories.

In the early 1900s, Albert Einstein stunned the scientific community when his special theory of relativity (1905) and general theory of relativity (1915) proved that Newton's gravitational law produced significant errors in situations of unusually strong gravity or high velocities. However, Einstein's work did not invalidate Newton's law, but rather represented a modified or improved version that was accurate under more extreme conditions. This example helps highlight the fact that scientific theories and hypotheses can be improved and modified through continued testing. Although it is rare, well-established theories are sometimes rejected when they fail to explain new data, or when a better explanation is presented. Perhaps the most well-known example is how the scientific community completely abandoned the Earth-centered theory of the solar system during the 1500s. A Sun-centered theory, based on the earlier work of Nicolaus Copernicus, eventually gained acceptance because it provided a much simpler explanation for the known movements of the planets.

Science and Society

Over the centuries scientists have accumulated a vast amount of knowledge regarding the natural world. Highly trained scientists in today's specialized fields are, of course, the ones who best understand the details of this knowledge, whereas all of society benefits from the results. Consider how advances in medical research have led to an increase in doctors specializing in fields such as cardiology, neurology, gynecology, and dermatology. Today if a person has a medical problem he or she can go to a specialist for a more accurate diagnosis and more sophisticated treatment. This, in turn, greatly increases the chance of being cured of a serious illness. Also consider the scientists with specialized knowledge on the complex interactions between hurricanes and the atmosphere and oceans. This area of science has progressed to the point now where sophisticated computer models routinely make projections of where hurricanes will make landfall, thereby saving large numbers of lives.

Although scientists are able to make useful predictions based on well-established theories, it is important to realize that there is almost always some degree of uncertainty associated with any prediction. The amount or degree of uncertainty usually depends on the nature of the process and the amount of error involved in making measurements. Take, for example, how scientists use Newton's laws of motion and gravity to predict, with great confidence and accuracy, the future position of the planets as they orbit the Sun. These laws were used to land a spacecraft on Saturn's moon Titan in 2005 and on a comet in 2014, which were impressive feats considering the individual crafts traveled seven and ten years through space before meeting up with their respective targets orbiting around the Sun. Also consider how hydrologists can predict with great certainty that deforestation will lead to increased flooding. Due to the sporadic nature of flood events, however, predicting an actual flood can only be done using statistical probabilities.

Nearly everyone in a modern society benefits from science, but problems can arise when people have a poor understanding of science and then take the benefits for granted. For example, when you enter a friend's number into your cell phone you probably never think about all the science behind the wireless technology that makes your call possible. Likewise, when we put gas in our vehicles or enjoy a warm house, few of us think about how the science of geology enables oil companies to locate petroleum deposits hidden deep within the Earth. Or, we may not consider how it took years of medical research before surgeons could perform open-heart surgeries that extend the lives of our loved ones. Although society reaps great benefits from science, history is full of examples where new knowledge met considerable resistance. Much of this resistance can be attributed to the tendency of knowledge to create change. In some instances scientific advances present society with new moral and ethical questions that must be addressed, as is the case with stem-cell research and its potential to develop cures for different diseases. Other times people feel their religious views are being threatened by science, as with the theory of evolution and how it explains the development of the biosphere. Perhaps the most common reason people resist scientific knowledge is that it often triggers changes that threaten established economic interests within a society. Pollution controls on coal-burning power plants, elimination of lead in gasoline, and restriction of ozone-depleting refrigerant gases are but a few examples where science identified a clear threat to human health, but corrective measures were strongly opposed by business and political interests.

It is not surprising then to find science thrust into public debates whenever new information runs counter to the interests of economic, political, or religious groups. A common reaction from interest groups is to try to discredit scientific information by referring to it as "only a theory." This reaction, of course, takes advantage of the common misperception that a scientific theory is just speculation or an educated guess. Another well-used tactic is to create doubt about the science by saying "scientists are not certain." The implication here is that the science is untrustworthy, conveniently ignoring that all scientific work by its very nature has some level of uncertainty. Perhaps the best example is how the tobacco industry, for over 30 years, used scientific uncertainty to effectively sow doubt regarding the link between smoking and lung cancer. Today we see the words "theory" and "uncertainty" again being misused in an effort to convince the public that the threat of global warming is a hoax (Chapter 16).

Another common tactic by interest groups is to create doubt by putting forth nonscientific work whose results run counter to the legitimate work of scientists. Here the nonscientific work is simply labeled "scientific," which is usually effective because most people find it difficult to distinguish between good, solid science and so-called *junk science*. The key difference is that good scientific explanations are always capable of being proved false and are consistent with *all* the data, not just selective data that fit a particular viewpoint. Another measure of good scientific work is whether it has passed the so-called *peer-review process* before being published. Here scientists submit their work to scientific organizations so it can undergo rigorous scrutiny, not just by any scientists, but by leading experts within a particular field. In this process, papers are published and presented to the public only if their conclusions are supported by physical data and by methods that can be verified.

Finally, because scientific knowledge is so vital to society, scientists often find themselves trying to educate policymakers on important environmental issues, many of which are politically charged. Although science itself is objective and nonpolitical, scientists commonly speak out publicly so that policymakers can base decisions on the best information available. Since today's students will be the decision makers of tomorrow, it is especially important for you to understand not only the science behind environmental issues, but also how science itself operates.

FIGURE 1.7   This house erupted into flames when a volcanic lava flow moved through a residential neighborhood on the big island of Hawaii. Lava flows have been taking place on the big island for the past 700,000 years and have resulted in the formation of the island itself (the other Hawaiian islands are much older). Although lava flows are a natural geologic process that creates habitable living space, they also pose a serious risk to human life and property. USGS

Environmental Geology

Environmental problems related to geology generally fall into one of two categories: *hazards* and *resources*. We will define a **geologic hazard** as any geologic condition, natural or artificial, that creates a potential risk to human life or property. Examples include earthquakes, volcanic eruptions, floods, and pollution. Some geologic processes, such as the eruption of volcanic lava (Figure 1.7), present a clear hazard to society. Ironically, what we consider as geologic hazards are often processes that play an important role in maintaining our habitable environment. Volcanic eruptions, for example, are as old as the Earth itself and have been instrumental in the development of the atmosphere and oceans. Also interesting is the fact that human activity can affect certain types of geologic processes, increasing the severity of an existing hazard and making it more costly in terms of the loss of life and property. A good example is the use of engineering controls to minimize flooding in one area, which often end up increasing the flooding somewhere else. Human interference in natural processes also commonly produces unintended consequences, as in the loss of wetlands. By destroying wetlands, humans inadvertently disrupt the food web within critical ecosystems, which ultimately results in the loss of sport and commercial fisheries.

Pollution is considered a hazard because it directly impacts human health and the ecosystems on which we depend. Although pollution can occur naturally, human activity is by far the most common cause (Figure 1.8). An example is metallic mercury, which cycles through the biosphere after being released during the natural breakdown of certain types of minerals. Mercury tends to accumulate in wetlands due to the acidic and oxygen-poor conditions there, but is then periodically released into the atmosphere when droughts allow fires to sweep through dried-out wetlands. Since the Industrial Revolution, humans have been releasing mercury into the environment by burning vast quantities of coal, which are ancient swamp deposits that naturally contain mercury-bearing minerals (Chapters 13 and 15). Consequently, the amount of mercury in the biosphere is now significantly higher than natural background levels. Mercury is a problem because it bonds with carbon atoms and creates highly toxic compounds capable of moving through the aquatic food chain and into humans. Of particular concern are pregnant or nursing women who eat mercury-contaminated fish and unknowingly pass the toxic compounds on to their children. This results in children with elevated levels of mercury, who run a higher risk of developing severe and irreversible brain damage. A somewhat different form of pollution is the emission of greenhouse gases (e.g., carbon dioxide) from the burning of fossil fuels, which is contributing to the problem of global warming (Chapter 16). Although greenhouse gases are natural and have helped regulate Earth's climate system for millions of years, the volume of gases being released into the atmosphere by humans is disrupting the atmosphere's natural equilibrium, which in turn threatens to disrupt the planet's entire ecosystem. This type of pollution is global in nature and has the potential to threaten all of humanity.

FIGURE 1.8   Pollution of air and water resources is a serious problem that affects both natural ecosystems and our quality of life, particularly human health. (Left) © Steve Cole/Getty Images; (Right) © Doug Menuez/Getty Images

The other major area of environmental geology relates to **earth resources**, which include water, soil, mineral, and energy resources. Resource issues generally involve society trying to maintain adequate supplies and minimizing the pollution that results from resource extraction and disposal of waste by-products. Freshwater and soil resources are the most critical to society since they form the basis of our agricultural food supply (Figure 1.9). As soils continue to be lost through erosion and water supplies become stretched to their limit, rapid population growth may soon outstrip world food production. Depletion of soil and water resources therefore is potentially one of humanity's greatest challenges. Moreover, by removing the natural vegetation from the landscape in order to grow food and expand urban areas, we inadvertently create a problem known as *sediment pollution.* When soils are left exposed, excessive amounts of sediment wash off the landscape and into our natural waterways. This destroys the natural ecology of streams and fills the channels with sediment, leaving them more prone to flooding.

In addition to soil and water resources, modern society is also highly dependent on nonrenewable supplies of energy and minerals. Mineral resources (Chapter 12) are critical because they provide most of the raw materials used in building our modern infrastructure. Of particular importance are iron for making steel, copper for electrical systems, and limestone for making concrete. Despite being nonrenewable, some mineral resources such as limestone, sand, and gravel are so abundant that their supplies are essentially inexhaustible. In contrast, there are some minerals with very specific and critical applications whose supply is so limited that they are considered to be of strategic importance. Examples include chromium and cobalt minerals needed to produce high-performance jet engines for military aircraft. Equally important are the energy resources that power the industrial, transportation, commercial, and residential sectors of an economy. Crude oil is especially critical because it is the primary source of our transportation fuels, and it serves as the raw material for making plastics and agricultural chemicals. Because oil is truly the lifeblood of modern societies, one of our major challenges is replacing our dwindling oil supplies with alternative sources of energy (Chapters 13 and 14).

FIGURE 1.9   Our ability to use water and soil resources in conjunction with fossil fuel–based fertilizers and pesticides has resulted in high rates of food production. Depletion of these resources threatens society's long-term ability to feed our growing population.
(top) © Reed Kaestner/Corbis; (bottom) © PhotoLink/Photodisc/Getty Images

Environmental Problems and Time Scales

One of the key factors that influences the way in which humans respond to environmental problems is the nature of the geologic processes involved and the time scales over which they operate. Some geologic hazards such as earthquakes, for example, happen suddenly and have unmistakable consequences that we easily recognize. In areas where earthquakes are common, society typically takes steps to minimize the future loss of life and property damage. This usually involves constructing buildings that resist ground shaking and having emergency personnel better equipped and trained for earthquake disasters. Pollution, in contrast, is a problem that may occur gradually over many years and whose effects on human health are not so readily apparent. Because the effects take place gradually, society often delays taking corrective measures until the consequences become more severe and noticeable.

With respect to geologic hazards that occur suddenly and in random or sporadic manner, society's response is governed in part by the frequency at which the events recur relative to the human life span. Generally speaking, once those with a living memory of a hazardous event begin to die off, the generations that follow tend to forget or become complacent about the hazard and its

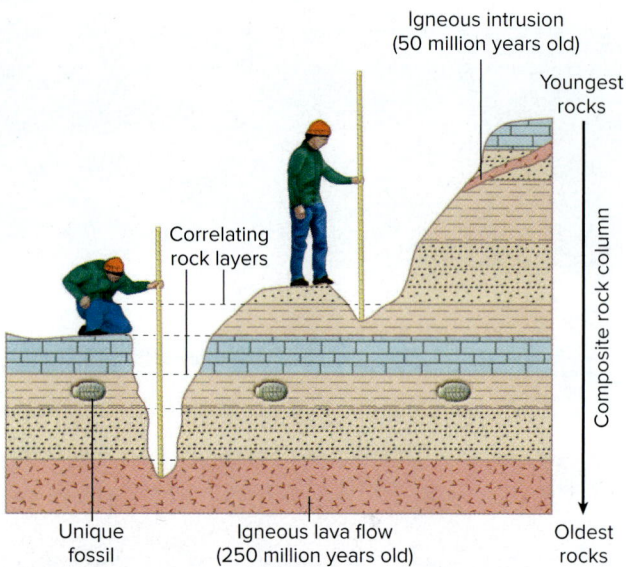

Igneous intrusion
(50 million years old)

Youngest rocks

Correlating rock layers

Composite rock column

Unique fossil

Igneous lava flow
(250 million years old)

Oldest rocks

FIGURE 1.10   In a sequence of undisturbed sedimentary rocks, the oldest rocks always lie on the bottom. By correlating specific layers at different exposures across a wide area, geologists can establish the relative age of an entire rock column. Should igneous rocks which formed from magma be present, radiometric dating can be used to determine an age range for the sedimentary column.

consequences. On the other hand, if a hazard happens frequently enough, we are less complacent and tend to take steps to reduce the risk. A good analogy is how people commonly carry an umbrella or wear rain gear in areas where it rains frequently. Hardly anyone is going to bother taking similar precautions in desert climates since the chance of getting caught in the rain is quite low. In much the same way, humans tend to become complacent about geologic hazards that recur on the order of decades or centuries. Because geologic processes often operate on time scales unfamiliar to humans, we need to briefly explore the concept of geologic time.

Geologic Time

During the late 1800s and early 1900s, geologists began systematically studying sections of sedimentary rocks exposed at the surface. Sedimentary rocks (Chapter 3) are unique in that they are derived from the erosion and weathering of older rocks and then deposited in layers. These rocks are important because they hold clues to the environmental conditions and life-forms present at the time the rocks were deposited. As indicated in Figure 1.10, geologists studying the exposed parts of a sequence of sedimentary rocks will typically record the composition of each layer and describe any fossils it may contain. They also determine the **relative age** of each layer using the *law of superposition,* which is based on the principle that the bottom layers were deposited first, and are therefore the oldest. Also note in Figure 1.10 how individual layers can be correlated across valleys from one exposure to another. By studying sedimentary sections that correlate and overlap with one another, geologists have been able to construct large, composite sections called *rock columns* (Figure 1.10). Together, sedimentary rock columns from around the world have provided scientists with a reliable record of Earth's environmental changes and the evolutionary history of its plants and animals. This worldwide rock record has also led to the development of the **geologic time scale**, which classifies all rocks according to their relative or chronological age. From Figure 1.11 one can see that the geologic time scale uses various names to subdivide Earth's rock record into progressively smaller time intervals. Perhaps the most familiar interval is the Jurassic period of the Mesozoic era, whose rocks contain dinosaur fossils.

   Although early geologists understood that the rocks on which the geologic time scale is based represent an enormous amount of time, they had no way of quantifying the actual or **absolute age** of the rocks in terms of years. What they needed was to find some characteristic of the rocks that forms at a steady and reliable rate. By knowing the rate they could then measure time, similar to a using a stopwatch. In the early 1900s, scientists discovered that as uranium atoms undergo radioactive decay they are transformed into lead atoms at a dependable rate; the term *half-life* describes the time required for half of the radioactive atoms in a given sample to decay into stable atoms (Chapter 15). Because nearly all igneous rocks contain uranium-bearing minerals, scientists could now use uranium's known decay rate (i.e., half-life) to calculate the number of years that passed since the original magma (molten rock) solidified into igneous rock. All that was needed was some means of measuring the amount of uranium and lead atoms in a rock—older rocks will contain progressively more lead atoms. By the 1950s, instruments called *mass spectrometers,* which measure the ratio of different atoms, were refined to the point that lead-uranium dating became quite precise. This was highly significant because igneous rocks often come into contact with and cut across sedimentary sequences (see Figure 1.11). Therefore, sedimentary rocks within the geologic time scale could now be assigned absolute dates in terms of years.

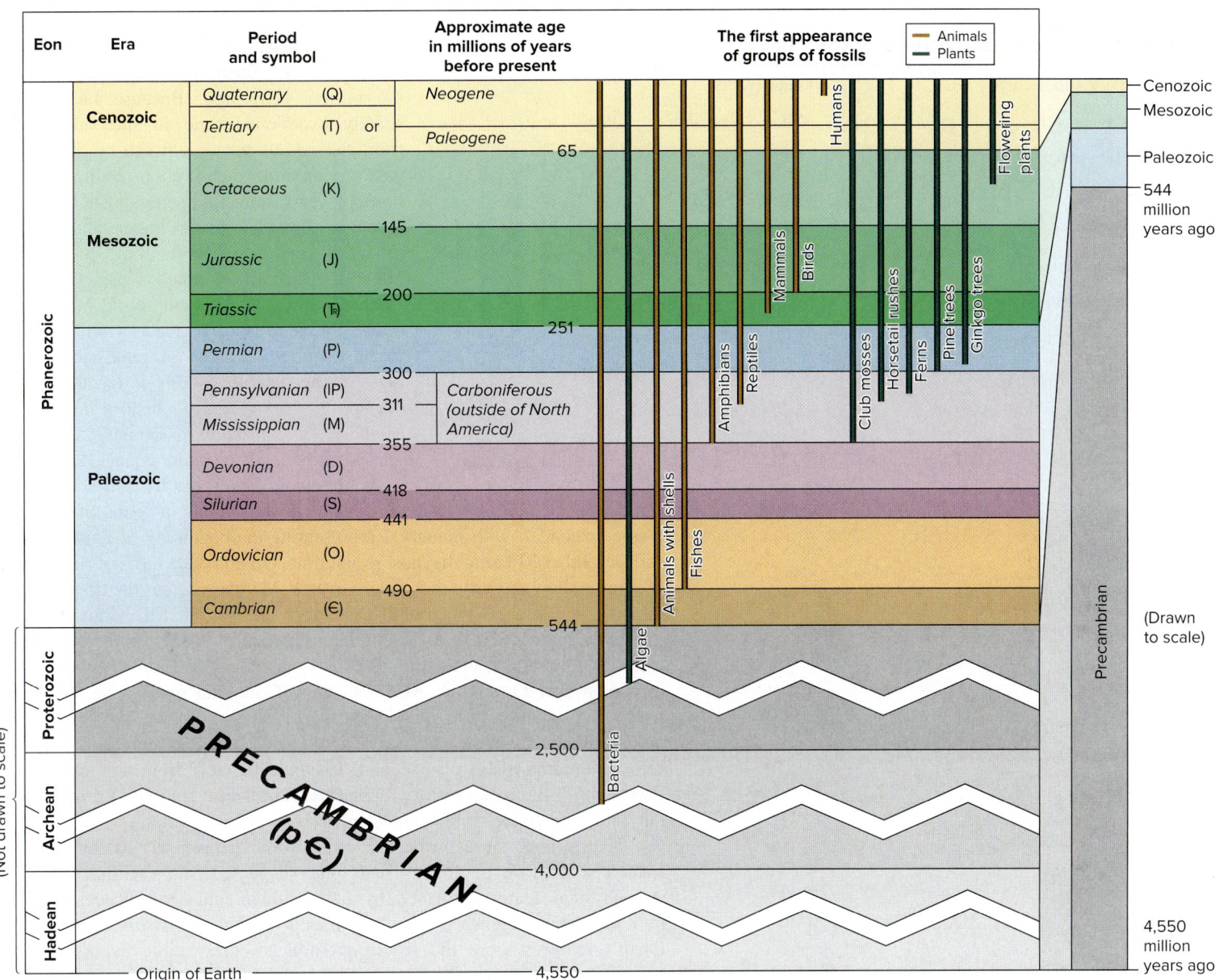

FIGURE 1.11   The geologic time scale was first developed by determining the relative age of sedimentary rocks from around the world. Radiometric dating was later used to establish the absolute dates of the various divisions within the scale. This 4.6-billion-year-old rock record holds the key to understanding the evolution of plants and animals as well as the changes in Earth's environment over time.

Note that lead-uranium dating is a specific type of **radiometric dating**, which is the general term applied to absolute dating techniques involving any type of radioactive element and its decay product. Moreover, different radioactive elements decay at different rates, which allows scientists to obtain reliable dates for events ranging anywhere from thousands to billions of years old. Elements that decay rapidly are used to date younger events, and those with slower decay rates are better suited for older events (see Chapters 14 and 15 for details on radioactive decay).

TABLE 1.1   Milestones in Earth's 4.6-billion-year history as represented on a compressed calendar year consisting of 365 24-hour days.

Beginning of Earth history	January 1
Oldest surviving rocks	Middle February
Oldest fossils—single-cell cyanobacteria	Early March
First fossils of animals with hard body parts	Middle October
First dinosaur fossils	December 11
Last dinosaur fossils	December 26
First modern human fossils	23 minutes before midnight, December 31
Egyptian civilization	35–14 seconds before midnight
Roman civilization	18–11 seconds before midnight
Columbus arrives in North America	3.5 seconds before midnight
Past 20 years	0.14 seconds before midnight

Another important outcome of radiometric dating is that scientists were able to determine that the Earth solidified approximately 4.6 billion years ago. Because humans personally experience time in intervals ranging from seconds to decades, it can be difficult for us to grasp time measured in millions or billions of years, something geologists refer to as **geologic time**. To better understand geologic time, imagine someone handing you a dollar bill every second, 24 hours a day. In order to reach a million dollars you would have to stay awake for 11.6 days. To reach a billion dollars would require 32 years of continuous counting. Another useful analogy is to compress all 4.6 billion years of Earth's history into a single calendar year consisting of 365 days. This means that a 24-hour day would equal 12.6 million years of actual time. Table 1.1 lists some important historical milestones in terms of this compressed calendar, with January 1 representing the beginning of Earth's history and December 31 being the most recent time. On this scale the first rudimentary life-forms show up in the rock record in early March, whereas the first dinosaurs do not come into existence until December 11. Note that the dinosaurs ruled for nearly 200 million years, which represents only 15 days on the compressed calendar. In contrast, the entire 200,000 years of modern human existence correspond to just 23 minutes. Also notice how Columbus reaches North America a mere 3.5 seconds before the clock strikes midnight on New Year's Eve. Even more striking is how the life span of a 20-year-old student represents only fourteen hundredths (0.14) of a second! It should be clear from this analogy that Earth's 4.6-billion-year history represents an immense amount of time, and that our human presence on the planet has been exceedingly brief.

Determining the age of the Earth through radiometric dating also confirmed something geologists had long suspected, namely that our planet's physical features formed by both sudden and slow processes. Prior to radiometric dating, a popular concept called *catastrophism* held that Earth's features were the result of sudden, catastrophic events, such as earthquakes, floods, and volcanic eruptions. Once the true age of the Earth was established, a concept known as *uniformitarianism* was used to explain how most of the planet's features are formed by slow processes acting over long periods of time. While geologists recognized that some features do indeed form by catastrophic processes, features such as deep canyons and thick sedimentary sequences could be more easily explained by very slow processes. For example, suppose that sediment being deposited at the mouth of a river accumulates at a rate of only one millimeter per year. If this continued for just a million years, the result would be a sediment sequence 1,000 meters or 3,280 feet thick. Likewise, given a sufficient amount of geologic time, slow rates of erosion can carve deep canyons and wear down entire mountain ranges. The Grand Canyon, shown in Figure 1.12, is an excellent example of how slow processes can do tremendous work over time. Because of the geologic time scale and absolute dating techniques, geologists now know that the sedimentary sequence within the Grand Canyon accumulated over hundreds of millions of years. The area was later uplifted along with the Rocky Mountains by forces within the Earth, causing the Colorado River to begin cutting downward into the sequence of sedimentary rocks. The canyon we see today formed over the past 5 to 17 million years by the slow downcutting of the river combined with mass wasting processes (Chapter 7) that have acted to widen the canyon.

Environmental Risk and Human Reaction

We will define an **environmental risk** as the chance that some natural process or event will produce negative consequences for an individual or society as a whole. The level of risk is based on the probability of an event taking place and its expected consequences, which can be expressed with the following relationship:

$$risk = (probability\ of\ an\ event) \times (expected\ consequences)$$

The matrix in Table 1.2 illustrates how the risk level is a function of probability and severity of consequences.

An example of an environmental risk is an earthquake in San Francisco, California. Here the risk is considered high because there is a high probability of a large earthquake taking place in the next 30 years, and the damage is expected to be severe. Another example is an asteroid hitting the Earth. Although small fist-sized asteroids routinely strike our planet (i.e., high probability), the actual risk is low since the potential damage is negligible. If an asteroid a mile in diameter were to strike the Earth, the damage would be catastrophic, but the risk can still be considered low since the probability of such a strike in the near future is extremely low. The differences in probability in this example are due to the fact that there are very few mile-sized asteroids compared to fist-sized ones. Therefore, when evaluating risk, one should not simply focus on the severity of the consequences without considering the probability of the event itself.

This leads us to the concept of *risk management,* which involves taking steps to identify and reduce, or *mitigate,* a specific risk. Before any action can be taken, it is of course necessary to identify the threat itself. This requires that the scientific method be used to understand those aspects of the physical world that pose a threat to society. Once a threat is identified, the risk can be mitigated by applying science and engineering to: (a) reduce the probability of an event taking place; and (b) minimize the impact or consequences. For example, driving a car puts people at high risk of being injured or killed in an accident. You can greatly reduce this risk by driving defensively, which lowers the probability of an accident. You can also minimize the potential consequences (injuries) by wearing a seat belt and driving a car with air bags. Another example is how society mitigates the risk of flooding by constructing dams and levees (lowers probability), and passes zoning laws that limit development on floodplains (minimizes consequences). Throughout this textbook we will focus on how science can be used to identify environmental risks, and then develop solutions to lower the probability of an event and minimize its consequences.

One of the key factors that affects how people respond to environmental risk is the nature of different geologic processes and the time scale over which they operate. Natural processes can be classified as

FIGURE 1.12   The Grand Canyon is an example of how slow processes can create dramatic features when given a sufficient amount of time. The thick sequence of sedimentary rocks in the canyon required hundreds of millions of years to accumulate. Later, as the area was uplifted by forces within the Earth, the Colorado River began cutting downward into the sedimentary section. The canyon we see today is the result of between 5 and 17 million years of downcutting by the river and mass wasting processes, which have widened the canyon.
© John Wang/Getty Images

TABLE 1.2   **Matrix showing the level of risk, which is based on the probability that an event will occur versus its severity. Mitigation efforts are more likely to take place as the level of risk increases.**

		Consequences →	
	Minor	**Moderate**	**Severe**
Near Certain	Low	Medium	High
Likely	Low	Medium	Medium
Rare	Low	Low	Low

(Probability ↑)

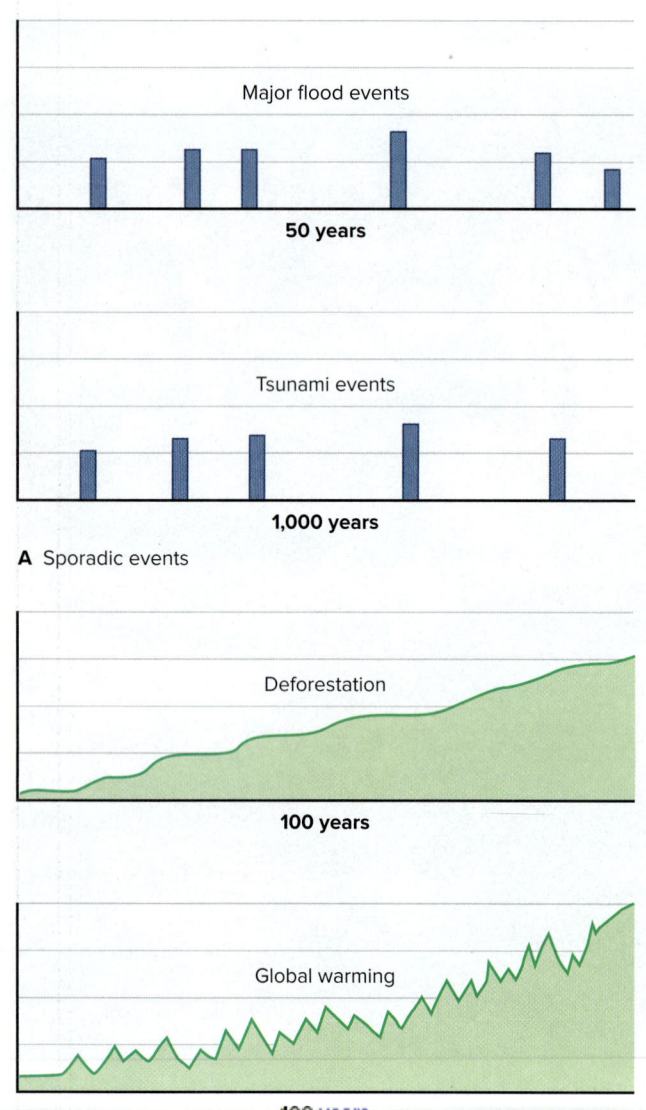

A Sporadic events

B Incremental changes

FIGURE 1.13 Natural processes generally operate in either a sporadic (A) or incremental manner (B). When sporadic events occur infrequently, it becomes more likely that a human generation will pass between events, causing people to be complacent about the risk. With incremental processes, people are typically slow to recognize environmental problems because the changes from year to year are small. Most difficult to recognize are incremental problems where upward and downward swings mask the overall change.

being either incremental or sporadic. An *incremental process* is one that generates small changes with time, such as the forces that are uplifting and eroding the sedimentary rocks in the Grand Canyon. Although incremental forces operate continuously, the rates of change are so small that their impact on the day-to-day lives of people is inconsequential. On the other hand, the incremental loss of topsoil due to erosion is occurring at such a rapid rate that significant amounts of soil have been lost in a matter of decades. This is a serious environmental issue since it reduces society's ability to grow food (Chapter 10). Therefore, if the rate of an incremental process is high enough, it is more likely that undesirable changes will occur within the life span of humans.

In contrast, *sporadic processes* are those that take place somewhat randomly as discrete events. Examples include floods, volcanic eruptions, earthquakes, and landslides. Sporadic processes are commonly referred to as *hazards* when they produce dramatic and sudden changes that humans consider undesirable. Events with particularly severe consequences are typically called *disasters* or *catastrophes*. Although sporadic processes occur randomly, they generally repeat in somewhat regular intervals, with the frequency depending on the process itself. For example, floods occur with greater frequency than do volcanic eruptions. Also important is how small-magnitude events are more common than large, catastrophic ones. Consider how minor floods occur more frequently than do major floods—similar to how your chances of winning a small daily lottery are much higher than those of winning a million-dollar jackpot.

The fact that geologic processes operate differently is important because of the way this difference influences how humans respond to environmental risks. To illustrate this point consider the sporadic nature of major floods and tsunamis shown by the graphs in Figure 1.13A. Notice that the floods occur frequently enough that a person is likely to experience several major floods within his or her lifetime. The relatively high risk of flooding helps explain why people living along rivers have historically reduced their risk by building settlements on the highest ground possible. On the other hand, major tsunamis (Chapters 5 and 9) occur much less frequently, which means several generations may pass between events such that people have no living memory of the previous tsunami. In this case some people may be completely unaware of the danger, whereas others are aware but feel that the benefits of living on flat ground next to the sea outweigh the risk. A tragic example of this phenomenon is the massive tsunami that swept into coastal villages around the Indian Ocean in 2004. Over 230,000 people lost their lives when they suddenly found themselves in harm's way with no means of escape (Figure 1.14).

Finally, environmental problems associated with incremental processes are particularly challenging to address since they can be hard to recognize. Take for example the process of deforestation (Figure 1.13B), where the amount of forest lost each year through logging can be so small that many people do not notice the overall change taking place. Because the forest looks pretty much the same as it did the year before, people tend to believe that things are "normal"—a phenomenon some refer to as *creeping normalcy*. By the time the forest is gone and erosion has washed the topsoil into nearby stream channels, there may be no one with a living memory of the environment that once existed. Even more difficult to recognize are slow incremental processes that vary naturally from year to year, such as the global warming example in Figure 1.13B. Here natural swings in temperature from year to year can easily mask the incremental change taking place. It was because of natural temperature swings that scientists were slow to recognize that Earth is currently in a warming trend.

A   B

FIGURE 1.14   Before and after photos of a coastal city in Indonesia that was completely obliterated by the 2004 tsunami. The low frequency of tsunamis means that several generations may pass between events, thereby lulling people to live in harm's way. People either become ignorant of the hazard or choose to accept the risk as they understand that the probability of such an event is low.
(a–b) © Digital Globe/Getty Images

Earth as a System

It is clearly in society's best interest to avoid environmental problems and minimize their impacts, regardless of whether a particular problem is one we create ourselves (e.g., pollution) or cannot prevent (e.g., earthquakes). A key component is having an appreciation of the time scale over which natural processes operate as this helps us identify problems, and allows us to make better risk management decisions. From the field of geology we now understand that Earth's history is one of constant change where everything evolves, including the atmosphere, oceans, continents, and, of course, plants and animals. In addition to an appreciation of geologic time, another important lesson is that humans are part of a complex natural system, and that our actions impact the very environment in which we live and depend on. For example, modern societies have grown and prospered because of our ability to modify the landscape for growing food, extracting natural resources, and constructing cities. These activities, however, have also produced unintended consequences that are highly undesirable, such as increased flooding, pollution, and destruction of wildlife habitats. By using science to understand how the Earth operates as a large system, we can learn to minimize our environmental problems and avoid creating new ones.

Recall that science strives to understand how the natural world operates, which in turn has led to many remarkable achievements. As our knowledge of the world has grown more detailed, we have seen the emergence of specialized fields within science, namely mathematics, physics, chemistry, biology, and geology. Today there are subdisciplines within each of these fields where specialists study rather narrow aspects of the physical world in great detail. In geology there are many specialized fields, including the study of minerals (mineralogy), rocks (petrology), ancient life (paleontology), chemical reactions within the earth (geochemistry), movement of groundwater (hydrogeology), and Earth's internal structure (geophysics). Many of these subdisciplines are *interdisciplinary* in nature, which means they are a combination of two or more scientific fields. For example, geochemistry is the study of both geology and chemistry.

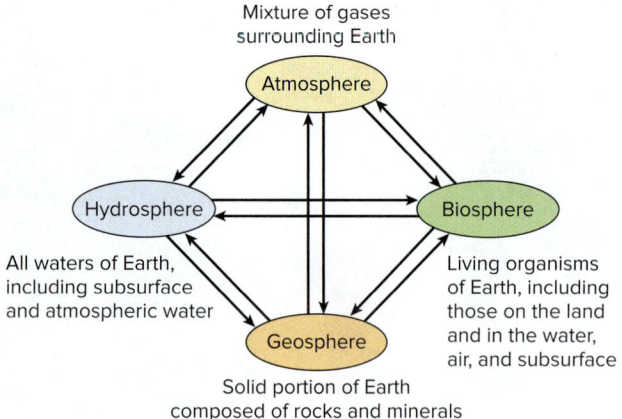

Mixture of gases
surrounding Earth

Atmosphere

Hydrosphere

Biosphere

All waters of Earth,
including subsurface
and atmospheric water

Living organisms
of Earth, including
those on the land
and in the water,
air, and subsurface

Geosphere

Solid portion of Earth
composed of rocks and minerals

FIGURE 1.15   The Earth system is composed of four major subsystems that interact in a highly complex and integrated manner, where changes in one subsystem affect the others.

The downside of specialization has been that scientists sometimes lose sight of nature as a whole. In recent years, however, there has been a growing interest in studying the interconnections that exist between major scientific fields. For instance, fisheries biologists have learned that individual fish species breed or spawn in very specific locations or habitats within river systems. These unique habitats are often found to be related to the type of sediment in the riverbed and to zones of groundwater discharge (springs), both of which are controlled by geologic processes. Scientists are learning that the natural world is highly interrelated, where individual processes depend on one or more other processes. Moreover, the entire Earth is now seen as operating as a single system where all natural processes are interconnected in one way or another. Scientists also now understand that Earth's entire history is one of constant change where everything evolves.

Today this concept that all natural processes are interrelated and in a constant state of change has led to a new, comprehensive field of study called **Earth systems science**. In this field the Earth is seen to be operating as a dynamic system made up of four major subsystems or components: the *atmosphere, hydrosphere, biosphere,* and *geosphere* (solid earth). As illustrated in Figure 1.15, these subsystems are an integral part of the overall system and continually interact with one another. The Earth system is said to be *dynamic* because when one component undergoes a change it almost invariably induces a change in one or more of the other subsystems. A good example is the damming of the Columbia River and its tributaries in North America (Figure 1.16). Groups that originally promoted the dams emphasized that the project would bring cheap electrical power and jobs to the Pacific Northwest. Plus, the vast amounts of water to be stored behind the dams would make large-scale agricultural production possible throughout the region. Unfortunately the dams also disrupted the natural hydrology of the region, resulting in the collapse of the salmon fisheries within the biosphere. The once-thriving fisheries that had sustained indigenous populations for thousands of years are now all but gone. This disappearance has resulted in a loss of fishing-related jobs and a host of other changes for the small communities throughout the region. By modifying the regional hydrology to achieve certain benefits, humans set in motion changes within the system that produced some rather undesirable consequences.

In some cases human activity produces changes that ripple through the entire Earth system. A good example is clearing the land through deforestation. The sequence of satellite images in Figure 1.17 shows the tremendous loss of forest in parts of South America in just 25 years. The photo from Indonesia in Figure 1.18 gives a more immediate view of the devastating environmental impact of deforestation, particularly in areas of more rugged terrain. What was once a dense forest with a diverse array of plants and animals is now relatively devoid of life. Moreover, the lack of forest cover will lead to landslides and extensive soil loss. In terms of the Earth system (see Figure 1.15), this means that deforestation within the biosphere has a major impact on the geosphere. In addition, as the excess sediment washes off the landscape and begins filling nearby stream channels, it destroys the habitat of aquatic species within those streams and disrupts their entire ecosystems. The filling of stream channels with sediment also impacts the hydrosphere component

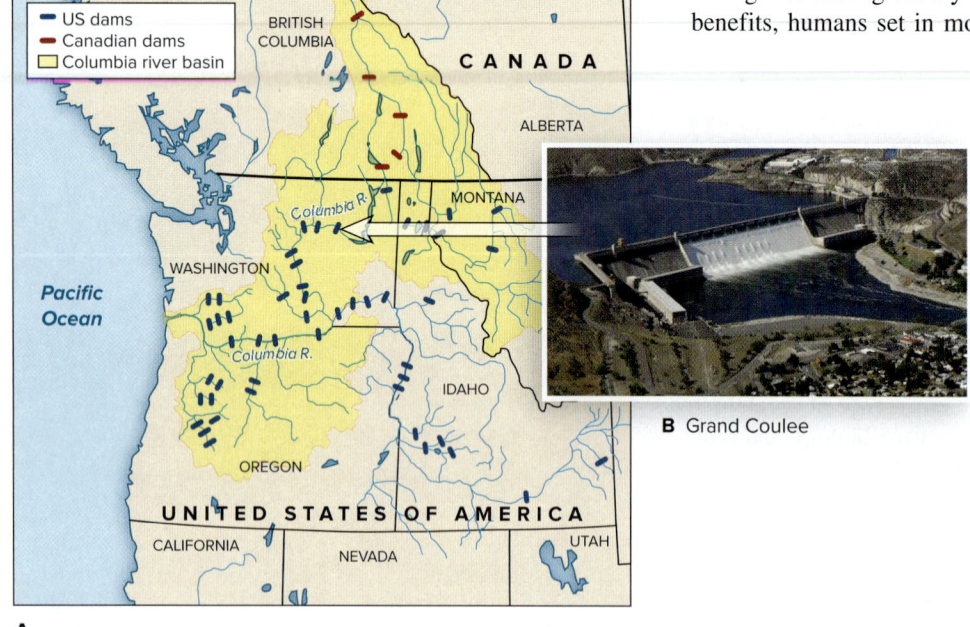

- US dams
- Canadian dams
- Columbia river basin

BRITISH COLUMBIA

CANADA

ALBERTA

Pacific
Ocean

Columbia R.

MONTANA

WASHINGTON

Columbia R.

IDAHO

OREGON

UNITED STATES OF AMERICA

CALIFORNIA

NEVADA

UTAH

A

B Grand Coulee

FIGURE 1.16   The construction of dams along the Columbia River system, such as the Grand Coulee Dam shown here, provides electrical power and water to large areas of the Pacific Northwest. The dams also prevent the migration of salmon, which has led to the collapse of the region's salmon fishing industry and traditional way of life.
(b) US Bureau of Reclamation

1975   1992   2000

FIGURE 1.17   Sequence of false-color satellite images showing deforestation near Santa Cruz de la Sierra, Bolivia, from 1975 to 2000. Areas of forests and dense vegetation are shown in deeper shades of red, whereas areas cleared of trees are in lighter shades.
(all) USGS

of the Earth system by leaving less room for the flow of water, which increases the chance of floods during heavy rain events. Finally, we need to consider what happens to the biosphere itself. During deforestation much of the wood is converted into lumber, but a significant portion is simply burned, releasing carbon dioxide gas into the atmosphere. Deforestation therefore contributes to the problem of global warming since carbon dioxide is known to trap some of the heat Earth radiates out into space (Chapter 16). The contribution to global warming is compounded by the fact that the trees that were removed had been an important means of removing carbon dioxide from the atmosphere.

Throughout this text we will examine numerous environmental issues in which scientists use the concept of Earth as a system to first identify the underlying cause of a problem, and then develop effective solutions based on our knowledge of how the system operates.

The Earth and Human Population

When humans modify the natural environment, it is usually done with good intentions so as to achieve some positive benefit. However, our activity almost always impacts one or more natural processes, generating secondary effects that ripple through different parts of the Earth system. The result is a series of unintended consequences that ultimately affect the lives of people and other living organisms that inhabit the planet. There are certain geologic processes though, such as earthquakes and volcanic eruptions, that humans have no ability to control.

FIGURE 1.18   Photograph of a deforested landscape in Sumatra, Indonesia. This once resource-rich landscape has been seriously degraded by deforestation. In addition to destroying the diverse ecosystem of plants and animals, deforestation typically leads to landslides and excessive soil erosion, causing streams to become choked with sediment. The filling of stream channels results in collapsed aquatic ecosystems and increased flooding during heavy rains.
© Bagus Indahono/epa/Corbis

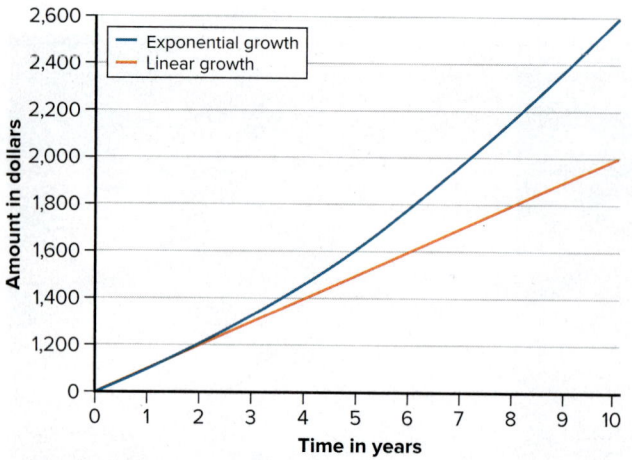

FIGURE 1.19 Plots showing how an initial sum of $1,000 responds to a linear growth rate of $100 per year versus an exponential (nonlinear) rate of 10% per year.

Regardless of whether an environmental problem is a result of our own actions or is a natural hazard we cannot control, the focus in this text will be on how humans interact with the geologic environment. It should be obvious that as human population continues to increase there will be greater interaction between people and the environment. For example, as population grows, more people will be living in areas at risk of floods, earthquakes, volcanic eruptions, landslides, and hurricanes. More people also mean greater resource depletion, pollution, and habitat destruction. Because human population plays such a key role in environmental issues, we will begin by taking a closer look at population growth.

Population Growth

We are all familiar with things that increase over time, such as gas prices, traffic congestion, and college tuition. Our interest here is in *how* things increase. Scientists classify growth rates as being either linear or nonlinear, meaning their graphs will plot as a straight line or a curve, as shown in Figure 1.19. **Linear growth** occurs when the amount added over successive time periods remains the same. In other words, if 10 is added to the total one month, then 10 more is added the next month, and so on. When you plot the growing total against time, the result is a straight line with a constant slope. Nonlinear or **exponential growth** is when the amount added over successive time increments keeps increasing. Adding 10 to the total one month, 15 the next, and then 25, would be considered exponential growth, generating a graph where the slope increases with time.

Because the slope keeps getting steeper, exponential growth leads to much greater increases over time compared to linear growth. For example, the data used in the plots shown in Figure 1.19 are listed in Table 1.3. Notice how the initial sum of $1,000 grows linearly at $100 per year. After 10 years this $1,000 would grow to $2,000. If the money grew exponentially at a fixed percentage of 10% per year, then after 10 years the total would be $2,593 rather than $2,000. Note that the difference occurs because with exponential growth you keep adding a percentage of a *growing* total. Thus as the total grows, so too does the amount you add over successive time intervals.

There are many natural processes that expand exponentially, but the most important in terms of our discussion is human population. From Figure 1.20 you can see that throughout most of history the population of the modern human species, known as *Homo sapiens,* grew quite slowly. Then, around the 1700s, population growth accelerated rapidly in response to increased industrialization and advances in medicine that brought about a decline in death rates. Notice in the graph that it took our species until 1830 to reach a population of 1 billion, which represented a time span of nearly 200,000 years. Incredibly, the population then doubled and reached 2 billion in only 100 years. It doubled again to 4 billion in just 45 years and is now projected to reach 9.5 billion by 2050. This leads to an obvious question: Can our population continue to expand exponentially, or is there a limit to how many people Earth can hold?

Limits to Growth

The idea of there being a limit to the number of humans that Earth can support was first proposed in 1687 by Antoni van Leeuwenhoek, a Dutch naturalist who estimated an upper limit of 13.4 billion.

TABLE 1.3 Calculations showing how an initial sum of $1,000 responds to linear and exponential growth rates. Each interval lists the beginning and ending totals as well as the yearly increase. Note how the yearly increase is constant under linear growth but keeps expanding under exponential growth.

	Linear Growth	Exponential Growth
End of year 1	$1,000 + $100 = $1,100	$1,000 + $100 = $1,100
End of year 2	$1,100 + $100 = $1,200	$1,100 + $110 = $1,210
End of year 3	$1,200 + $100 = $1,300	$1,210 + $121 = $1,331
End of year 4	$1,300 + $100 = $1,400	$1,331 + $133 = $1,464
End of year 5	$1,400 + $100 = $1,500	$1,464 + $146 = $1,610
End of year 6	$1,500 + $100 = $1,600	$1,610 + $161 = $1,771
End of year 7	$1,600 + $100 = $1,700	$1,771 + $177 = $1,948
End of year 8	$1,700 + $100 = $1,800	$1,948 + $195 = $2,143
End of year 9	$1,800 + $100 = $1,900	$2,143 + $214 = $2,357
End of year 10	$1,900 + $100 = $2,000	$2,357 + $236 = $2,593

He came up with this number by first calculating the inhabited land area of the world at the time, and then assumed that the maximum population it could support would have the population density of Holland. This upper limit seemed reasonable since Holland was importing many of its resources in order to support its high population density. Then in 1798 a British political economist named Thomas Malthus recognized that human population growth was exponential, whereas increases in food production were linear. Malthus concluded that unless population was brought under control, it would outstrip food supply and lead to a future of poverty and recurring famine. Famine, after all, was nature's way of keeping the population of animals in line with their available food supply.

Malthus's ideas on population limits have been highly influential over the years, leading to various predictions that human population would collapse in a catastrophic manner. What his population model failed to take into account, however, was the ability of humans to increase food production at an exponential rate through technology and innovation. The driving force behind this modern production increase has been the use of fossil fuels, particularly crude oil and natural gas (Chapter 13). Oil has made the use of mechanized farm equipment possible, along with irrigation systems that depend on diesel-powered pumps. Equally important has been the use of synthetic fertilizers and pesticides produced from oil and gas. In more recent years the development of genetically modified grains (rice, corn, wheat) has also helped increase agricultural yields. World population, therefore, has continued to expand, rather than collapse as Malthus's model predicted. The result has been that cities grow ever larger and spread across the landscape in a process known as *urbanization* or *urban sprawl* (Figure 1.21).

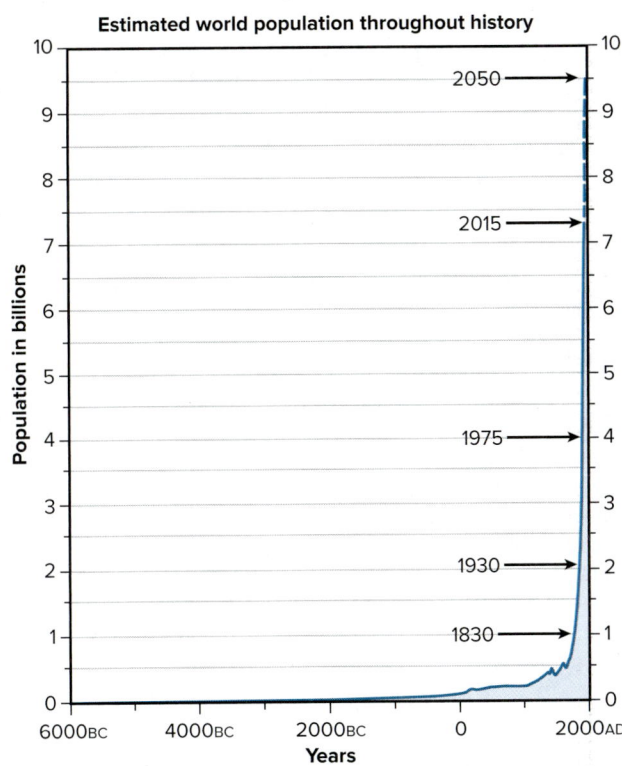

FIGURE 1.20   Graph showing exponential growth of world population throughout history. Population is projected to reach 9.5 billion by the year 2050.

A   1984   **B**   2007

FIGURE 1.21   Satellite images of Las Vegas, Nevada, showing rapid urban growth over a 23-year period. This rapid growth has increased the demand for water from nearby Lake Mead, which is the region's principal water supply. Water levels in the lake have fallen significantly in recent years due to long-term drought and may limit Las Vegas's future growth.
(a–b) USGS

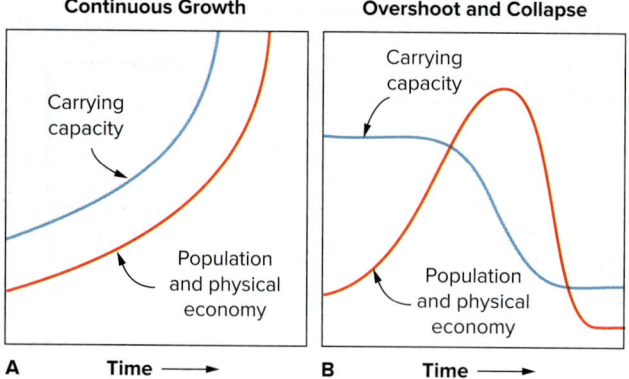

FIGURE 1.22   If Earth's carrying capacity could continue to increase (A), it would allow human population and economic activity to grow indefinitely. Since many of Earth's resources are finite, carrying capacity is fixed (B) and will begin to decline as population and economic expansion consume progressively more resources. Eventually growth exceeds the carrying capacity, causing the economy and population to crash.

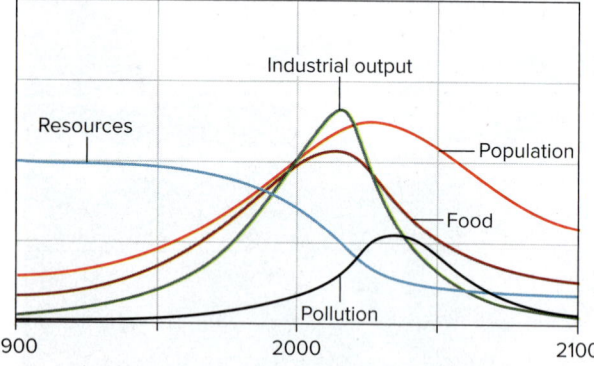

FIGURE 1.23   Modeling results showing relationships among population growth, industrial capital, food production, resource consumption, and pollution should the world continue under the policies and consumption patterns in place since 1900. Once population exceeds available resources, industrial output peaks as more and more capital is diverted to extract the remaining resources. This eventually leads to decreases in food production and population.

The idea that population growth would eventually collapse gained strength again in the early 1970s when an international group of scientists and economists examined the relationships among population growth, industrial capital, food production, resource consumption, and pollution. In this analysis they developed a computer model to study the ways in which human population and economic growth would respond as civilization approaches Earth's ability to support its inhabitants, often referred to as *carrying capacity*. For example, the scenario in Figure 1.22A shows Earth's carrying capacity continuing to increase such that it allows for infinite population and economic growth. Although humans have succeeded in increasing Earth's carrying capacity through increased food production, this has required massive amounts of investment capital and fossil fuels to produce the necessary mechanical equipment, fertilizers, irrigation systems, and so forth. Since fossil fuels are a finite resource, this infinite growth model is simply not feasible. In contrast, the scenario in Figure 1.22B has a carrying capacity that is initially fixed, but begins to decline at an accelerating rate as the expanding population consumes the finite resources. Contributing to the decline is the loss of renewable resources, such as fisheries and forests, which at some point are consumed faster than they can be replenished. Population and economic growth will then exceed or *overshoot* the planet's declining carrying capacity. Once overshoot occurs, the population and economy eventually crash and then stabilize at a level that is consistent with the system's new carrying capacity.

When running the model under different rates of resource consumption and policy changes, the researchers found that overshoot and collapse was the most likely outcome. For example, the results shown in Figure 1.23 are based on the scenario in which the world continues with the basic policies and consumption patterns that have been in place since 1900. Notice that once population exceeds the available resources, industrial output soon peaks. This occurs since population growth and economic expansion cause more and more capital to be diverted for extracting the remaining resources. Keep in mind that as time progresses, the remaining resources become more expensive and difficult to extract. The resulting diversion of capital makes it impossible to maintain the growth in industrial output. This means fewer factories and consumer goods, but most importantly, fewer farm machines and equipment. As food production and consumer goods decline, a decline in population and economic activity soon follows.

While many people have been skeptical over the years that population and economic growth will someday cease, there is now abundant evidence that infinite growth is not possible. For example, topsoils, which are critical for food production, are being lost at an alarming rate worldwide due to agricultural practices that leave soils exposed to erosion (Chapter 10). Another critical problem is that water supplies are being stretched to the limit in many regions, leaving little room for expanding irrigation (Chapter 11). Equally troublesome is that world production of cheap crude oil has declined, forcing us to rely on more expensive oil, which has led market instability and wide fluctuations in price (Chapter 13). It seems reasonable to expect that if we continue to rely on finite fossil fuels as our primary energy source, then at some point food production will begin to decline. The result would be progressively higher food prices and higher death rates as many people would be unable to afford the higher costs.

Sustainability

Since human civilization seems to be on a path toward eventual collapse, a critical question is how can we avoid exceeding Earth's carrying capacity? This leads us to the concept of **sustainability**, which is where a system or process can be maintained for an indefinite period of time. With respect to humans living within the Earth system, the term *sustainable society* is used to describe a society that lives within Earth's capacity to provide resources such that the resources remain available for future generations. The concept of sustainability is something that is readily observed in both nature and in our daily lives. Consider deer, whose population if left unchecked will quickly grow to the point where they outstrip their food supply. Starvation and disease then follow, resulting in a population crash. Another example is a person's finances. Suppose you make $50,000 a year, but spend $60,000. This deficit spending is not sustainable as it will eventually lead to a level of debt where the interest makes it impossible to pay off. To avoid bankruptcy you would have to earn more money and/or spend less by changing your lifestyle. There are no other choices. The same relationship exists between society and Earth's limited ability to provide resources. Therefore, the only options for humans are to level out population growth and change our consumption habits, or suffer the consequences of living beyond Earth's means.

Suppose that the human race is unable to control its exploding population and ends up overwhelming the planet's ability to provide resources. The consequences would likely vary among nations due to differences in population growth rates, level of economic development, and types of resources being consumed. From Figure 1.24 one can see that there are considerable differences in population growth rates between developing nations and those that are more developed, with higher living standards and more consumer goods. Note how the population in developing countries is presently growing very rapidly compared to that in developed countries (e.g., the United States, members of the European Union, and Japan). The difference is partly due to developing nations having high birth rates that tend to match their high death rates, which creates a pyramid-shaped population distribution as illustrated in Figure 1.25. Here large numbers of people are in age groups of potential child-bearing years. As these nations begin to develop, improvements in sanitation, health care, and food supply cause the death rate to decline, whereas the birth rate remains high. The result is rapid population growth. As development continues and living standards improve, couples commonly decide to have fewer children. Population growth then begins to slow. In highly developed countries, the population tends to stabilize when birth rates finally equal death rates—in some cases population actually declines because the number of births is less than the number of deaths. Geographers refer to the change in population growth rates associated with development as a *demographic transition.*

Because of the vastly different growth rates, the population of less-developed countries is projected to rise dramatically, whereas the population in more-developed nations will decline (see Figure 1.24). In fact, the population of

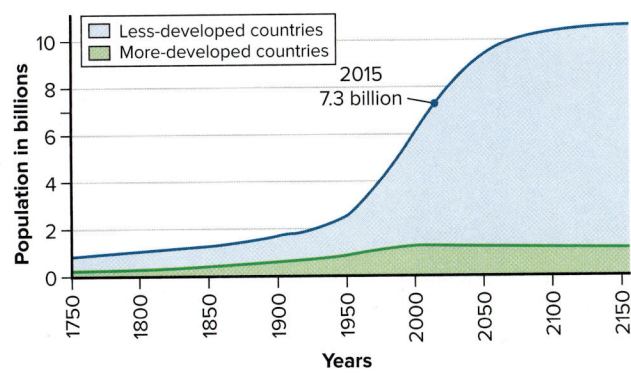

FIGURE 1.24   Graph showing world population growth and projected trends in both developed and developing countries.

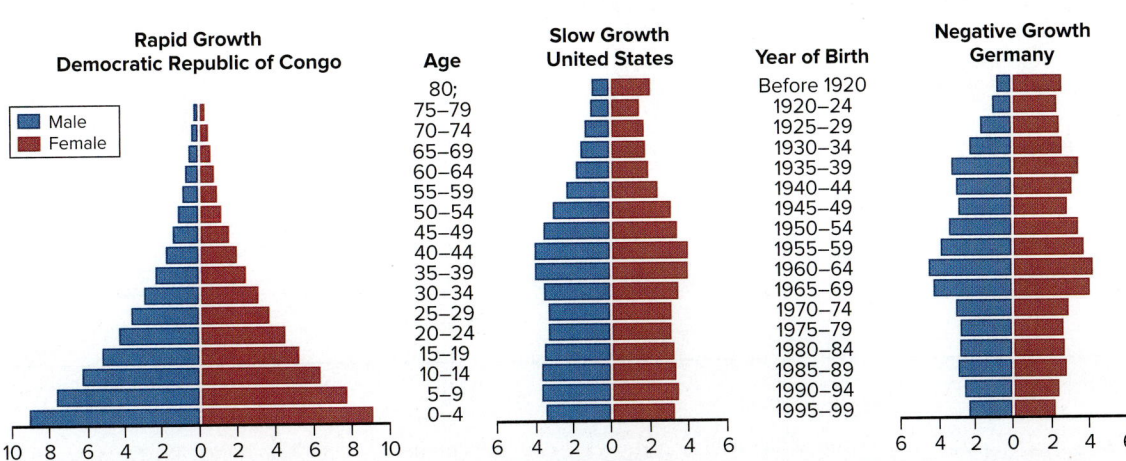

FIGURE 1.25   As nations become more developed, they usually undergo a demographic transition where death rates initially fall, followed later by declining birth rates. This results in a period of explosive population growth, gradually slowing to a point where population becomes steady, or in some cases actually declines.

some European countries has already started to decline, making further economic expansion more difficult due to a surplus of retirees and a shortage of workers. The rapid population growth of less-developed countries, on the other hand, will make it increasingly more difficult for them to obtain adequate supplies of food and water. Even if these countries could afford to import food, world food production will still be limited by the availability of Earth's soil and water resources. To help illustrate this point, notice the uneven distribution of the world's population shown in Figure 1.26. Here we see that the areas with low population density lie in regions with desert and/or polar climates. People historically have tended to avoid living in these climatic zones due to the lack of liquid water and difficulty of growing food.

Some people believe that developing nations will be able to make it through the demographic transition by increasing food production in a similar manner as developed nations have done, namely through mechanization, irrigation, synthetic fertilizers and pesticides, and genetically modified grains. With increased economic prosperity the high birth rates in these countries should fall, at which point global population would stabilize and become sustainable. The problem is that as developing nations go through the transition and begin to modernize, they naturally increase their per capita (per person) consumption of resources. This increased consumption places additional demands on Earth's finite mineral and energy resources, threatening the living standards of the developed countries.

FIGURE 1.26 Map showing global population density in 1994. Note how vast regions that are sparsely populated correspond to desert and polar climates where food production is limited or nonexistent.

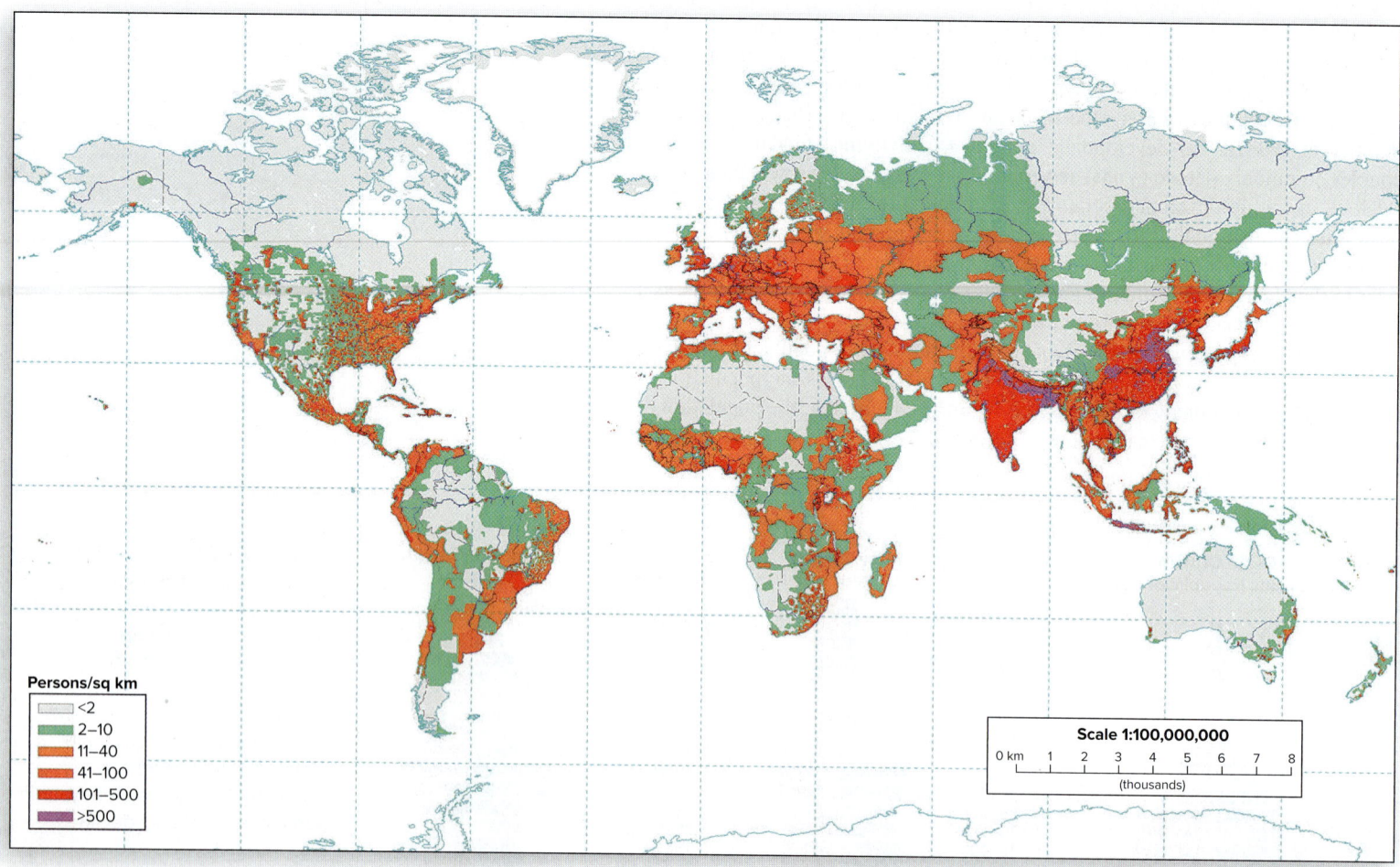

Persons/sq km
- <2
- 2–10
- 11–40
- 41–100
- 101–500
- >500

Scale 1:100,000,000
0 km 1 2 3 4 5 6 7 8
(thousands)

For example, China's rapid industrialization is currently placing greater and greater demand on the world's dwindling supplies of inexpensive crude oil (Chapter 13), causing market instabilities and wide fluctuations in price. This in turn could help destabilize the world economy and cause living standards to decline.

The key to sustainability, therefore, is not just Earth's total population, but also humanity's per capita consumption rate of resources. More people naturally means greater demand on Earth's resources, but when per capita consumption rates increase along with population, the depletion of resources will follow a non-linear downward path, as shown in Figure 1.27. Consider for a moment that if China's entire population were to achieve the present living standards of developed nations, humanity's impact on the planet would double. If all of Earth's inhabitants were to attain these higher living standards, then demand on resources would increase 14-fold. Achieving a sustainable society at this level of consumption would be virtually impossible.

Ecological Footprint

It should be apparent that one way to view sustainability is from the perspective of how many people Earth can feed with its existing water and soil resources. Another is the number of people living at developed world standards that can be supported by the planet's mineral and energy resources. We can also look at sustainability in terms of maintaining the present-day biosphere of the Earth system. A particularly useful concept here is the idea of an **ecological footprint**, which is simply the amount of *biologically productive* land/sea area needed to support the lifestyle of humans. The idea behind an ecological footprint is that every person requires a certain portion of the biosphere for extracting the resources that he or she needs and for absorbing the waste that is generated. Remember, we depend on the biosphere for its ability to purify water, provide forest resources, and regulate oxygen and carbon dioxide levels in the atmosphere. Simply put, humans could not survive without the ecosystems that make up the biosphere.

Biologists have estimated that the ecological footprint for all of humanity is currently 2.6 hectares per person (1 hectare is about the size of soccer field). Citizens living in developed countries naturally have a much larger footprint than the global average due to their higher rates of per capita consumption. For example, the Swiss average is 5 hectares per person and the British 4 hectares, whereas Americans require a staggering 7 hectares. The Chinese average only about 2.5 hectares per person, but since China is undergoing rapid industrialization, this average is expected to rise as its citizens purchase more and more consumer goods. If all the people in developing counties were to achieve the living standards of the United States, Europe, and Japan, then humanity's ecological footprint would obviously be far greater than the current average of 2.6 hectares per person. According to the Global Footprint Network, a nonprofit organization, humanity's current ecological footprint is already estimated to be over 50% larger than what the planet can support. This means humans are already consuming Earth's renewable resources faster than they can be replenished by natural ecosystems. Based on this footprint analysis, many people have concluded that we have gone beyond the ecological limits of the planet and that the present state of humanity is not sustainable. This view is consistent with the overshoot and collapse scenario illustrated in Figures 1.22 and 1.23.

Environmentalists generally believe the solution is for humanity to stabilize its population and reduce its per capita consumption of resources through conservation. Otherwise we will have to suffer the consequences of living beyond the ability of Earth to support civilization. The collapse of the society on Easter Island (Case Study 1.1) is perhaps the best example of people who lived unsustainably and eventually destroyed the very ecosystem on which they depended.

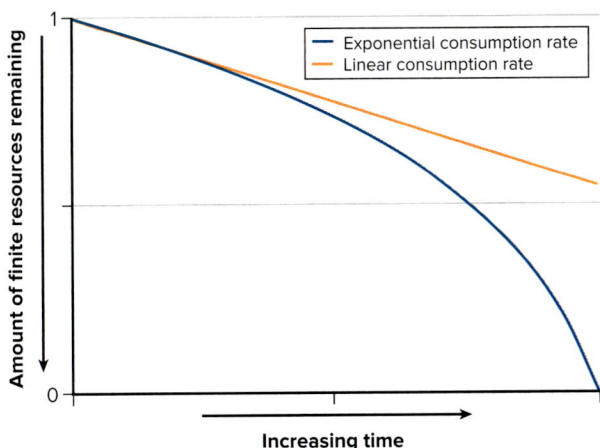

FIGURE 1.27   When an expanding population begins to increase its per capita consumption rate of a finite resource such as crude oil, resource depletion will accelerate and follow a nonlinear downward path.

Environmentalism

Environmental awareness in the United States really began in the 1960s and 1970s as a grassroots movement, driven in large part by widespread water and air pollution. One of the sparks for this environmental movement was Rachel Carson's classic 1962 book, *Silent Spring*. Her book helped awaken both the public and scientists to the fact that Earth's complex web of life is sensitive to environmental change, particularly pollution. The basic problem in the United States was that industries had historically been free to discharge their waste by-products into the atmosphere and nearby water bodies. However, as people recognized that pollution was fouling their air and drinking water as well as their beaches, fishing holes, and other recreational sites, they began to demand change. Eventually federal laws, such as the Clean Air Act (1970) and Clean Water Act (1972), were enacted and forced industries to properly dispose of their waste (Chapter 15). Businesses could no longer freely dump waste into the environment and pass the cleanup and health costs on to society. Proper waste disposal now had to be treated as a business expense.

Another defining event in the environmental movement came in 1968 when Apollo astronauts heading to the Moon provided humans with views of the Earth no one had ever seen before (Figure 1.28). People around the globe were struck by both the beauty and isolation of our planet in the darkness of space. It also caused many to start thinking of humanity as a single race, surviving on a fragile oasis in space, rather than as different nationalities all in competition with one another. This new perspective helped us see how Earth operates as a system, providing the necessary resources that make our lives possible. Moreover, it helped us understand how we could damage this system and overuse its limited resources to the

FIGURE 1.28   Photographs of planet Earth taken by Apollo astronauts helped humans understand that Earth behaves as a system and that we depend on the system for survival.

(Left) © PhotoLink/Getty Images; (Right) © Brand X/PunchStock RF

point where the planet would become less hospitable for humans. Should this occur, we would be stuck on this island in space with nowhere else to go, similar to the Easter Islanders in the vast Pacific (Case Study 1.1).

The environmental regulations passed in the United States since the 1970s succeeded in eliminating the most visible and obvious forms of pollution, leading many people to think pollution is no longer a problem. More subtle forms of pollution, however, still exist and pose a threat to the health of both humans and ecosystems in the biosphere. Another consequence of the environmental regulations was that the federal government began exercising its authority over individual and state property rights in order to protect the health of society as a whole. This issue of personal property rights, combined with the perception that pollution is no longer a problem, has contributed to a backlash against the environmental movement in recent years. Today various groups and individuals portray environmentalists as "wackos," "tree huggers," or "ecoterrorists" who feel that plants and animals are more important than people. This is simply not true. Environmentalism is not just about saving owls in a forest or fish in a river; it is about saving humans from themselves. Forests and wetlands, for example, are important not simply because they contain interesting plants and animals, but because they provide people with clean water and help regulate our climate. If we end up destroying Earth's ecosystems, then humans would find it difficult to survive.

The biggest environmental issue facing the human race is sustainability. The question is whether we will be able to use Earth's limited resources in a sustainable manner, or will population growth outstrip the planet's ability to support us? The answer to this question will depend on whether we can control our population and reduce per capita consumption of resources through conservation. Achieving sustainability will also require many societies to change from constant economic growth to nongrowth, similar to what is taking place today in some developed countries with stable populations. The problem is that many nations define economic success in terms of increased numbers of new homes and jobs and expanded factory production and trade. These economic indicators all depend on greater numbers of people consuming greater amounts of Earth's resources. Both science and common sense tell us that it is not possible to have permanent economic growth on a planet whose resources are finite. David Brower, a leading environmentalist for over 50 years, often spoke on the subject of sustainability and the rapid pace at which we have been consuming Earth's resources since the Industrial Revolution. John McPhee described Brower's thoughts on this subject in the 1971 book *Encounters with the Archdruid:*

> Sooner or later in every talk, Brower describes the creation of the world. He invites his listeners to consider the six days of Genesis as a figure of speech for what has in fact been four billion years. On this scale, a day equals something like six hundred and sixty-six million years, and thus "all day Monday and until Tuesday noon, creation was busy getting the earth going." Life began Tuesday noon, and "the beautiful organic wholeness of it" developed over the next four days. "At 4 p.m. Saturday, the big reptiles came on. Five hours later, when the redwoods appeared, there were no more big reptiles. At three minutes before midnight, man appeared. At one-fourth of a second before midnight, Christ arrived. At one-fortieth of a second before midnight, the Industrial Revolution began. We are surrounded with people who think that what we have been doing for that one-fortieth of a second can go on indefinitely. They are considered normal, but they are stark, raving mad. . . . We've got to kick this addiction. It won't work on a finite planet. When rapid growth happens in an individual, we call it cancer."

It is certainly not pleasant to think of the human race as being some type of disease on the planet, but we are most certainly having an enormous impact on the environment. We have eliminated, and continue to eliminate, large numbers of species by destroying their habitat. Some of these species have been around for over 250 million years, a time when dinosaurs began roaming the Earth. Our impact may become so great that the planet will simply no longer be able to provide sufficient resources, forcing the human population to decline to more sustainable levels. Anti-environmentalists often state that we do not need to worry because the Earth is simply too large for us to destroy, and in a sense, that is true. However, it is an undeniable fact that Earth is an interactive system that is responding to our actions. A very real concern is that we could disrupt this system to the point where the climate is no longer hospitable for humans. While the task of creating a sustainable society will certainly not be easy, it is possible because we are a species that has been given the gift of being able to make intelligent choices. Our ability to make choices that impact our future relationship with Earth's environment is perhaps best summed up in the following excerpt from the August 18, 2002, issue of *Time* magazine:

> For starters, let's be clear about what we mean by "saving the earth." The globe doesn't need to be saved by us, and we couldn't kill it if we tried. What we do need to save—and what we have done a fair job of bollixing up so far—is the earth as we like it, with its climate, air, water and biomass all in that destructible balance that best supports life as we have come to know it. Muck that up, and the planet will simply shake us off, as it's shaken off countless species before us. In the end, then, it's us we're trying to save—and while the job is doable, it won't be easy.

SUMMARY POINTS

1. Geology is the study of the solid earth. Environmental geology is the study of how humans interact with the geologic environment, particularly with regard to geologic resources and hazards.

2. Scientists develop hypotheses in order to *explain* phenomena in the natural world that can be observed or measured. All hypotheses must be falsifiable and are considered valid as long as they remain consistent with all existing data. Supernatural explanations are not scientific because they're not falsifiable.

3. A theory describes the relationship between several different hypotheses, and thus provides a more comprehensive explanation of the natural world. Scientific laws describe natural phenomena in which the relationship between different data occurs regularly and with little deviation.

4. Environmental problems related to geology generally fall into one of two categories: hazards and the use of resources. Humans sometimes make these problems worse due to a lack of scientific understanding or appreciation for the time scale in which natural processes operate.

5. Some geologic processes operate in a sporadic manner, producing dramatic and sudden changes that can be disastrous for humans living nearby. The more frequent these hazards, the more likely people will take steps to minimize their risk. Environmental problems associated with incremental processes are challenging for society because they can be hard to recognize.

6. Earth is a complex system made up of several subsystems: atmosphere, biosphere, hydrosphere, and geosphere. A change or disruption within one of the subsystems invariably leads to changes in one or more of the others. Scientific knowledge of how the Earth system operates can help society solve existing environmental problems and avoid creating new ones.

7. Humans are part of the biosphere and are therefore an integral part of the Earth system. The way in which we interact with the Earth system can have a profound impact on the very environment upon which we depend.

8. Geologic hazards and resource issues become more pronounced as human population continues to grow. Exponential population growth exacerbates both these problems.

9. As less-developed nations begin to develop, the death rate declines while births remain high, producing rapid population growth. As living conditions improve, the birth rate declines and approaches the death rate. Countries that go through this demographic transition tend to have stable or negative population growth rates.

10. Modern agricultural practices have allowed world food production to keep pace with population growth. Food production and population will reach a maximum due to the worldwide loss of topsoils, limited water supplies, and depletion of finite mineral and energy resources needed for fertilizers, pesticides, and mechanized farm equipment.

11. As developing nations modernize and increase their per capita consumption rates, they place exponentially greater demands on Earth's limited resources. Earth could not sustain all of humanity living at developed-country standards.

12. Sustainability can also be viewed in terms of the amount of biosphere each person requires for resources and waste disposal. Biologists estimate humanity's current ecological footprint is 50% larger than what the planet can support, thus the present state of humanity is not sustainable.

13. In order for humans to live in a sustainable manner, many environmentalists believe humanity needs to stabilize its population and reduce per capita consumption of resources through conservation.

KEY WORDS

absolute age   14
earth resources   13
Earth systems science   20
ecological footprint   27
environmental geology   6
environmental risk   17
exponential growth   22
geologic hazard   12

geologic time   16
geologic time scale   14
geology   6
historical geology   6
hypothesis   8
law   9
linear growth   22
multiple working hypotheses   8

physical geology   6
radiometric dating   15
relative age   14
scientific method   8
sustainability   24
theory   9
tragedy of the commons   4

APPLICATIONS

Student Activity

Look around your house/apartment/dorm room, and make a list of five different materials that relate to geology. For example, do you have a granite countertop? Slate floor or pool table? Salt in your kitchen? Drywall (made from gypsum)? Metal objects? Plastic items (made from petroleum)? Indicate those items that can be recycled. If you currently do not recycle, describe what would cause you to start recycling.

Critical Thinking Questions

1. Why do environmental scientists stress the importance of maintaining the health of Earth's ecosystems? Why is this important to you personally?
2. How does our human perception of time differ from geologic time? Why is it important to understand the difference?
3. What is a sustainable society, and why is it important? How can you personally help?

Your Environment: YOU Decide

US Dept of Agriculture, National Resources Conservation Service, Lynn Betts

In this chapter you learned how Earth consists of an atmosphere, biosphere, hydrosphere, and geosphere, all of which operate together as a highly complex system. Because humans are an integral part of this system, our activities have had a profound effect on the planet in modern times. For example, consider how the natural environment has been changed by farming, logging, mining, and construction activities. You also learned about Easter Island and how its expanding population began consuming resources in an unsustainable manner, ultimately leading to the society's collapse.

Describe whether or not you agree with the viewpoint that what happened on Easter Island is a small-scale example of how humanity is living in an unsustainable manner on Earth. Provide one or more reasons to support your answer.

Chapter 2

Earth from a Larger Perspective

LEARNING OUTCOMES

After reading this chapter, you should be able to:

▶ Understand how the nebular hypothesis explains the formation of the solar system and how it accounts for the orbital characteristics of the planets and moons.
▶ Describe our solar system and the size of the Earth relative to the size of the solar system as well as to the size of our galaxy and the universe.
▶ Explain how extremophile bacteria are related to the origin of life on Earth and how they relate to the extraterrestrial search for life.
▶ Understand the concept of habitable zones and why complex animal life that may exist elsewhere will likely be restricted to such zones.
▶ Explain what mass extinctions are and be able to name some of their possible triggering mechanisms.
▶ Understand how scientists came to appreciate the serious nature of comet and asteroid impacts, and describe the steps being taken to reduce the risk.

In modern times humans have sent machines into space in order to study the Earth as well as the solar system and universe. This knowledge helps us to better understand the Earth system and gives us a larger perspective from which to view environmental problems. The image shown here is centered over Mexico in North America and was taken by NASA's polar-orbiting satellite called VIIRS.

Introduction

© StockTrek/Getty Images

At first glance one may question why a textbook on environmental geology includes a chapter on what is beyond planet Earth. This chapter was included in part because our planet operates within an astronomical or cosmic environment that has a major influence on the Earth system and the environment in which we live. Consider how the Sun generates wave energy (e.g., visible light, infrared, ultraviolet) that warms our planet and drives the climate system along with the biosphere and hydrosphere. Although the amount of energy produced by the Sun has been fairly steady over much of Earth's history, subtle variations are known to produce significant changes within the Earth system. Another important astronomical force is the Moon's gravitational field. As the Moon orbits the Earth, its gravity plays a major role in creating ocean tides, causing water and nutrients to move within the coastal environment in a cyclic manner. Because the coastal environment serves as the nursery grounds for a large number of marine species, the Moon-induced tides are critical to the ecosystem of the oceans. Finally, it is important to note that other planets in the solar system can also influence the Earth system. For example, the gravitational fields of the planets occasionally alter the trajectories of asteroids and comets such that they begin to cross Earth's orbit. This creates the potential for large impacts, whose consequences could be catastrophic for the present-day biosphere and life as we know it.

In addition to understanding the external forces that affect the Earth, this chapter is intended to provide students with a better sense of how humanity fits into the larger scheme of things, namely the universe. Recall that one of the key themes in Chapter 1 was that humans are a small, but very influential part of the Earth system. Another key concept was that humans have been on the Earth for only a small fraction of the planet's 4.6-billion-year history. Having an appreciation for the Earth as a system and for the vastness of geologic time is important if we are to effectively address the environmental problems facing humanity. In this chapter we will go a step further and view the Earth from an even larger perspective, namely our cosmic environment, consisting not only of the solar system but the entire universe.

Humans have long sought this larger perspective by studying the stars and planets in the night sky, pondering the nature of our very existence. However, for most of history the only tools we had for learning what was beyond our planet were our naked eyes and the ability to reason. The telescope was a major technological advancement that provided many answers, but as is typical in the process we call science, this new knowledge led to new questions. The development of powerful rocket engines eventually allowed humans to escape Earth's gravity and send spacecraft to distant parts of our solar system, including the landing of probes on several planets and moons. Many consider the human exploration of the Moon (Figure 2.1) to be our single greatest achievement. In addition to probes and landing craft, scientists now have the orbiting Hubble space telescope along with an array of sophisticated ground-based telescopes (Figure 2.2). Together these telescopes enable us to peer into the farthest reaches of the universe with clarity that was unimaginable a mere 50 years ago. These instruments collect data from more than just visible light, revealing strange and incredible phenomena within the universe (Figure 2.3).

FIGURE 2.1   Humans have long sought answers to what exists beyond planet Earth. Shown here is *Apollo 17* astronaut Harrison Schmitt exploring the Moon in 1972.
NASA

A

B

FIGURE 2.2   Sophisticated space and ground-based instruments have allowed scientists to peer into the deepest reaches of the universe and gather data in ever-greater detail. Photo (A) is of the Hubble space telescope taken from space shuttle *Discovery*; photo (B) shows part of an array of 27 radio telescopes in Socorro, New Mexico, used to study everything from black holes to planetary nebula.
(a) NASA; (b) NRAO/AUI/NSF

In this chapter we will briefly explore some of the answers science has provided regarding Earth's place in the universe, along with the extraterrestrial forces that help shape our planet. We will begin by examining how our Sun and its system of planets formed and why Earth is the only place where life is known to exist. From there we will take a look at the relationship of our solar system to the other stars in the universe. We will end by examining some of the solar system hazards that Earth faces. Hopefully this larger perspective will give you a better understanding of the environmental challenges facing humanity, and a greater appreciation for the fact that Earth's environment is both rare and highly susceptible to change.

FIGURE 2.3   Modern telescopes collect data on many different types of phenomena. Shown here is a composite image from X-ray, radio, and optical telescopes showing the collision of a spiral galaxy and a black hole. A band of dust and gas is bisected by opposing jets of high-energy particles ejected away from the supermassive black hole in the center. X-ray data is shown in blue, optical data in orange and yellow, and radio data in green and pink.
X-ray (NASA/CXC/M. Karovska et al.); Radio 21-cm image (NRAO/VLA/J. Van Gorkom/Schminovich et al.), Radio continuum image (NRAO/VLA/J. Condon et al.); Optical (Digitized Sky Survey U.K. Schmidt Image/STScI)

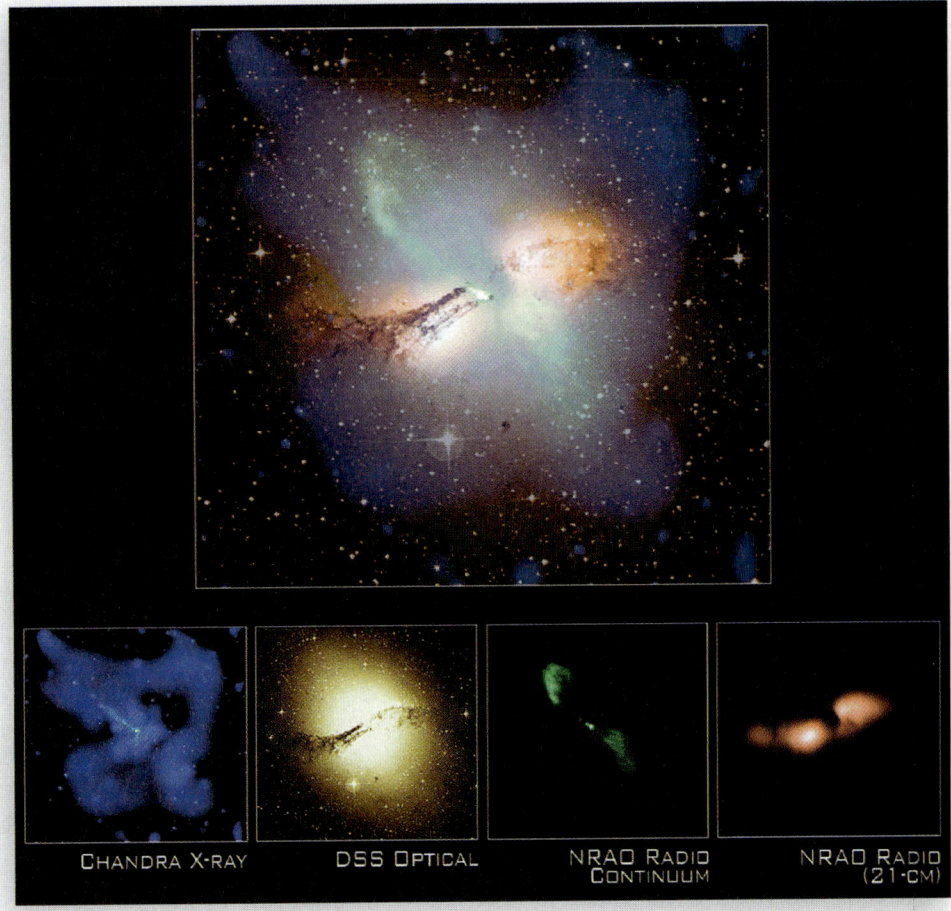

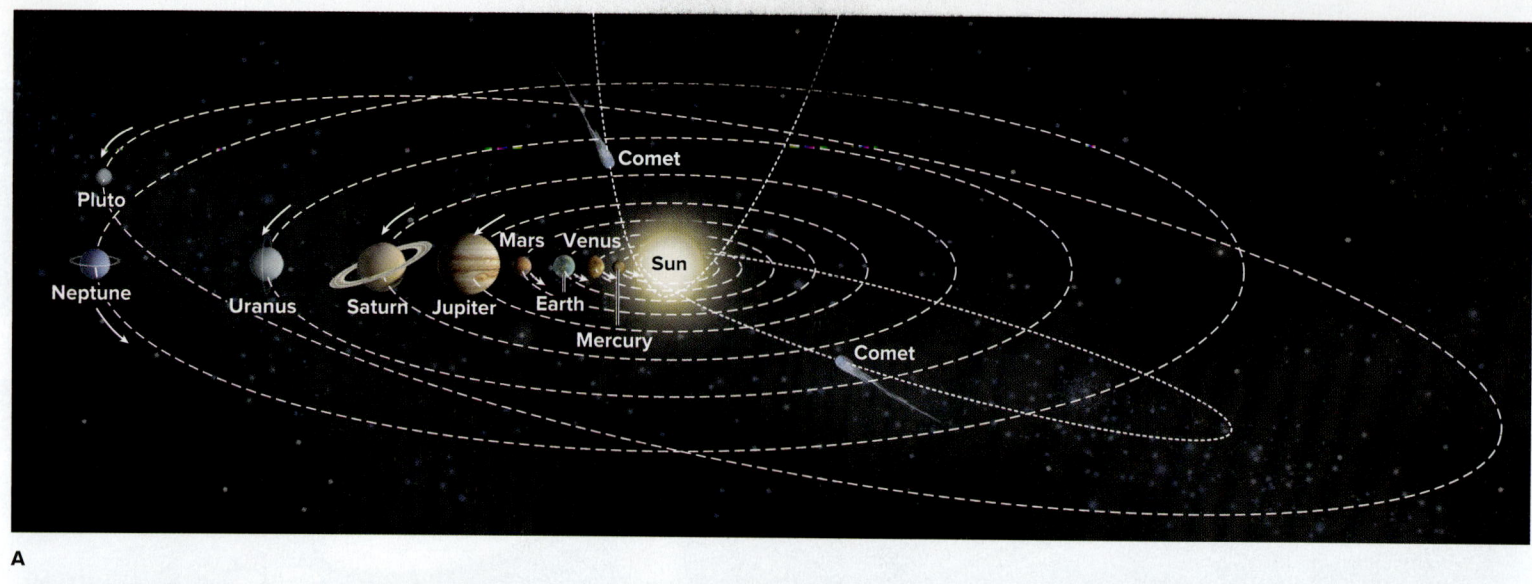

A

B

FIGURE 2.4 (A) Planets within the solar system and their orbital paths around the Sun—not drawn to scale (asteroids and asteroid belt not shown). (B) When the orbital paths are drawn to scale, one can begin to understand the vast distances between those planets out beyond Mars.

FIGURE 2.5 By compressing the distances between the planets, it becomes possible to view the relative size of the Sun and planets at the proper scale. Note how much larger the outer (gas) planets are compared to the rocky (terrestrial) planets of the inner solar system. The dwarf planet, Pluto, is shown on the far right. Images are actual photos.

Lunar and Planetary Lab NASA

Our Solar System

Before discussing the universe, we need to learn something about Earth's neighbors in the solar system. Our solar system consists of eight planets (Pluto has been reclassified as a dwarf planet), more than 100 named moons, a belt of asteroids, and millions of comets, all of which orbit the Sun (Figure 2.4). It is important to realize that Figure 2.4A was not drawn to scale. This was done because the distance between each of the planets is so great that it is impossible to show their correct size and orbits all in the same illustration. However, if only portions of the orbits are drawn to scale, as in Figure 2.4B, then we can get a better sense for the vast distances between the planets, particularly those beyond Mars. In order to get an accurate view of the relative size of the planets, it is necessary to compress their distances as seen in Figure 2.5. Here the vast differences in size between the inner and outer planets becomes quite striking (Pluto again is an exception). Perhaps even more impressive is Figure 2.6 showing the Sun and the planets all at the same scale. From this perspective we can see that the Earth is very small compared to the Sun.

The Sun

The Sun (Figure 2.7) is an immense sphere composed mainly of hydrogen and helium atoms along with relatively small amounts of the other elements. Similar to all stars, the Sun has an extremely hot and dense center surrounded by

FIGURE 2.6   When all the solar system bodies are shown at the same scale, one can see that the Sun is enormous compared to the planets, particularly Earth and the other rocky planets of the inner solar system.

Earth

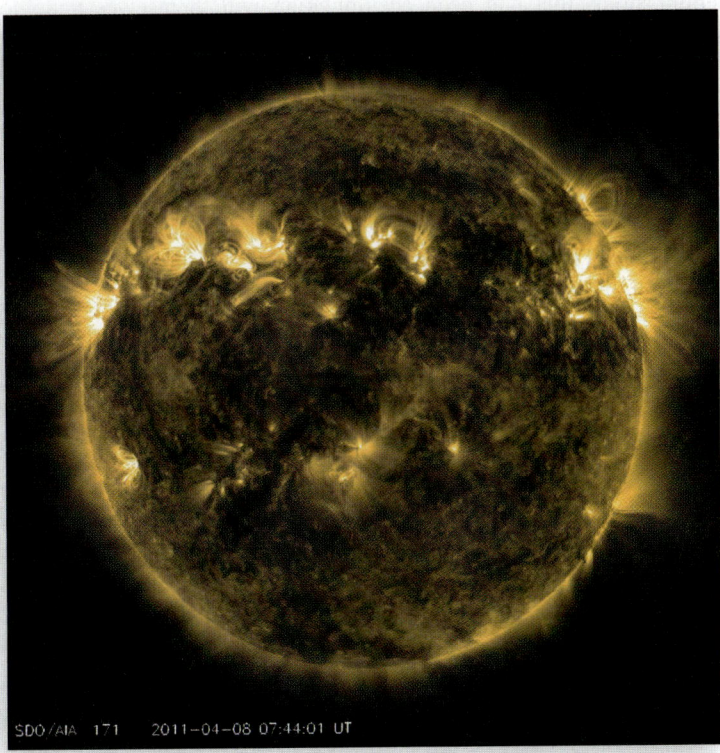

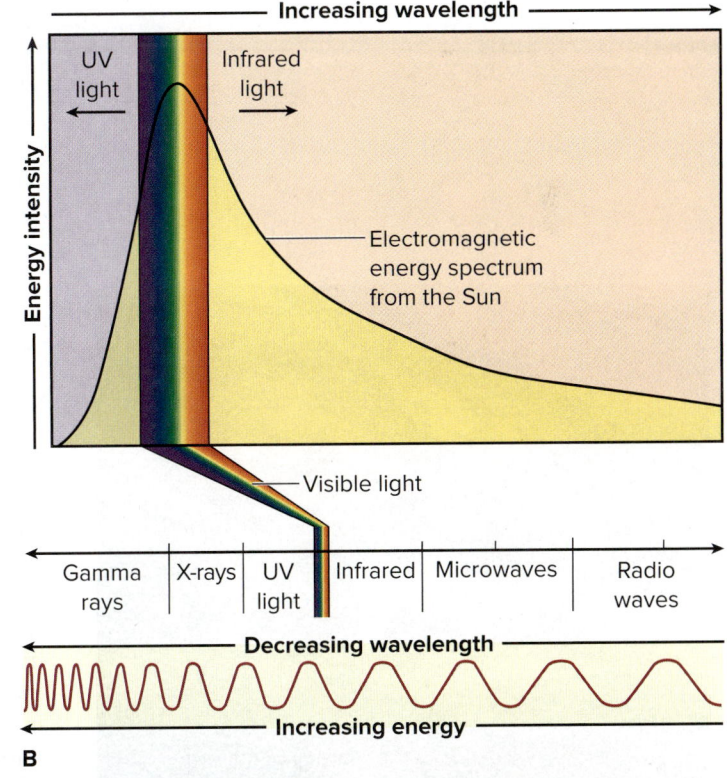

FIGURE 2.7   The Sun's immense gravity causes hydrogen atoms to undergo nuclear fusion (A) and form helium atoms—image taken in ultraviolet light and color coded to appear gold. This nuclear reaction also releases a continuous spectrum of wave energy (B), known as the electromagnetic spectrum. Note that the peak energy output from the Sun lies in the visible part of the spectrum. Electromagnetic energy travels outward from the Sun and provides much of the energy that drives the Earth system.
(a) Courtesy of NASA/SDO and the AIA, EVE, and HMI science teams

Jupiter (Figure 2.4B). Scientists believe that during the formation of the solar system the strong gravitational influence of Jupiter prevented material in the asteroid belt from developing into a planet. It is also believed that gravitational disturbances and/or collisions with other asteroids cause some asteroids to leave their normal orbits, placing them on a collision course with planets and moons within the solar system. Note that the term *meteoroid* is used to describe a body of rock and metal that is smaller than a planet or asteroid. Such a body is called a *meteorite* if it passes through Earth's atmosphere and strikes the ground. When scientists date meteorites using radiometric dating techniques (Chapter 1), they find that all fall between 4.5 to 4.7 billion years old. The 4.6-billion-year age of the Earth was determined based on both the radiometric dates of asteroids and the oldest surviving rocks on our planet. In fact, scientists now believe that the planets and moons within the solar system all formed during this same 4.5- to 4.7-billion-year period.

The Moon

As indicated earlier, Earth's Moon plays a very important role in the Earth system. In particular, the Moon's gravitational field has a strong effect on ocean tides, which, in turn, influence important processes where the marine and terrestrial (land) environments meet (Chapter 9). These coastal processes are vital to the food web that supports the ecosystem of the oceans. Also important is how the Moon's gravity acts to minimize the amount of movement (i.e., wobble) in Earth's axis as the planet spins or rotates. Minimizing this wobble in the axis has reduced the seasonal extremes between summer and winter. This has produced a more stable climate system where complex life-forms have had the necessary time to evolve. Were it not for the Moon, it is quite likely that complex life-forms would not have been able to evolve on planet Earth.

Scientists have learned a great deal about the Moon in modern times from various space and ground-based studies. For instance, the false-color image shown in Figure 2.11 was created from data collected by the *Galileo* spacecraft as it headed toward Jupiter in 1992. The different colors in this image represent rocks of different chemical composition, which, in turn, reflect different terrain, namely the rugged lunar highlands and low-lying lunar seas, or maria. Most significant is the fact that when rocks from the *Apollo* moon landings were brought back to Earth, radiometric dating techniques proved that the age of the Moon was similar to that of the Earth, around 4.5 billion years old. As we will examine next, this was an important piece of information that helped scientists refine their hypotheses as to the origin of the Moon.

Origin of the Solar System

Recall from Chapter 1 that science operates by developing multiple hypotheses, each of which is capable of explaining the data (facts) that have been collected from various observations

FIGURE 2.11   False-color image of the Moon taken by the *Galileo* spacecraft. Image processing created a color scheme that reflects compositional variations in the rocks near the Moon's surface. Reddish colors generally correspond to older rocks that make up the more rugged lunar highlands, whereas the bluish colors represent the younger rocks making up the flat lava flows of the maria, or lunar seas.
NASA

and experiments. The testing of hypotheses through additional observations and experiments will cause some hypotheses to be refined and others thrown out. Ultimately this rigorous testing process leads to scientists having greater confidence in the hypotheses that survive. In this section we want to examine the hypothesis scientists have developed for explaining how the solar system formed.

The Nebular Hypothesis

In the previous section we learned that the Earth, Moon, and asteroids are all roughly the same age (around 4.6 billion years old), thus indicating a common origin. In addition to age data, there are several important observations that must be accounted for in any explanation of how the solar system formed. First, all of the planets revolve around the Sun in the same counterclockwise direction (as viewed from above) and have regular orbits that are nearly circular (Figure 2.4). Second, the Sun and most of the planets rotate (spin) about their axes in the same counterclockwise direction—Venus is the exception as it rotates clockwise. Nearly all of the moons in the solar system also spin on their axes and orbit their respective planets in a counterclockwise manner. Last, all the planets and their moons lie in what is called the *solar plane,* which is a plane that coincides with the Sun's equator.

Most astronomers agree that these data and observations are best explained by the **nebular hypothesis,** where all objects in the solar system formed from a rotating cloud of dust and gas called a *nebula*. Note that various nebular hypotheses were first proposed in the 1600s and have been refined over the years as new data were collected. The basic idea is that the solar system formed when an exploding star (supernova) disturbed a nebular cloud of dust and gas, composed primarily of hydrogen and helium along with smaller amounts of other elements. As illustrated in Figure 2.12, this disturbance from a supernova caused the nebula to begin collapsing in on itself due to the gravitational attraction between the dust and gas particles. As the nebula continued to contract, it also began to spin, causing it to flatten into the shape of a disc (Figure 2.12B). This contraction resulted in higher temperatures and pressures in the center of the nebula, whereas the spinning motion caused the outer portions of the disc to thin. Because the thinner portions of the disc were farther away from the hot central region, they cooled down first. Eventually the temperatures became low enough for liquids to condense and solids to crystallize. In the outer reaches of the disc, it became cold enough for liquid water, ammonia (NH_3), and methane (CH_4) to turn into ice.

Once the solid materials formed within the disc, gravitational attraction caused the individual particles to clump together into larger masses, a process called **accretion.** Eventually accretion created larger bodies called *planetesimals,* which tended to form in a plane within the swirling nebula (Figure 2.12C). During this period the temperature and pressure within the center of the nebula increased to the point where hydrogen atoms began to undergo nuclear fusion, and a new star was born.

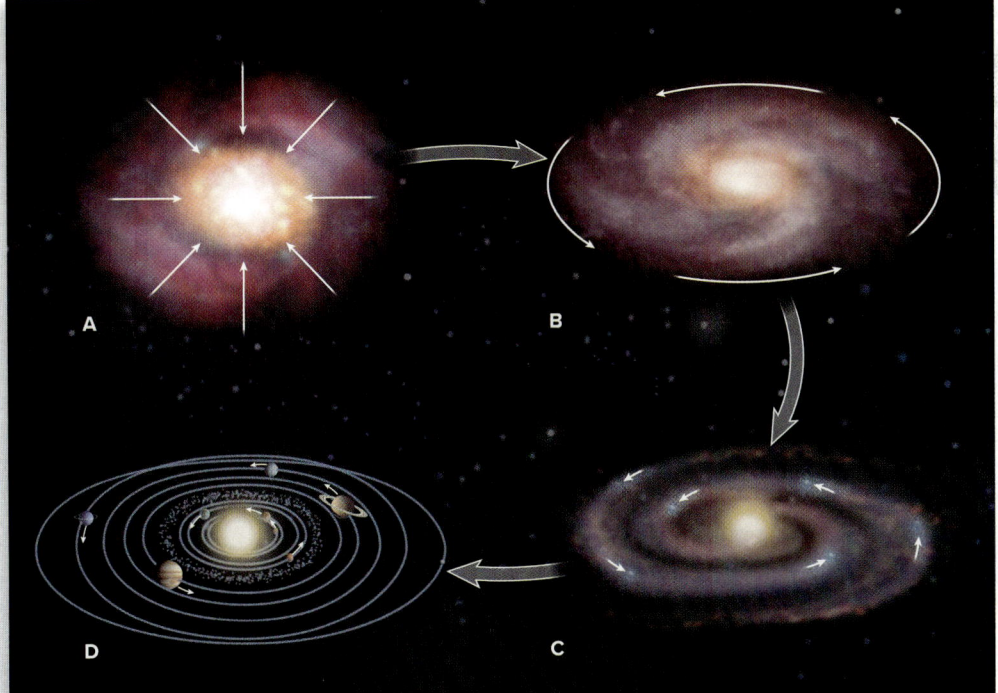

FIGURE 2.12   Illustration showing how the solar system evolved from a nebular cloud. Collapsing nebula (A) increases in density and begins to rotate. Continued collapse (B) results in nuclear fusion and the formation of a star, whereas the rotation forces the nebula to take on the shape of a disc. Planetesimals develop by accretion of particles (C), while solar radiation drives off remaining parts of the nebula. When the debris is cleared, what is left are planets (D) that lie in the same plane around the Sun and revolve and rotate in the same counterclockwise manner.

FIGURE 2.13 Artist's illustration of what a young planetary system might look like as planets clear their orbits of debris. Electromagnetic radiation and streams of charged particles emitted from the newly formed star act to clear the remaining dust and gas from the system.
© David Hardy/www.astroart.org/STFC

FIGURE 2.14 The leading hypothesis for the origin of the Moon involves a Mars-sized impact (A) early in Earth's history. Some of the ejected material went into orbit around the Earth where it underwent accretion (B), forming the Moon. Such a giant impact is also believed to have caused the young Earth to melt.

As electromagnetic radiation and gases began flowing outward from the Sun, the ice and lighter elements that had collected on solid objects within the nebula began to either evaporate or melt. Gradually the innermost planetesimals lost their lighter and more volatile constituents, which ended up recondensing back into solids farther out in the new solar system.

Around the time the Sun was born, the planetesimals had become large enough that their gravitational fields started attracting smaller bodies from nearby orbits. During this early stage of accretion the planetesimals grew very rapidly, with some eventually becoming **planets,** whose gravity dominates their individual orbital zones (Figure 2.12D). While the planets were clearing their orbits of debris, it is believed that the frequency of impacts was quite high. Moreover, some of the bodies that the young planets were attracting must have been rather large, resulting in tremendous impacts. As time progressed, the solar system was largely swept clear of debris, causing a dramatic decrease in the frequency and size of impacts. Figure 2.13 is an artist's illustration showing what the early solar system probably looked like during the heavy bombardment period when the planets were clearing out their orbits. Note how the leftover dust and gas is being cleared from the inner solar system by electromagnetic radiation and charged particles streaming outward from the newly formed Sun.

Finally, a key aspect of the nebular hypothesis is how it may also explain the origin of Earth's Moon. The current leading hypothesis is that the Moon formed early during the heavy bombardment period when the Earth experienced a Mars-sized impact as it was clearing its orbit. This proposed impact, illustrated in Figure 2.14, was so large that Earth remelted, and huge amounts of debris were ejected out into space. Some of this debris fell back onto the Earth, but much of it went into orbit around the Earth, where it underwent accretion

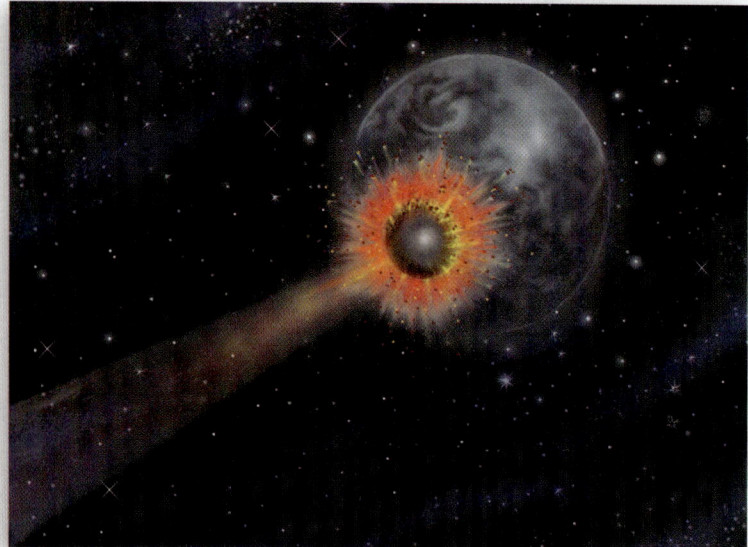

A

B

and eventually formed the Moon. This hypothesis not only explains the nearly identical age of the Moon and Earth, but also why the Earth has such a disproportionally large moon compared to other planets in the solar system. Most significant is that this giant impact created a moon that was large enough to have affected Earth's climate and ocean tides, thereby helping to produce an ideal environment for the evolution of complex life.

How Reliable Is the Nebular Hypothesis?

At this point we should take a look at how well the nebular hypothesis explains the known facts about the solar system. As noted earlier, this hypothesis explains why nearly all the planets and moons revolve (orbit) and rotate (spin) around the Sun in the same counterclockwise manner. These bodies also all lie within a plane that coincides with the Sun's equator. For those planets and moons that do not fit this general pattern, their deviation can be accounted for by giant impacts early in the solar system history—similar to that proposed for the origin of the Moon. For example, Venus rotates about its axis very slowly and in the opposite (clockwise) direction from the other planets. The best explanation is that Venus experienced a large, glancing impact such that it actually reversed its direction of spin. Similarly, a massive impact could explain why Uranus rotates on its side rather than in an upright position. Several moons in the outer solar system actually orbit their planets in the opposite (clockwise) sense. This can only be explained if the moons formed elsewhere and were then captured by the gravity of the planets they now orbit.

Direct evidence for the nebular hypothesis and the intense bombardment associated with accretion is the heavily cratered surfaces we see today on some of the planets and moons. Examples include Mercury and Earth's Moon (Figure 2.15). By analyzing the landforms and density of the craters, scientists have been able to show that the majority of the impacts occurred early in the solar system's history. This of course supports the accretion concept. Further evidence is that Venus and Earth have relatively few craters, which is consistent with the fact that weathering and erosion processes on these planets would have long ago erased most of the original impact craters. Finally, radiometric dating has shown that the Earth, Moon, and asteroids all solidified around the same time.

Perhaps the most dramatic evidence for the nebular hypothesis comes from recent data astronomers have gathered using powerful space and ground-based telescopes. For example, over 3,000 planets have been confirmed to be orbiting other stars. In some cases discs of dust have been observed around stars, which likely represent the early stage of planetary development when planetesimals begin to form. In addition to proving the existence of other planets, the Hubble telescope has provided us with vivid images of new stars being born in towering clouds of dust and gas (Figure 2.16). This seems to indicate that most stars form in a similar manner to our Sun and that planetary systems are a somewhat common occurrence. It is also quite possible that there are planets similar to Earth,

A   B

FIGURE 2.15 Heavily cratered surfaces of Mercury (A) and the Moon (B). A lack of an atmosphere along with weathering and erosion processes on these bodies has preserved the record of the heavy bombardment that took place early in the solar system history.
(a) NASA/Johns Hopkins University Applied Physics Laboratory/Carnegie Institution of Washington; (b) NASA

FIGURE 2.16 Photo from the Hubble space telescope showing new stars forming in the gas and dust clouds of the Eagle Nebula.
© Brand X/PunchStock RF

Gas and dust: The 6-trillion-mile-high fingers, made of dust and hydrogen gas, turn blue when they are hit by ultraviolet radiation

Star eggs: Clumps of hydrogen inside the pillars called Evaporating Gaseous Globules, or EGGs, eventually hatch into stars

FIGURE 2.17   View looking down toward the central core of a clockwise-rotating galaxy. Our Sun lies on the outer band of the Milky Way Galaxy, which is similar to the spiral galaxy shown here. Such galaxies are estimated to contain hundreds of billions of stars.
European Southern Observatory

where conditions are favorable for life, ranging from bacteria to highly evolved and intelligent life-forms. If this is indeed the case, then it brings up the interesting question of how many Earth-like planets there might be in the universe. In the next section we will examine this issue by getting a better feel for the number of stars in the universe.

Other Stars in the Universe

To anyone who has gazed into the night sky, there obviously are many stars in the universe. Less obvious is the fact that the stars are not distributed uniformly throughout the universe, but rather are found in large groupings called **galaxies.** Figure 2.17 shows an example of what astronomers call a *spiral galaxy.* Our Sun lies within a spiral galaxy known as the Milky Way Galaxy, which is estimated to contain 200 to 400 billion stars. Because numbers this large are difficult to comprehend, it is helpful if we use an analogy. If you counted stars at a rate of one per second, it would take 32 years to count to 1 billion, and 3,200 years to reach 100 billion. Also interesting is that parts of the Milky Way can be seen with the naked eye on dark and clear nights. As shown in Figure 2.18, this spinning disc of stars appears as a cloudy band that arcs across the night sky. Because our solar system is located out on a spiral arm of the Milky Way, what we are seeing is an edge-on view looking toward the center of the galaxy. Another interesting aspect of the Milky Way is that it takes light approximately 100,000 years to travel across the galaxy, where a *single* light-year equals 5,870 billion miles (9,450 billion km). To help grasp what this means in terms of size, consider that the average distance between the Earth and Sun is a mere 93 million miles (150 million km). Equally amazing is the fact that it takes our Sun about 230 million years to make one full revolution around this spinning disc. The Milky Way, therefore, dwarfs the Earth and our solar system on a scale that is almost beyond comprehension.

Although it may be difficult for us to comprehend the scale of our galaxy, it is even harder to grasp the size of the universe. For example, in Figure 2.18 you can see what appear to be stars in the parts of the sky away from the Milky Way. The inset image shows that when astronomers focus the powerful Hubble telescope on this background, what looked like stars with the naked eye turn out to be galaxies. Because this view into deep space represents such a tiny fraction of the night sky, it tells us that there must be a very large number of galaxies in the universe. Another important point is that the galaxies, as seen by the Hubble telescope, are so far away that the light we see has been traveling for more than 10 billion years. In fact, the light is so old that many of the stars in these galaxies have burned out and no longer exist. This means that we are actually looking backward in time and seeing the galaxies as they existed 10 billion years ago. Astronomers believe that what we are seeing is the state of the galaxies shortly after the origin of the universe itself.

The scientific explanation astronomers and physicists have developed to explain the origin of the universe is called the **big bang theory.** This theory, first proposed in 1927 by a Belgian priest, states that all matter in the universe had at one time existed at a single point. Approximately 14 billion years ago, this matter then began to expand outward in all directions, and has been expanding ever since. Supporting evidence for the big bang theory came in 1929 when the

FIGURE 2.18 As seen from Earth, the Milky Way Galaxy appears as a cloudy band of stars across the night sky. The inset shows a deep view of space taken by the Hubble space telescope, where what appeared as stars with the naked eye are actually distant galaxies of various shapes and colors. This view represents a very small portion of the night sky, about the width of a dime located 75 feet away. Chris Angelos and Richard Nolthenius; (inset) Courtesy of STSCI

astronomer Edwin Hubble, for whom the Hubble telescope was named, proved that all the galaxies are moving away from one another. Logic tells us that if the galaxies are continually moving outward, similar to the surface of an expanding balloon, then if we go backward in time they must have been closer together. Figure 2.19, which shows a slice of the universe, illustrates how the galaxies have been expanding outward from a central point since the big bang. Note that the Hubble and other telescopes view distant objects as they once existed, thus we are able to look backward in time.

Additional evidence for the big bang theory came in 1964 when two astronomers inadvertently discovered microwave (electromagnetic) radiation coming from deep space. They found it odd that this radiation was not coming from a single source, such as a star, but rather from wherever they pointed their instrument into the farthest reaches of space. What these scientists had discovered was the electromagnetic background radiation that formed during the initial expansion of the universe (Figure 2.19). In simple terms, they found direct evidence for the big bang itself. The discovery of this background radiation not only led to a Nobel Prize, but caused the big bang theory to gain widespread acceptance within the scientific community.

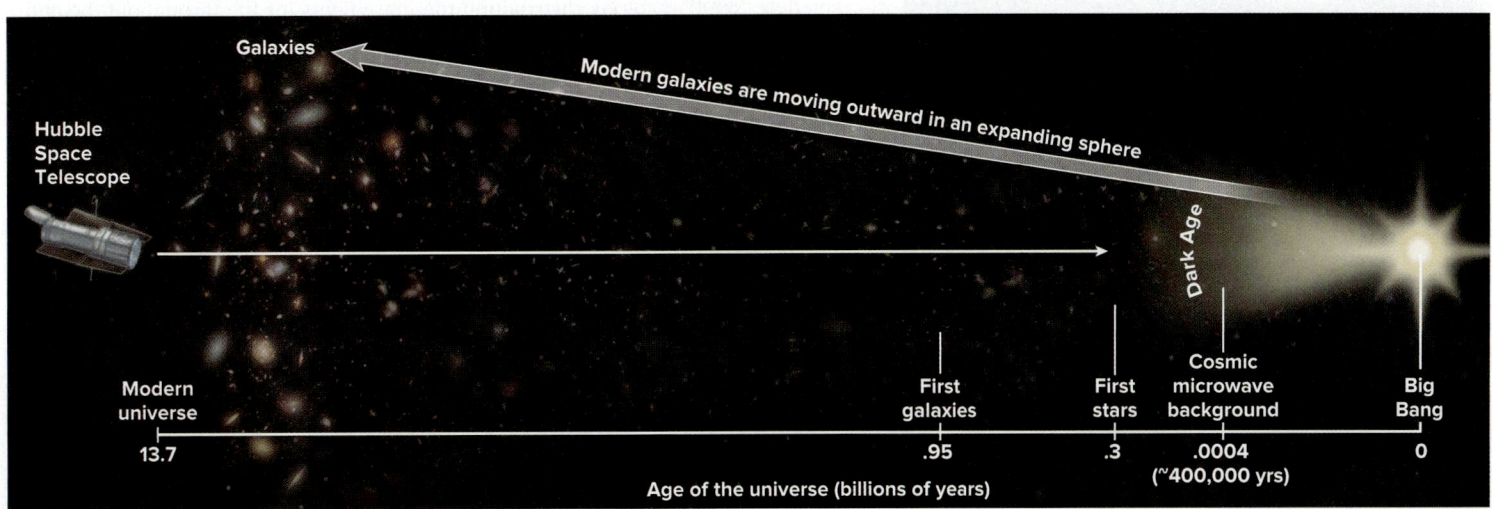

FIGURE 2.19 Conceptual diagram illustrating the big bang and the origin of the universe. Galaxies we see today have been rushing outward as an expanding sphere. The Hubble and other powerful telescopes are able to observe distant features as they appeared nearly 13 billion years ago.

Does Life Exist Beyond Earth?

Perhaps the greatest mystery in all human history is the question of how life began here on Earth. Also fascinating is whether life exists elsewhere (i.e., extraterrestrial life), and if it does, how common it is in our galaxy and the universe. Although this topic is beyond the scope of this book, we will take a brief look at it so you can gain a better appreciation of why Earth's environment is so special. From this discussion we can also learn how Earth's geology has played an important role in creating our environment and its incredible array of life. We will begin by briefly examining what scientists have learned about the origin of life on Earth and what may exist beyond our planet.

Life on Earth

While studying some of Earth's oldest surviving rocks, paleontologists have discovered evidence of bacteria in rocks as old as 3.6 billion years. Such ancient bacteria prove that life began very early in Earth's 4.6-billion-year history. Moreover, life began at a time when the planet's atmosphere and climate were considerably different from today. Earth's environment, in fact, was so different that it would have been inhospitable for most of today's life-forms. Interestingly, biologists today are finding what are called **extremophile bacteria,** which thrive under extreme conditions that would be lethal to nearly all of Earth's other life-forms. Extremophile bacteria are commonly found in places such as ancient Antarctic ice, superhot vents on the seafloor, and rocks located deep underground. Based on evolutionary changes recorded in the geologic rock record (Chapter 1), most biologists now believe that the complex plant and animal life we see today evolved from extremophile bacteria. One possibility are the extremophile bacteria that thrive in the oceans along hot water vents associated with volcanic activity. Long periods of geologic time would have led to progressively more complex organisms and photosynthesis, ultimately creating an oxygen-rich atmosphere and the diverse animal and plant life we see today.

The question of how nonliving material actually developed into the first primitive forms of bacteria (i.e., the origin of life itself) will likely remain a mystery. Recall that scientists now understand that Earth and all the other planets and their moons underwent an intense bombardment of asteroids and comets early in our solar system's history. Of particular interest are comets because they are known to be composed mostly of water and organic (i.e., carbon-based) compounds, both of which are critical ingredients for life as we know it. It is quite possible that comets acted as "seed" material, distributing the ingredients for life throughout the entire solar system. Life therefore could have developed in those places with favorable conditions. An analogy would be if one were to take grass seed and spread it over a wide area. While the seeds might germinate in a variety of places, they would truly flourish only in those areas with the right combination of soil, water, and nutrients. In terms of our solar system, Earth clearly presented the most favorable conditions for life to thrive and evolve into complex plants and animals.

Habitable Zones

The only life-forms we know of today are those found on Earth, and all of them require the presence of liquid water (H_2O). This is why many scientists believe the key to finding life beyond planet Earth is to first find liquid water. Consequently, the term **habitable zone** has been defined as that relatively narrow zone around a star where orbiting planets would have surface temperatures that would allow liquid water to exist. Figure 2.20 illustrates the concept of a habitable zone around a star. For planets located too close to their star, water near the surface would vaporize and be lost, which is what scientists think happened on Mercury and Venus. On the other hand, water will remain frozen on planets located too far from their star. Earth, of course, is ideally located within the Sun's habitable zone.

A

B

FIGURE 2.20 Habitable zones are those regions of space where conditions are believed to be most favorable for the development of life. Such zones can be defined in terms of areas where liquid water can exist around individual stars (A), and also areas within a galaxy (B) where there is an abundance of heavier elements but fewer cosmic hazards.
(b) © Brand X/PunchStock RF

Note however that the position of the habitable zone will vary depending on the size and energy output of a given star. This means that when the energy output and size of a star change as it goes through its natural life cycle, the position of the habitable zone will shift accordingly. As our Sun evolves and slowly grows in size over the next several billion years, its habitable zone will move out beyond Earth's orbit. Life on planet Earth then will eventually cease to exist. Finally, note in Figure 2.20 that a habitable zone can also be defined for an entire galaxy. The centers of galaxies are considered less favorable locations for life because they contain more hazards (exploding stars, intense radiation, black holes, etc.). Also less favorable are the extreme outer parts of galaxies where there are fewer stars capable of producing the heavier elements necessary for life.

One problem with the concept of habitable zones is the assumption that all life-forms require liquid water as a solvent (i.e., a fluid capable of dissolving different substances). It is certainly possible that life-forms exist in the universe that are based on some solvent other than water, such as ammonia (NH_3), sulfuric acid (H_2SO_4), or an organic solvent like formamide (CH_3NO). Another problem is that life, particularly extremophile bacteria, could exist out beyond a habitable zone as long as a planet or moon has an internal heat source similar to the Earth (Chapter 4). Remember, comets probably seeded the entire solar system with water and organic compounds believed to be essential for life. Thus, if a moon or planet had an internal heat source, then it could contain liquid water and life far beyond the solar system's main habitable zone. After all, life on our planet may have originated on the seafloor around volcanic vents. Such volcanic activity is powered not by the Sun, but by Earth's internal heat. The geology of a planet or moon then may play an important role in determining whether life develops.

Many scientists believe that extraterrestrial life is most likely to be found in the form of bacteria associated with liquid water. One promising place is Saturn's moon Enceladus, which is covered with water ice (Figure 2.21). In 2005 scientists with NASA's *Cassini* mission first detected geysers of water vapor, ice, salts, and organic material erupting from Enceladus. Then, in 2012, scientists using the Hubble telescope observed plumes of water vapor blasting into space from Jupiter's moon Europa (Figure 2.22). It is hypothesized that because Jupiter and Saturn are large and have strong gravitational fields, tidal forces alternately push and pull on their moons' rocky interiors. As the moons orbit their respective planets, this flexing generates frictional heat that has caused liquid oceans to form beneath the ice-covered surfaces. These oceans could contain extremophile bacteria and other life-forms despite being far outside of our Sun's habitable zone.

NASA is currently developing plans to send a probe to collect samples directly from Europa's water plumes and look for signs of life. Another likely candidate for extraterrestrial life is Mars. Since 2004, NASA robots have explored the surface of Mars and have identified rocks that were deposited in water. Future missions will search for telltale visual and chemical clues of life in these rocks (Case Study 2.1).

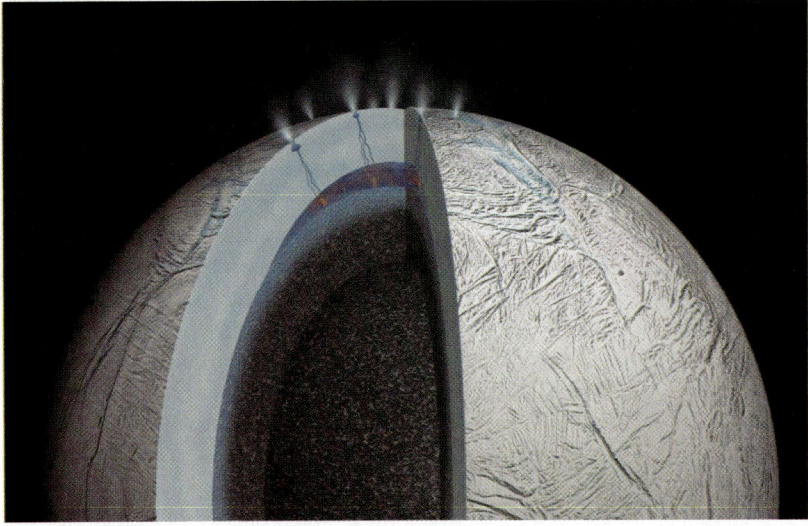

FIGURE 2.21   Cutaway view of Saturn's moon Enceladus showing plumes of water, dissolved salts, and organic material being blasted into space. These plumes indicate the presence not only of a liquid ocean between the moon's rocky interior and ice-covered surface, but also of possible hydrothermal vents that provide energy and nutrients for extremophile bacteria and other life-forms. NASA/JPL-Caltech

FIGURE 2.22   Jupiter's moon Europa is also covered with water ice and has geysers ejecting water into space. Its icy surface consists of broken slabs with grooves and ridges (inset), indicating a liquid ocean between this highly dynamic surface and the moon's rocky interior. Because of Europa's internal heat, this ocean could contain primitive bacteria and other life-forms. Galileo Project/NASA; (inset) NASA

Searching for Life on Mars

One of humanity's most intriguing questions is how life developed on Earth. This naturally leads us to ask whether the biology that governs life here operates elsewhere. In other words, does life exist beyond Earth? Answering these questions is important not just because we are curious about our own existence, but because it helps us better understand the Earth system and its ability to support complex life.

Since liquid water is a powerful solvent that transports dissolved nutrients needed for Earth's living organisms, NASA's initial strategy has been to find where water exists in the solar system. Although liquid water has been found on the moons of Jupiter and Saturn, the focus historically has been on Mars, which is much closer and more Earth-like. The first solid evidence of flowing water came in 1971 when a NASA satellite orbiting Mars sent back photos showing numerous stream channels on the planet's surface. Some of the channels originate in ancient impact craters, which indicates that the craters were once filled with water. After much research, scientists now believe that Mars was much warmer and wetter 4.5 to 3.7 billion years ago, resulting in a liquid ocean that covered nearly 20% of the planet's surface (Figure B2.1). Although most of the original water is thought to have escaped into space as Mars's atmosphere began to thin, new data indicate that liquid water remains stored in rocks below the surface. Because water is critical to life as we know it, Martian rocks could contain evidence of primitive bacteria. Consequently, the

United States and other nations have sent spacecraft to study Mars more carefully, with the initial goal of verifying that some of Mars's rocks had been deposited in a water-rich environment. Future missions then would examine those rocks for signs of life.

In 1997 NASA put a more sophisticated satellite into orbit around Mars in order to create detailed surface maps. The mapping revealed tantalizing evidence of layered rocks that were likely deposited in liquid water. To verify this hypothesis, NASA landed two spacecraft and examined the rocks directly. One of the landing sites was a large impact crater with a clearly defined stream channel leading away from the crater. The rocks at the other landing site were believed to contain large amounts of the iron oxide mineral known as hematite (Fe_2O_3). On Earth, this particular form of hematite is typically associated with significant quantities of water. Then, in 2003, NASA landed robots, called rovers, at each site. The rovers were highly mobile and equipped with instruments for taking photos and determining the composition of minerals making up the rocks and sediment. The rover at the crater site discovered significant quantities of magnesium sulfate ($MgSO_4$). This mineral is actually a salt composed of magnesium (Mg^{2+}) and sulfate (SO_4^{2-}) ions, which often combine on Earth to form a solid when saline water undergoes evaporation. Similar types of so-called evaporite minerals were found at the other landing site, along with numerous nodules the size of BBs. Since the nodules were composed of hematite (Fe_2O_3), scientists concluded that they formed similarly to ones on Earth, namely by growing in concentric layers over time as iron precipitated out of mineral-rich water. Based on this and other evidence, it was concluded that the rocks were deposited in a salty sea that extended for miles in every direction.

In 2012 NASA sent a new, more sophisticated rover called *Curiosity* to Mars. As shown in Figure B2.2, *Curiosity* landed at the base of Mount Sharp, located in what is known as Gale Crater. Along the bottom edge of the crater the rover discovered sedimentary rocks that formed from wind-blown sand as well as much coarser-grained sedimentary rocks. The coarse-grained rocks (Figure B2.3) are similar to those found on Earth and could only have been deposited by running water, due to the large size of the grains. On Mount Sharp itself, *Curiosity* discovered rocks with cross-bedding (Figure B2.4), which forms when either wind or moving water causes sediment layers to become oriented at an angle to the main layers. However, the cross-bedding found in these rocks is indicative of sediment that was carried into a lake by flowing water. Scientists have concluded that Mount Sharp itself consists of sedimentary rocks that were deposited in an ancient lake, whereas the surrounding rocks were mostly deposited on land, and many of them have subsequently been removed by erosion. In addition to the evidence for water, *Curiosity* has discovered nitrogen compounds along with organic molecules. This means that during its ancient wet period, Mars basically had the same ingredients necessary for life as did Earth. Because geologists find evidence of bacterial life in Earth rocks as old as 3.5 billion years, it is quite possible that we may find similar evidence on Mars.

Building on the success of its *Curiosity* mission, NASA has announced plans to launch a more advanced rover to Mars in 2020.

FIGURE B2.1 Data indicate that Mars was much warmer and wetter 4.5 to 3.7 billion years ago, resulting in an ancient ocean that covered nearly 20% of the planet's surface. Many of the impact craters were also filled with water.

NASA/NOAA/GSFC/Suomi NPP/VIIRS/Norman Kuring

A

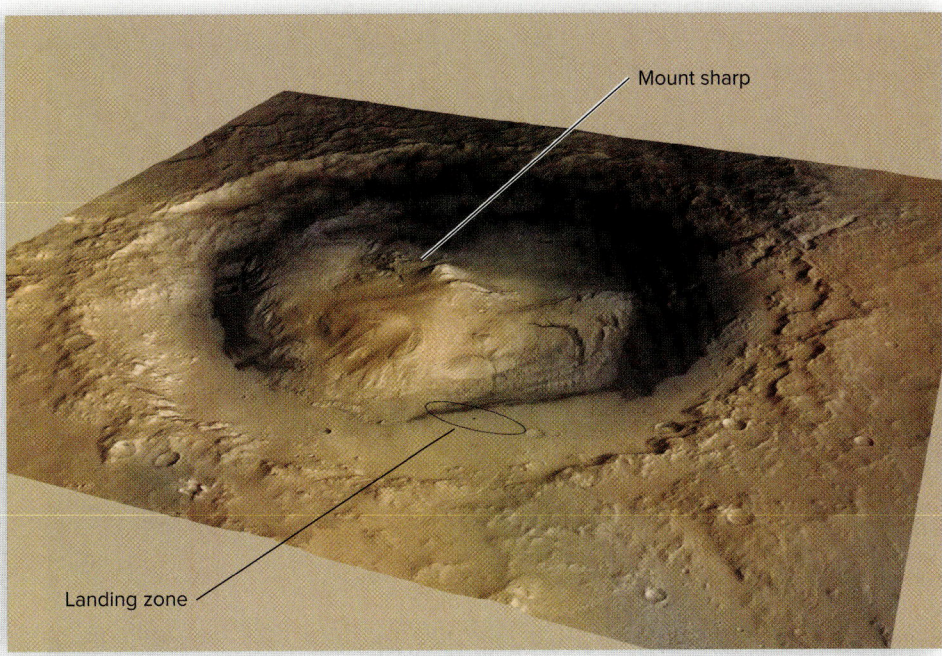

Mount sharp

Landing zone

B

FIGURE B2.2 In 2012 NASA's robotic rover named *Curiosity* (A) landed at the base of Mount Sharp in Gale Crater (B)—whose diameter is 96 miles (155 km). Since landing, the rover has sampled rocks and sediment on and around Mount Sharp. Results show that the mountain represents sedimentary rocks that were deposited in a former lake. The surrounding rocks were mostly deposited on land, and many have subsequently been removed by erosion.
(a) NASA/JPL-Caltech/MSSS; (b) NASA/JPL-Caltech/ESA/DLR/FU Berlin/MSSS

1 cm
Mars Earth 1 cm

FIGURE B2.3 Coarse-grained sedimentary rock from Gale Crater (left) with a similar rock found on Earth (right). Based on the fact that the gravel fragments are rounded and too large to be transported by wind, geologists conclude that the fragments were carried by flowing water.
NASA/JPL-Caltech/MSSS and PSI

centimeters
0   5   10   15   20   25   30

FIGURE B2.4 Sedimentary rock from Gale Crater showing cross-bedding, which forms when wind or moving water causes sediment layers to become oriented at an angle. By analyzing the way the layers accumulated here, geologists can tell that the sediment was transported by running water as it entered a lake.
NASA/JPL-Caltech/MSSS

One of the key objectives of this mission will be to search for *bio-signatures,* which are telltale visual or chemical signs of life, both past and present. Should such evidence be found, it would strongly support the hypothesis that primitive life-forms are relatively common throughout our galaxy and universe. This discovery would also increase the likelihood that intelligent life exists someplace other than Earth. Perhaps most important, such a discovery might help humans better appreciate how rare and fragile life here on Earth is, and in turn encourage better stewardship of our own environment.

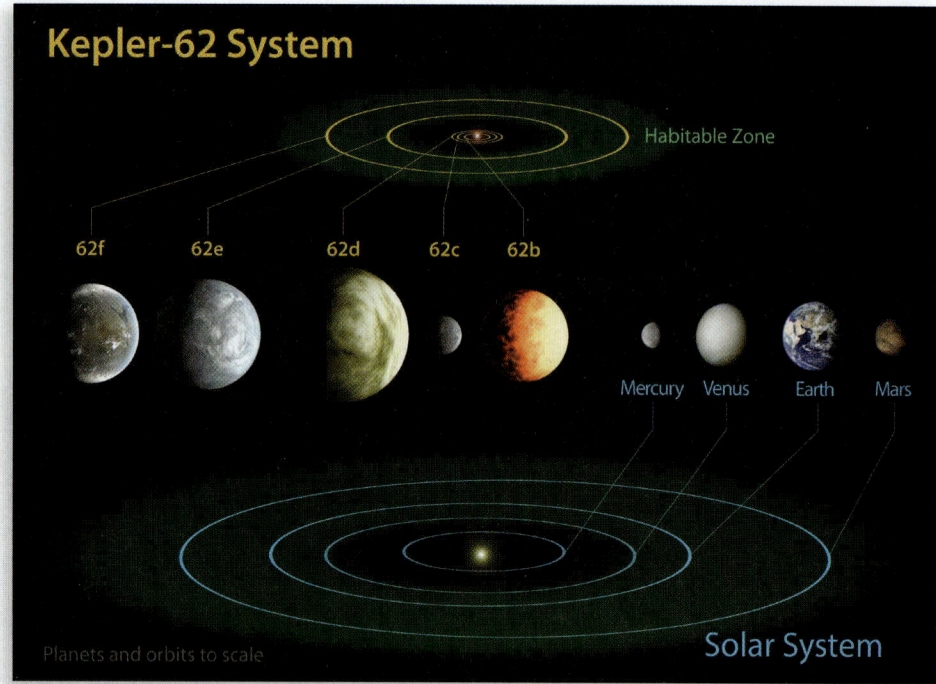

Kepler-62 System

Habitable Zone

62f   62e   62d   62c   62b

Mercury   Venus   Earth   Mars

Planets and orbits to scale

Solar System

FIGURE 2.23   Illustration comparing the five-planet system known as Kepler-62 to the planets in our inner solar system. Two of the planets, 62f and 62e, orbit within the habitable zone of the system's star, which is two-thirds the size of our Sun and only one-fifth as bright. Note that planet 62f completes its orbit every 267 days and is only 40% larger than Earth.
NASA Ames/JPL-Caltech

Possible Intelligent Life

Discovering evidence of life beyond the Earth would certainly be a historic achievement, but even greater would be finding evidence of intelligent life. It is commonly thought that intelligent life would be restricted to the main habitable zone around a star. This is based on the assumption that more complex life-forms not only require liquid water, but also land-masses on which to live. The idea is that intelligent life would evolve as it did on Earth, namely from marine organisms that eventually adapted to conditions on land. Such land-based or terrestrial life would need liquid water on the surface, which, in turn, would require the planet to lie within the star's main habitable zone. Thus, most scientists believe that any search for intelligent life should focus on stars with planets orbiting within their respective habitable zones.

Although we do not know if intelligent life exists elsewhere, scientists can get some idea of its likelihood by estimating the number of habitable planets. In 2009 NASA launched the *Kepler* mission and placed a space telescope into Earth orbit to obtain a clearer and more continuous view of space. Unlike the orbiting Hubble telescope, which was designed to look into deep space, the *Kepler* system uses a very high resolution digital camera to survey a relatively small area containing just 150,000 stars in our Milky Way galaxy. As illustrated in Figure 2.23, the goal was to document the existence of other planetary systems and estimate the number of Earth-sized planets within habitable zones. As of 2016, the Kepler telescope had identified more than 4700 possible planets, 2300 of which were verified. Of all the possible planets, over 1300 were found to be similar in size to planet Earth. Most important is the discovery that 297 possible planets are orbiting within the habitable zones of their respective stars. From these data, astronomers have determined that about 20% of the stars similar to our Sun, or slighter smaller, contain Earth-like planets in habitable zones.

Considering that there are between 200 and 400 billion stars in the Milky Way, astronomers now estimate that there about 40 billion Earth-like planets capable of supporting life in our galaxy. Keep in mind that this is just for the Milky Way, and there are approximately 100 to 200 billion galaxies in the universe! If one assumes that each galaxy averages several hundred billion stars, then there should be something on the order of 100,000 billion billion or 10^{23} stars in the universe—a number so large it is nearly impossible to comprehend. Based on the *Kepler* data, astronomers now estimate that there are about 4,000 billion habitable planets in the universe.

In recent years an idea called the **rare earth hypothesis** has been proposed, which contends that life is relatively common throughout the universe, but complex animal life similar to Earth's is likely to be exceedingly rare. The basic premise, as described in the book *Rare Earth* by paleontologist Peter Ward and astrobiologist Donald Brownlee, is that an Earth-like planet needs to maintain a stable environment over a long period of geologic time so that complex life can evolve from more primitive forms. Below are some of the factors these scientists believe to have been critical in helping Earth maintain a stable environment so that animal life, including humans, could evolve.

1. Energy output from the Sun has remained fairly steady, allowing Earth to have a more stable climate.
2. Earth's internal heat and plate tectonics (Chapter 4) help regulate the amount of carbon dioxide (CO_2), which traps heat and warms the atmosphere.

By controlling CO_2 levels, these processes act as a thermostat to regulate atmospheric temperature (Chapter 16).

3. Jupiter's large size has helped clear asteroids and comets from the region near Earth's orbit, thereby reducing the number of large, catastrophic impacts that would alter the global climate.

4. The Moon's gravity has reduced the wobble in Earth's axis, which has provided Earth with a more stable climate system.

Obviously, no one knows how many Earth-like planets with complex animal life actually exist, so it is possible we are indeed alone. Although the distances and potential number of planets make it impractical for us to explore them directly, we can scan the stars of our own galaxy, listening for possible radio signals from highly advanced civilizations. The assumption is that other advanced civilizations would be broadcasting radio waves into space in the same manner humans have been doing since the invention of the radio. Starting in the 1960s, various groups have been listening for such extraterrestrial signals using radio telescopes. Although NASA started a program in 1993, it ended less than a year later when Congress eliminated its funding. Today a privately funded organization known as the SETI (Search for Extra-Terrestrial Intelligence) Institute is actively searching within our galaxy.

In the end, it seems reasonable to conclude that Earth-like planets in the habitable zone of stars are fairly common, but those with complex and intelligent life are very rare. Due to the sheer size of the universe, we may never learn whether intelligent life exists elsewhere. While Earth may be just a small planet orbiting an ordinary star in the vastness of space, it is our one and only home. When we view Earth from the perspective of space, as shown in Figure 2.24, we can get a better sense that our environment is not only very special, but quite fragile. Despite all our efforts at trying to understand the universe and our place in it, what really counts is how well we take care of our home and the environment that makes our lives possible.

FIGURE 2.24   Earth in the vastness of space. Image taken by *Apollo 8* astronauts on their way to the Moon.
NASA

Solar System Hazards

The Earth is often thought of as a self-contained system composed of the atmosphere, hydrosphere, biosphere, and geosphere (Chapter 1). However, as noted earlier, our planet operates within a cosmic environment that has a major influence on the Earth system. For example, recall how the Earth depends on solar radiation and gravitational forces within the solar system. We can think of the Earth then as a system that interacts with even larger systems, namely the solar system and our own galaxy. In this section we will briefly examine some natural hazards that originate outside of the Earth system. Our focus will be on electromagnetic radiation and the impact of comets and asteroids. Interestingly, these extraterrestrial hazards appear to have played an important role in the evolution of life on Earth.

Electromagnetic Radiation

Although electromagnetic radiation streaming outward from the Sun is the critical energy source that drives Earth's biosphere and climate system, it can also pose a hazard to living organisms. Of particular concern is radiation in the

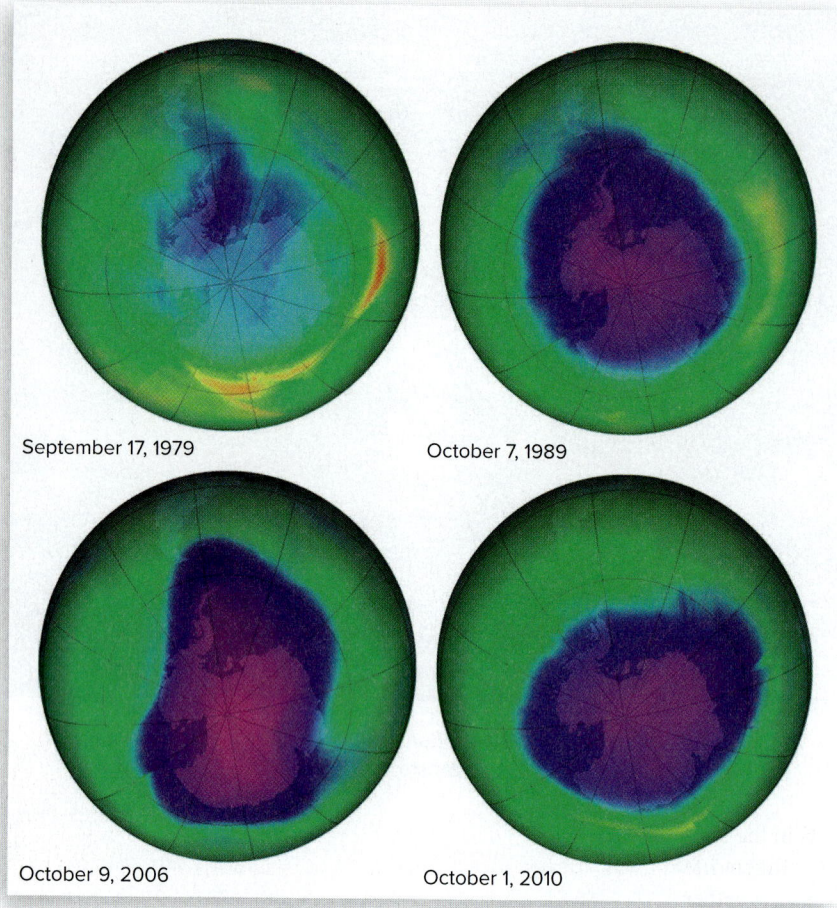

September 17, 1979

October 7, 1989

October 9, 2006

October 1, 2010

FIGURE 2.25   Satellite maps showing the concentration of ozone (O_3) in the upper atmosphere—blue indicates areas where concentrations are the lowest and green the highest. By the 1980s society's use of ozone-destroying gases had caused a hole to develop in the ozone layer over the Antarctic each winter. A 1987 international treaty phased out the use of these gases, causing ozone levels to stabilize. NASA Ozone Hole Watch

ultraviolet (UV) portion of the spectrum (see Figure 2.7) as these higher-energy wavelengths can damage the cell tissue of carbon-based life-forms. Fortunately, Earth's upper atmosphere contains a thin layer with a relatively high concentration of a type of oxygen molecule called *ozone* (O_3), which naturally absorbs much of the incoming UV radiation—the oxygen molecules we breathe are composed of two oxygen atoms (O_2). This thin layer rich in ozone molecules, called the *ozone layer,* acts as a protective shield for Earth's biosphere. In fact, the ozone layer has been shielding Earth's biosphere from UV rays for billions of years, making it possible for carbon-based organisms to evolve into the diverse life-forms we see today.

In the 1930s humans began using chlorine and fluorine-based gases, called *chlorofluorocarbons,* as a coolant for refrigeration and air conditioning systems. Chlorofluorocarbons, commonly called CFCs, also became quite popular in various commercial applications, such as propellants in cans of spray paint and hair spray. Then, in 1974, scientists discovered that ozone molecules (O_3) chemically break down into free oxygen (O_2) in the presence of CFCs and UV radiation. They warned that CFCs released from human activity would slowly make their way to the upper atmosphere and cause the ozone layer to become dangerously thin, a problem referred to as **ozone depletion.** It was not until 1985 that researchers from the British Antarctic Survey proved that ozone levels were actually declining. Additional research showed that the combination of air currents and cold temperatures over the Antarctic produced the most severe ozone depletion, as much as 60% during the spring. This annual thinning has become known as the *ozone hole* (Figure 2.25). It was also discovered that while ozone depletion is most severe over the polar regions, it is actually a global problem. Over the United States, for example, ozone levels had fallen as much as 5–10%.

Ozone depletion resulting from human activity was soon recognized as a serious health threat to people and the entire biosphere. Excess exposure to UV radiation in humans, for example, is known to cause skin cancer and eye cataracts. Left unchecked, ozone depletion could impact the entire food web within the biosphere. Because this threat was serious and global in nature, an international agreement was reached in 1987, called the *Montreal Protocol,* in which nations agreed to phase out the production and use of CFCs. As a result of this agreement, ozone concentrations in the atmosphere stabilized by 2000, and today appear to be slowly increasing. Scientists have projected that global ozone levels will return to 1980 levels by 2050, whereas it will take until 2070 for the hole over the Antarctic to fully recover.

Another radiation hazard that originates in space is known as a **gamma-ray burst,** which is a short-lived burst of very high-energy waves from the gamma-ray portion of the electromagnetic spectrum (see Figure 2.7). Gamma-ray bursts were first detected in the 1960s by orbiting U.S. spy satellites. More detailed studies have found that bursts average about one per day, each lasting on the order of fractions of a second to several minutes. Astronomers now know that at least some gamma-ray bursts originate in distant galaxies from stars that explode violently as they reach the end of their life cycle. Gamma rays are more hazardous to living organisms than ultraviolet (UV) rays because they possess much more energy. The good news is that Earth's ozone layer absorbs both gamma and UV rays. Unlike UV rays, however, gamma rays by themselves can destroy ozone molecules. Fortunately, gamma rays lose considerable amounts of energy during their long journey from distant galaxies.

FIGURE 2.26 Illustration showing how a burst of gamma rays from a nearby exploding star could destroy much of Earth's ozone layer. This would allow the biosphere to be exposed to intense ultraviolet (UV) radiation, causing many species to die. The food chain could then begin to collapse, leading to a mass extinction. NASA

This has allowed the production of ozone by natural Earth processes to keep pace with the steady rate of ozone loss. Earth therefore has been able to maintain a protective ozone layer for eons of geologic time.

In recent years scientists have considered the frightening possibility of a gamma-ray burst originating not in some distant galaxy, but from a nearby star in our own galaxy. Such a scenario is illustrated in Figure 2.26. Based on modeling results, astronomers estimate that if a gamma-ray burst were to originate from a star 6,000 light-years away and last just 10 seconds, half of Earth's ozone would be destroyed. The greatly thinned ozone layer would then immediately allow a steady stream of UV rays to pass through the atmosphere and strike the surface. The radiation would be expected to kill off large numbers of terrestrial species as well as many species living in shallow waters in lakes and oceans. The loss could become so great that the food web would reach a tipping point and collapse, leading to a **mass extinction** where large numbers of species go extinct in a relatively short period of time. As shown in Figure 2.27, mass extinctions show up in the fossil record as abrupt, major declines in the number of species. Scientists generally agree that these die-offs are related to significant changes in Earth's environment that are global in nature. Although gamma-ray bursts associated with stars exploding nearby are relatively rare, most astronomers believe the probability is high that one or more has occurred during Earth's 4.6-billion-year history. This does not mean that all mass extinctions are related to gamma-ray bursts. In fact, there are other mechanisms quite capable of triggering a global change in Earth's environment. One such mechanism is a large asteroid or comet impact, a topic we will explore next.

Asteroid and Comet Impacts

Earlier you learned that the heavily cratered surfaces of Mercury and the Moon are evidence of the heavy bombardment of comets and asteroids that took place early in the solar system's history. Also interesting is the fact that the surfaces of some planets and moons have areas where the crater density is high, yet in other areas craters are relatively sparse. Because the initial cratering should have been fairly well distributed, scientists conclude that the places we see today with few craters must have been resurfaced with younger rocks. Examples include the lava-filled basins on the Moon (Figure 2.28), which show relatively light cratering compared to the heavily impacted upland areas. Based on radiometric dating (Chapter 1) of different rock samples returned from the Moon, scientists have been able to confirm that the lightly cratered lava basins are considerably younger than the heavily impacted areas. From this evidence it is now known that the intense bombardment period ended about 3.6 billion years ago, a time when the solar system was approximately 1 billion years old.

Of considerable interest is the fact that the end of the heavy bombardment, 3.6 billion years ago, roughly coincides with the first evidence of microbial life in the geologic rock record. Many astronomers and astrobiologists interpret this to mean that the conditions on Earth during the intense bombardment period were too harsh for life to develop. Note that during this period comets are believed to have been impacting planets and moons throughout the solar system. Since comets contain carbon-based compounds and water, this means that they would have seeded the entire solar system with the essential ingredients

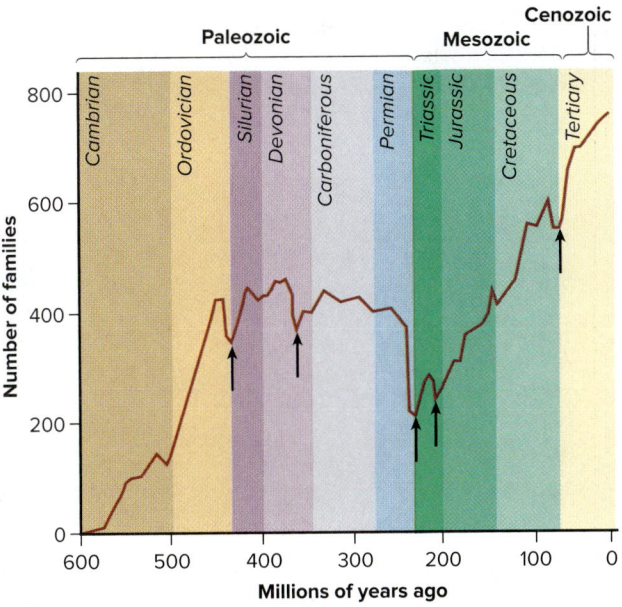

FIGURE 2.27 Graph showing the number of species recorded in the fossil record over the last 600 million years of geologic time. Mass extinctions are marked by periodic and sudden decreases in the numbers of families of species. Scientists believe these die-offs are related to environmental changes that are global in nature.

FIGURE 2.28   The dark areas on the Moon consist of flat-lying lava deposits where the cratering density is significantly lower than the surrounding highlands. The difference in crater density combined with radiometric dating of Moon rocks proves that the heavy bombardment period ended when the solar system was about 3.6 billion years old, which is also when primitive life-forms first show up in Earth's rock record.
NASA

for life. Once the heavy bombardment phase ended, life could have developed any place in the solar system with suitable conditions. Earth happened to have the most ideal conditions because of its location in the Sun's habitable zone. It is ironic that the early bombardment may have led to the development of life on Earth, but was then followed by occasional impacts with global consequences, triggering some of the mass extinctions found in the geologic record (Figure 2.27). In this section we will explore how scientists came to realize that impacts are an important geologic process and why they still present a serious hazard to the Earth.

Discovering the Impact Threat

Prior to the Moon landings there was considerable debate among geologists whether the Moon's numerous craters were the result of impacts or volcanic activity. The scientific debate was reasonable at the time because volcanic craters are quite common on Earth. Some geologists even questioned whether the 50,000-year-old crater in the Arizona desert (Figure 2.29) was formed by an impact, especially since a large meteorite had never been found there. While geologists knew that asteroids continue to strike the Earth, hardly anyone could imagine one large enough to make a crater of this size. Moreover, it was thought that large asteroids would break up into relatively small pieces due to the stresses encountered upon hitting Earth's atmosphere.

The mystery of the Arizona crater, called Meteor Crater, or Barringer Crater, was solved in the 1960s when a geologist named Gene Shoemaker mapped the rock layers at the site. He found that the orientation or structure of the rocks,

FIGURE 2.29   Meteor Crater in the Arizona desert formed approximately 50,000 years ago when a 150-foot (45-m) asteroid struck the Earth. The crater is 0.75 mile (1.2 km) in diameter.
USGS

FIGURE 2.30 Approximately 190 impact sites have now been identified on Earth. Note that the number of impacts varies across the globe, in part because of accessibility and the age of the rocks exposed at the surface. Also note that very few impacts have been found in offshore areas. A 214-million-year-old impact structure in Canada (inset) is believed to have had an original crater 50 to 60 miles (75–100 km) in diameter, formed by an asteroid over 3 miles (5 km) in diameter. Some scientists believe this event may be linked to a mass extinction in which nearly 60% of all species on the planet were lost.
Reto Stockli, NASA Earth Observatory; (inset) USGS

which were once flat-lying, was identical to that of craters formed by nuclear explosions at the U.S. government's test site in Nevada. After determining the rock structure of Meteor Crater, Dr. Shoemaker and two other scientists found a rare, high-pressure mineral called *coesite* within the crater. This was significant because the only other place coesite had ever been found was in nuclear craters. Since there is no natural Earth process capable of generating the pressure necessary to create coesite, the only possible explanation was that Meteor Crater had formed from a large impact. The reason a large meteorite had never been found there could be explained by the meteor's tremendous speed, which would have caused it to completely disintegrate upon impact—similar to a high-velocity bullet hitting a solid object. What remains then is a relatively large crater and small asteroid fragments scattered about the site.

Although Dr. Shoemaker's work and the Apollo Moon landings ultimately proved that impacts occurred on both the Moon and Earth, it took years for the scientific community to recognize that large impacts continue to take place today. Most scientists thought that after the heavy bombardment period, Earth's orbit had been swept clear of large debris; hence large impacts today would be highly unlikely. This view eventually changed as geologists began finding impact craters in the more recent parts of the geologic rock record (Chapter 1). By looking for coesite and the characteristic rock structure associated with impacts, geologists have currently identified approximately 190 impact craters on Earth (Figure 2.30). The actual number is undoubtedly much higher than this, partly because most of the asteroids or comets would have struck in the oceans, making the craters extremely difficult to detect. Weathering and erosion processes (Chapter 3) would also have erased or buried the evidence for many of those that struck land.

Despite the fact that Earth's crater record is very incomplete, it still contains evidence of very large impacts in the past. Of particular interest are those craters known to have formed within the last 500 million years, a time when complex life-forms began to flourish. A large impact during this period could have led to a mass extinction and affected the evolution of life. Aside from the

FIGURE 2.31   Rock outcrop (A) near Trinidad, Colorado, that contains the boundary between rocks of the Cretaceous and Tertiary periods of the Mesozoic and Cenozoic eras, respectively. Dinosaurs were dominant during the Mesozoic, but went extinct when a large asteroid struck the Earth. Fallout from the impact created an iridium-rich clay layer (B) found around the world, marking the boundary between the Cretaceous (K) and Tertiary (T) periods, often referred to as the K/T boundary.
(a–b) Eleanor Camann

obvious blast and possible tsunamis, the most critical effect of a large asteroid or comet impact would be the debris ejected from the crater. Recent studies have shown that massive amounts of material would be ejected into space, which would then generate frictional heat as it reentered the atmosphere. This would create a global heat pulse lasting several minutes that would be hot enough to kill many types of land animals. The hot debris falling back to the land surface would also ignite wildfires on a global scale, sending large volumes of soot into the atmosphere. This soot, combined with fine debris from the impact, would stay suspended in the atmosphere for a considerable period of time, reducing the amount of sunlight reaching the surface. As global temperatures quickly fall, the environment would become inhospitable for many of the species that had survived the initial heat pulse. The result would be a mass extinction.

The Mesozoic/Cenozoic Extinction Event

Perhaps the most famous of all mass extinctions was the one that ended the reign of the dinosaurs about 65 million years ago (see Figure 2.27). This event was first recognized over 100 years ago as geologists noted that, worldwide, rocks of the same relative age record a dramatic change in fossil content. Most fascinating was that the underlying (older) rocks proved that dinosaurs had once been the dominant type of animal, whereas mammals were the dominant species in the overlying (younger) rocks. The younger rocks also showed that the dinosaurs went extinct rather abruptly. Because this change in animal life was so significant and recorded worldwide, it was used to separate what geologists call the Mesozoic and Cenozoic eras of the geologic time scale (Chapter 1). Altogether there are three eras (Paleozoic, Mesozoic, and Cenozoic), which represent the entire 550-million-year history of complex life on our planet. Geologists have always been keenly interested in discovering what caused the dinosaurs to go extinct at the end of the Mesozoic, leaving mammals to dominate the planet. This key boundary, shown in Figure 2.31a, is commonly called the *Cretaceous (K) and Tertiary (T) boundary,* which reflects the names of finer subdivisions within the Mesozoic and Cenozoic eras.

After many years of extensive study, geologists have learned a great deal about the demise of the dinosaurs and rise of mammals. It is now known that for about 20 million years prior to the Cretaceous/Tertiary boundary, dinosaurs were becoming less diversified as individual species became more specialized. Geologists generally believe that some type of global environmental change then occurred, making it impossible for these highly specialized species to survive. Interestingly, the rock record shows that nearly 75% of *all* species on the planet at this time became extinct—terrestrial and marine combined. In the oceans approximately 90% of the plankton were lost, which almost certainly led to a major collapse within the marine food web. The question is what triggered such a massive die-off. Most ideas center on some type of global change in Earth's climate, triggered perhaps by a large impact or series of large volcanic eruptions, or even the movement of continental landmasses over time. Another possibility is a nearby gamma-ray burst that depleted Earth's ozone layer, allowing intense UV radiation to reach the surface.

A significant breakthrough in the mystery of the dinosaur extinction came in the 1980s when physicist Luis Alvarez and his son Walter, a geologist, were studying rocks in Italy that straddled the Cretaceous/Tertiary boundary. Here they discovered a thin layer of clay within the sequence of limestone rocks. Moreover, this clay layer marked a distinct change in the fossil content of the rocks, suggesting it represented the top of the Cretaceous (Mesozoic). Laboratory analysis revealed that the clay layer was 65 million years old and contained unusually large amounts of the element iridium. This was highly significant because iridium is extremely rare in Earth rocks, but relatively common in most meteorites.

The iridium-rich clay therefore represented the first direct evidence that an asteroid or comet impact coincided with the Cretaceous/Tertiary extinction event. The Alvarezes concluded that a large asteroid or comet hit the Earth, creating a huge dust cloud that led to a mass extinction.

After the discovery of the iridium layer, scientists began taking a closer look at rocks around the world where the Cretaceous/Tertiary boundary is exposed. Not only was a layer of iridium-rich clay found at many sites (Figure 2.31b), but it consistently yielded radiometric dates around 65 million years old. This left little doubt that the clay layer represented the atmospheric fallout of a global dust cloud. Moreover, the clay at some sites also contained the mineral coesite, providing conclusive proof that the dust cloud resulted from a large impact 65 million years ago. By mapping the location of coesite and other geologic data, scientists were able to determine the general area where the asteroid made impact.

The chance of finding the actual crater, though, was rather small considering so much of Earth's impact record is missing. But then, in 1990, a large crater was found whose location and 65-million-year-old age were consistent with the other data. This crater, called *Chicxulub,* is 112 miles (180 km) wide and located along Mexico's Yucatán peninsula (Figure 2.32). In 2010, after years of debate as to whether the timing of the Chicxulub crater is consistent with the iridium layer, a team of nearly 40 scientists completed a thorough review of all the data. They found that the debris ejected from a single crater was unique and compositionally linked to both the iridium-rich fallout layer and the Chicxulub impact site. From this they concluded that a large comet or asteroid had indeed crashed into the Earth at Chicxulub, altering the global environment such that it caused the mass extinction found in the geologic rock record 65 million years ago. The large and highly specialized dinosaurs would have been rather susceptible to the intense heat pulse and subsequent global cooling described earlier. In contrast, mammals at the time were mostly small, burrowing creatures that would have had a much easier time surviving the dramatic change in the environment. We humans then owe our current position at the top of the food chain to the environmental changes brought on by a large asteroid impact 65 million years ago. This brings up an interesting question: Might we humans eventually succumb to the same fate as the dinosaurs, undergoing a mass extinction brought on by a global change in the environment?

FIGURE 2.32   The Chicxulub impact crater on the Yucatán peninsula in Mexico is one of the largest found on Earth, and the impact that formed it is believed to be a major factor in the extinction of the dinosaurs. The crater measures 112 miles (180 km) across and was formed by an asteroid estimated to be 6.2 to 9.3 miles (10–15 km) in diameter. A computer-generated gravity map (inset) indicates that the crater may be much larger, nearly 185 miles (300 km) across. In 2016, researchers began drilling into the inner ring of the crater to examine the rocks and learn details of how life changed after the impact.
(insert) Dr. Virgil L. Sharpton, University of Alaska Fairbanks

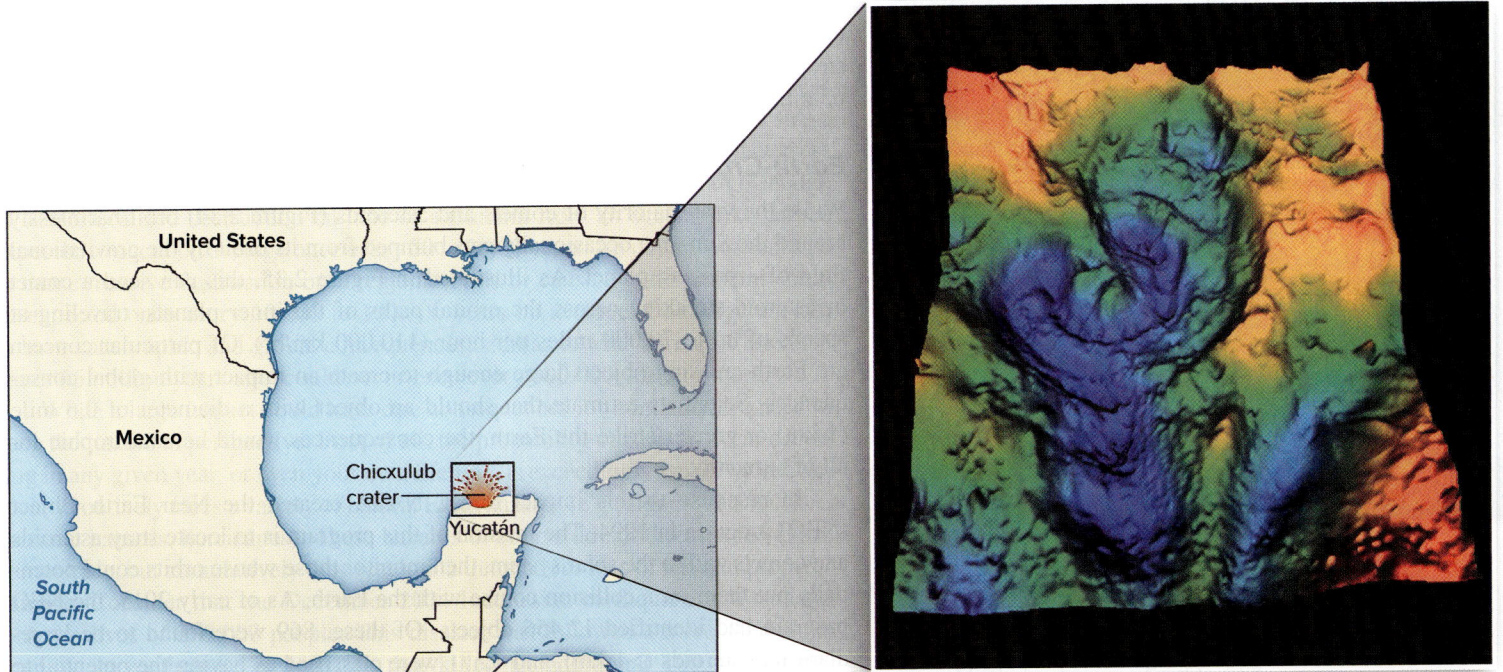

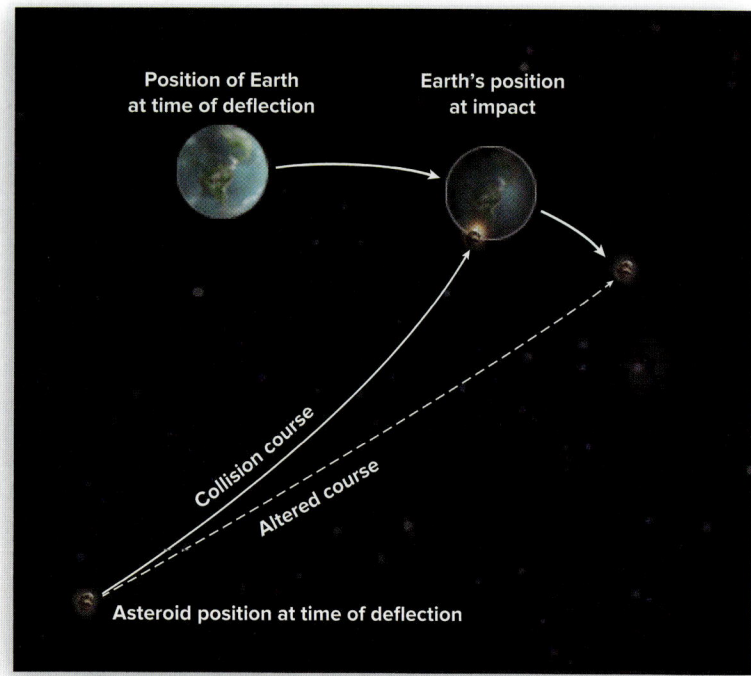

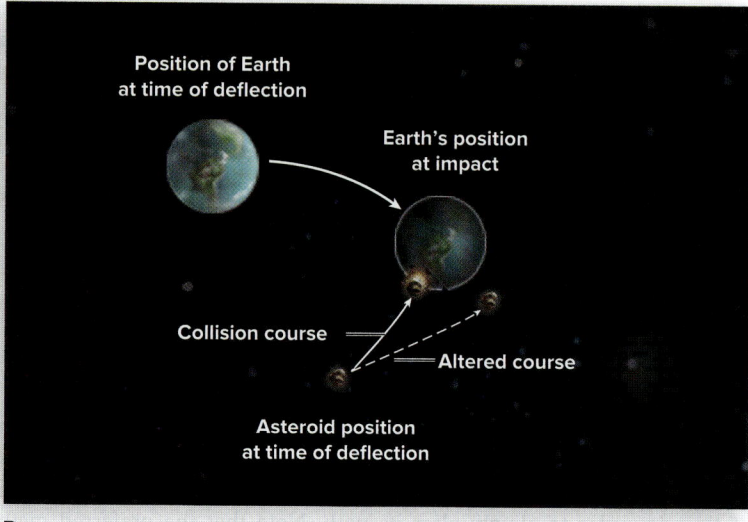

FIGURE 2.36   Technology already exists that would allow humans to alter the trajectory of an Earth-crossing object so that it would miss the Earth. The ideal situation (A) is for a spacecraft to intercept the object while it is still far from the Earth. In this situation only a slight change in its trajectory would be required to make it miss the Earth. If the object is intercepted too close to the Earth (B), then the degree to which it would have to be deflected would be beyond the capacity of humans.

Although it is important that society takes steps to reduce the risk of a major impact, we must also reduce the risk of those hazards that pose problems on a routine basis. Ironically, humans are powerless to prevent natural disasters associated with earthquakes and volcanoes, but have the ability to prevent a large impact, a hazard capable of ending human civilization.

SUMMARY POINTS

1. Knowledge of the solar system and universe can help us to better understand the Earth system and provide a larger perspective from which to view our environmental problems.

2. Nuclear fusion of hydrogen atoms in the Sun produces electromagnetic radiation that travels outward, eventually striking the planets and moons within the solar system. This energy is what drives critical processes in Earth's biosphere, hydrosphere, and atmosphere.

3. The nebular hypothesis describes how the solar system formed and accounts for the orbital characteristics of the planets and moons. Here a cloud of dust and gas condensed and slowly began to rotate, flattening into a disc in which material accreted into planets. The immense gravity at the center caused nuclear fusion of hydrogen, creating the Sun.

4. Our solar system lies in the Milky Way Galaxy along with hundreds of billions of other stars. Powerful telescopes have shown there to be hundreds of billions of other galaxies in the universe. The big bang theory accounts for the fact that the universe is continuing to expand outward.

5. The oldest forms of life recorded in Earth's rocks are traces of extremophile bacteria. Scientists believe that such ancient bacteria could have formed elsewhere in the solar system wherever the temperature would allow liquid water to exist.

6. Complex animal life in other solar systems is likely to be restricted to the habitable zones around stars where liquid water can exist and where the climate is stable for long periods of geologic time.

7. Mass extinctions recorded in Earth's geologic record are believed to be related to changes in the global environment. Possible triggers for such global change include large volcanic eruptions, comet or asteroid impacts, and nearby gamma-ray bursts.

8. Geologic studies have shown that large asteroids and comets have repeatedly struck the Earth and the other planets. A worldwide iridium-rich clay layer and a large crater in Mexico are strong evidence that a large impact triggered a mass extinction that ended the age of the dinosaurs 65 million years ago.

9. After witnessing a large impact on Jupiter in 1994, scientists realized that the Earth is still at risk from being hit by a stray asteroid or comet. NASA then began a program to map Earth-crossing objects to detect those on a potential collision course with Earth. With advanced warning, humans may be able to deflect the object, thus avoiding a collision.

10. A large impact with Earth could be devastating for humanity, but the probability of it occurring in any given year is exceedingly small. On the other hand, natural Earth hazards such as earthquakes and volcanoes happen relatively frequently, and their cumulative effect in terms of deaths and property loss is enormous.

The iridium-rich clay therefore represented the first direct evidence that an asteroid or comet impact coincided with the Cretaceous/Tertiary extinction event. The Alvarezes concluded that a large asteroid or comet hit the Earth, creating a huge dust cloud that led to a mass extinction.

After the discovery of the iridium layer, scientists began taking a closer look at rocks around the world where the Cretaceous/Tertiary boundary is exposed. Not only was a layer of iridium-rich clay found at many sites (Figure 2.31b), but it consistently yielded radiometric dates around 65 million years old. This left little doubt that the clay layer represented the atmospheric fallout of a global dust cloud. Moreover, the clay at some sites also contained the mineral coesite, providing conclusive proof that the dust cloud resulted from a large impact 65 million years ago. By mapping the location of coesite and other geologic data, scientists were able to determine the general area where the asteroid made impact.

The chance of finding the actual crater, though, was rather small considering so much of Earth's impact record is missing. But then, in 1990, a large crater was found whose location and 65-million-year-old age were consistent with the other data. This crater, called *Chicxulub,* is 112 miles (180 km) wide and located along Mexico's Yucatán peninsula (Figure 2.32). In 2010, after years of debate as to whether the timing of the Chicxulub crater is consistent with the iridium layer, a team of nearly 40 scientists completed a thorough review of all the data. They found that the debris ejected from a single crater was unique and compositionally linked to both the iridium-rich fallout layer and the Chicxulub impact site. From this they concluded that a large comet or asteroid had indeed crashed into the Earth at Chicxulub, altering the global environment such that it caused the mass extinction found in the geologic rock record 65 million years ago. The large and highly specialized dinosaurs would have been rather susceptible to the intense heat pulse and subsequent global cooling described earlier. In contrast, mammals at the time were mostly small, burrowing creatures that would have had a much easier time surviving the dramatic change in the environment. We humans then owe our current position at the top of the food chain to the environmental changes brought on by a large asteroid impact 65 million years ago. This brings up an interesting question: Might we humans eventually succumb to the same fate as the dinosaurs, undergoing a mass extinction brought on by a global change in the environment?

FIGURE 2.32   The Chicxulub impact crater on the Yucatán peninsula in Mexico is one of the largest found on Earth, and the impact that formed it is believed to be a major factor in the extinction of the dinosaurs. The crater measures 112 miles (180 km) across and was formed by an asteroid estimated to be 6.2 to 9.3 miles (10–15 km) in diameter. A computer-generated gravity map (inset) indicates that the crater may be much larger, nearly 185 miles (300 km) across. In 2016, researchers began drilling into the inner ring of the crater to examine the rocks and learn details of how life changed after the impact.
(insert) Dr. Virgil L. Sharpton, University of Alaska Fairbanks

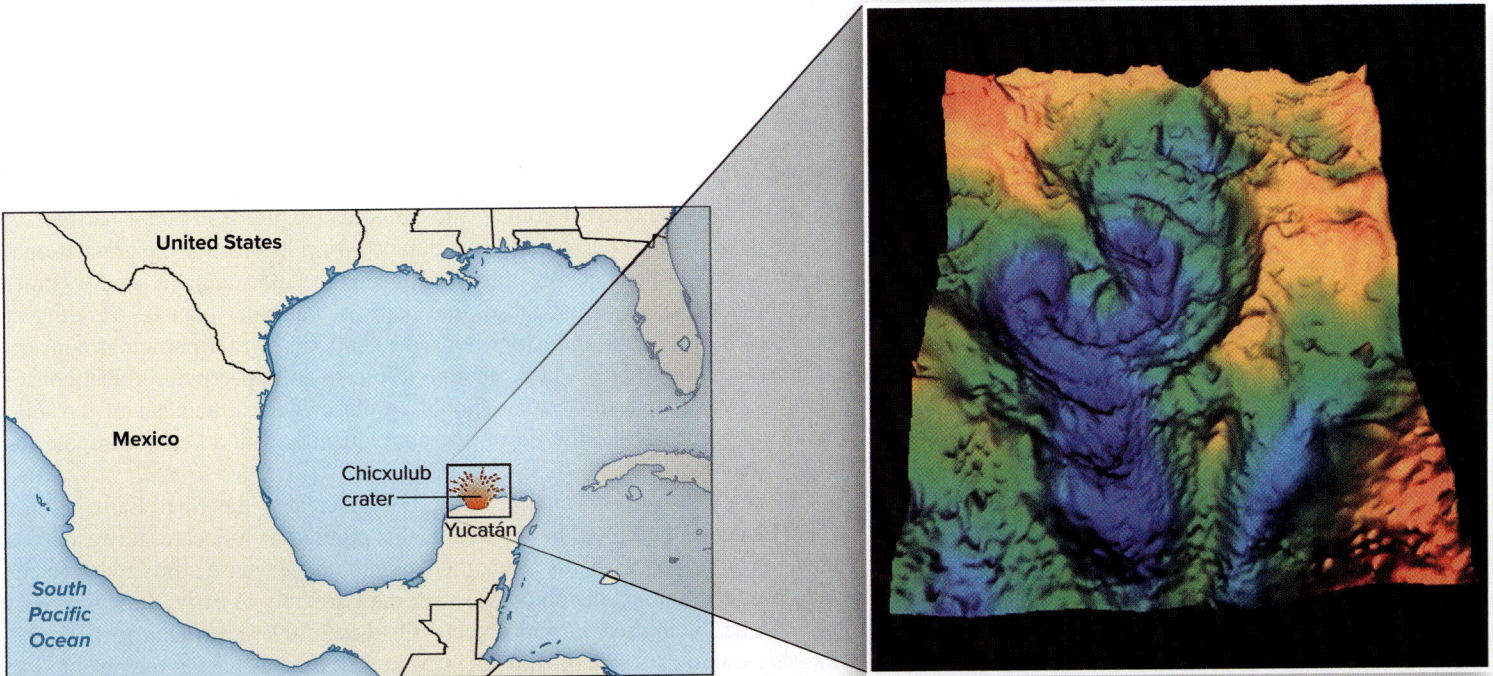

FIGURE 2.33   Image (A) taken by the Hubble space telescope of the broken fragments of comet Shoemaker-Levy 9 prior to the 1994 impact with Jupiter. An infrared image (B) showing the impact plumes made by several of the comet fragments.
(a–b) NASA

Recent Impact on Jupiter

Shortly after the Chicxulub crater was discovered, scientists were fortunate enough to actually witness a major impact on Jupiter in 1994. The only reason this extremely rare event was known in advance was because of the continued research of Gene Shoemaker. Based on his earlier work, Shoemaker hypothesized that the solar system still contained large objects, perhaps a mile or more in diameter, whose paths sent them streaking across Earth's orbit. If such Earth-crossing comets and asteroids truly existed, then they would pose a serious threat to life here on Earth. In the early 1980s Dr. Shoemaker, his wife Carolyn, and colleague David Levy began using a small telescope with a wide viewing area to map stray objects orbiting the inner solar system. In 1993 they discovered a comet (Figure 2.33a), which during a previous pass by Jupiter had been broken up into more than 20 major pieces by the planet's strong gravitational field. Shortly after their discovery it was determined that the comet, named Shoemaker-Levy 9, was on a collision course with Jupiter. When the broken-up comet began impacting Jupiter in July 1994, nearly every major telescope in the world was trained on the planet to record this historic event. Images of the impacts (Figure 2.33b) both awed and shocked the scientific community. It was immediately obvious that had even one of the larger fragments struck the Earth, the result would have been catastrophic. This was a dramatic illustration that what had happened to the dinosaurs 65 million years ago could happen to us.

Earth-Crossing Asteroids and Comets

While the vast majority of comets and asteroids (Figure 2.34) orbit harmlessly around the Sun, one occasionally gets bumped from its orbit by the gravitational field of a passing planet. As illustrated in Figure 2.35, this can send a comet or asteroid streaking across the orbital paths of the inner planets, traveling at speeds of up to 70,000 miles per hour (110,000 km/hr). Of particular concern are Earth-crossing objects large enough to create an impact with global consequences. Scientists estimate that should an object with a diameter of 0.6 mile (1 km) or greater strike the Earth, the consequences would be catastrophic for all of humanity.

In response to this impact threat, NASA created the Near Earth Object (NEO) program in 1994. The mission of this program is to locate stray asteroids and comets within the solar system, then monitor those whose orbits could potentially put them on a collision course with the Earth. As of early 2015, the NEO program had identified 12,456 objects. Of these, 869 were found to be large-diameter asteroids (>1 km), and 1,571 were classified as having the potential to strike the Earth. Fortunately, none of the larger objects located so far are projected to make impact with Earth in the foreseeable future. However, the paths of Earth-crossing objects can change as they orbit the solar system and encounter the gravitational fields of the planets (Figure 2.35). NASA therefore continues to monitor the orbits of objects that cross Earth's orbit.

If it is determined that a large comet or asteroid is going to collide with Earth, the next step would be to have a spacecraft intercept the object in an attempt to alter its trajectory such that it misses the Earth. Recent international mission probes have rendezvoused with comets and asteroids, gathering data that would be useful in a future attempt at deflecting an object—the European Space Agency actually landed a probe on a comet in 2014 as part of this international effort (Figure 2.34). Although various methods have been proposed for altering the trajectory of a comet or asteroid, most involve using existing technology. One approach is to simply crash a probe directly into the object so that the energy of the impact changes the object's trajectory. Others involve some means of exerting a small force over a long period of time in order to slowly alter its path. Note that the preferred method in Hollywood movies is to vaporize an asteroid or comet using a nuclear weapon. In reality this would be a poor choice as even our most powerful nuclear weapons would likely result in the object being broken up into several fragments. As was the case with Shoemaker-Levy 9 (Figure 2.33), large fragments can still cause considerable damage upon impact.

Finally, a critical aspect in any mission to deflect an asteroid or comet is to have enough advance warning so that spacecraft can rendezvous with the object before it gets too close to the Earth. As illustrated in Figure 2.36, if the deflection attempt is made when the object is far away, then all it takes is a slight trajectory change for it to miss the Earth. However, if the object gets too close, then the required trajectory change would be so large that it would be beyond our capability. In this case a catastrophic collision could not be avoided. One of the reasons NASA began mapping stray asteroids and comets in the first place was to have enough time to mount a mission before an object gets too close.

Impact Risk

Recall from Chapter 1 that in managing environmental risk, one needs to consider both the probability of a undesirable event taking place and its potential consequences. When it comes to the risk of impacts, tons of material strike Earth's atmosphere every day, but the particles are so small that most burn up harmlessly in the atmosphere. The concern, of course, is a large impact whose consequences would be regional or global in extent. Fortunately, the probability of such a strike is low because there are comparatively few large asteroids and comets that cross Earth's orbit. For example, NASA scientists estimate that an impact large enough to cause localized destruction on land or a tsunami from an ocean strike can be expected to occur on average once every 50 to 1,000 years. An object large enough to cause massive destruction on a regional basis or cause a giant tsunami could be expected once every 10,000 to 100,000 years on average. One large enough to have catastrophic and global consequences, similar to what formed Chicxulub crater 65 million years ago, could occur once every 100,000 years or more.

While a large impact could have dire consequences for humanity, we need to keep the hazard in perspective. First of all, the probability of such an event occurring in any given year, or even your own lifetime, is exceedingly small. Moreover, society routinely faces a variety of natural Earth hazards (e.g., floods, earthquakes, volcanic hazards, hurricanes). These hazards of course do not threaten the entire human race, but cumulatively they result in a tremendous loss of life and property damage. Consider that the December 2004 Indonesian earthquake and tsunami took the lives of 230,000 people. This disaster was followed a mere 10 months later by an earthquake in Pakistan that killed an additional 100,000 people.

FIGURE 2.34   Photo of comet 67P/Churyumov-Gerasimenko taken by the European Space Agency's *Rosetta* spacecraft as it orbited the comet in 2014. *Rosetta* also landed a probe on the comet's surface, taking numerous measurements and analyzing the comet's composition. Comet 67P measures about 2.5 by 2.0 miles (4.1 by 3.2 km) along its widest dimension.
ESA/Rosetta/MPS for OSIRIS Team MPS/UPD/LAM/IAA/SSO/INTA/UPM/DASP/IDA

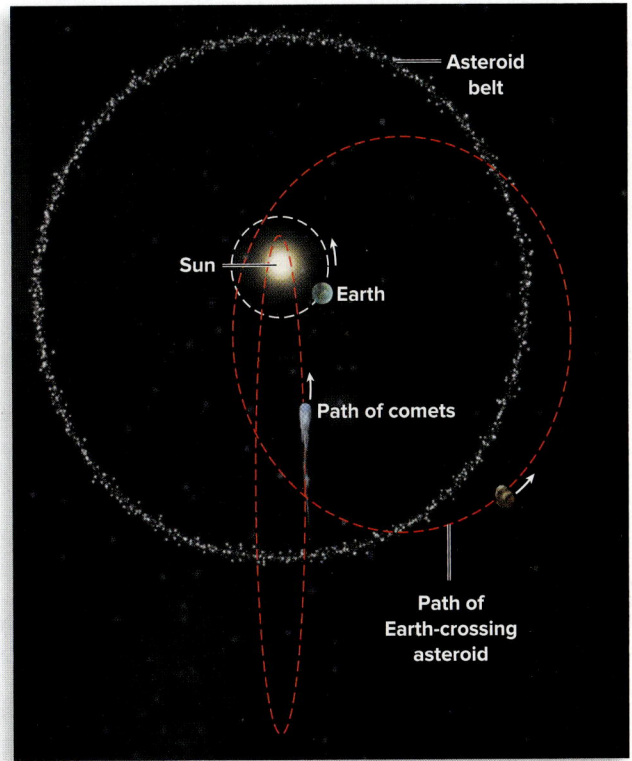

FIGURE 2.35   Asteroids and comets can be bumped from their normal orbits around the Sun by the gravitational effect of passing planets, sending them across the orbital paths of the planets. Of particular concern are comets and asteroids that cut across Earth's orbit.

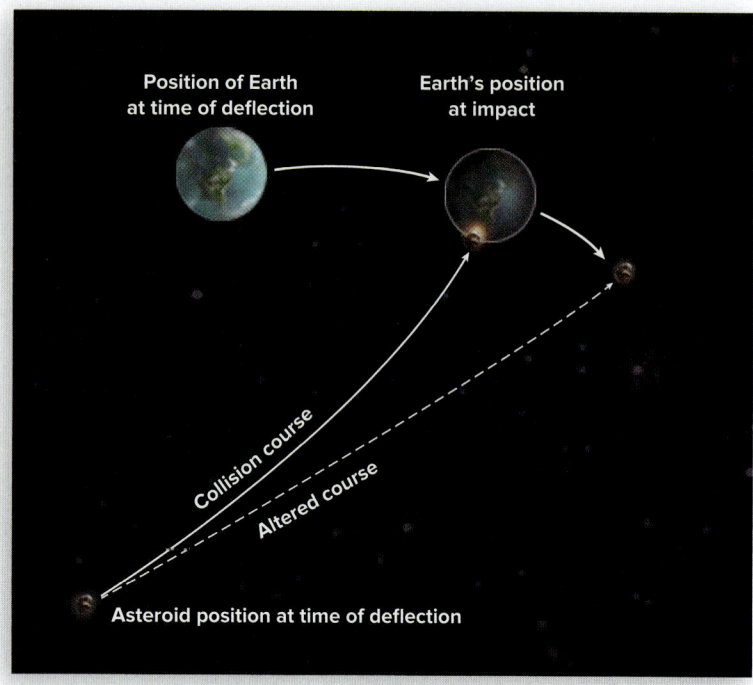

A

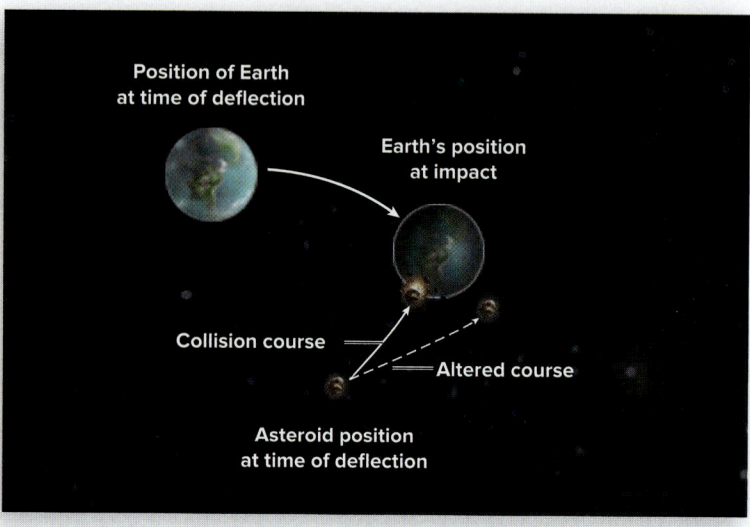

FIGURE 2.36   Technology already exists that would allow humans to alter the trajectory of an Earth-crossing object so that it would miss the Earth. The ideal situation (A) is for a spacecraft to intercept the object while it is still far from the Earth. In this situation only a slight change in its trajectory would be required to make it miss the Earth. If the object is intercepted too close to the Earth (B), then the degree to which it would have to be deflected would be beyond the capacity of humans.

Although it is important that society takes steps to reduce the risk of a major impact, we must also reduce the risk of those hazards that pose problems on a routine basis. Ironically, humans are powerless to prevent natural disasters associated with earthquakes and volcanoes, but have the ability to prevent a large impact, a hazard capable of ending human civilization.

SUMMARY POINTS

1. Knowledge of the solar system and universe can help us to better understand the Earth system and provide a larger perspective from which to view our environmental problems.

2. Nuclear fusion of hydrogen atoms in the Sun produces electromagnetic radiation that travels outward, eventually striking the planets and moons within the solar system. This energy is what drives critical processes in Earth's biosphere, hydrosphere, and atmosphere.

3. The nebular hypothesis describes how the solar system formed and accounts for the orbital characteristics of the planets and moons. Here a cloud of dust and gas condensed and slowly began to rotate, flattening into a disc in which material accreted into planets. The immense gravity at the center caused nuclear fusion of hydrogen, creating the Sun.

4. Our solar system lies in the Milky Way Galaxy along with hundreds of billions of other stars. Powerful telescopes have shown there to be hundreds of billions of other galaxies in the universe. The big bang theory accounts for the fact that the universe is continuing to expand outward.

5. The oldest forms of life recorded in Earth's rocks are traces of extremophile bacteria. Scientists believe that such ancient bacteria could have formed elsewhere in the solar system wherever the temperature would allow liquid water to exist.

6. Complex animal life in other solar systems is likely to be restricted to the habitable zones around stars where liquid water can exist and where the climate is stable for long periods of geologic time.

7. Mass extinctions recorded in Earth's geologic record are believed to be related to changes in the global environment. Possible triggers for such global change include large volcanic eruptions, comet or asteroid impacts, and nearby gamma-ray bursts.

8. Geologic studies have shown that large asteroids and comets have repeatedly struck the Earth and the other planets. A worldwide iridium-rich clay layer and a large crater in Mexico are strong evidence that a large impact triggered a mass extinction that ended the age of the dinosaurs 65 million years ago.

9. After witnessing a large impact on Jupiter in 1994, scientists realized that the Earth is still at risk from being hit by a stray asteroid or comet. NASA then began a program to map Earth-crossing objects to detect those on a potential collision course with Earth. With advanced warning, humans may be able to deflect the object, thus avoiding a collision.

10. A large impact with Earth could be devastating for humanity, but the probability of it occurring in any given year is exceedingly small. On the other hand, natural Earth hazards such as earthquakes and volcanoes happen relatively frequently, and their cumulative effect in terms of deaths and property loss is enormous.

KEY WORDS

accretion 43
asteroids 41
big bang theory 46
comets 41
electromagnetic radiation 40
extremophile bacteria 48

galaxies 46
gamma-ray burst 54
gas giants 40
habitable zone 48
mass extinction 55
nebular hypothesis 43

ozone depletion 54
planets 44
rare earth hypothesis 52
terrestrial planets 40

APPLICATIONS

Student Activity　　Look up at the sky on a dark, clear night. (If you live in an urban area, take a drive into the country on a clear night). Allow your eyes to adjust to the dark and look up into the sky. You should be able to see the Milky Way. It goes from east to west almost directly above your head. It looks like a bright, but cloudy band across the sky. Here we are actually seeing an edge view of our galaxy and one of its spiral arms!

Critical Thinking Questions

1. The nebular hypothesis explains how our solar system formed. Describe several lines of evidence that support this hypothesis. Why do scientists think that other planetary systems formed in a similar manner?
2. The big bang theory describes the origin of the universe. What is the most convincing evidence that the entire universe actually did exist at a single point approximately 14 billion years ago?
3. Extremophile bacteria are able to live and thrive under extreme conditions. Why are they important in the search for extraterrestrial life?

Your Environment: YOU Decide　　Should humanity continue exploring space in order to better understand the Earth system and the origin of life, or should we focus our resources on other problems? Provide one or more reasons to support your answer.

NASA/NOAA/GSFC/Suomi NPP/VIIRS/Norman Kuring

Chapter 3

Earth Materials

LEARNING OUTCOMES

After reading this chapter, you should be able to:

▶ Explain how different types of atoms are created and how they are assembled into minerals and rocks.

▶ Describe why the rock-forming minerals are so important to the study of geology.

▶ Define the basic way in which each of the major rock types forms.

▶ State the relationship between weathering, erosion, transportation, and deposition, and identify the different types of sedimentary rocks.

▶ Explain why some rocks are more susceptible to chemical weathering than others.

▶ Describe the basic paths rocks can take through the rock cycle in response to different geologic processes.

▶ Explain how rocks can be used to interpret ancient environments.

The Earth is composed of different types of rocks and minerals, which are commonly sculpted into striking landscapes by weathering and erosion processes. Shown here are iron-stained sedimentary rocks in Monument Valley, located in Arizona and Utah. These rocks were deposited in an ancient sea and have since been eroded into buttes by running water and wind. Rock and mineral deposits not only let geologists interpret Earth's history, but also serve as important raw materials used in modern societies.

Introduction

In this chapter we will take a closer look at minerals and how they form the different types of rocks that make up the solid earth. It is important for you to have a working knowledge of rocks and minerals because nearly every topic in environmental geology in some way or another involves these materials. For example, most of the material used to build the infrastructure of modern societies and produce consumer goods comes from rock and mineral resources (Chapter 12). To meet the high demand, enormous quantities of iron, copper, aluminum, and other metallic-bearing minerals must be removed from the solid earth and processed into pure metal. Large amounts of nonmetallic minerals are also extracted and used to produce key materials such as concrete, glass, ceramics, and drywall (i.e., sheetrock). In addition to raw materials, rocks provide critical supplies of coal, oil, and natural gas (Chapter 13). Without these energy resources, modern society as it exists today would simply grind to a halt.

Knowing about rocks and minerals is also important when it comes to our food and water supplies. The crops farmers grow depend on fertile soils derived from the breakdown of rocks and minerals, a process that requires water percolating into the subsurface over long periods of time. In addition to producing soil, infiltrating water accumulates in porous layers of rock and sediment, which humans extract and use for a multitude of things. One key use of this groundwater has been to irrigate crops to increase food production in areas where rainfall is limited. Moreover, chemical reactions between minerals and water affect the very quality of the water we drink. In some areas water supplies are naturally contaminated with heavy metals or radioactive gases, creating a potential human health hazard.

The chemical and physical properties of rocks and minerals also have a profound influence on many types of geologic hazards, as well as on the landscapes that we inhabit. For example, the physical strength of rocks and minerals helps determine the amount of energy a rock body can accumulate before it fails, at which point the stored energy is released in the form of vibrating waves that we call an *earthquake.* Therefore, the more energy rocks can store, the stronger the earthquake (Chapter 5). The inherent strength of rocks and minerals also helps determine the stability of steep slopes, and hence is important in mass wasting hazards such as landslides and rock falls (Chapter 7). Lastly, we can see from Figure 3.1

A

B

FIGURE 3.1   Earth's different landscapes reflect the underlying geology and erosion history of an area. A key control in landscape development is the way different rocks respond to weathering and erosion processes. Igneous rocks of Yosemite Valley (A) have fairly uniform chemical and physical properties, and thus weather to produce more rounded shapes. The properties of layered sedimentary rocks in the Grand Canyon (B) vary significantly, generating a landscape with a distinctive stair-step pattern.
(a–b) © Jim Reichard

that different landscapes are due in part to the different properties of the underlying rocks. Geologists often like to point out that our scenic landscapes are a reflection of interesting rocks and their geologic history.

Simply put, rocks and minerals are fundamental to nearly every interaction between people and the geologic environment. Consequently, the purpose of this chapter is to provide you with a basic knowledge of these materials so you can develop a better understanding of the various topics discussed in this text.

Basic Building Blocks

In this section we will explore the different building blocks that make up the solid Earth. In particular we want to examine how rocks are composed of individual mineral grains, which themselves are made up of even smaller particles called atoms. We will begin at the smallest scale, namely the particles that make up individual atoms.

Atoms and Elements

From chemistry we know that all matter is composed of individual *atoms,* where tiny *electrons* orbit around a nucleus of much larger particles called *protons* and *neutrons.* Atoms containing the same number of protons are referred to as *elements.* As illustrated in Figure 3.2, the simplest types of atoms are those of the element hydrogen, in which a single electron orbits around a proton. Progressively heavier elements in the periodic table. located on the last text page, have a nucleus with additional protons and neutrons along with a corresponding increase in the number of electrons. For example, all carbon atoms contain six protons and between six and eight neutrons. By definition, the number of protons in the nucleus of a particular element is fixed, whereas the number of neutrons can vary. Scientists use the term *isotopes* to refer to the different combinations of protons and neutrons found in atoms of a particular element. Figure 3.3 illustrates how the number of neutrons varies among the different isotopes of the element hydrogen, which always contain just one proton. Some isotopes are naturally unstable and transform themselves into new elements, emitting electromagnetic radiation in a process referred to as *radioactive decay.* Later we will see how isotopes are important in age dating, nuclear energy, and Earth's climate record.

Also recall from basic chemistry that electrons have a negative electrical charge and protons are positively charged, whereas neutrons have no charge. Thus when atoms have an equal number of electrons and protons, they are considered to be electrically neutral. However, some elements are more stable when they can acquire an extra electron; others become more stable after giving off an electron. When this occurs, atoms will have either a positive or negative charge, at which point they are referred to as **ions**—the actual charge depends on the number of electrons gained or lost. Examples of common ions include sodium (Na^+), iron (Fe^{3+}), chlorine (Cl^-), and oxygen (O^{2-}).

An obvious question to ask is, how do electrons, protons, and neutrons get assembled into the various configurations that make up the elements in the periodic table? In other words, how do the different elements actually form? Recall from Chapter 2 that hydrogen atoms, which are the simplest, undergo nuclear fusion in stars to produce helium atoms with two protons and two neutrons in the nucleus. Eventually the hydrogen fuel in a star diminishes to the point where fusion starts converting helium into progressively heavier atoms such as carbon, oxygen, iron, and nickel. As some stars approach the end of their life cycle, they begin ejecting this material out into space. The different types of atoms (i.e., elements) then

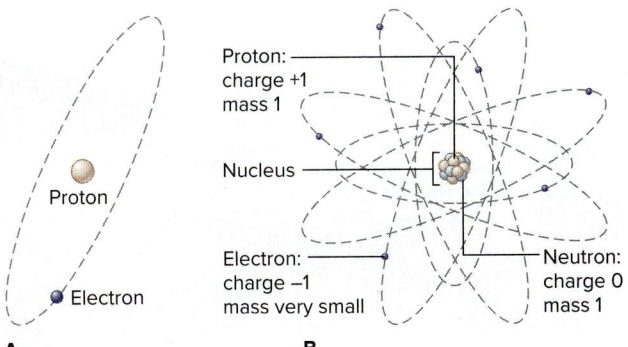

FIGURE 3.2   In a simplified view of atoms, negatively charged electrons orbit a nucleus composed of much larger protons (positive charge) and neutrons (neutral). The simplest types of atoms are of the element hydrogen (A), with a single electron orbiting a proton. Each succeeding element in the periodic table contains an additional proton and varying numbers of neutrons, thereby making those elements heavier. A carbon atom (B) contains roughly the same numbers of neutrons and electrons as it does protons.

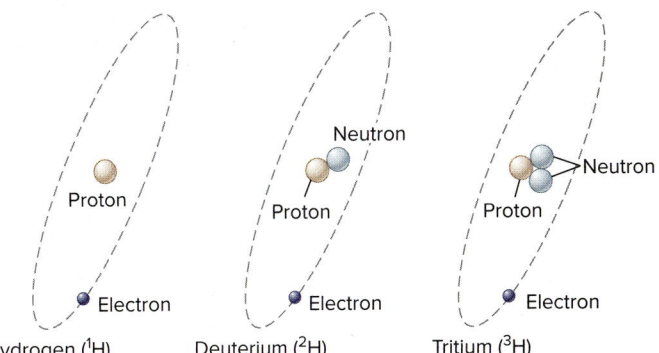

FIGURE 3.3   Illustration showing the three isotopes of the element hydrogen. The lightest isotope, common hydrogen, contains only a single proton in the nucleus. Deuterium and tritium are progressively heavier isotopes that have one and two neutrons, respectively, along with a single proton. A superscript next to the chemical symbol (i.e., ^1H, ^2H, ^3H) represents the number of protons and neutrons in the nucleus, which indicates the mass of the atom.

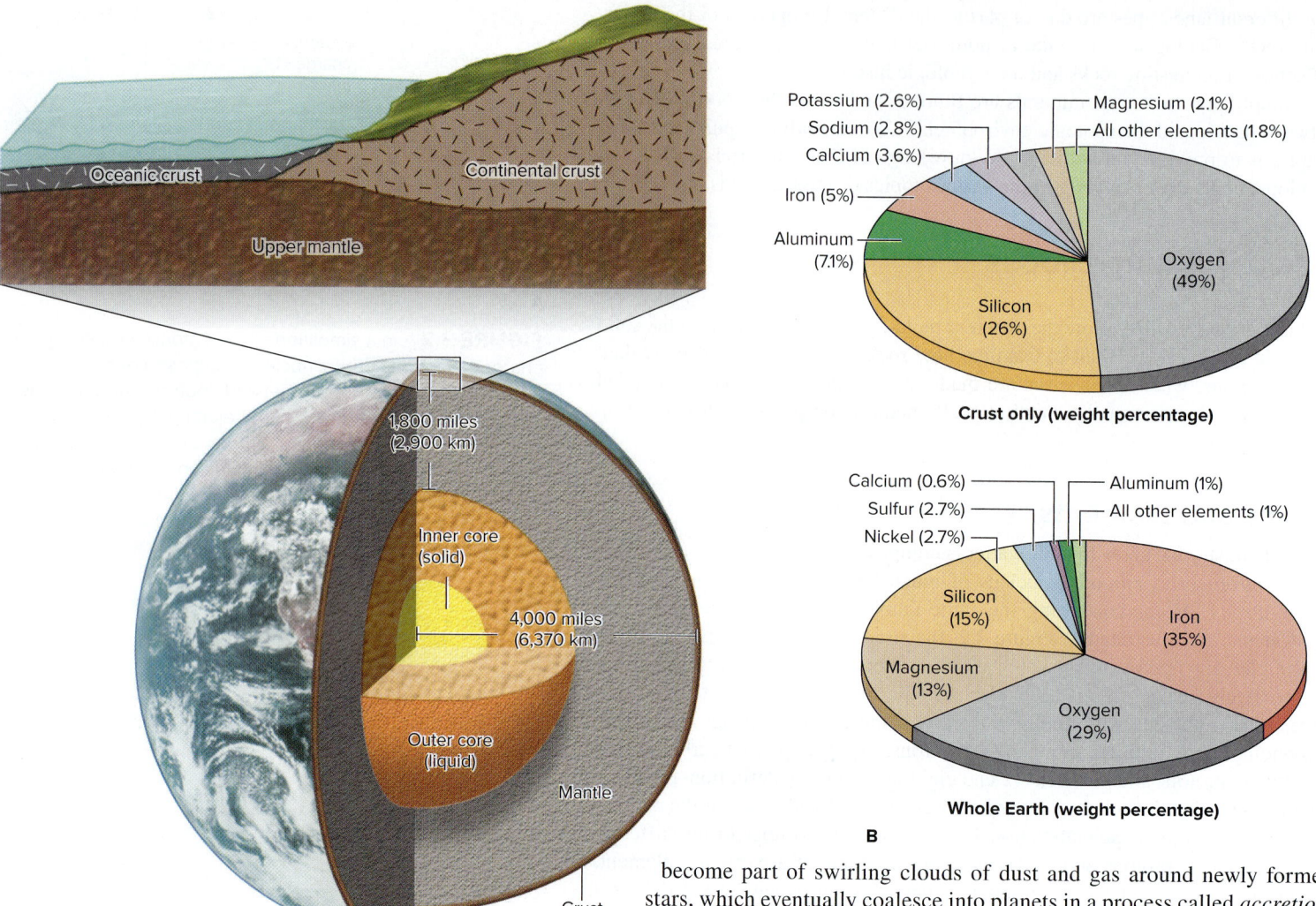

Potassium (2.6%)
Sodium (2.8%)
Calcium (3.6%)
Iron (5%)
Aluminum (7.1%)
Magnesium (2.1%)
All other elements (1.8%)
Silicon (26%)
Oxygen (49%)

Crust only (weight percentage)

Calcium (0.6%)
Sulfur (2.7%)
Nickel (2.7%)
Aluminum (1%)
All other elements (1%)
Silicon (15%)
Magnesium (13%)
Oxygen (29%)
Iron (35%)

Whole Earth (weight percentage)

Oceanic crust
Continental crust
Upper mantle

1,800 miles (2,900 km)
Inner core (solid)
4,000 miles (6,370 km)
Outer core (liquid)
Mantle
Crust

A

B

FIGURE 3.4   Earth has a layered structure (A) consisting of the core, mantle, and crust. Geologic processes have caused the heaviest elements to become concentrated in the core over time, whereas the lighter elements have tended to accumulate in the crust. Pie diagrams (B) show that the crust is largely composed of oxygen and silicon atoms, but in the planet as a whole, iron atoms are the most abundant.

become part of swirling clouds of dust and gas around newly formed stars, which eventually coalesce into planets in a process called *accretion.* This means that all the different elements we see on Earth, from those making up rocks to our own human bodies, were created by nuclear fusion in the interior of stars.

Of considerable interest in our study of rocks and minerals is the relative abundance of the elements found on Earth. Although there are around 90 naturally occurring elements in the periodic table, one can see from Figure 3.4 that just a few elements make up most of the Earth. If we consider the Earth as a whole, iron atoms account for 35% of the planet by weight, followed by oxygen at 29%. On the other hand, if we look at just the outermost portion of Earth called the *crust,* oxygen atoms make up an incredible 49% of the crust by weight, whereas the iron content decreases to around 5%. This compositional difference between the whole Earth and the crust is due to internal processes that have caused the planet to become layered. These processes tend to cause heavier elements to migrate toward the center of the planet, called the *core,* whereas lighter elements tend to accumulate in the crust; the intermediate layer is called the *mantle.* Note that Earth's internal processes and layered structure will be described in more detail in Chapter 4.

Minerals

Atoms combine to form solids, liquids, and gases, but our interest here is in a particular type of solid known as a mineral. A **mineral** is a naturally occurring inorganic (non-carbon-based) solid composed of one or more elements, where the individual atoms have an orderly arrangement called a *crystalline structure.* In this

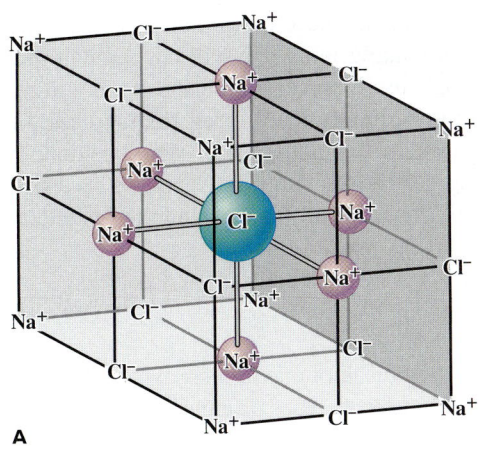

A

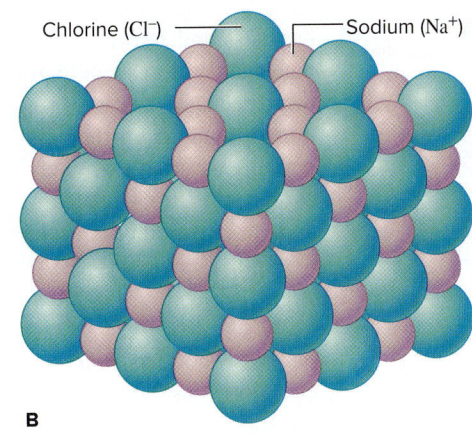

Chlorine (Cl⁻) —— Sodium (Na⁺)

B

FIGURE 3.5   All minerals have an internal structure and a definite chemical composition where the atoms are arranged in a set pattern that repeats itself in a three-dimensional manner. In view (A) the distance between atoms has been exaggerated to illustrate the fact that the angles and distances within the crystal structure are fixed. Also note that the surface of each atom represents the outermost shell or cloud of the atom's electrons.

structure the atoms are in a fixed pattern that repeats itself in a three-dimensional manner as illustrated in Figure 3.5. Solids such as glass and plastic, for example, are not minerals because their atoms are not arranged in a crystalline structure; such solids are referred to as noncrystalline or *amorphous*. Note that different minerals can have different crystalline structures that vary in terms of the angles and distances between atoms. In any given mineral, however, the structure is fixed and always the same. This holds true for minerals that form on bodies other than the Earth, such as those found on Mars and the Moon.

In addition to having a crystalline structure, all minerals have a definite chemical composition. What this means is that only certain elements, and in certain proportions, are allowed into the crystalline structure. For example, the chemical formula for the mineral pyrite (fool's gold) is FeS_2; thus, only iron (Fe) and sulfur (S) atoms are in the structure, and there are exactly two sulfur atoms for every iron atom. However, some minerals form what geologists call a *continuous series,* which is where two or more types of atoms freely substitute for one another. A good example is the mineral series called *plagioclase feldspar*— $(Ca,Na)(Al,Si)AlSi_2O_8$—where calcium (Ca) and sodium (Na) substitute for each other in varying proportions. Although minerals are very specific as to the type and number of atoms allowed into their structure, atoms of similar size and electrical charge will occasionally be allowed to substitute. This results in some minerals having minor impurities.

Interestingly, mineralogists have identified well over 4,000 different minerals on Earth, with each having a unique combination of crystalline structure and chemical composition. This means that if two minerals happen to have the exact same internal structure, then their chemical composition must be different. For example, the minerals calcite ($CaCO_3$) and siderite ($FeCO_3$) have the same crystalline structure, but siderite has iron atoms in place of calcium atoms within the structure. In other cases minerals may share the exact same chemistry, but differ in their crystalline structure—as with calcite and aragonite (also $CaCO_3$).

A key point here is that physical properties of individual minerals, such as melting point, hardness, and density, are all controlled by their internal structure and chemical composition. Therefore, because each mineral has a unique combination of structure and chemistry, each has a corresponding unique set of physical properties. For example, graphite (common pencil "lead") and diamond are two minerals composed entirely of carbon (C) atoms, but they have different crystalline structures (Figure 3.6). Despite their being composed of the same type of atoms, the structural difference results in these two minerals having vastly different physical properties. Diamond is the hardest known substance,

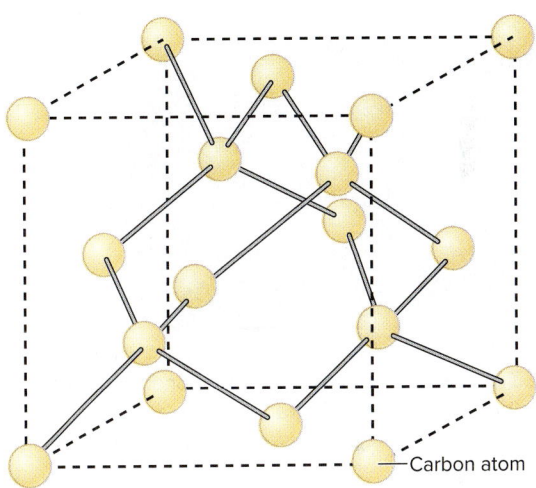

— Carbon atom

Structure of diamond

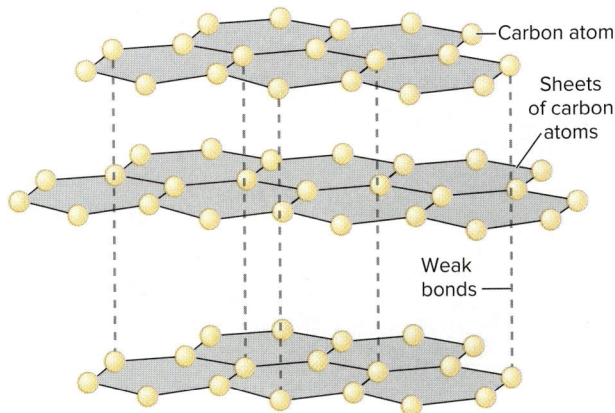

— Carbon atom

Sheets of carbon atoms

Weak bonds —

Structure of graphite

FIGURE 3.6   Although both diamond and graphite are composed entirely of carbon atoms, they have different crystalline structures. Diamond's structure helps make it the hardest known substance, whereas the weak bonds between the sheets of carbon atoms in graphite make it one of the softest minerals.

A

1 cm

B   1 cm

FIGURE 3.7 Rocks are commonly composed of multiple types of mineral grains like the granite in (A), but some contain only a single type of mineral grain, as in the quartzite shown in (B).
(a–b) © J. Geisler

FIGURE 3.8 Minerals can grow crystal faces, as in these quartz crystals, provided there is sufficient space for the faces to develop.
© Doug Sherman/Geofile

whereas graphite is one of the softest. Humans have taken diamond's extreme hardness and put it to practical use in making tools that will cut through *any* material. With respect to graphite, people use it as a dry lubricant and in writing tools because it is very soft. Interestingly, humans have been finding uses for minerals based on their physical properties for thousands of years, beginning perhaps with the extremely sharp edges of broken flint. In Chapter 12 we will explore this link between the physical properties of minerals and their practical applications in modern society.

Rocks

A **rock** is defined as an aggregate or assemblage of one or more types of minerals. Most rocks are composed of several different types of mineral grains bound together, similar to the granite shown in Figure 3.7A. However, there are some rocks that consist almost entirely of grains of one type of mineral. A good example is the rock known as quartzite (Figure 3.7B), where grains of the mineral called quartz are bound together. In addition to mineral composition, another important property of rocks is *texture,* which refers to the way the mineral grains themselves are arranged. The texture of rocks varies depending on such things as the size (coarse, fine, or mixed) and shape (round, angular, tabular, and elongate) of the individual mineral grains. In some cases the grains are oriented in a specific direction, giving a rock what geologists refer to as a *fabric,* similar to how fibers within cloth are oriented parallel to one another. Later in this chapter you will see that texture provides geologists with important clues as to the rock's origin and history.

One of the ways in which mineral grains combine and form rocks is when molten rock, called **magma,** begins to cool. Magma itself forms deep within the Earth where the temperature and pressure cause chemical bonds in minerals to begin breaking down. As the bonds break, the individual atoms within the minerals are released, forming a complex mass of positively and negatively charged ions (e.g., Na^+, Fe^{2+}, Cl^-, SO_4^{2-}). When magma encounters conditions that allow it to cool, the ions will begin to arrange themselves back into the crystalline structure of individual minerals. The mineral grains will then continue to grow as ions are selectively removed from the magma. Minerals also form when ions crystallize out of aqueous (i.e., water) solutions that are rich in dissolved ions, in which case the crystallization process is referred to as *precipitation.* These solutions are often present around magma bodies where the minerals precipitate at relatively high temperatures. Other types of minerals precipitate from cooler waters associated with shallow groundwater systems. In either case, if the precipitating mineral grains have sufficient room to grow, as in a cave or small void space, then they can develop crystal faces similar to the example in Figure 3.8. Note that when precipitation occurs in open bodies of water, mineral grains will settle out over time, creating layers of sediment. The mineral halite (NaCl), or common table salt, forms in this manner.

Finally, new minerals also form when older minerals undergo chemical and physical transformations. This process is quite common when changes in temperature and pressure deep within Earth's crust force minerals to recrystallize—here ions slowly move between individual grains within the rock body. Minerals are also transformed into new ones during chemical reactions involving water and oxygen molecules associated with Earth's surface environment. This process, referred to as *weathering,* is one of the primary ways that clay minerals form. Humans have been taking advantage of this weathering process for thousands of years to make pottery from clay minerals.

Rock-Forming Minerals

Geologists classify minerals based primarily on the type of negatively charged ion in the mineral's crystalline structure—this electrical charge is balanced by positively charged ions. For example, *sulfide* minerals all contain sulfur ions bonded to positively charged ions like lead (PbS), zinc (ZnS), and iron (FeS_2). *Carbonate* minerals all have positive ions bonded to the carbonate ion (CO_3^{2-}), *oxide* minerals contain the oxygen ion (O^{2-}), and *sulfates* contain the sulfate ion (SO_4^{2-}). The *silicate* class contains the greatest number of minerals, in which the basic building block is the silicate ion (SiO_4^{4-}). Figure 3.9 illustrates how the silicate ion itself creates a variety of complex chain and sheetlike structures, all of which can be bonded together by positive ions.

Although there are over 4,000 known minerals on Earth, only a dozen or so make up most of the rocks in Earth's outermost layer known as the *crust* (see Figure 3.4). Geologists generally refer to the common minerals listed in Table 3.1 as **rock-forming minerals.** Note that the table contains several mineral groups, each of which includes different, but closely related minerals. Also note how a majority of the rock-forming minerals are silicates (i.e., contain the SiO_4^{4-} ion), which is why silicon and oxygen atoms make up 75% of Earth's crust by weight. While many of the remaining 4,000 or so minerals are relatively rare and insignificant in terms of the overall volume of rocks, they are often concentrated by geologic processes. Because many of the more rare minerals have important applications in modern society, such localized concentrations can be of considerable economic value—a topic covered in detail in Chapter 12.

TABLE 3.1 Some of the more common minerals that make up most of the rocks in Earth's crust. For a listing of chemical symbols see periodic table, located on last text page.

Mineral Name or Group	Class—Important Negative Ion	Important Positive Ions
Olivine ($[Mg,Fe]_2SiO_4$)	Silicate (SiO_4^{4-})	Mg, Fe
Pyroxene Group (e.g., augite)	Silicate (SiO_4^{4-})	Fe, Mg, Ca
Amphibole Group (e.g., hornblende)	Silicate (SiO_4^{4-})	Ca, Na, Mg, Fe, Al
Plagioclase Feldspar Group	Silicate (SiO_4^{4-})	Ca, Na, Al
Potassium Feldspar Group (e.g., orthoclase)	Silicate (SiO_4^{4-})	K, Al
Mica Group (e.g., muscovite and biotite)	Silicate (SiO_4^{4-})	K, Fe, Mg, Al
Quartz (SiO_2)	Silicate (SiO_4^{4-})	n/a
Clay Minerals Group (e.g., kaolinite and illite)	Silicate (SiO_4^{4-})	K, Ca, Na, Mg, Fe, Al
Calcite ($CaCO_3$)	Carbonate (CO_3^{2-})	Ca
Dolomite ($CaMg[CO_3]_2$)	Carbonate (CO_3^{2-})	Ca, Mg
Gypsum ($CaSO_4 \cdot 2H_2O$)	Sulfate (SO_4^{2-})	Ca

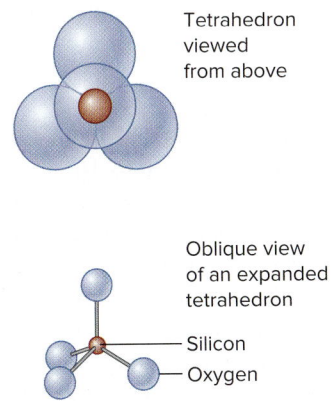

Tetrahedron viewed from above

Oblique view of an expanded tetrahedron

— Silicon
— Oxygen

A

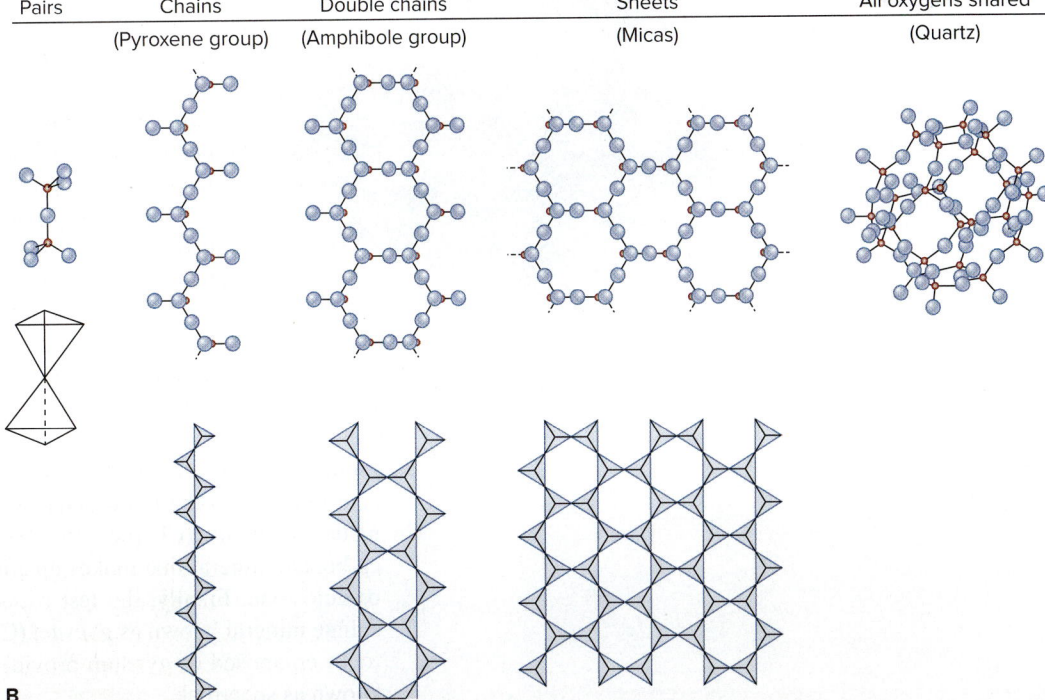

Pairs Chains (Pyroxene group) Double chains (Amphibole group) Sheets (Micas) All oxygens shared (Quartz)

B

FIGURE 3.9 Minerals in the silicate class all have the silicon-oxygen tetrahedron as their basic building block, which can be linked together in the various ways shown here.

FIGURE 3.10   Some of the more important rock-forming minerals: (A) olivine, an iron and magnesium-rich silicate believed to be compositionally similar to the minerals making up the mantle; (B) feldspars, a mineral group that makes up the largest percentage of crustal rocks; (C) quartz, a very abundant mineral in continental rocks and sediment; and (D) micas, a group of common platy minerals.

(a–d) © J. Geisler

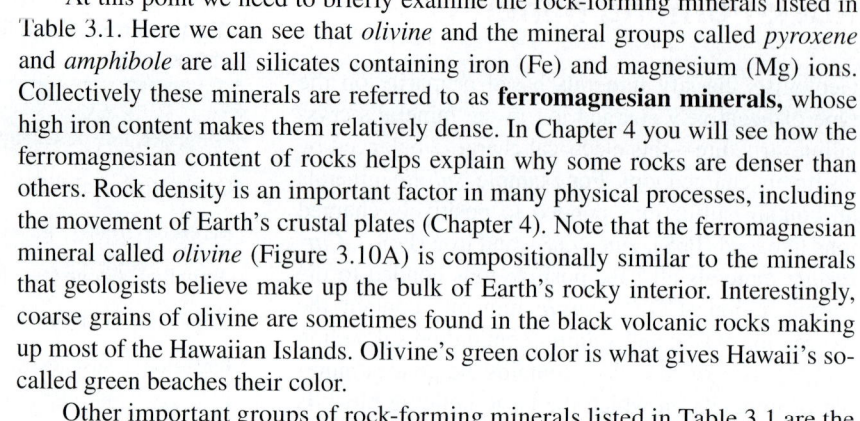

A

B

C

D

At this point we need to briefly examine the rock-forming minerals listed in Table 3.1. Here we can see that *olivine* and the mineral groups called *pyroxene* and *amphibole* are all silicates containing iron (Fe) and magnesium (Mg) ions. Collectively these minerals are referred to as **ferromagnesian minerals,** whose high iron content makes them relatively dense. In Chapter 4 you will see how the ferromagnesian content of rocks helps explain why some rocks are denser than others. Rock density is an important factor in many physical processes, including the movement of Earth's crustal plates (Chapter 4). Note that the ferromagnesian mineral called *olivine* (Figure 3.10A) is compositionally similar to the minerals that geologists believe make up the bulk of Earth's rocky interior. Interestingly, coarse grains of olivine are sometimes found in the black volcanic rocks making up most of the Hawaiian Islands. Olivine's green color is what gives Hawaii's so-called green beaches their color.

Other important groups of rock-forming minerals listed in Table 3.1 are the silicate minerals rich in aluminum (Al) that are called plagioclase and potassium feldspar, often simply referred to as **feldspars** (Figure 3.10B). Together, feldspar and ferromagnesian minerals make up the bulk of Earth's crust. Of particular importance is the fact that when these minerals are exposed to acidic solutions (low pH), such as natural rainwater, they readily break down and are transformed into **clay minerals,** a group of silicate minerals also rich in aluminum. Because feldspar and ferromagnesian minerals are so abundant, clays end up being a major component of the soils and sediment found blanketing the surface. The formation of clays has been critical to the development of complex life on Earth because the base of the food chain, vegetation, is highly dependent on the clay content in soils (Chapter 10). Another common group of silicates that are often transformed into clays includes the platy minerals known collectively as *micas* (Figure 3.10D). Last there is **quartz** (SiO_2), a silicate mineral composed entirely of silicate ions (SiO_4^{4-}). Quartz (Figure 3.10C) is commonly found in significant quantities along with feldspars in the crustal rocks underlying much of Earth's continental landmasses. Unlike feldspars though, quartz is chemically resistant to acidic solutions, and is therefore found in great abundance along with clays in soil and sediment. In fact, much of the sand-sized sediment we find near Earth's surface is composed of small, rounded grains of quartz. Note that pure deposits of quartz are used as the raw material for making glass, whereas clay minerals serve as the raw material for ceramics.

Table 3.1 also lists several rock-forming minerals that are based on carbonate or sulfate ions rather than the silicate ion. The carbonate mineral called **calcite** ($CaCO_3$) is the dominant mineral in an important group of rocks called *limestone,* many of which contain shell fragments composed of calcite produced by marine organisms. Calcite-bearing rocks serve as the raw material for making cement and concrete. A key property of calcite, and calcite-bearing rocks, is the tendency to dissolve in acidic water. For example, when groundwater travels through fractures in limestone, it can slowly dissolve the rock to form passageways and caverns. It turns out that calcite's ability to dissolve creates some unique environmental problems. One problem is when underground caverns collapse and form sinkholes (Chapter 7), and another is when acid rain causes the rapid deterioration of monuments and concrete structures made of calcite (Chapter 15). Note that *dolomite* is a magnesium-bearing carbonate mineral that makes up a fairly common type of rock called dolomite, or dolostone. Finally, the last rock-forming mineral listed in Table 3.1 is the sulfate mineral known as *gypsum* ($CaSO_4 \cdot 2H_2O$). Gypsum is important because rocks composed of gypsum provide the raw material for making drywall, also known as sheetrock.

In the following sections we will explore how rock-forming minerals are assembled into one of the three major types of rocks: *igneous, sedimentary,* or *metamorphic.* Because a thorough review of the classification of rocks is beyond the scope of this text, we will focus on only the most common types of rocks.

Igneous Rocks

Rocks that form when minerals crystallize from a cooling body of magma (molten rock) are referred to as **igneous rocks**—magma that breaches the surface is called *lava.* In general, minerals with the highest melting points crystallize first, which means that only the elements needed for those particular minerals are removed from the cooling magma. As additional minerals crystallize, a mass of interlocking crystals begins to form such that the remaining magma is forced to reside in the space between the mineral crystals—similar to how water occupies the spaces in a glass full of ice. A solid rock eventually forms when the last minerals crystallize from the remaining liquid. Note that a rock consisting of interlocking mineral crystals is said to have a *crystalline texture.* In general, the more time the crystals have to grow before the magma solidifies, the larger the grains become.

Geologists classify igneous rocks based on their texture and chemical composition. Igneous rocks are referred to as *extrusive* if their texture is so fine that individual mineral grains are too small to be seen without the aid of a microscope. Coarse-grained igneous rocks are called *intrusive* and have mineral grains that are clearly visible with the naked eye. Extrusive rocks are fine-grained because they represent magmas that make it to the surface and cool rather quickly, on the order of perhaps several years (Figure 3.11). A cooling period this short gives individual mineral grains very little time to grow. In contrast, intrusive igneous rocks are coarse-grained since they represent magmas that solidify deep within the crust, taking as long as a million years or more to cool. Note that when molten rock breaches the surface, the lava sometimes cools so quickly that the atoms making up the melt do not have time to organize into a crystalline structure. This results in the formation of a noncrystalline solid called *volcanic glass,* often given the rock name *obsidian.* Figure 3.12 illustrates the relationship between the grain size of igneous rocks and the cooling history of magma.

FIGURE 3.11 A lava lake developed inside the crater of Hawaii's Kilauea volcano. When lava is exposed to the surface environment, it cools quickly, resulting in fine-grained igneous rocks.
R.L. Christiansen/USGS

FIGURE 3.12 Volcanic glass (A) is an extrusive rock that cooled from magma so fast that the atoms were not able to establish a crystalline structure and form minerals. Other extrusive rocks such as basalt (B) cool slowly enough that small mineral crystals are able to develop, but are too small to be visible with the naked eye. Intrusive rocks like granite (C) cool much more slowly, allowing mineral grains to grow to the point that they are clearly visible.
(a–c) © Jim Reichard

A Obsidian   1 cm   **B** Basalt   1 cm   **C** Granite   1 cm

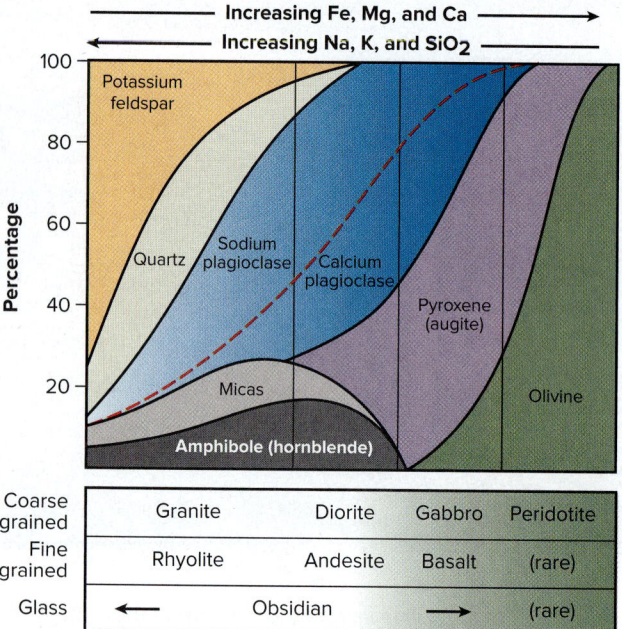

Increasing Fe, Mg, and Ca →

← **Increasing Na, K, and SiO₂** →

	Coarse grained			
	Granite	Diorite	Gabbro	Peridotite
Fine grained	Rhyolite	Andesite	Basalt	(rare)
Glass	←	Obsidian	→	(rare)

FIGURE 3.13   The classification of igneous rocks is based on texture and mineral composition. Of the various rock types, granites and basalts are the most common igneous rocks in Earth's crust.

In addition to texture, igneous rocks are classified based on their relative abundance of iron and magnesium-rich (ferromagnesian) silicate minerals. As illustrated in Figure 3.13, *peridotite* is the name given to intrusive (coarse-grained) igneous rocks composed entirely of ferromagnesian minerals such as olivine and pyroxene—extrusive rocks with this composition are quite rare. The next category includes the extrusive rock called **basalt** and its intrusive equivalent known as *gabbro.* Both of these rocks contain plagioclase feldspar along with lesser amounts of ferromagnesian minerals than in peridotite. As the ferromagnesian content further decreases, amphibole, mica, and plagioclase feldspars increase, creating intrusive rocks called *diorite* and extrusive rocks called *andesite.* The last category in the chart consists of **granite** (intrusive) and *rhyolite* (extrusive), which contain relatively few ferromagnesian minerals, but are rich in quartz and potassium feldspar.

Of the seven igneous rock types described above, basalt and granite were highlighted because they are the most abundant igneous rocks in Earth's crust. In Chapter 4 you will learn how the oceanic portions of Earth's crust are composed largely of basaltic rocks, whereas the continental crust is dominated by rocks of granitic composition. Moreover, the differing ferromagnesian content of basalts and granites creates important density differences that influence the movement of Earth's crustal plates.

Weathering Processes

Before describing the next major rock type, we need to examine how rocks are physically and chemically broken down in the process known as *weathering.* As noted earlier, weathering processes generate loose material that eventually forms sedimentary rocks and the soils that blanket the landscape. To begin, consider what happens to rocks that are exposed to Earth's surface environment. For many rocks Earth's surface is a rather hostile environment. Here they are exposed to liquid water, atmospheric gases, biologic agents, and relatively large fluctuations in temperature, all of which tend to cause rocks to slowly disintegrate and decompose over time. Of considerable importance is the fact that weathering rates are highly dependent on the types of minerals in a particular rock. Another key factor is climate, which, in turn, controls the temperature and availability of water. In the following sections we will explore the two basic types of weathering: *physical* and *chemical.*

Physical Weathering

Scientists use the term **physical weathering** to describe those processes that cause rocks to disintegrate into smaller pieces or particles by some mechanical (i.e., physical) means. The most common way this takes place is when water within rocks freezes and thaws in a repetitive manner. As water freezes, it increases in volume by about 9%, something you can readily observe by freezing a bottle of water. When water freezes in a confined space, this volume increase can exert as much as 30,000 pounds per square inch (2,100 kilograms per square centimeter) on its surroundings. Freezing then can literally break rocks apart because the amount of pressure involved far exceeds the strength of even the hardest and most durable rocks. Note that when water freezes within the internal (pore) space of rocks, the mechanical action generally affects only the outer surface of the rock. However, if the rocks contain planar openings called *fractures,* the effects can extend much deeper into the rock mass. As illustrated in Figure 3.14, a process called *frost wedging* occurs when a fracture filled with water undergoes repetitive freeze and thaw cycles, causing the expanding ice to act as a wedge that eventually breaks the rock in two. Frost wedging is most effective in fractured rocks and in climates where water is abundant and temperatures commonly range above and below the freezing point of water.

Freezing water is not the only mechanical means by which rock fractures can be wedged open. For example, fractures are sometimes filled with fluids rich in dissolved ions. Under the right conditions minerals can precipitate from these solutions, and as the crystals grow, they exert sufficient pressure to wedge fractures open. Wedging also takes place when plant roots, particularly tree roots, move down into a fracture and expand in size as the plant grows. In addition to widening existing fractures, some mechanical processes actually create new fractures. For example, some deeply buried rocks develop fractures when erosion strips away the overlying rocks, thereby reducing the weight or pressure on the rocks below. This decrease in pressure allows the rocks to literally expand and begin to fracture. Fractures can also form when surface rocks experience large fluctuations in daily temperature, causing them to expand and contract on a daily basis. Eventually this repetitive expansion and contraction weakens the rocks to the point that they begin to fracture. Regardless of how fractures form, they result in large rock masses being broken down into smaller particles (Figure 3.14). This process also greatly increases the surface area of the rock where chemical reactions can occur. Therefore, the physical breakdown into smaller particles can greatly enhance the chemical weathering of rocks.

Chemical Weathering

The term **chemical weathering** refers to the decomposition of minerals in rocks via chemical reactions. Chemical weathering can be thought of as a process where minerals decompose into simpler compounds and where individual ions are released into the surrounding environment. In general, minerals become susceptible to chemical weathering when they encounter an environment different from the one in which they formed. For example, an igneous rock like granite forms deep below the surface in a relatively closed environment where temperature and pressure are stable. When erosion strips away the overlying material, the granite can become exposed to the surface environment. In this new environment certain minerals within the rock may no longer be chemically stable, particularly in the presence of atmospheric gases and liquid water. These minerals then will begin to chemically decompose. Because the water molecule plays an essential role in many chemical reactions, climate and the availability of water are key factors in chemical weathering. Also important is the climate's temperature range, since chemical reaction rates typically increase with temperature. In general, chemical weathering is most pronounced in warm climates where rainfall is plentiful, and in rocks with minerals that are chemically unstable under such climatic conditions.

Although there are over 4,000 known minerals on Earth, there are only three basic types of chemical reactions that break down minerals: *dissolution, hydrolysis,* and *oxidation/reduction.* Figure 3.15 provides example reactions to help illustrate the chemical changes that are involved in each type of reaction (names of chemical symbols are listed in the periodic table). From this figure one can see that in **dissolution** reactions there is no solid product on the right side of the equation (try not to let the chemistry intimidate you as this is actually quite simple). What this means is that the original mineral will completely dissolve or disassociate in water, leaving only individual ions (charged atoms) in the solution. A familiar example is how halite (i.e., common table salt) completely dissolves in water. When you pour halite into a pan of water, the chemical bonds

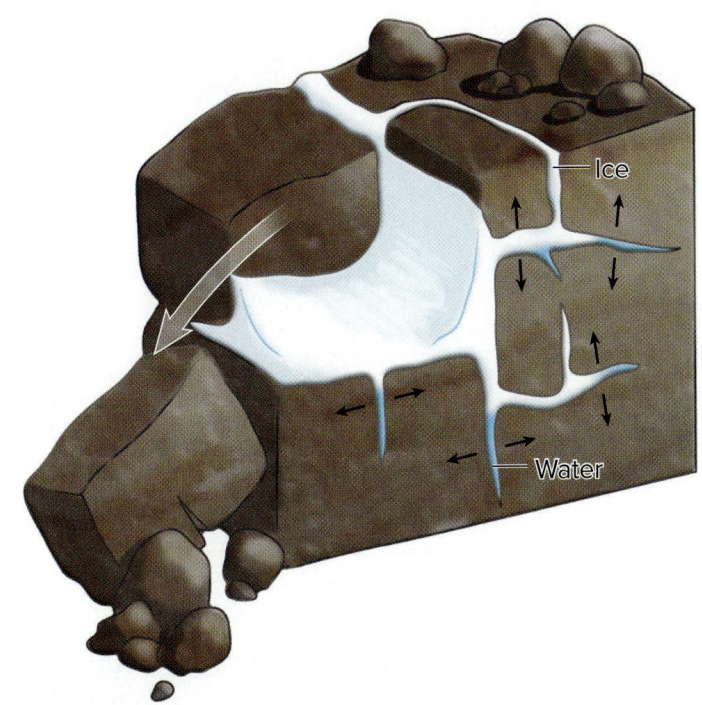

A

B

FIGURE 3.14 One of the ways physical weathering occurs is when water repeatedly freezes and expands within a fracture (A), slowly wedging the rock into smaller pieces. This breakage causes the surface area of the rock body to increase dramatically, thereby increasing the area where chemical weathering can take place. Photo (B) shows a large slab of rock slowly being wedged away from a rock body in British Columbia, Canada.
(b) Marten Geertsema, Geologic Survey of Canada

within the mineral begin to break, thereby releasing positively charged sodium (Na^+) ions and negative chlorine (Cl^-) ions into the solution. These dissolved ions (i.e., salts) are what give the water its salty taste.

The other dissolution example in Figure 3.15 is the reaction for calcite ($CaCO_3$). The dissolution of calcite is geologically significant because calcite is the primary mineral in limestone, a fairly abundant rock type. Note how this reaction requires the presence of an acid in order for calcite to dissolve. This requirement means that calcite will dissolve faster in more acidic waters. There are many

FIGURE 3.15
Important types of chemical reactions involved in the chemical weathering of minerals. Example reactions are shown for a few of the more common minerals that undergo chemical decomposition.

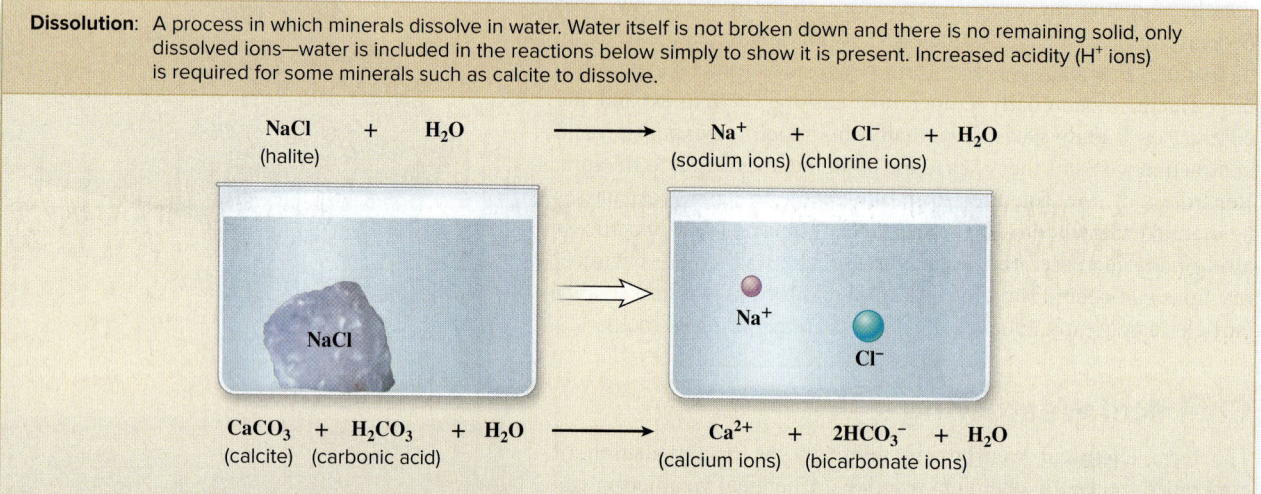

Dissolution: A process in which minerals dissolve in water. Water itself is not broken down and there is no remaining solid, only dissolved ions—water is included in the reactions below simply to show it is present. Increased acidity (H^+ ions) is required for some minerals such as calcite to dissolve.

$$NaCl + H_2O \longrightarrow Na^+ + Cl^- + H_2O$$
(halite)   (sodium ions) (chlorine ions)

Na^+

Cl^-

$$CaCO_3 + H_2CO_3 + H_2O \longrightarrow Ca^{2+} + 2HCO_3^- + H_2O$$
(calcite) (carbonic acid)   (calcium ions) (bicarbonate ions)

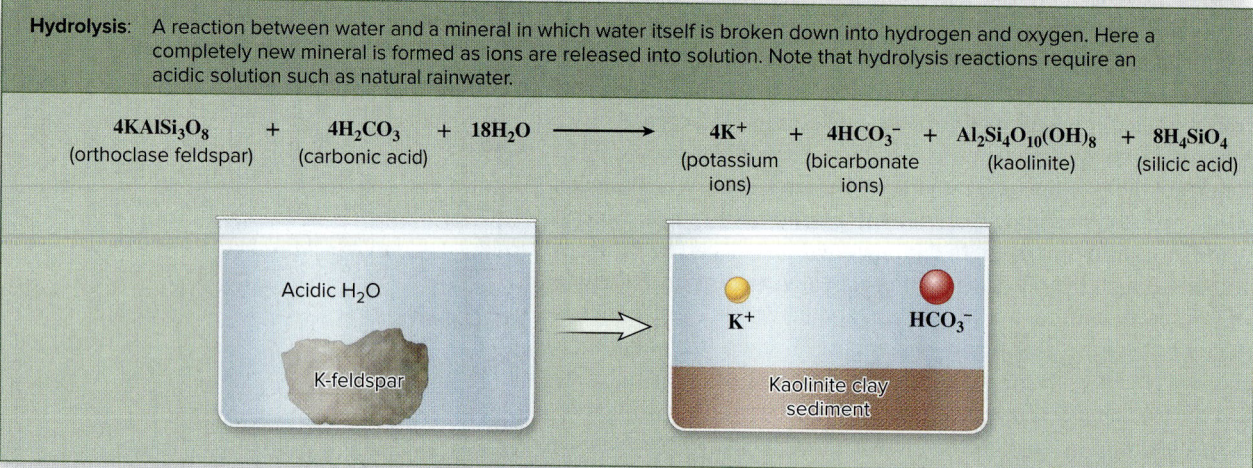

Hydrolysis: A reaction between water and a mineral in which water itself is broken down into hydrogen and oxygen. Here a completely new mineral is formed as ions are released into solution. Note that hydrolysis reactions require an acidic solution such as natural rainwater.

$$4KAlSi_3O_8 + 4H_2CO_3 + 18H_2O \longrightarrow 4K^+ + 4HCO_3^- + Al_2Si_4O_{10}(OH)_8 + 8H_4SiO_4$$
(orthoclase feldspar) (carbonic acid)   (potassium ions) (bicarbonate ions) (kaolinite) (silicic acid)

Acidic H_2O

K-feldspar

K^+

HCO_3^-

Kaolinite clay sediment

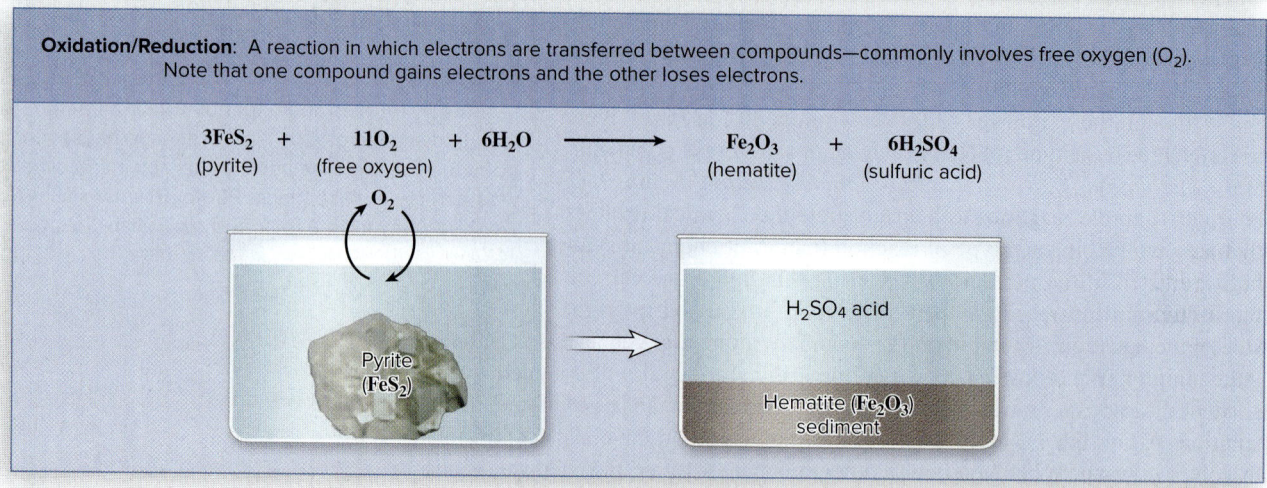

Oxidation/Reduction: A reaction in which electrons are transferred between compounds—commonly involves free oxygen (O_2). Note that one compound gains electrons and the other loses electrons.

$$3FeS_2 + 11O_2 + 6H_2O \longrightarrow Fe_2O_3 + 6H_2SO_4$$
(pyrite) (free oxygen)   (hematite) (sulfuric acid)

O_2

Pyrite (FeS_2)

H_2SO_4 acid

Hematite (Fe_2O_3) sediment

types of acids, but one of the most common is the carbonic acid that forms when water reacts with atmospheric carbon dioxide (CO_2). Because CO_2 is well mixed in the atmosphere, rainwater is naturally acidic. Note that unusually acidic rainwater forms from sulfur dioxide (SO_2) and is primarily a problem caused by human activity (Chapter 15). The key point here is that since rainfall is naturally acidic, calcite-rich rocks (Figure 3.16A) will dissolve more rapidly in climates where rainfall is more plentiful.

The second type of reaction shown in Figure 3.15 is known as **hydrolysis,** which differs from simple dissolution in that water molecules actually take part in the chemical reaction. For example, notice in the reaction for the potassium feldspar mineral called orthoclase how the H_2O molecule itself is broken down to create other chemical compounds. Another important difference is that in hydrolysis the original mineral does not simply dissolve away, but is transformed into a new mineral—often called a *secondary mineral* or *weathering product.* In the feldspar example you can see how potassium ions (K^+) are released into the water and the original mineral is transformed into *kaolinite*—one of the clay minerals in Table 3.1. Similar reactions transform calcium- and sodium-rich feldspars into clay minerals, but release calcium (Ca^{2+}) and sodium (Na^+) ions instead. Because feldspar minerals are so abundant in crustal rocks, the breakdown of these minerals by hydrolysis is the primary means by which sodium, potassium, and calcium ions are released into Earth's surface environment. Finally, it is worth noting that the decomposition of feldspars requires acidic solutions. Again, atmospheric CO_2 plays a critical role in generating naturally acidic rainfall, which makes the breakdown of feldspars and calcite possible.

The last type of reaction listed in Figure 3.15 is called **oxidation/reduction,** where individual atoms exchange electrons. When an atom gains or loses electrons, its chemical properties change such that it can then form an entirely new mineral or compound. Perhaps the most familiar oxidation/reduction reactions are those in which iron objects Figure 3.16B) begin to rust (i.e., oxidize). In the presence of water and free oxygen (O_2), iron atoms in minerals can exchange electrons with oxygen atoms, resulting in the formation of an *iron oxide* mineral. Iron oxide minerals are usually reddish, yellowish, or brownish in color. There are a large number of iron-bearing minerals that are susceptible to oxidation/reduction reactions, but the most abundant are the ferromagnesian silicates common in rocks like basalt. Because these minerals are susceptible to hydrolysis and oxidation/reduction reactions, the chemical weathering of ferromagnesian-rich rocks generates sediment consisting of iron-rich clays and iron-oxide minerals.

Another important oxidation/reduction reaction shown in Figure 3.15 is the one for the iron mineral called pyrite—more commonly known as *fool's gold.* Pyrite is of considerable interest in environmental geology because it is found in a wide variety of rock types and oxidizes very rapidly. Note in the chemical reaction for pyrite how it reacts with atmospheric oxygen (O_2) and water to form the bright red, iron oxide mineral known as hematite. Consequently, soils derived from rocks containing even small amounts of pyrite are typically bright red due to the presence of hematite. Also note how the sulfur (S) from pyrite reacts with water to form sulfuric acid. You will see in Chapter 12 how the mining and processing of rocks containing pyrite is the primary way in which humans release sulfuric acid into the environment. Sulfuric acid is particularly strong and causes significant damage to both the environment and human structures.

A

B

FIGURE 3.16 Materials made of the mineral calcite, such as this tombstone (A), readily break down by dissolution reactions with acidic rainwater. The iron found in automobiles (B) undergoes oxidation/reduction reactions and forms a variety of secondary minerals collectively known as iron oxides

FIGURE 3.17   Sediment that forms by the weathering of rocks is normally transported to some other site where it is deposited. Given the right conditions, the deposit may turn into sedimentary rocks. Example photos show massive sandstone rock (A) that represents ancient sand dune deposits, active wind transport of sediment (B), and a wind-blown sand deposit (C).

(a) © Jim Reichard; (b) © MarketPlace/Media Bakery; (c) © Royalty Free/Corbis

A Coconino sandstone, Grand Canyon

Sedimentary Rocks

When rocks are exposed to Earth's surface environment, they naturally undergo physical weathering and are mechanically broken down into smaller rock fragments and mineral grains, collectively called **sediment.** While this is occurring, minerals within the rocks that are susceptible to chemical weathering start decomposing. This generates secondary minerals (e.g., clays) and releases electrically charged ions, which attach themselves to water molecules. Both physical and chemical weathering therefore produce sediment. A different process known as **erosion** occurs when sediment and ions are removed from a given area. This takes place whenever rock or sediment is chemically dissolved, physically picked up, or mechanically worn down by the slow abrasive action of moving sediment particles. The way in which Earth materials actually move from one location to another is called **transportation,** which involves some combination of gravity, running water, glacial ice, and wind. During transportation, material is removed from areas of higher elevation and carried to low-lying areas, where it accumulates in a process referred to as **deposition.** In the case of dissolved ions, they are carried away by flowing water and end up either in a body of surface water or as part of the groundwater system (Chapter 11).

Given enough geologic time, the combination of weathering, erosion, and transportation results in entire mountain ranges being broken down into sediment and ions and then deposited elsewhere. In this section we are interested in how sediment grains and ions are reassembled to form what geologists call **sedimentary rocks** (Figure 3.17). We will examine the two basic types of sedimentary rocks: those fconsisting of weathered rock and mineral fragments and those composed of new mineral grains that chemically precipitate from dissolved ions. Sedimentary rocks are particularly important in environmental geology because they often contain void spaces between the individual sediment grains. These void spaces are capable of storing significant quantities of freshwater and petroleum, which, of course, represent two of society's most critical resources. Sedimentary rocks also serve as the raw materials for making glass, concrete, drywall (sheetrock), and a host of other items of great importance. Consequently, the information on sedimentary rocks that follows will be quite useful in understanding the topics of water resources, mineral and rock resources, and petroleum supplies (Chapters 11, 12, and 13, respectively).

B Sandstorm approaching a town in Africa

C Death Valley, California

Detrital Rocks

Perhaps the most common type of sedimentary rocks are those known as *detrital sedimentary rocks,* which consist of preexisting rock and mineral fragments, called *detritus,* cemented together to form a solid rock (because geologists refer to the granular nature of this material as having a *clastic texture,* the term *clastic* is sometimes used in place of detrital). As illustrated in Figure 3.18, the process of turning sediment into detrital rocks begins when sediment is deposited in low-lying areas, reaching thicknesses of hundreds or even thousands of feet. Depositional sites include lakes, shallow seas, and near-shore areas of the oceans. Here water naturally fills the void or *pore spaces* that exist between sediment grains. As the sediment is buried progressively deeper, the weight of the overlying layers causes the individual grains to rearrange themselves such that the sediment becomes more compact. With increased depth of burial both the sediment and its pore water are exposed to higher temperatures and pressures. In this environment new minerals begin to precipitate from the dissolved ions within the pore water. The new mineral matter acts to bind or cement the sediment grains together, thereby forming layers of solid rock. It is through this process of *burial, compaction,* and *cementation* that loose sediment is gradually transformed into sedimentary rock sequences (Figure 3.18).

Of considerable interest to geologists are the size and composition of the sediment itself. Earlier you learned that the bulk of Earth's crust is composed of ferromagnesian, feldspar, and quartz minerals. When rocks containing these minerals are exposed to chemical weathering, the ferromagnesian and feldspar minerals are transformed into clay minerals. Quartz, on the other hand,

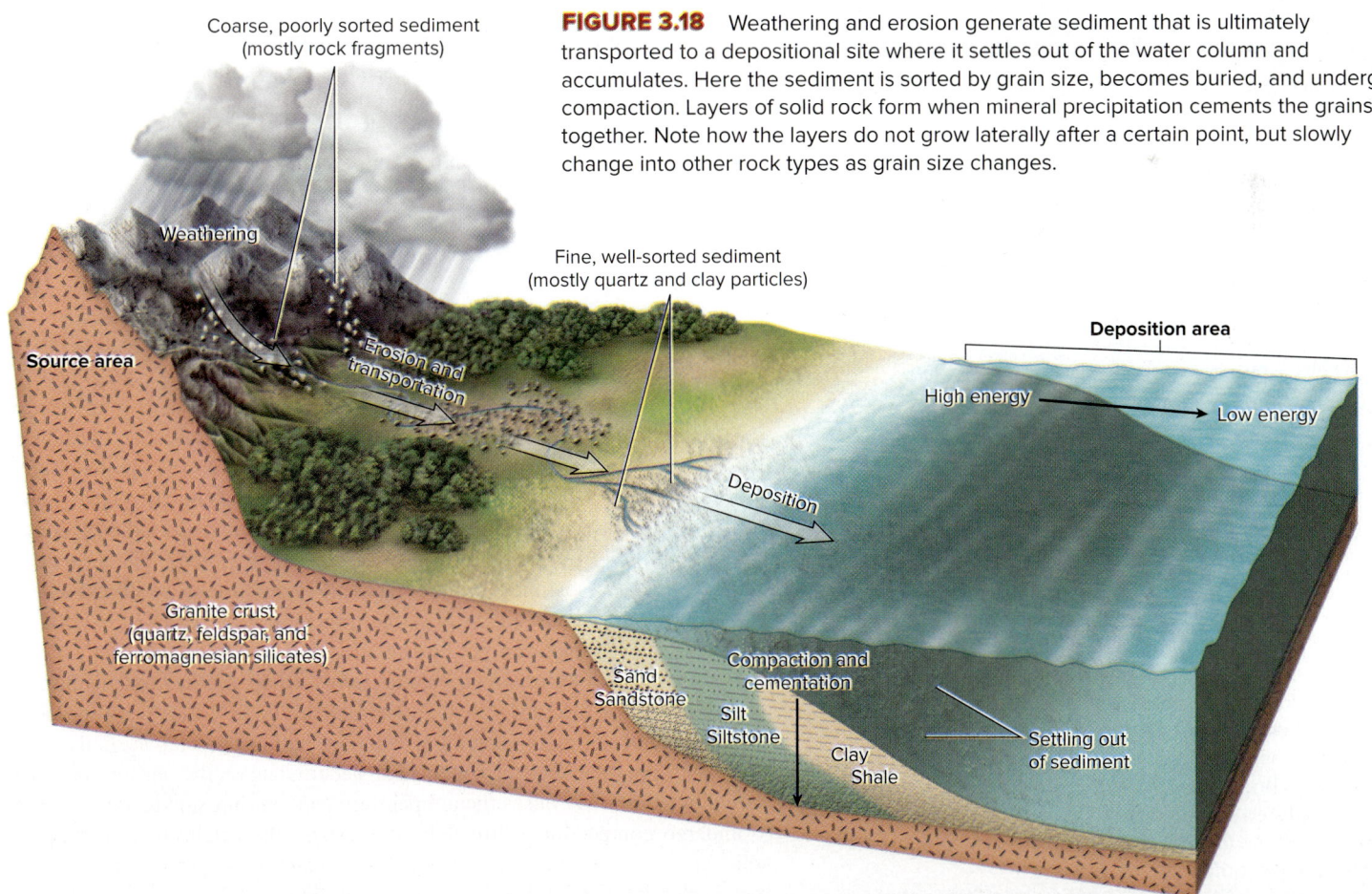

FIGURE 3.18 Weathering and erosion generate sediment that is ultimately transported to a depositional site where it settles out of the water column and accumulates. Here the sediment is sorted by grain size, becomes buried, and undergoes compaction. Layers of solid rock form when mineral precipitation cements the grains together. Note how the layers do not grow laterally after a certain point, but slowly change into other rock types as grain size changes.

Coarse, poorly sorted sediment
(mostly rock fragments)

Weathering

Fine, well-sorted sediment
(mostly quartz and clay particles)

Source area

Erosion and transportation

Deposition area

High energy   Low energy

Deposition

Granite crust
(quartz, feldspar, and
ferromagnesian silicates)

Compaction and
cementation

Sand
Sandstone

Silt
Siltstone

Clay
Shale

Settling out
of sediment

TABLE 3.2   Common detrital sedimentary rocks. Classification is based on grain size and shape.

Particle Diameter	Sediment	Rock Name
> 2 mm	Gravel	Conglomerate (rounded) or breccia (angular)
1/16 to 2 mm	Sand	Sandstone
1/256 to 1/16 mm	Silt	Siltstone
< 1/256 mm	Clay	Shale

A

B

FIGURE 3.19   Detrital sedimentary rock called conglomerate (A) consists of coarse rock and mineral fragments, which represent sediment that is young and has not traveled far. As the transport distance increases, feldspar and ferromagnesian minerals in the fragments break down into clay particles, whereas quartz remains unaltered and tends to dominate the grain size called sand. Photo (B) shows a sandstone rock composed almost entirely of quartz grains.

(a) © Norris Jones; (b) © Charlie Jones; (inset) Courtesy of Stan Celestian

is extremely resistant to chemical weathering, and thus remains largely unaltered. Consequently, when sediment is transported, particularly by running water, the rock fragments not only get progressively smaller due to mechanical action, but the ferromagnesian and feldspar minerals turn into clay minerals. Ultimately the rock fragments break apart during transportation, liberating individual quartz grains and clay particles. As time increases and sediment travels farther from its source area, the overall grain size and number of rock fragments will decrease (Figure 3.18). Given enough distance and time, the sediment will eventually be dominated by individual grains of quartz and clay minerals. It will also become better sorted, meaning the fragments are closer to being all the same size.

When sediment reaches its final resting place, such as the mouth of a river, it is sorted one last time. Here differences in water energy, related to stream velocity or wave action, will sort sediment particles based largely on grain size (Figure 3.18). In the case of sediment consisting of sand, silt, and clay-sized particles, the larger sand grains are deposited in areas where the water energy is relatively high. Progressively finer silt and clay particles are deposited farther out as energy continues to decrease. In this way, fairly uniform layers of sand, silt, and clay are deposited. Note in Figure 3.18 how the grain size of an individual layer is not constant, but becomes finer as one moves laterally in the direction of lower energy.

Because of the way sediment is naturally sorted by size, geologists classify detrital sedimentary rocks primarily on grain size. As can be seen in Table 3.2, sedimentary rocks dominated by fine, clay-sized particles are called **shale**, those consisting of silt particles are called *siltstone*, and those with more coarse sand grains are called **sandstone**. Rocks composed of gravel-sized particles are referred to as *conglomerate* or *breccia*, depending on whether the particles are angular or have been rounded during transportation. In general, sediment that has been transported great distances will form sandstones and siltstones dominated by quartz grains, whereas shales are composed largely of clay particles. This differentiation results from the fact that quartz is highly resistant to chemical weathering, but feldspar and ferromagnesian minerals decompose into clay minerals. On the other hand, conglomerates and breccias typically consist of various rock and mineral fragments, representing sediment that had not traveled very far, hence had less exposure to chemical weathering. The examples in Figure 3.19 illustrate the relationship between transport distance and sediment composition and size.

Chemical Rocks

The other major class of sedimentary rocks is referred to as *chemical sedimentary rocks,* which form when dissolved ions precipitate from water to form new mineral grains. Table 3.3 lists some of the more common chemical rocks—note that coal is composed of altered plant material rather than mineral matter, thus it is not a chemical precipitate. The most common chemical sedimentary rocks are those called **limestone,** which consist chiefly of the mineral calcite ($CaCO_3$). Although calcite can precipitate inorganically from seawater or groundwater, most calcite forms when marine organisms remove calcium (Ca^{2+}) and carbonate (CO_3^{2-}) ions from seawater and then biochemically precipitate their own hard parts or protective shells. As these organisms die, their skeletal remains can accumulate on the seafloor and become buried, which slowly raises the temperature and pressure and causes the material to undergo compaction. Ultimately, this causes the calcite to recrystallize into solid limestone. Should the original fossil material be preserved, the resulting rock is classified as a *fossiliferous limestone* (Figure 3.20).

TABLE 3.3   A list of some of the more common chemical sedimentary rocks. The classification for these rocks is based on mineral composition. Note that coal is often listed as a chemical rock despite the fact it consists of altered plant material and does not form by chemical precipitation.

Composition	Rock Name
Calcite ($CaCO_3$)	Fossiliferous limestone
Calcite ($CaCO_3$)	Crystalline limestone
Calcite ($CaCO_3$)	Chalk
Dolomite ($CaMg[CO_3]_2$)	Dolomite
Microcrystalline quartz (SiO_2)	Chert
Gypsum ($CaSO_4 \cdot 2H_2O$)	Rock gypsum
Halite ($NaCl$)	Rock salt
Altered plant material	Coal

FIGURE 3.20   Many marine organisms create body parts made of calcite by extracting dissolved ions from seawater. Their skeletal remains can accumulate on the seafloor over time to form fossiliferous limestone, like the rock shown here.
© Charlie Jones

In some cases the burial and recrystallization process destroys any visible evidence of the original fossil material, producing a rock consisting of a mass of interlocking calcite grains that geologists call a *crystalline limestone*. The crystalline texture of this type of limestone is similar to that of igneous rocks where individual mineral crystals interlock as they grow. Note that under the right conditions calcite grains will chemically react with magnesium ions (Mg^{2+}) in seawater to form the mineral called dolomite ($CaMg[CO_3]_2$), resulting in a rock called *dolomite*—some geologists use the term *dolostone*.

Although calcite can precipitate inorganically, the most common mechanism is carried out by marine organisms that thrive in what are known as reef systems. As shown in Figure 3.21, reefs typically form in shallow environments where the water is relatively free of suspended sediment. This lack of sediment causes the water to be rather clear, which, in turn, allows direct sunlight to penetrate deeper into the water column. Such conditions can occur close to shore in areas where very little sediment is being transported by streams into the marine environment.

FIGURE 3.21   Fossiliferous limestone typically forms where the water column is free of suspended sediment (A), allowing calcite-producing marine organisms to thrive. Limestone forms in shallow seas beyond the point where sediment settles out to form detrital rocks (B) or in near-shore areas where there is minimal sediment influx (C).
(a) © Cavan Images/Getty Images

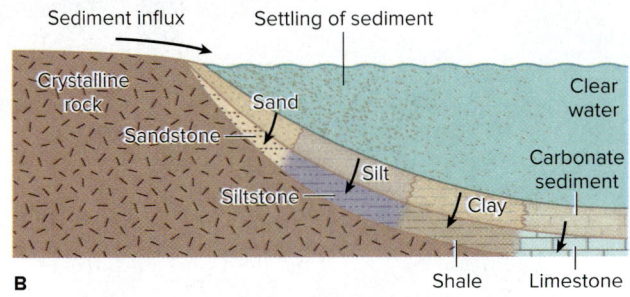

FIGURE 3.22 The exposed limestone rocks of the Guadalupe Mountains in Texas are approximately 250 million years old and were once part of an extensive marine reef system.

© John Karpovich/University of Virginia

Relatively clear water can also be found farther offshore, beyond the point where clay particles are settling out of the water column. A good example of a major deposit of reef limestone is the Guadalupe Mountains in Texas (Figure 3.22). This limestone sequence, approximately 250 million years old, represents part of an extensive reef system that was over 400 miles (650 km) in length. Finally, there is a special type of fossiliferous limestone geologists call *chalk* (Table 3.3), consisting of tiny calcite shells made by single-celled plants and animals. The White Cliffs of Dover, England, are perhaps the most famous example of an extensive chalk deposit. Because of its fine-grained texture and low hardness, chalk has historically been used as a writing tool.

The rock known as *chert* is commonly found as nodules within beds of limestone and as discrete layers, and has long been used by humans for making stone tools (Figure 3.23). Chert is quite dense and composed of extremely fine, interlocking crystals of quartz (SiO_2)—often referred to as microcrystalline quartz (Table 3.3). Most chert is thought to originate from tiny marine organisms that biochemically secrete shells made of SiO_2 rather than calcite. After burial, the SiO_2 shell material recrystallizes into a more compact and dense form. Because chert comes in a variety of forms and colors, it has been given specialized names, such as *flint, jasper,* and *agate.* Since chert is quite hard and forms razor-sharp edges when broken, it has played an important role throughout human history for making stone tools, including arrowheads and implements for scraping animal hides.

Another important group of chemical sedimentary rocks form when a body of water evaporates such that the concentration of dissolved ions becomes so great that minerals begin to precipitate. The newly formed minerals then settle out of the water column, creating layers of sediment on the seafloor or lakebed (Figure 3.24). Rocks that form in this manner are referred to as *evaporites,* and include valuable mineral deposits of halite (common table salt—NaCl) and gypsum ($CaSO_4•2H_2O$). Because different evaporite minerals precipitate under specific chemical conditions, individual minerals typically form in different locations within a body of water (Chapter 12). In this way relatively pure deposits of *rock salt* are able to form, consisting almost entirely of the mineral halite. Layers of nearly pure gypsum, known as *rock gypsum,* form in a similar manner. Note that rock salt is used as a source of sodium and chlorine in the chemical industry and as a deicing agent for roadways, whereas gypsum is the raw material used in making drywall (sheetrock).

Finally, **coal** is a sedimentary rock composed of altered plant remains, which originally accumulate in swamps where oxygen-poor conditions help preserve the organic matter. As additional plant material accumulates, the underlying material undergoes compaction and is transformed into what is known as *peat.* Should the peat become even more deeply buried, temperature and pressure can increase to the point where the organic matter undergoes chemical and physical changes. This process results in a more concentrated form of carbon we call coal. Because coal gives off considerable amounts of heat during combustion and is quite abundant, modern societies have found it to be a useful and inexpensive source of energy. The origin and use of coal will be described more thoroughly in Chapter 13 on fossil fuels.

FIGURE 3.23 Chert is commonly found associated with sedimentary rock deposits. Because of its hardness and ability to create sharp edges when broken, chert has been used for thousands of years to make stone tools, such as these arrowheads.

© Getty Images; (inset) © Steve Cole/Photodisc/Getty Images

Metamorphic Rocks

In the previous section you learned that when sediment is buried, it undergoes compaction and experiences increases in temperature and pressure. This triggers physical and chemical changes that transform the sediment into rock. Similarly, when rocks are placed in a new environment, they often undergo physical and chemical changes that transform them into new rock types. In some cases the temperature and pressure become high enough that certain minerals within the rocks become chemically unstable, at which point they begin transforming into new minerals. This results in an altered rock that contains a new assemblage of minerals that are chemically stable under higher levels of temperature and pressure. Geologists use the term *metamorphism* to describe the process where rocks are altered by some combination of heat, pressure, and fluids, producing what are called **metamorphic rocks.** Metamorphism may cause rocks to become so highly altered that the original rock type can no longer be recognized. Other times the temperature may become so high that the metamorphic rocks begin to melt, forming magma that eventually cools into igneous rocks.

Metamorphism occurs in two basic types of geologic environments. One involves an increase in heat, and the other an increase in both heat and pressure—reactive fluids are common in both environments. As illustrated in Figure 3.25, *contact metamorphism* occurs when magma comes into contact with preexisting rocks. Here just the addition of heat causes minerals to recrystallize into larger grains and/or be transformed into new and more stable minerals. Because any type of rock (igneous, sedimentary, or metamorphic) can undergo metamorphism, the degree to which a particular rock is altered is highly dependent on the original minerals it contains. Also important is the amount of heat the minerals are exposed to and the presence of reactive fluids in the metamorphic environment. For example, in Figure 3.25 one can see that the effect of the magma coming into contact with igneous rock is minimal since granitic minerals are generally stable at higher temperatures. On the other hand, sedimentary rocks often contain minerals that are very susceptible to being altered during contact metamorphism. In the case of limestone composed of calcite, the mineral grains typically recrystallize into larger grains of calcite, forming the metamorphic rock known as *marble.* Similarly, when a quartz sandstone undergoes metamorphism, the quartz grains recrystallize and form a more coarse-grained rock known as *quartzite.*

The other major type of metamorphism is known as *regional metamorphism,* which occurs when deeply buried rocks are exposed to elevated levels of both temperature and pressure. As with all metamorphic environments, the minerals that are unstable under the new conditions will either recrystallize or be transformed into more stable minerals. However, as illustrated in Figure 3.26, the increased pressure during regional metamorphism also causes elongate and sheetlike (i.e., platy) minerals to reorient themselves in a parallel manner. This parallel realignment of minerals gives the rocks a **foliated texture,** similar to how wood fibers align themselves in a parallel manner to form sheets of paper. For example, when shales undergo regional metamorphism, their clay minerals are transformed into platy minerals of the mica family, forming a fine-grained and highly foliated rock called **slate** (Figure 3.27A). In the case of granites composed primarily of quartz, feldspar, and ferromagnesian minerals, regional metamorphism produces a strongly

FIGURE 3.24 Death Valley in California once held a freshwater lake that eventually evaporated, causing the concentration of dissolved ions to become so great that minerals began to precipitate. Shown here is an evaporite deposit of mostly rock salt (halite) covering the valley floor.
© William Perry/age fotostock; (inset) © Doug Sherman/Geofile

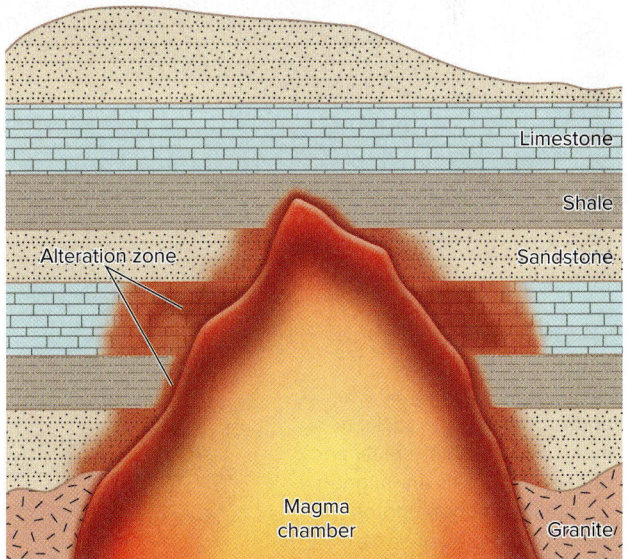

FIGURE 3.25 When magma comes into contact with rocks, the increased heat can cause minerals to recrystallize into larger grains and/or be transformed into more stable minerals. The width of the metamorphic alteration zone depends on how susceptible the original minerals are to higher levels of heat.

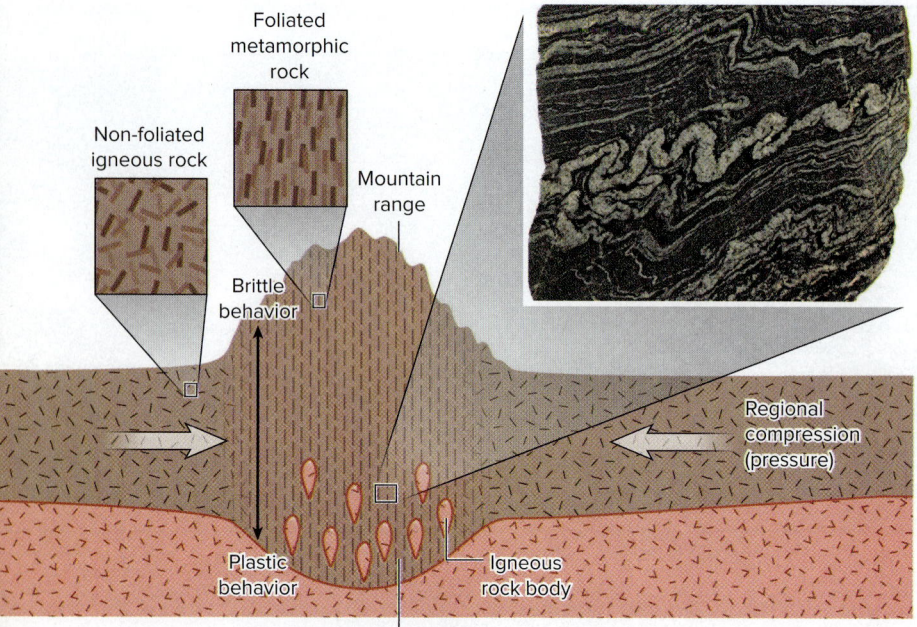

FIGURE 3.26   Regional metamorphism commonly occurs when deeply buried rocks are subjected to compressive forces. Elevated levels of both heat and pressure cause minerals within the rocks to recrystallize or be transformed into more stable minerals. The directed pressure forces elongated and platy minerals to become aligned, giving the rock a foliated (layered) texture. At higher levels of heat and pressure, rocks may begin to deform by flowing in the solid state (plastic flow) as opposed to fracturing in a brittle manner. At high enough temperatures the rocks can begin to melt and form magma.
(inset) © Siim Sepp/Alamy

foliated rock known as *schist,* which is dominated by micas and recrystallized quartz grains (Figure 3.27B). If the temperature and pressure become high enough, the minerals will start to separate into dark and light bands, forming a rock called *gneiss* (Figure 3.27C). Note that when rocks are exposed to increasing levels of temperature and pressure, they tend to become less brittle and develop fewer fractures. Instead of fracturing, the rocks begin to deform in a plastic manner and literally begin to flow, as is evident in the deformed gneiss shown in Figure 3.26.

The Rock Cycle

Based on what you learned about rocks in this chapter, it should be clear that geologic processes can take any type of rock and transform it into an entirely different type. Igneous rocks, for example, can undergo weathering and erosion and be transformed into layers of sandstone and shale. These sedimentary rocks can later be altered by heat and pressure and become metamorphic rocks, which might even undergo melting and form magma. The magma, of course, would eventually cool and create igneous rock. Geologists refer to this recycling of rocks from one rock type to another as the **rock cycle.** As shown in Figure 3.28, the rock cycle is commonly represented in terms of a flow chart in order to illustrate the various ways in which rocks can be transformed. This simple flow chart is a helpful tool for understanding the fairly complex ways in which rocks can be recycled.

Imagine if we could follow the possible paths that an igneous body of granite might take through the rock cycle in Figure 3.28. One path would be for the granite to become exposed at the surface. Here weathering and erosion processes would slowly break the rock down into sediment and dissolved ions. If the sediment were to undergo compaction and cementation after being transported to a depositional site, it would develop into detrital sedimentary rocks. Likewise, some of the ions might precipitate into minerals and form chemical sedimentary rocks. Should these new sedimentary rocks later become exposed to high levels of heat and pressure, they would be transformed into metamorphic rocks. However, should the heat and pressure become too intense, the metamorphic rocks would begin to melt and form magma. The magma of course would eventually cool and form igneous rocks, thereby completing the cycle.

From Figure 3.28 you can see that our original body of igneous rock could take a different path through the rock cycle; namely, it could stay buried and never be exposed to weathering and erosion. In this case the igneous rock could either remain unaltered for eons of time or undergo metamorphism and be transformed into metamorphic rocks. Should the heat and pressure continue to increase, these rocks would begin to melt and ultimately turn back into igneous rocks. However,

FIGURE 3.27

The increased pressure associated with regional metamorphism gives rocks a foliated texture where platy and elongated minerals are aligned in a parallel manner. Photos showing examples of some of the more common types of foliated metamorphic rocks.
(a–c) © J. Geisler

the metamorphic rocks do not necessarily have to melt, but instead may stay buried and experience repeated episodes of metamorphism. The rocks could also be uplifted and become exposed to weathering and erosion, and hence turn into sedimentary rocks. Note that sedimentary rocks themselves can undergo weathering and erosion and be recycled back into sedimentary rocks. Ultimately then, any of the three rock types can either be: (a) exposed at the surface and transformed into sedimentary rocks; (b) remain buried and unaltered; (c) go through metamorphism; or (d) experience extreme metamorphism and begin to melt and form magma. The particular path a rock body takes through the rock cycle is entirely dependent upon changes in its geologic environment.

A key question then is why do rocks encounter different geologic environments within the rock cycle? In other words, why do some rocks remain buried and subjected to metamorphism, yet others become uplifted and exposed to weathering and erosion? The answer remained a mystery until the 1960s when geologists developed the *theory of plate tectonics*. This new theory explains how Earth's crust is broken up into rigid slabs or plates, which are set in motion by forces associated with heat generated in the interior of the planet. These internal forces create the conditions for metamorphism and cause parts of the crust to be uplifted, forming mountains. Plate tectonics not only plays a key role in the rock cycle, it also affects nearly every aspect of the Earth system (Chapter 1). Because understanding plate tectonics is central to the study of environmental geology, the subject will be covered in greater detail in Chapter 4.

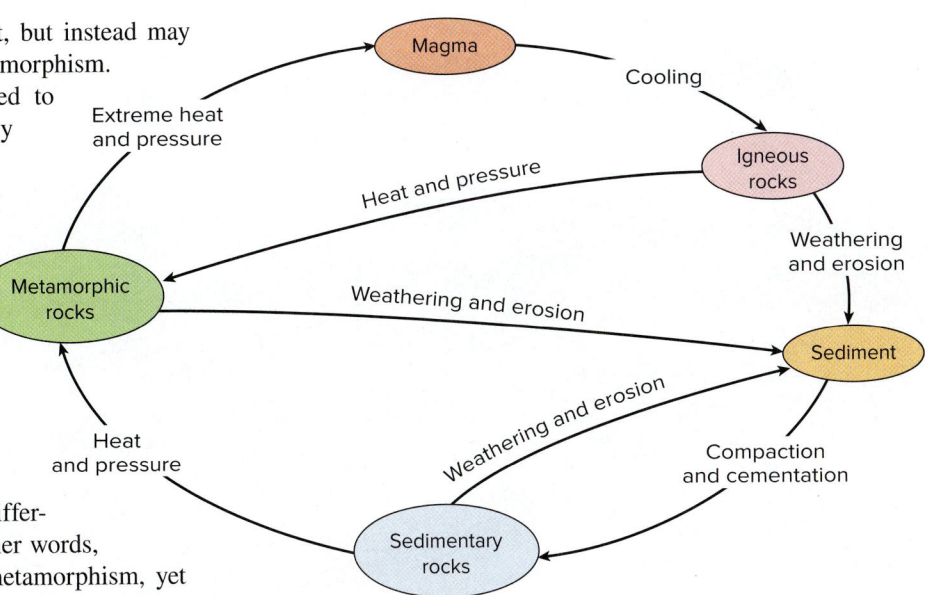

FIGURE 3.28   The rock cycle explains how various geologic processes can cause rocks to be transformed into different types of rocks. The geologic processes that operate within the rock cycle ultimately cause the rocks within Earth's crust to be recycled over time.

Rocks as Indicators of the Past

In Chapter 1 you learned how geologists constructed the geologic time scale by classifying rocks based on their relative ages—absolute ages were later obtained by radiometric dating. Because many sedimentary rocks contain fossils, the geologic rock record provides us with a vast history of life on our planet. The rock record also contains valuable evidence on an array of important topics, such as the movement of continents over time, locations of ancient mountain ranges, climatic changes, asteroid impacts, and more. In this section we will briefly explore how you can take what you learned about rocks in this chapter and make some basic interpretations about Earth's past.

Suppose you find a rock that is strongly foliated and shows evidence that it has been deformed and flowed in a plastic manner. Based on what you learned earlier, the foliation and plastic deformation tell us that this rock was once deeply buried and under high levels of temperature and pressure. In other words, it must have experienced regional metamorphism. The exposed rock in the cliff face shown in Figure 3.29, for example, represents a rock body that at one time was deeply buried and subjected to regional metamorphism. This rock is not only highly foliated, but has large veins of light-colored igneous rocks cutting through the entire rock mass. We can conclude therefore that the temperature and pressure must have reached the point where this mass of metamorphic rock underwent partial melting and formed magma. By using radiometric dating techniques we could also determine when the igneous veins cooled into solid rock.

While igneous and metamorphic rocks tell us a great detail about the history of Earth's interior, sedimentary rocks provide specific information about the surface environment. For example, because sediment and dissolved ions interact with the atmosphere and hydrosphere, scientists are able to interpret past climatic

FIGURE 3.29   The Gunnison River in Colorado has exposed this ancient body of metamorphic rock, which contains numerous veins of light-colored igneous rock. These igneous bodies within the metamorphic rock prove that the temperature during metamorphism was high enough to generate magma.
© Doug Sherman/Geofile

conditions based on the composition of sedimentary rocks. In the case of evaporite deposits such as rock salt (halite) and gypsum, we can infer a warm and arid environment due to the fact that formation of these rocks requires very high evaporation rates. The presence of coal in a sedimentary sequence tells us the environment must have been warm and humid because coal forms from thick accumulations of organic material in swamps. Sedimentary rocks also contain features that form during deposition, which provide valuable information about the environment. For example, sand that is transported by wind has a characteristic type of layering called *cross-bedding* (Figure 3.30A). Because large masses of blowing sand are found in arid climates where there is a lack of vegetation, cross-bedding in ancient sandstones is indicative of an arid environment. Similarly,

A Zion National Park, Utah

FIGURE 3.30 Features preserved in sedimentary rocks hold important clues as to the environment where the original sediment was deposited. The cross-bedding of layers in (A) is the result of windblown sand being deposited in shifting sand dunes. The ancient mudcracks in (B) developed in clay-rich sediment deposited in a shallow lake that periodically dried up.

(a–b) © Jim Reichard

B Glacier National Park, Montana

FIGURE 3.31 A 400-million-year-old fossiliferous limestone from the Great Lakes region in North America proves that life flourished in the marine environment that once existed in the area.
© Jim Reichard

FIGURE 3.32 Image of Mars taken from an orbiting spacecraft, showing what appear to be sedimentary rocks and an ancient shoreline.
NASA/JPL/Malin Space Science Systems

when *mudcracks* (Figure 3.30B) are found in layers of shale, they tell the story of a shallow lake that periodically evaporated, allowing the muddy lakebed to dry out and develop these characteristic cracks.

Sedimentary rocks also provide key environmental information through the fossils they contain (Figure 3.31). Although land animals can migrate and live in a variety of environments, land plants and many marine organisms live in very specific environments. A good example is corals, which thrive in clear, warm water that is also rich in microscopic organisms on which the corals feed. Other marine organisms live only in the turbulent beach zone, and some attach themselves to rocks in the intertidal zone. Still others crawl around on the seafloor in shallow waters, and some have even adapted themselves to the great pressure found in the deep ocean. Most fossils, therefore, reveal at least some information about the environment in which the original organisms once lived. In general, sedimentary rocks and their fossils tell us whether the sediment was deposited in fresh, brackish, or salt water and also provide some indication of the water depth and temperature (Case Study 3.1).

Finally, because Earth and other bodies in the solar system formed from the same solar nebula, one could expect that many of the minerals and rocks found on Earth might also be found on the other bodies. When Apollo astronauts brought samples back from the Moon, most of the rocks turned out to be nearly identical to the basalts and their coarse-grained equivalents found here on Earth. Robotic craft on the surface of Mars have likewise found basaltic rocks, but have also discovered sedimentary rocks that were deposited in a water-rich environment (Chapter 2). This means that water and gases in Mars's atmosphere had at one time caused these basaltic rocks to break down into sediment and ions through the processes of weathering and erosion. Supporting this interpretation are images of Mars's surface from orbiting spacecraft (Figure 3.32), showing what appear to be extensive deposits of sedimentary rocks and ancient shorelines. This means that we can use our knowledge of geologic processes and rocks not only to interpret past environmental conditions on Earth, but on other planets as well.

Early Earth History as Told by Our Oldest Rocks

Geologists have been able to unravel Earth's long history by learning what different rock types and fossils are able to tell us about past environments. One of the most fundamental questions in geology is, what was the Earth like right after it formed? According to the nebular hypothesis, scientists believe that when the newly formed planets were clearing their orbits during the heavy bombardment period, Earth experienced a Mars-sized impact (Chapter 2). Although questions remain, there is considerable evidence that the Moon formed from the debris of this massive impact. Based on radiometric dating of asteroids and rocks from Mars and the Moon, geologists have calculated that Earth developed a solid crust composed of silicate minerals 4.56 billion years ago. Despite the fact that geologists have examined rocks from every corner of the world, the oldest ever found is only 4.0 billion years old. This means that Earth's original crustal material did not survive the heavy bombardment period, leaving a span of 560 million years about which little is known. The original rocks therefore must have either remelted to form new rocks, or been destroyed by weathering and erosion and transformed into sedimentary rocks.

To help unravel the mystery, geologists have focused their attention on the silicate mineral called *zircon* ($ZrSiO_4$), which is found in igneous rocks of all ages. Zircon forms as tiny crystals when magma cools into igneous rocks. As the crystals grow, radioactive uranium and thorium atoms commonly substitute for some of the zirconium atoms in the mineral's crystalline structure. Because the uranium decays into lead atoms at a steady rate, scientists are able to use radiometric dating techniques (Chapter 1) to determine the age of individual zircon crystals. Although zircon crystals form in igneous rocks, they are also found in sedimentary and metamorphic rocks. What happens is that

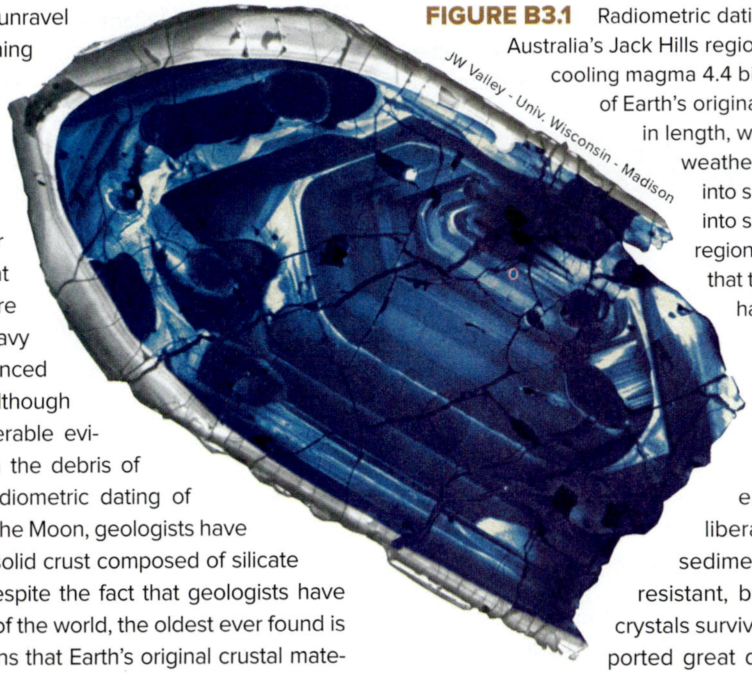

FIGURE B3.1 Radiometric dating of this zircon crystal, from Australia's Jack Hills region, indicates that it crystallized from cooling magma 4.4 billion years ago and was once part of Earth's original crust. The crystal, only 0.4 mm in length, was liberated from the crust by weathering and erosion and incorporated into sediment, which was then transformed into sedimentary rock before undergoing regional metamorphism. This crystal tells us that the hydrologic and rock cycles must have been active early in Earth's history.
JW Valley - Univ. Wisconsin - Madison

as igneous rocks break down due to physical and chemical weathering, the tiny zircon crystals are liberated and become incorporated in sediment. Because zircon is extremely resistant, both physically and chemically, the crystals survive intact while the sediment is transported great distances. The zircon crystals then become part of detrital sedimentary rocks. Interestingly, the crystals are so resistant they can even survive the high temperatures and pressures associated with regional metamorphism. Therefore, zircon crystals can remain intact while sedimentary rocks are transformed into metamorphic rocks.

Because zircon forms in igneous rocks and has the ability to survive Earth's tectonic rock cycle, we know that when zircon crystals are found in sedimentary or metamorphic rocks, the crystals must be older than the rocks themselves. By radiometric dating of zircon crystals in some of the oldest metamorphic rocks on Earth, geologists have been able to find individual crystals as much as 4.4 billion years old (Figure B3.1). This means that we now have direct evidence of the early crust that not only survived the heavy bombardment period, but formed only 160 million years after the Earth itself. Moreover, oxygen isotope data from the zircon crystals indicate that liquid water was associated with the magma that formed the crystals. What is also interesting is that the zircon crystals had to first be liberated from the original igneous rocks by weathering processes, and then deposited as sediment that was later transformed into sedimentary rock. For this to occur, Earth must have had an active hydrologic cycle along with an atmosphere and liquid ocean very early in its history (Figure B3.2). In addition, we

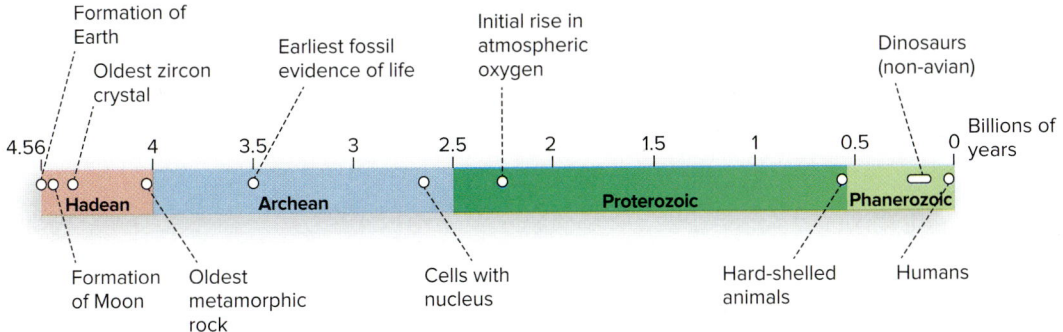

FIGURE B3.2 Illustration showing where the Jack Hills zircon crystal falls along Earth's 4.6-billion-year timeline. Other major events shown here are based on evidence contained in Earth's rock record along with rocks from the Moon and Mars. Note that the majority of the fossil record lies within the most recent 500 million years of the Phanerozoic Eon.

—Continued

find metamorphic rocks as much as 4.0 billion years old, which tells us that plate tectonics and the rock cycle were both operating quite early in Earth's history.

Based on the study of zircon crystals and Earth's oldest rocks, geologists now understand that the early Earth had conditions suitable for life as it cooled to form a crust and developed an atmosphere and liquid ocean. These studies also tell us that the basic Earth system we see today has been operating in a similar manner for most of the planet's 4.6-billion-year history. Finally, by understanding Earth's ancient past, we know that planets in habitable zones around other stars are likely to develop atmospheres, oceans, and active rock cycles, all of which were key elements in making life possible here on Earth.

SUMMARY POINTS

1. Different types of atoms (elements) are created in stars by the nuclear fusion of hydrogen. These atoms can then be ejected into space and incorporated into planets.

2. Minerals are inorganic solids composed of one or more types of atoms that are arranged in a crystalline structure. Each mineral is unique and has a unique set of physical properties. Rocks are simply assemblages of one or more types of minerals.

3. Oxygen, silicon, and aluminum make up over 80% of Earth's crust by weight. Of the more than 4,000 known minerals, only a dozen or so make up the bulk of Earth's crustal rocks. These minerals, referred to as the rock-forming minerals, are assembled into three basic rock types.

4. Igneous rocks form when magma begins to cool, which allows different types of ions to be incorporated into the crystalline structures of minerals. As the cooling period increases, mineral grains have more time to grow, producing an igneous rock with a coarser texture.

5. Physical weathering at the surface breaks rocks down into smaller fragments called sediment. Minerals within rocks that are susceptible to chemical weathering then decompose into ions and/or are transformed into new minerals.

6. Erosion and transportation take place when sediment is removed from an area by wind, water, or ice—ions move primarily by flowing water. The sediment is ultimately transported from its source area to a depositional site where it accumulates.

7. Sedimentary rocks form when accumulated sediment undergoes compaction and cementation, or when dissolved ions precipitate to form new minerals.

8. Metamorphic rocks form when preexisting rocks are altered by heat, pressure and/or chemically active fluids. Contact metamorphism occurs when rocks are altered by the heat from a magma body, whereas regional metamorphism involves both heat and directed pressure.

9. The continual transformation of rocks from one rock type to another is referred to as the rock cycle. The rock cycle is ultimately driven by plate tectonics and Earth's internal heat.

10. Because each rock type forms under certain geologic conditions, rocks provide scientists with a record of past events and environmental conditions on Earth.

KEY WORDS

basalt 74	feldspars 72	magma 70	rock-forming minerals 71
calcite 72	ferromagnesian minerals 72	metamorphic rocks 83	sandstone 80
chemical weathering 75	foliated texture 83	mineral 68	sediment 78
clay minerals 72	granite 74	oxidation/reduction 77	sedimentary rocks 78
coal 82	hydrolysis 77	physical weathering 74	shale 80
deposition 78	igneous rocks 73	quartz 72	slate 83
dissolution 75	ions 67	rock 70	transportation 78
erosion 78	limestone 80	rock cycle 84	

APPLICATIONS

Student Activity Go to a local cemetery and see how many different types of rocks you can identify among the various headstones (i.e., gravestones or tombstones). The most common igneous rock types used for headstones are granite and gabbro, whereas metamorphic headstones are typically made of gneiss or marble. In some cases you may find an older headstone made of sedimentary sandstone.

Critical Thinking Questions

1. The different elements in the periodic table ultimately are produced when hydrogen and helium atoms undergo nuclear fusion reactions in the interior of stars. How did all these different elements get from the interior of stars and come together to form the Earth?

2. If two minerals have the exact same chemical formula or composition, then what can you infer about their crystalline structure? Are their physical properties likely to be different or the same?

3. The rock cycle indicates how rocks can be transformed from one type to another. In other words, older rocks are recycled into new ones. How can an older sedimentary rock be transformed into a new sedimentary rock without first becoming a metamorphic rock?

Your Environment: YOU Decide

© Image Ideas/PictureQuest

In this chapter you learned how geologists study the environments in which modern rocks form, and then use this information to interpret Earth's history based on ancient rocks. Features such as mudcracks, cross-bedding, and fossils are particularly useful as they provide very specific information about past environments.

How reliable do you think it is to use rocks as a means of studying Earth's history? Likewise, do you think it is reasonable to use rocks to help unravel the history of other planets in our solar system? Explain.

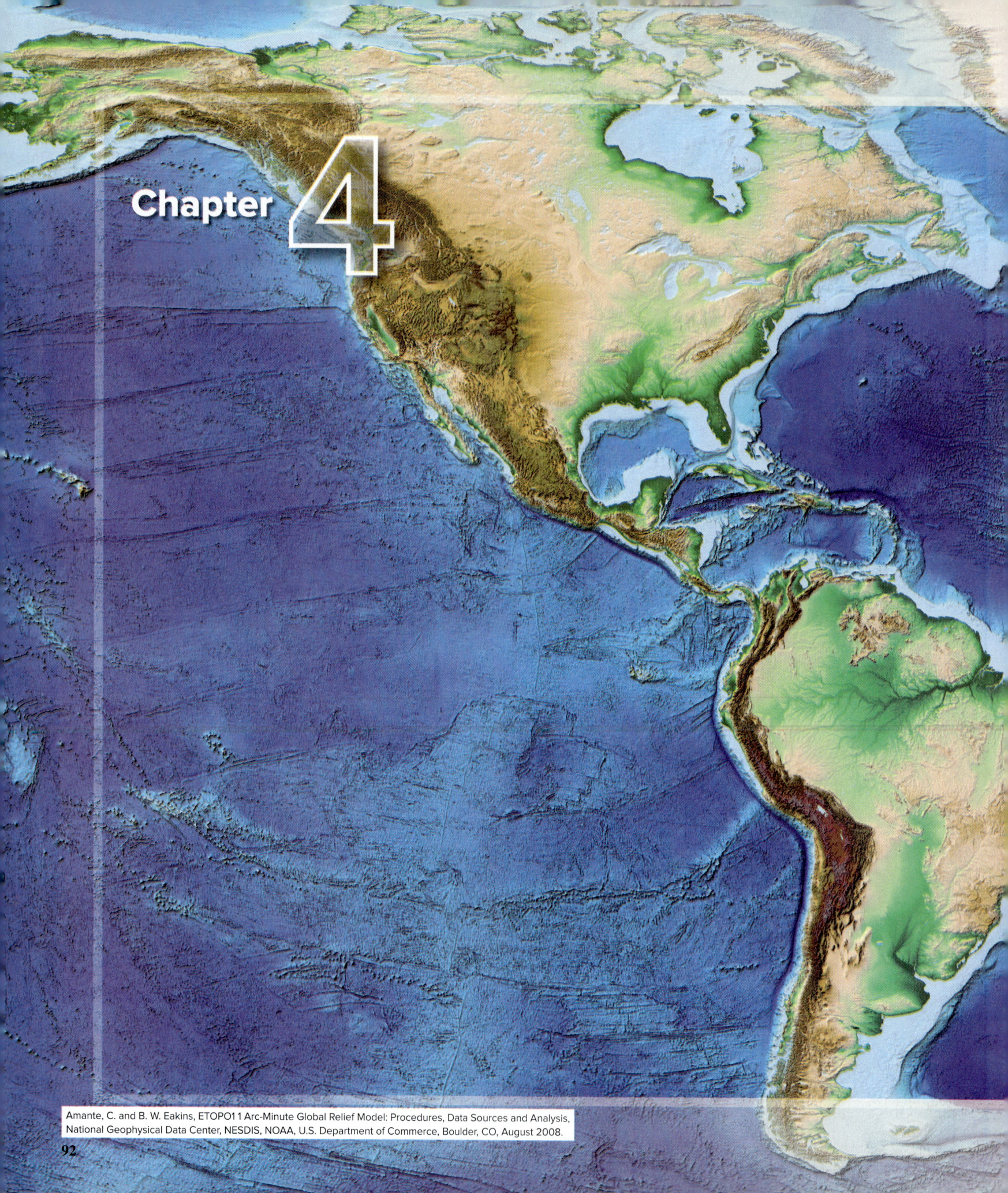

Chapter 4

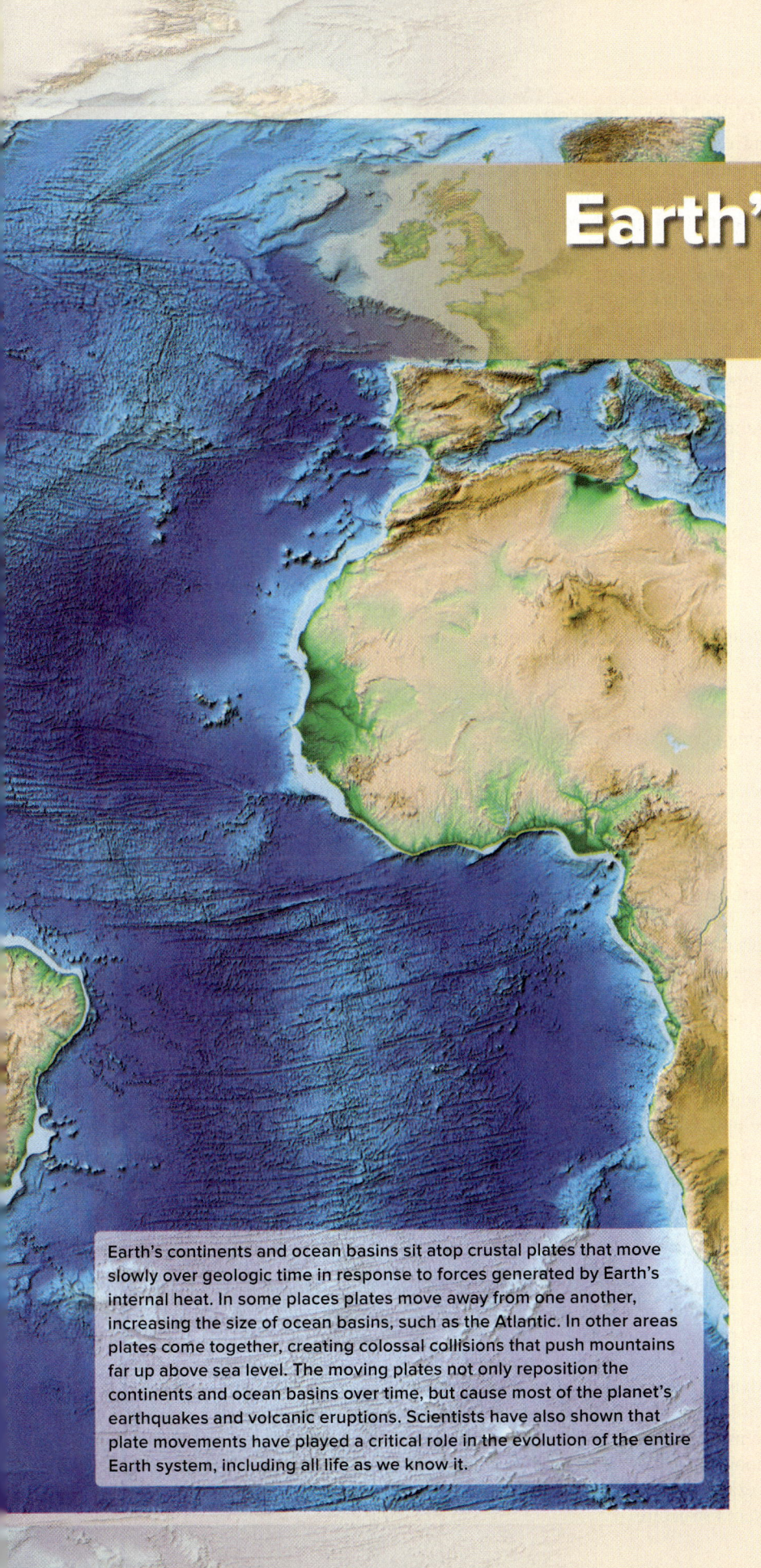

Earth's Structure and Plate Tectonics

LEARNING OUTCOMES

After reading this chapter, you should be able to:

▶ Describe the three different forces that deform rocks, and explain what happens to rocks when they are deformed beyond their elastic limits.

▶ Characterize the different layers making up Earth's internal structure, and describe the basic way in which scientists have determined this structure.

▶ Explain the difference between oceanic and continental crust, and describe the difference between the lithosphere and asthenosphere.

▶ List the two sources of Earth's internal heat, and explain how this heat helped create the planet's layered structure and is driving its system of moving tectonic plates.

▶ Describe how scientists have been able to confirm that seafloor spreading and subduction processes are taking place, and explain how this confirmation proved continental drift.

▶ List the major types of plate boundaries, and describe the types of surface features that develop at each one.

Earth's continents and ocean basins sit atop crustal plates that move slowly over geologic time in response to forces generated by Earth's internal heat. In some places plates move away from one another, increasing the size of ocean basins, such as the Atlantic. In other areas plates come together, creating colossal collisions that push mountains far up above sea level. The moving plates not only reposition the continents and ocean basins over time, but cause most of the planet's earthquakes and volcanic eruptions. Scientists have also shown that plate movements have played a critical role in the evolution of the entire Earth system, including all life as we know it.

Introduction

© StockTrek/Getty Images

Unlike Mars, Mercury, and the Moon, the Earth is a very active and restless planet where earthquakes and volcanic eruptions are relatively common. In addition, vast amounts of freshwater evaporate from Earth's oceans and fall over the landmasses, making it possible for life to flourish in the terrestrial biosphere. Much of the energy that drives this interactive and dynamic system, called the *Earth system* (Chapter 1), comes from the solar radiation that streams out from the Sun. The other primary energy source is the heat contained within our planet. It is this internal heat that sets rock masses in motion, causing large-scale metamorphism and uplift of the land surface. This uplift produces rugged and mountainous terrain, thereby exposing greater amounts of rocks to the surface environment. At the surface, weathering and erosion processes work to lower the elevation of the landscape, generating sediment that eventually accumulates and forms sedimentary rocks. Earth's internal heat therefore is a major driving force in transforming rocks from one type to another, referred to as the *rock cycle* (Chapter 3). The purpose of this chapter is to explore the critical role that Earth's internal forces have in shaping our environment.

Humans have long been aware that Earth's landscape is highly varied, from broad flat plains to rugged mountains that are virtually impassible. As with other aspects of our physical world, we learned to use the process known as *science* to explain how different landscapes form. This resulted in a new area of study called geology, which focuses on the different types of rocks and sediment that all landscapes are built from. From the study of rocks, early geologists were able to develop hypotheses that explained how volcanic mountains were built by rising magma. However, they found it difficult to explain the presence of mountains such as the Appalachians, Himalayas, and Alps, whose strongly deformed rocks show few signs of volcanic activity. Not until the 1960s did geologists come to understand that Earth's outer layer, or *crust,* is broken up into rigid plates that are in motion because of forces associated with the planet's interior heat. This concept of moving plates eventually became known as the **theory of plate tectonics.**

The theory of plate tectonics has enabled modern geologists to explain how deformed mountain ranges represent giant collision zones between crustal plates, some the size of continents. Plate tectonics not only provides a simple and elegant explanation for the rock cycle and why the land rises, it explains the occurrence of most earthquakes and volcanic eruptions. The theory also explains how internal forces can cause landmasses to break up and begin drifting apart, eventually becoming separated from one another by an ocean. Biologists have discovered that the great diversity of plant and animal life is a direct result of both evolutionary processes and the movement of landmasses over time. Similarly, climatologists and oceanographers have found that the reconfiguration of the continents and ocean basins creates important changes in the circulation patterns of the oceans and atmosphere. These changes have altered the global climate system over eons of geologic time.

Finally, it should be noted that plate tectonics is referred to as a *theory* rather than a *hypothesis* (Chapter 1) because it acts as a unifying framework that explains a wide variety of natural phenomena. It is no exaggeration to say that plate tectonics has revolutionized the way in which scientists view our planet. The solid earth is no longer seen as a rigid mass of rock, but rather a dynamic system that affects nearly every aspect of the Earth system. Plate tectonics is not just central to the study of geology, but also represents one of the most profound advances in modern science. Because plate tectonics is so important, we will devote this entire chapter to the subject. But first, we need to examine some background material on the deformation of rocks and Earth's layered structure.

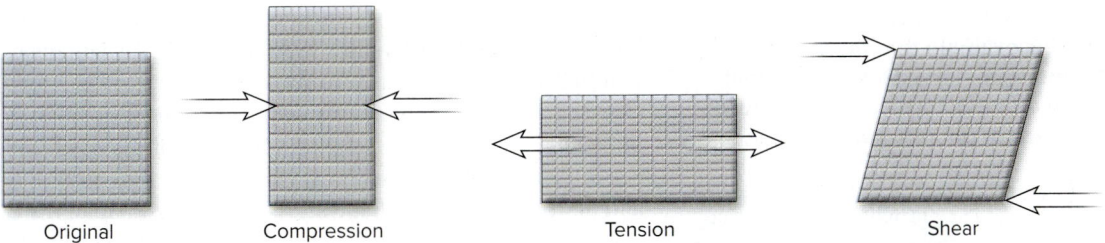

Original   Compression   Tension   Shear

FIGURE 4.1   A two-dimensional representation of the deformation (strain) that would result from different types of stress acting on a square.

Deformation of Rocks

For rocks to become deformed, they must be acted upon by some type of force, or *stress*. Deformation involves some change in shape or volume, technically known as *strain*. As illustrated in Figure 4.1, there are three basic types of stress that can result in deformation. **Compression** pushes on rocks from opposite directions, causing them to be shortened as if they were put in a vise. **Tension** pulls on rocks from opposite directions, resulting in the rocks becoming stretched or lengthened. Finally, **shear** occurs when rocks are being squeezed in an uneven manner, causing them to become skewed such that different sides slide or move in opposite directions.

People are often surprised to learn that rocks near the surface are *elastic,* meaning that when a force (stress) acting on them is removed, the rocks return to their original shape. Common examples of elastic materials include rubber bands and tree limbs. All elastic materials have what is called an **elastic limit,** which is the point beyond which they no longer behave elastically and deformation becomes *permanent*. For example, if the wind forces a tree limb to bend (deform) beyond its elastic limit, the limb will break or snap, which of course is permanent. In the case of rocks exceeding their elastic limit, permanent deformation occurs in one of two ways. As illustrated in Figure 4.2, one way is by fracturing, in which case the rocks are referred to as being *brittle*. The other way is by flowing, where the rocks are called *ductile*. For example, a glass rod is considered brittle because if it bends beyond its elastic limit, it will fracture and snap in two. On the other hand, a steel rod is ductile because if its elastic limit is exceeded, it will literally flow and develop a permanent bend. When it comes to rocks that are ductile, geologists often refer to the deformation as *plastic deformation.*

Permanent deformation by plastic flow

Permanent deformation by fracturing

FIGURE 4.2   Rocks will deform elastically up to a point, beyond which deformation becomes permanent. Ductile materials deform permanently by flowing plastically, whereas brittle materials fracture. Rocks near the surface are typically brittle and will fracture, but when buried, the higher temperatures and pressures cause them to become ductile and deform plastically.
Top: Courtesy of Phil Dombrowski/Bottom: © Jim Reichard

TABLE 4.1 Approximate temperature and confining pressure at selected depths below the Earth's surface. Note in this example that there is a 50-fold increase in pressure, whereas temperature increases only 10-fold. Estimates based on average crustal density of 2.8 g/cm³ and temperature gradient of 2.5°C/100 meters; pressure in units of pounds per square inch (psi) and megapascal (MPa).

Depth	Approximate Temperature	Confining/Overburden Pressure
1,000 feet (305 m)	74°F (23°C)	1,210 psi (8.4 MPa)
10 miles (16 km)	790°F (420°C)	64,000 psi (440 MPa)

The reason glass and steel rods have different elastic limits and deform differently is because of their composition and internal structure. In the case of rocks, their texture and mineral composition help determine their elastic limits and whether they deform in a brittle or ductile manner. Other key factors include temperature, pressure, and length of time over which the stress acts. For example, if we applied enough heat to a glass rod, it would become ductile; hence, it would undergo permanent deformation by flowing rather than fracturing—a *fracture* is simply a planar opening or break. On the other hand, if we kept the glass rod at room temperature, but wrapped it tightly with a fabric, the fabric would confine or add pressure to the rod such that its elastic limit would increase. Many materials will experience greater deformation under a relatively small force (stress) acting over a long period of time compared to a stronger, but short-duration force. Note that when rocks deform, they often slide past one another along a fracture plane, at which point the fracture is called a **fault.** All faults involve some type of slippage or movement, whereas fractures do not.

The factors that affect how rocks are deformed are important to our discussion because as one descends toward the center of the planet, rocks experience progressively higher temperature and pressure. What happens is that subsurface rocks have to bear the weight of the overlying column of rocks, creating what geologists refer to as *overburden* or *confining pressure.* Confining pressure is similar to the pressure you feel when diving to the bottom of a pool—the difference being that one is from the weight of water and the other due to the weight of the rocks. When rocks are under greater confining pressure, their elastic limit increases, just as the fabric wrapped around the glass rod made it stronger. Keep in mind that because of Earth's internal heat, temperature also increases along with confining pressure. Consequently, rocks near the surface are under little confining pressure and tend to be quite brittle, so they will fracture when subjected to a stress that exceeds their elastic limit. As can be seen in Table 4.1, temperature and pressure increase rapidly with depth; hence, rocks that are deeply buried readily become ductile and deform plastically. Examples of rocks that have undergone brittle and plastic deformation are shown in Figure 4.2.

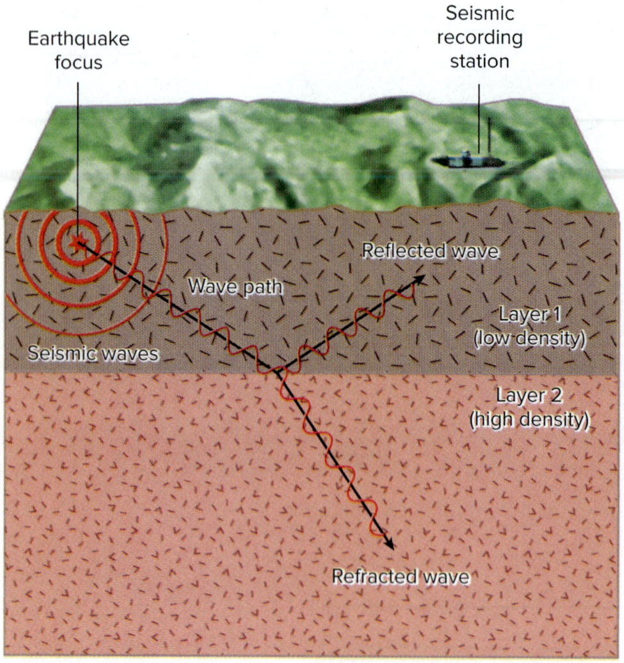

FIGURE 4.3 Seismic waves generated by earthquakes and human-made explosions will reflect and refract when encountering layers of different density. Recording instruments measure the waves that return to the surface, enabling scientists to determine the depth of different layers all the way to Earth's core.

Earth's Interior

As one descends deeper into the Earth, not only do temperature and pressure change, but so too does the composition of earth materials. Recall from Chapter 3 how Earth's crustal rocks are made up of a relatively small number of minerals called the *rock-forming minerals.* Because most rock-forming minerals contain the silicate ion (SiO_4^{4-}), oxygen and silicon atoms account for 49% and 26% of the crust by weight, respectively. However, if we consider the Earth as a whole, iron becomes the dominant element at 35% by weight, followed by oxygen at 29%. This means that Earth must have a layered structure and that the lighter elements are more abundant in the outermost layers. Likewise, Earth's interior must be denser than the outer shell we call the crust.

Scientists today know with certainty that Earth is layered because of the way earthquake waves, also called *seismic waves* (Chapter 5), change velocity and direction as they travel through the planet's interior. As indicated in Figure 4.3, when seismic waves encounter layers of different density, their velocity changes, causing the waves to both reflect and refract (bend)—similar to how light reflects and bends when passing from air into a pool of water. By measuring refracted and reflected seismic waves that return to the surface, scientists have been able to locate the boundaries between different materials deep within the Earth. Seismic studies have also shown that the Earth has four major layers that vary in composition and physical properties. Beginning at the surface, the major layers are the *crust, mantle, outer core,* and *inner core.* Interestingly, the deepest well ever drilled reached 7.6 miles (12.3 km), which is a mere pinprick considering that the center of the Earth is nearly 4,000 miles (6,400 km) deep. If one compared the Earth to an apple, this deep well would not even have penetrated the apple's skin! Therefore, were it not for the study of seismic waves, Earth's internal structure would have remained a mystery.

Earth's Structure

From Figure 4.4 one can see that the Earth consists of a metallic sphere surrounded by a rocky shell mostly composed of silicate minerals. The center of the metallic sphere is solid and referred to as the **inner core,** which is surrounded by a shell of molten metal called the **outer core.** In order to account for the overall known density of the Earth, scientists hypothesize that both the inner core and outer core are composed of an iron-nickel alloy. When the density of this alloy is combined with the density of the silicate shell, it produces an average that is consistent with the known density of the whole Earth. This nickel-iron hypothesis is also supported by the fact that many of the asteroids that strike the Earth are composed of these two metals. Because asteroids represent leftover material from the formation of the solar system (Chapter 2), it follows that Earth's inner sphere is an alloy of iron and nickel. Scientists have also discovered that certain types of seismic waves, called *shear waves,* do not penetrate the outer core, whereas *compressional waves* do. Because laboratory experiments show that shear waves are unable to pass through liquids, but compressional waves can, scientists conclude that the outer core is molten and the inner core is solid.

Earth's metallic center is surrounded by an 1,800-mile (2,900-km) thick rocky shell called the **mantle,** which is composed of iron-rich silicate minerals (Figure 4.4). Mantle rocks are subjected to very high temperatures and pressures because they lie at such great depths below the surface. This makes mantle rocks rather susceptible to plastic deformation as described in the preceding section. Note that as you descend progressively deeper into the Earth, temperatures eventually reach the melting points of different silicate minerals. However, pressure increases with depth at a faster rate than temperature (Table 4.1). Because increases in pressure raise the melting point of minerals, this means that despite the high temperatures deep within the mantle, the tremendous pressure generally keeps the silicate minerals from reaching their melting points (similar to how water in a pressure cooker is forced to boil at a higher temperature). This effect explains why the mantle is not molten, but instead made of solid rock that is very hot and subject to plastic flow.

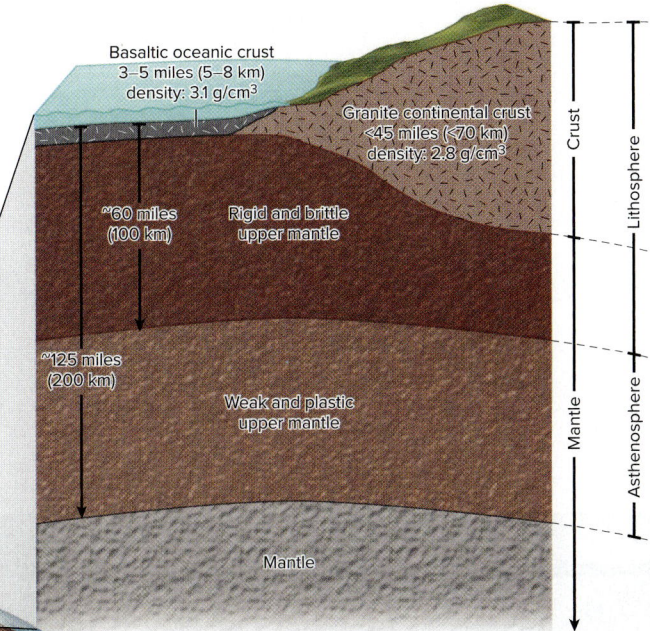

FIGURE 4.4   The Earth has a layered structure consisting of a high-density metallic core surrounded by a less-dense rocky shell of silicate minerals, called the mantle. Near the top of the mantle the silicate minerals are close to their melting points, creating a weak and plastic zone called the asthenosphere. The outermost silicate shell is called the crust, which has the lowest density of all the layers. Geologists refer to the crust and uppermost mantle as the lithosphere since they behave as a single, rigid slab that moves over the asthenosphere. Note that the lithosphere in continental areas contains granitic crust, whereas basaltic lithosphere lies beneath the oceans.

Near the top of the mantle is a zone called the **asthenosphere,** where the combination of temperature and pressure puts silicate minerals close to their melting points. This makes the asthenosphere a rather weak layer that is also prone to melting, particularly in areas where the rocks are disturbed by a localized decline in pressure, or by the introduction of water-rich material. One of the reasons the asthenosphere is important in geology is because it is a likely source for the magma that rises toward the surface, forms igneous rocks, and causes volcanic eruptions (Chapter 6). The asthenosphere is also important because it is a very weak layer, which allows it to flow and deform plastically when acted upon by Earth's internal forces. Note that above the asthenosphere, the temperature and pressure become low enough that the mantle rocks are more rigid and brittle.

Earth's outermost layer is called the **crust,** which consists primarily of silicate-rich rocks whose density is even lower than those in the underlying mantle. One of the more striking features of the crust, shown in Figure 4.4, is that it is exceedingly thin compared to the 4,000-mile (6,400-km) depth of the planet. Also striking is how the composition and thickness of the crust vary. Under the ocean basins the crust is made of more dense, basaltic-type rocks that maintain a fairly uniform thickness of around 3 miles (5 km). Under the continents, however, crustal rocks have an overall composition similar to granite and can reach thicknesses as much as 45 miles (70 km). Because both the mantle and the crustal rocks above the asthenosphere are relatively brittle and rigid, they effectively behave as a single layer, which geologists refer to as the **lithosphere.** The fact that the lithosphere is brittle and rigid is important because this allows forces within the Earth (compression, tension, or shear) to break the lithosphere up into individual slabs called **tectonic plates.** Earth's internal or *tectonic forces* then cause these rigid plates to glide on top of the weak and easily deformed asthenosphere. We will take a closer look at the actual movement of tectonic plates in a later section of this chapter.

Earth's Magnetic Field

The core is of considerable importance to life on Earth because of its role in creating the planet's strong magnetic field. As indicated in Figure 4.5, scientists think that Earth's magnetic field results from the circulation of metallic ions within the outer core—creating a magnetic force field similar to that around a familiar bar magnet. It is generally believed that the fluid nature of the outer core allows the inner core to rotate slightly faster than the rest of the planet. This, in turn, causes the electrically charged metallic ions in the outer core to circulate, essentially making the Earth a giant electromagnet. For over 700 years humans have made use of Earth's magnetic field for navigational purposes using compasses. Interestingly, biologists now hypothesize that certain types of migrating animals (e.g., birds and turtles) also use the magnetic field to navigate. Most important of all is how the magnetic field acts as a critical shield that helps block out harmful radiation streaming out from the Sun (Figure 4.5). Earth's core then has likely played an important role in making Earth's environment hospitable and in the evolution of life as we know it. This interaction

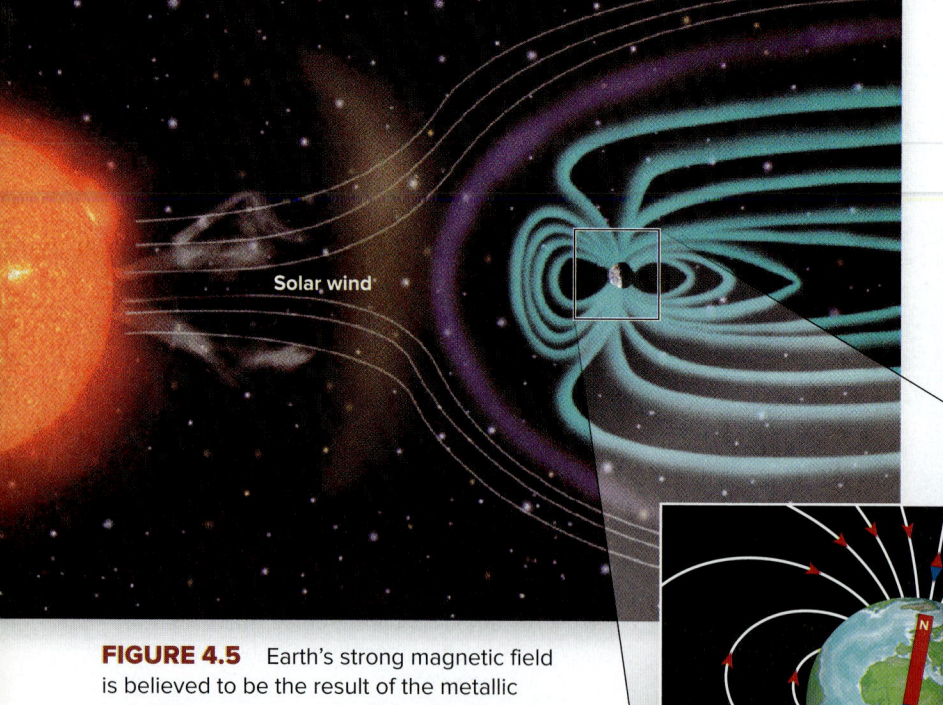

FIGURE 4.5 Earth's strong magnetic field is believed to be the result of the metallic ions in the outer core circulating as the planet rotates. The magnetic field is important to the biosphere because it helps shield the Earth from harmful radiation streaming from the Sun.
National Geophysical Data Center/NOAA

between the solid earth and the biosphere is another example of how the entire Earth operates as a system (Chapters 1 and 2).

Earth's Internal Heat

It is well established that the Earth radiates more heat energy into space than the planet receives from the Sun. Moreover, temperature is known to increase as you go deeper into the planet. This leads to the obvious question: What is the source of Earth's internal heat? Scientists originally thought that this heat was *residual,* meaning it was left over from the formation of the planet 4.6 billion years ago. When the young Earth was accreting large amounts of material from the solar nebula (Chapter 2), gravity caused material within the planet to undergo compaction due to the increasing overburden (confining) pressure. As the material compressed, individual particles reoriented themselves, which generated frictional heat—similar to heat generated by rubbing your hands together. The planet was also gaining heat energy during this early period from the impacts themselves. Scientists now believe that at some point the early Earth was mostly molten except for a relatively thin crust.

Temperatures within the young planet would naturally have been higher in the center and cooler near the crust where heat was radiating outward into space. Geologists refer to the increase in temperature with depth as the **geothermal gradient,** which averages about 75°F/mile (25°C/km). Naturally the geothermal gradient today is much lower since the planet has cooled considerably over the last 4.6 billion years. Interestingly, in the mid-1800s physicists attempted to calculate Earth's age based on the difference between the modern geothermal gradient and the gradient believed to have existed when the planet first formed. Assuming Earth was initially molten, they calculated it would take approximately 20 to 40 million years of cooling for the planet to reach its present geothermal gradient. Then in the early 1900s scientists discovered that the nuclei of certain types of atoms, particularly uranium (U), thorium (Th), and potassium (K), give off heat as they emit energetic particles in a process called *radioactive decay.* This means that minerals containing *radioactive elements* must be generating considerable amounts of heat deep within the Earth. The fact that radioactive heat is being added to the residual heat means that the planet must be cooling at a much slower rate compared to what had been calculated earlier. The Earth therefore had to be much older than 20 to 40 million years. By using radiometric dating techniques, described in Chapter 1, scientists now calculate the age of the Earth to be about 4.6 billion years old. Recall that this older age is consistent with many other lines of geologic evidence found on both the Earth and Moon.

Scientists today believe that early in Earth's history, it became molten as it continued to gain heat energy from a combination of internal compression (friction), impacts, and radioactive decay. During this period the planet was also starting to cool through a process called *conduction,* where the heat was transferred through the atmosphere and into space. Because of its molten state, Earth was also able to cool by *convection* as molten matter transported heat toward the surface where some of it would be lost via conduction through the atmosphere. During convection, the hottest material near the center of the planet becomes less dense, causing it to rise toward the surface—similar to how a hot-air balloon rises. Once near the surface this material loses some of its heat, at which point it becomes denser and starts to sink. As illustrated in Figure 4.6, **convection cells** represent this circular movement of heat and matter, which is driven by temperature-induced changes in density.

The convection process (Figure 4.6) is important because it explains how planets can have a layered structure. For example, when the Earth was mostly molten, the high density of iron and nickel would cause these elements to sink and accumulate in the core. On the other hand, lighter elements such as oxygen and silicon would tend to rise and form a crust made of silicate minerals. As the planet

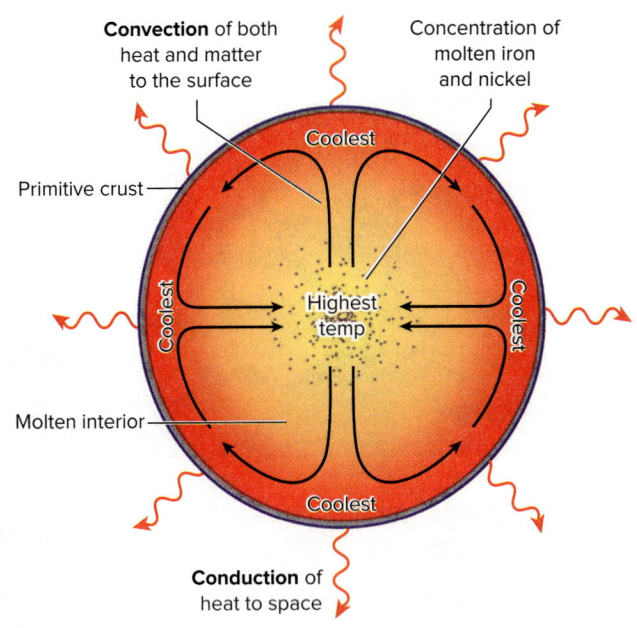

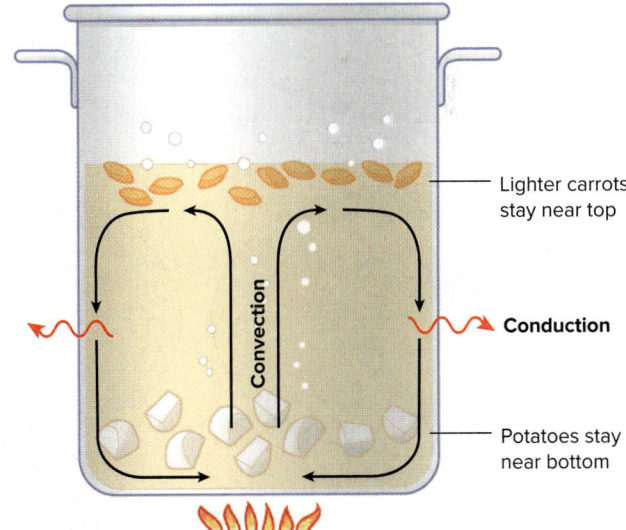

FIGURE 4.6 Soon after Earth formed, enough heat was generated to cause the planet to become molten, with only a thin crust. Convection cells developed and began transporting both heat and matter to the surface, where some of the heat was lost to space by conduction—similar to that of a pot of boiling soup. Earth soon developed its layered structure as molten iron and nickel sank to the core and lighter elements tended to rise, eventually forming the crust and mantle.

FIGURE 4.7   The distribution of unique plant and animal fossils on different continents (A) supports the idea of a single supercontinent. The matching fossil assemblages between South America and Africa were first noted by Alfred Wegener. The supercontinent called Pangaea (B) was originally proposed by Wegener, who suggested that the continents slowly drifted to their present position. Modern data show that Pangaea began to break up approximately 225 million years ago.

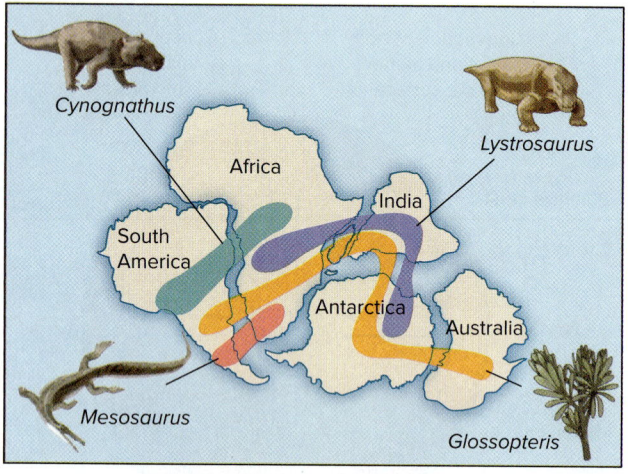

A

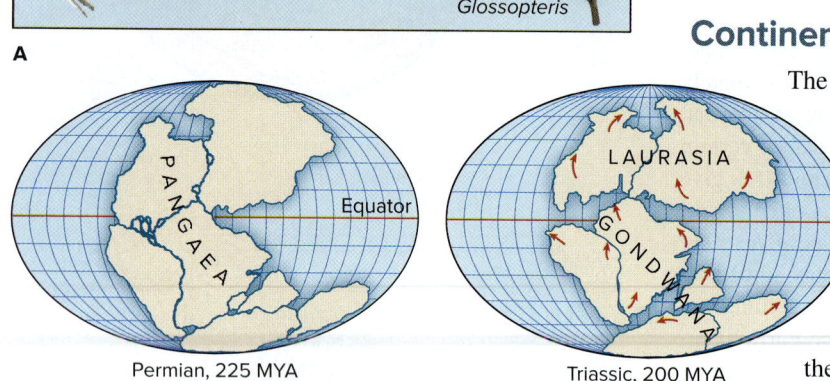

Permian, 225 MYA   Triassic, 200 MYA

Jurassic, 135 MYA   Cretaceous, 65 MYA

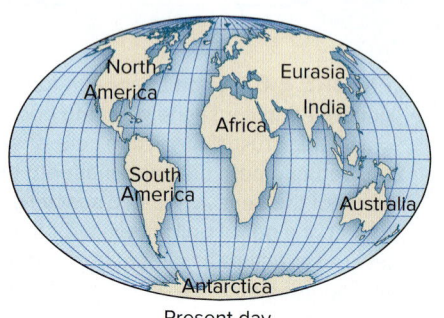

B   Present day

continued to cool, iron-rich silicates would eventually crystallize and form the mantle. This process of convection and separation of elements and minerals based on density nicely explains Earth's layered structure. Although the mantle is now solid rock, its temperature remains hot enough for it to undergo plastic flow. Scientists generally believe that convection cells continue to operate within the mantle, slowly bringing iron-rich silicate rocks up from great depths, where they cool and then sink back into the mantle. This convective motion in the mantle may help explain why lithospheric plates are slowly moving on top of the asthenosphere. The movement of tectonic plates will be our focus for the remainder of this chapter.

Developing the Theory of Plate Tectonics

Recall that plate tectonics is a unifying theory that revolutionized the way scientists view our planet. The geosphere is no longer seen as a rigid mass of rock, but rather a dynamic system with crustal plates that slide on top of a semimolten zone in the upper mantle. Scientists now think that even the mantle is being constantly overturned by deep convection cells. In short, the theory of plate tectonics has become central to the study of geology and represents one of the most profound advances in modern science. Although this theory is now widely accepted, its basic principles had been rejected by most geologists prior to the 1960s. In this section we will briefly examine how this relatively new concept was developed, and how it finally became accepted by the scientific community.

Continental Drift

The idea that Earth's continents actually move is thought to have first been suggested in 1596 by a Dutch mapmaker who proposed that the Americas had been torn away from Africa and Europe by some unknown catastrophic event. Later in the 1850s an American writer, noting how closely the shorelines of South America and Africa fit together (Figure 4.7), also suggested that the continents had been ripped apart in a violent and sudden manner. Then in 1910 an American geologist named Frank Taylor published a paper proposing that the continents were once joined, but then separated and reached their present position by slowly plowing through the ocean basins—an idea which later became known as the *theory of continental drift.* Taylor's paper drew little attention until 1922 when a book written by a German meteorologist named Alfred Wegener was translated into several languages, including English. In this new work Wegener had taken Taylor's concept of continental drift and actually supported it with some rather compelling geologic evidence. Continental drift then quickly aroused the interest of the geologic community.

Like others before him, Wegener was intrigued by how South America and Africa fit together like pieces of a jigsaw puzzle. However, Wegener was able to show that the two continents, now separated by the Atlantic Ocean, held similar sequences of rocks and unique plant and animal fossils that matched up remarkably well (Figure 4.7A). This fossil evidence and the fact that coal had been found in Antarctica strongly suggested that the continents were at one time located in vastly different climatic zones. Because this evidence was difficult to explain other than by continental drift, Wegener proposed an ancient supercontinent called *Pangaea,* which broke apart into separate continents that slowly began to drift apart (Figure 4.7B). Although Wegener's argument was quite compelling, it met considerable opposition within the geologic community. Most scientists rejected Wegener's idea of continental drift largely because of his inability to provide a plausible explanation as to *how* the continents actually moved.

Despite the fact no one could come up with a mechanism to explain how continents could move, additional evidence for continental drift continued to be found. For example, geologists discovered evidence for glaciation on continents where the climate is presently hot and dry (Figure 4.8). This odd occurrence could be accounted for had the continents broken up and drifted into warmer climatic zones. In 1958, geologists studying the continental margins of Africa and South America found that the true edges of the continents were below sea level, thus they actually extended farther offshore. When the two continents were reassembled based on the edge of their undersea margins rather than their coastlines, the fit was now almost perfect (Figure 4.9). Other lines of evidence were soon uncovered that also supported Wegener's idea of continental drift, eventually leading to the more comprehensive theory of plate tectonics.

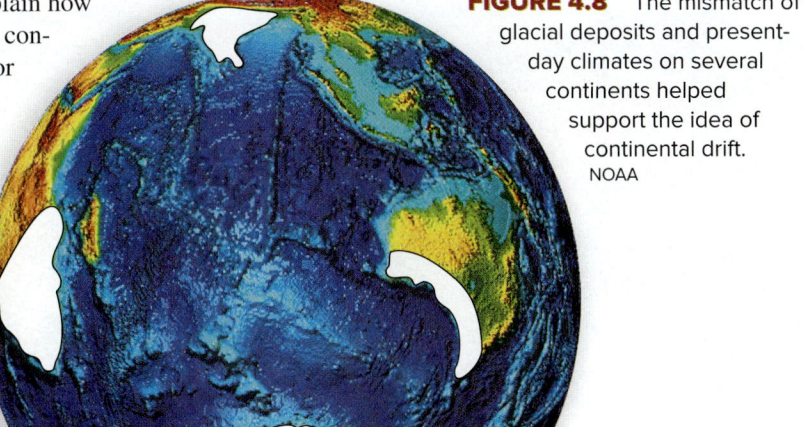

FIGURE 4.8   The mismatch of glacial deposits and present-day climates on several continents helped support the idea of continental drift. NOAA

Mapping the Ocean Floor

Prior to the 1800s very little was known about the ocean floor for the simple reason it was inaccessible to humans. This meant that nearly two-thirds of Earth's surface was completely unknown. The first real data came from depth measurements, called *soundings,* whereby a weighted line was dropped to the seafloor. Based on sounding surveys, the U.S. Navy published a depth map or *bathymetric map* in 1855, showing the presence of a submarine mountain in the middle of the Atlantic Ocean. The ability to gather bathymetric data greatly increased after World War I with the development of *sonar.* Here a ship could measure depth along a line, called a *profile,* by bouncing sound waves off the ocean floor and recording the time it takes for the waves to return—the sound is the familiar "ping" you hear in movies.

An international effort during the 1950s led to a great number of sonar surveys that gave scientists their first detailed look at the entire ocean floor. The previously mapped mountain in the mid-Atlantic was now shown to be part of a chain of submarine mountains referred to as **mid-oceanic ridges.** As illustrated in Figure 4.10, these mid-oceanic ridges form an extensive network that circles nearly the entire globe. The detailed mapping also revealed the existence of **ocean trenches,** which are narrow, steep-sided depressions running parallel to land-masses. Ocean trenches reach depths as great as 36,000 feet (11,000 m) below sea level, which is deeper than Mount Everest is high! Modern maps of Earth's surface (Figure 4.10) show that the ocean floor is not the flat and featureless plain it once was assumed, but rather a complex surface containing long mountain chains and extremely deep canyons.

Magnetic Studies

Certain rocks, igneous basalts in particular, contain appreciable amounts of the iron mineral known as magnetite (Fe_3O_4). Magnetite has the rather rare property of being magnetic, meaning it is naturally attracted to a magnet. When a magnetite crystal cools below 1,075°F (580°C), called its *Curie point,* its iron atoms orient themselves parallel to Earth's magnetic field lines shown earlier in Figure 4.5. Grains of magnetite then act as tiny compasses frozen in time, provided that metamorphism has not heated the grains above 1,075°F. This property has led to the development of a new field of study called *paleomagnetism,* in which geologists use magnetite-rich rocks (e.g., basalts) as a means of recording changes in Earth's magnetic field over time.

Early paleomagnetic studies found that the magnetite grains in many ancient rocks had reversed *polarity,* meaning the magnetic field lines pointed in the opposite direction compared to Earth's current magnetic field. In other words,

FIGURE 4.9   A near-perfect fit of Africa and South America was obtained when the two landmasses were reassembled using the edge of their continental shelves as opposed to their coastlines.

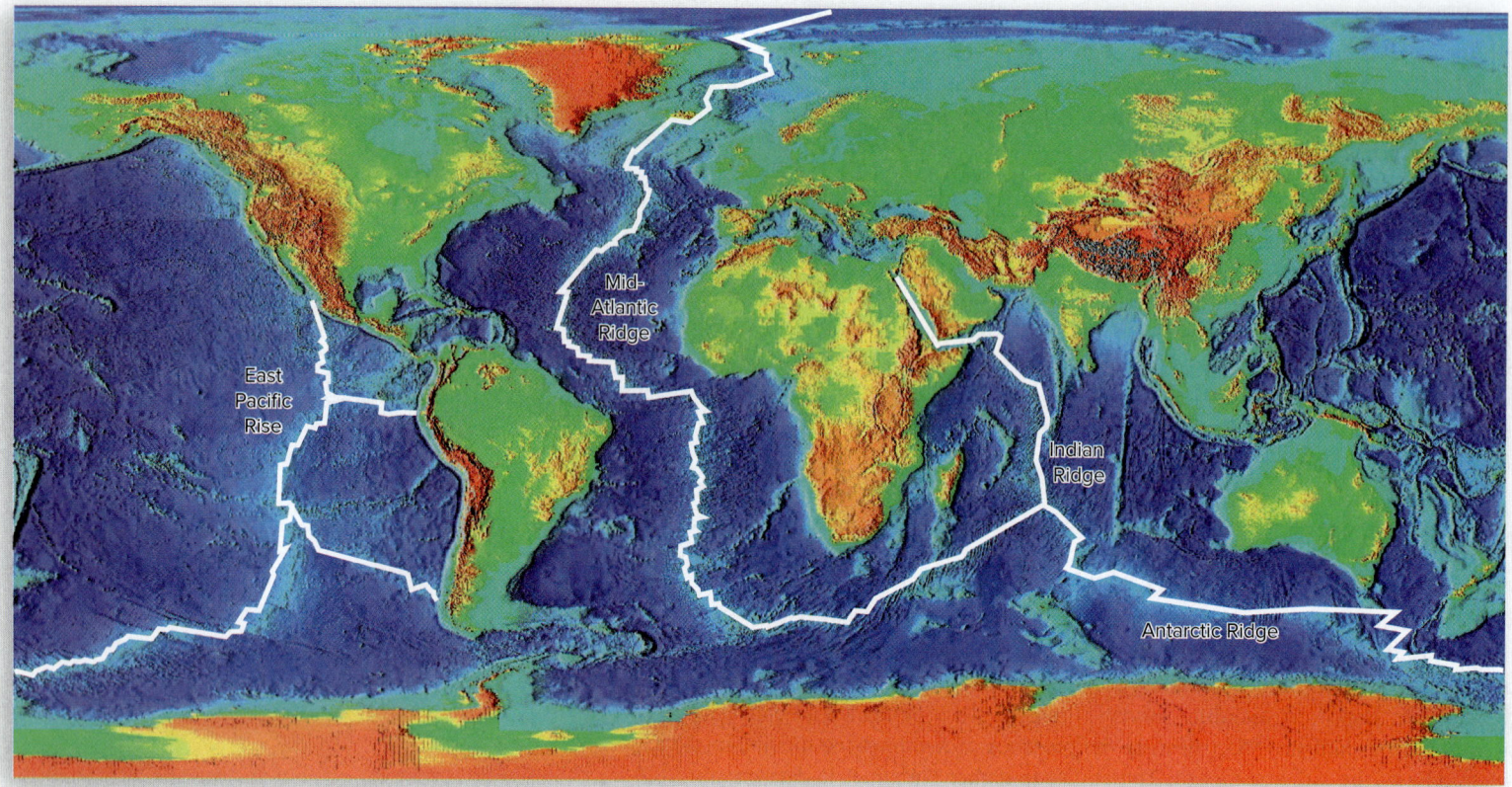

FIGURE 4.10   Modern map showing the topography of both the land and seafloor. One of the striking features of the oceans is the extensive network of mid-oceanic ridges that circle the globe, shown here in white. The oceans also contain narrow trenches that reach depths of nearly 7 miles (11 km).
NOAA

what was "north" in these rocks actually points toward today's magnetic South Pole. Consequently, geologists classify magnetic rocks as having either normal or reversed polarity. The reason Earth's magnetic field reverses itself is not well understood, but it is generally believed to be related to periodic changes in the circulation of metallic ions within the outer core.

Up until World War II paleomagnetic studies had largely been restricted to the continents; hence no one had ever systematically studied the magnetic properties of basaltic rocks known to exist on the seafloor. During the 1950s ships began measuring the magnetic properties of the seafloor by towing an instrument called a *magnetometer* along profile lines. Seafloor maps were constructed that revealed alternating bands of normal and reverse polarity. As illustrated in Figure 4.11, these mysterious bands extended for great distances and were oriented parallel to mid-oceanic ridges. One of the keys to understanding this strange striping pattern was that the bands were of different widths and were symmetrical on either side of a ridge. This meant the patterns were mirror images, and so each individual stripe could be paired up with an identical stripe on the opposite side of a ridge. Another key observation was that the rocks along the ridge crest show normal polarity, and that they become progressively older as one moves away from the ridge in either direction. Geologists soon recognized that the magnetic properties of the seafloor were somehow related to the formation of mid-oceanic ridges.

In order to explain this new oceanographic data, geologists developed a hypothesis known as **seafloor spreading,** where mid-oceanic ridges represent weak zones along which magma erupts to form new oceanic crust. According to this hypothesis, ocean ridges spread or open up over time and are filled with new magma, as shown in Figure 4.11. This process accounts for the symmetrical pattern of magnetic reversals and the fact that ridge crests all contain rocks with magnetic signatures consistent with Earth's present magnetic field. Seafloor spreading also explains why rocks get progressively older farther away from the ridge crests. The hypothesis gained widespread acceptance in 1968 when a research vessel equipped with a drilling rig systematically collected actual rock samples over much of the Atlantic basin. Radiometric dating of the basaltic rocks (Figure 4.12) showed that the entire Atlantic seafloor gets progressively older on either side of the mid-oceanic ridge. Moreover, the oldest parts of the seafloor were found to be 200 million years old, which is quite young compared to 4.0-billion-year-old rocks found on the continents. This was conclusive proof that seafloor spreading was acting like a conveyor belt, carrying newly formed rocks away from the mid-oceanic ridges.

Seafloor spreading presented a problem because if new oceanic crust is constantly being formed, then either the Earth must be growing in size or some unknown process must be destroying crustal rocks. Because the age of the Atlantic seafloor (Figure 4.12) proved that the oceanic crust is expanding, geologists reasoned that perhaps this growth was being compensated for by the destruction of crustal material under the Pacific. This would prevent the Earth from getting larger since part of the Pacific basin would be decreasing in size at the same time the Atlantic basin grew larger. Scientists felt that the most likely place where crustal material is being destroyed is in the deep ocean trenches located around the perimeter of the Pacific Ocean. This idea would eventually be confirmed through the study of earthquake (i.e., seismic) waves, a field that had also revealed the Earth to have a layered structure as described earlier.

Location of Earthquakes

We will explore earthquakes in greater detail in Chapter 5, but for now we can define an earthquake as the release of energy that occurs when rocks are deformed beyond their elastic limit, causing them to rupture. This energy then travels outward in all directions in the form of *seismic waves*—the *epicenter* is the point on the surface that directly overlies where rocks deep underground ruptured and released their stored energy. In the 1960s a global network of seismic recording stations enabled seismologists to map the location of earthquake epicenters around the world. In the map in Figure 4.13, one can clearly see that earthquakes do not occur randomly, but rather are concentrated in distinct zones. Perhaps most striking is just how well the epicenters correlate with the location of mid-oceanic ridges (Figure 4.13A). This can be explained by magma rising vertically up from the mantle, buckling the crust and thereby forming the ridge. The force of the rising magma also causes rocks to deform beyond their elastic limit, at which point they rupture and release stored energy in the form of seismic waves.

In Figure 4.13B you can see there are also places where epicenters are clustered in relatively wide zones, which happen to coincide with mountain ranges and ocean trenches. When seismologists performed detailed studies near ocean trenches, they found that the earthquakes were originating in an inclined zone that extended several hundred miles down into the mantle. This inclined zone was interpreted to be caused by the collision of two lithospheric plates, generating earthquakes as an oceanic slab of lithosphere sinks down into the mantle (Figure 4.13B). This process of lithospheric slabs descending into the mantle is referred to as **subduction,** whereby the oceanic slabs eventually undergo partial melting and are incorporated into the mantle. Subduction can explain the occurrence of certain types of earthquakes, and provides a mechanism by which oceanic crust is destroyed. In addition, scientists believe that ocean trenches themselves are the result of the descending slabs forcing the lithosphere to buckle or downwarp. In the next section you will see that melting of the descending slabs also explains the volcanic activity commonly found along ocean trenches.

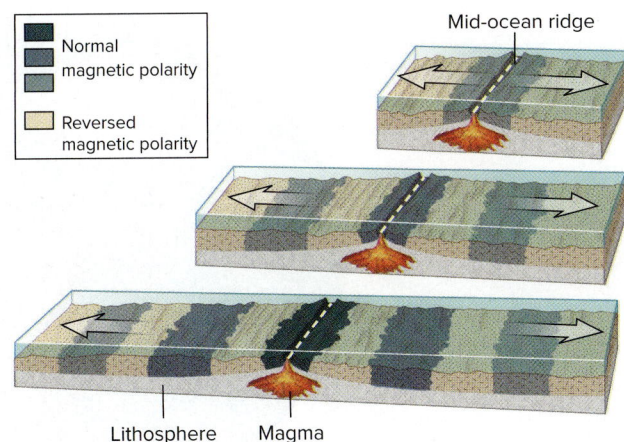

FIGURE 4.11   Illustration showing the development of magnetic striping by seafloor spreading. Magma rises up through mid-oceanic ridges and then cools to form basaltic rocks whose magnetite grains record the orientation and polarity of Earth's magnetic field. As the seafloor continues to spread and new rocks form, a symmetrical pattern of reverse and normal polarity develops on opposite sides of the ridge.

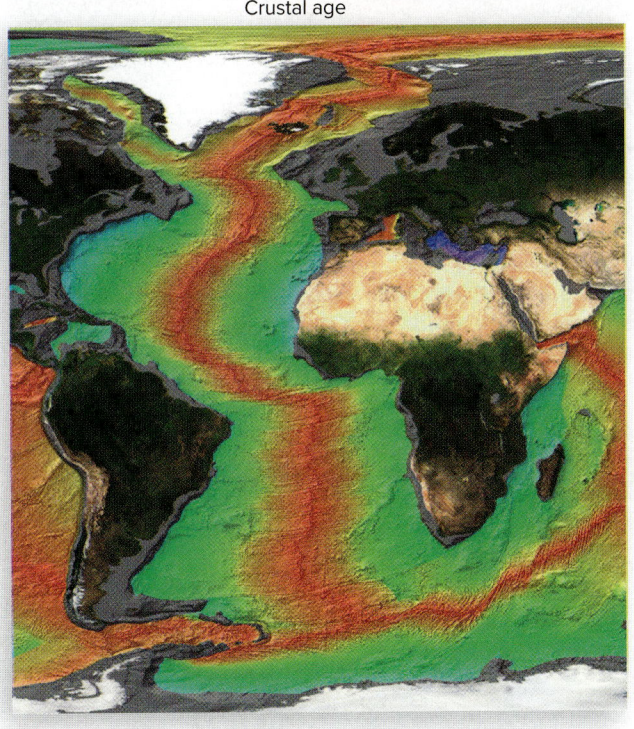

Crustal age

Millions years before present

0   20   40   60   80   100   120   140   160   180   200   220   240   260   280

FIGURE 4.12   Map showing the age in years before present (BP) and topography of the Atlantic seafloor. Note how the seafloor gets progressively older away from the mid-oceanic ridge, with the oldest seafloor being about 200 million years old.
NOAA

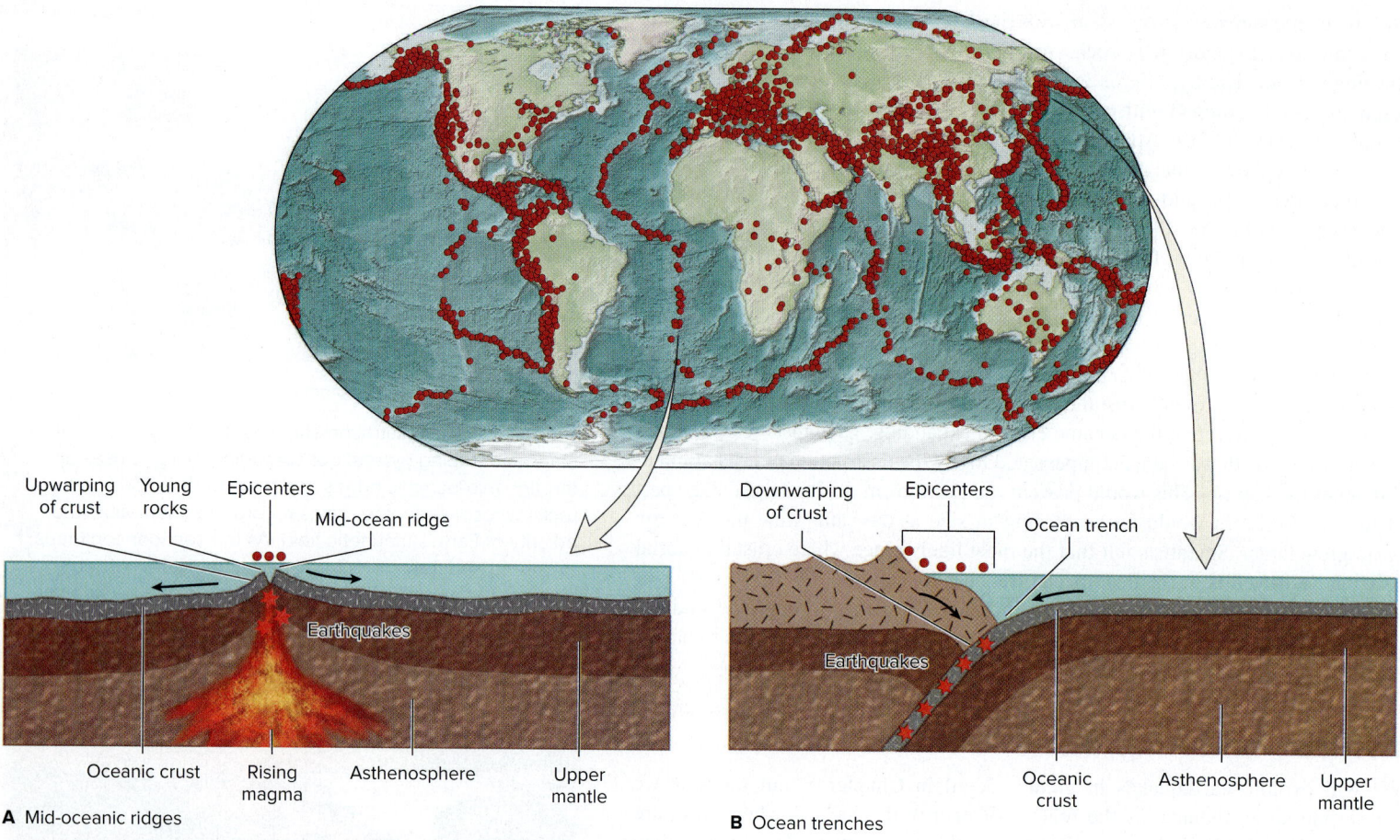

Upwarping
of crust

Young
rocks

Epicenters

Mid-ocean ridge

Earthquakes

Oceanic crust

Rising
magma

Asthenosphere

Upper
mantle

A Mid-oceanic ridges

Downwarping
of crust

Epicenters

Ocean trench

Earthquakes

Oceanic
crust

Asthenosphere

Upper
mantle

B Ocean trenches

FIGURE 4.13 Map showing the location of earthquake epicenters from 1963 to 1998. Note how the earthquakes are not randomly distributed. Inset (A) illustrates how rising magma beneath mid-oceanic ridges generates earthquakes whose epicenters lie in a relatively narrow zone at the surface. Inset (B) shows how epicenters near ocean trenches are spread over a wider area due to the way the subducting slab generates earthquakes in an inclined zone.

In the end, seismic studies provided convincing evidence that magma is rising up from the asthenosphere at mid-oceanic ridges, forming new oceanic lithosphere. These studies also indicated that oceanic plates are being destroyed along ocean trenches as lithospheric slabs undergo subduction. It was now quite clear that the Earth was not expanding, but rather the Atlantic seafloor was growing at the expense of the seafloor under the Pacific.

Polar Wandering

The last major line of evidence supporting continental drift came as a result of paleomagnetic studies involving continental rocks. Recall that magnetite-rich rocks record the orientation of the magnetic field at the time the rocks form. Data from these paleomagnetic studies were used to make maps showing the location of the magnetic North Pole at different times in the geologic past. Scientists were at first baffled when the location of the pole took on a strange wandering path, which was referred to as *polar wandering* (Figure 4.14A). Even more intriguing was that each of the continents showed a different polar wandering path. The separate paths made no sense since the Earth has only one magnetic North Pole at any given time. As illustrated in Figure 4.14B, the puzzle was solved when scientists took into consideration the possibility that the continents had been moving rather than the poles. Sure enough, when seafloor spreading data were used to reposition each of the continents at different times in the geologic past, paleomagnetic data showed a common location for the pole. Therefore, the combination of paleomagnetic and seafloor spreading data provided convincing proof that Wegener's concept of continental drift was indeed correct. Even the most skeptical geologists now agreed that a supercontinent had once existed, and then subsequently broke apart and began moving in different directions as separate landmasses.

Plate Tectonics and the Earth System

Although it took nearly 50 years of gathering data and testing various hypotheses, by the late 1960s scientists had proven that Wegener's basic idea of continental drift was correct. Radiometric dating and paleomagnetic studies of oceanic crust left no doubt that seafloor spreading was actually taking place at mid-oceanic ridges. Moreover, earthquake studies showed that magma rises up from the mantle at ocean ridges to form new crust, while at the same time older crust is being destroyed through subduction at ocean trenches. By unraveling the mystery of polar wandering, geologists also proved Wegener's concept that a supercontinent had once existed, which later broke up and began drifting apart. Finally, it was the clustering of earthquake epicenters that allowed geologists to define the boundaries of the rigid slabs of lithosphere, now called *tectonic plates.*

Perhaps the most important outcome of this long process of gathering data and testing different hypotheses was that scientists were able to show that continental drift, seafloor spreading, and subduction are all connected to one another. This resulted in the development of a single, unifying theory that explained the interrelationship between these different processes. Geologists referred to this new and more comprehensive theory as the *theory of plate tectonics,* a name that originates from the Greek word *tekton,* meaning "builder." Tectonics is an appropriate term since Earth's major surface features are literally built by the planet's internal forces and moving plates. These tectonic forces help drive the rock cycle (Chapter 3) by causing metamorphism and raising mountains above sea level. Weathering and erosion processes then act to wear the uplifted terrain back down to sea level, ultimately transforming the weathered material into sedimentary rocks. Plate tectonics therefore is a critical component of the Earth system that has helped make life as we know it possible. We will begin this section by taking a brief look at tectonic plates and the types of interactions that occur at plate boundaries.

Types of Plate Boundaries

Based largely on seismic data, scientists have determined that Earth's rigid lithosphere is broken up into seven major plates and several smaller ones as shown in Figure 4.15. Note that some of the plates, such as the Pacific, are covered almost entirely with oceanic (basaltic) crust, whereas others are covered by both continental (granitic) and oceanic crust. For example, notice how the eastern edge of the North American plate does not stop at the Atlantic shoreline, but rather extends all the way out to the oceanic ridge in the middle of the Atlantic. Keep in mind that plate boundaries are usually not sharp, distinct features, but rather are zones that are defined by concentrated earthquake activity.

Another key aspect of plate tectonics is that Earth's internal forces cause these rigid slabs to move over the weak asthenosphere. A useful analogy is how wind or water currents will cause broken slabs of ice to move over a water body. In some areas the currents will cause individual slabs of ice to collide, generating compressive forces along the boundary where they meet. This causes the ice to buckle and fracture, forming so-called pressure ridges. In other places a slab may be pulled in opposite directions, generating tensional forces that cause it to fracture and break in two. The movement of Earth's lithospheric plates is similar in that the tectonic forces are most prevalent along the boundaries of plates. It is

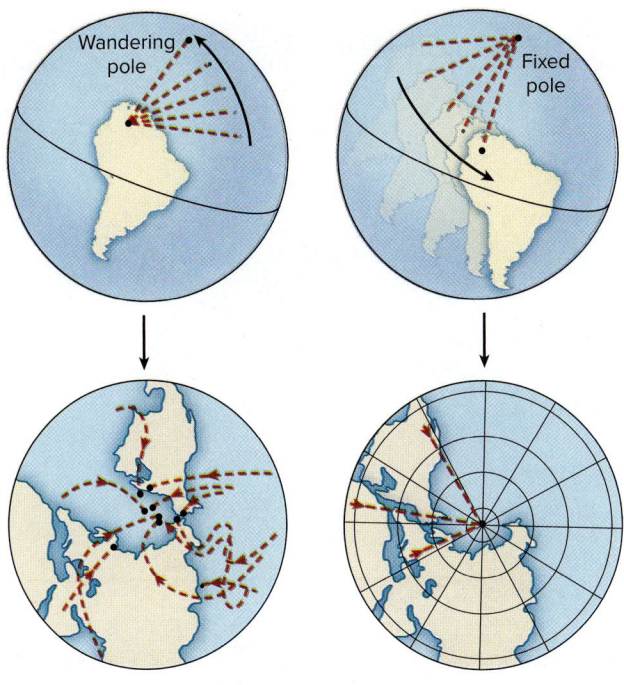

A Fixed continents   **B** Moving continents

FIGURE 4.14   When paleomagnetic studies assumed that the continents remained fixed (A), the position of the magnetic North Pole appeared to wander over time. Moreover, each continent showed a different wandering path for the pole. When seafloor spreading data were used to reposition the continents at different times in the geologic past (B), a single location for the pole emerged.

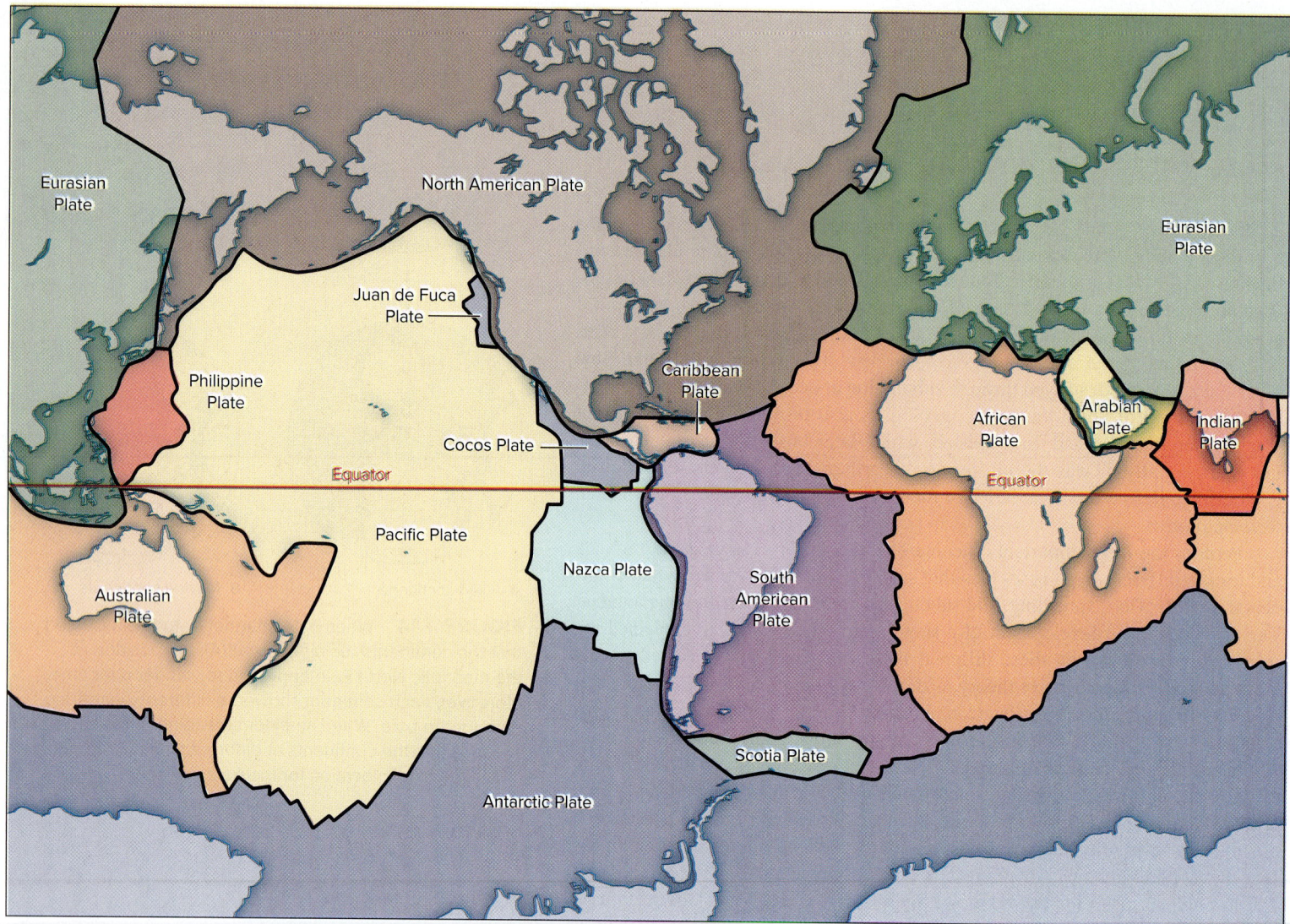

FIGURE 4.15   Map showing the distribution of lithospheric plates. Note how the North American plate is covered by both continental and oceanic crust, whereas the Pacific plate is covered entirely by oceanic crust.

along these plate boundaries that we find the vast majority of earthquakes, volcanic eruptions, and mountains ranges. Therefore, our focus will be on the way plates interact with each other at their boundaries.

The movement of Earth's tectonic plates generates three basic types of forces at the plate boundaries, namely *compression, tension,* and *shear* (see Figure 4.1). Depending on the relative motion of the plates and the dominant type of force being generated, plate boundaries can be placed into one of the three categories shown in Figure 4.16. A **divergent boundary** is dominated by tension forces and involves two plates moving away from one another. For example, a mid-oceanic ridge defines a divergent boundary because it marks where two plates are moving in opposite directions. In contrast, a **convergent boundary** is under compression and is where two plates are moving toward each other. An example is a subduction zone in which the plates collide, forcing one plate to slide under the other. Finally, when two plates slide past each other and are dominated by shear forces, the boundary is called a **transform boundary.** Although we have yet to describe a geologic example of a transform boundary, the concept can be illustrated by imagining two cars moving toward each other. If the cars were to have a glancing

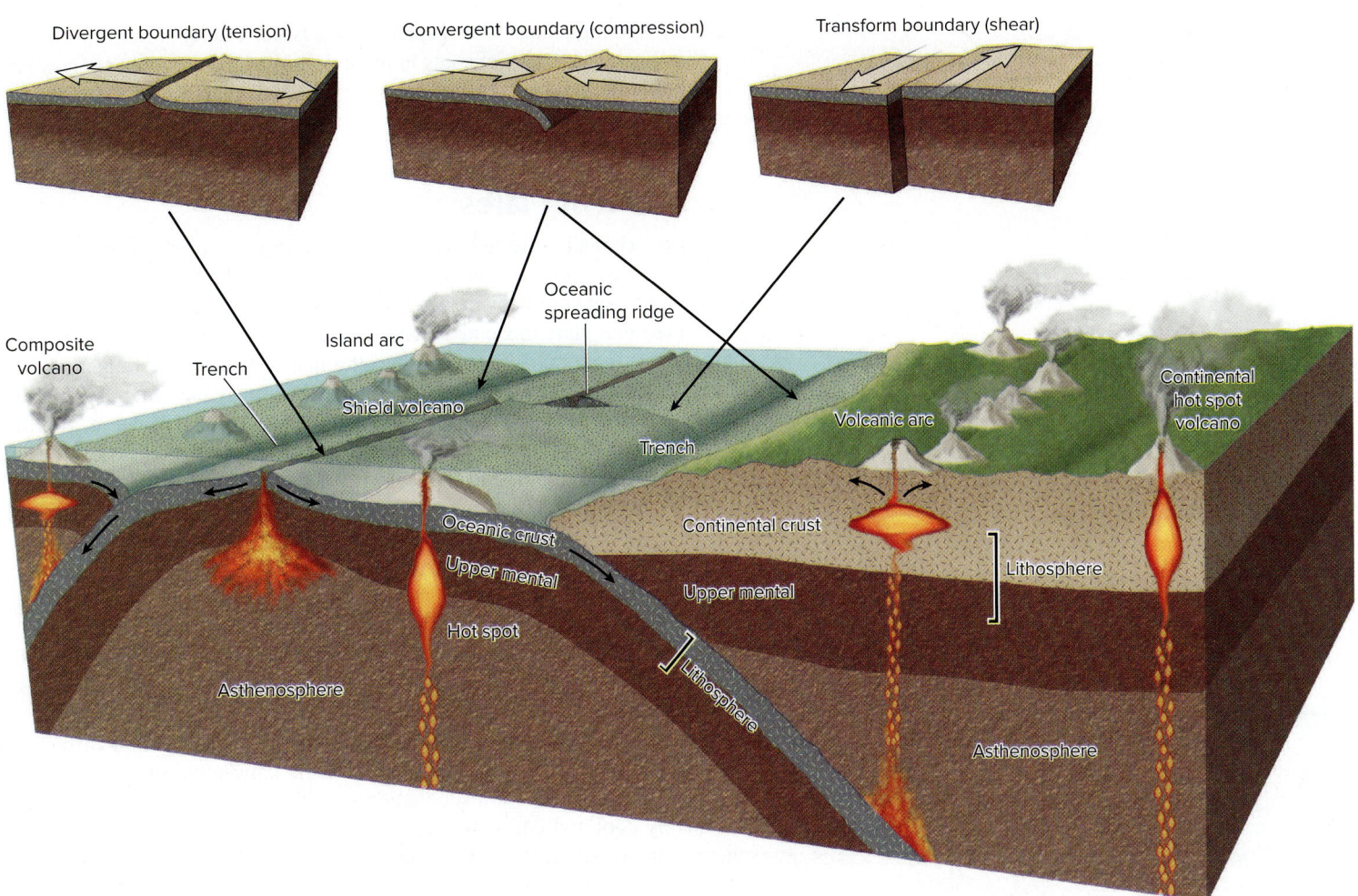

Divergent boundary (tension)   Convergent boundary (compression)   Transform boundary (shear)

Composite volcano

Trench

Island arc

Shield volcano

Oceanic spreading ridge

Oceanic crust

Upper mental

Hot spot

Asthenosphere

Trench

Continental crust

Upper mental

Lithosphere

Volcanic arc

Continental hot spot volcano

Lithosphere

Asthenosphere

FIGURE 4.16   Illustration showing how divergent, convergent, and transform plate boundaries are under tension, compression, and shear forces, respectively. Note how mid-oceanic ridges and mountain chains are features that form parallel to plate boundaries. Volcanic hot spots occur away from plate boundaries and are believed to be related to hot plumes of material that rise from deep within the mantle.

impact, as in a side-swipe, it would create shear forces on the cars and be analogous to a transform boundary. A head-on crash would generate mostly compression forces, thereby representing a convergent boundary.

One of the key features of Figure 4.16 is how oceanic lithosphere is created along mid-oceanic ridges. Here we see magma being generated in the asthenosphere, which then rises into the mid-oceanic ridge and cools to form basaltic rock. As the two plates diverge away from the ridge system, fresh magma fills the gap, creating new basalt. Over time, this results in a layer of basalt being deposited on top of the rigid rocks of the upper mantle—recall that the combined crust and upper mantle is called the *lithosphere.* Notice that while seafloor spreading is taking place, the opposite end of the oceanic plate is being destroyed as it descends into the asthenosphere at a subduction zone. As a slab descends and starts to melt, it generates magma that rises through the overlying plate and forms a volcanic mountain range. This rising magma at convergent boundaries commonly interacts with continental rocks, which are relatively rich in silica (SiO_2). This interaction produces andesitic and granitic magmas that are richer in SiO_2, but contain fewer ferromagnesian minerals compared to basaltic magmas (Chapter 3). Therefore, the oceanic lithosphere that is destroyed at convergent boundaries ends up producing magma, which forms new continental rocks. At transform boundaries, however, magma does not form since the plates simply grind past one another. Also notice in Figure 4.16 how magma rises from deep within the mantle at so-called *hot spots,* creating volcanic landmasses in the middle of tectonic plates. The Hawaiian Islands, for example, lie over a mantle hot spot. We will discuss hot spots in more detail in Chapter 6, but for now simply note that the majority of volcanic activity takes place along divergent and convergent plate boundaries.

Movement of Plates

Geologists originally rejected the concept of continental drift because Wegener could not come up with a mechanism to adequately explain how the plates move. Although the exact mechanism is not fully understood even today, scientists generally agree that plate movements result from the complex interaction between convection cells within the mantle and the plates themselves. Remember that deep within the mantle the temperature and pressure are extremely high, which allows the silicate rocks to flow as they undergo plastic deformation. As illustrated in Figure 4.17, hot mantle material tends to rise toward the surface where it begins to lose heat, causing its density to increase. Because of the increased density the rocky material eventually begins to sink, creating large-scale convection cells within the mantle. Notice that the lithospheric plates are believed to be moving in the same direction as the convection cells. Keep in mind, however, that the asthenosphere is near its melting point and is very weak. Therefore, it is unlikely that the convective motion in the mantle is very effective in moving tectonic plates lying on top of the weak asthenosphere. Scientists have concluded that additional forces must be operating to help drive the plates over the asthenosphere.

Modeling studies have shown that the density and elevation differences of tectonic plates play an important role in the convective movement within the mantle. For example, at a subduction zone the descending lithospheric slab is relatively cool and dense compared to the hot material within the asthenosphere. This density difference causes the slab to sink, which tends to drag or pull the rest of the plate into the subduction zone, a process called *slab pull* (Figure 4.17). In addition, the rising mantle material at divergent boundaries pushes the lithosphere upward, forming an elevated ridge system. This creates what is known as *ridge push,* where gravitational forces cause the elevated rocks to literally push the oceanic plate downward over the sloping surface of the asthenosphere. An example of ridge push can be seen in the snow-covered car in Figure 4.18. In this case gravity caused the elevated section of snow on the windshield to push downward on the sloping surface of the glass. This forced the entire layer of snow to slide over the hood of the car. Note that this movement was facilitated by unfrozen water on the car's surface, which acted as a weak layer similar to the asthenosphere.

Surface Features and Plate Boundaries

In this section we will take a closer look at the different types of features and processes that occur at the three types of plate boundaries. We want to focus our attention on plate boundaries because this is where earthquakes, volcanic eruptions, and the uplift of mountains take place. These processes, of course, are important in the study of environmental geology because they represent natural hazards. As you read through this section, it will be helpful to refer back to Figure 4.19 as it shows the boundaries of Earth's major plates and the directions in which the plates are moving.

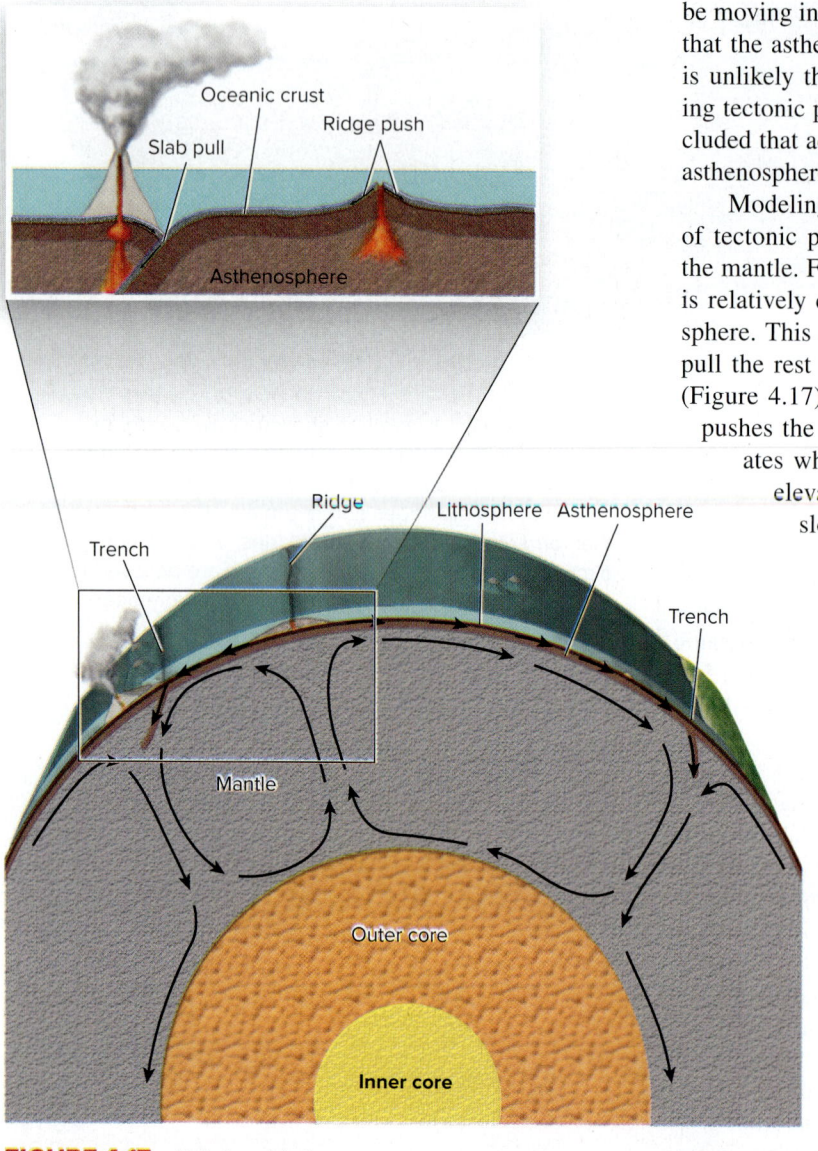

FIGURE 4.17   Relationship between mantle convection cells and plate boundaries. Note how the plates move in the direction of the convection cells.
© Jim Reichard

Divergent Boundaries

Earlier you learned that mid-oceanic ridges are divergent boundaries where seafloor spreading is taking place. Here the tectonic plates on opposite sides of a ridge move away from each other like giant conveyor belts. The sequence of events in Figure 4.20 shows how scientists believe such spreading centers originate. The process begins when a convection cell causes mantle material to rise, forcing the asthenosphere and lithosphere (crust and upper mantle) to buckle upward. This process also creates a zone of low pressure within the asthenosphere, which causes it to start melting and form magma. The upward force of the mantle plume not only causes the lithosphere to buckle, it also creates a considerable amount of tensional stress within the brittle plate. Eventually the plate begins fracturing, which typically leads to the development of faulting in a stair-step fashion, forming a down-dropped featured called a **rift valley,** or *graben* (Figure 4.20B). The rising magma results in basaltic rocks forming on the valley floor. As the basalt-filled valley continues to widen, the entire rift can become flooded with seawater, at which point a linear-shaped ocean is born.

The mid-Atlantic ridge (see Figures 4.10 and 4.19) is perhaps the most well-studied site of seafloor spreading. With a length of nearly 10,000 miles (16,000 km), this ridge system is actually the world's longest mountain range. Based on the age of the basaltic rocks making up the Atlantic seafloor and their distances from the spreading center, geologists calculate that the seafloor is spreading at an average rate of 2.5 centimeters per year. Although this may seem quite slow to humans, if this rate continued for another million years of geologic time, the Atlantic Ocean would be 15 miles (25 km) wider. Likewise, if we could somehow reverse the plate motion and go backward in time, the ocean basin would get smaller. Were we to go back 200 million years,

FIGURE 4.18   A thick layer of snow on this car provides an example of the ridge push mechanism. Gravity caused the snow to push downward on the sloping surface of windshield, forcing the rest of the snow to slide over the hood. Note that the snow layer behaved as a rigid plate, allowing it to buckle, forming a fold. Also note that unfrozen water was present on the surface of the car, creating a weak layer that facilitated the movement of the overlying snow.
© Jim Reichard

FIGURE 4.19   Map showing the types of movement taking place at the boundaries of Earth's major plates.

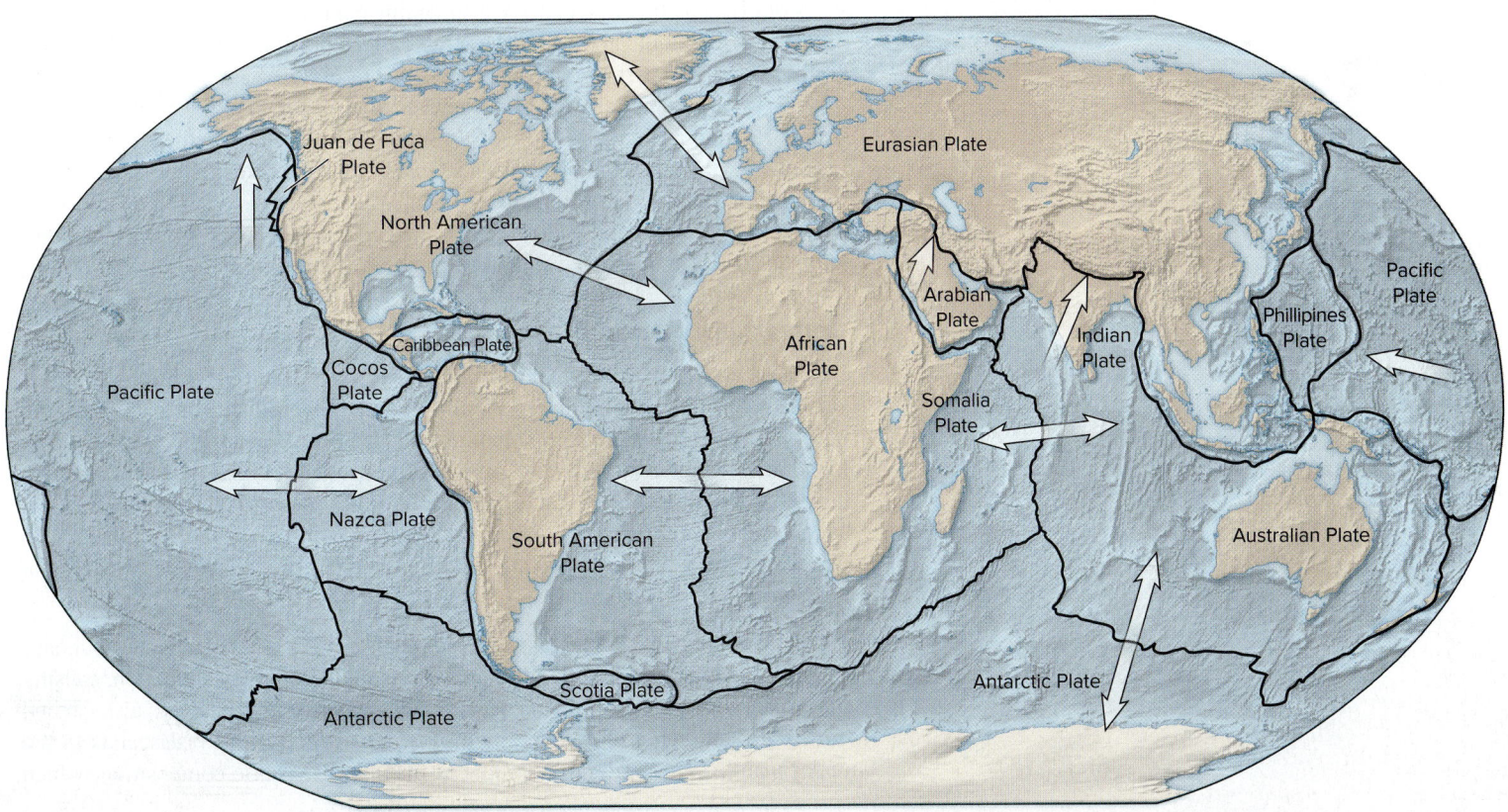

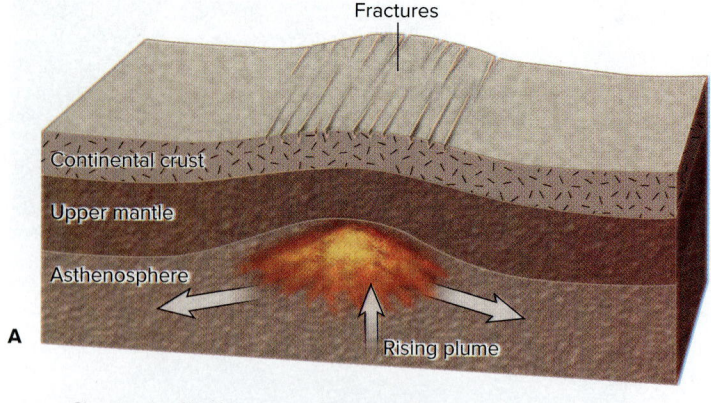

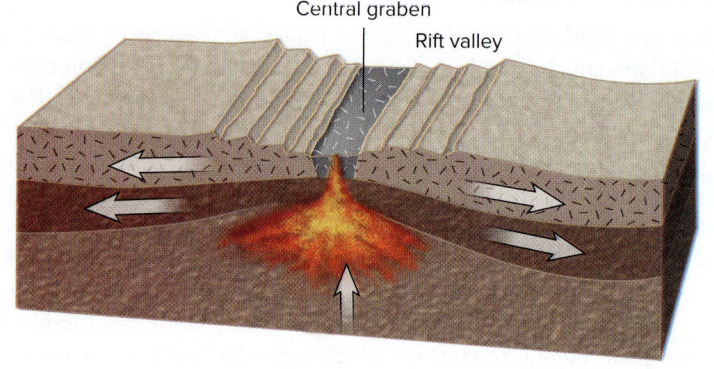

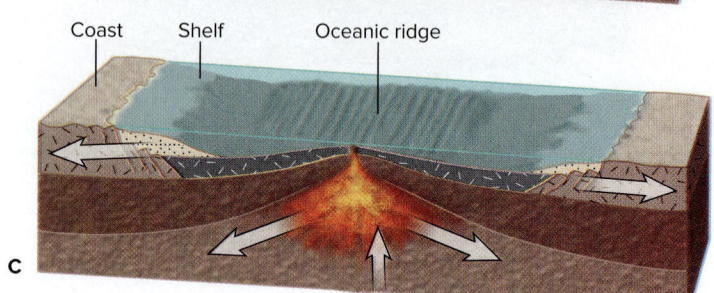

FIGURE 4.20 Sequence of events showing the development of a mid-oceanic ridge system. A rising convection cell (A) pushes upward against the brittle lithosphere, creating tensional forces that cause the plate to fracture. This also creates a zone of lower pressure, causing melting of the asthenosphere and formation of magma. Continued tension (B) leads to faulting and the development of a rift valley where rising magma forms layers of basaltic rock. As the rift widens, it becomes flooded (C), forming a new ocean with a spreading center in the middle.

the Atlantic would be a narrow sea bounded by the African and American continents, which were just starting to drift away from each other.

One of the fascinating things about geology is that we can study tectonic features in different stages of development. For instance, the mid-Atlantic ridge and ocean basin show us what an active spreading center can do given 200 million years of geologic time. To observe a spreading center in its initial stages of development, we can look at Africa, where the eastern portion of the continent is currently being torn apart by tensional forces. From Figure 4.21 one can see that the Arabian peninsula has already broken away from Africa, creating a rift valley that is now flooded and called the Red Sea—this is similar to how the Atlantic once looked. Moving south we find Africa's Great Rift Valley where the rifting is even more recent. Should the spreading continue, this valley will also flood and form a narrow sea. Note that the Great Rift Valley contains some of the world's oldest hominid fossils and is also well known for its volcanic activity.

Convergent Boundaries

Unlike the tensional forces that form rift valleys and spreading centers, compressional forces associated with convergent boundaries produce complex mountain ranges. Moreover, different combinations of oceanic and continental plates can be found at convergent boundaries. This, in turn, produces different types of mountains. As shown in Figure 4.22, the various types of convergent boundaries are as follows: *oceanic-oceanic, oceanic-continental,* and *continental-continental.* In this section we will briefly explore the different types of collisions and the type of landforms that are generated.

Oceanic-Oceanic   When two oceanic plates collide at a convergent boundary (Figure 4.22A), one of the basalt-covered slabs will undergo subduction, causing the crust to buckle and form an ocean trench. Water-rich sediment is also carried downward and acts to lower the melting point of the slab as it descends in the upper mantle. This commonly produces magma of basaltic composition, which then rises up through the buckled plate to form a string of volcanic islands called an **island arc.** The Aleutian Islands off the Alaska mainland (Figure 4.23A) are a good example of a relatively young island arc system. Over time, the subduction process not only results in the growth of the islands, but will produce *andesitic* magmas. Recall from Chapter 3 that andesite is an igneous rock that contains more SiO_2 (silica) but less iron and magnesium than basalt. One way andesite forms is when a subducting slab of basaltic crust becomes hot enough that it undergoes *partial melting.* Because the minerals in basalt with the lowest melting points also happen to be the richest in SiO_2, partial melting generates andesitic magmas that are relatively rich in SiO_2. Andesite can also form when a rising body of basaltic magma causes silicate minerals in the surrounding rocks to melt. This material is then incorporated into magma, enriching it in SiO_2—a process called *assimilation.*

Over time, the processes of partial melting and assimilation produce crustal rocks that are richer in SiO_2, but less dense compared to the basaltic crust being destroyed in the subduction zone. This means that as island arcs get older, they not only become larger, they generally contain more andesite rock. The islands of Japan (Figure 4.23B) are a good example of a relatively large, more mature island arc system composed of crustal rocks of somewhat lower density.

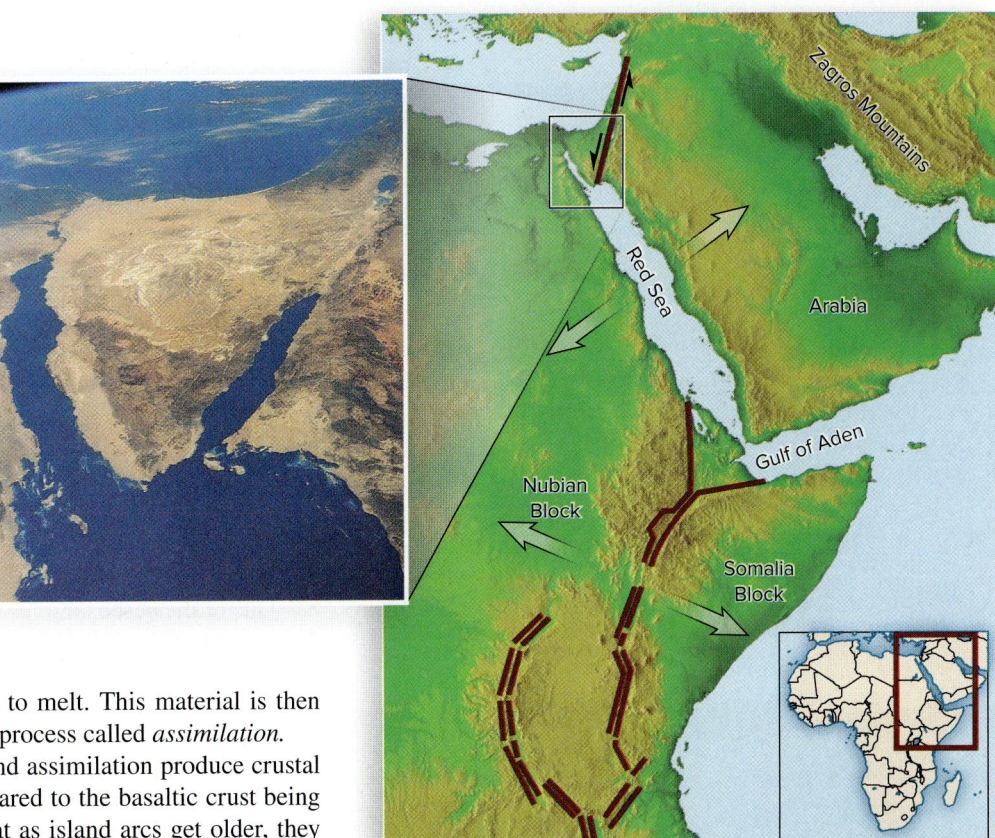

FIGURE 4.21 Map showing the location of a young spreading center in eastern Africa. The northern portion of the rift zone has opened to a point where it has flooded, forming the Red Sea and Gulf of Aden. To the south is the much younger Great Rift Valley. Inset is a satellite photo showing the Sinai peninsula located at the top of the Red Sea. This peninsula is bounded on the left by the spreading center and by a transform boundary on the right.
(inset) NASA

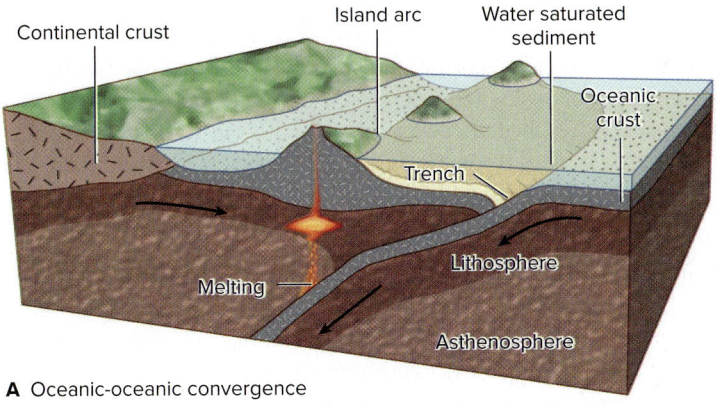

A Oceanic-oceanic convergence

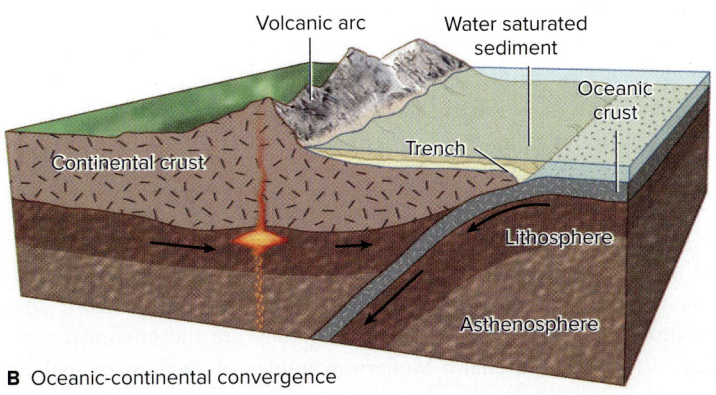

B Oceanic-continental convergence

FIGURE 4.22 Different types of convergent plate boundaries produce different landforms. The collision of two oceanic plates (A) results in subduction and the formation of a volcanic island arc. When oceanic and continental plates collide (B), subduction leads to a continental arc system that contains more SiO_2-rich rocks. When two continental plates collide (C), there is no volcanic activity since neither plate undergoes subduction. Here the collision produces a highly deformed mountain range called a suture zone.

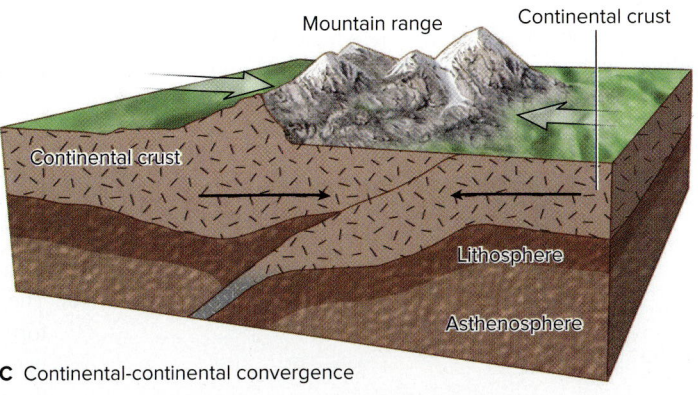

C Continental-continental convergence

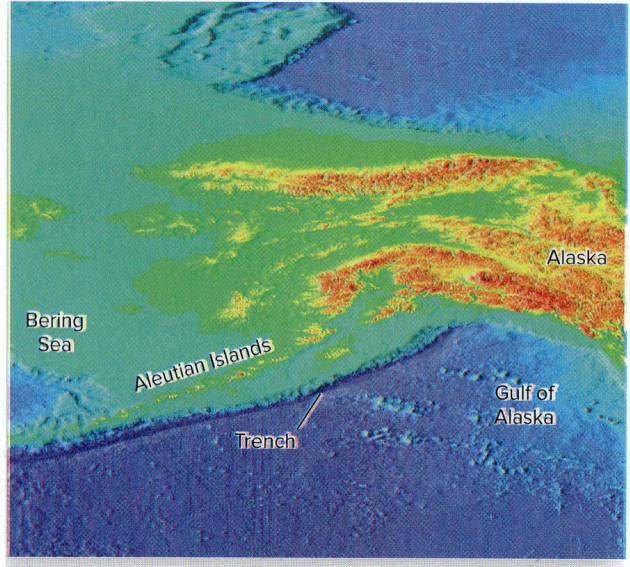

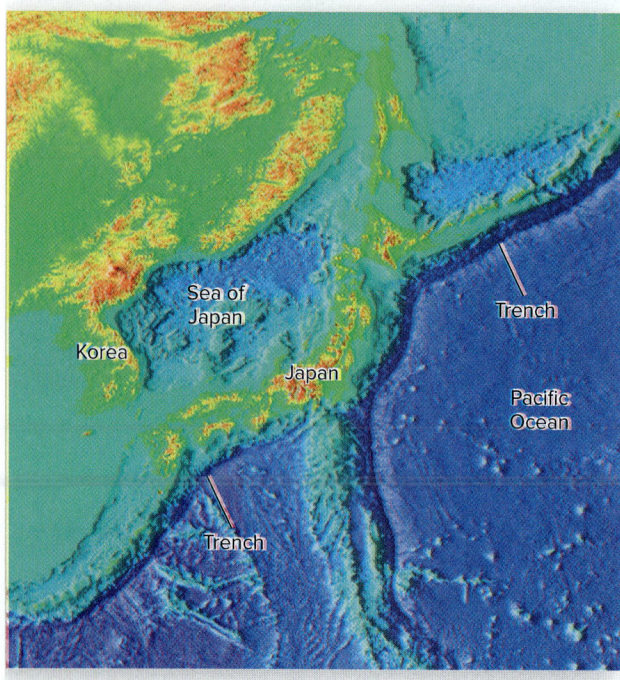

FIGURE 4.23   The Aleutian Islands off the Alaskan mainland (A) are a narrow string of volcanic islands that represent a young island arc system. The much larger islands of Japan (B) represent a more mature island arc. Note the deep ocean trench in both examples, caused by the subducting plate on the ocean side of the island arc.
(a–b) National Geophysical Data Center/NOAA

Oceanic-Continental   Convergent boundaries involving the collision of oceanic and continental plates (Figure 4.22B) also lead to subduction and volcanic activity. In this case, however, the resulting string of volcanoes forms on the continental plate, and is therefore referred to as a **continental arc** as opposed to an island arc. Recall that continental (granitic) lithosphere is considerably less dense than oceanic plates because the rocks are more enriched in SiO_2 and contain relatively few ferromagnesian minerals. Since the oceanic and continental plates are moving over the asthenosphere, when they collide, the more buoyant continental plate overrides the denser oceanic slab. The oceanic plate then undergoes subduction and provides the water that helps initiate melting in the upper mantle. Here the magma must rise through a much thicker continental plate as opposed to a thin oceanic slab. This means the magma is likely to assimilate even more SiO_2-rich material into the melt as it ascends to the surface. The result is a magma that forms igneous rocks called *granite* and *rhyolite,* which contain few ferromagnesian minerals but are rich in quartz (Chapter 3).

One of the best examples of an oceanic-continental plate boundary is along the western coast of South America (Figure 4.24). Here nearly the entire edge of the continent is converging with an oceanic plate, forming an ocean trench and a continental arc system known as the Andes Mountains. This impressive mountain range contains numerous volcanoes and has peaks over 22,000 feet (6,700 m) above sea level. Moreover, the Andes are approximately 5,500 miles (9,000 km) in length, which is greater than the distance from New York City to Rome, Italy. Another example of a continental arc is the Cascade Range, located in the Pacific Northwest region of the United States. Note that Mount St. Helens, which erupted in 1980, is one of the better known volcanoes in the Cascades.

It is important to emphasize that both island and continental arcs are built by magma that is derived from the subduction of oceanic lithosphere. This process generates igneous rocks that become part of the overlying plate, ultimately producing landmasses that are more granitic in composition. The continued recycling of oceanic plates and creation of new continental rocks explains why oceanic crust is generally less than 200 million years old. It also explains why the continents contain ancient rocks as much as 4.0 billion years old. Subduction zone studies have not only helped scientists unravel Earth's history, they have also led to a better understanding of the geologic hazards associated with earthquakes and volcanic eruptions (Chapters 5 and 6).

Continental-Continental   The last type of convergent boundary is where two continental plates collide (Figure 4.22C). Because both lithospheric plates in this case are composed of low-density granitic material, making them rather buoyant, neither plate will undergo subduction. Instead the two plates will literally plow into each other, forcing rocks to push vertically above sea level and also down into the asthenosphere. In this setting the rocks are under so much compressive force that they deform beyond their elastic limits. This causes the rocks near the surface to fail mostly by fracturing and folding, whereas more deeply buried rocks undergo plastic deformation. As illustrated in Figure 4.25, the result is a linear mountain chain consisting of intensely folded and faulted rocks, most of which have undergone regional metamorphism (Chapter 3). Geologists often refer to such mountain belts as *suture zones* because they act as a bond between two landmasses that have come together.

Numerous examples of ancient suture zones can be found around the world, indicating that plate tectonics has been operating since Earth was quite young. The Appalachian Mountains of North America represent a suture zone that formed 250 million years ago, thus what we see today are the erosional remnants of a once lofty mountain chain. Modern examples of such zones include the Himalayas in Asia and the Alps in Europe. The formation of the Himalayas is of considerable interest to geologists because the collision of India with Asia is a relatively recent event, geologically speaking. By studying the Himalayas, scientists have learned a great deal about the formation of ancient suture zones.

As illustrated in Figure 4.26, India was at one time a separate landmass surrounded by oceanic crust that was moving away from a spreading center as if on a conveyor belt. For millions of years the oceanic plate on the leading edge of India was being subducted under the Asian continent, thereby creating a volcanic arc system. Then about 40 to 50 million years ago the two continental landmasses started to collide, and the modern Himalayas began to rise. Subduction of the oceanic plate eventually stopped, causing most of the volcanic activity to cease. This collision continues today and is actively raising the Himalayas at a relatively rapid rate of more than 1 centimeter per year. Similar to the uplift of most other mountains, the rise of the Himalayas is not necessarily smooth and continuous, but rather occurs by sudden movements related to major earthquakes. Note that the height of all mountains, regardless of their tectonic setting, is controlled by the difference in the rates of uplift and erosion. Mountains naturally rise when uplift is greater than erosion, and they become lower when erosion is greater than uplift. Climate also plays a key role because erosion rates are affected by the amount of precipitation and the formation of glacial ice.

Transform Boundaries

The last of the three major types of plate boundaries we need to consider are *transform boundaries,* which are where *shear* forces dominate between two plates. In this setting the plates simply grind or slide past one another along what are called *transform faults.* These faults were first identified in the 1960s when

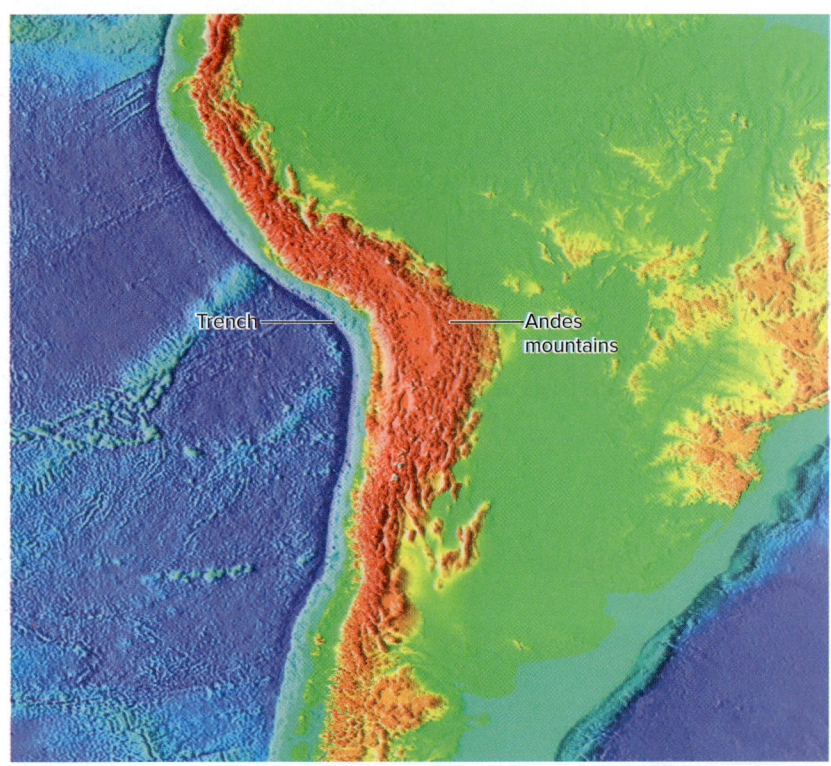

FIGURE 4.24 The Andes Mountains are a continental arc system associated with an oceanic-continental boundary. The boundary of these plates is marked by the position of the ocean trench. The Andes are the result of continued convergence and subduction of the oceanic plate, which has produced volcanic activity and numerous earthquakes. NOAA

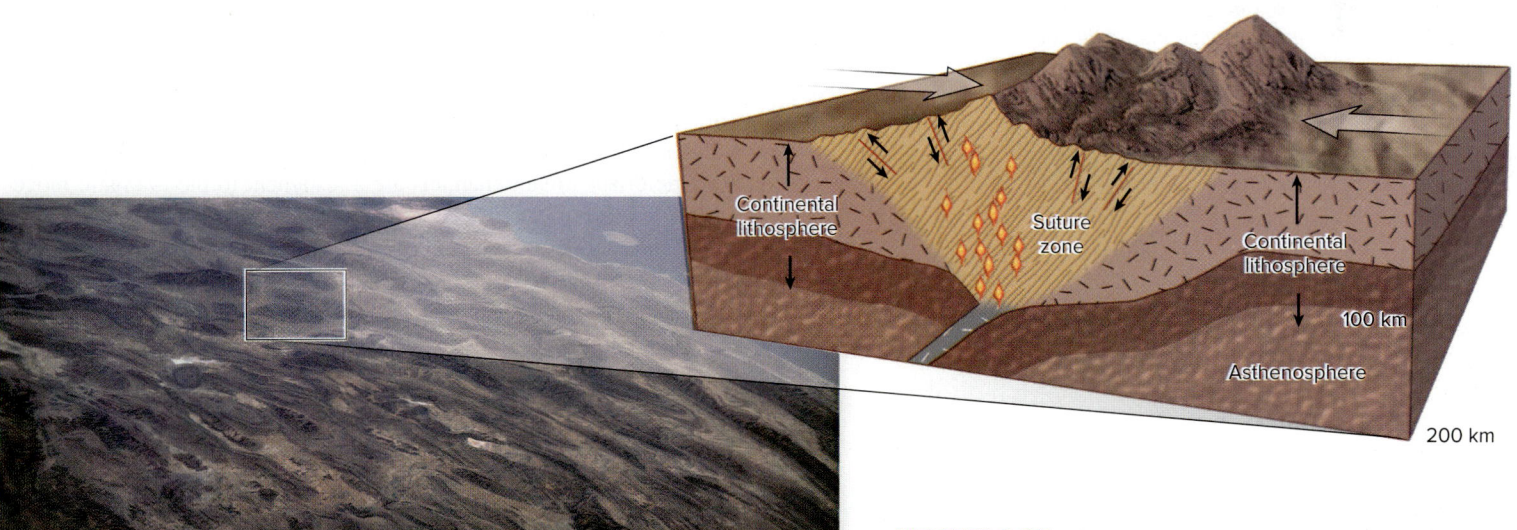

FIGURE 4.25 Subduction does not occur along convergent boundaries involving two continental plates composed of relatively low-density material. Instead, the collision creates a thick zone of highly faulted and deformed mountains where the rocks undergo regional metamorphism. Photo shows the folded Zagros Mountains in Iran. NASA

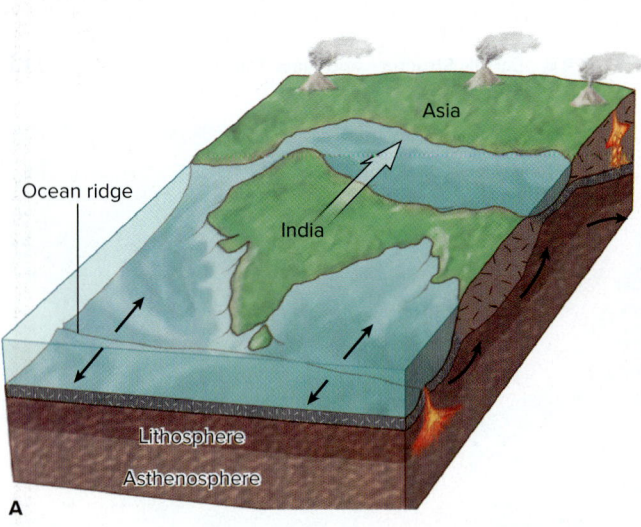

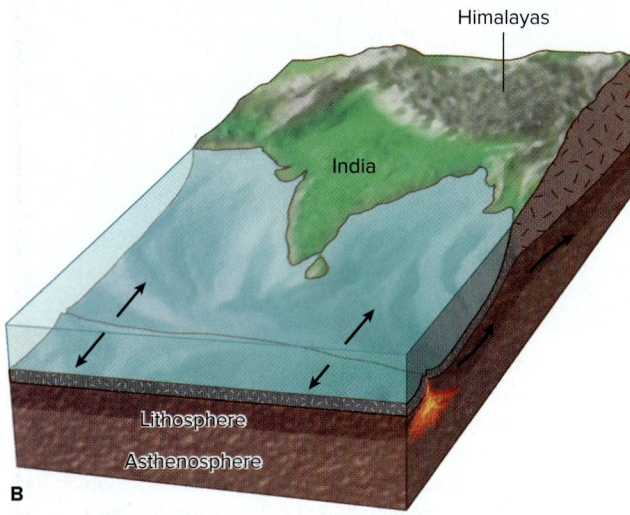

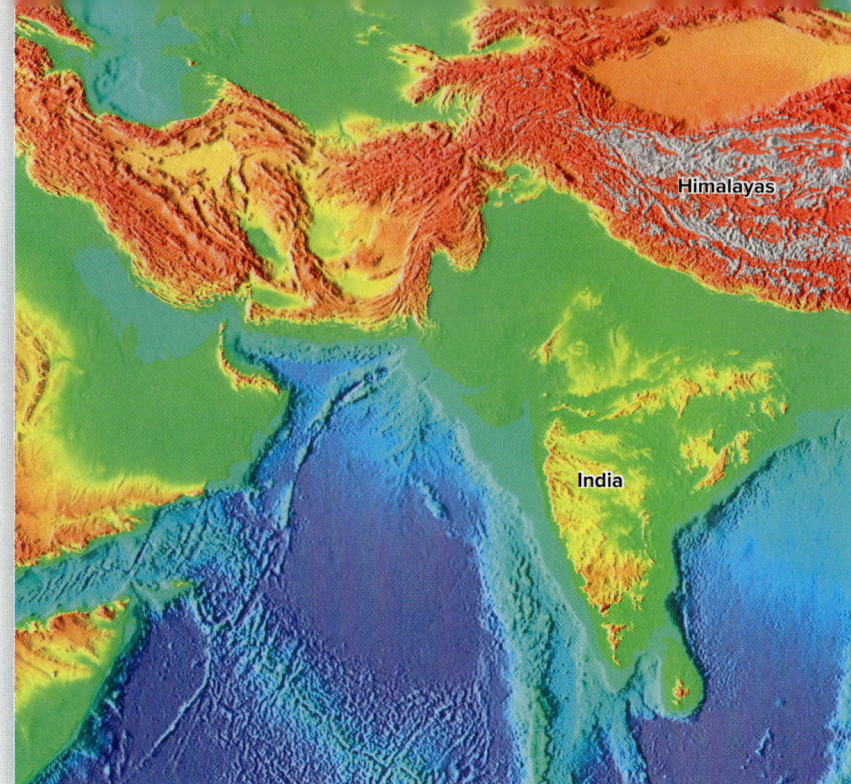

FIGURE 4.26 The Himalaya Mountains are a modern example of the collision between two continental plates. As India moved toward the Asian landmass, the subduction of oceanic crust created a volcanic arc (A). After the two landmasses collided, the subduction zone eventually shut down and volcanic activity ceased. The collision continues to cause uplift and deformation today (B).
National Geophysical Data Center/NOAA

scientists noticed that mid-oceanic ridges are commonly offset, creating a zig-zag pattern as shown in Figure 4.27. Notice in the areas where the ridges are offset, how seafloor spreading produces a shearing motion along the various transform faults. Because the plates are moving in a horizontal manner, there is no subduction and magma does not form. The lack of subduction and volcanic activity means that lithospheric material is neither being created nor being destroyed. In Chapter 5 will examine how the shearing forces that develop along transform faults can result in major earthquakes.

Because of their association with spreading centers, most transform faults are found within oceanic plates. Consequently, the vast majority of transform plate boundaries lie beneath the oceans, and relatively few are found on continents. An example of a transform boundary is the San Andreas fault in California, which is well known because it happens to be on land and generates powerful earthquakes in highly populated areas. Notice in Figure 4.27 how the San Andreas is actually a long transform fault that connects two rather distant spreading centers. The photo in Figure 4.28 shows a site north of Los Angeles where the San Andreas is well exposed. Here the offset stream channels are a testament to the dramatic movement that has taken place relatively recently along the fault. Interestingly, at this site you could stand on one tectonic plate, then walk a short distance and be on another plate that is literally moving in the opposite direction. This lateral movement along the San Andreas means that Los Angeles, which sits on the Pacific plate moving to the northwest (Figure 4.27), will eventually meet up with San Francisco as it moves to the southeast on the North American plate. Keep in mind that it will take millions of years for the cities to meet. Also, contrary to common folklore, these plate movements are not going to cause Los Angeles to suddenly slide off into the ocean in some giant landslide!

Finally, notice in Figure 4.27 that as one goes north of San Francisco, the plate boundary changes from a transform to a convergent boundary. Moreover, the presence of a subduction zone and volcanic arc (Cascade Range) tells us that the boundary along the Pacific Northwest involves the convergence of an oceanic plate with the continental plate. As a result, residents of the Pacific Northwest face both volcanic and earthquake hazards, whereas the primary hazard in Southern California is one of earthquakes. We will explore both earthquake and volcanic hazards in considerable detail in Chapters 5 and 6.

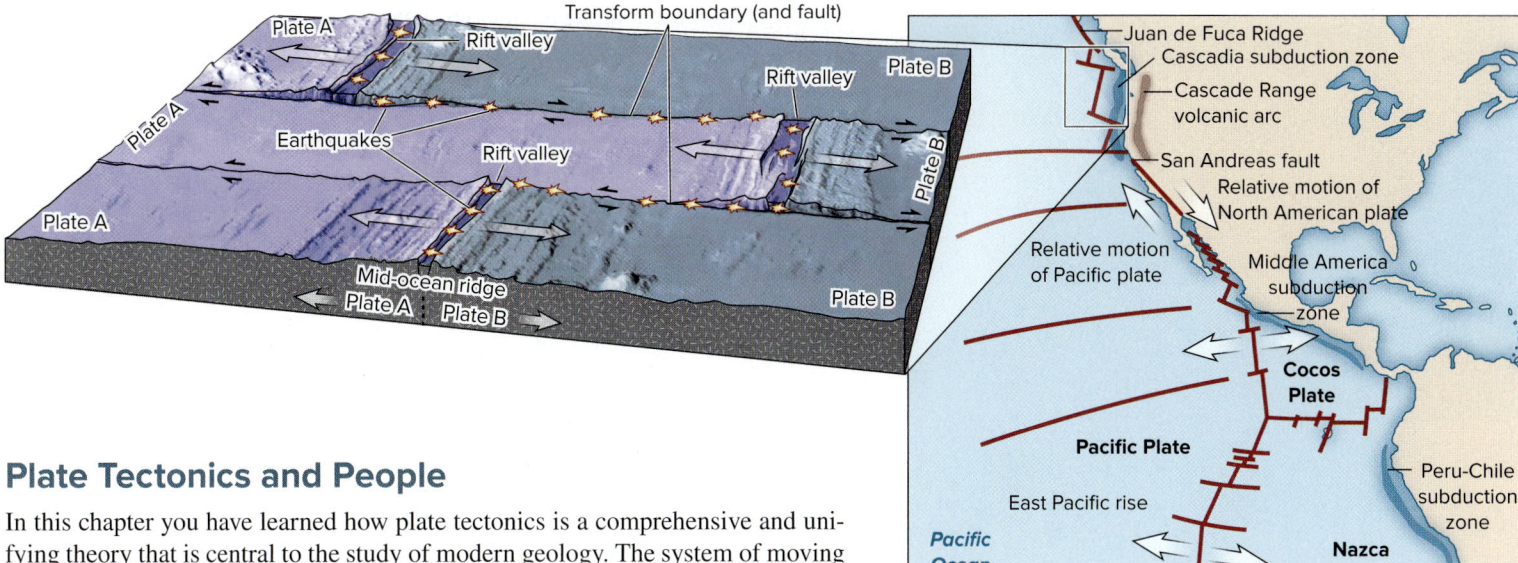

FIGURE 4.27   The San Andreas fault is a transform boundary associated with the East Pacific Rise spreading center. The San Andreas forms the boundary between the Pacific and North Amercian plates. Note that just north of San Francisco, the western edge of North America changes from a transform to a convergent boundary that includes a subduction zone and volcanic arc.

Plate Tectonics and People

In this chapter you have learned how plate tectonics is a comprehensive and unifying theory that is central to the study of modern geology. The system of moving plates not only is responsible for driving the rock cycle, but also creates the major surface features of our planet. However, the importance of plate tectonics goes far beyond geology. Because Earth acts as a system, we should expect that anything as significant as plate tectonics will have an important impact on other parts of the system, namely the atmosphere, hydrosphere, and biosphere. In this final section we will briefly touch on some of the ways in which plate tectonics affects the lives of humans and modern societies. After all, the focus of this textbook is on the interaction between environmental geology and people.

Natural Hazards

There are many geologic processes that create natural hazards for humans and society. In this text we will focus on the hazards associated with earthquakes, volcanic eruptions, and the downslope movement of earth materials (i.e., landslides), all of which are clearly related to plate tectonics. As you have learned in this chapter, the majority of earthquakes and volcanic eruptions occur along plate boundaries as does the uplift of mountain ranges, which generates steep slopes and the potential for landslides. Therefore, the level of risk people face from these hazards depends on where they live relative to Earth's plate boundaries.

If we consider just earthquakes and volcanic eruptions, we find that the risk varies according to the type of plate boundary. The relative risk is indicated in Table 4.2. For example, although earthquakes can happen far from a plate

TABLE 4.2   Relative risk of earthquake and volcanic hazards at the different types of plate boundaries.

Tectonic Plate Setting	Relative Risk of Earthquake Hazards	Relative Risk of Volcanic Hazards
Divergent	Moderate	Moderate
Convergent: oceanic-oceanic	Major	Major
Convergent: oceanic-continental	Major	Major
Convergent: continental-continental	Major	None
Transform	Major	None

FIGURE 4.28   Aerial photo showing the San Andreas fault in Southern California, which forms the boundary between the North American and Pacific plates. Note how the stream channel has been offset due to recent movement along this transform plate boundary.
USGS

boundary, most occur along boundaries where rocks are subjected to tremendous amounts of stress. Because rocks can store a great deal more energy when under compression or shear forces compared to tension, major quakes are much more common along convergent and transform boundaries (Table 4.2). Volcanic hazards, on other hand, exist only in areas where rocks deep in the subsurface are able to melt and form magma. The formation of magma then is largely restricted to divergent boundaries and those convergent boundaries involving subduction. The reason volcanic hazards are ranked higher along convergent settings in Table 4.2 is because subduction zones typically generate magmas that cause more explosive and violent eruptions.

Natural Resources

Earth provides not only all the food and water we need to survive, but also the mineral and energy resources that make modern societies possible. Plate tectonics is tied to our food and water supplies and crop production because it influences Earth's climate and precipitation patterns. In many dry regions of the world people depend almost entirely on rivers whose water comes from melting snow. This snow, in turn, accumulates in mountains that had been uplifted far above sea level by tectonic forces. For example, millions of people who depend on such meltwater live in the desert regions below the Himalayas of Asia, the Andes of South America, and the Rockies of North America. Plate tectonics also influences the formation and fertility of the soils in which we grow our crops.

In addition to being dependent on tectonic processes for water and food, modern societies depend on a vast array of mineral and energy resources to supply the goods and services we enjoy. As you will learn in Chapters 12 and 13, mineral and fossil fuel deposits are not randomly located around the world, but rather form under very specific geologic conditions related to the rock cycle and plate tectonics. Consequently, some countries have the good fortune of being located in an area whose tectonic history has left them with valuable mineral and fossil fuel deposits. For example, Saudi Arabia is quite wealthy because of its enormous oil deposits, whereas many nearby African countries remain desperately poor due in part to a lack of mineral or energy resources.

FIGURE 4.29 The shifting position of continents affects Earth's climate by altering the circulation of heat and water in the oceans. Climate is also affected by the uplift of mountains and by the rates at which heat-trapping gases are released by volcanic activity and absorbed in the rock cycle.

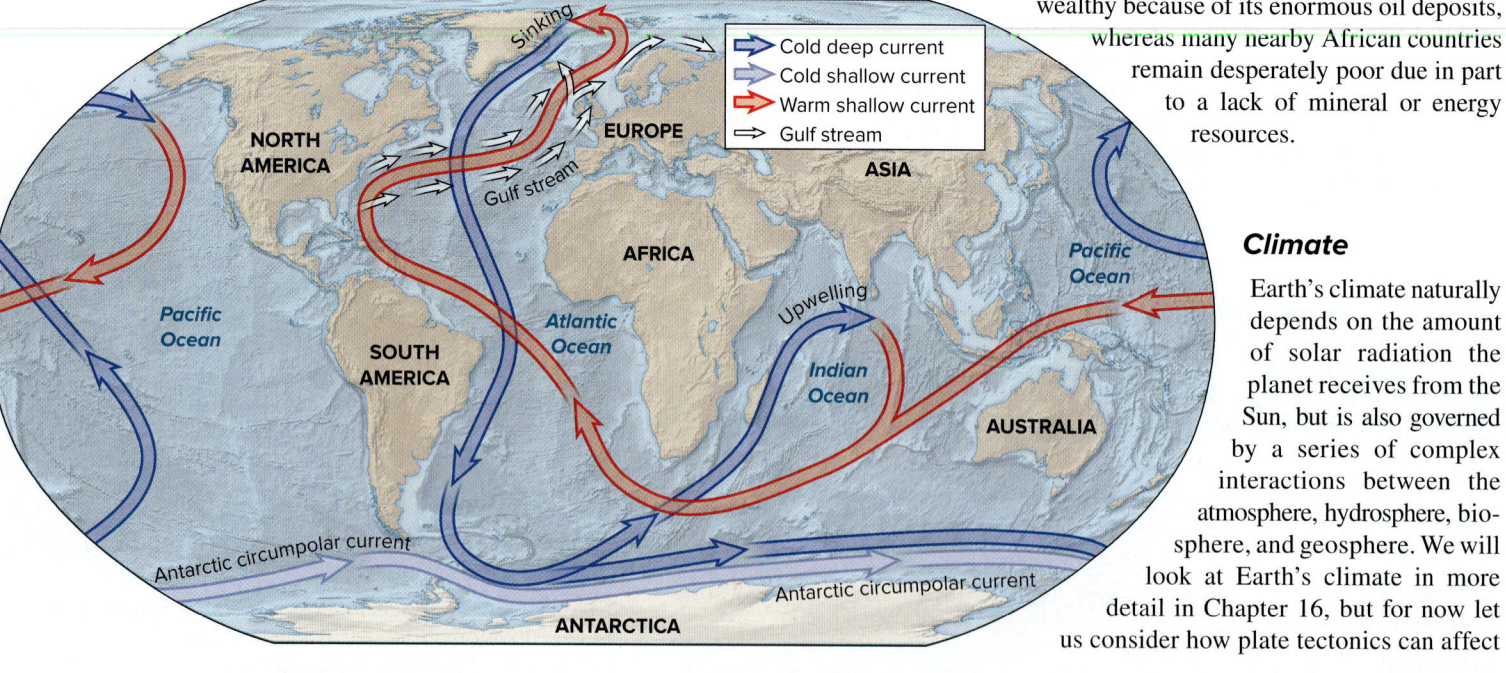

Climate

Earth's climate naturally depends on the amount of solar radiation the planet receives from the Sun, but is also governed by a series of complex interactions between the atmosphere, hydrosphere, biosphere, and geosphere. We will look at Earth's climate in more detail in Chapter 16, but for now let us consider how plate tectonics can affect the climate over the course of geologic time. For example, the major ocean currents shown in Figure 4.29 transport vast quantities of both heat and water, thus playing a significant role in regulating Earth's climate. As the positions of the continents change over geologic time due to continental drift, so too does the pattern of ocean circulation and Earth's climate. The shifting of tectonic plates also produces large mountain ranges (e.g., Himalayas and Andes) that have a significant impact on precipitation patterns around the globe. Of course, the climate will also change on individual continents as they change latitude by slowly drifting toward either the poles or equator.

In addition to atmospheric circulation patterns, Earth's global climate is influenced by the rates at which volcanic gas and ash are emitted into the atmosphere. For example, volcanic ash has a cooling effect on the climate because airborne particles reduce the amount of solar radiation that can strike Earth's surface. Sulfur dioxide (SO_2) gas from volcanic eruptions has a similar effect and contributes to global cooling. In contrast, carbon dioxide (CO_2) emitted by volcanoes causes the planet to warm as this gas traps some of the heat that is radiating out into space. Therefore, varying rates of volcanic activity can have complex and long-term impacts on Earth's climate system. The climatic effects from long-term changes in the rates of mountain-building activity are also complex. As tectonic uplift increases, so too does erosion and the subsequent formation of limestone ($CaCO_3$) in the rock cycle (Chapter 3). The deposition of limestone is important because it is the primary way the Earth system removes carbon dioxide (CO_2) from the atmosphere. Because carbon dioxide is a greenhouse gas that traps some of Earth's heat radiating into space, the formation of limestone has prevented a buildup of carbon dioxide and a runaway greenhouse effect. An example of a runaway greenhouse is the planet Venus, where surface temperatures today exceed 800°F (425°C).

Development of Life

There is little doubt that one of the most important aspects of plate tectonics is the crucial role it has played in the evolution of life on our planet. As mentioned in Chapter 2, many scientists now believe life may have originated around volcanic vents on the seafloor early in Earth's history. One of the reasons for this is that modern extremophile bacteria are found thriving in the hot gases and fluids discharging from seafloor vents along spreading centers. Whatever the exact origin, complex life is known to have evolved and eventually colonized the landmasses.

Once terrestrial (land) life was established, the combined effects of continental growth, continental drift, and evolutionary processes led to the great diversity and abundance of life-forms we see today. As you learned in this chapter, subduction allows continents to grow at the expense of oceanic lithosphere, and mountain building creates various landforms at different elevations. The increase in land area over time allowed terrestrial populations to increase, and the more varied landscape led to greater species diversification. Diversification increased further when the supercontinent of Pangaea began to break up around 225 million years ago. As the fragmented landmasses began to drift apart, terrestrial life-forms were allowed to evolve in isolation, leading to even greater diversification (Case Study 4.1). The increased abundance and diversity of life resulting from plate tectonics proved to be of considerable importance when life had to reestablish itself after periodic mass-extinction events (Chapter 2). In the end, one can safely say that if it were not for plate tectonics, life as we know it would not exist.

Biogeography and Plate Tectonics

In the mid-1800s a naturalist named Alfred Wallace spent considerable time observing and collecting various plants and animals on islands throughout Southeast Asia. As he sailed along the chain of volcanic islands known today as Indonesia (Figure B4.1), Wallace was struck by the sudden and dramatic change in the types of birds between two islands that were only 20 miles (30 km) apart. Moreover, he noted that the plants and animals on the islands to the west of this break were clearly related to one another and shared similar characteristics to species found on the continent of Asia. Sailing in the opposite direction toward the east, Wallace found a similar pattern of related species on different islands, but here they all shared common characteristics with those living on the Australian continent. It was as if the two halves of this island chain had life-forms that had come from opposite sides of the world.

In 1859 Wallace published his maps showing *different biogeographic regions,* each of which defined a region where plants and animals were related, but distinctly different from those in other regions. Interestingly, his work was similar to that of Charles Darwin, with both scientists coming to the same, but independent, conclusions regarding evolution and natural selection. Darwin's work, however, received more recognition in part because he was better known and because he published his work before Wallace. In honor of Wallace's contribution, the sharp boundary he mapped in Indonesia is still referred to as the *Wallace Line.* Modern biogeographers have found that the Wallace Line is more of a zone consisting of several different lines, with each line representing the biogeographic boundary of more specific groups of plants and animals.

Another interesting point is that Wallace was never able to provide a biological explanation for the differences he observed, but instead suggested that geologic processes were somehow involved. As the study of geology advanced, it became apparent that many of the islands in the region were once joined by land bridges that developed during glacial periods when sea levels were much lower. When the landmasses were interconnected, various species would be forced to compete with each other; in other words, they evolved together. Eventually sea level would rise and submerge the land bridges, allowing species on the various islands to once again evolve under conditions of geographic isolation. This, of course, nicely explains why many islands in Southeast Asia today contain different species that share common ancestors. However, it does not account for the sharp break at the Wallace Line where species on either side do not share a common ancestry.

The nature of the Wallace Line itself remained a mystery until the development of the theory of plate tectonics in the 1960s. Scientists were then able to show that the Wallace Line coincides with a deep ocean trench and associated subduction zone. Here the Asian and Australian continents are slowly getting closer together along a convergent plate boundary. Most critical though was the fact that regardless of how low sea level might fall, land bridges between the islands could never have formed across the deep trench. The trench therefore permanently isolated the plant and animal communities on either side, forcing them to evolve separately. The Asian and Australian continents may be getting closer together, but the ocean trench remains an effective migration barrier.

Perhaps the most familiar example of animals forced to evolve in isolation across this ocean trench are the mammals known as *placental* and *marsupial.* The offspring of placental mammals develop internally (e.g., humans), whereas marsupial young develop externally in a pouch (e.g., kangaroos). It turns out that marsupials are the dominant type of mammal to the east of the Wallace Line, particularly on the Australian continent. On the other hand, placental mammals are dominant west of the Wallace Line, which includes the Asian, African, and American continents.

FIGURE B4.1 The Wallace Line in Southeast Asia represents the boundary between distinctly different groups of plant and animal species. This boundary coincides with a deep ocean trench that forced species to evolve in isolation from one another on opposite sides of it.

SUMMARY POINTS

1. Rocks deform elastically when they are subjected to tension, compression, or shear forces. If they deform beyond their elastic limit, they either fail by fracturing (i.e., are brittle) or flow plastically (i.e., are ductile).

2. With increasing depth, rocks are exposed to greater levels of temperature and pressure, making them more ductile and prone to plastic deformation.

3. Earth's internal structure is known through the study of how earthquake (seismic) waves refract and reflect at boundaries between materials with different properties. The planet has a layered structure consisting of the crust, mantle, outer core, and inner core. Density of these layers increases going downward, reflecting an increase in iron content.

4. Oceanic crust is relatively thin and composed of basalt, whereas continental crust is thicker and composed of granitic rock. The crust and upper mantle act as a single rigid layer and together are called the lithosphere. The lithosphere is broken up into rigid plates that move over the weak, semimolten layer called the asthenosphere.

5. Decay of radioactive elements within the Earth generates heat and helps create a large temperature difference between the core and the crust. This sets in motion large convection cells that transport both heat and plastic mantle material toward the surface.

6. The rising and sinking motion of convection cells explains how Earth obtained its layered structure and the movement of lithospheric plates. Spreading centers develop along rising parts of convection cells, whereas subduction zones correspond to areas where cells descend down into the mantle.

7. Seafloor and seismic studies have confirmed that new oceanic crust forms at mid-oceanic ridges and is eventually destroyed along subduction zones to form continental crust. This explains why ocean crust is younger than continental crust.

8. The three basic types of plate boundaries are defined by the dominant forces that exist along boundaries: divergent (tension), convergent (compression), and transform (shear).

9. Major surface features develop along plate boundaries and include ocean ridges, ocean trenches, rift valleys, island arcs, volcanic arcs, and complex mountain belts. The majority of earthquakes and volcanic eruptions also occur along plate boundaries.

10. In addition to geologic processes, plate tectonics plays a central role in the Earth system by affecting the atmosphere, hydrosphere, and biosphere. For humans, plate tectonics is important because it creates hazards (earthquakes, volcanic eruptions, landslides), regulates our climate, distributes natural resources, and was important in the development of life.

KEY WORDS

asthenosphere   98
compression   95
continental arc   112
convection cells   99
convergent boundary   106
crust   98
divergent boundary   106

elastic limit   95
fault   96
geothermal gradient   99
inner core   97
island arc   111
lithosphere   98
mantle   97

mid-oceanic ridges   101
ocean trenches   101
outer core   97
rift valley   109
seafloor spreading   102
shear   95
subduction   103

tectonic plates   98
tension   95
theory of plate tectonics   94
transform boundary   106

APPLICATIONS

Student Activity Go to the website of a major news organization, such as the *New York Times,* the *Wall Street Journal,* or the BBC, and make a list of all the stories you can find that are in some way related to plate tectonics. Then, briefly state how the topic relates to plate tectonics. For example, earthquakes and volcanoes are easy to relate, but if you think about it, mineral, energy, and water resources and even climate change can be linked to plate tectonics.

Critical Thinking Questions

1. Rocks that are brittle will deform by fracturing when they exceed their elastic limit, but if they become ductile, they will deform by flowing plastically. Can you think of an everyday material that can go from being brittle to being ductile, and thus go from fracturing to flowing plastically when deformed?

2. Large-scale convection cells develop in the mantle due to differences in density caused by Earth's internal heat. Can you think of an example of density-driven convection circulation that takes place in Earth's surface environment?

3. Some disaster movies have shown lava erupting in downtown Los Angeles, California. Given the fact that Los Angeles is situated along a transform plate boundary, is a volcanic eruption there very likely? Explain.

Your Environment: YOU Decide In this chapter you learned how the theory of plate tectonics was finally developed in the 1960s, based on data gathered from numerous geologic studies over many years. These data include maps of the seafloor, magnetic studies, earthquake locations, and similar rock and fossil sequences on the continents.

Describe the piece of evidence you found to be most convincing in showing that tectonic plates are indeed moving on top of the asthenosphere. Be sure to explain why.

Amante, C. and B. W. Eakins, ETOPO1 1 Arc-Minute Global Relief Model: Procedures, Data Sources and Analysis, National Geophysical Data Center, NESDIS, NOAA, U.S. Department of Commerce, Boulder, CO, August 2008.

Chapter 5

Photo taken as a giant tsunami wave rolls ashore in Natori, Japan, on March 11, 2011. Waves as much as 130 feet (40 m) high formed when tectonic plates suddenly ruptured in a subduction zone located 80 miles (130 km) offshore. The first wave, shown here, arrived approximately 30 minutes after the quake, giving residents time to flee to safety. Even with public education and advanced planning, over 21,000 people perished. The tsunami also knocked out electrical power to the Fukushima nuclear plant, causing three of its six nuclear reactors to melt down, releasing radiation into the environment. Although tragic, lessons from this disaster may help minimize the risk along other subduction zones around the world.

Earthquakes and Related Hazards

CHAPTER OUTLINE

LEARNING OUTCOMES

After reading this chapter, you should be able to:

▶ Describe the elastic rebound theory, and explain how it relates to the occurrence and strength of earthquakes.

▶ Explain the basic difference between the various seismic waves and why some types are more damaging than others.

▶ Describe the difference between intensity and magnitude scales for measuring earthquakes.

▶ Explain the basic science behind making short-term and long-term earthquake predictions.

▶ Characterize the different types of secondary hazards associated with earthquakes.

▶ Describe the factors that can lead to increased ground shaking and structural damage.

▶ Characterize the basic ways in which society can reduce the loss of life and property damage from earthquakes.

Introduction

© StockTrek/Getty Images

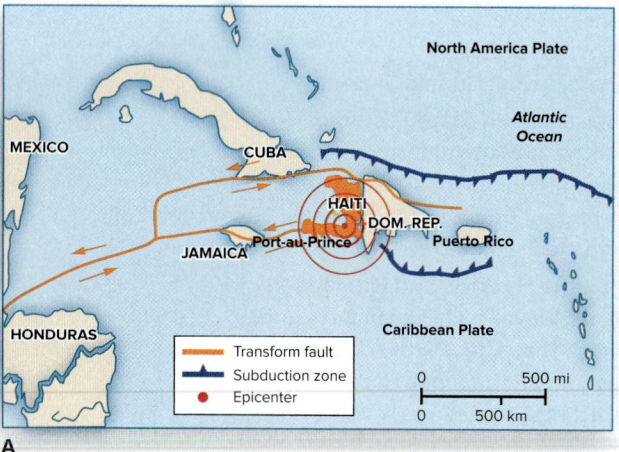

At around 5:00 p.m. in the afternoon of January 12, 2010, Earth's crust suddenly shifted beneath Port-au-Prince, the capital city of Haiti. The resulting earthquake unleashed waves of vibrational energy, causing buildings throughout this impoverished city to quickly collapse (Figure 5.1). The Haitian government estimated that 316,000 people were killed and over a million left homeless (independent groups estimate a much lower death toll, ranging from 47,000 to 158,000). In contrast, a slightly less powerful 1994 earthquake struck around 4:30 a.m. in densely populated Los Angeles, California, near the town of Northridge, causing extensive damage and 61 deaths. Although damage from the Northridge earthquake was considerable, it paled in comparison to the devastation in Haiti. Since both earthquakes released about the same amount of energy and occurred in populated areas, we need to ask why these two events had such vastly different outcomes.

The difference in this case was due almost entirely to preparedness. Engineers today know how to design and construct buildings capable of withstanding the violent ground shaking associated with powerful earthquakes. Earthquake-engineered buildings therefore are less likely to collapse and crush their occupants. In addition, emergency managers have learned how to coordinate medical personnel, police, and firefighters in the aftermath of a major earthquake. They also know the value of education programs designed to teach citizens basic steps for surviving an earthquake and minimizing damage to their property. Wealthy nations, like the United States, have the resources and organizational structure needed to prepare for a major earthquake and thereby reduce the risk. Unfortunately, Haiti is one of the poorest nations in the world and simply does not have the resources or political stability needed to institute building codes and effective emergency management plans.

In addition to financial and organizational resources, another key factor affecting the level of preparedness is the frequency of earthquakes in a given area. For example, in 2004 a massive earthquake occurred in a subduction zone along the coast of Indonesia (Figure 5.2), resulting in the deaths of approximately

FIGURE 5.1   In 2010, a major earthquake occurred along a transform plate boundary (A) beneath Port-au-Prince, Haiti, killing as many as 300,000 people and damaging or destroying approximately 90% of the buildings (B) in this capital city. The widespread death and destruction were largely due to the lack of building codes in this impoverished nation.
© AP Photo/Dario Lopez-Mills

A

B

FIGURE 5.2 The map in (A) shows the shorelines around the Indian Ocean impacted by the 2004 tsunami, which originated from a subduction zone earthquake off the coast of Indonesia. The massive ocean waves swept through low-lying coastal areas, killing an estimated 230,000 people. Before and after photos (B) of the once thriving community of Meulaboh, Indonesia, illustrate how entire communities were literally swept away by the tsunami. Note that concrete foundations mark the remains of buildings.
(a–b) © Digital Globe/Getty Images

230,000 people around the Indian Ocean, the majority of whom were literally swept away by a series of ocean waves known as a *tsunami*. During this earthquake the tectonic plates within the subduction zone suddenly shifted, displacing a large volume of seawater that triggered the tsunami. The ocean waves then traveled outward, taking the lives of people on distant shorelines where they never even felt the vibrations from the earthquake.

Then in 2011, another massive subduction zone earthquake occurred, but this time along the coast of Japan. Just as in Indonesia, this earthquake generated a giant tsunami that swept ashore, demolishing everything its path (see photo on opening page of chapter). Although the two earthquakes and resulting tsunamis were similar in size, the death toll of around 21,000 in Japan was far less than the

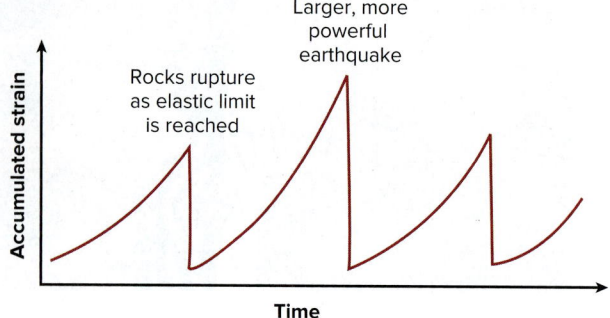

FIGURE 5.5   Graph showing how a steady tectonic force can cause earthquakes to occur in a repetitive manner. Here a rock body accumulates strain until its elastic limit is reached. The instant the rock ruptures, the strain is released as vibrational wave energy, which produces ground shaking. Note that more powerful earthquakes occur when greater amounts of strain are allowed to accumulate before the rupturing event.

The elastic rebound theory not only explains how earthquakes occur, but can account for why some areas experience repeated earthquakes. Recall from our discussion on plate tectonics (Chapter 4) that Earth's tectonic forces operate over long periods of geologic time. This means that when a rock body ruptures, the strain may be released, but the tectonic forces are still present. After each rupturing event, the strain slowly rebuilds until the elastic limit of the rock body is again reached, thereby producing repeated earthquakes as shown in Figure 5.5. Because faults and fractures represent weak zones within a rock body, rupturing generally takes place where the strain overcomes the frictional resistance along a fault or fracture. Since the rocks in tectonically active areas usually contain numerous faults, the sudden release of strain along one fault can alter the distribution of strain on the others. This redistribution of strain commonly produces a series of smaller earthquakes called *aftershocks*, which may continue to occur for days or weeks after the primary earthquake, often called the *main shock*.

Finally, the elastic rebound theory also tells us that the more strain a rock body accumulates, the more energy it will release when it finally ruptures. This naturally translates into greater vibrational energy, hence a larger earthquake. The key factors then that determine the size of an earthquake are the frictional resistance along fault planes and the elastic properties of the rock body. For example, imagine two faults under the same tectonic force. One fault offers little frictional resistance such that it slips fairly easily, whereas the other one does not slip easily because its resistance is high. In the first case the rock body will accumulate a relatively small amount of strain before the fault slips and releases the strain. On the other hand, a lot more strain must accumulate to cause slippage along the fault with more frictional resistance. This additional strain energy will translate into a larger earthquake.

In addition to faults, we also need to consider the elastic property of the rock itself. Recall that when rocks are more ductile (less brittle), they tend to accumulate less strain, and instead undergo plastic deformation (Chapter 4). This is important because an earthquake requires a rock body rigid enough to accumulate strain. It turns out that earthquakes do not occur deeper than 435 miles (700 km) below the Earth's surface because the rocks become so ductile that they deform only by plastic flow, and hence do not rupture.

Earthquake Waves

Today scientists understand that most earthquakes are caused by the buildup of strain associated with tectonic forces, and hence are often referred to as *tectonic earthquakes*. There are also many *magmatic earthquakes* that are caused when magma forces its way up through the crust (Chapter 6). However, not all earthquakes result from the slow buildup of strain; some also occur from the sudden transfer of energy during landslides, volcanic explosions, and meteor impacts. The vast majority of earthquakes though are tectonic or magmatic in nature, which explains why most earthquakes are found along plate boundaries (Chapter 4).

Since most earthquakes are restricted to plate boundaries, relatively few people in the world will ever experience an earthquake. However, most everyone has felt the ground shake from small *artificial* earthquakes in which the vibrational waves are generated from a human-derived energy source—explosions, passing trains, and heavy construction equipment all generate such waves. To help illustrate this point, imagine you were standing outside when a demolition company brought down a large building using small explosive devices. As the building collapsed, you would begin to feel a rumbling sensation in the ground. What happens is that as the building falls, it generates kinetic energy, which is then transferred to the ground when the debris makes impact. The energy then travels outward in all directions in the form of vibrational waves, eventually reaching the ground where you are standing.

A

FIGURE 5.2   The map in (A) shows the shorelines around the Indian Ocean impacted by the 2004 tsunami, which originated from a subduction zone earthquake off the coast of Indonesia. The massive ocean waves swept through low-lying coastal areas, killing an estimated 230,000 people. Before and after photos (B) of the once thriving community of Meulaboh, Indonesia, illustrate how entire communities were literally swept away by the tsunami. Note that concrete foundations mark the remains of buildings.
(a–b) © Digital Globe/Getty Images

B

230,000 people around the Indian Ocean, the majority of whom were literally swept away by a series of ocean waves known as a *tsunami*. During this earthquake the tectonic plates within the subduction zone suddenly shifted, displacing a large volume of seawater that triggered the tsunami. The ocean waves then traveled outward, taking the lives of people on distant shorelines where they never even felt the vibrations from the earthquake.

Then in 2011, another massive subduction zone earthquake occurred, but this time along the coast of Japan. Just as in Indonesia, this earthquake generated a giant tsunami that swept ashore, demolishing everything its path (see photo on opening page of chapter). Although the two earthquakes and resulting tsunamis were similar in size, the death toll of around 21,000 in Japan was far less than the

230,000 lives lost in the Indonesian event. Again, the question is, why were the results so different? In this case the primary reason is that tsunamis are relatively rare in the Indian Ocean, but quite frequent in the Pacific. The difference here is due to the Pacific Ocean being ringed almost entirely with subduction zones, which is where most tsunamis originate. Since the Indian Ocean has far fewer subduction zones, it has fewer tsunamis. Recall from Chapter 1 how humans are more likely to take steps to minimize a risk when the frequency of a hazard is high and its consequences more severe. The last major tsunami in the Indian Ocean happened several generations ago, which meant relatively few people were aware of the risk. Hence, most residents were caught completely unprepared. Japan, on the other hand, has a long history of devastating tsunamis. Both its government and people were well aware of the danger and took steps to minimize the risk. Although the Japan tsunami was highly destructive, were it not for the nation's high level of preparedness, the death toll would have been far greater.

Regardless of whether the danger is from collapsing buildings or massive ocean waves, earthquakes are a sober reminder that we are often at the mercy of natural forces over which we have little or no control. However, we can use science to better understand the nature of earthquakes and then develop ways of minimizing the hazards. In this chapter we will first explore how earthquakes occur, then discuss how science can be used to reduce the risk.

How Earthquakes Occur

Prior to modern science, most people believed earthquakes were random events; some even thought they were punishment from gods for evil or immoral behavior. By gathering data and testing various hypotheses, scientists have developed a theory based on the elastic properties of rocks that nicely explains how earthquakes occur. Recall from Chapter 4 that when rocks are placed under a force, also called *stress*, they can deform and change their shape or volume, a process known as *strain*. Interestingly, rocks are also considered to be *elastic*, meaning that if the force (stress) is removed, they will return to their original shape. Similar to a rubber band, rocks then are capable of releasing the strain they have accumulated and returning to an undeformed state (i.e., zero strain). All elastic materials have what is known as an **elastic limit,** which is the maximum amount of strain they can accumulate before either fracturing or undergoing plastic deformation. Moreover, when *brittle* materials reach their elastic limit, they undergo permanent deformation by fracturing, whereas *ductile* materials deform by flowing plastically.

This somewhat complex process of how rocks accumulate strain and deform can more easily be understood if we consider the example in Figure 5.3. Here the wooden rod represents more brittle rocks in the lithosphere, and the iron rod is analogous to ductile rocks that are more deeply buried. Notice that when a force is applied to the rods, they both bend and accumulate strain. If the force does not exceed their elastic limits, then this strain can be released once the force is removed, allowing the rods to return to their original shape. However, if we exceed the elastic limit of the iron (ductile) rod, it will flow plastically and become permanently deformed (i.e., bent). On the other hand, should the wooden (brittle) rod exceed its elastic

FIGURE 5.3 Iron and wooden rods will both bend (deform) and return to their original shape as long as their elastic limit is not exceeded. When an iron rod exceeds its elastic limit, it deforms permanently. When a wooden rod exceeds its limit, it will suddenly break by fracturing, releasing energy in the form of vibrational waves. Note that when the fractured rod breaks, the separate pieces rebound and become straight again.

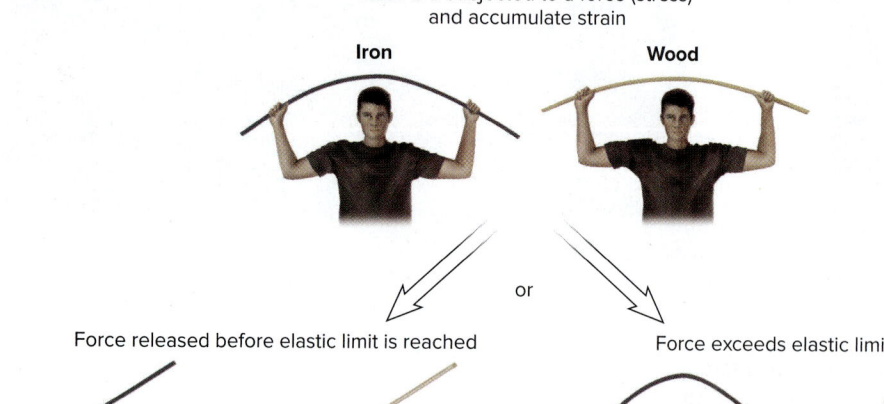

Rods are subjected to a force (stress) and accumulate strain

Iron   **Wood**

or

Force released before elastic limit is reached   Force exceeds elastic limit

Accumulated strain is released

Iron rod begins to deform plastically and continues to accumulate strain

Wooden rod breaks and accumulated strain is suddenly released as vibrational wave energy

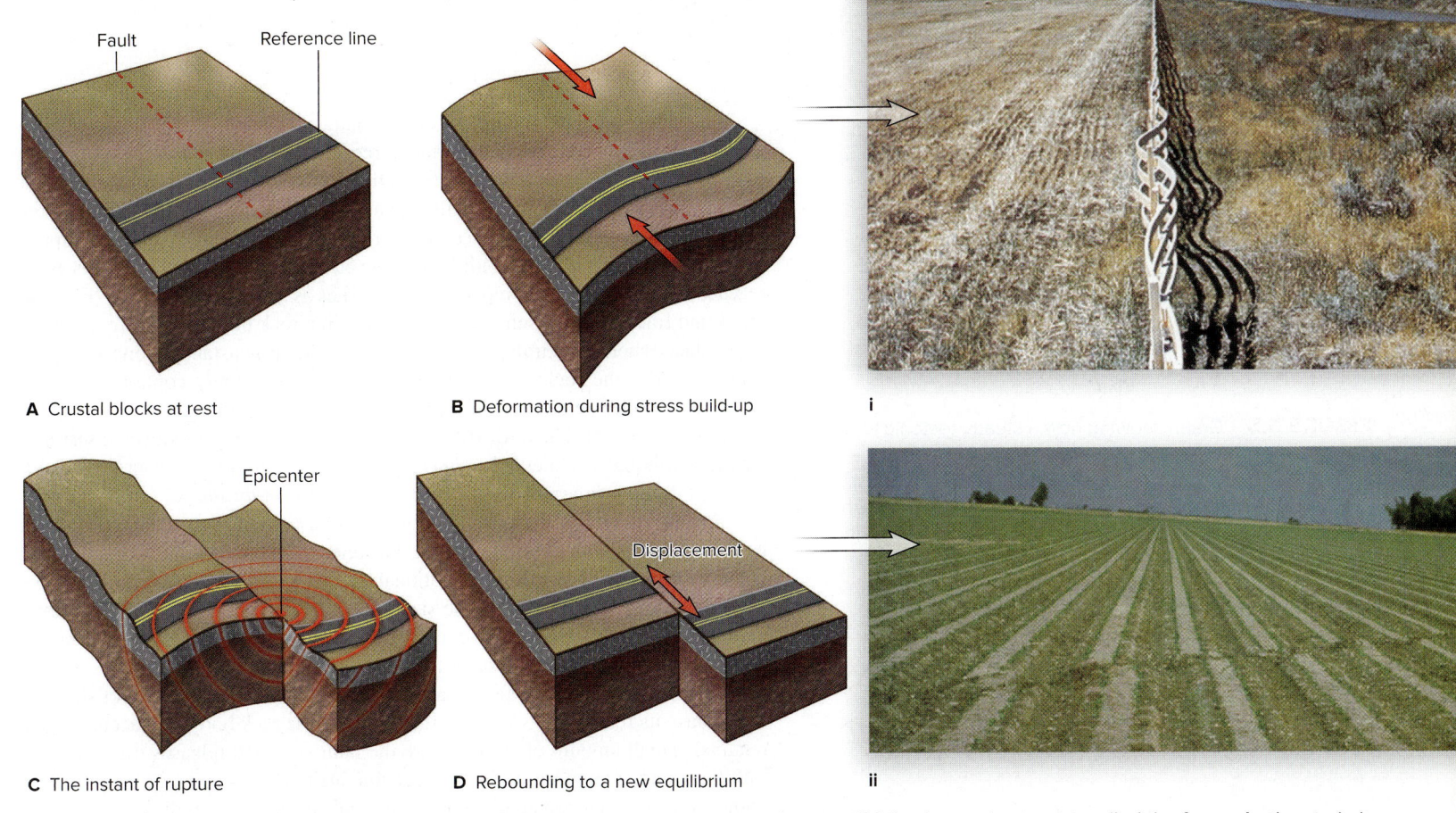

A Crustal blocks at rest

B Deformation during stress build-up

i

C The instant of rupture

D Rebounding to a new equilibrium

ii

FIGURE 5.4 When a rock body accumulates strain (A, B) and reaches its elastic limit, it will fail at its weakest point, called the focus. As the strain is suddenly released (C), waves of vibrational energy begin radiating outward in all directions from the focus, causing the ground shaking known as an earthquake. After the earthquake, the rock body becomes displaced (D) on opposite sides of the fault, but is no longer deformed because the strain has been released. (i) Buckled fence is evidence that the underlying rocks are accumulating strain. (ii) Displacement of rows in a farm field along a fault is evidence of a recent earthquake and that strain has been released.
(Top) USGS; (Bottom) Courtesy of National Geophysical Data Center, Boulder, CO

limit, it will suddenly fracture and break. At the moment the rod fractures, the accumulated strain is suddenly released and transformed into vibrational wave energy. This energy then quickly travels away from the fracture and moves down the two broken pieces. You can experience this yourself by bending a dried tree branch until it snaps, at which point you should feel a vibration in your hands as the energy waves travel down the wood.

Based on the relationship between stress and strain and the deformation of rocks, earth scientists have developed the **elastic rebound theory** that explains the occurrence of earthquakes. As illustrated in Figure 5.4, this theory holds that earthquakes originate when a force (stress) acts on a rock body, causing it to deform and accumulate strain. Eventually the rock reaches its elastic limit, at which point it suddenly *ruptures* or fails, releasing the accumulated strain. This sudden release of strain, lasting anywhere from several seconds to a few minutes, is transformed into vibrational wave energy that radiates outward, causing the ground to shake in what is called an **earthquake.** In addition, the release of energy generally begins at a point called the **focus,** also called the hypocenter, which is typically located where the rock is the weakest, such as along a preexisting fracture plane. While the strain is being released, rocks on either side of the fracture will begin moving in opposite directions, creating what geologists refer to as *displacement*. Note that when rocks are displaced along a fracture, the fracture technically becomes a *fault* (Chapter 4). Also notice in Figure 5.4 that after the earthquake, when the strain has been released, the displaced rock body is no longer deformed. It is worth pointing out that when rocks return to their undeformed state, scientists refer to this as *rebound*, hence the name elastic rebound theory.

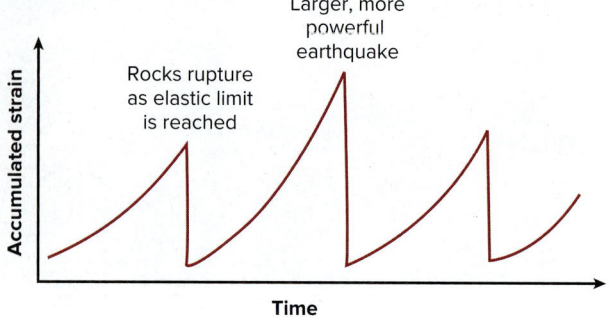

FIGURE 5.5   Graph showing how a steady tectonic force can cause earthquakes to occur in a repetitive manner. Here a rock body accumulates strain until its elastic limit is reached. The instant the rock ruptures, the strain is released as vibrational wave energy, which produces ground shaking. Note that more powerful earthquakes occur when greater amounts of strain are allowed to accumulate before the rupturing event.

The elastic rebound theory not only explains how earthquakes occur, but can account for why some areas experience repeated earthquakes. Recall from our discussion on plate tectonics (Chapter 4) that Earth's tectonic forces operate over long periods of geologic time. This means that when a rock body ruptures, the strain may be released, but the tectonic forces are still present. After each rupturing event, the strain slowly rebuilds until the elastic limit of the rock body is again reached, thereby producing repeated earthquakes as shown in Figure 5.5. Because faults and fractures represent weak zones within a rock body, rupturing generally takes place where the strain overcomes the frictional resistance along a fault or fracture. Since the rocks in tectonically active areas usually contain numerous faults, the sudden release of strain along one fault can alter the distribution of strain on the others. This redistribution of strain commonly produces a series of smaller earthquakes called *aftershocks*, which may continue to occur for days or weeks after the primary earthquake, often called the *main shock*.

Finally, the elastic rebound theory also tells us that the more strain a rock body accumulates, the more energy it will release when it finally ruptures. This naturally translates into greater vibrational energy, hence a larger earthquake. The key factors then that determine the size of an earthquake are the frictional resistance along fault planes and the elastic properties of the rock body. For example, imagine two faults under the same tectonic force. One fault offers little frictional resistance such that it slips fairly easily, whereas the other one does not slip easily because its resistance is high. In the first case the rock body will accumulate a relatively small amount of strain before the fault slips and releases the strain. On the other hand, a lot more strain must accumulate to cause slippage along the fault with more frictional resistance. This additional strain energy will translate into a larger earthquake.

In addition to faults, we also need to consider the elastic property of the rock itself. Recall that when rocks are more ductile (less brittle), they tend to accumulate less strain, and instead undergo plastic deformation (Chapter 4). This is important because an earthquake requires a rock body rigid enough to accumulate strain. It turns out that earthquakes do not occur deeper than 435 miles (700 km) below the Earth's surface because the rocks become so ductile that they deform only by plastic flow, and hence do not rupture.

Earthquake Waves

Today scientists understand that most earthquakes are caused by the buildup of strain associated with tectonic forces, and hence are often referred to as *tectonic earthquakes*. There are also many *magmatic earthquakes* that are caused when magma forces its way up through the crust (Chapter 6). However, not all earthquakes result from the slow buildup of strain; some also occur from the sudden transfer of energy during landslides, volcanic explosions, and meteor impacts. The vast majority of earthquakes though are tectonic or magmatic in nature, which explains why most earthquakes are found along plate boundaries (Chapter 4).

Since most earthquakes are restricted to plate boundaries, relatively few people in the world will ever experience an earthquake. However, most everyone has felt the ground shake from small *artificial* earthquakes in which the vibrational waves are generated from a human-derived energy source—explosions, passing trains, and heavy construction equipment all generate such waves. To help illustrate this point, imagine you were standing outside when a demolition company brought down a large building using small explosive devices. As the building collapsed, you would begin to feel a rumbling sensation in the ground. What happens is that as the building falls, it generates kinetic energy, which is then transferred to the ground when the debris makes impact. The energy then travels outward in all directions in the form of vibrational waves, eventually reaching the ground where you are standing.

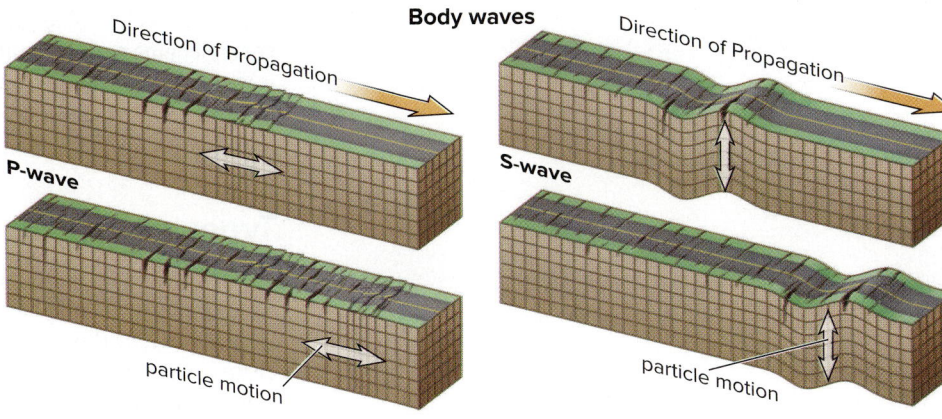

Body waves

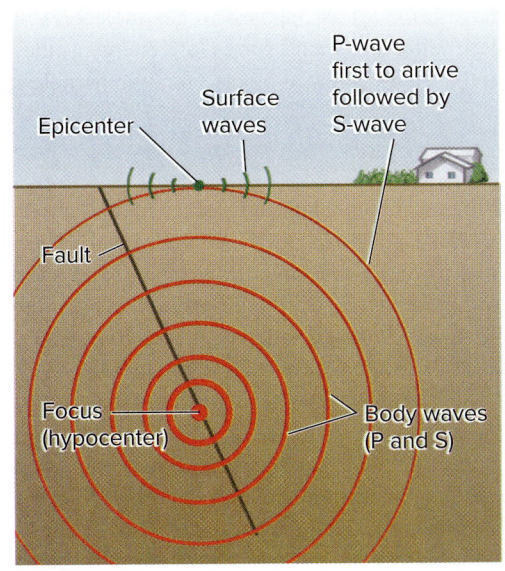

A

FIGURE 5.6 Body waves travel through the solid earth (A) and then generate surface waves upon reaching the land surface. Note how the direction of particle motion varies relative to the direction the wave type is traveling. P-waves are the least damaging and travel the fastest, thus are the first to arrive at any given location. Most of the damage to human structures results from the side-to-side and rolling motion of surface waves as illustrated by the apartment building (B) that collapsed during the 2008 earthquake in Sichuan, China. (b) © Chien-min Chung/Getty Images

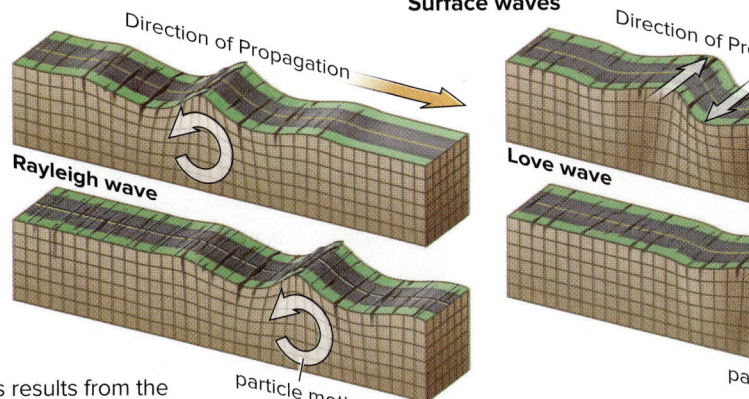

Surface waves

B

Regardless of their origin, we will refer to vibrational waves that travel through solid earth materials as **seismic waves,** as opposed to the less scientific term of earthquake waves. From Figure 5.6 you can see that in the case of tectonic or magmatic earthquakes, seismic waves originate at the point of rupture called the *focus*—also called the *hypocenter*. From there the waves travel outward in all directions, similar to how exploding fireworks expand outward as a sphere. A key point here is that as seismic waves travel away from the focus, their energy is spread out over an ever-increasing volume of rock; thus they become weaker. Consider what would happen if two earthquakes released the same amount of strain energy, but were located at different depths. Because wave energy steadily decreases with distance, we could expect that the earthquake with the more shallow focus would be more destructive than the deeper one.

Finally, note that the **epicenter** is the point on the surface that lies directly above the focus, which means it is also the closest point to where the strain energy was released (Figure 5.6). The epicenter then is not only the place where seismic waves *first* reach the surface, but more importantly it is where the waves contain the *most* energy—similar to being the closest to a bomb when it explodes. This is why the epicenter is commonly where the ground shaking is most severe, posing the greatest risk to humans and human-made structures.

Keep in mind that seismic waves are similar to other types of vibrational waves that transport energy, such as sound and water waves. The basic difference is the material (solid, liquid, or gas) that the waves must pass through. For example, the sounds we hear involve wave energy that originates at a point, then forces gas molecules to vibrate as the wave travels through the air. Interestingly, when a stone falls into a pond, we can actually *see* this process taking place. As the stone makes impact, its kinetic energy is quickly transferred to the water, at which point the energy begins to travel outward in all directions as visible waves. Here the passing waves cause water molecules to move (vibrate) as opposed to gas molecules or rock particles. As with seismic waves, water and sound waves lose energy as they travel away from their source due to the friction created by the vibrating particles.

Types of Seismic Waves

When a rock body is actually rupturing and being displaced along a fault, it generates seismic waves called **body waves** because they travel through Earth's interior, or body, as illustrated in Figure 5.6. When body waves reach the surface, the energy is not transferred to the atmosphere, but instead begins to move along the land surface in what are called **surface waves.** This is important in environmental geology because most of the damage in earthquakes is caused by surface waves. In this section we will take a closer look at seismic waves in order to understand why some types of waves cause more damage than others.

As with other forms of wave energy, body and surface waves can be classified based on the direction that particles within rocks vibrate relative to the direction the wave is traveling, or propagating. Consequently, there are two types of body waves shown in Figure 5.6. In **primary waves,** or **P-waves,** the particles vibrate in the same direction the wave is traveling, causing rocks to alternately compress and decompress with each passing wave. On the other hand, particles in **secondary waves,** or **S-waves,** vibrate perpendicular to the wave path, which creates a shearing (side-to-side) motion. Although primary and secondary waves travel along the same path, the compressional nature of the P-waves allows them to travel faster; hence they are the first to arrive at any given point. Later in this chapter you will see how the early arrival of P-waves serves as the basis for earthquake early-warning systems.

Similarly there are two basic types of surface waves, but here the motion is more complicated because of the way the waves interact with the ground surface. As indicated in Figure 5.6, the particles in *Rayleigh waves* move back and forth in the direction of the wave, *plus* they move up and down, creating a rolling motion similar to ocean waves. In *Love waves* the ground moves back and forth and side to side at the same time. In essence then, both types of surface waves force the ground to move in two different directions at the same time, something human structures are not normally designed to handle. Consequently, surface waves cause far more structural damage than do body waves. Later in this chapter we will explore how engineers make use of this information to minimize the amount of structural damage in an earthquake.

Measuring Seismic Waves

When body and surface waves pass through an area, anything that is firmly attached to the ground is going to be forced to move in the same direction that the waves are vibrating. However, loose objects like file cabinets and suspended light fixtures tend to remain stationary. This seemingly odd behavior is the result of **inertia,** which is the tendency of objects at rest to stay at rest unless acted upon by some force; objects in motion also tend to stay in motion because of inertia. For example, imagine you are standing on a rug when someone quickly pulls the rug out from under you. Because you are just standing (i.e., at rest), you will tend to not move with the rug, but instead fall straight to the ground. Inertia is important as it helps explain much of the structural damage caused by earthquakes.

Inertia also forms the basis for instruments called **seismographs,** which measure the ground motion during earthquakes. The first known seismograph was built in China in 132 BC, but it was not until 1889 that one was made that could actually record the ground movement—such a record is called a *seismogram*. As illustrated in Figure 5.7, early seismographs consisted of a pen attached to a suspended weight (mass) that sat over a rotating drum wrapped with paper, all of which was

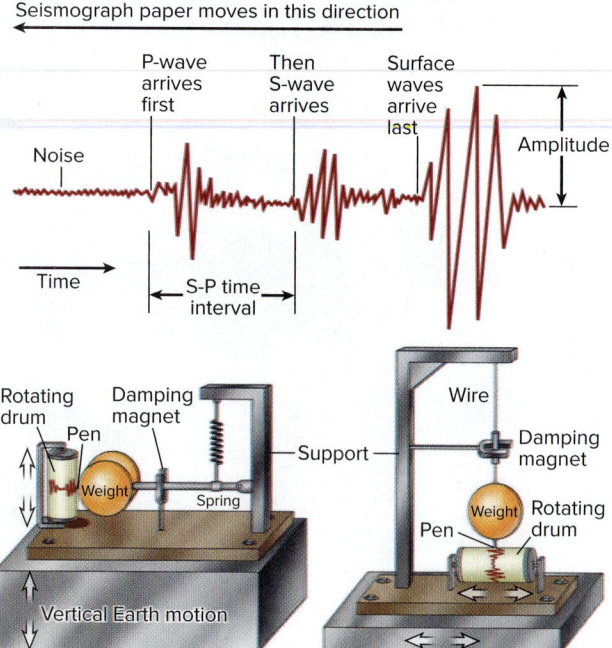

FIGURE 5.7 Seismographs operate on the principle of inertia, where a pen attached to a weight remains stationary whereas a rotating drum moves as the ground vibrates. The ground movements are recorded on the paper attached to the drum.

anchored to solid ground. When the ground is not shaking, which is most of the time, the pen simply traces a relatively straight line on the rotating paper. But when seismic waves pass through, the pen and attached weight remain stationary due to inertia, whereas the underlying paper and rotating drum vibrate or shake. Thus the ground motion is traced directly onto the paper, producing an accurate record of the successive waves that pass through the ground. You can simulate how this works by placing the tip of a pencil on a piece of paper, and then slowly pulling the paper toward you while keeping the pencil steady. Try it again, but this time move the paper from side to side as you pull.

From Figure 5.7 one can see that seismographs can be configured to record ground motion in both the vertical and horizontal directions. The seismogram (i.e., recording) is best viewed when the paper is removed from the drum and laid flat. Notice that the seismic waves from an earthquake arrive at a station in the following order: P-waves (primary waves), S-waves (secondary waves), and then surface waves. Also, note how the trace of the surface waves has the largest amplitude (height) and that of body waves the smallest. This means that surface waves create the greatest ground motion and body waves the least, which helps explain why surface waves cause the most structural damage. Finally, it should be mentioned that modern electronics have made mechanical seismographs obsolete. Instead of having drums anchored to the ground, the movement of the pen is now based on signals from a buried electronic sensor. Signals are also stored electronically and displayed on a computer screen as opposed to on paper.

Locating the Epicenter and Focus

As described earlier in this section, P- and S-waves are generated the instant a rock body begins to rupture. This means that the first set of P- and S-waves will leave the focus at the same time. But since the P-wave travels faster, the S-wave lags progressively farther behind as the waves get farther from the focus—similar to how two cars traveling at different speeds continue to get farther apart. Scientists have devised a means of locating the epicenter based on the time interval, or lag time, between the first set of P- and S-waves to arrive at different seismograph stations. As indicated in Figure 5.8, the distance from each station to the epicenter is determined from the stations' respective lag times. However, because the distance data do not indicate in which direction the epicenter is located, a minimum of three stations is required. By plotting the distances as circles around each station, the epicenter is the point where the circles all intersect.

Scientists have also developed methods for locating the actual focus within the subsurface. Although the methods are rather complex, the basic approach involves using multiple seismic stations to compare the time it takes P-waves to travel along different paths in the subsurface. This information is valuable because scientists can construct three-dimensional views of fault zones by plotting the location of focal points from multiple earthquakes. For example, Figure 5.9 shows a map and cross-sectional views from a study of the 1989 Loma Prieta earthquake, which occurred just south of San Francisco. From the map we can see that the epicenters of the main earthquake and numerous aftershocks are concentrated along the trace of the San Andreas fault zone, whereas the cross-sectional views show how the individual focal points are distributed throughout the subsurface. Of particular interest is the cross section BB' that cuts across the fault zone at a right angle. This view clearly shows that the fault plane is inclined at a steep angle.

By using seismic data to locate focal points and epicenters of earthquakes, geologists have a powerful tool for studying Earth's interior. In fact, it was through the analysis of seismic data that scientists were able to define the boundaries of tectonic plates (Chapter 4). From a hazard perspective, seismic studies are important as they help scientists understand how and where rock strain is being relieved along fault zones, which is useful in making long-term earthquake predictions.

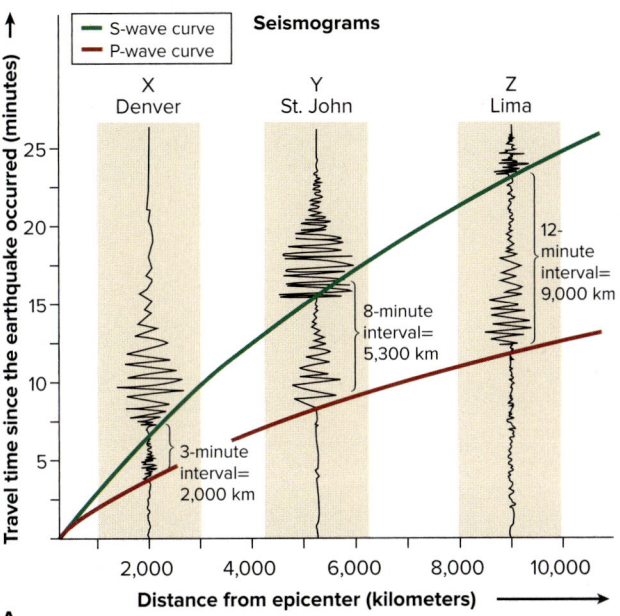

A

B

FIGURE 5.8 Locating the distance from a seismograph station to the epicenter makes use of the difference in arrival time between P- and S-waves from at least three stations. The distances from each station are then plotted as circles, with the epicenter located where the circles all intersect.

FIGURE 5.9   Plotting the epicenters of an earthquake and its aftershocks on a map allows geologists to determine the surface trace of a fault zone. The fault plane itself can be defined in the subsurface by plotting the focal points along different cross-sectional views. Note the cross section AA' is parallel to the fault trace and BB' is perpendicular. Here each focal point represents the place along the fault zone where rock strain had been relieved.

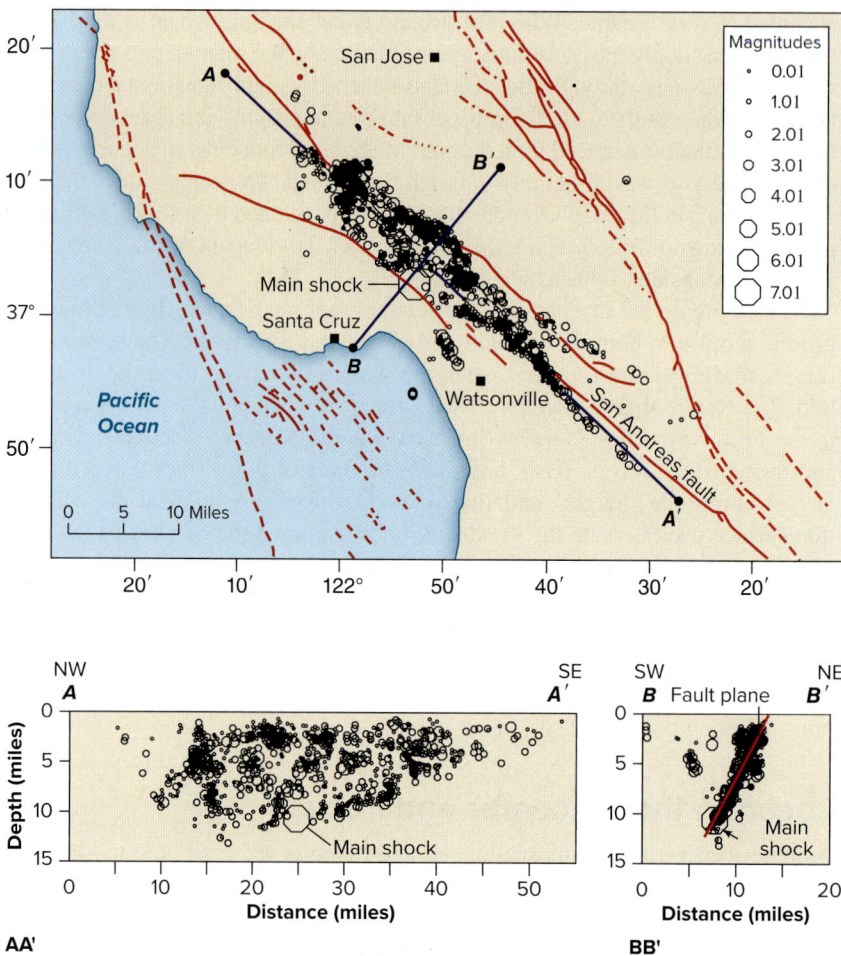

Measuring the Strength of Earthquakes

For thousands of years people have been both fearful of and fascinated by earthquakes due to their enormous destructive power and mysterious origin. Prior to the development of the modern seismograph, scientists had no means of quantifying the strength of earthquakes by direct measurements. All that was available were people's written accounts of the damage that had occurred and the sensations they felt during an earthquake. In this section we will explore the basic methods scientists use to measure the destructive power of earthquakes.

Intensity Scale

In 1902 an Italian seismologist named Giuseppe Mercalli developed a method for comparing both modern and historical earthquakes through the use of first-hand human observations during earthquakes. He created what is known as the **Mercalli intensity scale,** whereby earthquakes are ranked based on a set of observations most people could report objectively, particularly the type of damage sustained by buildings. A modified version of Mercalli's rankings and standardized observations is listed in Table 5.1. Note how the intensity scale ranks earthquakes from I to XII, with XII representing total destruction.

The way in which the intensity scale is employed is basically the same as taking a survey. Immediately after an earthquake, people throughout the region are asked to read the list of observations described on the scale (Table 5.1), and

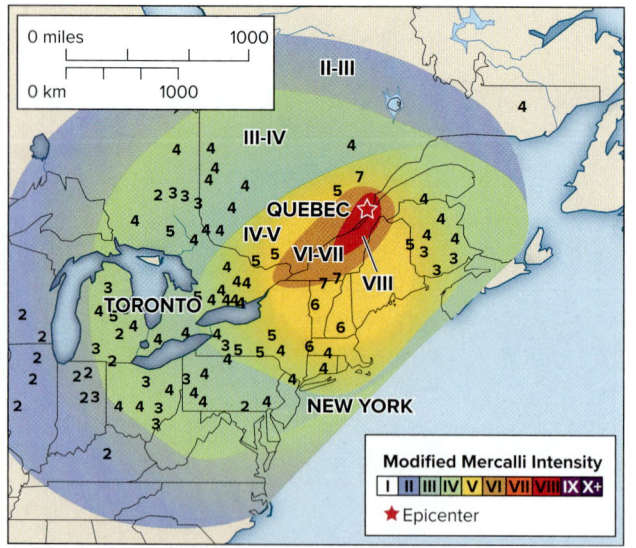

FIGURE 5.10   Mercalli intensity map of the 1925 Charlevoix-Kamouraska earthquake along the St. Lawrence River, in Quebec, Canada. Note the individual intensity rankings from the original survey.

TABLE 5.1   **Modified Mercalli intensity scale for classifying earthquake effects based on human observations.**

I. Not felt except by very few people under especially favorable conditions.
II. Felt only by a few people at rest, especially on upper floors of buildings. Delicately suspended objects may swing.
III. Felt quite noticeably indoors, especially on upper floors of buildings. Vibration like passing truck.
IV. Felt indoors by many, outdoors by few. Dishes, windows, and doors disturbed; walls make creaking sound. Sensation like heavy truck striking building.
V. Felt by nearly everyone. Some dishes, windows, and so forth, broken; a few instances of cracked plaster; unstable objects overturned.
VI. Felt by all; many frightened and run outdoors. Some heavy furniture moved; a few instances of fallen plaster or damaged chimneys. Damage slight.
VII. Everybody runs outdoors. Damage negligible in buildings of good design and construction; slight to moderate in well-built ordinary structures; considerable in poorly built or badly designed structures. Some chimneys broken.
VIII. Damage slight in specially designed structures; considerable in ordinary buildings, great in poorly built structures. Fall of chimneys, factory stacks, columns, monuments, walls. Heavy furniture overturned.
IX. Damage considerable in specially designed structures. Buildings shifted off foundations. Ground cracked conspicuously. Underground pipes broken.
X. Some well-built wooden structures destroyed; most masonry and frame structures destroyed. Ground badly cracked and rails bent. Landslides considerable along river banks and steep slopes.
XI. Few masonry structures remain standing. Bridges destroyed. Broad fissures in ground. Underground pipelines completely out of service. Rails bent greatly.
XII. Damage total. Waves seen on ground surfaces. Objects thrown upward into the air.

then pick the classification which best fits their experience. The individual rankings and locations are then plotted on a map and contoured so that similar rankings are grouped together, as shown in Figure 5.10. Although the intensity scale is a *qualitative* measure since it is based on subjective human observations, it still provides a useful means of comparing the strength of different earthquakes. Notice in Figure 5.10 how the intensity scale can be used to locate the epicenter in fairly populated areas with considerable accuracy. Prior to modern seismographs, being able to locate the epicenter represented a significant achievement.

Magnitude Scales

It was not until 1935 that Charles Richter and his colleague, Beno Gutenberg, devised a way of quantifying earthquake energy and ground motion based on the amplitude (height) of the waves recorded by seismographs. Here they analyzed the records of numerous earthquakes, ranking them according to the amplitude of the largest wave in each record. However, the amplitudes could not be directly compared since the distance between the earthquakes and seismic stations were different. Therefore, they applied a correction factor to the data to account for the different amounts of energy that were lost as the waves traveled to the seismic stations. Because there was such a large range in the amplitude data (i.e., some waves were very small and others quite large), they used a logarithmic scale and called it the *magnitude scale*. Their original magnitude scale, shown in Figure 5.11, is now known as the **Richter magnitude scale.**

As the number of seismograph stations around the world steadily increased, scientists eventually realized that results based on the Richter magnitude scale were not always consistent with one another, particularly for large-magnitude earthquakes. The problem is that the Richter scale was developed using a database with a somewhat narrow range of epicenter distances, and all the earthquake data was obtained from the same geologic setting, namely California. Today most scientists favor the **moment magnitude scale,** which is based on similar types of seismogram measurements as Richter's, but is more accurate over a wide range of magnitudes and geologic conditions. Despite the scientific community's preference for the moment magnitude scale, the Richter scale is still often used by the

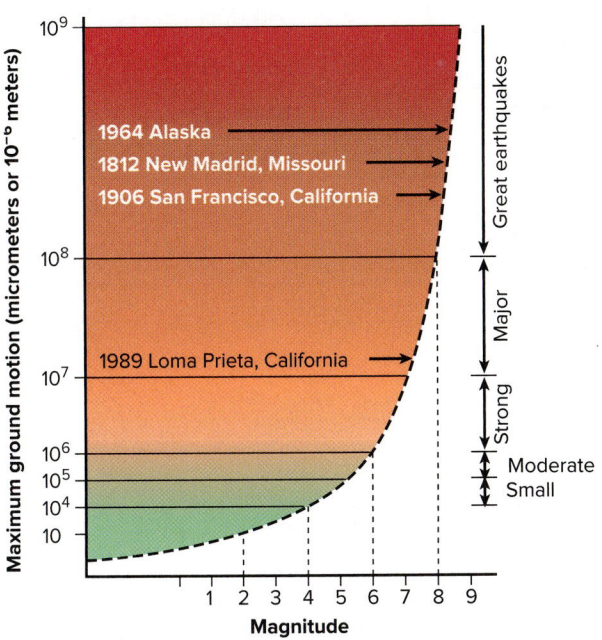

FIGURE 5.11   Graphic illustration of the exponential nature of the Richter magnitude scale, where each increase in magnitude of 1 represents a 10-fold increase in ground shaking. That is, a magnitude 8.0 quake has 10 times greater ground motion than a 7.0 quake, 100 times greater than a 6.0 quake, and 1,000 times greater than a 5.0 quake.

TABLE 5.2 Richter and moment magnitudes and corresponding death toll for selected earthquakes.

		Richter Magnitude	Moment Magnitude	Estimated Death Toll
1556	Shansi, China	~8	*	830,000
1811–1812	New Madrid, Missouri	~7.3–7.5	~7.8–8.1	*
1857	Fort Tejon, California	~7.6	~7.9	1
1886	Charleston, South Carolina	~6.7	~7.3	60
1906	San Francisco, California	8.3	7.8	3,000
1908	Messina, Italy	*	7.2	72,000
1923	Tokyo, Japan	*	7.9	143,000
1960	Chile, South America	8.5	9.5	5,700
1964	Anchorage, Alaska	8.6	9.2	131
1970	Chimbote, Peru	*	7.9	70,000
1976	Tangshan, China	7.6	7.5	650,000
1985	Mexico City, Mexico	*	8.0	9,500
1989	Loma Prieta, California	7.0	6.9	63
1990	Western Iran	*	7.4	50,000
1994	Northridge, California	6.4	6.7	61
1995	Kobe, Japan	6.8	6.9	5,500
2003	Bam, Iran	6.3	6.6	31,000
2004	Sumatra, Indonesia	*	9.1	230,000
2005	Kashmir, Pakistan	*	7.6	86,000
2008	Sichuan, China	*	7.9	88,000
2010	Port-au-Prince, Haiti	*	7.0	316,000
2010	Maule, Chile	*	8.8	507
2011	Honshu, Japan	*	9.0	21,000
2015	Lamjung, Nepal	*	7.8	9,000

Source: Data from U.S. Geological Survey, www.earthquake.usgs.gov.
*Data not available.

media when reporting earthquakes. In Table 5.2 you can compare the Richter and moment magnitudes for a select number of large, historical earthquakes. Note how magnitude itself is not necessarily a good predictor of death toll.

Magnitude and Ground Shaking

Magnitude scales are useful because they quantify the amount of ground motion during an earthquake, and also the energy that was released when the rocks ruptured. Interestingly, when the news media report the magnitude of an earthquake, the general public typically does not appreciate the logarithmic nature of the scale. People often think that the difference between a magnitude 5 earthquake and a magnitude 8 is simply 3 on a scale of 1 to 10. However, as can be seen in the exponential graph in Figure 5.11, Richter magnitudes are actually logarithms of the ground motion values found on the vertical axis. In other words, magnitude numbers are simply the exponents of the ground motion values. Notice on the graph that a magnitude 5 earthquake causes 10^5 (100,000) micrometers (0.1 meter) of

TABLE 5.3 Differences in ground motion (i.e., shaking) of selected earthquakes as determined by their differences in moment magnitudes.

Earthquakes		Difference in Moment Magnitude	Difference in Ground Motion
1906 San Francisco	M = 7.8	7.8 – 6.9 = 0.9	$10^{0.9} = 8\times$
1989 Loma Prieta	M = 6.9		
2011 Honshu, Japan	M = 9.0	9.0 – 7.8 = 1.2	$10^{1.2} = 16\times$
1906 San Francisco	M = 7.8		
2011 Honshu, Japan	M = 9.0	9.0 – 6.9 = 2.1	$10^{2.1} = 126\times$
1989 Loma Prieta	M = 6.9		

ground motion, whereas a magnitude 8 represents 10^8 (100,000,000) micrometers (100 meters) of movement. While there was only an increase of three units on the magnitude scale, the amount of ground motion increased by a factor of 10^3, or 1,000. Although a unit increase on the magnitude scale represents a 10-fold increase in ground motion, this corresponds to about a 30-fold increase in energy released at the focus—recall that the release of stored elastic energy is what causes the shaking in the first place.

To give you a better appreciation of what this all means, we will compare the two largest earthquakes in the San Francisco area in modern times, namely the 1989 Loma Prieta earthquake and the great earthquake of 1906. With a moment magnitude of 6.9, the 1989 Loma Prieta event was ranked as a strong to major earthquake with shock waves causing extensive damage as much as 60 miles (100 km) away from the epicenter. In comparison, the 1906 San Francisco earthquake had a moment magnitude of 7.8 and was considered to be a great earthquake. As shown in Table 5.3, we can easily calculate how much stronger the ground shaking was during the 1906 quake than the 1989 event by subtracting the two magnitudes (7.8–6.9) to get 0.9. Since magnitude represents the exponent on the ground motion value, we take 10 and raise it to the 0.9^{th} power ($10^{0.9}$), which comes out to an eightfold difference. In other words, the ground shaking in the 1906 earthquake was eight times greater than in the 1989 quake. For anyone who experienced the violent shaking of the 1989 Loma Prieta event, it might be difficult to imagine being shaken eight times harder.

Finally, we can take our example one step further and compare these two earthquakes to the 2011 earthquake and tsunami in Japan that took nearly 21,000 lives. With a moment magnitude of 9.0, the Japan earthquake ranks among the most powerful ever recorded. In Table 5.3 you can see that by doing the same type of calculations as before. The ground motion in the Japan earthquake was 16 times greater than in the 1906 San Francisco quake, and 126 times greater than in the Loma Prieta quake! Clearly, the 2011 Japan earthquake was a truly massive event whose power is difficult for humans to fully appreciate.

Earthquakes and Plate Tectonics

In Chapter 4 you learned that most earthquakes are not random events, but intimately tied to plate tectonics. For example, the maps in Figure 5.12 show that earthquake epicenters in the United States are clustered along the transform boundary in California and along the convergent boundaries off the coast of Oregon and Washington and Alaska. Another well-defined cluster is associated with tectonic uplift of the Rocky Mountains. However, there are also small clusters and isolated epicenters scattered throughout the continental interior. Geologists refer to earthquakes that occur far from a plate boundary or active mountain belt as **intraplate earthquakes.** Although less well understood than tectonic

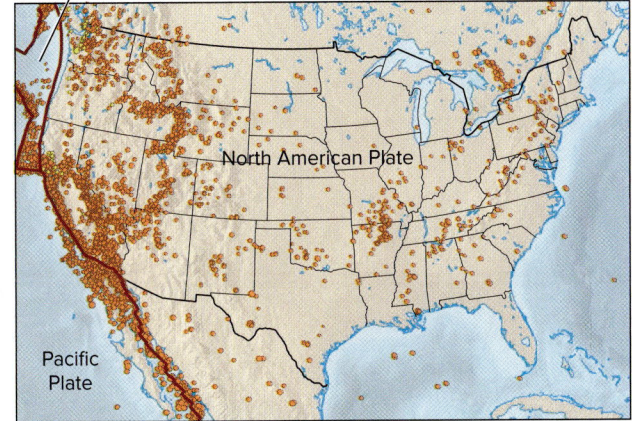

A

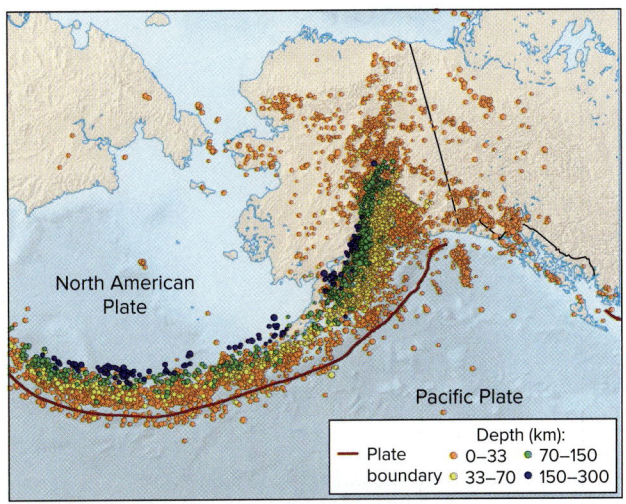

B

FIGURE 5.12 Earthquake epicenters in the United States from 1990 to 2000. Note that the majority of epicenters are associated with plate boundaries and active mountain belts; the remainder are called intraplate earthquakes as they occur in the interior of plates.

earthquakes, intraplate earthquakes are generally believed to be related to tectonic forces being transmitted through the rigid plates. This causes crustal rocks to slowly accumulate strain, which is then released along buried fault systems, producing earthquakes in the interior of continents.

In this section we will examine how the theories of plate tectonics and elastic rebound can help explain the location and frequency of most earthquakes. We will pay particular attention to the relatively small number of large-magnitude (>6) earthquakes since these are the ones that pose the greatest risk to people.

Earthquake Magnitude and Frequency

Seismologists estimate that several million earthquakes occur worldwide each year, with the vast majority going undetected because they are either too small or lie too far from an existing seismograph station (nearly all of the strong earthquakes are detected by an international monitoring network of over 150 seismograph stations). Interestingly, of the estimated several million earthquakes each year, only about 100,000 are large enough (magnitude >3) to be felt by people. From a hazard perspective, our concern is with the 150 or so *strong earthquakes* (magnitude 6 and higher) since these are the ones capable of causing significant damage.

At this point we need to examine how the theories of plate tectonics and elastic rebound can help explain where strong earthquakes occur, and why they are relatively rare in most parts of the world. From the elastic rebound theory we know that more powerful earthquakes are generated when rock bodies are able to accumulate greater amounts of strain energy before rupturing. The key here is the strength of the rock itself and the nature of the faults within a rock body. Recall from Chapter 4 that rocks are much stronger under a compressional force compared to a tensional force. This means that at convergent boundaries, where compressive forces dominate, rocks are able to accumulate much more strain before rupturing than at divergent boundaries, where tensional forces are dominant. Rocks can also accumulate considerable amounts of strain under the shear forces found along transform boundaries.

The other key factor in the ability of a rock body to store strain is the frictional resistance of the faults. In areas dominated by tensional forces, the friction along faults is naturally low, allowing them to slip in an almost continuous process known as *fault creep*. When a rock body experiences fault creep, it naturally cannot build up much strain, which helps explain why large-magnitude earthquakes generally do not occur at divergent boundaries. On the other hand, compressional and shearing forces at convergent and transform boundaries tend to generate high levels of friction on faults, creating the potential for large magnitude earthquakes. In fact, the resistance may become so great that faults become locked, allowing strain to build to the point where extremely powerful earthquakes are generated. We will now turn to some examples to illustrate the connection between tectonic setting and large-magnitude earthquakes.

Transform Boundaries—San Andreas Fault

The San Andreas fault, shown in Figure 5.13, is one of the few places in the world where a transform boundary is found on land. The San Andreas is actually a large transform fault that separates the Pacific and North American plates, but is often referred to as a *fault zone* due to the network of interlocking faults found on either side. This means that as tectonic forces cause strain to accumulate along

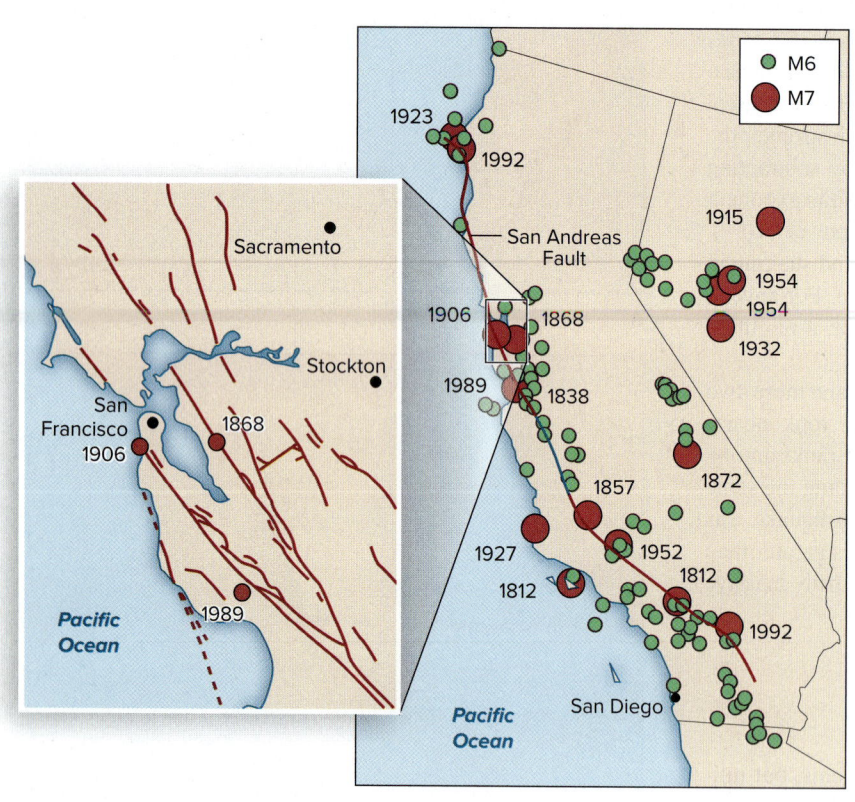

FIGURE 5.13   Map showing the location of magnitude 6 and 7 earthquakes along the San Andreas fault zone in California between 1800 and 1994. This transform fault is a boundary between the Pacific and North American plates, and also contains a network of interlocking faults (inset). Fault creep occurs along the blue segment of the main fault, which greatly limits the chance of earthquakes greater than magnitude 7.

the boundary, some of the strain is distributed among the different faults. As indicated by the location of epicenters in Figure 5.13, not only does the San Andreas fault occasionally slip, generating a strong earthquake, but other faults within the fault zone do so as well. Moreover, due to the interlocking nature of the faults, strain relieved along one fault can disrupt the delicate balance within the fault zone, triggering additional earthquakes on other faults.

From Figure 5.13, notice that the segment of San Andreas shown in blue is an area where fault creep prevents strain from building; hence there is a lack of major earthquakes (magnitude >7). Along those sections shown in red, the main fault tends to become locked, allowing strain to build to more dangerous levels. In fact, the largest earthquake ever recorded in California, an estimated magnitude 7.9, occurred in 1857 along the southern portion of the San Andreas. Because this section of the fault near Los Angeles has remained locked since 1857, the strain energy has continued to build. It is only a matter of time then before the fault ruptures and releases the strain, generating another major earthquake—Los Angeles residents often refer to such a quake as the "Big One." Unlike in the 1857 earthquake, however, Los Angeles is now a densely populated metropolis, which is why the damage and death toll from such a quake are expected to be high.

Convergent Boundaries—Cascadia Subduction Zone

In northern California the San Andreas fault moves offshore and the boundary of the North American plate changes from a transform (shear) setting to one of convergence (compression). At this point the North American plate starts to override a series of relatively small oceanic plates along what geologists call the *Cascadia subduction zone* (Figure 5.14). This subduction zone not only produces the volcanic arc (Chapter 4) known as the Cascade Mountain Range, it also generates **subduction zone earthquakes,** which form when an oceanic plate is overridden by another plate. Subduction zones are important because they are capable of generating extremely powerful and devastating earthquakes. For example, the ten largest earthquakes ever recorded were all subduction zone earthquakes, and of these, seven were magnitude 9 or higher. The possibility of such an earthquake along the Cascadia subduction zone is rather frightening since a magnitude 9 will unleash 32 times more energy than a magnitude 8. Keep in mind that throughout history, earthquakes in the magnitude 8 range have proven to be sufficiently powerful enough to level entire cities.

The reason subduction zone earthquakes are capable of releasing unusually large amounts of energy is partly due to the way the overriding plate buckles and becomes locked, as shown in Figure 5.14. Equally important is the fact that the descending oceanic plate is relatively cool, which makes the rocks more brittle and capable of accumulating more strain before rupturing. In addition to the intense ground shaking, some of this energy can be transferred to the ocean, creating tsunamis that reach heights of 100 feet (30 m) as they crash into coastal areas.

Although geologists have long been aware of the hazards associated with the Cascadia subduction zone, public awareness increased considerably after the massive magnitude 9.1 subduction zone quake and subsequent tsunami in Indonesia in 2004. This was followed by another horrific example in 2011 when a magnitude 9.0 earthquake was unleashed in a subduction zone along the coast of northern Japan. Similar to the earthquake in Indonesia, the Japan quake generated a giant tsunami that obliterated entire coastal communities. What is particularly worrisome about the Cascadia subduction zone is that a magnitude 9 quake and giant tsunami are likely to occur there also. The last major earthquake was in 1700, which means that for more than 300 years the strain has been building to dangerously high levels. Recent studies have found ample evidence of a large tsunami was associated with the 1700 event. To make matters worse, because there

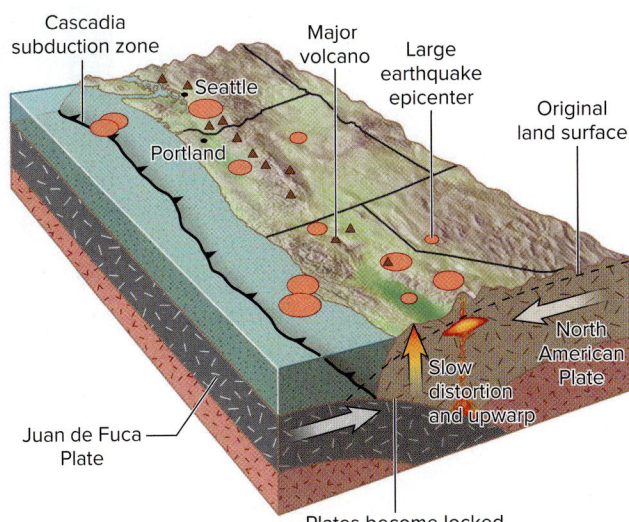

FIGURE 5.14 The Cascadia subduction zone along the Pacific Northwest is not only responsible for the volcanic activity in the Cascade Mountain Range, but also for considerable seismic activity—note the position of epicenters. Subduction zones are notorious for generating powerful earthquakes because of the way the plates lock and the ability of the rock to accumulate large amounts of strain.

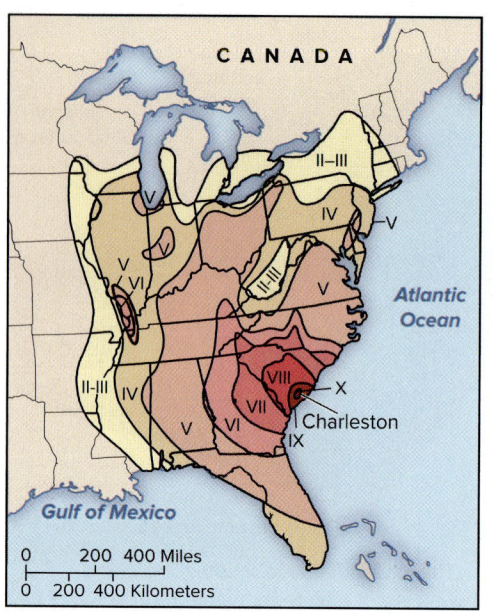

A

B

FIGURE 5.15 In 1886 an intraplate earthquake occurred near Charleston, South Carolina, causing severe damage. As indicated by a Mercalli intensity map, this earthquake was felt over a large portion of the eastern United States.
(b) USGS

FIGURE 5.16 The clustering of epicenters (most too small to be felt) shows the area of high seismic activity within the New Madrid seismic zone. Other geologic data define an ancient rift system that coincides with the seismic activity. Geologists believe that compressional forces within the continental plate cause strain to build within the rift, which eventually slips and causes earthquakes.

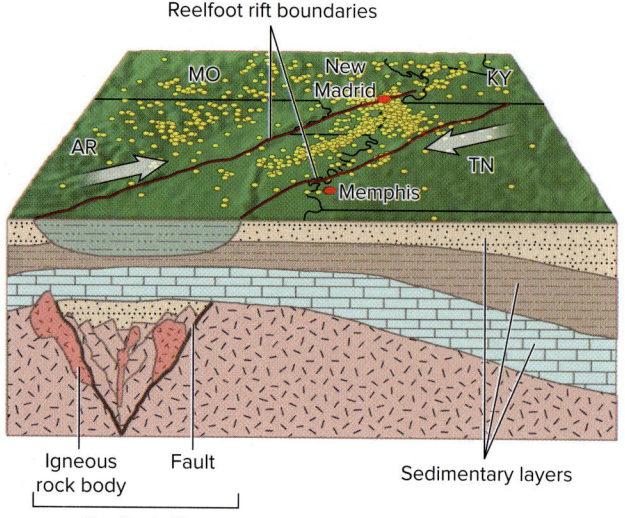

has not been a major earthquake in the Pacific Northwest in more than 300 years, many of the buildings have not been designed to withstand the shaking associated with a major earthquake. This leaves us with the frightening prospect of a giant tsunami crashing into coastal communities and populated urban areas whose buildings and other infrastructure are relatively unprepared for such an event.

Intraplate Earthquakes—North American Plate

Other high-risk areas in the United States are the New Madrid and Charleston seismic zones due to the fact they have a history of producing powerful intraplate earthquakes. In 1886 a strong earthquake occurred about 50 miles (80 km) outside of Charleston, South Carolina, causing 60 deaths and extensive property damage throughout the city and surrounding region. As indicated by the Mercalli intensity map in Figure 5.15, the earthquake was felt over most of the eastern United States. Amazingly, structural damage was reported as far away as central Alabama and central Ohio. Although this earthquake occurred prior to the development of modern seismographs, based on damage reports and other lines of evidence, scientists estimate its moment magnitude was 7.3. Modern studies have shown that this seismic zone is still active, as evident by the clustering of small earthquakes in three distinct areas west and north of Charleston. Based on seismic data, scientists have mapped the position of several faults buried beneath a thick sequence of sedimentary rocks. Despite the fact this region is far from a plate boundary, geologists believe the crust is still accumulating strain, which is then periodically released along buried faults.

In the case of the New Madrid seismic zone shown in Figure 5.16, geologists have been able to link modern earthquake activity there to faults associated with a large, buried rift system called the *Reelfoot rift*. This structure is over 500 million years old and is thought to be similar to the rift currently forming in East Africa, where tensional forces are literally tearing the continent apart (Chapter 4). Geologists now believe that compressional forces within the North American plate have reactivated ancient faults within the Reelfoot rift, generating a clustering of earthquake epicenters (Figure 5.16). Although the vast majority of these quakes are too small for people to feel, what concerns scientists is that during the winter of 1811–1812 a series of magnitude 8 earthquakes occurred in the New Madrid zone. These powerful earthquakes caused damage as far away as Washington, D.C., and Charleston, South Carolina. Closer to the epicenter the seismic waves triggered landslides along the Mississippi River, and in some areas, caused entire islands to sink beneath the river. Even more incredible, water waves developed on the Mississippi that were large enough to swamp some boats and wash others up onto dry land. Since the region was sparsely populated back in the early 1800s, structural damage and loss of life were minimal.

Today, of course, the New Madrid seismic zone is highly developed, including the nearby metropolitan areas of Memphis and St. Louis. Unfortunately, should a powerful earthquake occur again, people living in this region face the same danger as those in the Pacific Northwest in that relatively few buildings have been designed to resist the ground shaking. This lack of preparedness is a common problem in areas where powerful earthquakes are infrequent, lulling people into a false sense of security. Perhaps the most tragic example is the 1976 intraplate quake in Tangshan, China, where upwards of 650,000 people perished. This magnitude 7.5 earthquake occurred in a heavily populated area that was totally unprepared, largely because the people had no memory of a large quake ever taking place. Tangshan represents a sober lesson for cities located in areas with large but infrequent earthquakes.

On August 23, 2011, a magnitude 5.8 earthquake struck central Virginia, which served as a reminder to people in the eastern United States of the potential danger of intraplate quakes. This earthquake caused moderate to heavy damage to structures near the epicenter (Figure 5.17), whereas damage was relatively minor in the major urban areas of Richmond and Washington, D.C., located 38 and 84 miles (60 and 135 km) away, respectively. Scientists and engineers recognized that since few of the buildings were designed to withstand major ground shaking, the results would have been disastrous if the magnitude had been similar to the 1883 Charleston or 1811 New Madrid earthquakes. The question is, what is the best way of reducing the risk?

Earthquake Hazards and Humans

It should be clear that earthquakes are a natural consequence of Earth's shifting tectonic plates. Seafloor spreading, subduction, and uplift of mountain ranges are processes that generally do not occur in a smooth and continuous manner, but rather by the sudden release of strain energy and displacement along faults. Therefore, were it not for earthquakes and plate tectonics, Earth would not have the variety of landscapes or the biodiversity we see today. Simply put, life as we know it would not exist. Why then do we consider earthquakes a "problem"? After all, our ancestors managed to live for hundreds of thousands of years in the presence of earthquakes.

While earthquakes are an important part of the Earth system, the reality is they have always posed a number of hazards to humans. Prior to the development of agriculture and cities, the primary hazards people faced from earthquakes were landslides and tsunamis. While these hazards are significant, they are relatively minor compared to problems that developed once people started living and working in buildings, particularly as buildings became progressively taller and heavier over time. Another key factor is that exponential population growth has resulted in many more people living in earthquake hazard zones compared to the past.

In this section we will examine the different types of earthquake hazards, particularly those related to the failure of buildings and other human structures. This is important because structural failure is the leading cause of death and property damage in most earthquakes. We will also focus on how society can mitigate the risk of earthquakes by designing structures that are more resistant to ground shaking.

Seismic Waves and Human Structures

There is a common saying among seismologists that "earthquakes don't kill people, buildings do." You can perhaps better appreciate this statement by examining the collapsed structure in Figure 5.18. This structure was one of an

A

B

FIGURE 5.17 In 2011 a magnitude 5.8 intraplate earthquake occurred in central Virginia (A), 84 miles (135 km) from Washington, D.C. The quake was notable for damaging monuments in the nation's capital, but also caused heavy damage to buildings near the epicenter in Virginia. Photo (B) shows severe structural damage to a home located near the epicenter. (b) Matthew Heller, Virginia Department of Mines, Minerals, and Energy

estimated 5.4 million buildings that collapsed during the moment magnitude 7.9 earthquake in China in 2008, in which nearly 70,000 people died. Most victims in collapsed buildings die from being crushed, but some manage to survive in small void spaces within the piles of rubble. In major earthquakes, however, it is not uncommon for there to be numerous collapsed structures within a city, which completely overwhelms the available rescue personnel. There is also the constant threat of aftershocks causing the rubble to shift, endangering both rescuers and survivors. To make matters worse, rescue generally requires heavy equipment to remove the overlying debris, a process that may take days or weeks to accomplish. Sadly, most of the uninjured are never rescued from within the rubble, but die within a few days due to hypothermia or dehydration. Therefore, the most effective way of reducing the loss of life in an earthquake is to design buildings such that the chance of collapse is minimized.

In modern cities there are many different types of structures that may fail or become damaged in an earthquake, including homes, office buildings, factories, highways, bridges, and dams. Because all structures have mass, engineers must design them at a minimum to be strong enough to support their own weight against the force of gravity. Because gravity works in the vertical direction, structures are usually the strongest in the vertical direction. Engineers also design for horizontal (lateral) forces such as wind, but this is usually a minor consideration compared to the vertical load or weight. For most places this lack of structural strength in the lateral direction is not a problem, but it becomes one of critical importance in areas where strong earthquakes occur.

FIGURE 5.18 Although most people are killed in the total collapse of buildings, some survive in void spaces within the rubble. The problem is gaining access to the survivors before they die from their injuries, or from hypothermia or dehydration. Rescue is also made more difficult by the unstable nature of the rubble and the constant threat of aftershocks. Photos from the 2008 earthquake (magnitude 7.9) in Sichuan, China.
© Mark Ralston/AFP/Getty Images; (inset) © Jiang Yi/Xinhua Press/Corbis Images

Construction Design

Recall that when a rock body ruptures and releases its strain, it produces both body (P and S) waves and surface (Rayleigh and Love) waves. Of these, surface waves are the most destructive since they cause the ground to vibrate in a lateral direction, and at the same time, roll up and down like an ocean wave. The degree to which a structure withstands this violent shaking is highly dependent upon the way in which it was constructed. In the case of homes, most have a wood frame that either sits directly on a concrete slab or is built over a crawl space or a garage, like those shown in Figure 5.19. During an earthquake the base of the structure is forced to move laterally with the ground, but the top tries to remain stationary due to its inertia, thereby putting the structure under a lateral shearing force. Because most houses are not specifically designed for lateral shear, they offer very little resistance to ground vibrations and are damaged quite easily.

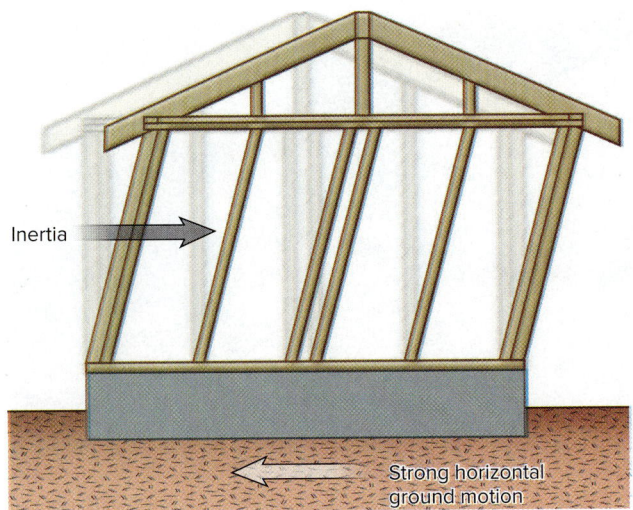

A

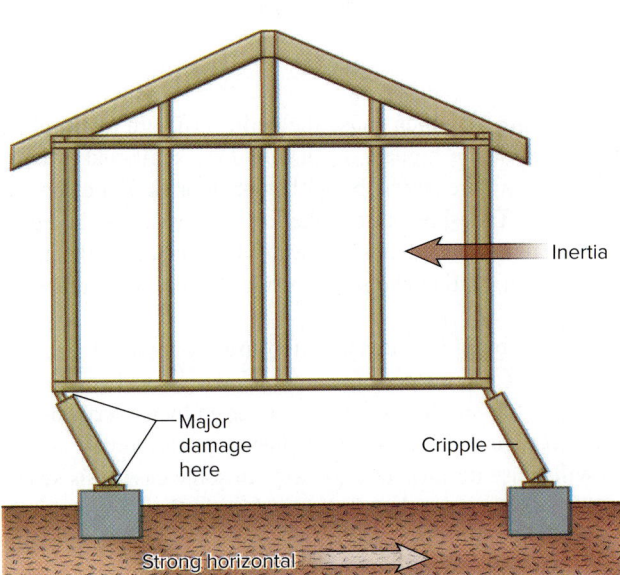

B

FIGURE 5.19 During an earthquake, buildings are subjected to lateral shear stress due to the horizontal ground motion and their own inertia. This lateral shear force causes structures built on slabs (A) to become skewed after an earthquake. Buildings with crawl spaces (B) or large open areas on the ground floor are inherently weak and prone to cripple-wall failure.

(a) NOAA/NGDC, E.V. Leyendecker, U.S. Geological Survey; (b) M. Mehrain, Dames and Moore/NOAA

Another important aspect of the two types of home construction shown in Figure 5.19 is the type of damage they suffer in an earthquake. In the case of a framed structure built on a concrete slab, the shearing motion typically leaves the building heavily skewed (Figure 5.19A). For a home built over a crawl space, the walls surrounding this open space commonly fail since they are the weakest part of the structure. In this case the house tends to stay intact while falling over into the open space (Figure 5.19B). Note that engineers refer to the collapse of a crawl space as a *cripple-wall failure*, but when the open space is taller the failure is called a *soft-story collapse*. Soft-story collapse is a common problem in commercial buildings where the first floor has considerable open space for parking or retail shopping.

Generally speaking, the most dangerous types of homes are those constructed of *unreinforced masonry* because they offer very little resistance to lateral shearing motion. The walls in this design are usually constructed of brick or stone bound together with mortar, as opposed to reinforced walls with internal supports of wood or steel—note that most brick homes in the United States have an internal wooden frame. Although unreinforced masonry walls have great load-bearing capacity, the mortar readily fails during an earthquake, leaving the wall in a weakened state

A   B

FIGURE 5.20   The mortar in unreinforced masonry walls such as these in Iran, 1990 (A) can easily fail during an earthquake. Oftentimes the entire structure crumbles, leaving a pile of rubble in which few survive. Shown in (B) is one of the many masonry homes to collapse during the 2015 Nepal earthquake (magnitude 7.8), which claimed nearly 9,000 lives.
(a) NOAA; (b) © Prakash Mathema/AFP/Getty Images

FIGURE 5.21   Floors in a multistory building can become detached from the supporting columns as the building sways due to lateral ground motion. This can lead to a total collapse, where the chance of survival is extremely remote. Notice in the photo from the 1985 earthquake (magnitude 8.0) in Mexico City, how the vertical column (center of photo) punched through the cascading floors as they fell.
Reinsurance Company, Munich, Germany/NOAA

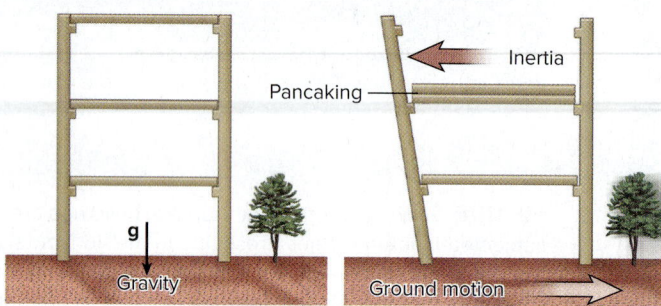

(Figure 5.20). Continued shaking may then cause the entire structure to crumble, crushing its inhabitants. Some of the highest death tolls from earthquakes have occurred in regions where homes were largely built of stone or brick. For example, most of the deaths in the 1976 Tangshan earthquake in China were attributed to homes having been built with unreinforced masonry walls. Compounding the problem in Tangshan was the fact the earthquake struck at night while residents were asleep in their homes.

With respect to multistory buildings, construction usually involves an interior skeleton made of steel or steel-reinforced concrete. Under normal conditions the entire weight of the building is easily supported by its vertical columns as shown in Figure 5.21. However, during an earthquake strong lateral forces will cause the structure to sway. In some cases this swaying motion becomes so great that floors within the building become detached from the columns. Once a floor becomes detached, it naturally falls onto the one below, which can cause additional floors to fail in a cascading manner that engineers call *pancaking*. The result is either a total or partial collapse of the structure in which few people survive.

Another important type of structural failure in earthquakes is the sudden rupture of steel-reinforced concrete columns (Figure 5.22). These types of columns are widely used for supporting highways, bridges, and buildings. However, they can fail when the swaying motion of the structure becomes so great that the columns, which are quite brittle, reach their elastic limit and literally explode. Once the concrete shatters, the entire structure can collapse since the steel-reinforcing rods alone are not capable of supporting the weight of the structure.

Natural Vibration Frequency and Resonance

Although all multistory buildings are flexible to some degree, a dangerous phenomenon can develop during an earthquake which causes a building's swaying motion to actually increase. A building swaying back and forth is technically considered to be vibrating—similar to how a guitar string vibrates. Moreover, the building will vibrate at a fixed frequency called its **natural vibration frequency;** frequency is the number of times the motion is repeated in a set amount of time.

As building height increases, the natural vibration frequency decreases—similar to how lengthening a guitar string will produce a note with a lower or deeper pitch (i.e., frequency). This means that in a city with multistory buildings of different heights, some will have a relatively low vibration frequency and others will vibrate at a relatively high frequency.

The problem occurs during an earthquake when the natural vibration frequency of a given building matches that of the seismic waves. The matching of frequency then leads to the phenomenon called **resonance,** whereby the amplitudes of the individual waves combine as shown in Figure 5.23A. Here the increased amplitude causes the building to sway even more violently, increasing the chance that its supporting structure will fail as previously described. In some instances the additional swaying motion from resonance can be so severe that a building can slam repeatedly into adjacent structures, as was the case for the two buildings in Figure 5.23B. However, because seismic waves have a relatively narrow range of frequencies, only those buildings whose vibration frequency falls within this range can experience resonance. It turns out that buildings around 10–20 stories high are the most susceptible to resonance. On the other hand, tall skyscrapers are unlikely to experience resonance as their vibration frequency is beyond the frequency range of most seismic waves.

Factors That Affect Ground Shaking

Strong ground shaking is clearly the principal cause of structural damage and loss of life in an earthquake. However, ground shaking creates several other hazards in addition to structural failure, topics we will explore in the following sections. But first we need to discuss why the level of shaking varies from earthquake to earthquake, and from one location to another.

Focal Depth and Wave Attenuation

Recall that magnitude and ground shaking are ultimately controlled by the amount of elastic energy that is released when rocks rupture. Moreover, the energy of the corresponding seismic waves steadily decreases as they travel away from the focus, a process referred to as **wave attenuation.** The reason

FIGURE 5.22 Brittle failure of steel-reinforced concrete columns occurs when swaying motion causes the columns to reach their elastic limit. Note that the steel rods themselves are not strong enough to support the weight of the structure. Photo from the 1999 earthquake (magnitude 7.6) in Taiwan. NOAA

FIGURE 5.23 When a building's natural vibration frequency matches the frequency of seismic waves, resonance can occur (A), causing it to sway more violently. Because vibration frequency varies with height, not all multistory buildings will experience resonance. In the photo (B) the shorter building on the far left experienced resonance and repeatedly slammed into the adjacent hotel, half of which then partially collapsed. Photo from the 1985 (magnitude 8.0) earthquake in Mexico City.
(b) C. Arnold, Building Systems Development, Inc./NOAA

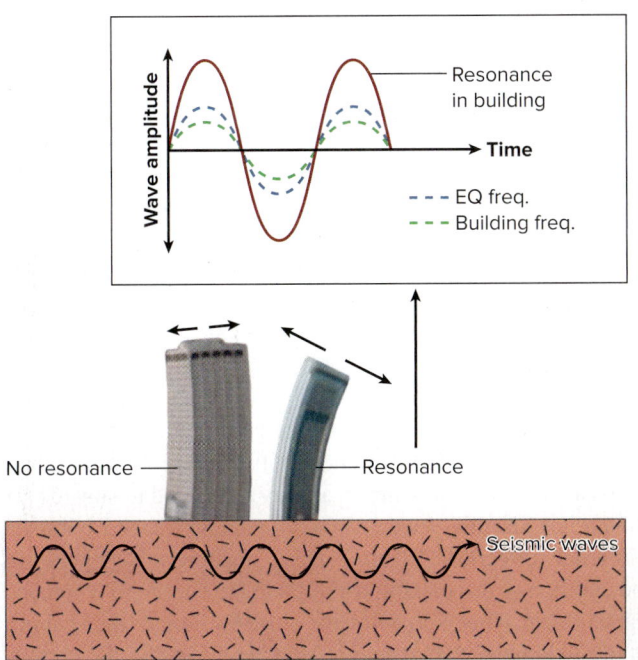

A

B

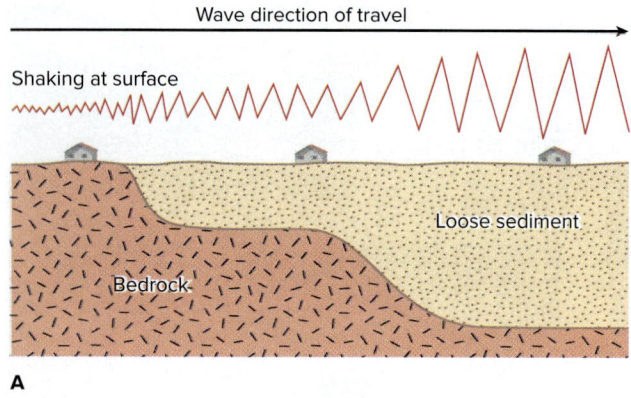

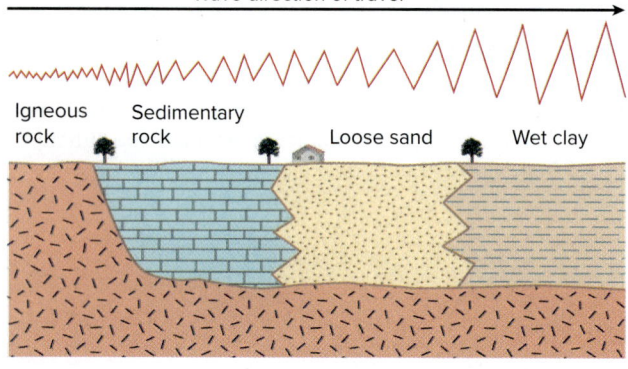

FIGURE 5.24   As seismic waves travel from bedrock into materials of lower density, their velocity decreases, which can cause the waves to amplify. Resonance in loose sediment (A) often leads to ground amplification and is most severe where the sediment is thicker. Ground amplification is also more severe in weaker materials (B), which offer less resistance to seismic waves.

ground shaking is typically the strongest at the epicenter is because this is the closest surface point to the focus, which means wave attenuation is at a minimum. Therefore, the level of ground shaking at any given location depends on the distance to the focus and the amount of energy released when the rocks ruptured. This explains why the most dangerous earthquakes tend to be those with a large magnitude *and* shallow focal depth. Keep in mind that it is quite possible for a shallow, low-magnitude quake to generate greater ground motion than a deep, high-magnitude quake.

Also of interest is the fact that seismic waves experience different amounts of wave attenuation, depending on the types of geologic materials the waves must past through. It turns out that loose materials and rocks of lower density will absorb more energy from passing seismic waves compared to rocks that are more rigid and dense. This means that in areas where the rocks are rigid, seismic waves are able to retain more of their energy as they travel. Because the waves undergo less attenuation, they have the potential to cause damage farther from the focus. A good example is the series of magnitude 8 earthquakes that struck the New Madrid (Figure 5.16) area during the winter of 1811–1812. The rigid rocks in this region were able to transmit the seismic waves efficiently and with little wave attenuation. In fact, the seismic waves retained enough energy to ring church bells as far away as Boston, Massachusetts.

Ground Amplification

Although seismic waves lose energy as they travel, their velocity remains relatively constant provided the density and rigidity of subsurface materials stays the same. However, in many regions the subsurface consists of a variety of rock types and loose sediment, forcing seismic waves to change velocity as they encounter the different materials. What is important here is that when seismic waves travel through weaker materials, they slow down and lose energy at a *faster rate*. This abrupt decrease in velocity causes wave amplitude to increase, creating a phenomenon known as **ground amplification.** To make matters worse, weaker materials can begin to vibrate at the same frequency as that of the seismic waves. This can lead to resonance, similar to that described in the previous section for buildings, but in this case it further increases ground amplification. One could expect then in Figure 5.24 for ground amplification to be most severe in areas with thick layers of loose sediment and where the surface materials are relatively weak.

Ground amplification is a serious problem because it greatly increases the level of shaking, which, in turn, increases the risk of structural failure. Due to the increased risk, scientists and engineers are keenly interested in identifying those areas with the greatest potential for ground amplification, as shown in the map of the Los Angeles area in Figure 5.25. Notice on the map how ground shaking is expected to be up to five times greater in the sediment-filled valleys (yellows and reds) than in the mountainous areas (purple) composed of solid rock. Naturally the areas of highest population density coincide with the more flat-lying terrain of the valleys, which unfortunately is where geologic conditions create the highest risk of ground shaking.

This brings up another problem, namely, how seismic waves behave in *sedimentary basins*, which are depressions in the crust filled with sediment and sedimentary rocks. As shown in Figure 5.26A, ground amplification can occur when seismic waves enter a basin and begin to amplify as they are forced to slow down in the sedimentary material. Seismic waves can also become trapped within a basin and undergo internal reflection (Figure 5.26B), creating a reverberating effect that extends the duration of the shaking—similar to the way *Jello* continues to vibrate after being set down. This is undesirable since the longer the shaking goes on, the greater the likelihood that structures will fail. Finally, the convex shape of a basin (Figure 5.26C) can cause waves to refract and merge, focusing

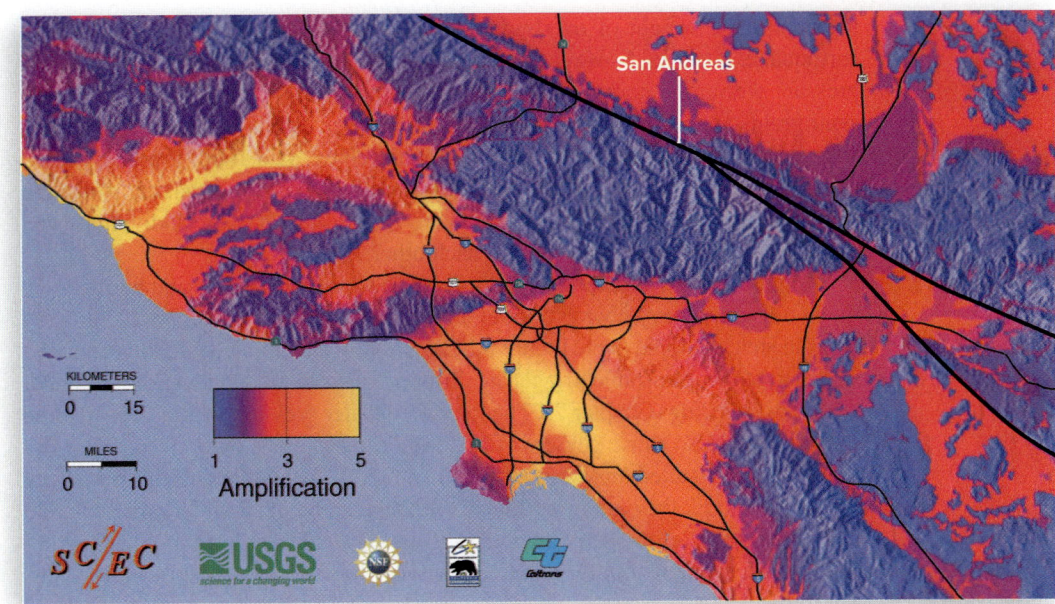

FIGURE 5.25 Map showing levels of ground amplification in the Los Angeles area as predicted by mathematical models. Note that areas of highest risk (red and yellow) are sediment-filled valleys, whereas the lowest risk is in the mountainous bedrock areas (purple). Edward H. Field, 2001 U.S. Geological Survey Open-File Report 01-164

their energy into localized areas, which then experience more intense shaking. All told, the combined effects of amplification, internal reflection, and focusing of seismic waves can cause ground shaking in a sedimentary basin to be as much as 10 times greater than in the surrounding areas composed of more dense and rigid bedrock.

Secondary Earthquake Hazards

Although the primary hazard associated with earthquakes is the failure of human structures, intense ground shaking often produces secondary hazards such as fires, landslides, and saturated ground that suddenly turns into a liquid. In addition to shaking-related hazards, subduction zone earthquakes can generate devastating tsunamis. In this section we will explore some of the secondary hazards associated with earthquakes.

Liquefaction

In areas where saturated conditions exist close to the surface (i.e., a high water table—Chapter 11), ground shaking can cause a phenomenon called **liquefaction,** where sand-rich layers of sediment behave as fluid. As illustrated in Figure 5.27A, compacted sand grains are normally in contact with one another, hence are able

FIGURE 5.26 Ground shaking commonly increases when seismic waves travel through sedimentary basins due to: (A) ground amplification occurring as waves slow down when encountering lower-density material within the basin; (B) seismic waves reflecting and becoming trapped within the basin, prolonging the shaking; and (C) refracted waves merging and focusing their energy.

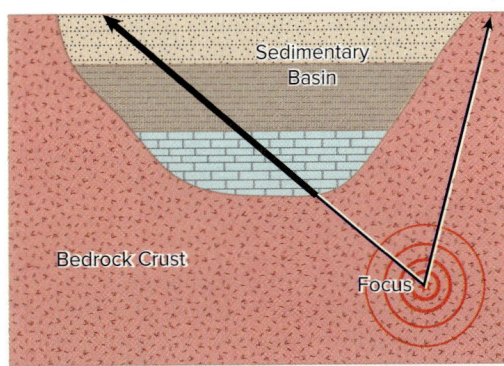

A Wave amplification

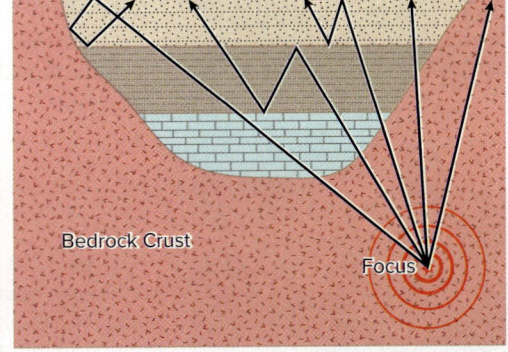

B Internal reflection

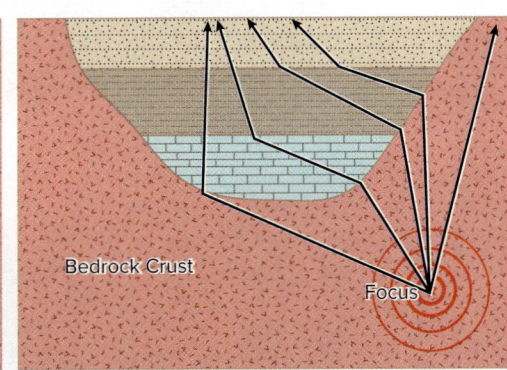

C Focusing by refraction

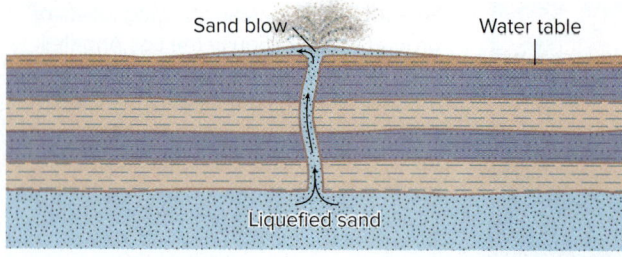

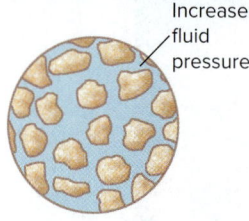

A   Before   During shaking

to support the weight of overlying sediment and human structures. During an earthquake, however, S-waves produce a shearing motion that increases the water pressure within the pore space of the sediment, thereby preventing the vibrating sand grains from making contact with one another. While the grains are in this suspended state, the saturated sediment will behave as a fluid, a process often referred to as becoming *liquefied* or *fluidized*. As soon as the shaking stops, the sand-rich material will return to behaving as a solid since the individual sand grains are able to make contact with each other.

Liquefaction is a serious problem because while subsurface sand layers are in the liquid state, heavy objects sitting on the surface are left unsupported, allowing them to sink or topple over (Figure 5.27B). In hilly terrain, liquefaction can cause slopes to become unstable, triggering different types of landslides geologists refer to as *mass wasting* (Chapter 7). The increased water pressure within the saturated sediment can also cause geysers of liquefied sand to erupt onto the surface, creating what are called *sand blows* (Figure 5.27A). Although sand blows do not present a hazard, they are important since they can be overlain by new sediment and become part of the geologic record. Buried sand blows have provided geologists with a valuable tool for dating ancient earthquakes associated with the New Madrid and Charleston seismic zones.

Disturbances of the Land Surface

In addition to ground shaking and liquefaction, earthquakes can cause disruptions to the land surface that damage buildings and other important infrastructure. For example, recall that when a fault ruptures, rocks on either side of the fault are *displaced* or offset. Obviously anything built across an active fault plane is at risk of being damaged when the fault slips and the ground becomes displaced, as with the bridge shown in Figure 5.28. Because of the potential for ground displacement, critical structures like dams, nuclear power plants, underground pipelines, hospitals, and schools should not be built across known faults.

As shown in Figure 5.29, earthquakes also create large open cracks called **ground fissures** over a wide area of the landscape (faults themselves do not open up like fissures). Ground fissures typically develop close to the surface in loose sediment where there is little resistance to the rolling and stretching motion associated with surface waves. Unlike ground displacement that occurs directly along the fault trace, open fissures have the potential to affect a greater number of structures because they develop over a much broader area. In addition to surface features such as roads and buildings, ground fissures also damage underground gas, electric, water, and sewage lines. Damage to transportation links and basic utilities can be particularly disruptive to the overall economy of a region. Moreover, the interruption of utility services creates additional hardships for people who are already struggling to cope in the aftermath of an earthquake. The loss of basic utilities can also lead to disease outbreaks due to a lack of sanitation (sewers) and clean water. Finally, ruptured gas lines pose a serious fire hazard, a topic we will explore separately in the next section.

Earthquakes also serve as one of the basic triggering mechanisms for *mass wasting*, which is where earth materials move downslope due to gravity (Chapter 7). Familiar forms

FIGURE 5.27 Liquefaction (A) occurs when ground shaking causes an increase in water pressure within sand-rich sediment. As individual sand grains lose contact with one another, the material behaves as a fluid and loses its ability to support the weight of overlying materials. This lack of support causes structures (B) to sink or topple over. Photo from the 1964 earthquake (magnitude 7.5) in Niigata, Japan.
(b) Courtesy W. Godden Collection, NISEE, University of California, Berkeley

FIGURE 5.28 This bridge was destroyed when an earthquake caused nearly 20 feet (6 m) of vertical displacement along a fault, which is marked by the newly formed waterfalls. Photo from the 1999 (magnitude 7.6) earthquake in Taiwan.
© Dr. Ross. W. Boulanger/University of California at Davis

of mass wasting include rock falls and mudflows. Although mass wasting events can occur over a variety of landscapes, they are largely restricted to hilly and mountainous terrain where steeper slopes are more common. Many slopes in steep terrain are inherently unstable, needing only the vibrations from an earthquake to fail and start sending material downslope. Particularly troublesome are slopes consisting of highly fractured rock and those covered with thick layers of loose material. One of the worst mass-wasting disasters occurred in Peru in 1970, when an earthquake triggered a rock and snow avalanche that killed an estimated 18,000 people. The earthquake itself was responsible for another 48,000 deaths.

Fires

As mentioned earlier in this section, underground gas lines are easily broken when surface waves roll through a city. All that is needed for a fire is an ignition source, which is often readily provided by sparks from downed power lines and damaged electrical lines in buildings (Figure 5.30). Gas-fed fires can be extremely difficult to extinguish, especially when blocked or damaged roads limit access to the fire. To compound the problem, once local firefighters get to a site, broken water mains may mean there is no water available to fight the fire. Often the sheer number of fires is so great that fire crews are simply overwhelmed. Such was the case in the 1995 earthquake near Kobe, Japan, where over 300 fires broke out, most of which were started by gas cooking stoves in residential homes. To make matters worse, fire-fighting efforts were greatly hampered by blocked streets and a lack of water resulting from an estimated 10,000 water-line breaks. In the end, over 7,000 buildings were destroyed by fire. Because of the lessons learned from Kobe and other recent earthquakes, some cities in earthquake-prone regions have built systems that will use seawater or underground storage reservoirs (cisterns) in future emergencies.

FIGURE 5.29   Ground fissures damaged this highway during the 1964 Alaskan earthquake (magnitude 9.2). Open fissures occur in unconsolidated sediment and can cause serious damage to highways, bridges, buildings, and underground utilities.
Karl Steinbrugge/Earthquake Engineering Research Center, University of CA, Berkeley

FIGURE 5.30   Fires are a common secondary hazard associated with earthquakes. Aerial panorama (A) of San Francisco showing the extensive damage caused by the fires that swept through the city after the 1906 (magnitude 7.8) earthquake. Photo (B) showing a gas-fed fire caused by a broken underground gas line. The spark for this fire probably came from the nearby electrical lines.
(a) Library of Congress Prints and Photographs Division; (b) USGS

A

B

Perhaps the best known example of an earthquake fire is the one that followed the great 1906 quake (magnitude 7.8) near San Francisco. Between 500 and 700 people perished and approximately 20% of the city was destroyed in this earthquake. Although the earthquake caused heavy structural damage, the fire that swept through parts of the city was responsible for 70% to 80% of all buildings that were destroyed (Figure 5.30A). Like Kobe nearly 100 years later, broken water mains made it nearly impossible to extinguish the initial fires, which then raged out of control for days. In a last-ditch effort to stop the spreading fire, firefighters resorted to dynamiting buildings to try to create firebreaks.

Tsunamis

A **tsunami** is a series of ocean waves that form when energy is suddenly transferred to the water by an earthquake, volcanic eruption, landslide, or asteroid impact. The majority of tsunamis, however, form during subduction zone earthquakes when tectonic plates abruptly move and displace large volumes of seawater. As illustrated in Figure 5.31, compressive forces cause the overriding plate to slowly buckle until the fault eventually ruptures, at which point the plate lurches upward as the strain is released. The kinetic energy from the vertical displacement is quickly transferred to the water, where it takes the form of wave energy that travels in both directions away from the subduction zone. Note that a tsunami is similar to seismic waves in that they originate at a site where energy is suddenly transferred. The basic difference is that the wave energy in a tsunami travels through water as opposed to rock. Because wave attenuation (rate of energy loss) in water is relatively small compared to rock, a tsunami will carry the energy for considerable distances. For example, tsunamis originating along the subduction zone of South America can travel across the entire Pacific Ocean and still have enough power to cause considerable damage in Japan, at a distance of over 10,000 miles (16,000 km).

Over the deeper parts of the ocean a tsunami will travel at great speed, around 450 miles per hour (725 km/hr), but its amplitude (height) is usually less than 3 feet. As the waves approach shore and enter shallower waters, they are forced to slow down, a process which causes them to become taller. When the waves finally crash ashore, they may be as high as 65 feet (20 m). Even taller waves can be generated in bays and inlets that tend to collect or funnel the waves. Here the energy becomes more focused, producing wave heights that can exceed 100 feet (30 m). What ultimately governs wave height though is the volume of rock along the subduction zone that displaces seawater. This, in turn, is controlled by the amount of vertical displacement on the fault and the horizontal distance over which the rupture occurs.

As an illustration of this process consider the devastating 2011 tsunami that crashed ashore in northern Japan. This event was triggered by an extremely powerful earthquake (magnitude 9.0) located in a subduction zone 80 miles (130 km) offshore (Figure 5.32). Here the overriding tectonic plate lurched upward 100 to 130 feet (30–40 m) over a 185 mile (300 km) long by 93 mile (150 km) wide swath of seafloor. The sudden displacement of such a large volume of rock produced an exceptionally large tsunami. As the waves moved into shallow waters, they naturally gained height before sweeping ashore. The largest recorded wave was an incredible 125 feet (38 m) high. Of the 21,000 people who perished in the earthquake, nearly 95% were killed by the tsunami. The tsunami was also responsible for destroying the Fukushima nuclear power plant. Engineers had built breakwaters to guard the plant against giant waves, but assumed a worst case scenario of a magnitude 8.6 quake and a tsunami 18.7 feet (5.7 m) high. Unfortunately, the 2011 earthquake generated waves as high as 46 feet (14 m) that reached the plant, easily washing over the breakwaters and destroying the emergency power supply. In the end, three of the plant's six reactors melted down, producing one of the world's worst nuclear disasters (see Chapter 14 for more details).

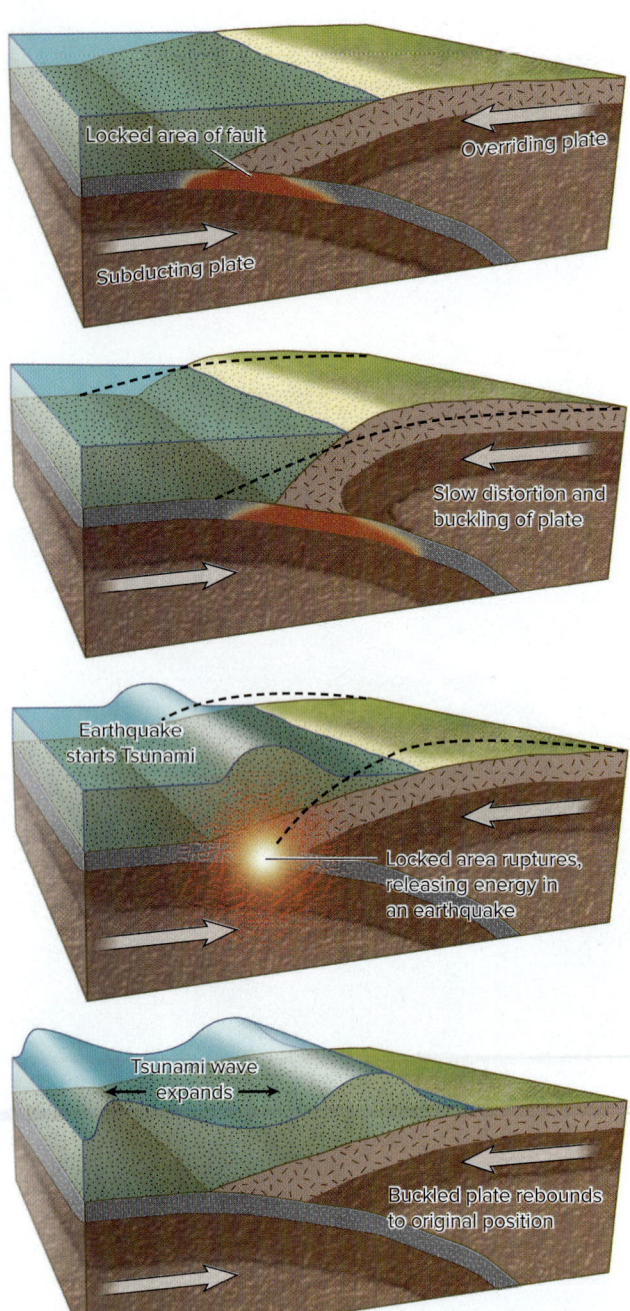

FIGURE 5.31 Tsunamis form at subduction zones when a buckled plate suddenly slips, displacing a large volume of seawater. Kinetic energy from this movement is transformed into wave energy, which then travels outward away from the subduction zone. As the waves approach shore they slow down, which causes an increase in wave height.

Case Study 5.1 describes how Japan's high level of preparedness greatly minimized the loss of life in the 2011 disaster. For more details on tsunamis, please see Chapter 9.

Predicting Earthquakes

Many lives certainly would be saved if scientists could predict earthquakes on a short-term basis, say within hours or days of the actual event. Despite knowing how earthquakes occur when a rock body accumulates strain beyond its elastic limit, seismologists still cannot predict just *when* the rock will rupture. Short-term predictions, therefore, are not yet a reality, and some seismologists believe they never will be. What we have been successful at is predicting earthquakes on a long-term basis using statistical probabilities. These long-term predictions are similar to weather forecasts in that they give the *probability* of an earthquake occurring within a given time period. In this section we will take a brief look at the current status of both short- and long-term earthquake predictions.

Short-Term Predictions

When strain energy is released along a fault, or series of faults, it usually takes place as a cluster of separate events. Small earthquakes called *foreshocks* will sometimes precede the major release of energy, known as the *main shock*, which can be followed by a series of less powerful *aftershocks*. In addition to the foreshocks, phenomena called **earthquake precursors** sometimes occur just prior to the main shock. Precursors include such things as changes in land elevation, water levels and dissolved gases in wells, and unusual patterns of low-frequency radio waves.

Earthquake precursors can largely be explained by the way strain accumulates in rocks on a microscopic scale. Here we can use the bending of a wooden rod as an analogy. As the wood accumulates strain and gets near its elastic limit, individual wood fibers will start separating and produce a subtle cracking sound. While this is happening, the volume of the rod increases slightly as air space develops between the fibers. Ultimately the strain becomes so great that the fibers are no longer able to resist the force, at which point the rod ruptures. Rocks behave similarly in that as strain accumulates, individual mineral grains begin to separate and develop microscopic cracks prior to rupturing. This model then can explain the occurrence of some foreshocks and the following earthquake precursors:

1. *Increase in foreshocks*—microcracks forming prior to complete rupture, or main shock.
2. *Slight swelling or tilting of the ground surface*—microcracks increasing the rock volume.
3. *Decreased electrical resistance*—water entering new void spaces is more conductive than surrounding minerals.
4. *Fluctuating water levels in wells*—water entering new cracks causes water levels to lower; levels rise when voids close again.
5. *Increased concentration of radon gas in groundwater*—new cracks allow the gas, a radioactive decay product of uranium, to escape from rocks and enter wells.
6. *Generation of radio signals*—changes in rock strain or movement of saline groundwater.

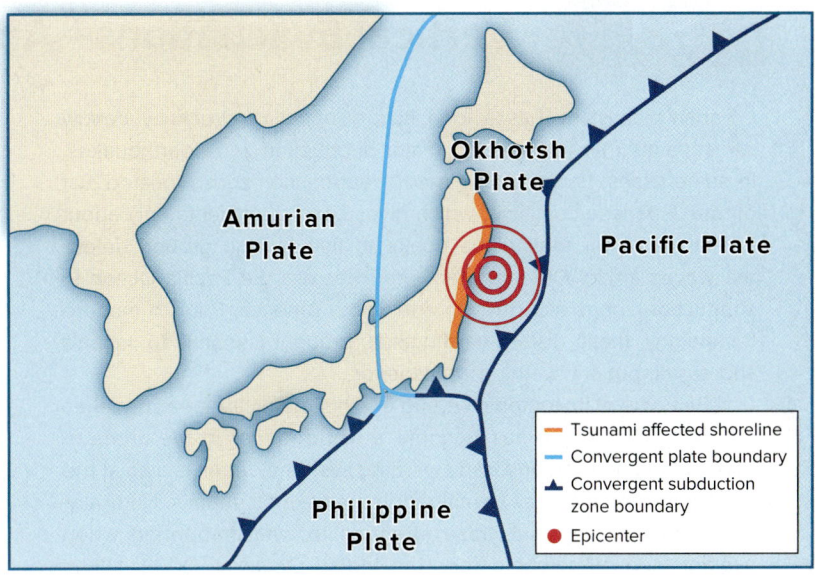

A

B

FIGURE 5.32   Map (A) showing the location of the 2011 subduction zone earthquake (magnitude 9.0) that struck Japan. During the quake the tectonic plates suddenly shifted, creating a giant tsunami that swept ashore. Aerial view (B) of a coastal community in northern Japan that was completely destroyed by the tsunami. Notice the ship left stranded on the two-story building in the center of the photo.
(b) U.S. Navy photo by Mass Communication Specialist 3rd Class Alexander Tidd

Vertical Evacuations—An Important Lesson from Japan

Japan's coastline has a long history of being struck by deadly tsunamis that form during major subduction zone earthquakes. In some cases, tsunamis originate in subduction zones located just off the Japanese coastline, which gives coastal residents only about 20 minutes after feeling the quake to flee to high ground before the waves arrive. Other tsunamis form across the Pacific Ocean in subduction zones along North and South America. Prior to modern technology, these distant earthquakes brought tsunamis to Japan's shores without any warning whatsoever.

Because of its tectonic setting and history of large earthquakes and tsunamis, Japan has become a world leader in the areas of seismic engineering and early warning systems. Japan is also at the forefront when it comes to mitigating the risk of tsunamis. Naturally, Japanese experts paid close attention to what happened when a giant tsunami swept across the Indian Ocean in 2004, killing nearly 230,000 people. Of those who perished, 168,000 had lived along the Indonesian coast, which was adjacent to the subduction zone where the tsunami originated. Japanese officials were quick to recognize that if a similar event occurred in a subduction zone off Japan's coast, the death toll could be horrific. Since Japan's coastal cities are located on flat terrain and have high population densities, there would not be enough time for most people to walk or run to higher ground before the waves struck. Driving to high ground would also be difficult since the roads leading away from the cities would likely experience gridlock, or be blocked by fallen debris from the earthquake. The only viable option would be to evacuate vertically.

After the Indonesian tsunami, Japan embarked on a program of establishing vertical evacuation sites throughout its coastal communities. Some sites consisted of specially built towers, whereas others made use of existing multistory buildings deemed strong enough to withstand the expected waves. All evacuation sites were clearly marked with signs, had stairwells leading to the top floor, and were designed to be accessible 24 hours a day. Other key elements included a public education program to inform residents of the location of the various evacuation sites and an early warning system so they would know when to evacuate.

When a magnitude 9.0 subduction zone quake struck on March 11, 2011, Japan was well prepared. The earthquake and tsunami early warning systems worked as planned, and upon receiving the warnings, the well-informed population took immediate action, many heading to vertical evacuation sites. As illustrated in Figure B5.1, coastal cities in northern Japan were decimated by the tsunami. In many areas the only structures still standing were the vertical evacuation sites. Although 21,000 people were killed in the disaster, 17,000 of whom were lost in the tsunami, it is clear that without Japan's high level of preparedness, the death toll would have been far higher.

The disaster that unfolded in Japan was closely followed by other nations with coastal residents at risk from a subduction zone earthquake and tsunami. In the United States and Canada, many of the coastal communities along the Cascadia subduction zone in the Pacific Northwest (see Figure 5.14) are located on flat, isolated spits of land that are extremely vulnerable to a tsunami. Some of these communities face the additional problem of large tourist populations during the summer months. Since it would be virtually impossible for most people to make it to distant high ground before the waves make landfall, many coastal communities are following Japan's example. They are coordinating with state and federal agencies to develop early warning systems and embarking on

One of the difficulties in using precursors as a predictive tool for earthquakes is a lack of consistent and reliable data. For example, with the exception of foreshocks, all of the precursors listed above can result from things other than the buildup of strain. Without sufficient data then, scientists cannot necessarily attribute a particular precursor to earthquake activity. Another problem is that the triggering of earthquakes is a highly complex process involving the interaction of many different factors, or variables. To make predictions even more difficult, individual earthquakes may involve different combinations of variables. This means that no two earthquakes are likely to have a similar set of precursors leading up to the main shock, making short-term earthquake predictions all but impossible at the present time.

Long-Term Predictions

Geologists have recognized that in seismically active areas, the chance of a large earthquake increases as more time passes since the last major event. This observation is consistent with the elastic rebound theory, where a persistent force will

A

B

FIGURE B5.1 Aerial view (A) showing the devastating effects of the 2011 tsunami on the city of Minami-Sanrikucho, in northern Japan. Here and in other coastal cities, few people were able to flee to high ground before the tsunami waves came ashore. Fortunately, many lives were saved because Japan had established vertical evacuation sites in coastal communities. As illustrated by the evacuation tower in (B), these sites were often the only structures still standing after the tsunami.

(a) © Junichi Sasaki/AFLO/Nippon News/Corbis; (b) U.S. Navy photo by Specialist 3rd Class Alexander Tidd

public education programs so that residents and tourists know how to respond when an alert is issued. Planners in the Pacific Northwest are also working with local communities to decide what type of vertical evacuation sites are best suited for their individual needs. Since constructing dedicated evacuation towers is expensive and often considered unattractive, planners are considering building earthen berms, or embankments. Berms are not only less expensive, but can be landscaped to make them aesthetically pleasing, and can also serve as public parks. Hopefully, the Pacific Northwest and other regions around the world that are at risk from tsunamis will have completed their preparations before the next major subduction zone earthquake.

cause a rock body to repeatedly accumulate strain and then rupture. Geologists use the term **seismic gap** to refer to those sections of an active fault where the strain has not been released for an extended period of time. Seismic gaps can be useful in predicting what areas are most likely to experience a large earthquake. For example, the map in Figure 5.33 shows the areas along the Alaskan subduction zone where strain has been released in modern times by major earthquakes. Based on the location of the ruptured areas, geologists have identified two seismic gaps where a major earthquake is most likely to strike. A similar situation exists along the Cascadia subduction zone in the Pacific Northwest. Here the entire subduction zone can be considered a seismic gap since a major earthquake has not taken place for nearly 300 years.

Although seismic gaps are good for indicating the potential of future earthquakes, statistical estimates are even more useful as they provide information on what to expect over a specific time period. One statistical approach involves the creation of *seismic hazard maps*, which incorporate information on past seismic activity, magnitudes, and displacement rates on faults. For example, the seismic hazard map of the United States in Figure 5.34 shows the potential strength of

Classification and Flow Characteristics of Volcanic Rocks

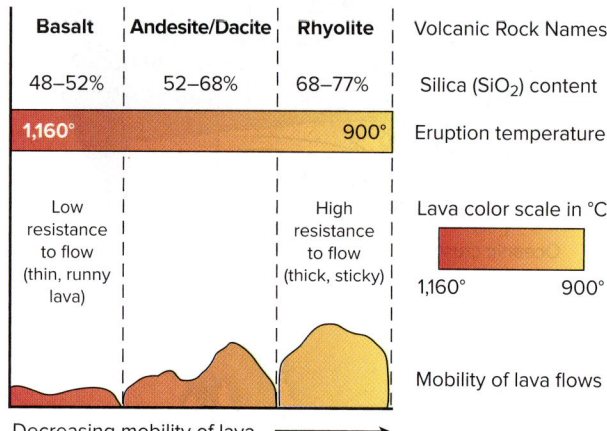

Basalt	Andesite/Dacite	Rhyolite	
Basalt	**Andesite/Dacite**	**Rhyolite**	Volcanic Rock Names
48–52%	52–68%	68–77%	Silica (SiO_2) content
1,160°		900°	Eruption temperature
Low resistance to flow (thin, runny lava)		High resistance to flow (thick, sticky)	Lava color scale in °C
			Mobility of lava flows

1,160°   900°

Decreasing mobility of lava ⟶

FIGURE 6.3   The viscosity of magma increases with increasing SiO_2 content and decreasing temperature. Basaltic magmas are the least viscous because they form in the upper mantle where the temperature is high and the SiO_2 content of the rocks is relatively low. Andesitic and rhyolitic magmas form at much shallower and cooler depths and under processes that cause these magmas to become enriched in SiO_2.

cool, which increases the internal friction within the magma. Therefore as the SiO_2 content increases, the ability of a magma to resist flowing also increases, a property called **viscosity.** A useful way of understanding viscosity is to think in terms of thick versus thin fluids. For example, water is considered to be a "thin" fluid with low viscosity because it has little internal friction; hence it flows quite easily. However, if you add flour or cornstarch to water, chemical bonds develop that increase the fluid's resistance to flow, making it more viscous. In a similar manner, silica is the key ingredient in magma that controls internal friction and viscosity. As shown in Figure 6.3, of the three types of magma, basaltic magma has the lowest viscosity primarily because it has the lowest SiO_2 content.

Figure 6.3 also illustrates the progressive decrease in eruption temperature among the three magma types, with basaltic being considerably hotter than rhyolitic magma. Part of the reason for this is that basaltic magmas generally form in the upper mantle where temperatures are naturally higher. On the other hand, the water present in subduction zones allows melting to take place at lower temperatures; hence the temperature range of andesitic magma is lower than that of basaltic. Rhyolitic magma has the lowest temperature range of all the types due to the fact that granitic rock within the crust is able to melt at even lower temperatures. The key point here is that as temperature *decreases,* the internal friction within magma *increases,* which means viscosity increases—similar to how maple syrup gets even thicker when it is cooled. Therefore, the reason rhyolitic magmas are the most viscous is because they have the highest SiO_2 content *and* the lowest temperature. In contrast, basaltic magmas are more fluid (Figure 6.4) because they contain around 25% less SiO_2 than rhyolitic magmas and are about 250°C (480°F) warmer.

In addition to individual ions (charged atoms), magma also contains various types of gas molecules that form when minerals within a rock body begin to melt. These gases remain dissolved until the magma gets near the surface, at which point the decreased pressure allows the gases to escape—similar to how carbon dioxide (CO_2) gas in soda remains dissolved until the can is opened. The type and amount of gases involved in this process largely depend on the material being melted. This is particularly important at subduction zones where water-rich sediment is incorporated into the molten material, generating andesitic magmas rich in water (H_2O) vapor. Rhyolitic magmas typically get their water vapor from the melting of continental crustal material. Table 6.1 compares the gas composition of a basaltic and an andesitic magma, illustrating how H_2O and CO_2 together make up close to 90% of all magmatic gases. We can also see from the table that water vapor is by far the most dominant gas in andesite magma. In the next section you will learn that the *amount* of dissolved gases is the ultimate control on the type of volcanic eruption.

Volcanic Eruptions

Recall from Chapter 3 that when magma stays buried deep within the crust, it slowly cools and forms *intrusive* igneous rock, but if it reaches Earth's surface environment, it cools quickly and forms *extrusive* or *volcanic* rock. A volcanic eruption can be defined as whenever magma breaks through Earth's surface. Note that once molten rock is on the surface it is referred to as *lava.* In this

FIGURE 6.4   How easily magma flows depends on its viscosity, which is controlled by its SiO_2 content and temperature. Shown here is high-temperature, silica-poor basaltic magma of relatively low viscosity. Photo taken during the 1989 eruption of Kilauea volcano, Hawaii.
J.D. Griggs/USGS

A

B

FIGURE B5.1 Aerial view (A) showing the devastating effects of the 2011 tsunami on the city of Minami-Sanrikucho, in northern Japan. Here and in other coastal cities, few people were able to flee to high ground before the tsunami waves came ashore. Fortunately, many lives were saved because Japan had established vertical evacuation sites in coastal communities. As illustrated by the evacuation tower in (B), these sites were often the only structures still standing after the tsunami.

(a) © Junichi Sasaki/AFLO/Nippon News/Corbis; (b) U.S. Navy photo by Specialist 3rd Class Alexander Tidd

public education programs so that residents and tourists know how to respond when an alert is issued. Planners in the Pacific Northwest are also working with local communities to decide what type of vertical evacuation sites are best suited for their individual needs. Since constructing dedicated evacuation towers is expensive and often considered unattractive, planners are considering building earthen berms, or embankments. Berms are not only less expensive, but can be landscaped to make them aesthetically pleasing, and can also serve as public parks. Hopefully, the Pacific Northwest and other regions around the world that are at risk from tsunamis will have completed their preparations before the next major subduction zone earthquake.

cause a rock body to repeatedly accumulate strain and then rupture. Geologists use the term **seismic gap** to refer to those sections of an active fault where the strain has not been released for an extended period of time. Seismic gaps can be useful in predicting what areas are most likely to experience a large earthquake. For example, the map in Figure 5.33 shows the areas along the Alaskan subduction zone where strain has been released in modern times by major earthquakes. Based on the location of the ruptured areas, geologists have identified two seismic gaps where a major earthquake is most likely to strike. A similar situation exists along the Cascadia subduction zone in the Pacific Northwest. Here the entire subduction zone can be considered a seismic gap since a major earthquake has not taken place for nearly 300 years.

Although seismic gaps are good for indicating the potential of future earthquakes, statistical estimates are even more useful as they provide information on what to expect over a specific time period. One statistical approach involves the creation of *seismic hazard maps*, which incorporate information on past seismic activity, magnitudes, and displacement rates on faults. For example, the seismic hazard map of the United States in Figure 5.34 shows the potential strength of

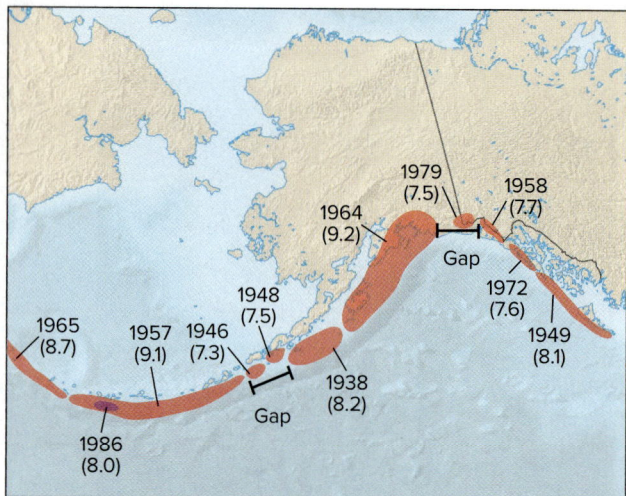

FIGURE 5.33 Map showing the location of seismic gaps along the Alaskan subduction zone. Note how the gaps represent areas where little of the strain along the plate boundary has been released by major earthquakes in modern times. Therefore, the next major earthquake is likely to occur in one of these seismic gaps.

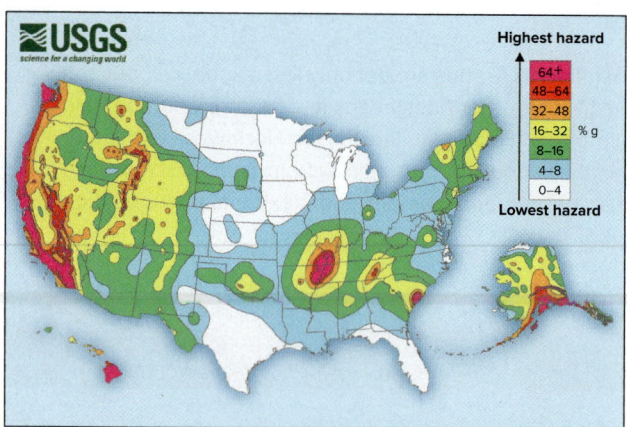

FIGURE 5.34 Seismic hazard map of the continental United States prepared by the U.S. Geological Survey. The map indicates the strength of the horizontal ground motion, relative to gravity, with a 2% chance of occurring over a 50-year period.

horizontal ground motion compared to the strength of gravity—commonly called *g-force*. The map colors represent different levels of ground motion that have a 2% probability of being exceeded in a 50-year period. For example, in the New Madrid seismic zone along the Mississippi River we see that the highest-order color (magenta) is found in the central portion of the seismic zone. Based on the map legend then, this area has a 2% chance over the next 50 years of experiencing lateral ground forces that are at least 64% or 0.64 times as strong as gravity. Despite their statistical nature, these types of projections can be quite useful to local governments in developing building codes that improve the ability of structures to resist lateral ground motion.

Statistical methods can also be used to estimate the probability of earthquakes with a certain *magnitude*. For example, the map in Figure 5.35A shows the percent probability of an earthquake of magnitude 6.7 or larger on individual faults within the San Francisco area between 2014 and 2043. Taken as a whole, seismologists estimate that this region has a 72% chance of a magnitude 6.7 or greater earthquake over the 30-year period. The basis behind these estimates is illustrated in the accompanying graph (Figure 5.35B). Notice how the level of seismic activity along the various faults in the San Francisco area abruptly decreased shortly after the great earthquake in 1906. Probabilities are then calculated from the frequency and magnitude of these earthquakes along with the amount of strain the rocks have accumulated.

One of the problems with earthquake probabilities is that historical records only go back a few hundred years, and the number of large earthquakes within this period is typically small. This tends to lower the confidence level of statistical projections. In recent years, however, geologists have been able to extend the record farther back in time by finding evidence for large earthquakes within layers of sediment. For example, radiometric dating techniques (Chapter 1) can be used to assign specific dates to buried sand blows and layers of sediment that have been offset, both of which form as a result of large earthquakes.

Although scientists' ability to predict earthquakes is improving, it is important to keep in mind that predictions can only be made for faults where data are available. This lesson was driven home in 1994 when a magnitude 6.7 quake struck the Los Angeles area along a "blind" fault previously unknown to geologists. Despite these limitations, scientists, engineers, and planners have made great progress over the past several decades in minimizing the risk from earthquakes, which is the topic of this chapter's final section.

Reducing Earthquake Risks

A common question students ask is whether scientists could prevent major earthquakes by somehow creating small earthquakes, thereby preventing strain energy from reaching dangerously high levels. Studies have shown that injecting water into the subsurface can initiate small earthquakes. Although a technique might someday be developed to prevent the buildup of strain, it would be of little use in areas where the strain is already at high levels. This leaves us with the fact that earthquakes are simply inevitable in certain tectonic settings. Our best solution is to reduce the risks associated with earthquakes by designing more earthquake-resistant structures. Another is to develop early warning systems so that critical preparations can be made before the shaking begins. We will begin by examining some of the engineering designs that have proven effective at minimizing structural damage.

Seismic Engineering

Over the last century engineers have learned how to design structures that are more resistant to the ground shaking associated with seismic waves. This has not only led to a reduction in structural damage and property losses, but it has also

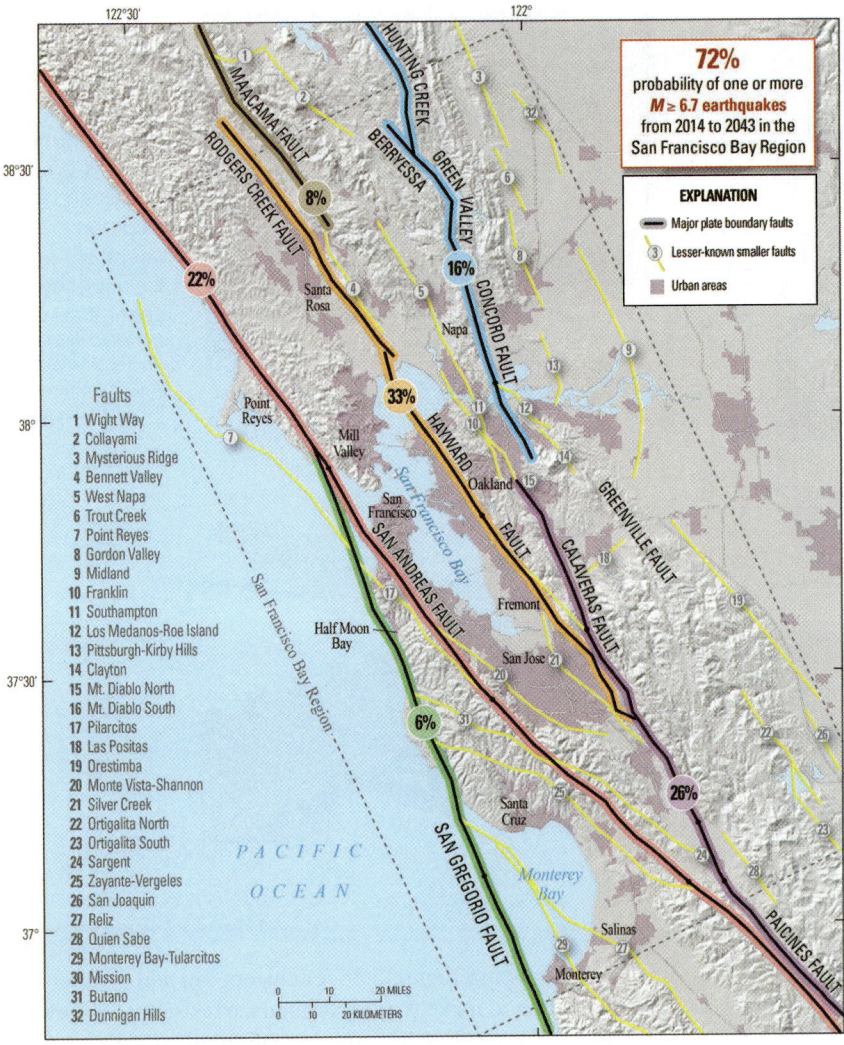

A

FIGURE 5.35   Map (A) showing earthquake probability estimates for individual faults within the San Andreas fault zone near San Francisco. Estimates are based on the frequency and magnitude of earthquakes over time as shown in the graph (B). Note in the graph how seismic activity abruptly decreased after the great quake in 1906. Overall, the region has a 72% chance of experiencing a 6.7 magnitude or larger quake between 2014 and 2043.
Courtesy of the U.S. Geological Survey. Used with permission

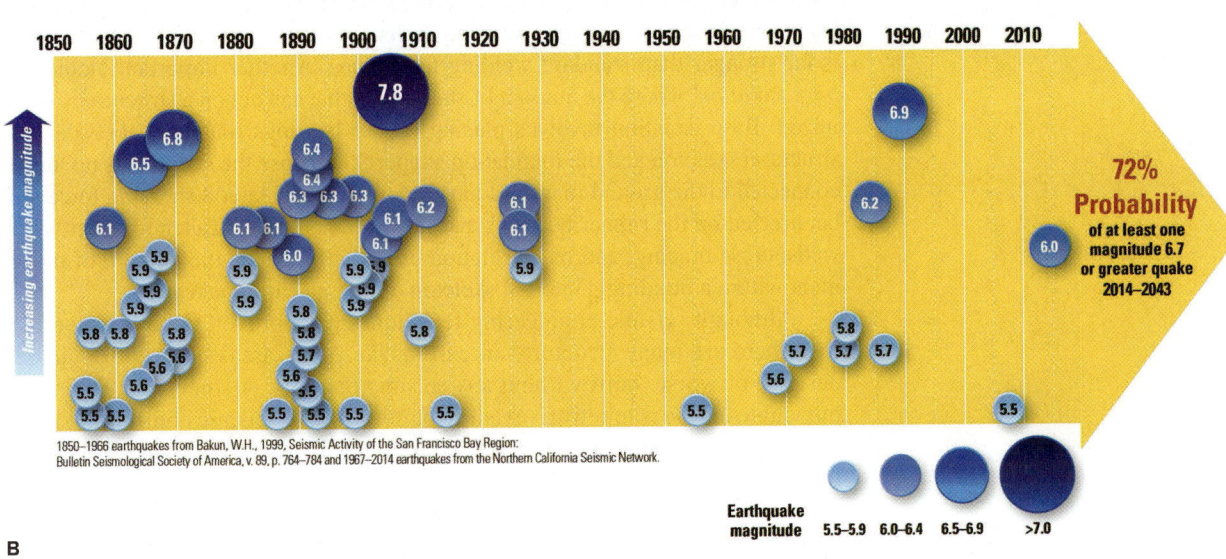

B

reduced the loss of life. The basic approach in this branch of engineering, known as *seismic engineering*, is to: (a) provide greater structural strength with respect to the lateral shearing forces generated during an earthquake; and (b) minimize the amount of shear force that can develop on the structure in the first place.

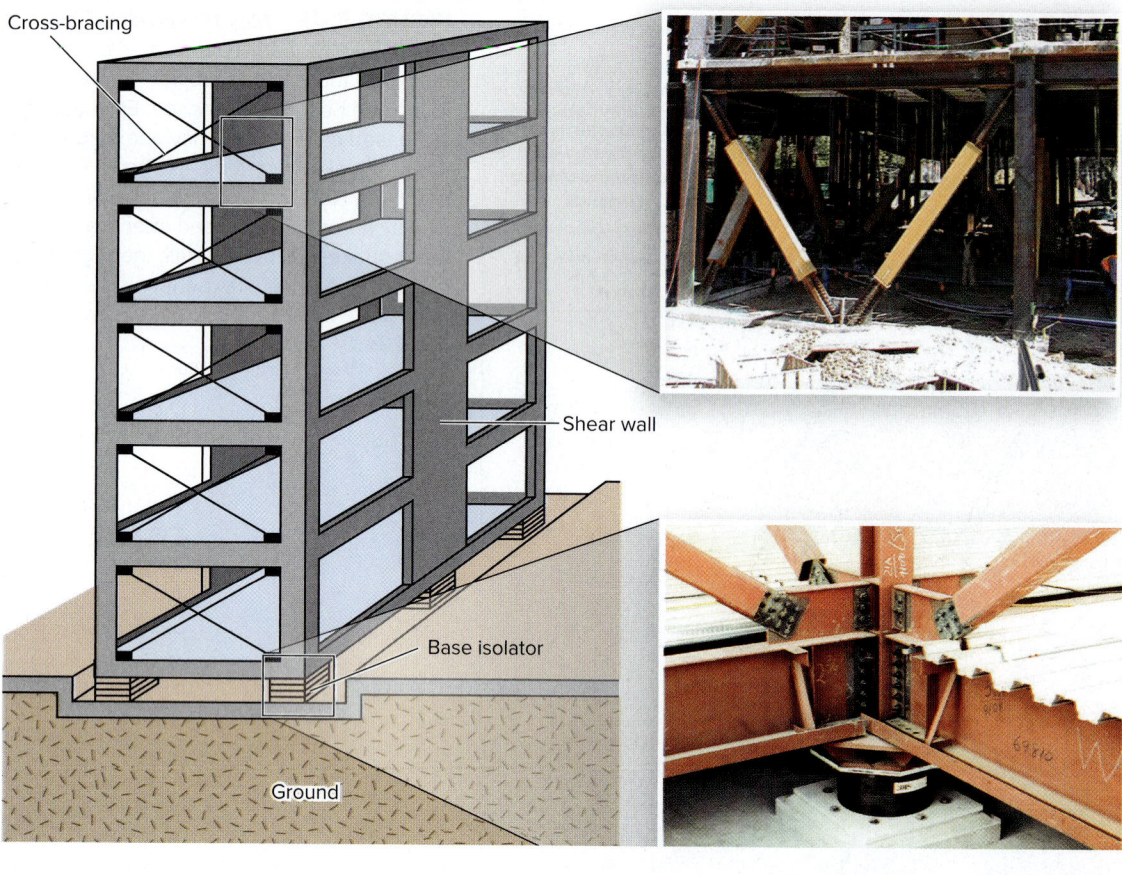

Cross-bracing

Shear wall

Base isolator

Ground

FIGURE 5.36 Seismic engineering involves adding elements to structures that reduce risk of damage during an earthquake. The addition of cross-bracing, shear walls, and bolted members gives the skeleton greater strength against lateral shear forces, whereas base isolation reduces the amount of shear that will act on the structure. (Top) Earthquake Engineering Research Center, University of CA, Berkeley; (Bottom) William Godden/Earthquake Engineering Research Center, University of CA, Berkeley

As shown in Figure 5.36, engineers can add a variety of structural elements to a building to make it more earthquake resistant. Of particular importance is the addition of cross-bracing and specially built *shear walls*, which give a building greater strength against the lateral shearing forces generated during an earthquake. Engineers have also learned that using bolts to connect structural steel members is far stronger than standard welding techniques. Another important element is *base isolation*, where the amount of shear force that can act on a structure is minimized. Base isolation involves placing rubber bearings (dampers) between the structural skeleton and the foundation supports. Because the skeleton is no longer connected to the ground in a rigid manner, more of the lateral shearing force will be directed on the rubber bearings and less to the building. Taken together, shear walls, cross-bracing, and base isolation can greatly reduce the amount of movement within a building's internal skeleton during an earthquake.

Although seismic engineering techniques have proven to be highly effective, there are many structures in earthquake-prone areas built with outdated designs, or worse, built without any seismic controls. A somewhat expensive, but viable option is to retrofit existing buildings with seismic controls as shown in Figure 5.37. Notice how an external frame with cross-braces can be attached to the outside of a building, providing strength against lateral shear. Rubber bearings can also be placed between the structural skeleton and its original foundation supports, thereby providing base isolation. Retrofitting projects are most common for those buildings whose continued operation would be critical in the aftermath of a major earthquake. Examples include emergency command and control centers, hospitals, and police and fire stations. Retrofitting is also becoming common in areas where people have only recently realized that they lie in an area with a history of major earthquakes. This typically occurs in places that were settled *after* the last major earthquake, thus no one ever saw much need to incorporate seismic engineering. Examples include the Pacific Northwest and New Madrid seismic zones (see the hazard map in Figure 5.34), where a powerful earthquake has not occurred in living memory.

Recall that steel-reinforced concrete columns are prone to brittle failure when they are forced to flex beyond their elastic limit during an earthquake. This presents a major problem since these types of columns are not only used in buildings, but as highway and bridge supports. Failure here can lead to serious disruptions in transportation networks. As shown in Figure 5.38, engineers have developed new designs in which the vertical steel rods are wrapped in a spiral fashion by a single rod. After the concrete is poured and allowed to harden, the entire column is then wrapped with a steel jacket or sleeve. The result is a support where the concrete is far less likely to fail in a brittle manner when forced to flex during an earthquake. Note that existing highway columns can be retrofitted with steel jackets, but nothing can be done to the reinforcing rods already set in concrete. Finally, notice in Figure 5.38 how the road deck can be tied to the columns using cabling devices.

Early Warning Systems

Although humans are unable to prevent earthquakes, or even provide short-term predictions, we do have the technology to create early warning systems that reduce the amount of damage and loss of life. Recall that when an earthquake begins, P- and S-waves leave the focus at the same time, but since P-waves travel faster, they arrive first at any given location. In addition, P-waves cause very little damage compared to the highly destructive S-waves and surface waves that follow. As illustrated in Figure 5.39, early warning systems involve placing sensors throughout a region that are designed to detect the first P-wave from an earthquake. Once a sensor detects a P-wave, it automatically sends an electronic signal to a processing center, which then quickly calculates the earthquake magnitude, epicenter location, and expected level of ground shaking across the region. Because seismic waves travel through shallow earth materials at 0.5 to 3 miles per second (0.8–4.8 km/sec), whereas electronic signals travel almost instantaneously, an alert can be sent out ahead of the destructive S-waves and surface waves (Figure 5.39).

Depending on the distance back to the epicenter, the early warning may range from a few seconds to a minute or more (beyond 1 minute the epicenter is usually far enough away that damage is minimal). A warning on the order of seconds may not seem significant, but it can be enough for people to seek shelter or pull off the highway and avoid accidents. It may also be enough time for doctors to suspend doing delicate surgery, or for people doing hazardous work to keep from being injured. An early warning can also allow utility and transportation networks to shut down in a controlled manner. For example, only seconds are needed for preprogrammed systems to close valves on gas lines, thereby reducing the risk of uncontrolled fires. Likewise, trains can be programmed to automatically stop. Electric utilities can also shut down critical control systems on electrical grids and at power plants. This not only prevents damage to the systems themselves, but allows electrical service to be restarted much more quickly.

Japan began operating the first early warning system in the 1980s, but it was used only for shutting down the country's high-speed rail system in the event of an earthquake. After a 1995 quake that killed 5,500 people and crippled the infrastructure of the city of Kobe, the Japanese government invested $600 million

A

B

FIGURE 5.37 Existing buildings can be strengthened against lateral shear by adding an external skeleton (A), whereas the amount of shear force can be reduced using base isolation (B).

(a–b) Earthquake Engineering Research Center, University of CA, Berkeley

A

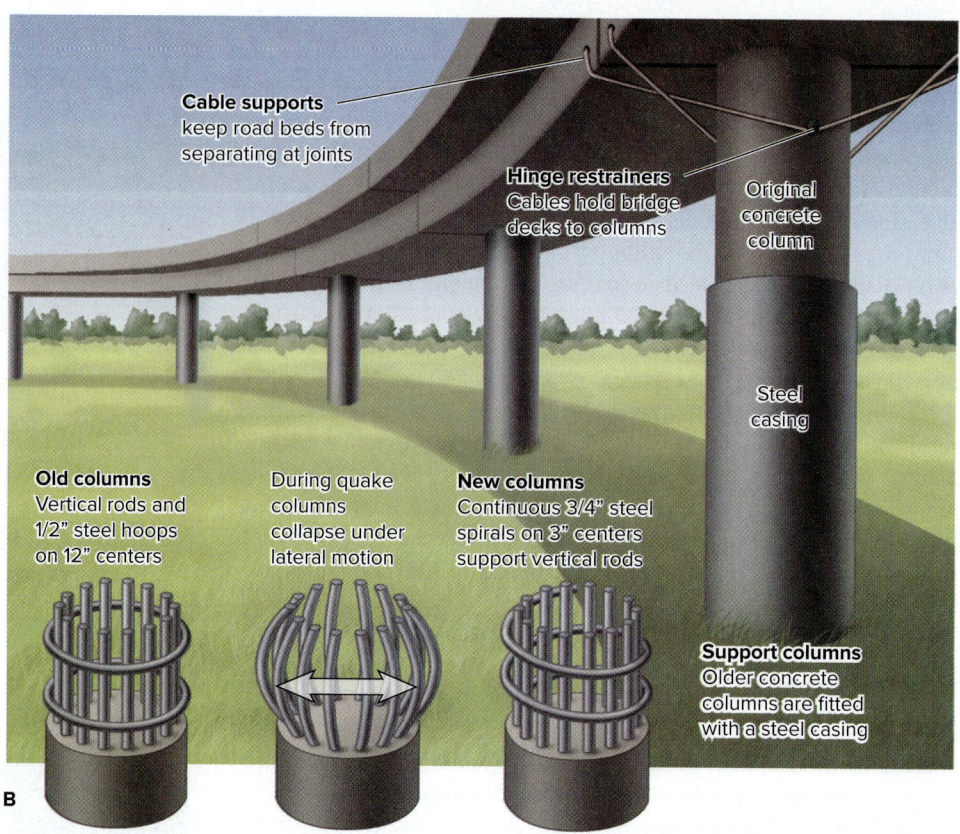

B

FIGURE 5.38   Lateral motion during an earthquake can cause steel-reinforced concrete columns to flex such that they reach their elastic limit, resulting in failure (A). One solution is to wrap the columns with a steel jacket (B) and use a spiral wrapping technique on the reinforcing rods, thereby reducing the chance of failure.

(a) Earthquake Engineering Research Center, University of CA, Berkeley

FIGURE 5.39   Earthquake early warning systems make use of motion sensors to detect the arrival of the first P-wave. Sensors then transmit electronic signals to a processing center that calculates expected ground shaking across the region. From there an alert is sent out ahead of the destructive S-waves and surface waves.

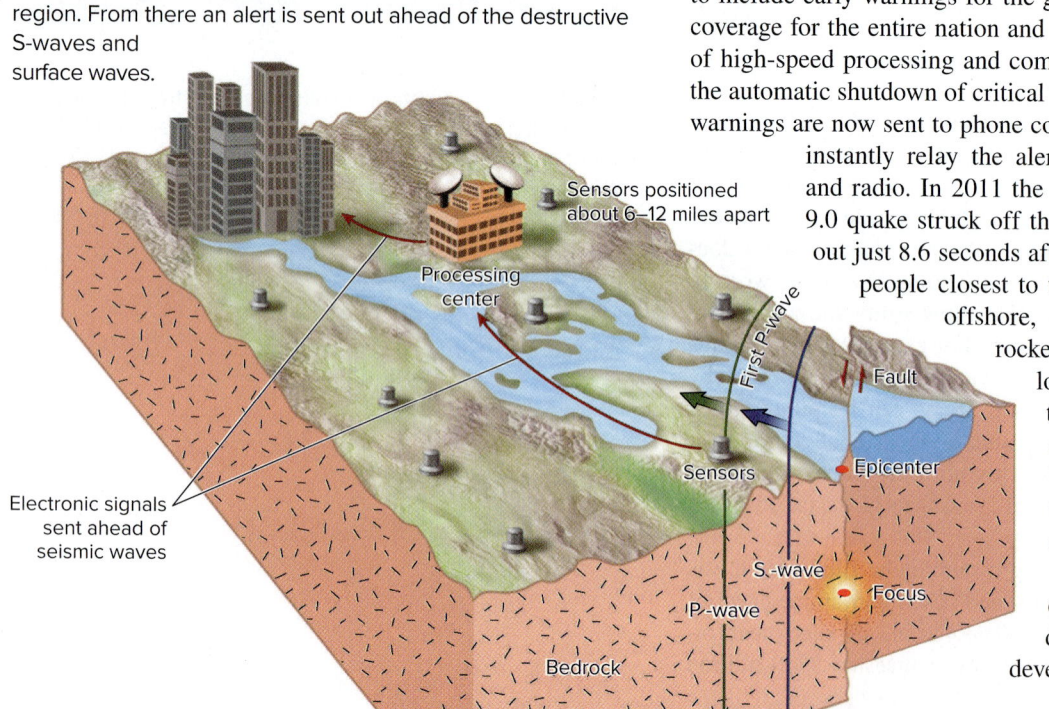

to improve and expand its early warning system. This new system was initially designed to provide early warnings for a select group of government agencies, businesses, and schools. In 2007, after years of testing, Japan expanded its system to include early warnings for the general public. The system currently provides coverage for the entire nation and consists of over 3,600 sensors and a network of high-speed processing and communication centers. In addition to triggering the automatic shutdown of critical utility and transportation systems, earthquake warnings are now sent to phone companies and broadcasting stations, who then instantly relay the alert to the public via cell phones, television, and radio. In 2011 the system was fully tested when the magnitude 9.0 quake struck off the coast of Japan. Automatic alerts were sent out just 8.6 seconds after the first P-waves were detected. This gave people closest to the epicenter, which was 80 miles (130 km) offshore, about 10 seconds warning before being rocked by the seismic waves. Residents of Tokyo, located 230 miles (370 km) from the epicenter, received the warning about 60 seconds ahead of the seismic waves. After the quake it became quite clear that the early warning system saved many lives and helped minimize damage to the nation's infrastructure.

In the United States, the U.S. Geological Survey (USGS) began working with academic, public, and private partners in 2006 to develop an early warning system for California.

By 2012 a prototype system, called *ShakeAlert*, was put in operation that made use of the state's existing network of 400 sensors. This system sends alerts on a test basis to select emergency response, public utility, and transportation networks in California. After 2 years of testing, the USGS developed an implementation plan that outlines the steps needed to complete the system and to begin issuing public alerts. If fully implemented, ShakeAlert would send announcements out via existing public emergency alert systems as well as smartphone applications and social media providers. For computer and smartphone users, a map will pop up on their screen showing the seismic waves as they approach the user's known position. In addition to the map, a voice would count down the time remaining before the waves reach that position.

As of 2016, the USGS was seeking $38 million to complete the ShakeAlert system and $16 million for annual operations. The U.S. Congress recently approved $8.2 million in funding for the program—compare this to the $600 million Japan invested for its national system. Once implemented in California, researchers hope to expand ShakeAlert into Oregon and Washington. Note that earthquake early warning systems are currently operating in parts of Mexico, Taiwan, Turkey, Romania, China, Italy, and Switzerland.

With respect to tsunamis, the United States began developing an early warning system in 1946 after a tsunami killed 170 people on the Hawaiian Islands. This system has been continuously improved and updated over the years, particularly after tsunamis struck Hawaii again in 1960, and the Oregon and northern California coasts in 1964. With the cooperation of numerous Pacific nations, the system now maintains a network of seismograph stations to detect earthquakes with the potential for generating a tsunami, and a series of deep-ocean buoys that can actually detect passing tsunami waves. When a potential threat is detected, an electronic message is sent over the network to various coastal centers around the Pacific. From there, an alert is sent to radio and television broadcasters, and emergency sirens and public address systems are activated in coastal communities. To increase the effectiveness of the system, a public education program was included so that when citizens hear the warnings, they know to immediately seek higher ground. Finally, in the aftermath of the 2004 tsunami that killed 230,000 people in the Indian Ocean, the United Nations led an effort to install an early warning system modeled after the one in the Pacific. The Indian Ocean system, which became operational in 2006, consists of seismograph stations, deep ocean buoys, and 26 communication centers in different countries. In 2012, a magnitude 8.6 subduction zone quake struck near the same area as the 2004 event. Fortunately, the quake did not generate a major tsunami, but the warning system was successful in providing an alert within 5 minutes to coastal communities closest to the epicenter.

Planning and Education

A key element in reducing the risk of earthquakes is to have all levels of society, from school-aged children up to emergency management officials, know what to do before, during, and after an earthquake. The first step is to determine the level or severity of risk in a given area. In the United States this task has largely been performed by the USGS through the construction of various hazard maps, particularly the seismic hazard map shown in Figure 5.34. Based on the hazard assessment, state and local government agencies will typically develop building codes that require appropriate levels of seismic engineering in buildings and other structures. The goal here is to reduce structural damage and the loss of life. Also important are local zoning ordinances that keep homes and key facilities, such as hospitals, schools, dams, and fuel storage areas, from being built across known faults.

In addition to taking steps to minimize structural damage, it is equally important that individual citizens, businesses, and local emergency services know exactly what to do both during and after an earthquake. Such planning is critical because unlike most other natural disasters, earthquakes occur without any warning. Individual citizens therefore need to know in advance how to seek protection in a variety of situations, such as being at work, walking down a sidewalk, or driving a car. Prior to an earthquake people should take steps that will minimize injuries and damage to their homes. This may include securing loose objects such as free-standing cabinets, water heaters, and other heavy objects. Family members should also plan on where to best seek shelter in different places within the house, and how to shut off gas and electricity when the shaking stops. Since emergency services will probably not be available, people need to be prepared to take care of themselves and family members for a minimum of several days after an earthquake. This means that emergency supplies of food, water, and medicine should be on hand at all times, and that family members should all have had first-aid training.

With respect to local government and emergency services, there are a number of things that can be done in advance of an earthquake that will help minimize property damage and the loss of life. For example, because water main breaks are all but inevitable, some municipalities have built underground storage systems so that water will be available for firefighting efforts. Planning can also help ensure that hospitals are equipped to treat large numbers of injuries, and that rescue crews have the specialized equipment needed for extracting people from collapsed buildings. Public utilities can make use of devices that detect P-waves and cause gas lines to shut down automatically, thereby reducing the threat of fire.

Although planning is critical, practice drills are important so that various professionals are more familiar with their assigned tasks in advance of an actual earthquake. Drills are also valuable in that they reveal flaws in the planning, which can then be corrected. Of course, practice drills are just as important for individual family members as they are for emergency personnel. Not surprisingly, in areas where earthquakes are frequent it is common to find high levels of earthquake planning and training as well as building codes that require seismic engineering controls. On the other hand, in places where large earthquakes occur less frequently, the level of planning and preparation is often quite low. Should you discover that you live in an area at risk from earthquakes and want to learn more on how to protect you and your family, just type "earthquake preparedness" into any Internet search engine. This should lead you to a number of governmental websites that provide detailed information on the subject.

SUMMARY POINTS

1. A rock body placed under a force will deform and accumulate strain. When deformation exceeds the rock's elastic limit, it will suddenly rupture, releasing the strain in the form of vibrational wave energy called seismic waves.

2. Seismic waves travel outward from the rupture point (focus) and cause the ground to vibrate or shake in what is called an earthquake. Because seismic waves lose energy as they travel, the greatest shaking on the surface is usually at the epicenter, which is directly above the focus.

3. Seismic waves are classified based on how they force rock particles to vibrate. Body waves (P- and S-waves) travel through solid earth and then generate surface waves when reaching the land surface. P-waves travel the fastest, but cause very little damage compared to the highly destructive S-waves and surface waves.

4. In tectonically active areas, earthquakes occur repeatedly because of the cyclic manner in which strain accumulates and is then released. Large earthquakes occur when rocks accumulate greater amounts of strain energy before rupturing.

5. The Mercalli intensity scale is a qualitative means of measuring the strength of an earthquake using human observations. Magnitude scales are quantitative measures of the amount of ground motion and are based on measurements taken from a seismograph.

6. Most large earthquakes occur along convergent or transform plate boundaries as rocks are stronger under compressional and shear forces compared to tension. Large earthquakes sometimes take place in the interior of plates when tectonic forces cause strain to accumulate along ancient faults.

7. Seismologists are currently unable to use precursor data to make reliable earthquake predictions on a short-term basis. More successful are long-term predictions using statistical probabilities that make use of historical earthquake frequency and magnitude.

8. Structural failure of buildings due to ground shaking is the single greatest cause of human deaths and property loss in earthquakes. Other related hazards include ground fissures, landslides, fires, and tsunamis.

9. Factors that affect the amount of structural damage due to shaking include magnitude, wave attenuation, resonance effects in buildings, ground amplification, and liquefaction.

10. Buildings and structural supports commonly fail during earthquakes if they do not have adequate strength against lateral shear force. Key elements in seismic engineering include cross-bracing, base isolation, and spiral-wrapped support columns.

11. Loss of life and property damage due to earthquakes can be reduced through seismic engineering controls, early warning systems, and better preparedness within communities.

KEY WORDS

body waves 128
earthquake 125
earthquake precursors 147
elastic limit 124
elastic rebound theory 125
epicenter 127
focus 125

ground amplification 142
ground fissures 144
inertia 128
intraplate earthquakes 133
liquefaction 143
Mercalli intensity scale 130
moment magnitude scale 131

natural vibration frequency 140
primary waves 128
resonance 141
Richter magnitude scale 131
secondary waves 128
seismic gap 149
seismic waves 127

seismographs 128
subduction zone
   earthquakes 135
surface waves 128
tsunami 146
wave attenuation 141

APPLICATIONS

Student Activity

Go to an area with lots of trees and find a small, dead branch or twig. Then see if you can find a downed tree that still has green leaves on its branches. Using a small kntife, trim off a branch that is about the same size as the dead one you collected earlier. With both hands, grasp the ends of the green (fresh) twig and try bending it slightly; then release the force so it goes back to its original shape. Try it again, only this time bend the green twig as far as you can. Did it break, or did it stay bent? Next, take the dry twig and bend it slightly. Does it go back to its original shape also? Now bend the dry twig until it breaks. Did you feel any vibrations in your hands?

Which stick has the higher elastic limit? Can you describe which stick behaves in a ductile manner and which exhibits brittle deformation? Explain how this relates to why some rocks deform by fracturing and others by plastic flow. When a rock body ruptures during an earthquake, are the rocks in a ductile or brittle state? Why?

Critical Thinking Questions

1. How are seismic and sound waves similar? How are they different?

2. Since earthquakes release the strain energy that accumulates in rocks, why do some fault segments experience earthquakes on a recurring basis?

3. Can an earthquake rank low on the Mercalli intensity scale, but at the same time rank relatively high on the moment magnitude scale? Describe the conditions necessary for this to occur.

Your Environment: YOU Decide

© AFLO/Mainichi Newspaper/epa/Corbis

In this chapter you learned that, with the exception of tsunamis, the loss of human life and property damage in an earthquake are directly related to the number of structures that fail. There is a lot that can be done to strengthen existing buildings and make new construction safer in earthquake-prone areas, but sometimes this does not take place.

Although there has been considerable research in the area of short-term earthquake predictions, we still cannot make such predictions. However, we have been successful in determining those areas that are at high risk from earthquakes. If you were in charge of granting research money, and had only a certain amount to allocate, would you prefer to fund earthquake prediction or improvements in earthquake resistance for both new and old buildings in earthquake-prone areas? Explain your reasoning.

Chapter 6

Volcanoes and Related Hazards

LEARNING OUTCOMES

After reading this chapter, you should be able to:

► Describe the different tectonic settings in which volcanoes form and the basic types of volcanoes and magma that are found in these settings.

► Explain why some magmas explode violently when they breach the surface and are exposed to atmospheric conditions.

► Describe what controls magma viscosity, and explain how viscosity affects the explosiveness of eruptions and the nature of lava flows.

► Explain why water is the key to explosive eruptions, and describe where it originates.

► List the various types of volcanic hazards; then make a separate list of those that can occur even when a volcano is not erupting.

► Describe the different monitoring tools geologists use to make eruption forecasts, and discuss how emergency managers use the forecasts to minimize the effects of eruptions.

Japan's Mount Fuji towering over Ashino Lake. Composite volcanoes such as this are well known for creating beautiful landscapes, but they also generate very powerful and violent eruptions that result in a variety of hazards for people.

Introduction

© StockTrek/Getty Images

As with earthquakes, people have long been fascinated by the destructive power of volcanoes and the various types of hazards they create. In addition to explosive eruptions and rivers of molten lava, volcanoes generate secondary hazards such as falling ash, mudflows, avalanches, and tsunamis. As with almost all natural hazards, however, people tend to become ignorant or complacent with respect to volcanic hazards, particularly whenever several human generations pass between major events. A tragic example where complacency and ignorance led to a horrible disaster is the 1985 eruption of Nevado del Ruiz, a 17,681-foot (5,389-m) volcano, that buried the Colombian town of Armero, located in the Andes Mountains of South America (Figure 6.1).

After nearly a year of minor activity by Nevado del Ruiz, explosive eruptions began in the afternoon of November 13, ejecting approximately 700 million cubic feet of hot rock and ash onto the volcano's snow- and ice-covered flanks. The eruption also released large quantities of water vapor, which quickly condensed and fell as rain. Together, the ejected material and heavy rains caused rapid melting of the snow and ice, forming mudflows that began flowing downslope into the network of streams around the volcano. As the stream valleys merged, the mudflows grew in size, eventually reaching heights of 130 feet (40 m) and speeds of over 30 miles per hour (50 km/hr). One of these massive mudflows swept down a steep and narrow river valley toward the town of Armero, a seemingly safe 46 miles (74 km) away from Nevado del Ruiz's summit. Armero though had been built on a fan-shaped sediment deposit at the mouth of the canyon, where the river empties out onto a broad valley (Figure 6.1). Around midnight, nearly two and a half hours after the eruption began, the giant mudflow reached the base of the canyon and deposited a new layer of sediment on the valley floor, burying most of the city. Of Armero's 28,700 residents, approximately 23,000 were killed and thousands more were injured or left homeless. It was the worst volcanic disaster in Columbia's history. However, much of the loss of life could have been avoided.

Armero is a grim reminder that ignorance and complacency can have serious consequences when it comes to the geologic environment (Chapter 1). Armero was founded by Spanish settlers who selected the site in part because it offered a flat area to build and a steady supply of water from the nearby river (Figure 6.1). Unfortunately, the site turned out to be not so ideal when Nevado del Ruiz erupted in 1595 and then again in 1845, killing hundreds of Spanish settlers when mudflows came roaring down the river valley. Just like its predecessors then, modern Armero was built on relatively fresh mudflow deposits. The major difference was that by 1985 population growth had resulted in many more people living in harm's way. Because 140 years had passed since the last mudflow, most of Armero's residents were either ignorant of the risk or had become complacent. Despite the growing level of volcanic activity leading up to the eruption, government officials and people in Armero tended to believe that the volcano was simply too far away to be much of a threat. Only the scientists seemed to appreciate the fact that the city was built on old mudflow deposits.

In this chapter we will focus on what science can tell us about the different hazards people face when living in volcanically active areas. Although our focus will be on hazards, keep in mind that volcanoes have also been responsible for making Earth habitable. Over eons of time, each volcanic eruption has released carbon dioxide and water vapor into the atmosphere. This carbon dioxide has been critical in maintaining the greenhouse effect that keeps Earth's average surface temperature above the freezing point of water (Chapter 4). It has also been the source of carbon that forms the basis of life as we know it. Equally important to life is the water that moves through the hydrologic cycle (Chapter 11). Volcanic eruptions even created the first landmasses, which were later colonized by plants and animals. Were it not for volcanoes, life as we know it would simply not exist.

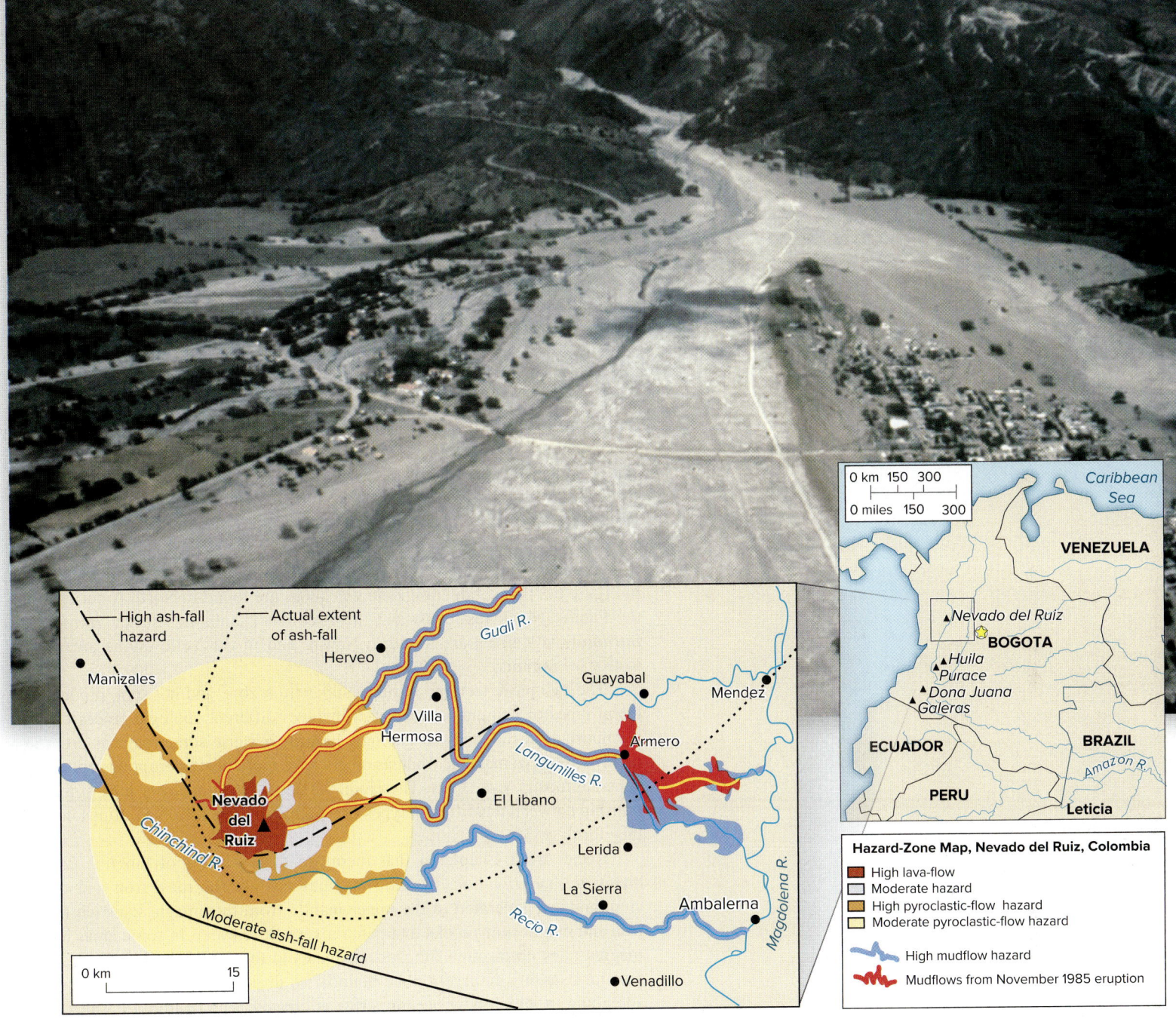

FIGURE 6.1 Three-quarters of the residents of Armero, Colombia, were killed when a massive mudflow came roaring down a canyon and emptied out onto the valley floor where the city was built. The mudflow formed when Nevado del Ruiz, located 46 miles (74 km) away, began to erupt, causing its glacial ice cap to melt.
Photo by R. Janda, Cascade Volcano Observatory/USGS

We will begin by taking a look at the basic science behind volcanic eruptions in order to better understand the hazards we face.

Nature of Volcanic Activity

In the previous chapters you learned that crustal rocks are continuously being transformed from one rock type to another in what geologists call the *rock cycle*. Moreover, a key part of the rock cycle is Earth's system of moving tectonic plates, which are driven by our planet's internal forces. It is primarily along the boundaries of these plates that we find molten rock called **magma** rising up through the lithosphere and eventually cooling to form igneous rock. The newly formed rock can then take a variety of paths through the rock cycle. Our interest in this chapter is the magma that makes it through the lithosphere and erupts at the surface, at which point it is technically called *lava*. This eruption of lava poses several hazards for us humans and, at the same time, produces a variety of landforms; the most familiar of these are what geologists call *volcanoes*. We will start this section by reviewing the relationship between the different types of magma and plate tectonics (Chapter 4).

Magma and Plate Tectonics

Although considerable volcanic activity takes place at divergent boundaries along spreading centers, the familiar cone-shaped landforms called *volcanoes* are mostly found along convergent boundaries, as shown in Figure 6.2. In fact, there are so many active volcanoes associated with the subduction zones around the Pacific plate that this relatively narrow belt is often referred to as the *Ring of Fire*. As described in Chapter 4, the magma for these volcanoes is generated when oceanic plates sink into the asthenosphere. Volcanoes are also found in the interior of plates where rising plumes (columns) of mantle material create what geologists refer to as **hot spots** (Figure 6.2). A hot spot will cause the overlying lithospheric plate to partially melt, creating blobs of magma that rise through weak zones within the plate. Here the type of magma that forms depends on whether the plate setting is oceanic or continental. Note in Figure 6.2 that subduction zone volcanoes are found in North America along the Pacific Northwest coast and the Aleutian Islands in Alaska. *Hot spot* or *intraplate* volcanoes are found in Hawaii and in the Yellowstone area of the Rocky Mountains.

The two basic tectonic settings, subduction zone and hot spot, produce different types of magma because each involves different geologic processes. This is important in the study of geologic hazards because some magmas erupt in a highly explosive manner, whereas others have more benign eruptions and tend to generate mostly lava flows. Therefore, we need to take a closer look at the relationship between the different types of magmas and the tectonic settings in which they form.

Recall from Chapter 4 that there are three classes of magma: *basaltic, andesitic,* and *rhyolitic.* Basaltic magmas contain the most iron and magnesium and are generated in the upper mantle from ferromagnesian-rich (rich in iron and magnesium) rocks that contain very little water. In some areas basaltic magma rises up through lithospheric plates to form hot spot volcanoes. Basalt also forms along divergent plate boundaries during the process called seafloor spreading in which new oceanic crust is created. The resulting basalt-covered lithospheric plates eventually descend into subduction zones where the abundant water acts to lower the melting point of the rocks. As shown in Figure 6.2, this subduction-zone melting at an oceanic-oceanic convergence can lead to basaltic magma that eventually erupts and forms volcanic arcs. However, when the rocks undergo *partial melting,* the magma tends to be andesitic as it contains less iron and magnesium but more SiO_2, a compound called *silica.* Remember that andesitic magmas also form by the process called *assimilation.* As a basaltic magma rises, it causes silicate minerals in the overlying plate to melt, enriching the magma in SiO_2 such that it becomes andesitic in composition. Andesitic magma is also common in oceanic-continental convergences where the magma has to rise through continental plates. Because continental crust is thicker and composed of granitic minerals with even higher proportions of SiO_2, assimilation can lead to rhyolitic magmas that are highly enriched in silica. Rhyolitic magmas also occur at hot spots where mantle plumes rise through thick continental plates.

From a hazard perspective, the varying amount of silica (SiO_2) in basaltic, andesitic, and rhyolitic magmas is important because it plays a key role in determining the explosiveness of volcanic eruptions. Silica is actually made up of SiO_4^{2-} ions that form strong chemical bonds as the magma begins to

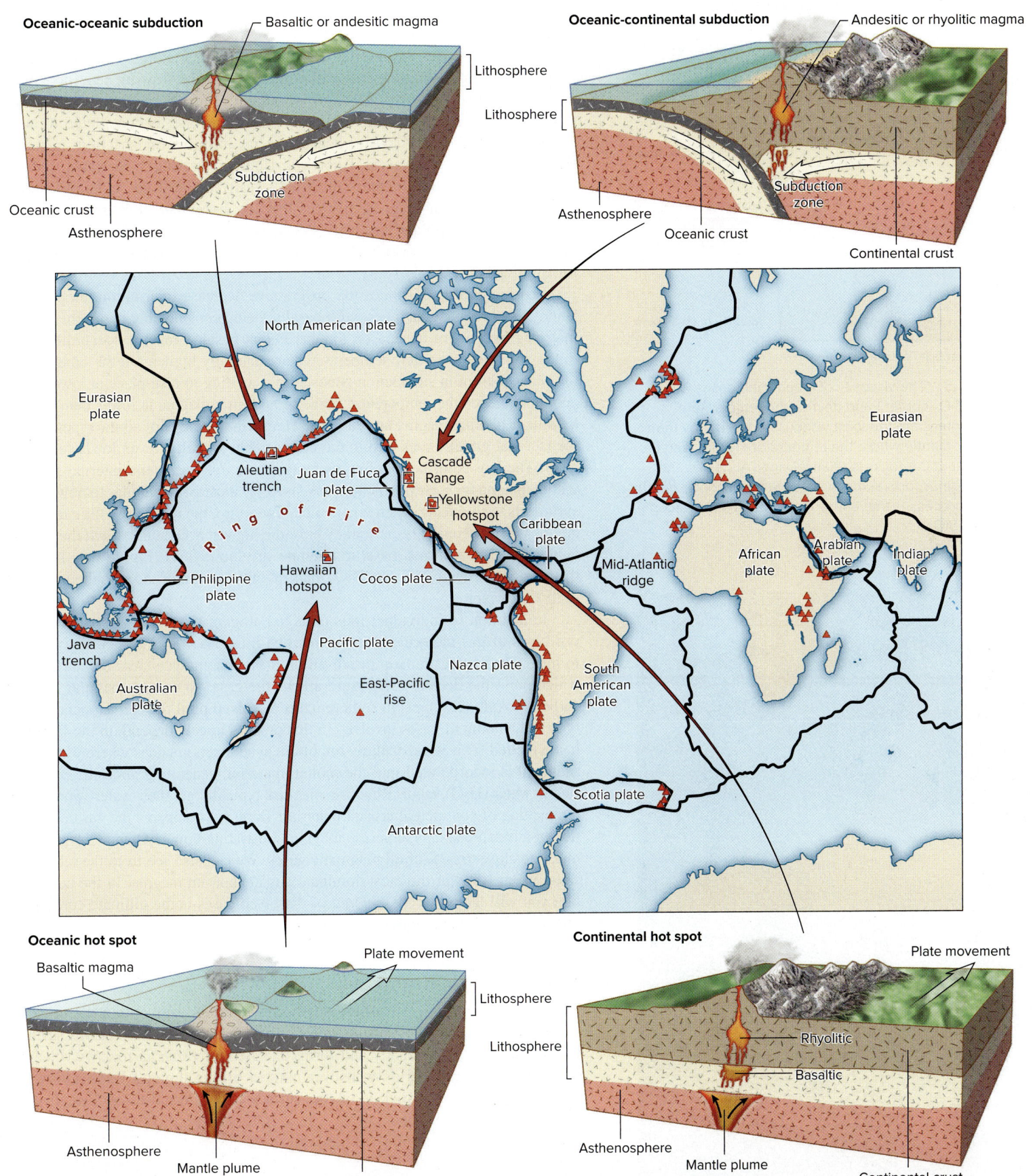

Oceanic-oceanic subduction
Basaltic or andesitic magma
Lithosphere
Lithosphere
Subduction zone
Oceanic crust
Asthenosphere

Oceanic-continental subduction
Andesitic or rhyolitic magma
Lithosphere
Lithosphere
Subduction zone
Asthenosphere
Oceanic crust
Continental crust

North American plate
Eurasian plate
Aleutian trench
Juan de Fuca plate
Cascade Range
Yellowstone hotspot
Caribbean plate
Eurasian plate
Ring of Fire
Philippine plate
Hawaiian hotspot
Cocos plate
Mid-Atlantic ridge
African plate
Arabian plate
Indian plate
Java trench
Australian plate
Pacific plate
East-Pacific rise
Nazca plate
South American plate
Scotia plate
Antarctic plate

Oceanic hot spot
Plate movement
Basaltic magma
Lithosphere
Lithosphere
Asthenosphere
Mantle plume
Oceanic crust

Continental hot spot
Plate movement
Lithosphere
Lithosphere
Rhyolitic
Basaltic
Asthenosphere
Mantle plume
Continental crust

FIGURE 6.2 Location of active volcanoes and their relationship to tectonic plate boundaries. Most volcanoes derive their magma from subduction zones or from the upper mantle in what are referred to as hot spots. Hot spot volcanoes in an oceanic setting produce basaltic magma, and those on continents generate more rhyolitic magma. Depending on the tectonic setting, subduction zone volcanoes will produce basaltic, andesitic, or rhyolitic magmas.

Classification and Flow Characteristics of Volcanic Rocks

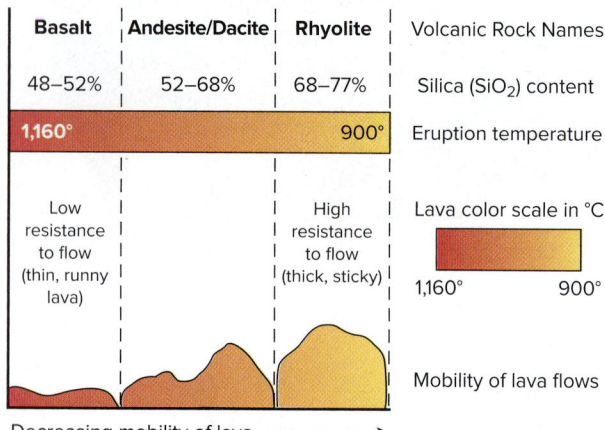

Basalt	Andesite/Dacite	Rhyolite	Volcanic Rock Names
48–52%	52–68%	68–77%	Silica (SiO₂) content

| 1,160° | | 900° | Eruption temperature |

| Low resistance to flow (thin, runny lava) | | High resistance to flow (thick, sticky) | Lava color scale in °C |
| | | 1,160°   900° | |

| | | | Mobility of lava flows |

Decreasing mobility of lava ⟶

FIGURE 6.3   The viscosity of magma increases with increasing SiO₂ content and decreasing temperature. Basaltic magmas are the least viscous because they form in the upper mantle where the temperature is high and the SiO₂ content of the rocks is relatively low. Andesitic and rhyolitic magmas form at much shallower and cooler depths and under processes that cause these magmas to become enriched in SiO₂.

FIGURE 6.4   How easily magma flows depends on its viscosity, which is controlled by its SiO₂ content and temperature. Shown here is high-temperature, silica-poor basaltic magma of relatively low viscosity. Photo taken during the 1989 eruption of Kilauea volcano, Hawaii.
J.D. Griggs/USGS

cool, which increases the internal friction within the magma. Therefore as the SiO₂ content increases, the ability of a magma to resist flowing also increases, a property called **viscosity.** A useful way of understanding viscosity is to think in terms of thick versus thin fluids. For example, water is considered to be a "thin" fluid with low viscosity because it has little internal friction; hence it flows quite easily. However, if you add flour or cornstarch to water, chemical bonds develop that increase the fluid's resistance to flow, making it more viscous. In a similar manner, silica is the key ingredient in magma that controls internal friction and viscosity. As shown in Figure 6.3, of the three types of magma, basaltic magma has the lowest viscosity primarily because it has the lowest SiO₂ content.

Figure 6.3 also illustrates the progressive decrease in eruption temperature among the three magma types, with basaltic being considerably hotter than rhyolitic magma. Part of the reason for this is that basaltic magmas generally form in the upper mantle where temperatures are naturally higher. On the other hand, the water present in subduction zones allows melting to take place at lower temperatures; hence the temperature range of andesitic magma is lower than that of basaltic. Rhyolitic magma has the lowest temperature range of all the types due to the fact that granitic rock within the crust is able to melt at even lower temperatures. The key point here is that as temperature *decreases,* the internal friction within magma *increases,* which means viscosity increases—similar to how maple syrup gets even thicker when it is cooled. Therefore, the reason rhyolitic magmas are the most viscous is because they have the highest SiO₂ content *and* the lowest temperature. In contrast, basaltic magmas are more fluid (Figure 6.4) because they contain around 25% less SiO₂ than rhyolitic magmas and are about 250°C (480°F) warmer.

In addition to individual ions (charged atoms), magma also contains various types of gas molecules that form when minerals within a rock body begin to melt. These gases remain dissolved until the magma gets near the surface, at which point the decreased pressure allows the gases to escape—similar to how carbon dioxide (CO_2) gas in soda remains dissolved until the can is opened. The type and amount of gases involved in this process largely depend on the material being melted. This is particularly important at subduction zones where water-rich sediment is incorporated into the molten material, generating andesitic magmas rich in water (H_2O) vapor. Rhyolitic magmas typically get their water vapor from the melting of continental crustal material. Table 6.1 compares the gas composition of a basaltic and an andesitic magma, illustrating how H_2O and CO_2 together make up close to 90% of all magmatic gases. We can also see from the table that water vapor is by far the most dominant gas in andesite magma. In the next section you will learn that the *amount* of dissolved gases is the ultimate control on the type of volcanic eruption.

Volcanic Eruptions

Recall from Chapter 3 that when magma stays buried deep within the crust, it slowly cools and forms *intrusive* igneous rock, but if it reaches Earth's surface environment, it cools quickly and forms *extrusive* or *volcanic* rock. A volcanic eruption can be defined as whenever magma breaks through Earth's surface. Note that once molten rock is on the surface it is referred to as *lava.* In this section we will look at how the gas content and viscosity of magma determine whether eruptions take place in an explosive or nonexplosive manner.

When rock begins to melt, droplets of magma form and eventually coalesce into larger blobs, which then rise because their density is lower than that of rock. Should enough magma accumulate, it can form a zone of molten material called a **magma chamber,** as illustrated in Figure 6.5. Because this accumulation occurs at considerable depth, the magma chamber must support the weight of the overlying column of rock, referred to as *confining* or *overburden pressure* (Chapter 4). The key point here is that this confining pressure gives molten material within the magma chamber a tremendous amount of fluid pressure. Therefore, as the pressurized magma rises and encounters less confining pressure, it tries to expand by pushing outward on the surrounding rock. Should the magma get close enough to the surface, its fluid pressure may be sufficient to force open fractures and faults. This effect not only creates small earthquakes, but can also provide the magma with a pathway to the surface. If the overlying rocks at some point are no longer capable of containing the fluid pressure, then significant amounts of magma can make its way to the surface, resulting in a volcanic eruption.

Once magma reaches the surface, it is free to expand since the only confining pressure now is atmospheric pressure (Figure 6.5). Because the confining pressure is now negligible, the dissolved or compressed gas within the magma forms bubbles and escapes into the atmosphere. However, should the pressurized magma contain significant amounts of dissolved gas, this rather sudden encounter with the atmosphere will allow large volumes of gas to rapidly decompress in an explosive manner—similar to what happens if you shake a bottle of champagne and then pull the cork. In some eruptions the volume of gas is so great that this decompression results in a cataclysmic explosion that literally blows much of the existing volcano skyward. For example, consider that when 1 cubic meter of rhyolitic magma, with just 5% dissolved water vapor, moves from being confined at depth to the surface environment, it will expand to 670 cubic meters! On the other hand, when magma contains relatively small amounts of gas, like a can of soda gone flat, it will simply flow out onto the surface as a lava flow. These two basic eruption styles, explosive and nonexplosive, are nicely illustrated in Figure 6.6.

It should now be clear that subduction zone volcanoes tend to erupt in an explosive manner because their andesitic magmas usually contain large volumes of dissolved gas—mostly water vapor. Hot spot volcanoes located in the interior of continental plates also erupt violently, as their rhyolitic magmas normally contain abundant water vapor. In addition to the dissolved gas content, the more viscous nature of andesitic and rhyolitic magmas contributes to the explosiveness of eruptions by making it more difficult for the decompressing gases to escape. In contrast, hot spot volcanoes in oceanic settings typically produce low-viscosity basaltic magma with relatively small amounts of gas; hence they generally do not erupt violently. Note that any volcano that erupts in an area where water is abundant in the subsurface, as in oceanic settings, is capable of generating a large steam explosion.

From Figure 6.6 one can see that during explosive eruptions most of the lava is ejected into the atmosphere, where it cools into different-sized particles. Because these eruptions also create pulverized rock, geologists use the term **pyroclastic material** when referring to the combination of solidified lava and pulverized rock particles—though some geologists prefer the term *tephra*. The coarsest pyroclastic fragments quickly fall from the sky and are deposited on the volcano itself. The finest pyroclastic material, called **volcanic ash,** can travel hundreds or even

TABLE 6.1   Percent breakdown of gases escaping from magmas at a hot spot and a subduction zone volcano. Water vapor and carbon dioxide are typically the most abundant gases, but notice the dominance of water vapor in the andesitic magma commonly found at subduction zones. Also note how the andesitic magma is relatively cool compared to the basaltic magma.

Volcano	Kilauea Summit (Hawaii)	Mount St. Helens (Washington)
Magma Type	Basaltic	Andesitic
Tectonic Style	Hot spot	Subduction zone
Temperature	1,170°C	802°C
H₂O	37.1	91.6
CO₂	48.9	6.64
SO₂	11.8	0.21
H₂	0.49	0.85
CO	1.51	0.06
H₂S	0.04	0.36
Other	0.16	0.28

Source: R.B. Symonds, Rose, W.I., Bluth, G., and Gerlach, T.M., 1994, "Volcanic gas studies: Methods, results, and applications," in Carroll, M.R., and Holloway, J.R., eds., *Volatiles in Magmas: Mineralogical Society of America Reviews in Mineralogy,* 30: 1–66.

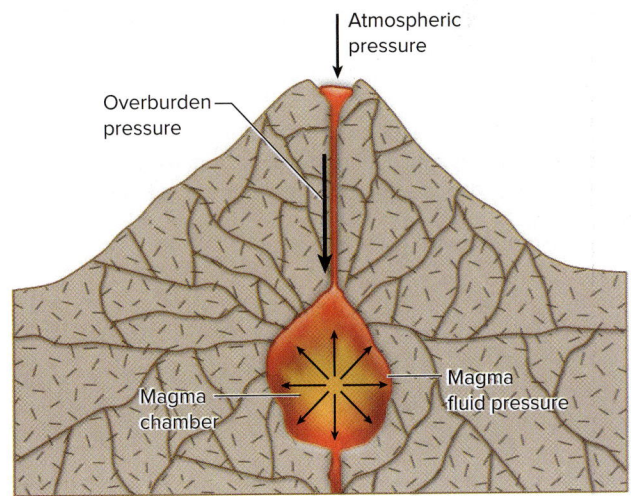

FIGURE 6.5   Molten rock eventually accumulates in what is called a magma chamber. The weight of the overlying column of rock creates overburden (confining) pressure, which is offset by the magma's fluid pressure. As magma continues to rise and encounters less overburden, the fluid pressure is able to open fractures, creating possible pathways to the surface. At the surface the pressurized magma is allowed to expand freely.

A Mount St. Helens, Washington

B Kilauea, Hawaii

FIGURE 6.6 Volcanic eruptions occur when pressurized magma breaches the surface. Explosive eruptions (A) are associated with more viscous and gas-charged magmas in which the dissolved gases rapidly decompress, ejecting rock and ash into the atmosphere. Nonexplosive eruptions (B) are associated with hot fluid magma containing less dissolved gas; in which case the eruption generates lava fountains and lava flows.
(a) USGS (b) J.D. Griggs/U.S. Geological Survey

thousands of miles before falling back to Earth's surface. In some eruptions ash can reach the upper atmosphere and circle the entire globe.

Volcanic Landforms

In this section we will briefly examine the types of landforms commonly created during volcanic eruptions. This is important because volcanic landforms can provide information as to the relative proportion of lava and pyroclastic material in ancient eruptions, which then tells us something about the gas content and viscosity of the magma itself. Using such information geologists can assess the potential hazards humans may face from volcanoes whose last eruption was prior to human settlement.

Lava Flows

When lava flows out onto the land surface, it eventually cools and solidifies into an igneous rock body called a **lava flow.** The shape and thickness of lava flows vary widely due to differences in the viscosity and volume of lava being extruded and the slope of the terrain over which it flows. For example, Figure 6.7 illustrates how low-viscosity basaltic magma can travel long distances and spread into thin sheets in areas where the terrain is relatively flat. On the other hand, more viscous andesitic lava flows are generally thicker and do not travel as far.

Low viscosity

High viscosity

A Basalt, Puʻu Oʻo, Hawaii

B Andesite, Lassen, California

C Rhyolite, Long Valley, California

FIGURE 6.7   The shape and thickness of a lava flow depend upon the magma's viscosity and the slope of the land surface. Low-viscosity basaltic lava flows (A) tend to travel greater distances and spread out into thin sheets in areas where the terrain is more relatively flat. Andesitic lava is more viscous and creates thicker flows (B) that travel relatively short distances. Highly viscous rhyolitic lava (C) hardly flows at all, but rather builds into a mound-shaped feature called a lava dome.

(a) Photograph by J.D. Griggs, USGS Photo Library, Denver, CO (b) Michael Clynne USGS and USDA (c) USGS

Rhyolitic lavas are so viscous that they hardly flow at all, which causes the lava to build into steep-sided mounds called **lava domes.** Although highly viscous lavas do not present much of a threat to distant human settlements, a lava dome can act like a plug as it begins to solidify. This can allow pressure to build in the magma chamber and result in a more explosive eruption.

In some geologic settings large volumes of basaltic lava will flow onto the surface along large fracture zones, creating vast deposits geologists call *continental flood basalts*. These basaltic magmas are believed to originate within the mantle, then rise through the lithosphere and erupt onto the land surface. Although this process is similar to the formation of hot spot volcanoes, the volume of magma here is so great that the lava flows occur on a more continuous basis as opposed to periodic eruptions. A good example of flood basalts is the extensive sequence located along the Columbia River in the Pacific Northwest region of the United States (Figure 6.8).

Volcanoes

In simple terms a **volcano** is the accumulation of extrusive materials around a vent through which lava, gas, or pyroclastics are ejected (Figure 6.9). Sometimes the vent is a fault or fracture, resulting in a *fissure eruption* whereby materials are extruded and accumulate in a linear fashion. Other times the opening is a single pipelike feature called a *central vent*, creating the familiar cone-shaped deposit most people know as a volcano. There are three basic types of volcanic cones we need to consider: *cinder, shield,* and *composite*.

Cinder cones are relatively small features that form when lava is ejected into the air and cools into pyroclastic material called *cinders*, which then fall and accumulate around the vent. In contrast, **shield volcanoes** are exceptionally large landforms composed primarily of basaltic lava flows. Because of the low viscosity of basaltic magma, these lava flows can travel considerable distances from

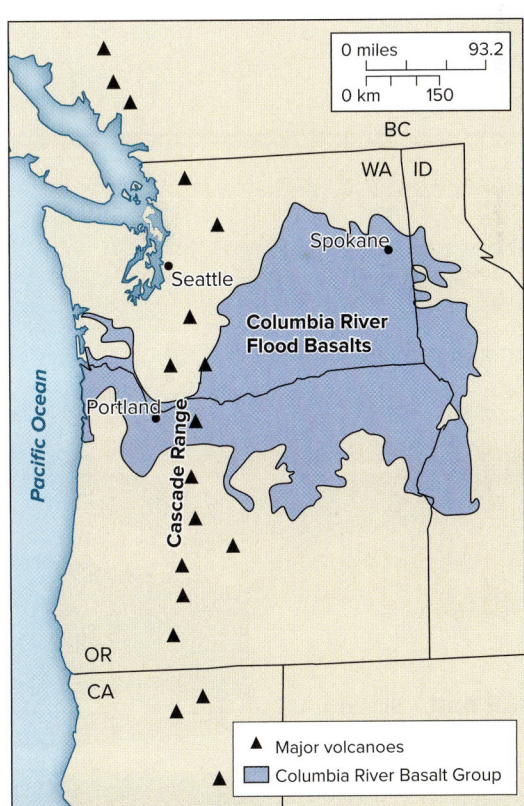

FIGURE 6.8   Extensive lava flows, called flood basalts, form when large volumes of highly fluid basaltic magma are extruded, usually along fracture zones. The map shows the extent of the Columbia River flood basalts in the Pacific Northwest region of the United States. These extensive flows reach a thickness of over 2,000 feet (600 m). The Cascade Range is a volcanic arc that contains subduction zone volcanoes.

A Pu'u O'o, Hawaii, 1986

B Mount Etna, Sicily, Italy, 2001

FIGURE 6.9   A volcanic vent can coincide with a fault or fracture, resulting in a linear extrusion known as a fissure eruption (A). In other cases the vent is a single opening whereby the ejected material creates the familiar cone-shaped feature known as a volcano (B).
(a–b) USGS

the vent and spread out over large areas. This results in shield volcanoes having a broad cross-sectional shape similar to a warrior's shield (Figure 6.10). From a hazard perspective, note that shield volcanoes generally do not erupt in an explosive manner because basaltic magma contains only small amounts of dissolved gas; thus these eruptions produce primarily low-viscosity lava flows. Although shield volcanoes have relatively gentle slopes, repeated lava flows over geologic time can allow them to grow to great heights. Perhaps the best example is Mauna Loa in Hawaii, which at 14,400 feet (4,400 m) above sea level is the largest shield volcano in the world. Because of its great mass, Mauna Loa has caused the sea-floor to become depressed or sag. Altogether, this volcano represents an enormous deposit that is nearly 56,000 feet (17,100 m) high. The reason Mauna Loa is so massive is because it sits over a mantle hot spot that provides a steady supply of magma (Case Study 6.1).

The other major type of volcano is known as a **composite cone,** also called a *stratovolcano,* which is cone-shaped with steep slopes consisting of alternating layers of pyroclastic material and lava flows (Figure 6.11). Composite cones are normally associated with viscous and gas-charged andesitic magmas that generally erupt in an extremely explosive and violent manner. In addition,

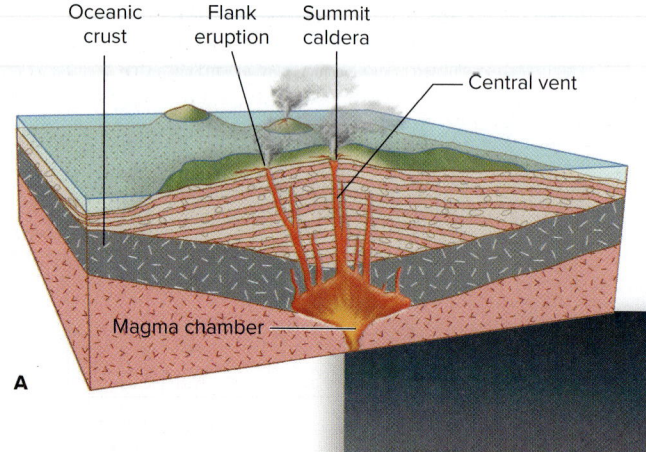

Oceanic crust   Flank eruption   Summit caldera   Central vent   Magma chamber

A

B

FIGURE 6.10   Shield volcanoes (A) are composed primarily of basaltic lava flows that accumulate over geologic time. Photo shows Mauna Loa (B) in the Hawaiian Islands, which sits over a hot spot, providing a steady supply of magma that has allowed it to grow to 14,400 feet (4,400 m) above sea level.
© J.S. Griggs/U.S. Geological Survey

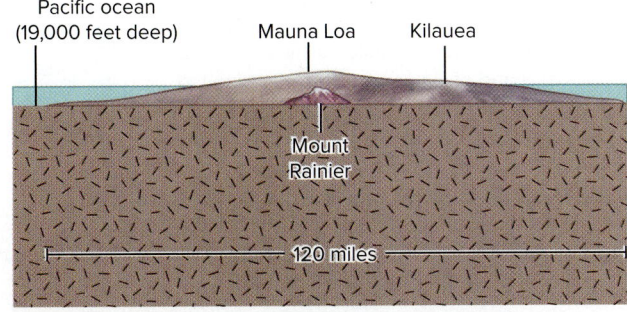

FIGURE 6.11 Composite cone volcanoes (A) are composed of alternating layers of pyroclastic material and lava flows. Viscous andesitic lavas tend to form short and thick flows that enable the volcano to maintain steep slopes and reach great heights. Composite cones are typically quite symmetrical, like Mount Fuji (B) in Japan.
© Image Plan/Corbis

these volcanoes are typically found along subduction zones, making up the high peaks in volcanic arc mountain ranges (Chapter 4). In the Andes Mountains of South America, for example, composite cones reach heights of over 20,000 feet (6,000 m) above sea level. The typical eruption sequence begins with an explosive event and the accumulation of pyroclastic material around the vent. This is followed by the extrusion of viscous andesitic lava that forms relatively thick and short lava flows, which then solidify on top of the pyroclastic layers. Because andesitic lavas are rather viscous and resist flowing downslope, the volcano is able to maintain steep slopes and thus attain great heights. Although composite volcanoes along subduction zones are indeed impressive landforms, they are actually quite small compared to shield volcanoes that develop over hot spots, like those in the Hawaiian Islands. Figure 6.12 helps illustrate the vast size difference between composite and shield volcanoes.

Craters and Calderas

During a volcanic eruption a circular depression called a **crater** will form around the vent where material is being ejected into the air (Figure 6.11A). Especially large craters can form in great explosive events that remove a volcano's summit. A **caldera** is also a circular depression, but one that forms *after* the eruption as rocks in the subsurface begin to collapse or subside (some geologists also refer to large explosive craters as calderas). Basically, a caldera forms when large volumes of magma are ejected from a shallow magma chamber, leaving it relatively empty. The roof of the magma chamber then becomes unsupported

FIGURE 6.12 Size comparison of one of the larger composite cones in the Cascade Range, Mount Rainier, with the shield volcanoes on the island of Hawaii (see Case Study 6.1).

Hawaiian and Yellowstone Hot Spots

Both Hawaii and Yellowstone are similar in that they are active volcanic hot spots related to mantle plumes. However, their eruption styles and associated hazards are entirely different because Hawaii lies in the interior of an oceanic plate whereas Yellowstone lies within a continental plate. The idea of mantle hot spots originated in the 1960s when geologists tried to explain why some volcanic centers have been active for long periods of geologic time. The Hawaiian Islands are particularly interesting because they are part of an exceptionally long chain of volcanic islands and seamounts (submerged volcanic remnants), shown in Figure B6.1. Moreover,

radiometric dating proved that the oldest islands in the chain are about 70 million years old and the islands get progressively younger toward the Hawaiian Islands, where three volcanoes are currently active on the big island of Hawaii. Geologists generally agree that a stationary mantle plume here has been generating basaltic magma in the overlying Pacific plate for the past 70 million years. Moreover, as the plate slowly moves in a northwesterly direction, individual shield volcanoes are cut off from their source of magma and become extinct, at which point new volcanoes develop over the plume. The shield volcanoes of Mauna Loa and Kilauea on the big island of Hawaii then are simply the latest in a long line of volcanoes to have been situated over the hot spot. Kilauea has been erupting almost continuously since 1983, and the youngest volcano, named Loihi, has not yet risen above sea level (Figure B6.1). Due to the fluid nature and low gas content of the basaltic magma, the primary hazard from these nonexplosive shield volcanoes is lava flows.

Yellowstone National Park is a hot spot located in a continental setting and is famous for its geothermal hot springs and geysers (Figure B6.2). To geologists, Yellowstone is also one of the largest calderas on the planet, measuring an impressive 53 by 28 miles (85 by 45 km) in diameter. In addition to geothermal activity, the Yellowstone caldera is seismically quite active; it produced a powerful (magnitude 7.5) earthquake in 1959 that killed 28 people. Although geologists have long known that Yellowstone's geothermal features are related to past volcanic activity, not until the 1990s did it become clear that

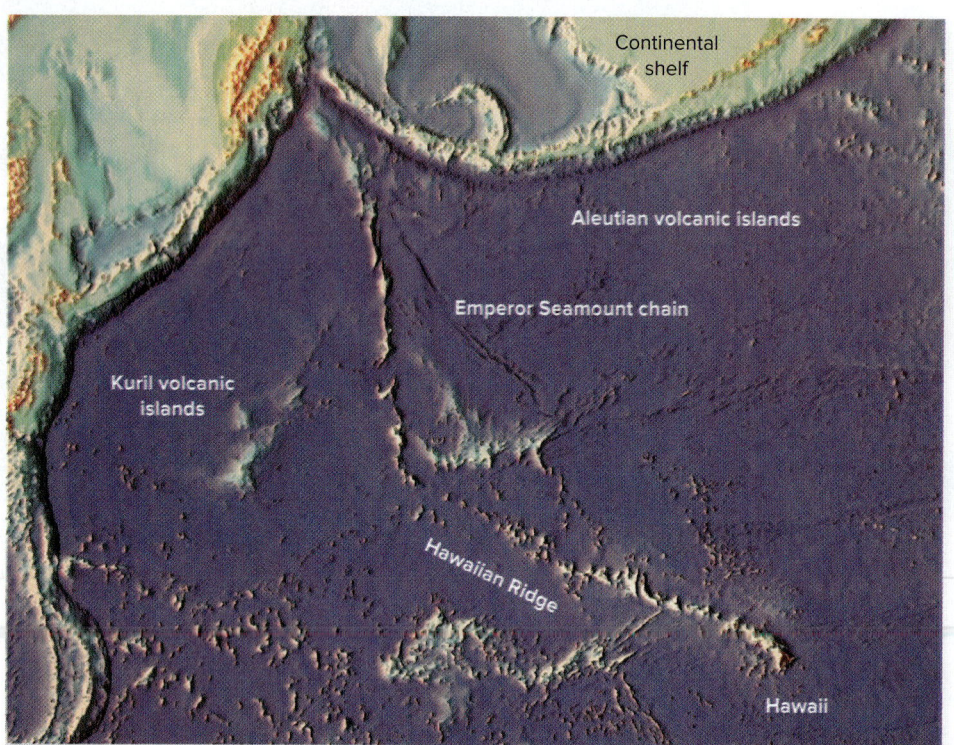

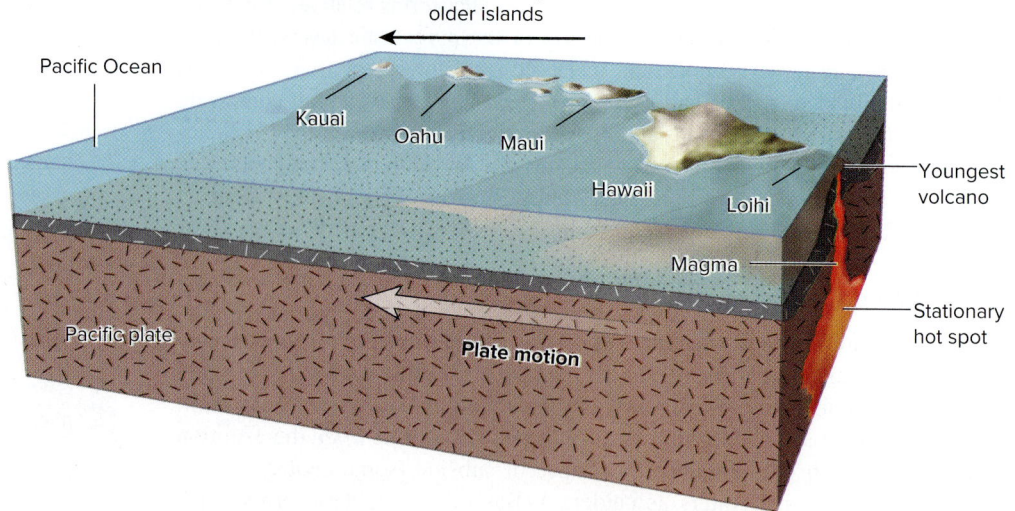

FIGURE B6.1 The Hawaiian Islands are part of a long chain of islands and submerged volcanic remnants known as the Hawaiian Ridge–Emperor Seamount chain. The active shield volcanoes on the big island of Hawaii are currently situated over a stationary mantle plume, which has been generating basaltic magma for approximately 70 million years while the Pacific plate slowly moves to the northwest.

this giant caldera sits over a mantle hot spot. Today, researchers have identified a series of calderas to the southwest of Yellowstone that get progressively older, as shown in Figure B6.2. Similar to the tectonic setting in Hawaii, the North American plate is slowly drifting over a hot spot that has been active for approximately 17 million years, with Yellowstone representing the site of the most recent series of eruptions.

Although the Hawaiian and Yellowstone hot spots share some common characteristics, Yellowstone is different in that the rising basaltic magma is able to melt continental crust of granitic composition, thereby producing highly explosive rhyolitic magma as illustrated in Figure B6.2. Eruptions at Yellowstone therefore tend to generate large amounts of pyroclastic material and comparatively few lava flows, which explains why massive shield volcanoes are not found there. Instead, a giant caldera developed at Yellowstone when huge volumes of magma were drained from the magma chamber after the last eruption around 640,000 years ago. Based on the size of the caldera and the extensive pyroclastic deposits surrounding it, scientists have concluded that the last major eruption was a truly cataclysmic event. If such an eruption were to happen today, it could have disastrous consequences within the United States and also likely affect the climate of the entire planet.

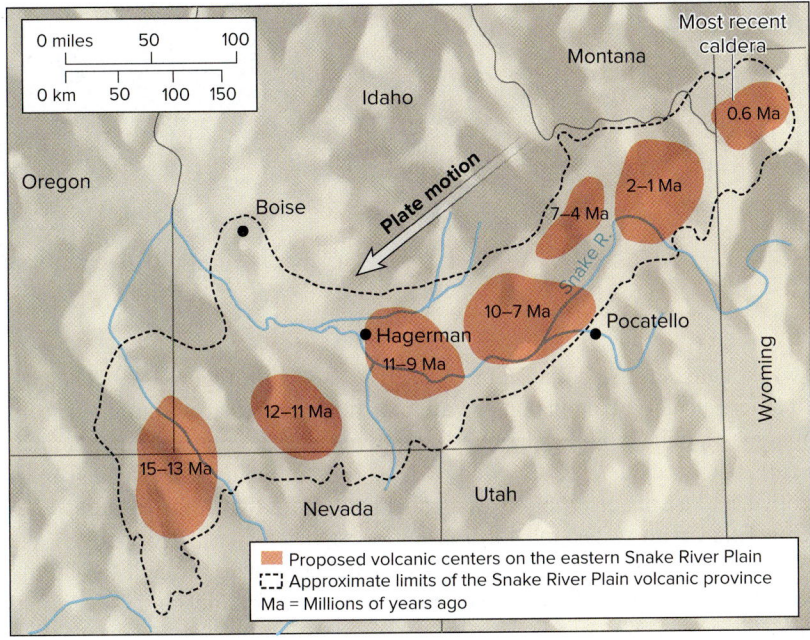

FIGURE B6.2 Map (A) showing the location of ancient calderas associated with the Yellowstone hot spot. Yellowstone National Park is the site of the most recent caldera, which formed after a colossal eruption 640,000 years ago. Cross section (B) showing how rising basaltic magma from the hot spot can cause melting of granitic crust, resulting in a rhyolitic magma chamber. The accumulation of new magma and release of hot fluids are believed to be responsible for recent earthquake activity and changing land elevations within the caldera.

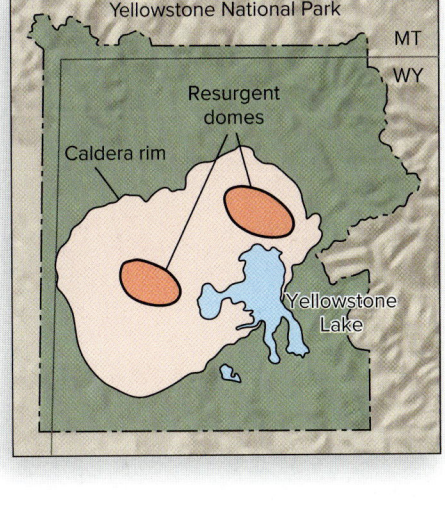

A **B**

FIGURE 6.13 Calderas form (A) when magma is ejected from a shallow magma chamber, leaving its roof unsupported and eventually causing it to collapse. Photo (B) shows the water-filled caldera of Crater Lake in the Cascade Range in Oregon.
(b) © Jim Reichard

Labels in figure A: Crater, Caldera

and weak, causing it to collapse in on itself as shown in Figure 6.13. The size of a caldera usually depends on the amount of magma that was drained from the magma chamber and the chamber's depth beneath the surface. For example, the Yellowstone caldera (Case Study 6.1) is exceptionally large, which indicates that its last major eruption was enormous. Note that some calderas later become filled with water and form scenic freshwater lakes, like Crater Lake in Oregon. In some cases, hot mineralized fluids circulate through the highly fractured rock and deposit valuable ore minerals containing gold, silver, and copper (Chapter 12).

Volcanic Hazards

Similar to their response to other natural hazards, the way in which humans react to potential volcanic hazards depends on the frequency of eruptions and the availability of habitable living space. For example, people have lived for thousand of years on the flanks of active volcanoes because these areas commonly provide critical resources in the form of fertile soils, lush forests, and freshwater. The availability of resources and the fact that several human generations may live between volcanic eruptions make the risk of living in such hazardous areas acceptable to most people. Moreover, when enough time passes between eruptions, people may become unaware of the risk altogether.

When it comes to volcanic hazards, people most often think of violent explosions and flowing lava, which can lead to the erroneous assumption that the only danger is in the immediate area around a volcano. While the area nearest an active volcano is indeed dangerous, there are a number of secondary hazards that extend outward considerable distances. A grim example is the mudflow described earlier that killed over 23,000 residents of Armero, Colombia, located a seemingly safe 46 miles (74 km) from the volcano. Other far-reaching hazards include the release of volcanic ash into the atmosphere and large tsunamis created during explosive eruptions or the sudden collapse of volcanoes. In addition, hazards such

TABLE 6.2   **Some notable volcanic catastrophes in human history and the primary hazard responsible for most of the deaths.**

Location	Date	Deaths	Volcanic Hazard
Vesuvius, Italy	79	3,600	Pyroclastic flow
Tambora, Indonesia	1815	80,000	Ash and short-term climate change (starvation from crop failure)
Krakatau, Indonesia	1883	36,000	Tsunami
Mount Pelee, Martinique (Caribbean)	1902	29,000	Pyroclastic flow
Nevado del Ruiz, Colombia	1985	22,000	Mudflow
Lake Nyos, Cameroon	1986	2,000	Asphyxiation by volcanic gases

Source: Data from the U.S. Geological Survey.

as mudflows, landslides, and deadly gases can pose serious risks even when a volcano is not erupting. Table 6.2 lists some of the more notable volcanic catastrophes in human history, illustrating the diverse nature of these hazards. In the following sections we will explore the different types of volcanic hazards in some detail.

Lava Flows

Perhaps the most familiar volcanic hazard is a *lava flow*, which occurs when magma reaches the surface and begins to move across the landscape. Because lava is a high-density fluid that is very hot, 1,650–2,100°F (900–1,150°C), nearly everything in the path of a lava flow will be either crushed, buried, or incinerated (Figure 6.14). As you might guess, the speed and distance that a flow travels are largely determined by the slope of the land surface, volume of lava being emitted at the vent, and its viscosity. Very fluid basaltic lavas, for example, move about a half mile per hour (1 km/hr) on gentle slopes and up to 6 miles per hour (10 km/hr) in steep terrain, and may travel as much as 30 miles (50 km) away from the vent. However, in Hawaii where fluid lavas are sometimes confined in steep channels that help retain the heat, flows have been clocked as high as 35 miles per hour (55 km/hr). On the other hand, more viscous andesitic lavas rarely travel faster than a few miles per hour or travel much more than 10 miles (16 km) from the vent. As discussed earlier, rhyolitic lavas are so viscous that they hardly flow and instead tend to form lava domes.

Lava clearly becomes a problem whenever people and their property are in the path of an advancing lava flow. If lava repeatedly erupts from the same vent or fissure, the path is somewhat predictable as it will follow the slope of the surrounding terrain. In areas with steep slopes, flows tend to be thicker since they are more likely to follow stream valleys, whereas in flat terrain lava tends to spread out and form broad sheets. Note that even the most fluid lava is much more viscous than water, which gives large-volume flows the ability to actually move up gentle slopes a short distance due to the height of the main lava body pushing from behind. Fortunately, most lava moves slowly enough that people can escape an advancing flow simply by walking away at a quick pace. However, as shown in Figure 6.14, permanent structures and certain land-use activities (agriculture, parks, etc.) are completely at the mercy of moving lava.

Various attempts have been made in modern times to stop or divert lava flows to protect valuable structures and property. In some cases explosives have been used as a means of disrupting moving lava so that it starts flowing in different directions. Ideally this causes the lava to spread out over a larger area and cool more rapidly, thereby forming a thinner

FIGURE 6.14   Lava flows cause considerable damage when they bury valuable real estate and infrastructure, such as the highway and personal property shown in (A). Flows are also destructive when they encounter combustible materials and cause them to catch fire, as in the case of Volcanoes National Park Visitor Center in Hawaii (B).
(a) USGS (b) J.D. Griggs, USGS

A Kalapana, Hawaii, 1986

B Volcanoes National Park, Hawaii, 1989

and shorter lava flow. Another method involves constructing an earthen barrier in the path of an advancing flow to divert it away from human structures or property. Finally, large volumes of water can be strategically sprayed on the leading edges of a lava flow, thereby creating a chilled zone of rock and a more continuous barrier. Of course, a key factor in any attempt to divert or stop a flow is the volume of lava being emitted. In many cases the volume is so large that the flow simply overwhelms any human effort to contain or divert it.

Perhaps the most famous diversion project was the successful 1973 attempt on the island of Heimaey in Iceland, where relatively slow-moving basaltic flows were threatening to block the harbor of the country's most important fishing port. Since unlimited seawater was readily available, pumping operations aimed at chilling the leading edge of the flows were able to begin within weeks of the eruption. After 19 miles (30 km) of pipe had been laid and pumping was performed continuously for nearly 5 months, the lava flows were finally stopped, and the vital fishing harbor was saved.

Explosive Blasts

The most spectacular volcanic processes, and also the most dangerous, are undoubtedly those associated with explosive eruptions of composite cone volcanoes (stratovolcanoes). Recall that the explosive power comes from highly compressed gases (primarily water vapor) dissolved within andesitic and rhyolitic magmas. When this type of magma breaches the surface and encounters the low-pressure conditions of the atmosphere, the dissolved gases will expand violently. Note that even gas-poor magmas can generate large steam explosions should they suddenly encounter a significant volume of groundwater or seawater.

Photographs like the one in Figure 6.15 can help us appreciate the immense power of volcanic eruptions. Here one can see how the rapid decompression of magmatic gases, including steam from ground or surface waters, can literally blow the summit of a composite cone to bits, leaving only a crater behind. One of the most powerful eruptions in recorded history was the 1883 eruption of the composite cone called Krakatau (or Krakatoa) in Indonesia, where seawater is believed to have entered the magma chamber, creating enormous steam explosions. The eruption of Krakatau was so violent that the 2,600-foot (792 m) mountain, along with most of the island, ceased to exist. Where this mountain once stood scientists found a hole in the seafloor over 1,000 feet (300 m) below sea level! While this colossal eruption ranks about fifth in terms of known volcanic explosions, it ranks first in the number of humans killed by a volcano. Krakatau's final series of explosions created several tsunamis, killing over 36,000 people on islands throughout the region.

Although explosive eruptions can generate tsunamis, the most common hazard results from the pyroclastic material and hot gases that get blasted away from a volcano. As the gases escape from the magma and rapidly decompress, they create a pressure wave or shock wave. This shock wave then pulverizes the rocks making up the volcano into smaller fragments, hurtling them outward at great speed along with superheated gases and blobs of lava (a similar phenomenon occurs during steam explosions when magma encounters seawater or groundwater). This expanding cloud of pyroclastic debris and hot gases will obliterate and incinerate everything in its path. For example, during the 1980 eruption of Mount St. Helens, dense forests of 200-foot (60-m) fir trees were completely swept away in a roughly 8-mile (13-km) radius around the volcano. From 8 to 19 miles away from the volcano, similarly giant trees were blown down like matchsticks as shown in Figure 6.16 (see Case Study 6.2 for more details on the eruption).

The 1980 eruption of Mount St. Helens provided a grim reminder of the serious risks people face living within the blast zones of composite cones along the Cascade Range in the Pacific Northwest. Although the Cascade volcanoes are quite active in terms of geologic time (Figure 6.17), the event at Mount St. Helens

FIGURE 6.15 Composite cones that have lost their summits are a testament to the enormous power of expanding gases in volcanic eruptions. Photo showing Mount St. Helens in 1982, after 1,300 feet (400 m) of its summit was lost during the 1980 eruption. Note the new dome growing within the crater.
Lyn Topinka/USGS

was the first major eruption of such magnitude since European descendants settled the region over 200 years ago. This has lulled people into living dangerously close to these ticking bombs. Although major cites like Seattle and Portland are located outside of the expected blast zones of their respective volcanoes, many smaller towns in the region are not. Other hazards, like mudflows and ash fall for example, extend far beyond the blast zone and threaten parts of the urban population centers.

Pyroclastic Flows

A **pyroclastic flow** is a dry avalanche consisting of hot rock fragments, ash, and superheated gas, all rushing down the side of a volcano at great speed. A typical pyroclastic flow consists of two parts: a tumbling mass of large rocks overlain by a turbulent cloud of finer material. Although highly destructive pyroclastic flows can form during explosive or nonexplosive eruptions, they are almost always associated with more viscous, SiO_2-rich magmas. As illustrated in Figure 6.18, one of the ways a pyroclastic flow can form is when a rising eruption column begins to collapse on itself, sending hot gases and fragmental material down the flanks of the volcano. Another way such a flow arises is during explosive blasts when hot fragmental material is sent rushing along the ground surface away from the volcano. Pyroclastic flows can also develop when gases escape rapidly from erupting lava, creating a frothing mass of semimolten material that rolls downslope. Lastly, flows often form in nonexplosive eruptions when a growing lava dome becomes unstable due to its slopes becoming too steep. This can lead

People

FIGURE 6.16 A lateral blast during the 1980 eruption of Mount St. Helens obliterated all trees and vegetation within about 8 miles of the volcano. From 8 to 19 miles away the blast leveled 200-foot (60-m) tall fir trees as shown here—note the people in the lower right for scale.
Lyn Topinka/USGS

Explosive Blast of Mount St. Helens

The 1980 eruption of Mount St. Helens in the United States generated considerable public attention despite the fact it was not particularly large in terms of explosive power, ash volume, or outpouring of lava. Even the death toll of 57 was relatively small compared to the tens of thousands killed in other volcanic eruptions (Table 6.2). However, Mount St. Helens was on the U.S. mainland, which made it easy for teams of scientists from the U.S. Geological Survey (USGS) and other institutions to monitor key indicators prior to the eruption, particularly seismic data (Chapter 5). Because the data allowed scientists to anticipate the eruption, many journalists and photographers were on hand to record the actual eruption. The dramatic imagery helped focus the world's attention on the hazards associated with explosive eruptions, particularly the blast generated by the expansion of gases within the magma.

Mount St. Helens's explosive blast was especially devastating because it was initially directed in a lateral direction as opposed to upward into the sky (Figure B6.3). This occurred due to the fact that, as the magma slowly began to rise within the vent, it created a bulge on one side of the volcano (Figure B6.4). As the bulge continued to swell, it caused the slope to become steeper, making the bulge less stable.

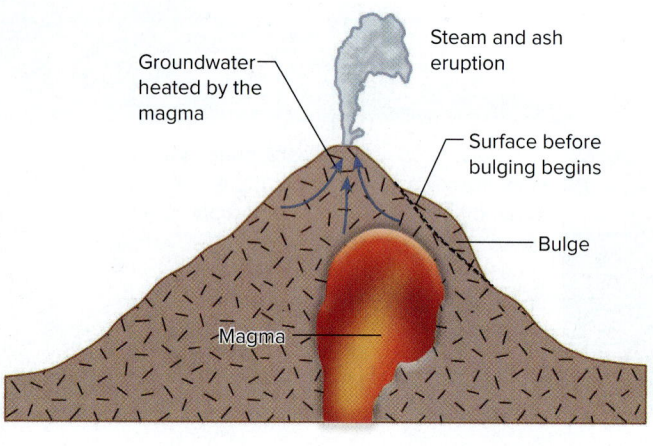

FIGURE B6.3 The eruption of Mount St. Helens was triggered by a small earthquake, which caused a landslide (A) over a bulge that had formed on the side of the volcano. Once the landslide removed the weight of this overlying rock (B), dissolved gases within the highly pressured magma were allowed to rapidly expand. Because of the location of the bulge, the initial blast was directed horizontally (C), devastating the landscape.
Original illustrations by T.R. Alpha, U.S. Geological Survey

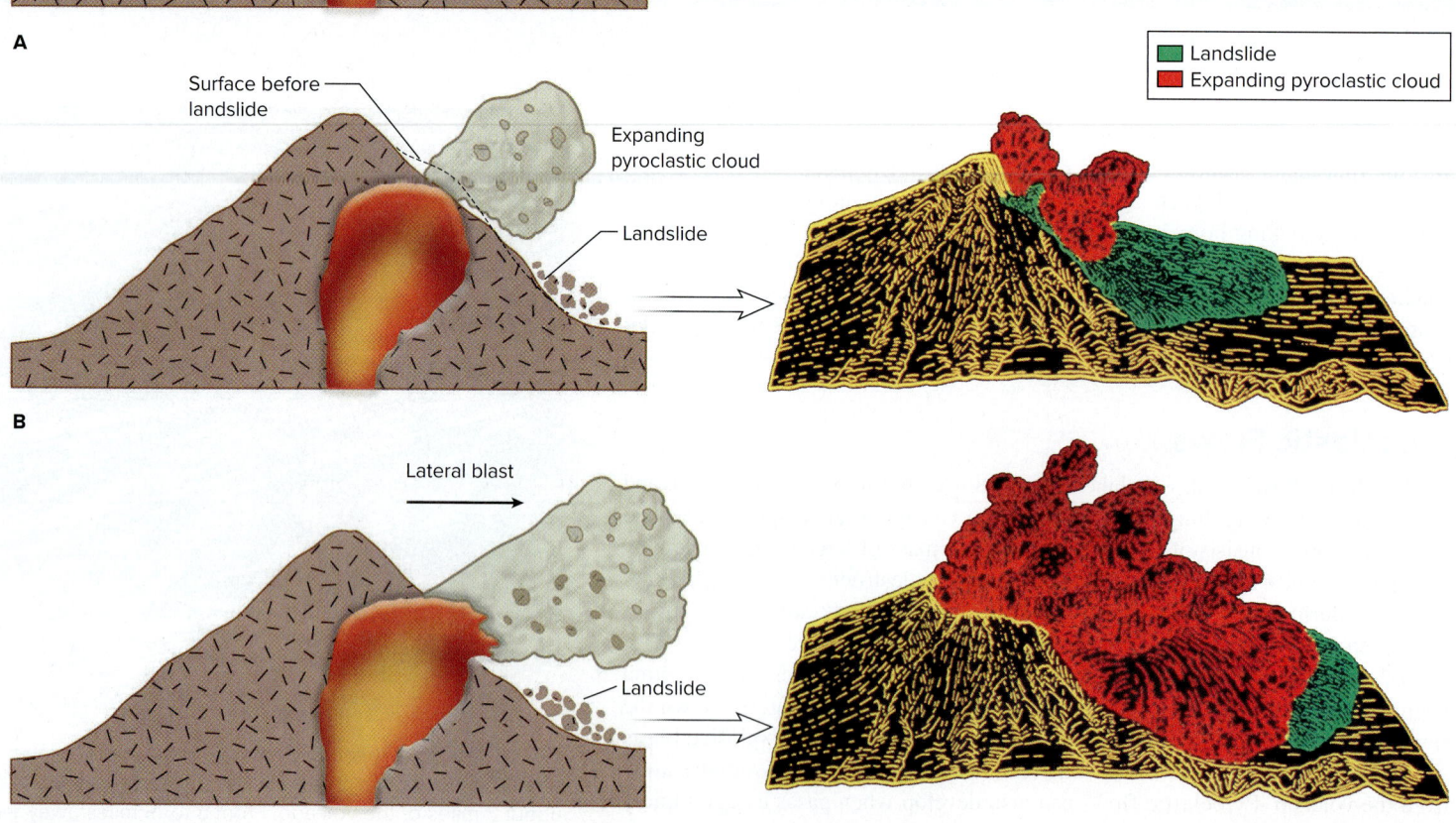

Ultimately, one of the many earthquakes generated by the rising magma caused the bulge and its oversteepened slope to collapse, creating a massive landslide. As the rock mass slid down the mountain, it suddenly removed much of the weight that had been containing the highly pressurized magma. Once the weight was removed, the dissolved gases began to rapidly expand, thereby creating a massive shock wave and explosion.

Among the people killed in the initial blast from Mount St. Helens was a geologist with the USGS named Dave Johnston. He was at a monitoring station located approximately 5 miles (8 km) from the summit. The site had been carefully chosen and deemed safe from the hot avalanches expected to take place in the pending eruption. Tragically, however, no one anticipated the lateral blast.

When the blast occurred, it hurled rocks as large as 9 feet (3 m) in diameter toward the monitoring station at an estimated speed of 670 miles per hour (1,080 km/hr). At this rate it took the cloud of hot ash and rocks approximately 30 seconds to reach the station. Of the 57 people killed in this eruption, most were on the side of the volcano that was devastated by the lateral blast. This tragedy caused scientists and emergency management officials to reconsider the size of the evacuation zones for similar volcanoes in the future.

For details about the eruption type "St Helens past present future" into any Internet search engine. If you want general information on the Cascade Range volcanoes, use "Cascades Volcano Observatory" in your search.

FIGURE B6.4   A massive bulge (A) developed on the northern face as magma within Mount St. Healens slowly pushed outward toward one side. It was here that an earthquake triggered a landslide, which then led to the lateral blast that blew out the side of the volcano (B).
(a) Peter Lipman/USGS (b) USGS

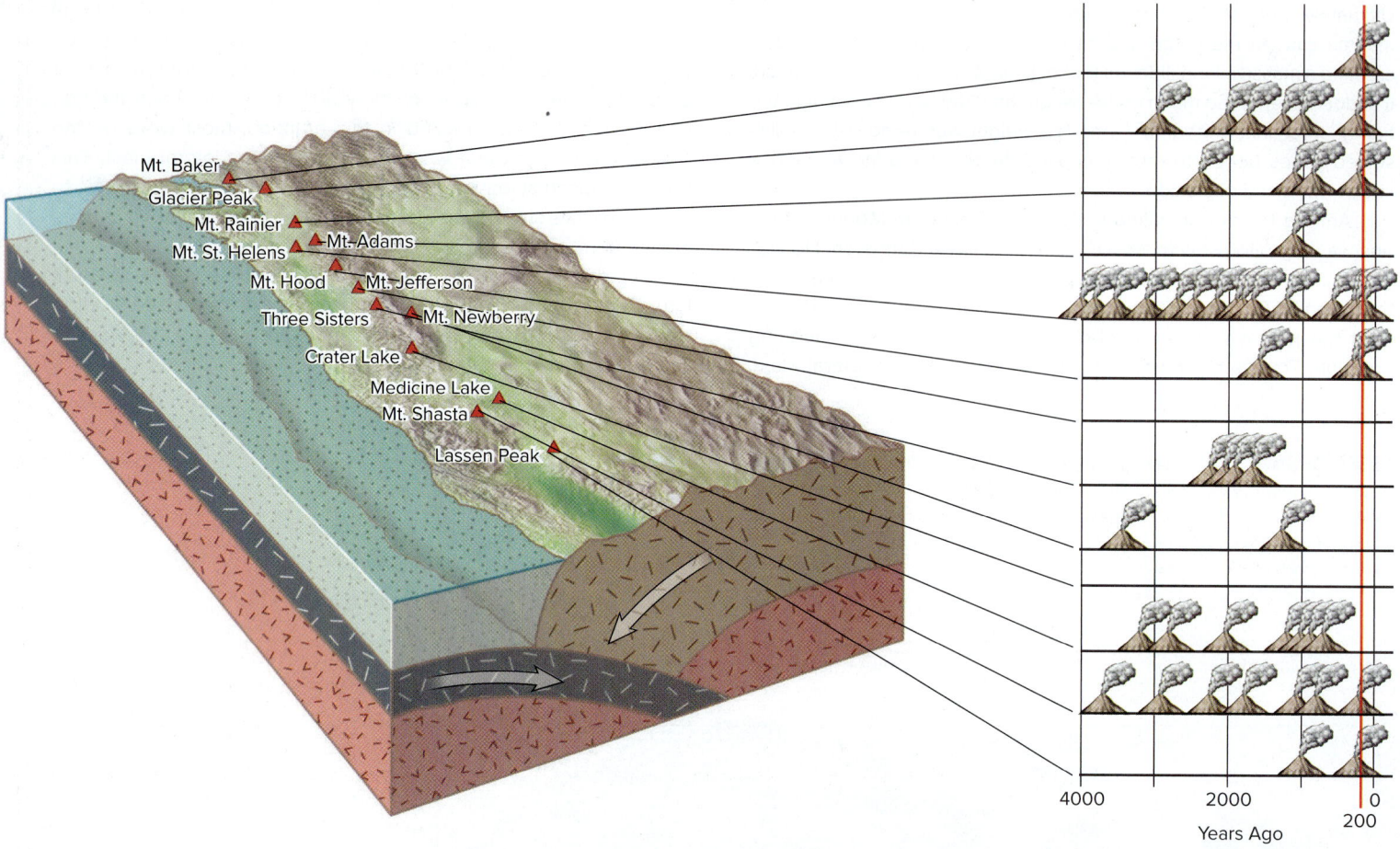

FIGURE 6.17   The Cascade Range in the Pacific Northwest contains active stratovolcanoes associated with a subduction zone. Most of these volcanoes have erupted in the recent geologic past; some like Mount St. Helens erupt more frequently.

to partial or total collapse of the dome, sending large volumes of hot rock tumbling down the volcano.

Pyroclastic flows are extremely destructive in part because of the sheer weight of the rocks and the fact that this material travels on average around 50 miles per hour (80 km/hr), but can reach speeds upward of 450 miles per hour (725 km/hr) depending on the steepness of the slope and relative amounts of gas and rock fragments. Adding to its destructive power is the high temperature of the gases within the flow, which ranges between 400°F and 1,500°F (200–700°C). Not surprisingly, a pyroclastic flow will obliterate or bury everything in its path and also incinerate any combustible materials. Like most avalanches these flows tend to funnel down narrow valleys on the flanks of volcanoes, then form a fan-like deposit at the base where the slope flattens out (Figure 6.19). Due to the infrequent nature of volcanic eruptions, people will often unknowingly place themselves in danger by using this relatively flat terrain for building settlements and cultivating agricultural crops—just as Armero, Colombia, was built on old mudflow deposits.

Perhaps the most famous example of a pyroclastic flow is the one that struck Pompeii, Italy, a thriving Roman resort city on the flanks of Mount Vesuvius. In 79 AD a series of pyroclastic flows rushed down from this stratovolcano, completely burying the city and its inhabitants. Although Pompeii is recognized today as having been destroyed by pyroclastic flows, this deadly hazard was largely unknown to scientists until 1902, when a series of flows devastated the city of St. Pierre on the island of Martinique (Figure 6.20). Residents of St. Pierre at the time knew that nearby Mount Pelée, a composite cone only 4 miles (7 km) away,

Explosive eruptions

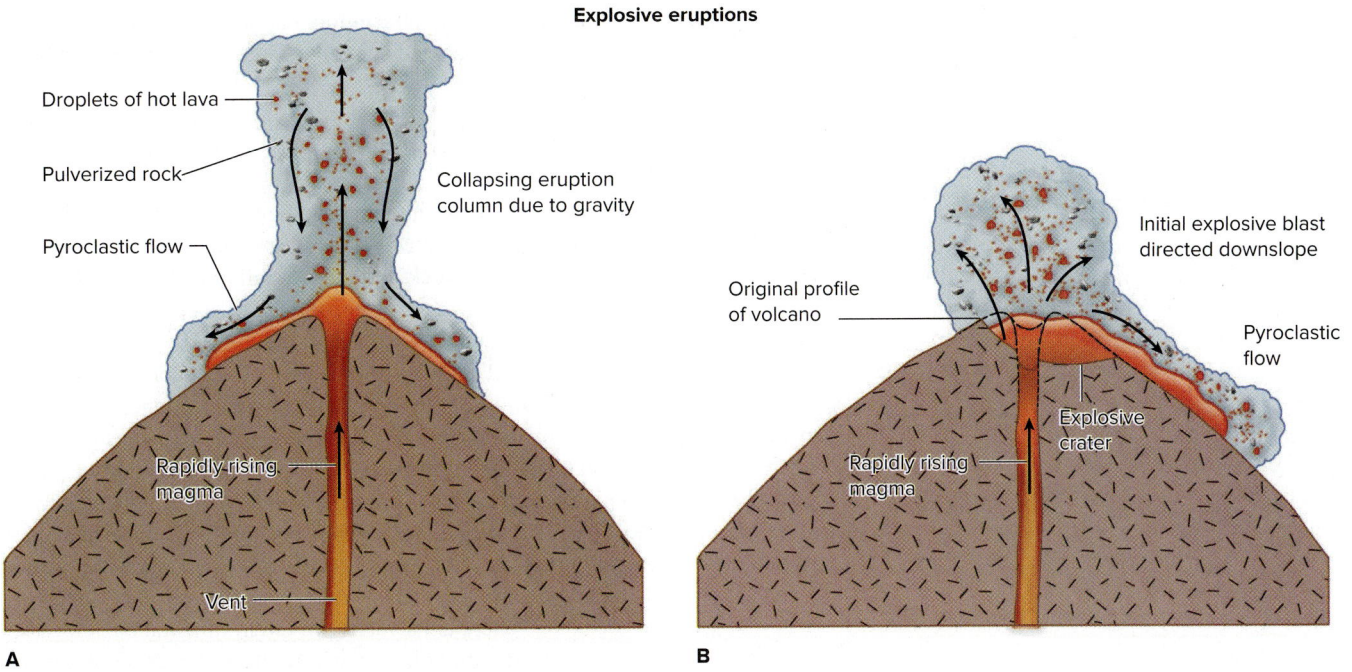

Non-explosive eruptions

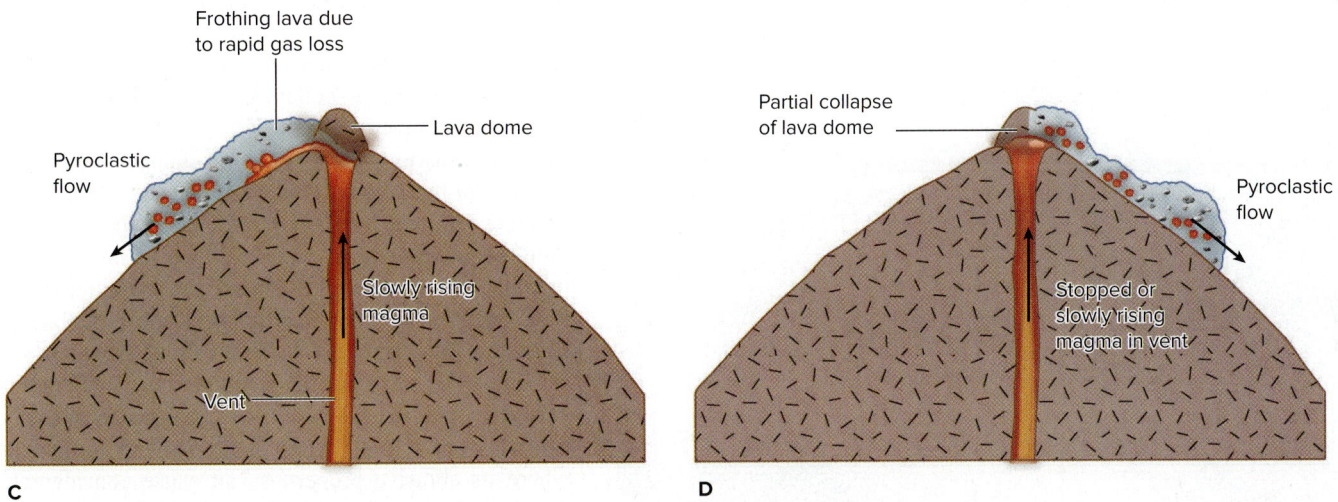

FIGURE 6.18   Pyroclastic flows are dry, hot avalanches where large rock fragments tumble along the ground surface and are overlain by a flowing cloud of finer fragments and droplets of lava. Mixed with these materials are superheated gases, creating a flow that will obliterate and incinerate everything in its path. The illustrations show some of the ways pyroclastic flows form during either explosive or nonexplosive events.

had previously been active, but eruptions were rare and none had caused serious damage. Consequently, people were not alarmed when minor eruptions began in April of 1902. Concern grew a few weeks later with the smell of sulfur gas followed by larger explosions and increasing amounts of ash falling on the city. By the morning of May 8, activity on the volcano had subsided and only a thin cloud of smoke was seen rising from the crater. Suddenly a burst of activity began around 8:00 a.m., resulting in an immense pyroclastic flow that rushed down the flanks of the volcano, engulfing the city and seaport in less than two minutes. The flow, consisting mostly of hot gases, incinerated nearly the entire population of 30,000 and instantly ignited everything that was combustible. Only two people from the city itself were known to have survived, along with a few sailors whose ships had been anchored offshore. Twelve days later another pyroclastic flow, one containing more rock fragments, entered the empty city and knocked down many of the walls in buildings that had remained standing. The once bustling seaport of St. Pierre simply ceased to exist.

Volcanic Ash

In explosive eruptions, parts of the volcano will disintegrate into smaller rock fragments and be blown skyward along with fine droplets of lava, which eventually cool into different-sized particles of rock and noncrystalline material called *volcanic glass* (Chapter 3). Moreover, the hot, expanding gases quickly carry this pyroclastic material into the atmosphere as shown in Figure 6.21, creating a towering eruption column up to 80,000 feet (25,000 m) high. The coarser fragments in this column naturally rain down closer to the volcano, whereas upper-level winds may transport finer particles around the globe before they are deposited. Our interest here is in the fine material geologists call *volcanic ash,* consisting of jagged rock and glass fragments less than 2 millimeters in diameter (2 mm is slightly larger than a pinhead). Due to its small size and loose nature, volcanic ash is often assumed to be soft and lightweight. This ash is actually extremely abrasive because it is made up of tiny rock and glass fragments, which also makes it fairly dense.

FIGURE 6.19 Repeated pyroclastic flows funneled down a steep valley, forming a large fan-shaped deposit at the base of Soufrière Hills volcano on the Caribbean island of Montserrat.
R, Hoblitt, USGS

There are a variety of environmental hazards associated with volcanic ash, such as inhalation of the fine particles, which is particularly dangerous to children or adults with cardiac or respiratory conditions. Other ash hazards occur when it falls from the sky (Figure 6.22) and forms a blanket-like deposit over the landscape. Here the additional weight of the ash can destroy crops and cause buildings to collapse. Ash can also ruin crops when it is not practical to wash it off before processing, or when it disrupts pollination and changes the acidity of soils. Pastureland covered with ash can force farmers to either sell their livestock or purchase expensive feed until the grasses can regenerate. Likewise, new sources of water may be needed to replace those contaminated with ash. Another serious problem occurs when ash is washed off the landscape by heavy rains, clogging rivers and streams and making flooding more frequent for years to come. Because ash is so fine-grained, it easily finds its way into all types of mechanical equipment where its abrasive properties can cause considerable damage. Since ash also conducts electricity, especially if wet, it can short-circuit electrical equipment and cause entire power systems to shut down.

In 1989 a new volcanic ash hazard was discovered when a commercial airliner flying from Amsterdam to Anchorage, Alaska, flew into what appeared to be a typical layer of clouds, causing all four of its engines to shut down. For nearly five minutes the pilots tried to restart the engines as the plane and its terrified passengers silently fell from the sky. Finally, just a few thousand feet above the ground, the pilots successfully restarted the engines and averted disaster. The plane and its 231 passengers safely landed in Anchorage, but the aircraft required four new engines at a cost of $80 million. What the pilots thought were normal-looking clouds turned out to be clouds of volcanic ash that had drifted 155 miles (250 km) after being ejected from Alaska's Redoubt volcano the previous day.

Later analysis showed that the high temperatures within the jet engines caused the ash to melt and resolidify on the engine blades and other moving parts, which ultimately led to engine failure. Because of this and similar encounters, the airline industry recognized the seriousness of this hazard and began working with various scientific institutions to develop an early warning system for pilots. These systems now include seismic monitoring of volcanoes and satellite imagery for tracking ash clouds so that pilots can avoid flying into these otherwise normal-looking clouds.

The potential hazard of volcanic ash for modern air travel gained worldwide attention in 2010 when Iceland's Eyjafjallajokull volcano sent plumes of ash drifting over northern and central Europe, shutting down major airports across the continent. Due to the integrated nature of the modern airline industry, the closing of major hubs in Europe created havoc on flight schedules around the world, stranding or delaying millions of travelers. During the shutdown, government and airline officials worked closely with scientists and engineers in trying to find

FIGURE 6.20 St. Pierre was once a thriving seaport on the island of Martinique in the Caribbean. Then in 1902 a pyroclastic flow raced down nearby Mount Pelée and incinerated the city and its entire population of 30,000.
(inset) Underwood & Underwood/Library of Congress

A Klyuchevskoy 1994, Kamchatka peninsula, Russia

FIGURE 6.21 Explosive eruptions can send a column of volcanic ash (A) as much as 80,000 feet (25,000 m) into the atmosphere, at which point it can get carried around the globe by upper-level winds. Airborne ash poses a serious threat to commercial air traffic, particularly routes over the North Pacific (B) where there is a high concentration of active composite cones. Ash itself consists of pulverized rock and glass fragments, some of which can be extremely fine.
NASA

181

Soufrière Hills volcano, 1997, Montserrat

FIGURE 6.22 Ash clouds can turn day into night. Deposits of ash often add excessive weight to buildings, damage moving parts in mechanical devices, and lead to mudflows during heavy rains. Ash also creates economic losses in agriculture and forestry.
M. Mangan, USGS, 1997

ways of safely operating passenger planes until the crisis was over. In addition to closely monitoring the drifting clouds of ash, the solution involved determining the ash concentrations in which jet engines could safely operate. After eight days, flights resumed intermittently based on the concentrations of ash in the clouds as they drifted over Europe. The crisis ended several weeks later when the volcano stopped emitting ash. Fortunately no aircraft were lost, but it was estimated that the airline industry suffered $1.7 billion in losses.

Upper-level winds can also carry volcanic ash and gases around the globe, creating a widely dispersed volcanic cloud. The amount of ash in some eruptions is so great that it cools the planet, in part by reducing the amount of sunlight reaching the surface. The 1991 eruption of Mount Pinatubo in the Philippines, which spewed 1.2 cubic miles (5 cubic km) of ash and 22 million tons of sulfur dioxide gas, provided scientists a good opportunity to study the effects of large eruptions on Earth's climate. A group of NASA scientists separated the effects of volcanic eruptions from natural climate fluctuations, and then compared the average response of the climate to the five largest eruptions this century. From this investigation they concluded that the Mount Pinatubo eruption resulted in a cooling of 0.5°F (0.25°C). Although the Mount Pinatubo eruption was significant, it was small compared to the 1815 eruption of Tambora in Indonesia where 7 cubic miles (30 cubic km) of ash were ejected into the atmosphere. The resulting ash and gas cloud from Tambora circled the globe, lowering global temperature as much as 5°F (3°C) and causing frosts during the summer months in parts of Europe and North America. The year of 1815 was long remembered as "the year without a summer." Most significant was the fact that the frosts had a devastating effect on crops, leading to an estimated 80,000 deaths by starvation.

In Figure 6.23 one can compare the volume of ash ejected by some of the eruptions discussed in this text along with other notable eruptions in the recent geologic past. Note the relatively small amount of ash emitted by Mount St. Helens in 1980 compared to the 1815 eruption of Tamboro. Even more amazing is Yellowstone's eruption 640,000 years ago (Case Study 6.1), which is estimated to have ejected 240 cubic miles (1,000 cubic km) of ash, an amount so large that it was left off the graph due to its scale.

Mass Wasting on Volcanoes

Geologists use the term *mass wasting* to refer to the downslope movement of earth materials due to the force of gravity. Although Chapter 7 is devoted entirely to the topic of mass wasting, in this section we will examine two types of mass wasting processes and associated hazards that are specific to volcanoes. Note that pyroclastic flows were discussed separately due to the fact that hot gases, rather than gravity, play a dominant role in transporting the volcanic debris in those events.

Volcanic Landslides

Whenever the steep flanks of a volcano become unstable, the underlying material can fail, resulting in the rapid downslope movement of rock, snow, and ice known as a **volcanic landslide**—also called a *debris avalanche*. Since composite cones have much steeper slopes than shield volcanoes, they are more likely to experience volcanic landslides. It is important to realize that these landslides do not necessarily take place during eruptions, but are often triggered by heavy rains or earthquakes, as was the case with the massive landslide just before the eruption of Mount St. Helens (Case Study 6.2). Another key factor is the movement of corrosive gases and groundwater within a volcano that can cause the feldspar-rich rocks to break down into much weaker clay minerals (Chapter 3). This breakdown weakens the rocks making up a volcano and tends to destabilize the slopes, thereby making it easier for an earthquake or heavy rain to trigger a landslide.

Volcanic Mudflows

Unlike a volcanic landslide, a **volcanic mudflow** (sometimes called a *lahar* or *debris flow*) is a mixture of ash and rock that also contains considerable amounts of liquid water. Because of their more fluid nature, volcanic mudflows tend to rush down stream valleys that lead away from a volcano, reaching speeds of 20 to 40 miles per hour (30–65 km/hr). The amount of ash and rock in the flow can be as much as 60–90% by weight, making some mudflows resemble a river of wet concrete. All this solid material causes mudflows to be fairly dense and viscous, which is why they are so much more destructive than ordinary river floods. In fact, mudflows are so powerful they can easily rip up trees and transport large boulders far downstream (Figure 6.24). The flow of debris also has a hammering or crushing effect on objects it encounters. As in the case of Armero, most of the debris within a mudflow will be deposited at the base of the volcano where the stream gradient abruptly flattens out, burying everything in its path. In addition to their destructive power, mudflows can travel considerable distances and occur even when a volcano is quiet; hence they can strike without warning.

Volcanic mudflows typically form whenever loose ash lying on the flanks of composite cones is picked up by running water. The water can originate from a heavy rain event or from rapidly melting snow and glacial ice. Rain-induced mudflows are common in eruptions that release large volumes of water vapor, which quickly condenses and falls as rain, carrying loose ash off into stream channels. In some areas with a humid climate, heavy rains may occur fairly frequently as part of normal weather patterns or be the result of tropical storms and hurricanes. Note that heavy rainfall events may produce mudflows long after the actual ash eruption. Such was the case around Mount Unzen in Japan in 1995, where mudflows destroyed thousands of homes more than a year after the ash was deposited.

Mudflows also form by the rapid melting of snow and glacial ice, which are commonly found on the summit of many composite cone volcanoes (Figure 6.25). Ice caps can hold tremendous quantities of water, and hence have the potential to produce exceptionally large mudflows. One of the ways this ice quickly melts is by coming into contact with pyroclastic flows, lava, or hot ash. Another mechanism is a sudden slope failure near a volcano's ice-capped summit, resulting in a massive landslide of rock and ice, as illustrated in Figure 6.26. As the debris travels downslope, some of the ice will begin to melt, transforming the landslide into a mudflow. This type of slope failure can occur when magma creates an unstable bulge or when gases and fluids within the volcano transform solid rock into much weaker clay minerals.

An area that is at serious risk of a mudflow is the Seattle-Tacoma metropolitan area, where nearly 3 million people live in the shadow of Mount Rainier (Figure 6.25). This 14,413-foot (4,393-m) composite cone contains more water in the form of glacial ice than all the other Cascade volcanoes combined. Moreover, geologic studies in river valleys leading up to Mount Rainier's summit have found evidence of repeated mudflows over the past 6,000 years—some are as recent as 500 years old. The scientific data also show that these

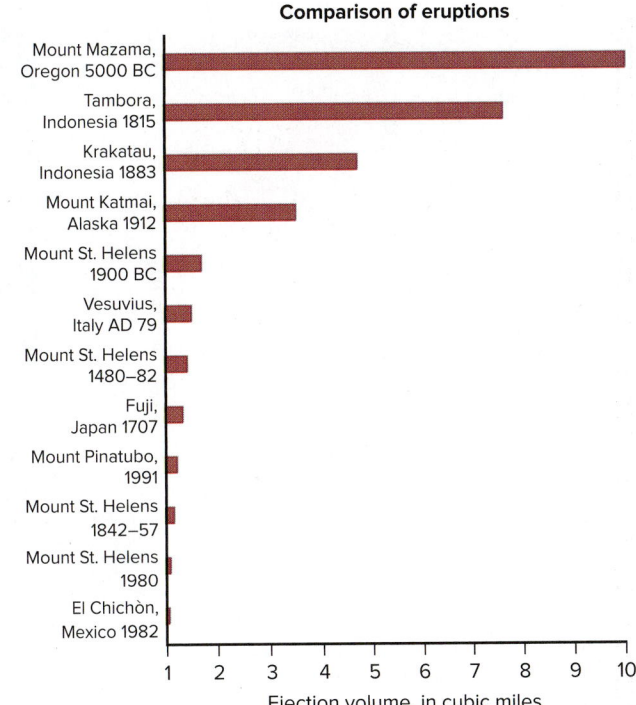

Comparison of eruptions

FIGURE 6.23 Graph comparing the volume of ash ejected in selected volcanic eruptions that have occurred in the recent geologic past.

FIGURE 6.24 These mud-stained trees clearly mark the height of a mudflow generated during the Mount St. Helens eruption—note the person for scale.
USGS

FIGURE 6.25 Mount Rainier in Washington state is an active composite cone whose snow- and ice-capped summit contains more water than all the other Cascade volcanoes combined. This volcano has a history of generating extremely large mudflows, as illustrated by the map showing ancient mudflow deposits in river valleys leading up to the volcano's summit. Many rapidly growing communities in the Seattle-Tacoma metropolitan area are at extreme risk as they are located within river valleys draining Mount Rainer.
Steve Brantley/USGS

flows were enormous, depositing debris in valleys all the way to Puget Sound, a distance of 55 miles (90 km). Of great concern to scientists and emergency management officials is that the geologic conditions that led to past mudflows are still present on Mount Rainier. In addition, due to continued growth of the Seattle-Tacoma metro area, large numbers of people have moved to communities nestled within the valleys leading up to Mount Rainier (Figure 6.25). Compounding the problem is the fact that geologists have recently found considerable amounts of clay minerals within the ancient mudflow deposits. This indicates that feldspar-rich rocks within the volcano have been transformed into clay minerals that destabilize the volcano's steep slopes. Consequently, it's quite likely that large portions of the volcano's summit periodically collapse, generating huge landslides that turn into mudflows as the ice rapidly melts. Keep in mind that Mount Rainier does not need to erupt for this to occur, meaning that residents would have no warning that giant mudflows are headed their way.

Because of the high mudflow risk to residents living in river valleys draining Mount Rainier, federal and state agencies have developed an early warning system. This system consists of a network of acoustic sensors in the upper reaches of the two most vulnerable valleys draining the volcano. The sensors are buried underground and designed to measure the characteristic ground vibrations of a passing mudflow. Signals from the sensors are constantly monitored by a computer system at an emergency management operations center. If a sensor records ground vibrations above some preprogrammed limit, the system automatically sends electronic alerts to television and radio stations, schools, and various other public and commercial facilities. In addition, warning sirens will be sounded that will give residents as much as 30–60 minutes to follow preplanned evacuation routes by vehicle or foot. This detection and warning system is particularly important since mudflows can occur even when the volcano is not erupting.

Volcanic Gases

When rocks begin to melt deep within the Earth, volcanic gases are produced and remain dissolved in the newly formed magma. Along subduction zones where water-rich sediment is incorporated into the melt, large volumes of H_2O gas become dissolved, making andesitic magmas highly explosive. Consequently, the most abundant volcanic gas is commonly water vapor (H_2O), followed by carbon

dioxide (CO_2), and sulfur dioxide (SO_2), which together account for over 95% of all volcanic gases. One of the reasons why a volcanic gas cloud is hazardous to humans is simply because it does not contain free oxygen (O_2). Therefore, should a volcanic cloud descend into a populated area, it will pose an asphyxiation (i.e., suffocation) risk to people. Because volcanic gases are typically quite hot, another life-threatening hazard is severe burns to skin and lung tissue.

The hazard from volcanic gas is normally the highest during or just before the eruption of magma since this is when the release of gas is most common. However, deadly gas clouds have been known to move down the flanks of a volcano in the complete absence of any volcanic activity. In 1986 clouds of carbon dioxide (CO_2) gas silently moved down the slopes of two water-filled craters in the Oku volcanic field in Cameroon, Africa. The result was the asphyxiation of over 2,000 people and thousands of head of livestock in a 12-mile (20-km) radius around the craters. Officials at first were baffled as to the cause since there was no indication of a volcanic eruption. What they did find, however, was that the once clear water within the craters was now laden with sediment. Scientists later determined that CO_2 gas had been accumulating in the thick sediment lying on the lake bottom. An unusual wind or cooling of the lake surface is believed to have triggered the sudden release of CO_2, which churned up the sediment and overturned the water within the lake as it escaped. Because CO_2 is heavier than air, the resulting gas cloud rolled down the flanks of the crater and into low-lying areas, causing people and livestock to quickly suffocate.

FIGURE 6.26   Large volcanic mudflows can form when a landslide develops beneath a glacial ice cap on a composite cone. Such landslides can be triggered when magma creates an unstable bulge in the mountainside, or when slopes are weakened by gases and hot fluids that turn solid rock into clay minerals.
(all) Michael Rymer, USGS

Tsunamis

In Chapter 5 you learned that earthquakes create tsunamis when the seafloor suddenly moves in a vertical direction, thereby displacing large volumes of seawater. Tsunamis also form when volcanoes explode violently in an oceanic setting. Due to the large number of volcanoes in the world along shorelines, volcanic tsunamis represent a threat that can strike coastal communities far from the volcano itself. As described earlier, the 1883 eruption of Krakatau in Indonesia is notorious for the deadly series of tsunamis created by the colossal eruption that obliterated most of the volcano.

Scientists have recently come to better appreciate the fact that large tsunamis can also form by volcanic landslides, partly because of geologic studies on the Hawaiian Islands. For example, the map in Figure 6.27 shows an extensive debris

A

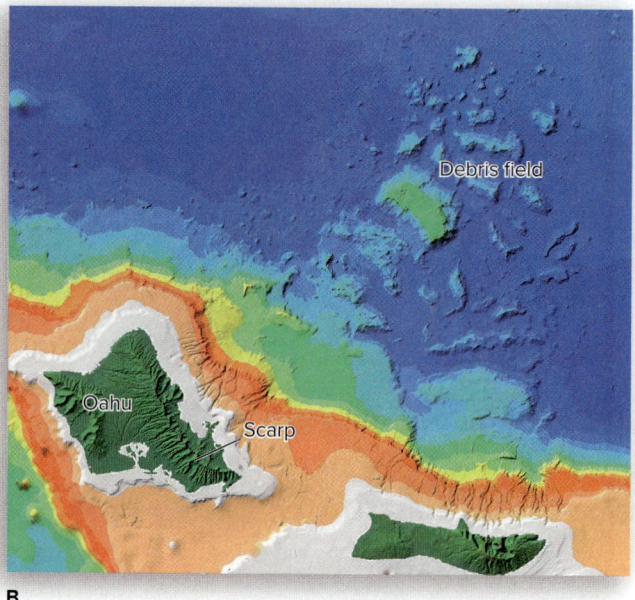

B

FIGURE 6.27   Large scarps (A) and an extensive debris field (B) offshore of the Hawaiian Islands point to an enormous landslide that had likely generated a large tsunami. The dark green areas on the map are above sea level, representing the islands of Oahu and Molokai.
(a) © Jenny B. Paduan

field on the seafloor north of the islands of Oahu and Molokai. In the photo one can see abrupt *scarps* (clifflike faces) on the islands themselves, which are also characteristic of landslides. This evidence strongly indicates that a massive slide started well above sea level, which would have displaced a tremendous amount of water, and, in turn, generated a very large tsunami. Such events appear to have occurred on nearly all of the Hawaiian Islands, implying that volcanic landslides may be a common process there. It is quite possible then that similar landslides may be common on volcanic islands in other places, creating a tsunami threat along unsuspecting coastlines where such waves are not known to have occurred in modern times.

Predicting Eruptions and Minimizing the Risks

Approximately 20% of Earth's population lives in areas at risk of a volcanic hazard. As noted throughout this text, people live in hazardous zones for many reasons, including fertile soils, limited availability of useable land, economic opportunity, scenery, and recreational value. However, another key factor is complacency. Remember that the time interval between volcanic events is often much longer than the life span of an individual human; many people therefore consider the risk to be low and worth taking. Although many generations may pass while an active volcano remains dormant, an eruption is basically all but inevitable. Human complacency is compounded by the fact that as population grows, greater numbers of people put themselves in harm's way, increasing the potential loss of life and property. For example, researchers at Columbia University used 1990 census data, satellite images, and geologic information to study the relationship between population patterns and volcanic hazards. Here they examined 1,410 volcanoes that have been active within the past 5,000 years, which is very recent in terms of geologic time. Of these, 457 volcanoes were found to have

A

B

FIGURE 6.28   False-color satellite image (A) showing Mount Vesuvius and the surrounding urban area of Naples, Italy. In 79 AD, Mount Vesuvius erupted and buried 3,600 residents in the Roman city of Pompeii and surrounding settlements. (B) Nearly 4 million people now live in the urban area of Naples.
(a) NASA/GSFC/MITI/ERSDAC/JAROS, and U.S. Japan ASTER Science Team; (b) © Author's Image/PunchStock

populations greater than 1 million living within a 60-mile (100-km) radius—only 311 active volcanic zones were found to be relatively uninhabited.

An example of a potential megadisaster site where people have been crowding into a known volcanic hazard zone is Naples, Italy. Here nearly 4 million people now live within 18 miles (30 km) of the summit of Mount Vesuvius, an active and very dangerous composite cone (Figure 6.28). For perspective, consider that this 18-mile radius is about the same size as the blast zone of Mount St. Helens (Case Study 6.2). Moreover, most of Naples has been built on ash and pyroclastic flow deposits laid down during previous eruptions. Recent geologic studies have shown that over the past 25,000 years Mount Vesuvius has experienced, on average, a major eruption every 2,000 years. The most famous eruption occurred in 79 AD when pyroclastic flows and volcanic ash entombed approximately 3,600 residents of the Roman city of Pompeii and surrounding settlements (Figure 6.28). Archaeological studies have recently found that an eruption in 1780 BC also killed thousands of people on the plains surrounding this composite cone. Geologic evidence shows that the 1780 BC eruption was considerably larger than the one that destroyed the Roman cities in 79 AD. Based on the distribution of ash and pyroclastic flow deposits, geologists estimate that Vesuvius unleashed a 900°F (480°C) pyroclastic flow that rushed along the ground at 240 miles per hour (385 km/hr), incinerating everything in its path—the pyroclastic material would have been hot enough to boil water as much as 10 miles (16 km) from the vent. Of major concern to scientists is the fact that based on its 2,000-year cycle, Mount Vesuvius is due for another major eruption.

Since more people worldwide are now living in volcanic hazard zones, an obvious question is, how can society reduce the risks? Clearly, there is nothing humans can do to prevent the hazards themselves. Although we have had limited success in diverting lava flows, the only thing we can do with respect to most volcanic hazards is to simply get out of the way. Unlike earthquakes, where buildings can be constructed to withstand the intense shaking, the forces involved in volcanic blasts, pyroclastic flows, and mudflows are beyond our capacity to protect ourselves. Moreover, there is the problem of the searing heat associated with many volcanic hazards. What society can do is take advantage of the fact that volcanic eruptions are almost always preceded by fairly reliable warning signs, such as swarms of earthquakes, topographic bulges, and escaping gases. The most effective way then of minimizing volcanic hazards is for scientists to monitor precursor activity and evaluate the potential for an eruption. Based on this information, emergency managers can decide whether to implement evacuation plans for those who are at risk.

Predictive Tools

In this section we will explore the different tools volcanologists use to assess the potential for an eruption. Note that when it comes to noneruptive hazards (mudflows and landslides), the effectiveness of some predictive tools is rather limited. The best approach for noneruptive hazards is to install sensors to detect the movement of earth materials as they begin to move downslope. Once a slide or flow is detected, a signal is sent to an early warning system that automatically alerts residents farther down the drainage system so they can quickly move to higher ground.

Geologic History

When evaluating potential volcanic hazards, one of the first things volcanologists do is try to learn how a particular volcano has behaved in the past. Because written accounts do not go back very far in terms of geologic time, scientists commonly make use of volcanic deposits since the sediment and rocks themselves hold clues to how they formed. This type of analysis is carried out by performing field studies and making maps of the different types of volcanic deposits found around a volcano. Based on the size, shape, composition, and layering characteristics of a deposit's particles, scientists can identify its origin (i.e., lava flow, mudflow, pyroclastic flow, ash fall, or landslide). By using a map to view the way various deposits are distributed around a volcano, geologists can get a pretty good idea of the hazards associated with past events. For example, it was the discovery of ancient mudflow deposits around Mount Rainier that led geologists to voice their concern about the safety of communities located on top of these deposits.

Topographic Changes

Another useful tool for assessing the potential for a volcanic eruption is monitoring changes in the topography, or shape, of a volcano. When rising magma begins to collect in void spaces near the surface, it forms a reservoir called a *magma chamber*. Because the confining (overburden) pressure is low near the surface, the presence of pressurized magma commonly causes the volcano to swell or inflate. After an eruption the pressure naturally decreases and the volcano will deflate. In cases where the magma simply moves, some parts of volcano will swell while others deflate. Therefore, by accurately surveying changes in the shape of a volcano over time, scientists can get an idea as to the position of magma within the volcano as well as the volume moving into the magma chamber. Today global positioning system (GPS) receivers are commonly used to survey topographic changes.

The conditions on some active volcanoes can quickly change, making it too dangerous for scientists to revisit and take measurements. In these situations electronic tiltmeters can be installed that accurately measure very small changes in slope, and then transfer the data in real time to a safe location. Tiltmeters work on the same principle as a carpenter's level, the major difference being that an electronic sensor is used to determine the position of a bubble in a fluid-filled container. By installing numerous tiltmeters at strategic locations around a volcano, scientists can observe, in real time, those parts of the volcano that may be inflating and those that may be deflating. This information can be quite useful in tracking the movement of magma within the volcano and assessing the potential for an eruption.

Seismic Monitoring

Since earthquake activity almost always increases as magma moves toward the surface, seismic monitoring (Figure 6.29) is an excellent tool for predicting eruptions. Recall from Chapter 5 that earthquakes occur when rocks reach their elastic limit, at which point they fail and the strain they had accumulated is suddenly released in the form of vibrational wave energy. Although most earthquakes result from forces associated with the movement of tectonic plates (i.e., *tectonic earthquakes*), strain also accumulates when rising magma forces its way through crustal rocks, creating what geologists call **magmatic earthquakes** (sometimes called *harmonic tremors*). Like all earthquakes, magmatic earthquakes cause rocks to become displaced along faults, which makes it easier for the magma to move upward. A key point here is that as magma pushes its way to the surface, the resulting earthquakes vibrate in a steady and rhythmic (i.e., harmonic) manner. In addition to rhythmic vibrations, magmatic earthquakes have relatively low magnitudes and occur in distinct swarms, which may last an hour or more and consist of tens to hundreds of small earthquakes (Figure 6.30). This stands in sharp contrast to more powerful tectonic earthquakes that take place very abruptly and last a minute or two at most.

Because tectonic and magmatic earthquakes have such vastly different characteristics, it is somewhat easy for seismologists to tell the difference between the two. Moreover, since magmatic earthquakes occur when magma is on the move, an increase in seismic activity under a volcano is a strong indication that an eruption is likely. In situations where people would be at risk in an eruption, seismologists typically place an array of portable seismographs (Figure 6.29) around a volcano. From the data they can determine the focal depth of each earthquake, and therefore monitor the position of the main magma body as it moves upward. Should the magma continue to move closer to the surface, the probability of an eruption will naturally increase.

Monitoring of Volcanic Gases

Recall that the dissolved gases in magma are mainly water vapor (H_2O), carbon dioxide (CO_2), and sulfur dioxide (SO_2). As a magma body gets closer to the surface, it becomes increasingly likely that some of the highly pressurized gases will escape through fracture systems and be released into the atmosphere (Figure 6.29). Volcanologists typically monitor the escaping gases on a fairly regular basis, and then look for changes in gas chemistry that may indicate a possible eruption. However, the use of this technique as a predictive tool is complicated by the fact that volcanic gases do not always originate from fresh magma moving up from depth. Other sources include heated groundwater and older magma left over from a previous eruption. Determining the source is important since leftover magma is likely to have lost much of its gas pressure, which means the chance of an explosive eruption is far less than if the magma is fresh and fully charged with gas. Should the primary source turn out to be heated groundwater, then the worst-case scenario would be a large steam explosion.

Fortunately, the origin of volcanic gases can be determined from careful chemical analysis of samples collected directly at the surface. For example, if results show the gases to be composed almost entirely of water vapor, then one can conclude that the primary source is heated groundwater. On the other hand, if the samples contain appreciable amounts of CO_2 and SO_2 in addition to water vapor, then the source is most likely magma. Finally, should the proportion of individual gases within a set of samples be significantly different from that in a set taken earlier, then it can be inferred that new magma is moving up from a

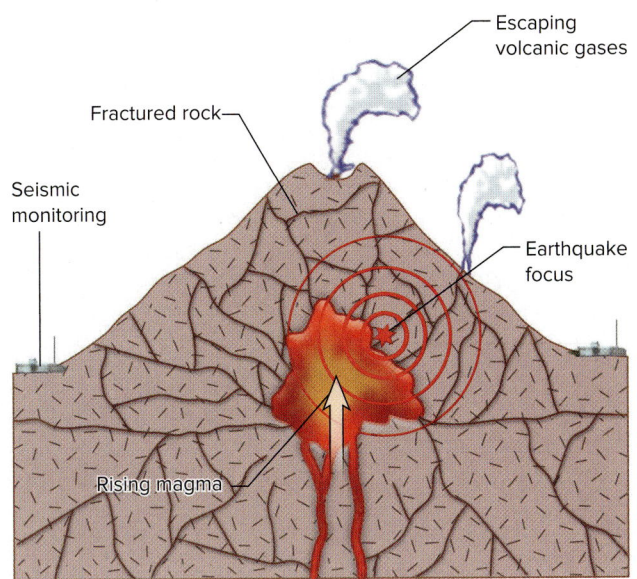

FIGURE 6.29 The monitoring of magmatic earthquakes and volcanic gases is key in predicting volcanic eruptions. Portable seismographs record the rhythmic vibrations of magmatic earthquakes and allow scientists to track the magma body as it pushes up through the crust. Measuring the chemistry of gas samples collected at the surface helps determine whether the magma is new, and hence potentially more explosive.

FIGURE 6.30 A seismograph recording showing numerous magmatic earthquakes taking place on Mount St. Helens.
© Jim Reichard

greater depth. This, in turn, means there is a strong possibility that a major eruption could occur. The interpretation that fresh magma is moving upward can be verified by plotting the depths of focal points obtained from seismic data.

Geophysical and Groundwater Changes

In addition to topographic changes, earthquakes, and release of volcanic gases, rising magma can also change the physical properties of rocks—called geophysical changes—as well as the temperature and chemical composition of groundwater. When measuring various rock and groundwater properties over a period of time, scientists look for changes that may indicate that magma is getting closer to the surface. For example, we would naturally expect the temperature of both the rocks and groundwater to increase as magma approaches the surface. Moreover, because magmatic gases and fluids commonly flow outward from the magma chamber along fracture systems, water samples from wells typically show an increase in acidity and sulfur content as magma gets closer to the surface. Finally, the electrical resistance of rocks will often change due to increased temperatures and circulation of conductive fluids and gases within the volcano.

Early Warning and Evacuation

The best way to minimize the risks associated with volcanic hazards is to use predictive tools to produce reliable eruption forecasts. The forecasts then provide an early warning so that officials can implement emergency response plans and allow people to safely evacuate. The ideal situation is to monitor a dangerous volcano with a variety of sensitive ground-based instruments that are capable of transmitting data to scientists in real time. Such a system allows baseline (i.e., background) data to be collected that can later be compared to data that arrive when activity on the volcano begins to increase. The ability to compare against baseline conditions is important as it helps scientists increase the reliability of eruption forecasts, which in turn provides more accurate early warning times for emergency managers.

Although early warning systems have proven to be very valuable, only a small number of the world's volcanoes that threaten populated areas are well monitored and have instruments with real-time data transmission capabilities. Most volcanoes are only lightly instrumented with seismographs whose sensitivity is not adequate for detecting the subtle earthquakes typically found during the earliest stages of an eruption. Even worse is the fact that some dangerous volcanoes are not being monitored at all. This means that when they do become active, scientists and emergency managers lose precious time gathering data and trying to assess the danger. Eruption forecasts in such cases tend to be less reliable. To help alleviate this problem, the USGS proposed a program for developing adequate monitoring capabilities on the 57 U.S. volcanoes that the agency identified as being undermonitored.

Ultimately, having adequate instrumentation on volcanoes that threaten populated areas is important because emergency managers need reliable forecasts when making a decision on whether to order an evacuation. For example, there simply may not be enough time for a successful evacuation should officials wait until scientists are more certain of an eruption. Such a delay could then lead to a large loss of life. On the other hand, should evacuation orders be given and an eruption not occur, then people will be less inclined to evacuate the next time, which may also lead to unnecessary deaths. Finally, keep in mind that the decision to evacuate is further complicated by the fact that evacuations are highly disruptive to the local economy and create serious hardships for individual citizens. Evacuations involve more than science; they include sensitive political and economic issues as well.

SUMMARY POINTS

1. Most of the world's active volcanoes are located along convergent and divergent boundaries between tectonic plates where magma is generated. Hot spot volcanoes occur in the interior of a plate where a plume of magma rises up from the mantle.

2. There are three basic types of magma: basaltic, andesitic, and rhyolitic (granitic). Basaltic magmas are the dominant type at divergent plate boundaries, whereas andesitic magmas are most common at convergent boundaries along subduction zones. Rhyolitic magmas form when continental crust becomes involved in the melting process at convergent boundaries and hot spots.

3. The viscosity of magma refers to its ability to resist flowing. Magma's viscosity increases with increasing SiO_2 (silica) content and decreasing temperature. Basaltic magmas are the least viscous, and rhyolitic the most viscous.

4. When rocks melt into magma, gases are formed and generally remain dissolved within the magma, with water (H_2O) vapor, carbon dioxide (CO_2), and sulfur dioxide (SO_2) being the most abundant gases. Highly gas-charged magmas contain mostly water vapor and generally form in subduction zones where water-rich sediment is involved in the melting process.

5. Magma deep within the Earth is under considerable pressure, forcing magmatic gases to remain dissolved within the magma. When magma breaches the surface and encounters atmospheric conditions, the compressed gases rapidly expand and can create an explosive eruption.

6. The type of landform that develops when magma reaches the surface largely depends on the relative proportions of lava, gas, and pyroclastics that are extruded. Explosive eruptions generate considerable amounts of pyroclastics, but relatively small volumes of lava. The opposite is true for nonexplosive eruptions.

7. Shield volcanoes have gentle slopes and usually form by the eruption of gas-poor, low-viscosity basaltic magmas, which tend to erupt in a nonexplosive manner. Composite cones have steep slopes and form during explosive eruptions of gas-rich, high-viscosity andesitic magmas.

8. Some volcanic hazards such as explosive blasts, lava flows, and pyroclastic flows typically occur in a relatively small radius around the volcano. Other hazards like mudflows, ash fall, and tsunamis can threaten people considerable distances from the volcano itself. Noneruptive hazards include volcanic landslides, mudflows, and deadly gases.

9. Explosive eruptions can have global consequences by ejecting large quantities of volcanic ash and gas into the upper atmosphere, creating a cooling effect over the entire planet.

10. As the human population continues to expand, greater numbers of people are living in areas subject to volcanic hazards, thereby increasing the potential loss of life and property.

11. Volcanic eruptions are normally preceded by precursor activity, such as topographic changes, swarms of earthquakes, and release of volcanic gases. Scientists can measure these precursors and make fairly reliable eruption forecasts. The forecasts then provide an early warning for implementing emergency response plans and evacuating people, thereby minimizing the effects of an eruption.

KEY WORDS

caldera 169

cinder cones 167

composite cone 168

crater 169

hot spots 162

lava domes 167

lava flow 166

magma 161

magma chamber 165

magmatic earthquakes 189

pyroclastic flow 175

pyroclastic material 165

shield volcanoes 167

viscosity 164

volcanic ash 165

volcanic landslide 182

volcanic mudflow 183

volcano 167

APPLICATIONS

Student Activity   Buy two small bottles of club soda or seltzer water. Do this activity outside. Without shaking, open one bottle and let it stand. Shake the second bottle and open immediately. What happens? There is dissolved gas in both, but the shaken one "erupts" more violently. Put the cap on the second bottle and shake it again, then open it back up. Was the "eruption" as violent as the first time? Shake the second bottle again and open. The intensity of the eruption should decrease significantly. This demonstration shows the energy that dissolved gases have in magma. More violent eruptions typically involve magma that contains higher amounts of dissolved gases.

Critical Thinking Questions

1. What happens along subduction zones that makes magmas erupt so violently compared to magmas that form along divergent plate boundaries?

2. With respect to lava flow hazards, why are basaltic magmas able to threaten developed areas much farther away from a volcano compared to andesitic magmas?

3. Mudflows are a common hazard associated with composite cone volcanoes. Why is our ability to predict volcanic eruptions of little use when it comes to mudflows?

Your Environment: YOU Decide   Given the fact that most volcanoes erupt rather infrequently and that scientists can now predict eruptions relatively accurately, has the risk been reduced to the point where you would be willing to live near an active volcano? Also, do you think that volcano monitoring is a good use of taxpayer dollars?

© Image Plan/Corbis

Chapter 7

Earth materials suddenly moved downslope and severely damaged these homes and roadway in Laguna Beach, California. This type of movement, called mass wasting, is a natural process that results from the pull of gravity on sloping surfaces composed of rock or sediment. Certain triggering events, such as heavy rains, earthquakes, and construction activity, can cause less stable slopes to suddenly fail. Buildings and other human structures in sloping terrain are inherently at risk from mass wasting processes.

Mass Wasting and Related Hazards

CHAPTER OUTLINE

Introduction

Slope Stability and Triggering Mechanisms
Nature of Slope Material
Oversteepened Slopes
Water Content
Climate and Vegetation
Earthquakes and Volcanic Activity

Types of Mass Wasting Hazards
Falls
Slides
Slump
Flows
Creep
Snow Avalanche
Submarine Mass Wasting

Subsidence
Collapse
Gradual Subsidence

Reducing the Risks of Mass Wasting
Recognizing and Avoiding the Hazard
Engineering Controls

LEARNING OUTCOMES

After reading this chapter, you should be able to:

▶ Characterize the main factors that affect slope stability as they relate to the balance between gravitational and frictional forces acting on a slope.

▶ Describe the different types of weakness planes found in earth materials, and explain how they can affect the stability of slopes.

▶ Explain the two ways in which water acts to destabilize a slope.

▶ Characterize the four main triggering mechanisms for mass wasting events.

▶ Explain the fundamental differences between the following types of mass wasting: falls, slides, flows, slump, and creep.

▶ Discuss why subsidence occurs, and identify the geologic conditions that lead to rapid or gradual subsidence.

▶ Describe the primary ways in which humans can reduce the risk of mass wasting.

FIGURE 7.3 Photo shows the Beartooth Highway traversing steep terrain northeast of Yellowstone National Park. Construction of the highway required excavating material from the hillside, which made the slopes even more prone to mass wasting—note the scars left by repeated movements of rock and sediment. This photo was taken immediately after a series of rockslides and mudslides had cut or blocked the road in 13 separate locations, requiring extensive and costly repairs.
(all) Montana Dept of Transportation

repairs. Cost and safety considerations are sometimes so high that they justify the building of a tunnel.

In the following sections we will explore the different types of mass wasting and the ways in which they occur. Of particular importance will be the consideration of how human activity can help trigger these events, which, in turn, lead to costly repairs and maintenance. Lastly, we will examine some of the more common engineering techniques used to stabilize slopes and minimize the hazards associated with mass wasting.

Slope Stability and Triggering Mechanisms

Every rock and mineral grain on our planet is under the influence of Earth's gravitational force, but some slopes are inherently less stable than others and are therefore more prone to mass wasting. One of the key factors affecting the stability of a slope is its steepness, which can range from almost horizontal to completely vertical. Figure 7.4 illustrates how the pull of gravity on a rock creates the force we call weight. Notice that on a horizontal surface (a zero slope) the rock's entire weight is directed vertically downward, making movement by gravity alone impossible. However, on an inclined surface a portion of the gravitational force, which itself is a constant, will act parallel to the slope. This means that on a sloping surface some of the rock's weight is directed downslope. As the slope increases, so too does the gravitational component acting parallel to the slope, which allows more of the rock's weight to be directed in the downslope direction.

Most of us know from experience that rocks and sediment grains usually remain stationary on hillsides despite the force of gravity. The reason for this is that frictional forces are keeping them in place, as illustrated in Figure 7.4. You can demonstrate this for yourself by placing a block of wood on a board, and then gradually lifting one end of the board until the block begins to slide. What happens is that the gravitational component in the slope direction continues to increase until it finally overcomes the frictional resistance, at which point the block starts to slide. The same thing happens on a hillside where rocks and sediment remain in place until something causes the gravitational force to become greater than the frictional forces. When this happens the movement can occur suddenly, such as in a rockfall, or be so slow that it is imperceptible to the human eye.

In order for us to better understand slope stability and the potential for mass wasting, we need to examine the factors that influence the gravitational and frictional forces on a slope. In addition to steepness, other key factors include the type of rock or sediment making up the slope, the presence of water or ice, and the amount of vegetative ground cover. We will also need to explore what geologists and engineers call **triggering mechanisms,** which are processes or events that reduce the frictional forces on a slope and/or increase the effect of gravity. Earthquakes, heavy rains, and removal of vegetation by wildfires are all common triggers for mass wasting. We will begin this section on slope stability by considering the types of material that make up slopes.

Nature of Slope Material

Some rocks are inherently so strong and homogeneous that they are able to form stable cliffs, such as the granitic rocks of Yosemite Valley, California, shown in Figure 7.5. Here slabs of rock are separated from the main rock body along

Mass Wasting and Related Hazards

LEARNING OUTCOMES

After reading this chapter, you should be able to:

► Characterize the main factors that affect slope stability as they relate to the balance between gravitational and frictional forces acting on a slope.

► Describe the different types of weakness planes found in earth materials, and explain how they can affect the stability of slopes.

► Explain the two ways in which water acts to destabilize a slope.

► Characterize the four main triggering mechanisms for mass wasting events.

► Explain the fundamental differences between the following types of mass wasting: falls, slides, flows, slump, and creep.

► Discuss why subsidence occurs, and identify the geologic conditions that lead to rapid or gradual subsidence.

► Describe the primary ways in which humans can reduce the risk of mass wasting.

Earth materials suddenly moved downslope and severely damaged these homes and roadway in Laguna Beach, California. This type of movement, called mass wasting, is a natural process that results from the pull of gravity on sloping surfaces composed of rock or sediment. Certain triggering events, such as heavy rains, earthquakes, and construction activity, can cause less stable slopes to suddenly fail. Buildings and other human structures in sloping terrain are inherently at risk from mass wasting processes.

Introduction

© StockTrek/Getty Images

In Chapter 4 you learned that Earth's moving tectonic plates generate long underwater mountain ranges and deep ocean trenches, creating considerable topographic *relief* or elevation differences in the ocean basins. Plate movements also result in parts of Earth's landmasses being elevated far above sea level. While plate tectonics gives the globe its vertical relief, gravity works in the opposite sense by moving rock and sediment from areas of high elevation to low-lying areas, thereby acting to fill in Earth's rough surface. Gravity alone can directly move rock and sediment downslope (i.e., downhill), as in rocks falling from a cliff, or it can indirectly move material through the action of flowing water, ice, and air. Geologists use the term **mass wasting** to describe the general process of earth materials moving downslope due only to gravity, whereas those materials carried by some secondary agent fall under the category of stream, glacial, or wind transport (Figure 7.1). Note that the terms *landslide* and *avalanche* are often used synonymously with *mass wasting*, but actually imply specific types of movement. In order to avoid confusion, only the term *mass wasting* will be used in this chapter to describe the general movement of material by gravity alone.

Okanagan Valley, British Columbia, Canada

FIGURE 7.1   Photo showing how gravity caused part of a hillside to slide downslope in a process known as mass wasting. Note how the rock and sediment slid into the stream valley, forming a dam and creating a temporary lake. Running water will then slowly transport the material downstream. Together with water, wind, and ice, mass wasting plays an important role in shaping the landscape.
British Columbia Ministry of Forests and Range/Kevin Turner.

Although mass wasting processes play an important role in many different aspects of the Earth system, our focus in this chapter will be on how they affect humans. The basic problem is that when a body of rock or sediment begins to move downslope, anything in the path of this material is in danger of being destroyed. People and human-made objects typically are no match for the forces generated by moving masses of earth materials, especially when the material is moving quickly. Mass wasting not only puts human lives at risk, but also threatens valuable buildings, transportation networks, and utility lines (water, sewer, and electric).

Similar to other natural disasters such as earthquakes and volcanic eruptions, some of the worst mass wasting disasters have occurred when people live and work in a hazard zone. For example, from the photo in Figure 7.2 one can see the effects of a mass wasting event in 1970 that buried parts of two cities in Peru, killing an estimated 18,000 people. Note how this event, which was triggered by an earthquake, created a fan-shaped deposit at the mouth of a canyon where the cities had been built on more flat-lying terrain. This disaster is very similar to the volcanic mudflow described in Chapter 6 that took the lives of 23,000 residents of Armero, Colombia. Both of these disasters involved mass wasting debris that swept into an area where large numbers of people were concentrated. Keep in mind that the cities were located in hazard zones because the sites at the mouths of canyons offered relatively flat terrain and an abundant water supply. Moreover, several generations had passed between major events; thus many people were either ignorant of or complacent regarding the risks they faced. In the end, population growth helped ensure that more people would be living in the hazard zone when earth materials came crashing down the valleys just as they had in the past.

While large-scale mass wasting events can have devastating consequences, they are somewhat rare. Much more common are small-scale movements, whose cumulative effects are quite substantial due to the fact that they are so numerous. For example, consider that in the United States alone, mass wasting is estimated to cause $1–2 billion in damage and 25–50 deaths, *each* year. One reason why small-scale events are so numerous is that mass wasting can occur even on moderate slopes, which means it is a potential problem over a significant portion of the landscape. In addition to being a common process that is geographically widespread, mass wasting is often inadvertently triggered by routine human activities. For example, excavating material from hillsides for construction purposes and harvesting timber and growing crops on sloping terrain are activities that tend to destabilize slopes, and therefore trigger mass wasting. Ironically, society allocates considerable sums of money and resources to address mass wasting problems resulting from routine human actions.

Perhaps the best example of the interaction between human activity and small, but numerous, mass wasting events is the construction and maintenance of highways. In order to connect towns and cities, highways naturally have to be built across a variety of different terrains. As shown in Figure 7.3, constructing a flat roadbed that winds up and down steep terrain requires that earth material be excavated from the hillsides. The problem is that removing this material only helps destabilize steep slopes that are already prone to mass wasting—note the scars in the photos where rocks and sediment have routinely moved downslope. This destabilization results in the ongoing need for costly maintenance and emergency

FIGURE 7.2 Aerial photo showing the source area for a rock and snow avalanche that destroyed much of the Peruvian city of Yungay in 1970—parts of Ranrahirca were also destroyed. Note how the cities were built at the mouth of a canyon that leads up to the source area on Mount Huascarán.
USGS

FIGURE 7.3 Photo shows the Beartooth Highway traversing steep terrain northeast of Yellowstone National Park. Construction of the highway required excavating material from the hillside, which made the slopes even more prone to mass wasting—note the scars left by repeated movements of rock and sediment. This photo was taken immediately after a series of rockslides and mudslides had cut or blocked the road in 13 separate locations, requiring extensive and costly repairs.
(all) Montana Dept of Transportation

repairs. Cost and safety considerations are sometimes so high that they justify the building of a tunnel.

In the following sections we will explore the different types of mass wasting and the ways in which they occur. Of particular importance will be the consideration of how human activity can help trigger these events, which, in turn, lead to costly repairs and maintenance. Lastly, we will examine some of the more common engineering techniques used to stabilize slopes and minimize the hazards associated with mass wasting.

Slope Stability and Triggering Mechanisms

Every rock and mineral grain on our planet is under the influence of Earth's gravitational force, but some slopes are inherently less stable than others and are therefore more prone to mass wasting. One of the key factors affecting the stability of a slope is its steepness, which can range from almost horizontal to completely vertical. Figure 7.4 illustrates how the pull of gravity on a rock creates the force we call weight. Notice that on a horizontal surface (a zero slope) the rock's entire weight is directed vertically downward, making movement by gravity alone impossible. However, on an inclined surface a portion of the gravitational force, which itself is a constant, will act parallel to the slope. This means that on a sloping surface some of the rock's weight is directed downslope. As the slope increases, so too does the gravitational component acting parallel to the slope, which allows more of the rock's weight to be directed in the downslope direction.

Most of us know from experience that rocks and sediment grains usually remain stationary on hillsides despite the force of gravity. The reason for this is that frictional forces are keeping them in place, as illustrated in Figure 7.4. You can demonstrate this for yourself by placing a block of wood on a board, and then gradually lifting one end of the board until the block begins to slide. What happens is that the gravitational component in the slope direction continues to increase until it finally overcomes the frictional resistance, at which point the block starts to slide. The same thing happens on a hillside where rocks and sediment remain in place until something causes the gravitational force to become greater than the frictional forces. When this happens the movement can occur suddenly, such as in a rockfall, or be so slow that it is imperceptible to the human eye.

In order for us to better understand slope stability and the potential for mass wasting, we need to examine the factors that influence the gravitational and frictional forces on a slope. In addition to steepness, other key factors include the type of rock or sediment making up the slope, the presence of water or ice, and the amount of vegetative ground cover. We will also need to explore what geologists and engineers call **triggering mechanisms,** which are processes or events that reduce the frictional forces on a slope and/or increase the effect of gravity. Earthquakes, heavy rains, and removal of vegetation by wildfires are all common triggers for mass wasting. We will begin this section on slope stability by considering the types of material that make up slopes.

Nature of Slope Material

Some rocks are inherently so strong and homogeneous that they are able to form stable cliffs, such as the granitic rocks of Yosemite Valley, California, shown in Figure 7.5. Here slabs of rock are separated from the main rock body along

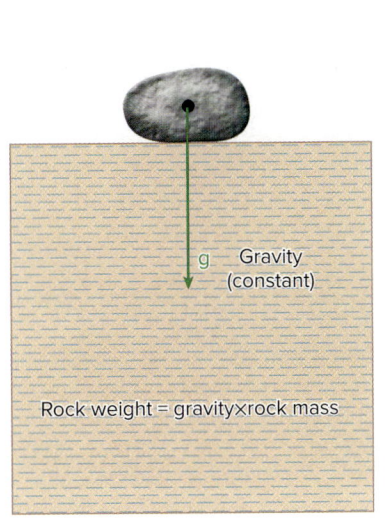

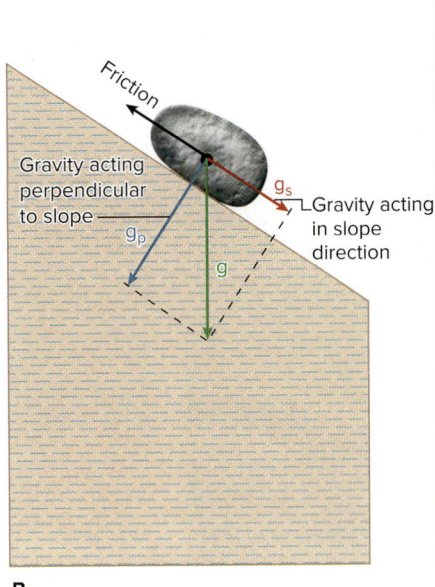

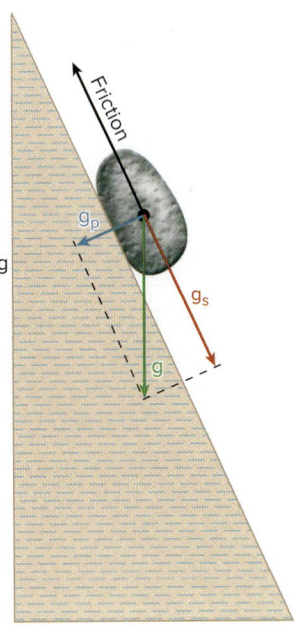

FIGURE 7.4   On horizontal (zero slope) terrain (A), the full weight of the rock due to gravity (g) is directed downward. On a hillside (B), part of the gravitational force (g_s) acts parallel to the slope, thereby directing some of the rock's weight downslope; the remaining component (g_p) acts perpendicular to the slope. On a steeper slope (C), more of the rock's weight is directed downslope, which requires greater friction to hold it in place.

fractures, producing walls that have remained nearly vertical for thousands of years. The walls are stable because the minerals in the massive granite are fairly resistant to chemical weathering, plus they form an interlocking network of crystals with great frictional resistance. In contrast to the towering cliffs in Yosemite, the alternating layers of sedimentary rocks in the Grand Canyon (Figure 7.6) are each composed of different types of minerals that vary greatly in their strength and ability to resist weathering. The weathering of these alternating layers of relatively weak and strong rocks creates a stair-step effect in the overall slope as opposed to a single massive cliff face. Note in Figure 7.6 that the steep, but relatively short, cliffs in the Grand Canyon form from individual layers of sandstone and limestone that are inherently strong and resistant to weathering. Between these resistant layers are easily weathered beds of shale, whose internal frictional resistance is much lower, which allows only gentle slopes to develop.

These examples illustrate that to have a nearly vertical cliff requires a material with strong internal friction, such as rock that is homogenous and composed of minerals that interlock and are resistant to weathering. Obviously, frictional forces within loose or unconsolidated sediments are generally lower than those in solid rock, making sediment more prone to mass wasting and less likely to form steep slopes. Depending on water content and the size and shape of the individual sediment grains, loose sediment typically does not form slopes greater than 35 degrees—geologists call this the *angle of repose*. Generally speaking, large angular fragments generate greater frictional forces, and are therefore capable of maintaining steeper slopes compared to small, well-rounded fragments. In terms of composition, sediments containing

FIGURE 7.5   This nearly vertical 2,000-foot (600-m) cliff has existed for thousands of years on Half Dome in Yosemite National Park due to the strength of the granite's interlocking mineral grains and resistance to weathering.
© Jim Reichard

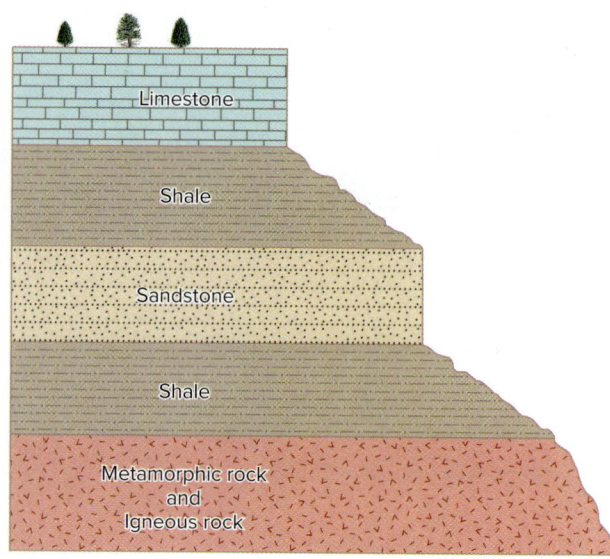

FIGURE 7.6 Differences in strength and resistance to weathering of sedimentary layers control the steepness of slopes in the Grand Canyon. Steep cliffs develop in resistant sandstones and limestones, whereas broad, gentle slopes form in the much weaker shales.
© Jim Reichard

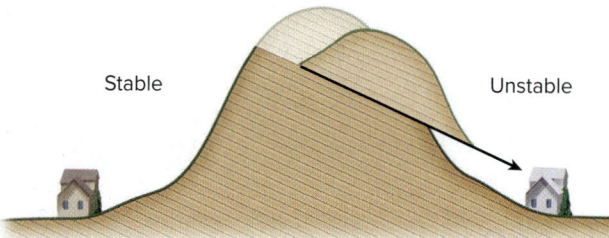

FIGURE 7.7 Planar surfaces such as bedding planes, faults, fractures, and foliation planes represent weaknesses within rocks that can greatly reduce slope stability. Particularly dangerous situations occur when these surfaces are inclined in the same direction as the slope, creating the potential for blocks of material to slide downslope.

considerable water and clay minerals tend to form the least stable slopes because they can behave as a plastic material (Chapters 3 and 4), which allows them to flow and spread out.

We also need to consider that earth materials commonly contain planar or sheetlike features that weaken them considerably. For example, both sediment and sedimentary rocks have *bedding planes*, which are horizontal surfaces that typically represent some change in the sediment during deposition. Sedimentary materials then tend to split or break along these planar surfaces. Should tectonic activity cause these materials to become inclined, slippage can occur along the bedding planes such that overlying sections of rock slide downslope, as shown in Figure 7.7. Faults and fractures represent planar breaks or openings and are typically found in most rock bodies (Chapter 3)—these features can even be found in semiconsolidated (partially cemented) sediments. As with bedding planes, masses of rock can slide along fracture and fault surfaces that are inclined in the same direction as the slope (Figure 7.7). Finally, foliation planes that commonly develop in metamorphic rocks (Chapter 3) can serve as sliding surfaces similar to bedding planes, fractures, and faults.

Oversteepened Slopes

As the steepness of a hillside increases, the component of gravity operating in the slope direction becomes greater, thereby increasing the potential for mass wasting. There are a number of ways, both natural and unnatural, whereby the steepness of a particular slope can change rather suddenly in terms of geologic time. Therefore, any activity that abruptly increases the slope and leads to mass wasting can be considered a triggering mechanism. Perhaps the most geologically important mass wasting trigger is the undercutting of stream banks (called *cutbanks*) caused by the natural migration of stream channels, as shown in Figure 7.8A. When a stream channel migrates and undercuts its bank, it creates a highly unstable overhang, which will eventually fall or slide into the channel. In fact, the primary way river valleys become wider over time is through mass wasting that is triggered by streams undercutting and destabilizing their banks.

In addition to natural processes, human activity often results in oversteepened slopes and costly mass wasting problems. For example, flat surfaces are required for the construction of roads, buildings, and parking lots. In sloping terrain this means that material must be excavated from hillsides as shown in Figure 7.8B to create a level surface. This removal of material weakens the slope significantly and greatly increases the potential for mass wasting. Note that this process can also trigger mass wasting below the leveled area, where the excavated material puts additional weight on the slope. Later in this chapter we will examine how engineers try to avoid these problems by building retaining walls above the cut and by hauling the excavated material away rather than placing it on the slope.

FIGURE 7.8 Both natural processes and human activity can destabilize a slope by increasing its steepness. (A) As a stream channel naturally migrates and cuts into the outside bank, it creates highly unstable overhang, which inevitably leads to mass wasting. (B) Construction of highways and buildings in hilly terrain requires that material be removed in order to create a level surface. This removal of material causes the slope to become oversteepened and susceptible to mass wasting.
(top) Courtesy of Myron McKee/River of Life Farm (bottom) Reproduced with the permission of Natural Resources Canada 2012 (Photo 2002-585 by Réjean Couture).

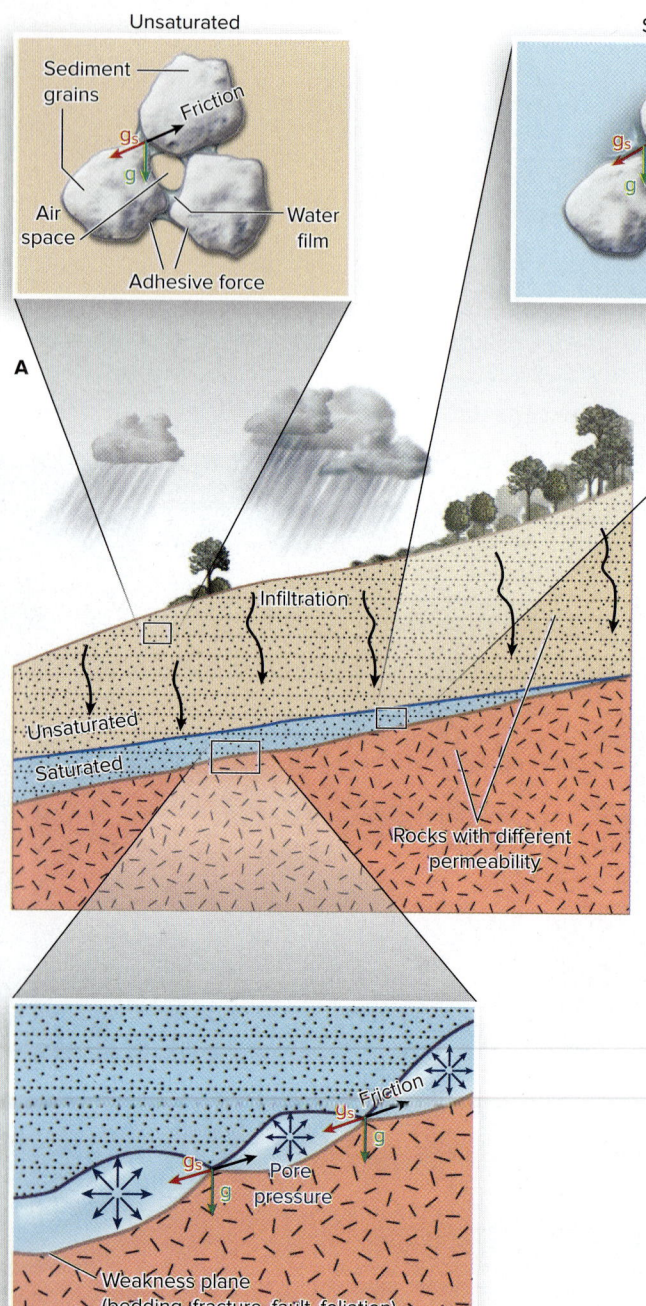

FIGURE 7.9 Rain or melting snow will infiltrate and eventually cause intergranular and planar voids in the subsurface to become saturated. The weight of the water causes the fluid or pore pressure within the saturated voids to increase (A), which reduces the friction between the solids. Downslope movement occurs when the frictional forces become less than the gravitational force in the slope direction. Note the enlarged view (B) shows the irregular nature of most planar surfaces.

Water Content

Another common mass wasting trigger arises when the water content of permeable earth materials increases and upsets the balance of forces existing on a slope. As illustrated in Figure 7.9, earth materials commonly contain voids, such as the *pore spaces* (Chapter 3) between individual grains as well as spaces along faults, fractures, bedding, and foliation planes. In the case of unconsolidated sediment, when a thin film of water forms within the pore spaces, the electrical attraction between the water molecules and mineral grains (adhesive forces) makes the sediment stronger. This effect, in turn, strengthens the slope. On the other hand, if too much water accumulates and the pores become saturated (Figure 7.9A), then the additional weight of the water tends to destabilize the slope. Here the excess water also reduces the frictional forces between the individual grains, thereby causing the material to become weaker. Water therefore can reduce the stability of a slope because it both increases weight *and* reduces friction in the direction of the slope. A good example is how relatively strong sand sculptures can be made at the beach by adding moderate amounts of water to the sand. However, if too much water is added, then the sculpture will collapse because of the additional weight and loss of friction between the grains.

The process whereby water destabilizes a hillside usually begins when rain or melting snow infiltrates the subsurface (Figure 7.9). As the infiltrating water moves downward through voids, it will eventually encounter a material with relatively low permeability. When this occurs, the voids in the overlying material will begin filling with water. The filling of voids is critical because when rock or sediment becomes saturated, the weight of the water generates fluid pressure or **pore pressure** that acts outward in all directions within the voids. Because part of this fluid pressure acts in the opposite direction of gravity, it reduces the weight of the earth material bearing down on the contact points between the solid grains and planar surfaces. This reduction in weight naturally decreases the frictional force within the materials, which makes the slope less stable. Consequently, if the saturated thickness within a rock or sediment layer keeps increasing, from heavy rains, for example, then the rising pore pressure can reduce frictional forces to the point where solid material begins to move downslope. This can be particularly dangerous when pore pressure reduces the friction along a weakness plane (Figure 7.9B), setting the entire mass of overlying material in motion.

Climate and Vegetation

The long-term average weather for a region is defined as *climate*, which is an important factor in slope stability because it ultimately determines how and when precipitation falls. Climate also determines the types of vegetation we see blanketing different slopes. The type of vegetative cover is important because it influences the fraction of rain or snow that infiltrates into the subsurface. For example, *humid climates* are characterized as having relatively consistent amounts of precipitation (rain and snow) distributed throughout the year (e.g., the Pacific Northwest and southeastern United States). Humid climates support fairly dense vegetation that tends to stabilize slopes as the plant roots help bind together loose particles of rock and sediment. However, during unusually large rainstorms or rapid snowmelts, dense vegetation will increase infiltration since it reduces the ability of surface water to move downslope. Excessive infiltration then adds significant weight to a slope and reduces friction through higher pore pressures. Although under normal conditions dense vegetation helps to stabilize slopes, during heavy and prolonged precipitation events it facilitates infiltration and leads to less stable slopes.

A much different situation exists in warm *arid climates* where relatively small amounts of precipitation produce sparse vegetation (e.g., the southwestern United States). Although rainfall is infrequent in these regions, when it does occur, it is often fairly intense. Intense rains combined with sparse vegetation in arid climates makes it easier for loose material to move downslope. Note that a similar situation can take place in humid climates where wildfires or logging activity suddenly remove vegetation from a hillside. Because of the rapid way in which the slope is destabilized, the loss of vegetation can be considered a triggering mechanism that initiates mass wasting. Once the vegetation is gone, mass wasting can be triggered by a rainfall event that would not otherwise cause the slope to fail. Vegetation, of course, will eventually reestablish itself, but the slope will remain unstable during the period of regrowth. Note that in California there is a fairly regular pattern of wildfires during the dry summer conditions, followed by winter rains that trigger mudslides, which explains some of the state's frequent mass wasting events. Finally, we should mention that in colder climates (e.g., in the Upper Midwest of the United States and in Canada), where freeze/thaw cycles are common, material can move downslope simply due to the expansion and contraction of water.

Earthquakes and Volcanic Activity

The ground vibrations associated with earthquakes and volcanic eruptions are another common type of triggering mechanism for mass wasting. For example, the massive landslide described earlier that took the lives of 18,000 people in Peru was triggered by a large (magnitude 7.9) earthquake. The 1980 eruption of Mount St. Helens (Chapter 6) also triggered an enormous landslide. Recall that these types of processes release tremendous amounts of energy, which are then transformed into seismic waves (Chapter 5) that travel outward in all directions. When seismic waves pass along the surface, the least stable slopes are the first to fail as the ground vibrations suddenly reduce the frictional forces within the slope. Clearly, stronger earthquakes and volcanic eruptions cause more slopes to fail due to the higher levels of ground shaking. Also note that passing seismic waves often cause surface materials to undergo liquefaction, which can immediately destabilize a slope and trigger a mass wasting event.

In addition to seismic waves, volcanic activity can trigger massive mudflows when lava or hot debris causes rapid melting of a volcano's snow and ice cap, as occurred in the Armero disaster described in Chapter 6. Gases and fluids commonly found within a volcano can also weaken the rocks to the point that the slope eventually fails. It is now believed that this internal weathering of rocks has played a major role in triggering the large landslides known to have occurred repeatedly on Mount Rainier (Chapter 6).

Types of Mass Wasting Hazards

As mentioned in the introduction to this chapter, mass wasting processes play an important role in the Earth system; hence they help shape the environment we humans depend on. However, these processes also pose serious risks to our safety and infrastructure, such as buildings, bridges, and highways. Mass wasting is a hazard because it involves large masses of earth material, whose downhill motion will obliterate, crush, or bury everything in its path. In some cases the material moves so slowly that the motion is imperceptible to humans, but still powerful enough to destroy or cause considerable damage. Perhaps most familiar, and most common, are situations where the sudden movement of material damages a section of highway, creating an immediate hazard to motorists and requiring difficult and costly repairs.

Because mass wasting presents substantial problems for society, geologists and engineers have devised different techniques and strategies for reducing the risk associated with these earth movements. The particular approach that is used largely depends on the type of mass wasting. For example, minimizing the effects of rocks free-falling from a cliff requires an entirely different approach than is used to keep wet, finer-grained sediment from flowing downhill. Scientists and engineers who study and try to mitigate mass wasting find it useful to classify the different types of movement. However, due to the number of variables involved, no single classification system has been devised that adequately describes all the different types of mass wasting processes. This has resulted in overlapping and somewhat confusing terminology.

In order to simplify our discussion of mass wasting, we will examine the different processes in terms of the type of movement, namely falling, sliding, flowing, and creeping. As the name implies, a *fall* involves material free-falling from a cliff or tumbling down a steep slope. On the other hand, a *slide* is where a body of material moves downslope along some surface, and in a *flow* the material moves with the consistency of a viscous fluid. The last type of movement is *creep*, which refers to the imperceptibly slow movement of loose material (earth or debris). By combining the kind of movement with four basic types of slope materials, namely *rock, debris, earth,* and *mud,* we get the simplified classification system shown in Figure 7.10. This system is well suited for our discussion that follows because the terminology is both informative and easy to understand. Note that some combinations of materials and movement are not represented in Figure 7.10 (e.g., mud fall, rock flow) either because they are rare or do not occur at all. Finally, a very common process called *slump* is not listed in this classification because it is a hybrid type of movement that involves both sliding and flowing.

Falls

A **fall** is the rapid movement of earth materials falling through air. As illustrated in Figure 7.11, *rockfalls*, sometimes called *topples*, involve relatively small amounts of material and begin when a slab or block of rock becomes detached

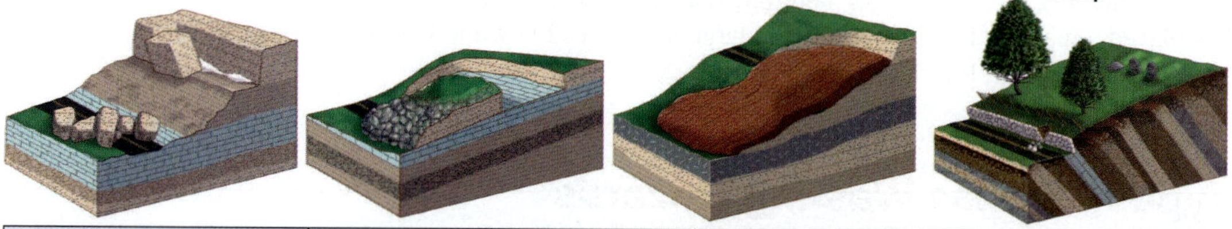

FIGURE 7.10 Mass wasting can be categorized based on the type of material involved and the way it moves downslope. Blank boxes in the chart indicate those combinations of materials and movement that take place rarely, or not at all.

Type of motion

	Fall	Slide	Flow	Creep
Rock (large blocks of solid rock) 	Rockfall	Rockslide		
Debris (mixture of rock, earth, plants, and mud) 	Debris fall	Debris slide	Debris flow	Creep
Earth (loose sediment, weathered rock fragments) 	Earth fall	Earth slide	Earth flow	Creep
Mud (mixture of water and finer-sized sediment)			Mudflow	

Type of material

from a steep wall of solid rock. This process typically includes the repeated freezing and thawing of water within fractures or other planar surfaces in a rock body. When liquid water freezes into ice, it naturally expands, and then acts to slowly wedge a block of rock away from the wall until the block becomes detached and begins to free-fall through the air. The falling block will usually crash into the cliff face before breaking up into smaller pieces as it hits the rocks at the base of the cliff that had fallen previously. These smaller pieces of rock will then bounce, roll, or slide on top of the rock pile, often dislodging individual rocks such that they too start to move downslope. Eventually the entire pile of rock attains a stable slope of between 30 and 35 degrees—referred to as the *angle of repose*. Over time this process produces a wedge-shaped deposit of rocks called a **talus pile,** which is a common feature found at the base of exposed rock bodies (Figure 7.11).

Rockfalls are an obvious hazard, but they are also an important geologic process that helps widen valleys and lower mountain ranges over long periods of time. Clearly this material must go somewhere; otherwise talus piles would simply grow until a valley is filled in. What normally happens is that the rocks at the bottom edge of the pile are slowly incorporated into streams or glaciers, which then carry the material away and deposit it elsewhere. Although most falling debris consists of solid rock, *earth falls* and *debris falls* are common in areas where migrating rivers undercut stream banks made of unconsolidated materials (refer back to Figure 7.8A). In this way earth falls and debris falls also help supply sediment to a stream.

Slides

As the name implies, a **slide** occurs when material moves in a sliding manner on some zone of weakness, such as along a bedding plane, fault, fracture, or foliation plane. Slides that involve masses of rock, earth, or debris (mixtures of rock and earth) are referred to as *rockslides, earth slides,* or *debris slides*, respectively. As illustrated in Figure 7.12, slides are more likely to take place when the zone of weakness, or weakness plane, is inclined in the same direction as the slope of the land surface. Also note how the material moves downslope over a relatively stable and undisturbed layer. In the case of rockslides, it is common for one or more large blocks to move as a single unit. However, in earth and debris slides the material above the zone of weakness is more likely to break up and form a jumbled mass as it moves downslope.

The potential for a slide is also high in areas where a permeable layer, such as sand or sandstone, overlies a less permeable layer of shale or clay (Figure 7.12). As described in the preceding section, when infiltrating water from prolonged rains or melting snow reaches a less permeable layer, it will begin filling up the pore spaces. This causes water to accumulate in the overlying layer, which leads to the slope becoming less stable. Here the extra water increases the weight acting in the direction of the slope, plus it raises the pore pressure such that the frictional forces are reduced along the contact between the two layers. In addition to the internal effects of water, the slope can be further destabilized by the undercutting action of a migrating stream or highway construction. Slides are a particularly serious problem for society because they commonly involve large volumes of material that move onto highways and into flat-lying areas where humans tend to live and build.

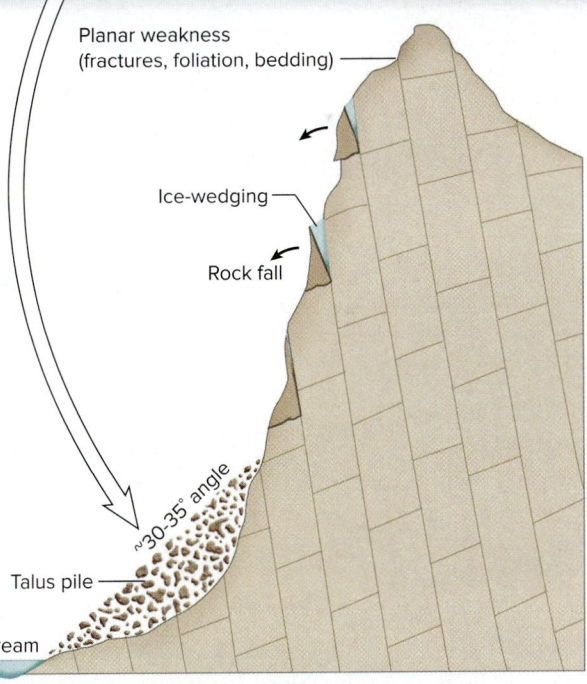

FIGURE 7.11 Rockfalls generally result from repeated freezing and thawing of water within fractures in steep exposures of solid rock. Expanding ice creates a wedging effect that eventually pushes a slab outward to the point where it free-falls. Over time this process creates a deposit of broken rocks at the base of the cliff called a talus pile. This material is then carried downstream by rivers and glaciers.
© Jim Reichard

FIGURE 7.12 Rockslides (A) consist of blocks of solid rock sliding on top of a zone of weakness such as a bedding or foliation plane, a fault, or a fracture. Earth and debris slides (B) are less coherent and tend to break up and move as a jumbled mass. Common triggering events for slides include streams undercutting their banks and infiltrating water that increases the pore pressure along less permeable layers.

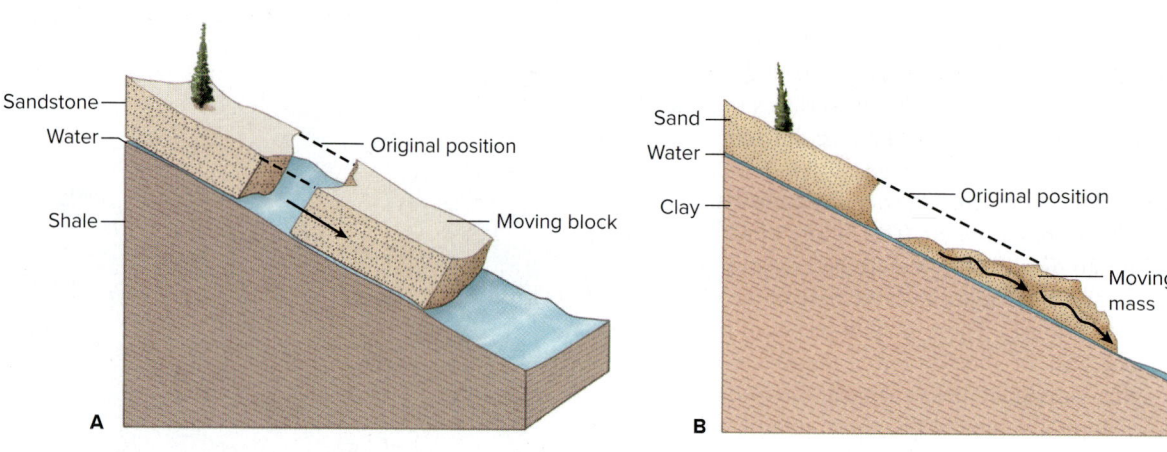

FIGURE 7.13 Photo showing the aftermath of the 2006 debris slide that killed over 1,100 people in the village of Guinsaugon on Leyte Island, Philippines. A period of unusually heavy rains is believed to have played a major role in triggering the deadly slide.
© Rick Guthrie

A tragic example of the hazards associated with slides occurred in 2006 on Leyte Island in the Philippines. Here over 1,100 people perished when a massive debris slide buried much of Guinsaugon, a village built in a valley at the base of the steep hillside shown in Figure 7.13. Material from the slide moved downward and covered nearly half the width of the valley, traveling as much as 2.4 miles (3.8 km) from the source area. Scientists later estimated that the slide reached a velocity of 78 miles per hour (126 km/hr), which made it nearly impossible for the residents to escape. The researchers also concluded that the primary triggering mechanism was a period of unusually heavy rain that ended four days prior to the slide.

Slump

As mentioned earlier in this section, a **slump** is a complex form of mass wasting where both sliding and flowing take place in unconsolidated material (earth and debris). From Figure 7.14 you can see that in the upper portion of a slump the material slides along a curved surface whose shape is similar to a spoon. At the very top of the slump is a curved and distinct scar called a *scarp*, which marks where the disturbed material becomes detached from the undisturbed material. Note how at the base of the slump, also called the *toe*, a prominent bulge develops. Because slumps generally occur in unconsolidated materials with a high water content, the bulging mass near the toe tends to flow (see the next section), giving the terrain a jumbled or disorderly appearance. Slumps are very problematic since once they form, additional movement is difficult to prevent, particularly if the toe is removed in order to clear debris away from a road or building. The mass within the toe actually provides support for the slide material lying farther up the slope. Therefore, when material is cleared away from the toe, the entire mass becomes destabilized and triggers additional slumping at the top, which then sends even more material downslope. The complex slump that occurred at La Conchita, California (Case Study 7.1), provides a good example of the problems associated with recurrent movement of this type.

Flows

In Chapter 4 you learned that when solid rock, deep within the Earth, is subjected to enough heat and pressure, it begins to flow and deform in a plastic manner. Here we will consider a **flow** to be a type of mass wasting where surface material moves in a continuous manner due to some external force. Take, for example, how gravity causes liquids, such as water and oil, to flow when they are on a sloping surface. A similar process occurs when loose (unconsolidated) material covering a hillside accumulates enough water that the material's internal friction is reduced, allowing it to behave like a fluid and start flowing downslope. The speed at which a flow moves depends on the type of material involved, the amount of water it contains, and the steepness of the slope. Perhaps the most common type of mass wasting involving flow is a *mudflow*, which is composed of mostly fine-grained sediment and enough water to enable it to move downslope. In mudflows the relative amounts of sediment and water may vary considerably, making some so viscous the entire mass flows rather slowly, whereas others are more fluid and resemble a turbulent river heavily overloaded with sediment.

Recall from Chapter 6 that volcanic mudflows (lahars) can form when heavy rains pick up ash lying on the flanks of composite volcanoes, and then carry it off into stream channels. Volcanic mudflows can also develop near a volcano's summit when glacial ice undergoes rapid melting. Note that mudflows form in nonvolcanic regions as well, particularly in rugged terrain where vegetation is sparse because of an arid climate or recent wildfires. Since sparse vegetation offers less protection against the impact of falling raindrops, heavy rain events can loosen significant amounts of fine sediment, making it easier for running water to carry it away. A key point here is that regardless of whether a mudflow begins on a volcanic or sparsely vegetated slope, the results are similar. From Figure 7.15 you can see that both cases involve large volumes of sediment that are picked up by water as it flows over a fairly rugged landscape. The resulting mixture of mud then moves into small stream channels, which progressively merge into a single channel that funnels the mudflow through a narrow canyon. Eventually the canyon empties out onto a relatively flat area, allowing the mudflow to spread and form a fan-shaped deposit. As was the case in the Armero disaster (Chapter 6), humans have historically put themselves at grave risk by building towns and cities on old mudflow deposits because of the flat terrain and steady water supply.

A *debris flow* is similar to a mudflow, but the primary difference is that it contains particles ranging in size from mud and sand to large boulders and trees (see Figure 7.10). Because of the wide variety of materials involved, debris flows naturally form in a greater number of geologic settings. For example, debris flows form on volcanic and sparsely vegetated slopes, consisting of both fine and coarse fragments as opposed to just fine particles. Water flowing down the slopes can then funnel this material (debris) into channels, sending it crashing downhill where it forms a fan-shaped deposit near the bottom of the slope. In addition, debris flows occur on heavily vegetated slopes that become unstable after taking on too much water, which reduces the internal friction. This situation is fairly common during unusually large rainfall events in more humid climates with steep terrain (Case Study 7.2). Examples include Pacific storms that move inland over rugged coastal areas and tropical storms and hurricanes that make landfall and move over parts of the Appalachian Mountains. Note that the news media typically use "mudslides" when referring to both mudflows and debris flows.

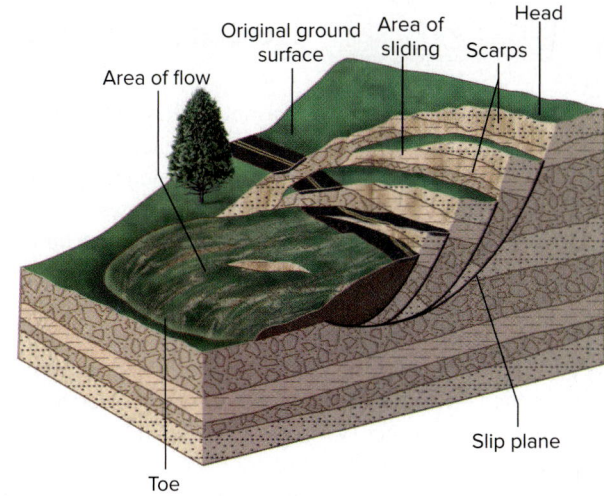

FIGURE 7.14 Slumps are complex events where material moves by sliding along a spoon-shaped surface near the top and then flows toward the bottom or toe. Note the distinctive scars called scarps at the top and the jumbled terrain near the toe.

Recurrent Mass Wasting at La Conchita, California

On January 10, 2005, a debris flow crashed into the small residential community of La Conchita, California, taking the lives of 10 people and completely destroying 23 homes and damaging 13 others. This disaster drew national news coverage in the United States, and then took on an added dimension when reporters learned that a similar event occurred at the same site 10 years earlier in 1995. Questions naturally arose as to why people would continue to live at the base of a hillside with a history of mass wasting. Other questions centered on why the local government allowed people to build there in the first place. Television interviews revealed that many of the local residents felt that living in a pleasant seaside community with easy access to the beach and wonderful ocean views was simply worth the risk. To understand why the local government allowed people to live

in a known hazard zone, it is helpful to examine the geology of the site and its historical human development.

La Conchita lies on a narrow strip of land located between the Pacific Ocean and a steep hillside (Figure B7.1). These bluffs consist of poorly consolidated layers of marine sediment that have recently been uplifted by tectonic activity. Modern geologic studies have shown that mass wasting has been taking place along these same bluffs for

FIGURE B7.1 Color infrared photo (vegetation in red) showing the town of La Conchita, California, located below bluffs overlooking the Pacific Ocean. The bluffs consist of uplifted marine sediments that are weak and prone to mass wasting after periods of heavy rain. The inset shows the area where movement occurred in 1995, and then again in 2005.
© Robert A. Larson, (inset) USGS

many thousands of years. In terms of human development, historical records as far back as 1865 note that a wagon trail along this section of the coast was plagued by masses of earth falling down from the bluffs. Later a railroad was built along the coastal strip, and it too experienced problems when earth movements buried the tracks in 1887 and 1889. In 1909 another event destroyed a work train, prompting the railroad company to try to reduce the hazard by removing part of the hillside using bulldozers. The idea was to create a relatively flat area that would collect earth material moving down from the bluffs, thereby minimizing the chance of any material reaching the railroad tracks. However, this operation also created level ground that attracted the attention of real-estate developers, who eventually purchased the property in 1924. A housing development soon emerged, with 330 lots with easy access to the ocean and an additional 47 lots located along the base of the bluffs themselves (Figure B7.1). La Conchita was thus born, but it was only a matter of time before earth materials would again start moving downslope, posing a threat to this new community.

Small-scale earth movements took place along the bluffs above La Conchita in 1988, 1991, and 1994; fortunately none were large enough to reach the town below. Then on March 4, 1995, a large section of the bluff moved several tens of meters downslope in a matter of minutes, taking the form of a combined slump and earth flow (Figure B7.1). Although this relatively slow-moving mass damaged or destroyed a total of nine homes, no one was injured. Geologic studies later determined that the event was triggered by unusually heavy rain that winter. For example, in the six-month period leading up to the event, this coastal region received nearly 30 inches (761 mm) of rain, which was nearly double the normal average of 15.4 inches (390 mm). Also significant is the fact that 24.5 inches (623 mm) of rain fell in January alone, a month where only 4.3 inches (108 mm) normally falls. In the end, scientists concluded that the primary triggering mechanism of the large slump and earth flow was a rise in pore pressure associated with the infiltrating water during the exceptionally heavy January rains.

A mere 10 years later a smaller event took place on January 10, 2005, claiming the lives of 10 people. Unusually heavy winter rains again served as the primary triggering mechanism, but in this case the movement took on different and more deadly characteristics. In 2005 the movement occurred at a much shallower depth and had the form of a fairly rapid debris flow, leaving residents no time to flee. Scientists later concluded that the differences in speed and volume between the 1995 and 2005 events were related to the depth that water had been able to infiltrate into the slope. From the graphs in Figure B7.2 one can see that the 2005 debris flow occurred right at the end of a 15-day period of heavy rain, whereas the 1995 event took place nearly a month after an extended period of heavy rain. This timing implies that water did not infiltrate as deep in 2005 before rising pore pressure reduced the frictional resistance within the slope materials to the point of failure; it also explains why the 2005 failure occurred at a much shallower depth and involved a smaller volume of material. Finally, the fact that water accumulated closer to the surface helps account for the 2005 debris flow being more fluid-like, giving it greater speed and less time for people to get out of the way.

Based on the geologic history and the scientific studies of mass wasting near La Conchita, one can expect that future earth movements will be triggered by periods of heavy rainfall. However, it is nearly impossible to predict, with any degree of certainty, exactly where and when movement will take place and whether it will be rapid or slow. The reason such predictions are difficult is because individual mass wasting events are somewhat unique, often depending upon complex subsurface conditions that are difficult to determine in advance. Further complicating any prediction is the knowledge that future movement could be triggered by an earthquake, which unlike periods of heavy rainfall provides absolutely no warning that movement may be imminent. What is known is that future mass wasting activity near La Conchita is almost certain to occur and that people will likely find themselves in harm's way.

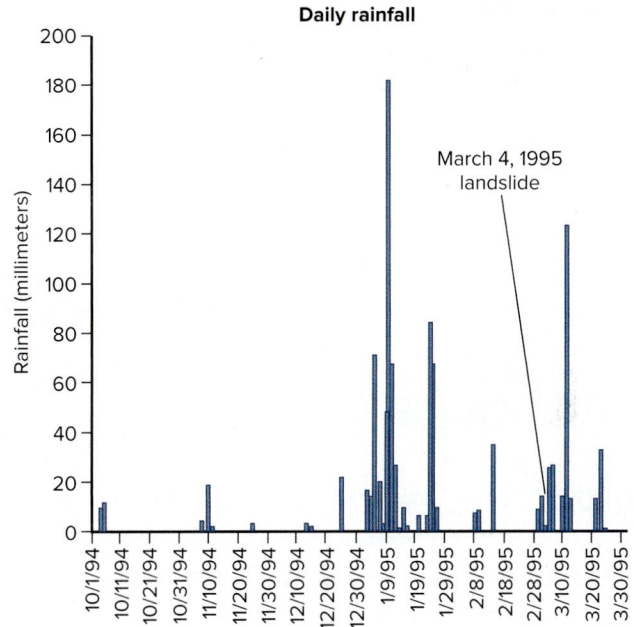

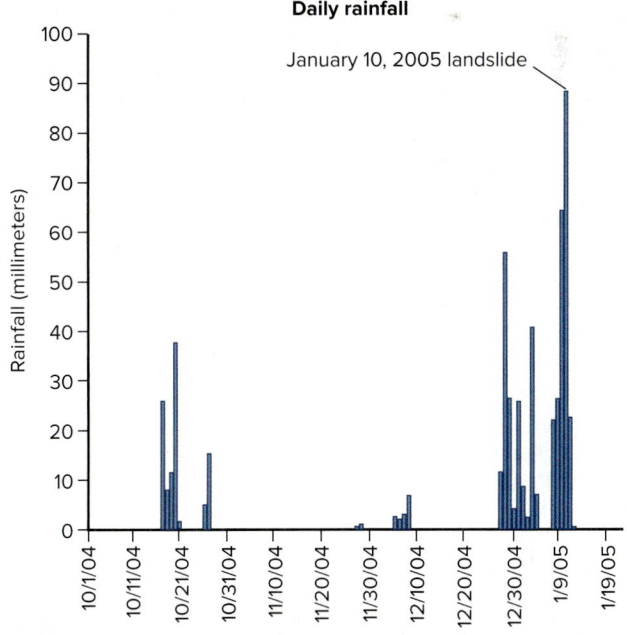

FIGURE B7.2 Graphs showing the different rainfall accumulation patterns that led up to the 1995 and 2005 mass wasting events at La Conchita.

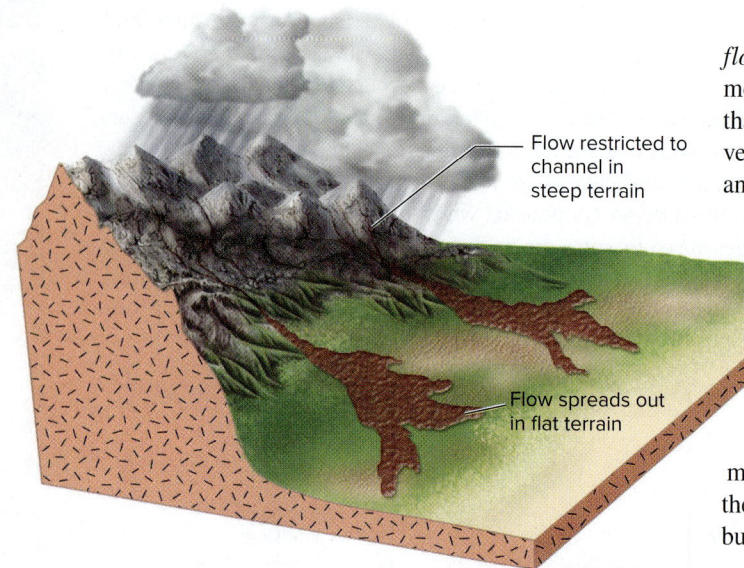

Flow restricted to channel in steep terrain

Flow spreads out in flat terrain

Unlike mud and debris flows where water picks up loose material, an *earth flow* involves a large section of an unstable hillside that flows downslope as a more coherent and viscous mass. In general, earth flows are common on slopes that are underlain by sediment richer in clay minerals and held in place by thick vegetation. Here the extensive root system helps bind the loose particles together, and therefore minimizing the potential for mudflows. However, during prolonged rains the infiltrating water can saturate the material, thereby adding weight to the slope and reducing the internal friction within the sediment. The presence of clay minerals is important in reducing friction because their *plasticity* (ability to deform by flowing) greatly increases as they take on water. A special type of earth flow, called *solifluction*, is common in cold climates, such as in northern Canada and Russia, where most of the ground remains permanently frozen—called *permafrost*. Problems occur when the uppermost zone thaws during the summer, which causes it to become saturated and begin flowing downslope over the frozen base. Solifluction can be a serious and costly problem for roads and buildings, especially since these features have the tendency to increase the rate of thawing, which, of course, can exacerbate the problem.

Creep

Creep is an exceptionally slow process where repeated expansion and contraction causes unconsolidated materials to move downslope. One of the ways creep takes place is by the freezing and thawing of water within the pore spaces of unconsolidated materials. Liquid water naturally expands as it freezes into ice, which forces the overall sediment volume to increase. When the ice melts, the volume decreases. Expansion and contraction also occur due to the cyclic wetting and drying of clay-rich materials. Here electrical forces allow the microscopic clay-mineral particles to alternately trap and release water molecules. During wet periods the clay particles take on water, causing the overall volume of material to expand. When the material dries, it will shrink—similar to a sponge that expands when it becomes wet and then shrinks as it dries.

The effect of repetitive expansion and contraction on unconsolidated material is illustrated in Figure 7.16. From the inset you can see that during expansion the individual particles move upward, perpendicular to the slope. But then, during contraction, gravity pulls the particles vertically downward. Over time this results in particles taking a zigzagging path downslope. Because the greatest motion occurs closest to the surface, this means that particles move more slowly as you go deeper in the sediment profile. It is this difference in speed that causes weathered layers of rock to bend downslope. Keep in mind that even the fastest creep rates are so slow that the resulting movement is imperceptible to humans. The rate of creep in loose materials depends on several factors, such as the amount of water and clay in the material, number of freeze/thaw and wet/dry cycles, slope steepness, and presence of plants with deep roots. Although creep is exceedingly slow, it exerts a nearly continuous force capable of damaging human structures, such as retaining walls, fences, buried utility lines, and aboveground utility poles (Figure 7.16). While creep may not be as dramatic as rockslides and falls, it is a widespread and costly problem.

FIGURE 7.15 Mudflows and debris flows typically form in areas where there is an abundance of loose sediment on relatively steep slopes. Water carries this material off the slopes, funneling it down steep channels where it forms a fan-shaped deposit near the base of the slope. Debris flows can also occur on heavily vegetated slopes when unusually heavy rains saturate loose material. Photo shows a 2007 debris flow that buried several homes and closed a highway near Clatskanie, Oregon.
Photo courtesy of Clatskanie, Oregon, PUD

Snow Avalanche

Mass wasting often involves snow rather than rock or sediment, in which case the process is usually referred to as a **snow avalanche** (Figure 7.17). In recent

Mass Wasting Tragedy at Oso, Washington

On March 22, 2014, a slump occurred near Oso, Washington, in which large blocks of glacial sediment broke loose from a forested plateau and slid down along a zone of weakness, generating a massive debris flow at the base of the slide. The highly mobile flow then rushed across the North Fork of the Stillaguamish River, burying a small subdivision and killing 43 people (Figure B7.3). The Oso slump (debris slide and flow) was the deadliest mass wasting event in U.S. history. Although there were numerous warning signs and predictions that movement would occur in this area, no one anticipated the full extent of the tragedy.

Based on aerial photographs from the 1930s, geologists mapped numerous slump deposits along the valley floor of the North Fork of the Stillaguamish River (Figure B7.4). Subsequent studies revealed almost ideal conditions for mass wasting within the valley due to the humid climate and underlying geology. During the last ice age, glaciers deposited a mix of well-sorted sand and clay layers throughout the region. Rivers, like the Stillaguamish, later cut down through these deposits, leaving a forested plateau standing 650 feet (200 m) above the valley floor. As the river naturally meanders (bends) and undercuts its banks, the slopes above become oversteepened and less stable (refer to Figure 7.8A). In addition to stream bank erosion, mass wasting in the valley is also triggered by increases in pore pressure that result from the high rates of rainfall. In areas underlain by sand, significant amounts of rainwater are allowed to infiltrate deep into the subsurface. The infiltrating water eventually encounters a clay layer, at which point the material above the clay starts to become saturated as groundwater levels rise (refer to Figure 7.9). The increased weight and pore pressure then lower the frictional resistance within the slope, increasing the likelihood that sandy materials will slide over the clay layers.

The first documented movement at the site of the Oso slump was based on aerial photographs taken in 1937. Geologists determined that the older event was caused by the Stillaguamish River migrating northwestward into previous slump deposits (Figure B7.4). The risk of future movement was considered low at the time since only a few homes had been built between the highway and the river, plus the older slump deposits were on the other side of the river. However, since the river was picking up sediment as it cut into the older slump deposits, fisheries were being negatively impacted. This led state officials to consider different engineering solutions. One option was to build a retaining wall to the prevent slump material from sliding into the river. The other was to reroute the river and disconnect the meander so it would no longer undercut the slope. Unfortunately, this second option had to be abandoned in 1959 when a developer built a 48-unit housing subdivision in the area where the river would have been rerouted (Figure B7.4). Engineering efforts were then focused on building various structures designed to prevent movement at the base of the old slump. Despite these efforts, in 1967, a 300 foot high by 800 foot long (90 m by 240 m) section of the slope collapsed and sent debris crashing into river, forming a temporary dam that caused flooding throughout the subdivision.

By the 1980s there was growing concern that logging on the plateau above the 1967 slump might help trigger renewed movement by letting additional rainwater enter more sensitive parts of the slope. After studies revealed that logging may have contributed to previous slides, the state began restricting the areas in which logging was permitted. Despite the logging restrictions and engineering efforts, parts of the 1967 slide were reactivated in 1988, 1996, and 2006, sending material crashing into the river. After the 2006 event caused renewed flooding within the subdivision, concerns were raised that future slides might actual cross the river and put residents in peril. Additional engineering controls were then put in place while officials examined the feasibility of buying out some of the property owners in the subdivision. Unfortunately, the buyouts never took place.

FIGURE B7.3 Aerial view of the Oso slump looking west. Notice the scarp at the top where water-rich glacial sediment slid along a zone of weakness, generating a highly mobile debris flow at the base (toe). The debris flow quickly crossed the river and buried 49 homes before reaching the other side of the valley. Note that a major highway was blocked along with the river, causing extensive flooding upstream.
Mark Reid, U.S. Geological Survey

—Continued

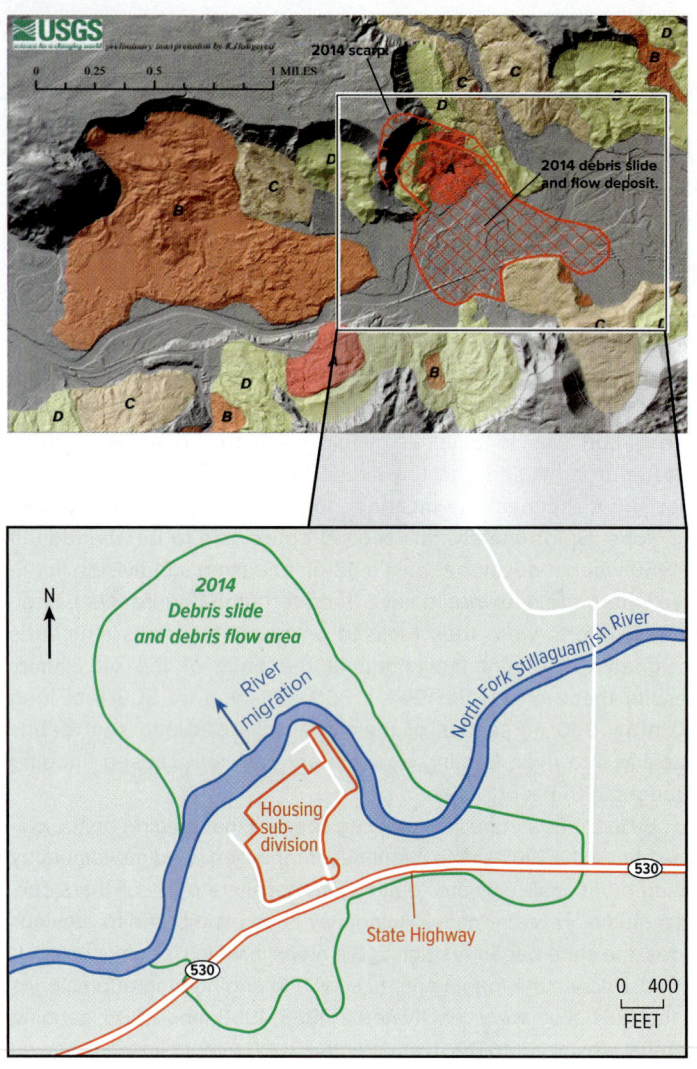

FIGURE B7.4 Elevation map showing mass wasting deposits along the Stillaguamish River valley. The 2014 Oso slump deposit and scarp are outlined in red, and the ages of older deposits are indicated by letters, with A being the most recent and D the oldest. Note the plateau area above the river valley and the steep scarps that mark the tops of the older slump deposits.

(left) Ralph A. Haugerud, U.S. Geological Survey.

After the massive 2014 slump near Oso killed 43 people, scientists and engineers undertook studies to better understand the cause of the tragedy. As with most mass wasting events, this debris slide and flow was ultimately triggered by an extended period of unusually heavy precipitation. In fact, the area was estimated to have received 12–16 inches (300–400 mm) of rain in the 21 days leading up to the event. Scientists concluded that excessive water not only added weight and increased pore pressure within the slope, but it also helped make the resulting debris flow exceptionally fluid. The highly fluid nature of the debris flow enabled it to cross the river and bury the homes in the subdivision. Interestingly, the flow was so mobile that some of the material actually reflected off the slope on the other side of the river valley (Figure B7.4).

In 2015 the governor of Washington state signed a bill passed by the legislature that expanded the state's hazard assessment program to include mapping areas at high risk of mass wasting. If the program is properly funded, it will allow residents to make more informed decisions as to where to live, and help local officials decide what areas to zone for new housing developments. Note that these types of zoning decisions are often complicated by property rights issues and the desire for continued development.

years the number of fatalities and property damage from snow avalanches has increased significantly as mountain resorts have expanded to accommodate the increasing number of people who enjoy winter sports. However, long before winter sports became popular, avalanches posed a serious hazard to railroad and highway traffic in mountainous areas where steep slopes and thick accumulations of snow are common. Under the right climatic and internal snow conditions, the snow may move downslope in a tumbling mass known as a *loose-snow avalanche*. This type of avalanche is relatively rare and accounts for only a small percentage of all snow avalanche-related deaths and property damage. Much more common is what is called a *slab avalanche*, where a coherent slab (sheet) of dense snow slides along a weakness zone within the snowpack. The key to understanding most snow avalanches is to know how slabs and weak layers form within a snowpack.

The development of slabs and weak layers is related to the way snow accumulates in mountainous terrain, and is then subjected to wide variations in temperature, sunlight, and wind. The climatic variations, combined with the weight of the overlying snow, cause individual snowflakes to recrystallize into a variety of more compact forms. During this process, layers of snow begin to bond to one another, forming more dense and coherent masses of snow. However, under certain weather conditions the bonding between two layers can be rather poor, resulting in the development of a weak zone or layer within the snowpack. A weak layer often forms, for example, when the surface of the snowpack becomes crusted with ice and is then overlain by a fresh layer of snow. As can be seen in Figure 7.17B, this produces thick slabs of coherent snow that are separated by thin, weak layers. Also important is the fact that all the layers are inclined in the same direction as the slope, which as you recall from Figure 7.7 is the least stable orientation. Clearly, movement is most likely to begin along a weak layer because this is where the frictional resistance is the lowest.

As with other types of mass wasting, a snow avalanche can be triggered by a natural process or human action that alters the balance between the weight acting in the slope direction and frictional forces along weak layers. Natural triggering mechanisms include heavy accumulations of fresh snow and surface melting that enables water to percolate downward and reduce the friction along a weak layer. Interestingly, snow avalanches are more common later in the winter partly because weak layers tend to form more frequently early in the snow season. This means that there are usually a greater number of weak layers in the lower parts of a snowpack. As snow continues to accumulate throughout the winter, the additional weight makes the slope progressively less stable.

With respect to snow avalanches triggered by humans, most occur when the additional weight from a skier or snowmobiler allows the weight of the snow to finally overcome the frictional resistance along a weak layer. When this happens, the overlying slab breaks free and quickly accelerates, reaching speeds of up to 80 miles per hour (130 km/hr); as it accelerates, the slab begins to shatter into smaller and smaller pieces. People who get caught in such an avalanche find themselves in a moving fluid whose density is less than a human body, thus they tend to sink into the flowing snow. In some cases people are able to literally

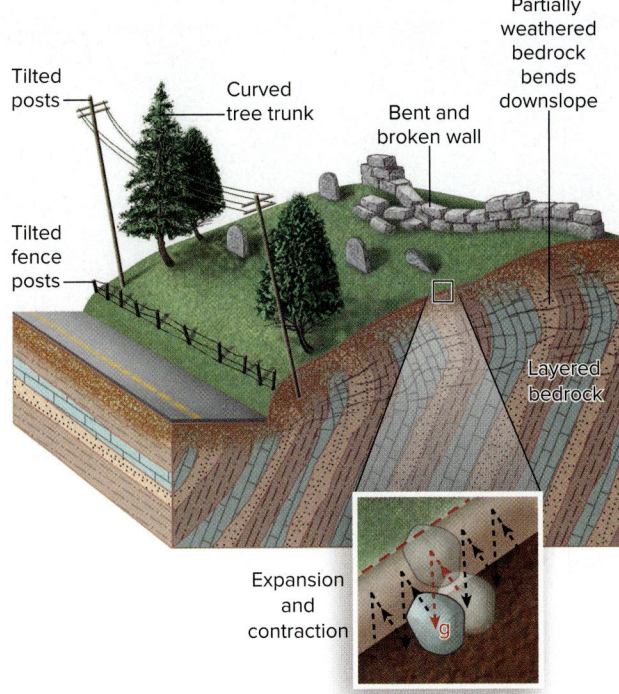

FIGURE 7.16 Creep is the extremely slow movement of unconsolidated materials caused by repeated expansion and contraction resulting from freeze/thaw and wet/dry cycles. The inset shows how expansion and contraction cause particles to take a zigzagging path downslope. Because the motion decreases with depth, the weathered rock layers appear to bend downslope. This movement can cause damage to a variety of human-made structures.

A 2008, San Juan Mountains, Colorado

FIGURE 7.17 Most snow avalanches occur in mountainous areas where weak layers form within the snowpack. Common triggers are heavy snowfall events and human activity that add weight to the slope and overwhelm the frictional forces along a weak snow layer. The avalanche in (A) began when a slab started sliding along a weak layer. Closer view (B) of a detached slab—note the skier for scale.
(a) © Mark Rikkers (b) © Toby Weed

B 2001, Logan Mountains, Utah

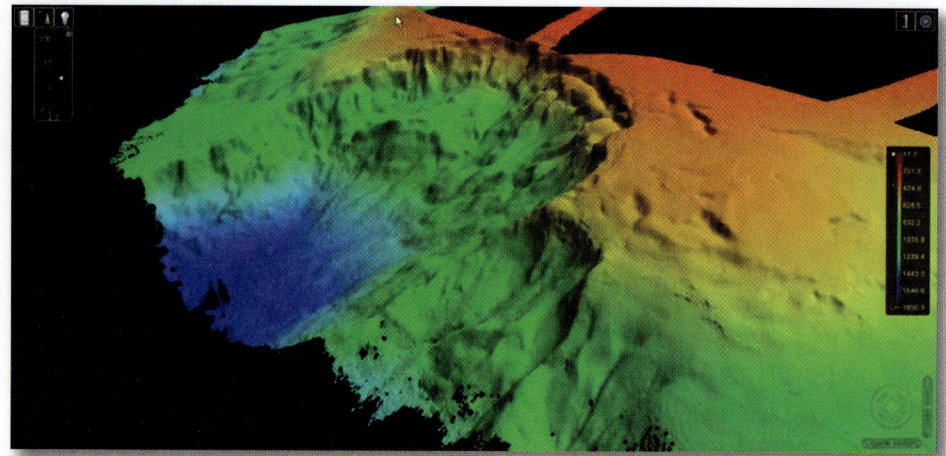

FIGURE 7.18   A sonar survey of the ocean floor near St. Croix Island in the Caribbean Sea reveals a complex submarine slump over 4.5 miles (7 km) across. Note the prominent scarp at the top of the slump.
NOAA

swim their way to the surface of the flow and stay there until it stops. Unfortunately, most people remain trapped within the flow. Once the fluidized snow stops, it takes on the consistency of concrete and escape becomes nearly impossible. Although the packed snow can be about 60–70% air, most people die from asphyxiation as carbon dioxide gas builds up around their mouths. In fact, over 90% of buried avalanche victims are found alive if dug out within 15 minutes, but only 20–30% will survive after 45 minutes. In this situation rapid rescue is obviously critical, which is why many skiers and snowmobilers now wear radio transmitters or beacons so they can be quickly located by rescuers.

Submarine Mass Wasting

Although hidden from human view, mass wasting also takes place on the seafloor, especially where sediment accumulates on slopes found along the edge of continental shelves. Here thick sediment sequences are naturally saturated and often contain gases, which helps reduce cohesion between the sediment grains. As sediment continues to accumulate on the shelf, the deposit pushes seaward. Eventually the sediment deposit begins to spill over the edge of the shelf, causing the slope areas to become progressively less stable. Mass wasting can then be triggered by an earthquake, hurricane, or passing tsunami, which causes movement within the sediment that reduces the cohesion or frictional resistance between the individual grains. Oceanographers today are obtaining the clearest pictures yet of the seafloor and are finding submarine slides, slumps, and flows similar to those found on land, but at much larger scales, such as the complex slump shown in Figure 7.18.

As oceanographers continue to study the offshore environment, they are discovering potential hazards associated with submarine mass wasting events. For example, should a submarine slope abruptly fail and involve a large volume of sediment or rock, the movement would displace an equal volume of seawater, creating the potential for a tsunami (Chapter 5). Just such an event is suspected to be responsible for a 55-foot (17-m) wave that struck the coast of Papua New Guinea in 1998. This tsunami originally was thought to have been triggered by an earthquake, but the source has now been traced to a large slump located 16 miles (25 km) offshore. Other slides and slumps that correlate to known tsunamis have also been found off the coasts of Peru and Puerto Rico.

In addition to tsunami hazards, submarine mass wasting can cause significant damage to offshore oil and gas pipelines lying on the seafloor. For example, in 2004 Hurricane Ivan roared up through the Gulf of Mexico, disrupting offshore oil production for months. After the storm, production facilities and over 10,000 miles (16,000 km) of pipeline lay damaged, largely due to submarine slides and flows. Overall, the damage caused a reduction in daily U.S. oil production by almost 500,000 barrels, representing nearly 15% of total U.S. production. Because this disruption came at a time when world petroleum supplies were already tight, the submarine mass wasting events in the Gulf of Mexico were a significant factor in keeping world oil prices relatively high throughout the latter part of 2004.

Subsidence

In this section we will examine the process known as **land subsidence,** which is the lowering of the land surface due to the closing of void spaces within subsurface materials. Since land subsidence occurs in a strictly vertical sense and does not require a slope, it is technically not a form of mass wasting. The reason subsidence is included in this chapter is because it fits the general theme of earth materials moving downward in response to gravity. In addition, introducing the topic here should provide you with a better understanding of the relationships between subsidence and coastal flooding (Chapter 9), groundwater withdrawals (Chapter 11), and extraction of mineral and energy resources (Chapters 12 and 13). Details on these relationships will be provided in the corresponding chapters.

The closing of void spaces that leads to subsidence is ultimately related to *overburden pressure*, which we defined in Chapter 4 as the weight of the overlying rock and sediment bearing down on subsurface materials. This means that a subsurface body of rock or sediment must be strong enough to support the weight of everything lying above it. Should something reduce the ability of subsurface materials to support this weight, pore spaces and open fractures tend to close. Subsurface materials then become more compact. Because most of the compaction takes place in the vertical sense, the end result is that the land surface naturally sinks or subsides. Under some geologic conditions subsidence takes place gradually, but in other situations it happens so suddenly that it is referred to as a *collapse*. In the following sections we will briefly explore both gradual and sudden (collapse) subsidence.

Collapse

Sudden or rapid land subsidence typically occurs in areas where unusually large void spaces are found in subsurface materials. Subsurface cavities, commonly called *caves*, can form by groundwater circulating through soluble rock, magma draining from a magma chamber, and underground mining. However, the vast majority of caves form when groundwater slowly dissolves away limestone rock, which is composed of the soluble mineral called calcite (Chapter 3). As illustrated in Figure 7.19, the key to this process is the way cavities are drained of water as a river slowly cuts downward and lowers the water table. When the

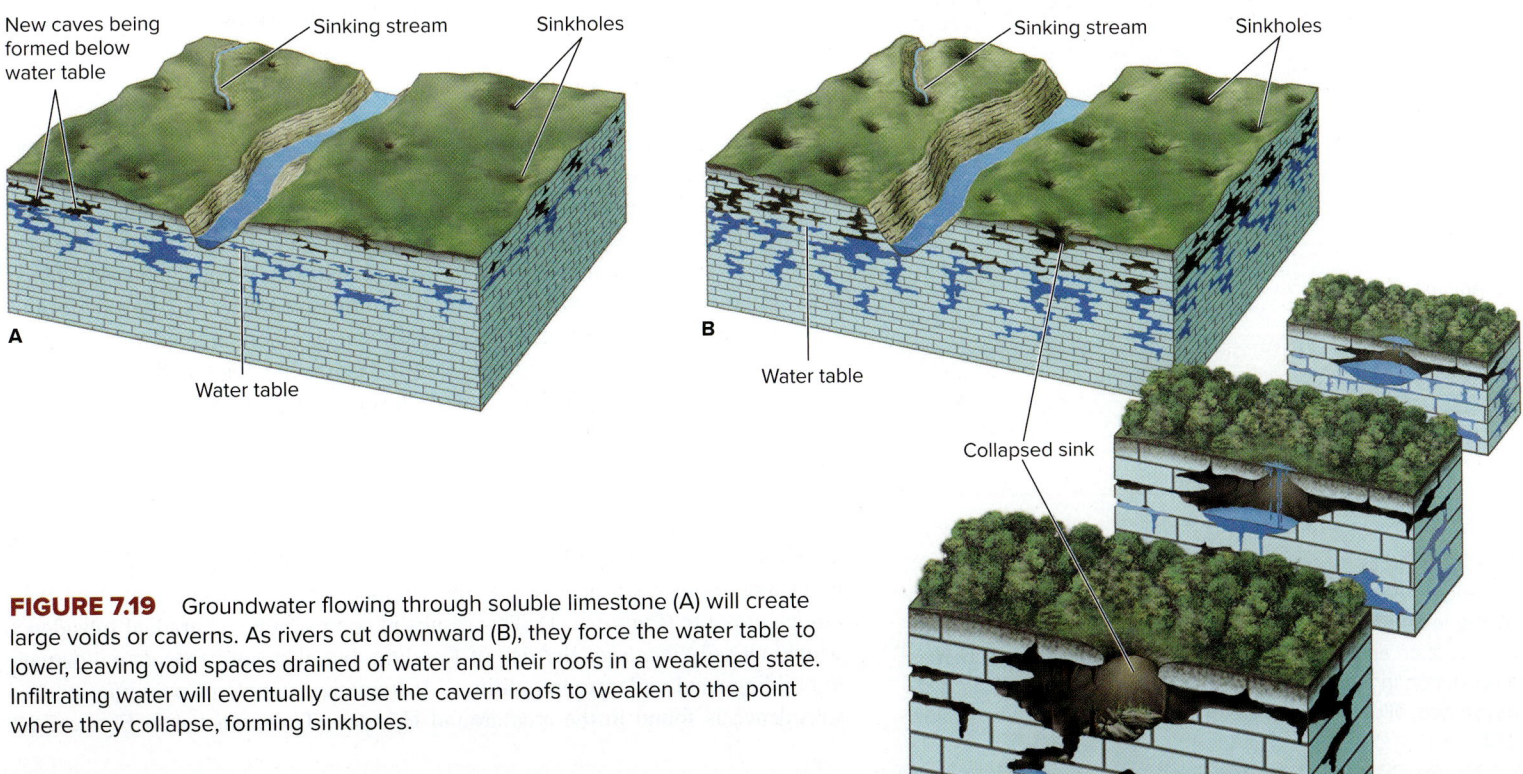

FIGURE 7.19   Groundwater flowing through soluble limestone (A) will create large voids or caverns. As rivers cut downward (B), they force the water table to lower, leaving void spaces drained of water and their roofs in a weakened state. Infiltrating water will eventually cause the cavern roofs to weaken to the point where they collapse, forming sinkholes.

Winter Park, Florida

FIGURE 7.20 The sudden collapse of sinkholes is normally associated with large cavities that form in layers of soluble rock relatively close to the surface. Sinkholes cause significant damage to buildings, highways, and utility lines. USGS.

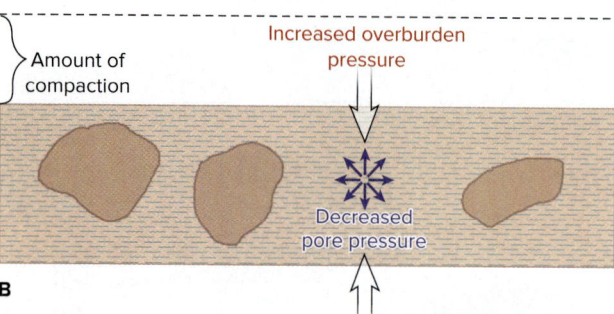

FIGURE 7.21 The amount of overburden pressure, or weight, that sediment grains must bear (A) is offset by the level of pore (fluid) pressure within the sediment. Compaction and subsidence can occur (B) whenever there is a reduction in fluid pressure, or when additional sediment is deposited, allowing more weight to bear down on the grains.

214

cavities are full, the water helps support the weight of the overlying rock, but once the water is gone, the roofs and walls must bear the full weight. At the same time infiltrating rainwater will continue to dissolve away the cavern roofs until they weaken to the point they suddenly fail and collapse, forming what geologists call **sinkholes.** In areas where limestone is relatively close to the surface, the number of sinkholes can be so great that the landscape takes on a pitted or cratered appearance (Figure 7.19). This type of landscape is often referred to as *karst* terrain.

The sudden collapse of sinkholes can be a widespread and serious problem in areas underlain by soluble limestone, as is the case for large portions of Florida and Kentucky. From the photo in Figure 7.20 one can see that any human structure which lies directly above a limestone cavity is at risk of being destroyed should the roof suddenly fail. The formation of sinkholes can also create serious and costly disruptions to transportation and utility networks. Although heavy rainfall events commonly trigger the collapse of limestone sinkholes, human activity can play an important role as well. For example, constructing a building directly over a large cavity may increase the overburden pressure (weight) beyond what the cavern roof can safely support. Another common factor is the lowering of the water table by large groundwater withdrawals (Chapter 11), leaving caverns empty and more prone to collapse.

Finally, void spaces created during underground mining for mineral resources or coal (Chapters 12 and 13) often collapse, thereby causing land subsidence. The collapse of mining voids that are relatively shallow will result in pits forming at the surface, which appear similar to limestone sinkholes. In the case of mining voids that are deeper, the collapsed space results in a more gentle sagging of the land surface.

Clearly, the best way for society to minimize the problems associated with collapses is to avoid constructing permanent and valuable structures over large cavities. Fortunately, the location of subsurface voids can be easily determined in advance, particularly in limestone terrain. In recent years ground-penetrating radar has been successfully used to locate voids that are relatively close to the surface. Should a cavity be found beneath an existing building, engineers can design systems that will prevent rainwater from weakening the cavern roof. This can be done by installing drains and covering the surface with impermeable material so as to minimize the amount of surface water that can infiltrate the void.

Gradual Subsidence

Gradual or slow land subsidence is usually associated with the compaction of pore spaces within a sedimentary sequence. As illustrated in Figure 7.21, compaction occurs in fine-grained sediment when overburden pressure forces individual clay mineral particles to become aligned in a parallel manner. The realignment causes the material to become more compact, which can lead to gradual land subsidence. This process occurs naturally in areas where new sediment is being deposited, such as in the Mississippi Delta. Here additional sediment increases the overburden pressure, causing compaction within the delta. In Chapters 8 and 9 we will take a closer look at how human activity has increased the rate of subsidence in the Mississippi Delta. This increased subsidence, in turn, has increased the risk of flooding along the Louisiana coast, including in New Orleans.

Perhaps the most common cause of slow subsidence is a reduction in pore pressure associated with the pumping of large volumes of water or oil from the subsurface. Since fluid pressure acts outward in all directions, any reduction in the amount of water or oil forces the sediment grains to bear more of the overburden pressure (Figure 7.21). This results in compaction and gradual subsidence which not only increases the risk of flooding, but also causes structural damage to buildings and underground utilities. A dramatic example of pumping-induced subsidence is found in the area around Houston, Texas, shown in Figure 7.22.

Here the subsurface withdrawal of water and oil has created a bowl-shaped depression that has subsided as much as 10 feet (3 m) since 1906 (some low-lying areas are now submerged). The decreasing land elevation has made it more difficult for streams to drain water off the landscape, thereby increasing the frequency and intensity of flooding.

Reducing the Risks of Mass Wasting

In the previous sections you learned that tremendous forces are generated when rock or sediment moves downslope. Unfortunately, people often put themselves in danger by building in areas susceptible to mass wasting. It is also common for normal human activity to serve as a triggering mechanism for mass wasting events. In order to reduce the risk of mass wasting, one of the most obvious things we can do is to simply avoid building or living in hazardous areas. This solution of course may not be feasible where people are already living in a hazard zone, or where there is a need for a certain activity, such as an important transportation link through mountainous terrain. If it is undesirable or impractical to avoid the hazard, then another option is to try to minimize the risk through engineering strategies. In some situations we may use control structures designed to help stabilize a slope, hence keeping the material from moving in the first place. Should engineers determine that it is not cost effective to try to stabilize a slope, a structure can be built that will help shield buildings and highways from any material that eventually moves downslope.

In this section we will examine the risk management approach in more detail, as well as some of the more common types of engineering controls.

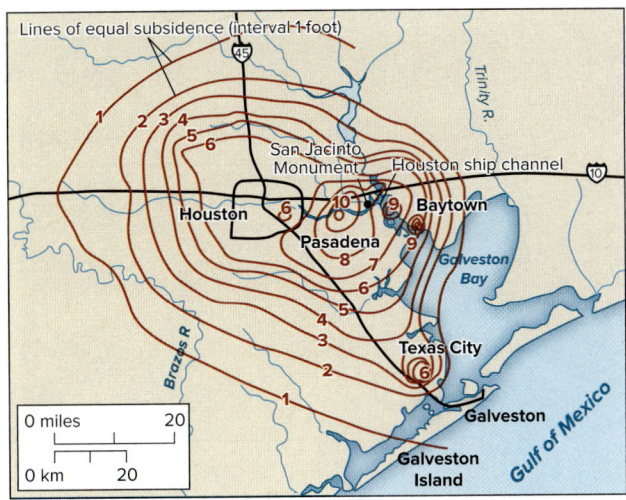

FIGURE 7.22 Gradual subsidence can take place when large volumes of water or oil are removed from the subsurface. Compaction occurs in clay-rich layers as water molecules between individual clay particles are slowly removed. Heavy withdrawals of water and oil in the Houston area have resulted in as much as 10 feet (3 m) of subsidence, creating a bowl-shaped depression that has led to serious flooding problems.

Recognizing and Avoiding the Hazard

The first step in minimizing the risks associated with mass wasting in a given area is to identify the slopes that are unstable. Because slopes tend to fail repeatedly, the easiest way to locate unstable slopes is to recognize the telltale signs of mass wasting. For example, boulders at the base of a cliff are a clear indication that rockfalls have taken place and will occur again in the future. Likewise, scarps and jumbled terrain point to past slump activity; curved or deformed walls are clear evidence of creep. These types of features can be seen in the field and on aerial photographs, enabling geologists to make hazard maps showing areas of past movement. In addition, by understanding the science behind mass wasting processes, geologists and engineers can identify slopes that have a high potential for failure, but for which movement has yet to occur. Key risk factors include oversteepened slopes, weakness planes oriented parallel to slopes, unconsolidated materials, and materials with a high water content. Note that hazard maps showing high-risk areas are often available through the U.S. Department of Agriculture, federal or state geological surveys, and local planning offices.

Once a high-risk slope is identified, steps can be taken to minimize human activity within the hazard zone, particularly that which may trigger a mass wasting event. Such steps commonly involve zoning laws designed to restrict development along with construction ordinances prohibiting the oversteepening of slopes. In general, it is much cheaper to avoid human activity in areas with unstable slopes than to design and construct engineering controls for preventing mass wasting. In situations where a hazard zone is simply unavoidable, detailed surveys can be performed to gather data on slope materials and conditions prior to the design and construction of an engineering solution. This preventive form of risk reduction is highly desirable since the cost involved in avoiding a mass wasting event is often far less than the potential property damage.

A

B

FIGURE 7.23
Retaining walls are commonly used both above and below highways to strengthen oversteepened slopes. The retaining wall in (A) helps stabilize an oversteepened slope created when part of the hillside was removed, whereas the wall in (B) is supporting fill material that was placed on the slope to create a flat area for the road.
(a) California Department of Transportation (b) © Jim Reichard

Engineering Controls

Engineering controls are a key element in most efforts to minimize the hazards associated with mass wasting. In general, engineering solutions involve using a number of different controls. The decision as to what combination of controls to actually use is based on the type of expected movement and the characteristics of the site itself. In some cases the goal is to prevent movement; in others it is simply to provide protection from movement that is all but impossible to avoid. If the objective is to prevent movement, then slope stability efforts can be maximized by increasing frictional forces within the slope, while at the same time decreasing the weight acting in the downslope direction. Consequently, most efforts aimed at increasing slope stability involve one or more of the following engineering controls.

FIGURE 7.24   Photo of a slide for which the material was removed from the highway. This removal reduced support for the remaining material above the road, thereby increasing the chance of continued movement. A retaining wall here would help provide support and prevent additional movement. Note the large amount of material placed below the road, forming a buttress that will help support the highway.
California Department of Transportation

Mendocino County, California

Retaining Walls

A **retaining wall** is an engineering structure made of steel, concrete, rock, or wood whose basic purpose is to strengthen oversteepened slopes. Retaining walls are commonly used whenever flat or level surfaces are needed in sloping terrain, such as for roadways, buildings, and parking lots. Recall from our earlier discussion that creating a level surface on a slope requires either removing (i.e., excavating) part of the slope or bringing in fill material and placing it on the slope (refer back to Figure 7.8). Either technique generates an oversteepened slope, which will likely begin to move unless it is supported by a retaining wall. For example, the photo in Figure 7.23A shows how a retaining wall was used to stabilize a slope above a highway, which was oversteepened when it was cut to make room for the road. In some situations, however, making this type of cut is not desirable because the amount of material above the road would be too great for a retaining wall to support. Engineers can then put fill material on the slope below the road and hold it in place with a retaining wall as shown in Figure 7.23B.

Although retaining walls are commonly used as a preventive measure, they can also be used to gain control over slides or flows that occur repeatedly. For example, Figure 7.24 shows an active slide for which the toe material has been removed from the highway. Because the toe normally provides support for the upper part of the slide, removing the toe material typically results in

renewed movement. A retaining wall above the road would provide additional support and decrease the likelihood of future movement. However, a retaining wall in this case may not be adequate depending on the amount of loose material that it has to support. A retaining wall is also not effective if the plane on which the material is moving lies beneath the wall, in which case the wall is likely to become part of the moving mass. Note in Figure 7.24 how engineers placed material below the roadway to form a buttress that helps support the highway from below, similar to a retaining wall.

Rock Bolts

Rockfalls or rockslides are common in areas where highly fractured rocks are exposed on steep slopes and where weakness are inclined in the same direction as the slopes. To minimize the potential for rockfalls or small-scale rockslides, **rock bolts** are commonly used to anchor loose rocks to more massive, solid bodies of rock (Figure 7.25). Installation requires that a hole first be drilled to some desired depth, after which a bolt is inserted and held in place by one of several different techniques. The most common anchoring technique is to use a tip that expands when the bolt is turned, forming a wedge that keeps the entire assemblage from backing out of the hole. Other types of rock bolts make use of cement grout or simple friction within the borehole to keep the bolts fixed in place. Rock bolts are used extensively along highways and rail lines where fractured rocks create a near constant threat of rockfalls. Bolts are also widely used for stabilizing walls and ceilings in tunnels and underground mines.

Controlling Water

In this chapter you learned that water plays an essential role in a number of mass wasting processes, particularly in acting as a triggering mechanism. As water accumulates in the subsurface, the increase in fluid pressure within pore spaces and along weakness planes reduces the frictional forces within slope materials. Water not only reduces friction, but also increases the weight acting in the direction of the slope. Consequently, a highly effective means of keeping a slope stable is to control or limit the amount of water that can accumulate in it. Here a common technique is to install a network of perforated pipes and/or gravel beds in order to drain water from within the slope. For example, Figure 7.26 illustrates how drains can be placed behind retaining walls to prevent the buildup of water. Note in this figure that another useful approach is to combine drains with features that prevent water from entering a slope in the first place. For example, a berm or channel can be used to divert surface water away from the upper portion of an unstable slope. Diversion is especially important in keeping water from flowing into open fractures at the top of slumps and slides. Under some circumstances it may even be necessary to cover large sections of a slope with impervious plastic sheeting to prevent water from infiltrating into unstable materials. Finally, the buildup of water in some situations can be minimized by limiting the amount of water being used for landscape or agricultural irrigation in areas above an unstable slope.

Terracing

As can be seen from the example in Figure 7.27, **terracing** involves creating a series of *benches* (flat surfaces) on a hillside—retaining walls are often used to support the oversteepened portions of the slope. The construction

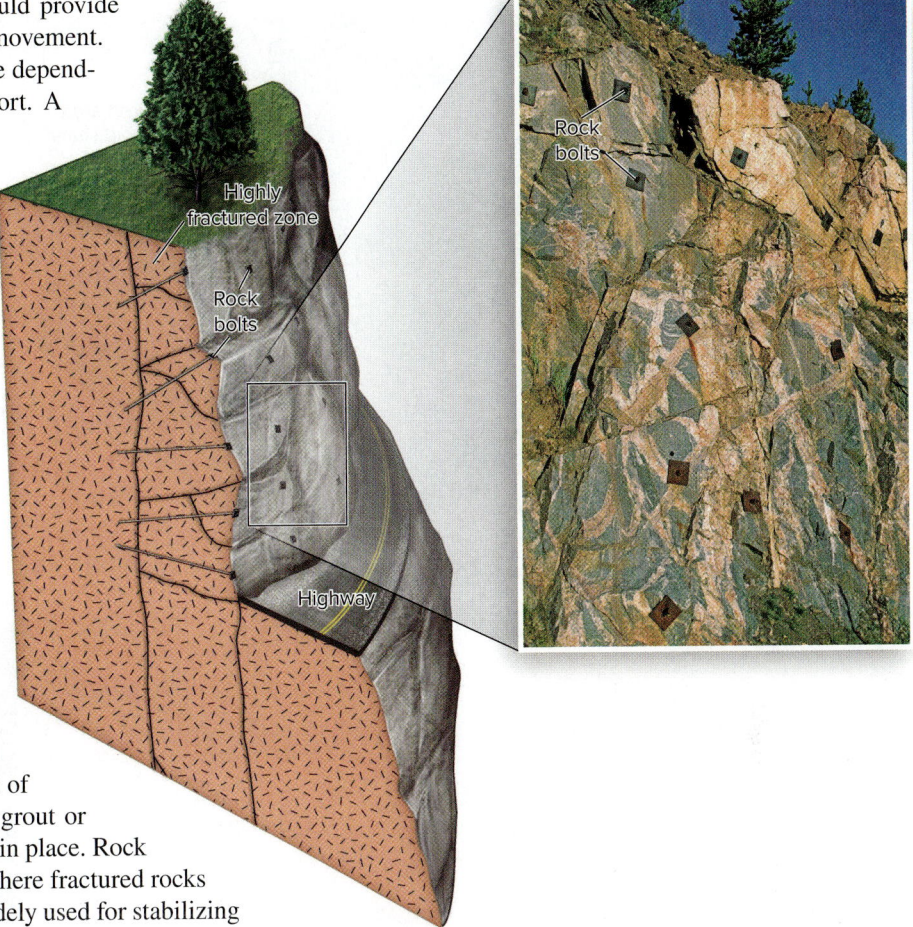

FIGURE 7.25 Rock bolts are used to attach loose slabs of rock to more massive, solid bodies of rock, thereby reducing the chance of rockfalls or small-scale rockslides along highways and in tunnels.
(inset) © Doug Sherman/Geofile

FIGURE 7.26 Drains can be used to prevent water from accumulating in porous earth materials, thereby minimizing weight in the slope direction and the buildup of pore pressure. Berms and channels can also be used to divert surface water away from unstable slopes. (inset) © King's Material, Inc.

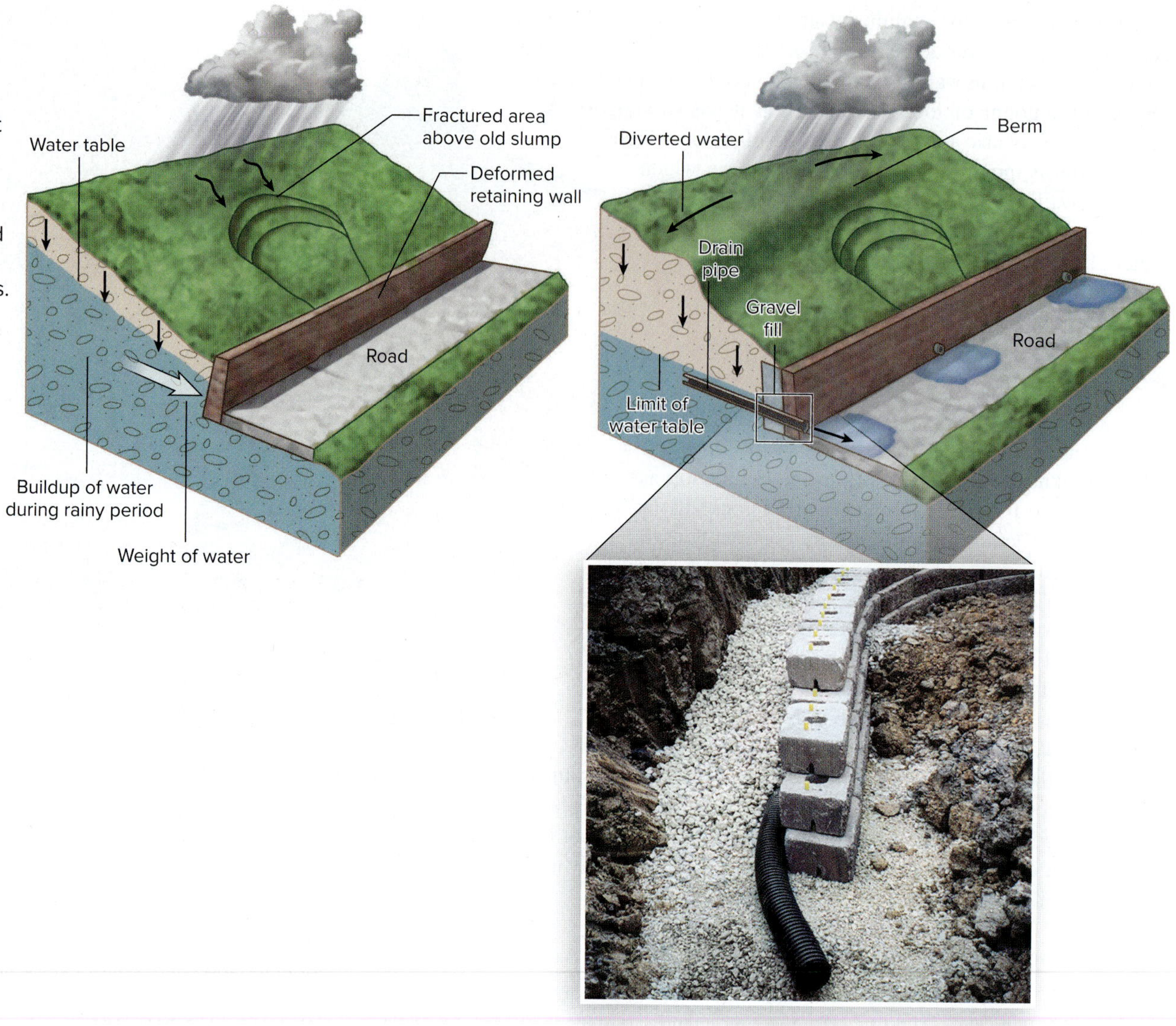

Water table · Fractured area above old slump · Deformed retaining wall · Road · Buildup of water during rainy period · Weight of water

Diverted water · Berm · Drain pipe · Gravel fill · Road · Limit of water table

FIGURE 7.27 Terraces were built into this road cut in order to keep rocks from falling onto the highway. By breaking the slope into shorter segments, terracing allows rocks to come to rest on the terrace as opposed to tumbling down the entire length of the slope. © Jim Reichard

Interstate 70, west of Denver, Colorado
Terrace step (bench)

of terraces is an ancient practice in parts of Asia and South America where the flat benches were used for growing crops in rugged terrain. Terracing is also common around homes built on slopes in order to create level areas for landscaping and recreation. In addition to creating level ground, terracing is effective in reducing the risk of mass wasting. For example, multiple terraces can be used to stabilize a slope in situations where it would not be feasible to construct a single, large retaining wall. In areas along highways where rockfalls are difficult to control with rock bolts, an alternative approach is to create a series of terraces by cutting into the rock face in a stepwise fashion (Figure 7.27). This not only decreases the overall steepness of the slope, but also breaks the slope up into shorter segments. When a rock does break loose, it can travel only a short distance before encountering a bench or step. The rock will most likely then come to a stop on the bench as opposed to tumbling down onto the highway.

Covering Steep Slopes

It is fairly common for slopes to become unstable when construction activity or wildfires remove the vegetation cover. The stability of a bare slope can be increased by covering it with new vegetation, crushed rocks, or a synthetic mesh or fabric in order to increase frictional forces within the slope. Vegetation is particularly effective in stabilizing slopes consisting of sediment because of the way plant roots help bind the loose material together. In many instances the roots will extend through the surface sediment and penetrate into fractures within the underlying rock, effectively anchoring the sediment. Generally speaking, vegetation with deeper root systems is more effective at stabilizing slopes. Regardless of root depth, however, should movement occur below the root zone, the vegetation will simply move downslope along with the underlying material. Vegetation also helps stabilize slopes by extracting water from the soil. The removal of water, in turn, helps limit the amount of deep infiltration and buildup of pore pressure, ultimately decreasing the potential for movement within the slope.

A

B

Depending on the slope material and steepness, grass alone may be sufficient, or in other cases it may simply serve as a temporary cover until permanent and more deeply rooted vegetation can take hold. Perhaps the most common way of establishing a blanket of grass is a technique called *hydroseeding*, where a slurry of seed, mulch, and fertilizer is sprayed onto a bare slope (Figure 7.28). However, on steeper slopes it becomes more of a challenge to reestablish vegetation because of the need for deep-rooted plants, which of course take more time to mature compared to grass. Compounding the problem is the fact that erosion will quickly form gullies on steeper slopes, and once a gully develops, it is especially difficult to reestablish vegetation. One solution is to fill a gully with new soil and then place deep-rooted plants along small terraces constructed in the hillside. The terraces greatly reduce erosion by decreasing the flow of water moving downslope, which gives the plants more time to grow.

FIGURE 7.28   Covering slopes with vegetation and synthetic materials increases friction within the slope, thereby decreasing the potential for mass wasting. Photo (A) shows workers applying a hydroseed mixture over a synthetic fabric, which had been draped over the sloped surfaces of a construction site. Photo (B) illustrates how the slopes soon became covered with vegetation. Note how large rocks were used to stabilize the slope in certain areas. (a–b) Profile Products

Reducing Slope Materials

Depending on the geological conditions at a site, it may be more cost effective to remove a dangerous slope rather than trying to stabilize it by some physical means. Figure 7.29 illustrates a situation where weakness planes are inclined in

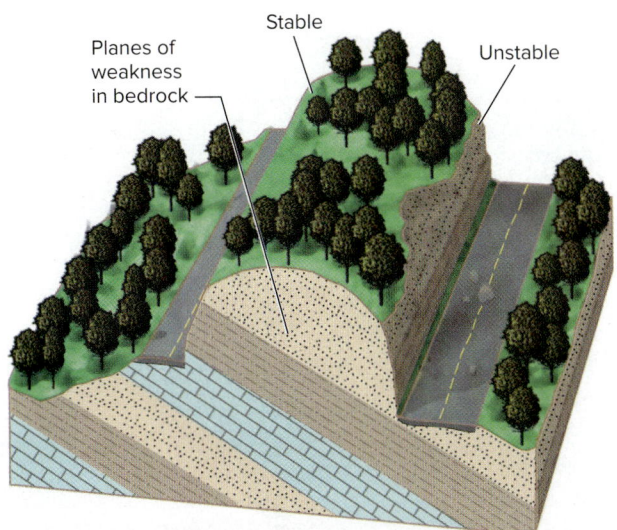

Planes of weakness in bedrock
Stable
Unstable

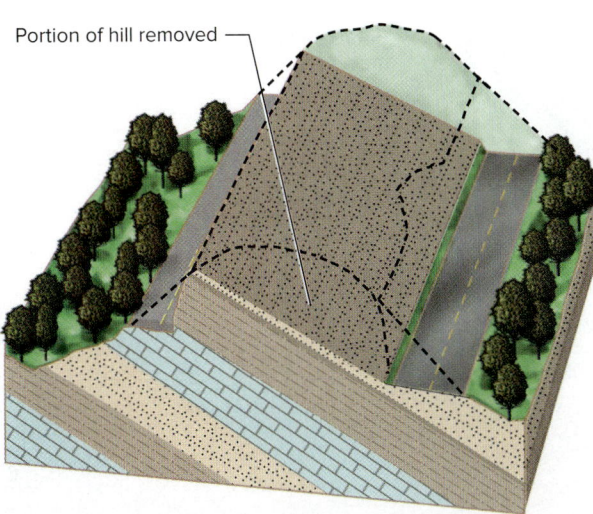

Portion of hill removed

FIGURE 7.29
Removal of a slope altogether is sometimes the most practical and cost-effective means of reducing a mass wasting hazard.

the direction of the slope, creating a serious hazard that can require continuous and costly engineering efforts to prevent a slide. Removing the entire slope has the advantage of eliminating both the hazard and the long-term costs associated with stabilization efforts. Another common situation where slope reduction or removal is a viable option is when it is not practical to use rock bolts to stabilize loose rocks above a highway. In some such cases hazard reduction may simply involve having highway engineers periodically dislodge threatening rocks in a controlled manner while traffic is stopped. Finally, note that slope reduction is the preferred method used by ski resort operators for minimizing the threat of snow avalanches. Here explosives are used to initiate snow avalanches so that dangerous accumulations of snow are removed from the surrounding slopes. Avalanche control programs at most ski resorts are typically quite effective, which accounts for the fact most snow avalanche fatalities occur outside the boundaries of resorts where such protection is not available.

Protective Structures

There are some situations where it is not cost effective to physically stabilize or remove materials from a dangerous slope. For example, some slopes may simply contain too much material to remove, while others may require an unreasonable amount of engineering in order to reduce the threat. An alternative approach is to build structures that keep people and/or infrastructure from coming into contact with material that moves downslope. One technique is to construct a retaining

FIGURE 7.30   In areas where it is not feasible to prevent mass wasting, buildings can be protected by constructing retaining walls (A) that will divert material, or by installing large barriers in valleys designed to trap debris (B). Shelters (C) are used to allow material to safely move over transportation routes.

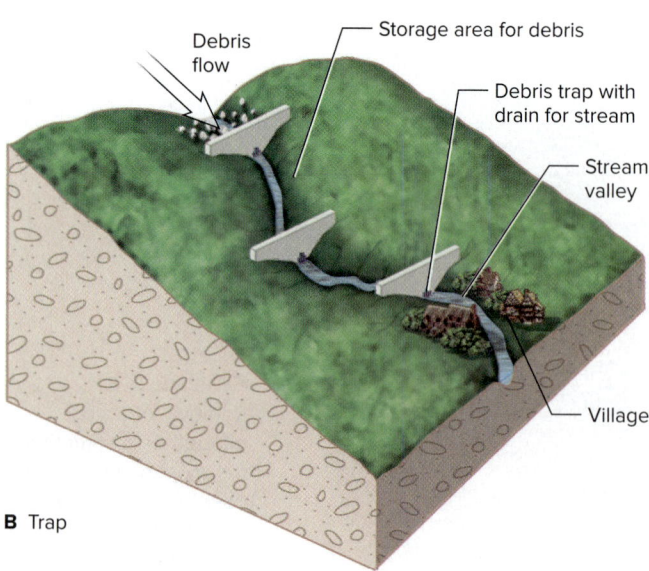

wall (Figure 7.30A) that is designed to divert material away from a building or group of buildings. To protect an entire village or town, large barriers made of reinforced concrete can be placed in a stream valley (Figure 7.30B) to trap material from flows or slides as it moves downslope. For highways and railroads, special shelters (Figure 7.30C) are used as a cost-effective means of shielding those sections where mass wasting occurs frequently. Shelters not only keep people from being injured, they protect the road or rail line from being damaged and therefore minimize costly repairs.

Perhaps the most common technique is to drape a heavy chain mesh over an entire rock exposure (Figure 7.31) in order to protect transportation routes from small- to medium-sized rockfalls. Alternatively, transportation routes can be shielded from rockfalls by constructing a tall, heavily reinforced fence parallel to a road or rail line. Finally, because mass wasting problems can lead to long-term repair and maintenance costs, it is sometimes more cost-effective to bore a tunnel (Figure 7.32). In this way troublesome sections of slopes near a highway or railroad can be bypassed altogether.

Lahaina, Maui, Hawaii

FIGURE 7.31   A heavy chain-link mesh was draped over a crumbling rock mass in Hawaii in order to keep small- to medium-size rocks from falling onto the roadway.
© Doug Sherman/Geofile

Wolf Creek Pass, Colorado

FIGURE 7.32   This tunnel was constructed as a long-term and cost-effective solution for protecting both the highway and motorists from earth materials moving downslope. Note that the old roadway (to the left of the tunnel) is no longer open to traffic.
© Jim Reichard

SUMMARY POINTS

1. Rock or sediment will begin to move downslope whenever the gravitational force acting on the material in the slope direction becomes greater than the frictional forces keeping the material in place.
2. The main factors affecting slope stability are the steepness of the slope, type of material, presence of water, existence of zones of weakness, and amount of vegetative ground cover.
3. A triggering mechanism is anything that destabilizes a slope by reducing the frictional forces on a slope or by increasing the effect of gravity. Earthquakes, heavy rains, removal of vegetation, and oversteepening of slopes are common examples.
4. Most rocks are not homogeneous, but rather contain zones of weakness such as bedding planes, fractures, faults, or foliation planes. Slopes are the least stable when these weakness planes are inclined in the same direction as the slope.
5. Water acts to destabilize slopes by adding weight and by reducing the internal friction between sediment grains and weakness planes. When materials become saturated, pore pressure will increase, which then reduces internal friction.
6. Mass wasting can be classified based on the type of movement (falls, slides, flows, and creep) and also on the material involved (rock, debris,

earth, and mud). Slump is a complex type of movement involving both sliding and flowing. Creep is imperceptibly slow movement due to repetitive expansion and contraction of sediment.
7. Snow avalanches are a form of mass wasting that pose a serious hazard along highways and for ski resorts in mountainous terrain.
8. Land subsidence occurs when void spaces within subsurface materials close. Rapid subsidence (collapse) takes place in limestone terrain and in mining areas where large underground voids are present. Gradual subsidence is common in areas where large water withdrawals reduce pore pressure, causing clay mineral particles to undergo compaction.
9. Reducing the risk of mass wasting requires that people first identify hazardous areas, then take steps to avoid these or implement engineering controls.
10. The goal of some engineering controls is to prevent movement by increasing slope stability. Efforts here focus on increasing frictional forces within slope materials and decreasing the amount of weight acting in the slope direction.
11. For mass wasting events that are either impractical or impossible to prevent, engineering efforts focus on protecting people and structures from the inevitable downslope movement of material.

KEY WORDS

creep   208
fall   202
flow   205
land subsidence   212
mass wasting   194

pore pressure   200
retaining wall   216
rock bolts   217
sinkholes   214
slide   203

slump   204
snow avalanche   208
talus pile   203
terracing   217
triggering mechanisms   196

APPLICATIONS

Student Activity   Take a drive around your town or countryside and make a list of the different engineering efforts you can find related to mass wasting. Also record the type of movement (e.g., rockfall, slump, creep) that you suspect to be involved. Even if you live in flat terrain, you can find retaining walls designed to prevent creep. You will probably be surprised by the number of examples you discover.

Critical Thinking Questions
1. Increased water content is one of the more common mass wasting triggers. Why does the actual mass wasting event sometimes take place as much as a month or more after a period of heavy rain?
2. Why is wet beach sand able to form steeper slopes than dry sand? How is this consistent with the fact that excess water can cause a slope to become unstable?
3. Mass wasting hazards are classified based on the type of movement (falls, slides, flows, and creep) and on the material involved (rock, debris, earth, and mud). However, not all combinations of movement and materials are seen in nature. Why do mud falls not occur?

Your Environment: YOU Decide

© AP Photo/Nick Ut

The logging business is a necessary part of modern life, but one of the industry's environmental impacts is mass wasting. When clear-cutting is done on very steep slopes, the vegetation cannot grow back fast enough to hold the soil in place. This has many consequences, including the fouling of streams with excess sediment and increased flooding. We all consume wood products, but as a society end up paying for the undesirable consequences of logging on steep slopes. Do you think there should be regulations that force logging companies to modify their practices of harvesting trees in steep terrain, or is it acceptable for the consequences and associated costs to be passed on to society?

Streams and Flooding

LEARNING OUTCOMES

After reading this chapter, you should be able to:

▶ Explain the various paths within the hydrologic cycle and the basic process that drives the cycle.

▶ Describe the connection between stream channels, drainage basin, and drainage divide.

▶ Define *flood*, and explain the basic way in which a flood develops.

▶ Characterize how streams transport sediment, and explain the concept of hydrologic sorting.

▶ Discuss the role that streams play in lowering the landscape down to base level.

▶ Explain how the severity and frequency of floods are measured.

▶ List and describe the various factors that affect the severity of floods.

▶ Describe the various human activities that have led to increases in flood frequency and severity.

▶ List and describe the ways in which humans have attempted to reduce the risk of flooding.

▶ Discuss how human attempts to reduce flooding can make flooding worse and lead to greater flood losses.

Streams play a critical role in the Earth system by transporting water and sediment across the landscape. As illustrated by this photo of the Yellowstone River in Yellowstone National Park, the abrasive power of running water and sediment allows streams to carve through solid rock. Streams also provide humans with supplies of freshwater and serve as means of transportation and sources of mechanical power. Because streams also periodically overflow their banks and cause flooding, humans have devised various ways of controlling them to both minimize flooding and increase water supplies.

Introduction

© StockTrek/Getty Images

The basic role of streams and rivers (larger streams) within the Earth system is to drain water off the landscape and to transport sediment. Periodically, however, a stream's ability to carry water is overwhelmed by the sheer volume of water flowing off the landscape. When this occurs a stream or river may overflow its banks and create what is known as a *flood* (Figure 8.1). Of all the natural hazards facing society, floods perhaps take the greatest toll in terms of human lives and property damage. For example, each year in the United States alone floods claim on average more than 90 lives and cause $8 billion in property damage. Although a single volcanic eruption, earthquake, or mass wasting event can result in tremendous losses, these events occur less frequently and are usually restricted to relatively small geographic areas. In comparison, rivers and streams are found everywhere on the landscape, and floods occur quite frequently in terms of human time scales. This frequency of occurrence combined with the fact that human activity tends to be concentrated along waterways puts large numbers of people and buildings at risk of flooding.

One of the reasons people have historically built settlements along rivers and streams is because of the basic need for water to survive. Rivers also served as important transportation corridors and sources of food (e.g., fish), and the adjacent low-lying areas provided rich topsoil where crops could flourish. As societies advanced, people eventually discovered that waterfalls and rapids were ideal sites for harnessing the energy of falling water; thus mills were built for grinding grain and cutting lumber. While there were many benefits to constructing settlements along waterways, there were also serious risks associated with periodic flooding. To reduce these risks, humans learned to build levees and dams that provide protection against floods. A consequence of these engineering controls, however, has been increased development in areas that people previously avoided due to frequent flooding. A prime example is the 2005 disaster in New Orleans, where levees and other engineering controls allowed the city to expand, but the controls eventually failed and resulted in a catastrophic flood with extensive property damage and loss of lives.

In this chapter our focus will be on the basic ways in which humans interact with the stream processes that transport water and sediment off the landscape. We will also examine how human modifications to the landscape have actually led to an overall increase in the frequency and severity of flooding.

Levee

FIGURE 8.1   Humans have historically built settlements along rivers to take advantage of the water supply, fertile soil, and ability to transport people and goods. This photo shows how engineering controls (levees) help keep the town of Trempealeau, Wisconsin, dry when the Mississippi River overflows its banks. However, the town would clearly flood if the engineering controls were to fail or become overwhelmed by an exceptionally large flood event.
Larry Robinson USGS

Role of Streams in the Earth System

Recall from Chapter 1 that the *hydrosphere, biosphere, atmosphere*, and *geosphere* are subsystems that interact with one another, forming the *Earth system*. A key part of this system is the *hydrologic cycle*, which is the constant transfer of water between the subsystems. The hydrologic cycle is ultimately driven by solar energy (electromagnetic radiation) that strikes Earth's surface. When liquid water is warmed by sunlight it undergoes *evaporation*, sending water vapor into the atmosphere. This vapor eventually cools and condenses into tiny water droplets, which fall back to the surface in the form of rain, snow, sleet, or hail, collectively called *precipitation*. In the following sections we will explore how water moves across the landscape on its journey through the hydrologic cycle.

Stream Discharge

Although nearly 75% of all precipitation falls directly into the oceans, the 25% that falls on Earth's landmasses can take several different paths through the hydrologic cycle. As illustrated in Figure 8.2, some of the precipitation that falls on the land surface moves downslope (under the force of gravity) in thin sheets in a process called **overland flow**—a good example is how water flows across a sloping parking lot. During an overland flow event the water will eventually move into a low area of the terrain and begin flowing as a discrete body or *stream*. The sinuous pathway that this flowing stream takes is called a *channel*. **Stream discharge** is the term used to describe the volume of water moving through a channel over a given time interval, commonly measured in units such as cubic feet per second (ft^3/s). Note that hydrologists often refer to water flowing through channels as *runoff*. Since water flows from areas of higher to lower elevation, most stream discharge (runoff) eventually makes its way to the oceans. Once in the ocean, the water undergoes evaporation and returns to the atmosphere, thereby completing its journey through the hydrologic cycle.

From Figure 8.2 we can see that some of the precipitation falling on the land surface infiltrates into the soil zone rather than flowing across the landscape. Once in the soil this water can return back to the atmosphere by evaporation or *plant transpiration* (process in which water is taken up by root systems and then released through plant leaves). However, when soils gain enough water, gravity is able to begin pulling it from the pore spaces. This allows infiltrating water to reach the water table and enter the saturated zone, at which point it is called **groundwater.** Groundwater moves slowly through subsurface materials (rocks or sediment) as it follows the slope of the water table, eventually discharging into a stream channel, lake, wetland, or ocean (see Chapter 11 for details). Hydrologists use the term **groundwater baseflow** to refer to groundwater that discharges into the surface environment (Figure 8.2). Note that unlike the sporadic input of water that streams receive from overland flow, groundwater baseflow is fairly continuous. Depending on the distance involved, once water reaches the groundwater system, it can take from a few days to thousands of years before discharging into a stream.

Hydrologists can learn a great deal about a river or stream from a *stream hydrograph*, which is simply a plot of discharge versus time. For example, the hydrograph in Figure 8.3 illustrates how the contributions of overland flow and groundwater baseflow affect stream discharge. The overland flow component typically occurs during heavy or prolonged rain events, creating a rapid input of water that generates a sudden rise or spike in stream discharge. Because it takes time for water to move across the landscape and into channels, **lag time** is the term used to describe the time difference between a rain event and the resulting discharge peak. Obviously, lag time varies depending on the distance between where the rain is falling and the particular channel where discharge is

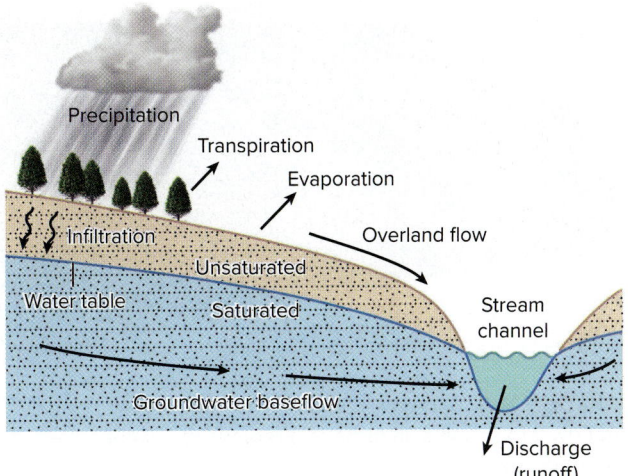

FIGURE 8.2   Precipitation falling on the land surface can take different paths through the hydrologic cycle. Some of the water moves as overland flow and enters directly into stream channels. Most of what remains infiltrates the soil zone where it can return to the atmosphere via evaporation and plant transpiration. If soil moisture becomes great enough, infiltrating water can reach the water table and then flow as groundwater until it discharges into a stream channel. Stream discharge (runoff) is simply the volume of water moving through a channel.

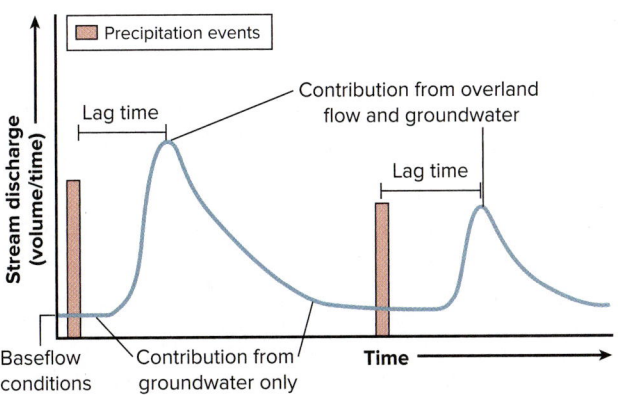

FIGURE 8.3   Stream hydrographs are useful in analyzing how stream discharge changes over time. Overland flow from heavy rain events creates spikes in discharge, which can be large enough to cause streams to overflow their banks. Note the lag time between rain events and peak discharge. In contrast, water that infiltrates and slowly makes its way to a stream through the groundwater system provides a steady supply of water called groundwater baseflow. Groundwater baseflow enables streams to keep flowing during periods of little or no rain.

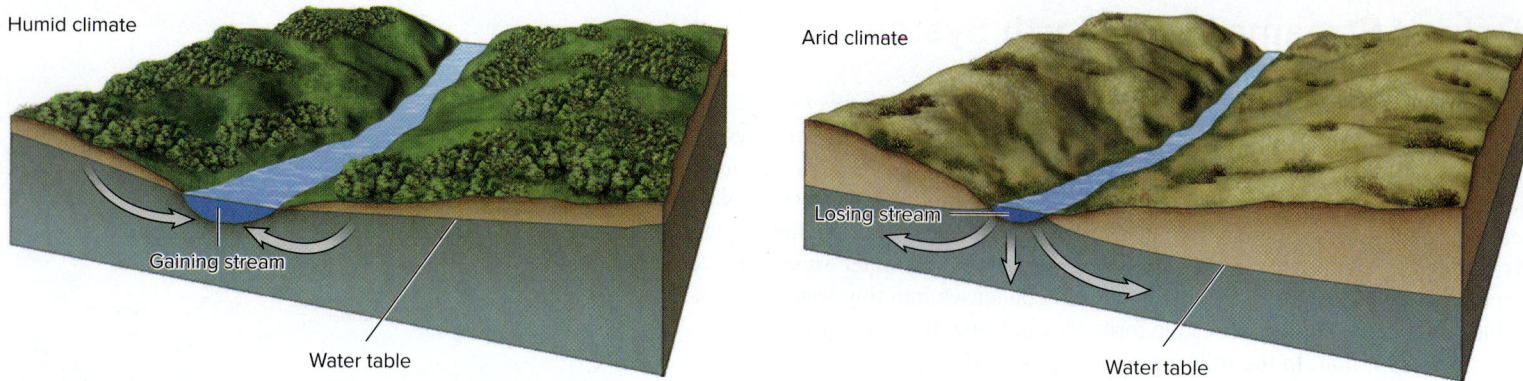

Humid climate

Gaining stream

Water table

Arid climate

Losing stream

Water table

FIGURE 8.4   Greater infiltration in humid regions causes the water table to be higher than the level of stream channels, forcing groundwater to flow into streams. In more arid regions the water table is commonly below the stream channels, causing streams to contribute water to the groundwater system.

being measured. The key point here is that overland flow is sometimes so great that the peak discharge is too large to be contained within the channel. This forces a stream to overflow its banks and flood low-lying areas next to the channel. Floods can also be caused by the rapid melting of snow, by ice jams, and by dam failures. Finally, notice in Figure 8.3 that between overland flow inputs, it is the continuous nature of groundwater baseflow that allows streams to keep flowing at some minimum level, often called *baseflow conditions*. During periods of little or no rainfall, groundwater baseflow is what keeps streams from going dry; thus it is critical to the health of stream ecosystems.

Although groundwater systems will be described in more detail in Chapter 11, we need to mention here how climate affects the relationship between streams and groundwater systems. In humid climates more water infiltrates to the water table because precipitation is more plentiful. As can be seen in Figure 8.4, this higher level of infiltration causes humid regions to have water tables that are higher than the stream channels, thereby forcing groundwater to flow into streams. Such streams are often referred to as *gaining streams*. In arid climates there is less deep infiltration, resulting in a water table that is below the level of most stream channels. Under these conditions the flow of water is reversed, meaning that water flows from streams into the groundwater system, and the streams are called *losing streams*. Because water is being lost, it is not uncommon for all but the largest rivers in arid regions to go completely dry during the summer months. Most of the water in these rivers typically comes from groundwater baseflow, melting snow, and rainfall in distant mountain areas.

Drainage Networks and Basins

Figure 8.5 illustrates that as a stream flows downhill, it eventually merges with other channels, where the smaller of any two merging channels is called a *tributary*. The result is a network of stream channels referred to as a *drainage system*, in which merging tributaries form progressively larger streams. Note that the term *river* is often applied to the larger stream that serves as the principal channel within a drainage system. You can also see in Figure 8.5 that the upper portion of the drainage system is called the **headwaters**, whereas the **mouth** is located in the lowest part of the system where a river empties into an ocean, lake, or another river. An important characteristic of drainage systems is the relatively large number of small tributaries in the headwaters. Headwater tributaries usually occupy somewhat narrow valleys and have relatively steep channels that are gently curved (as viewed from the air). As one gets closer to the mouth, the valleys tend to become wider and the steepness and number of channels decreases. In addition, the main channel begins to develop tight, S-shaped curves called *meanders*. Therefore, headwater streams are generally small and fast moving and

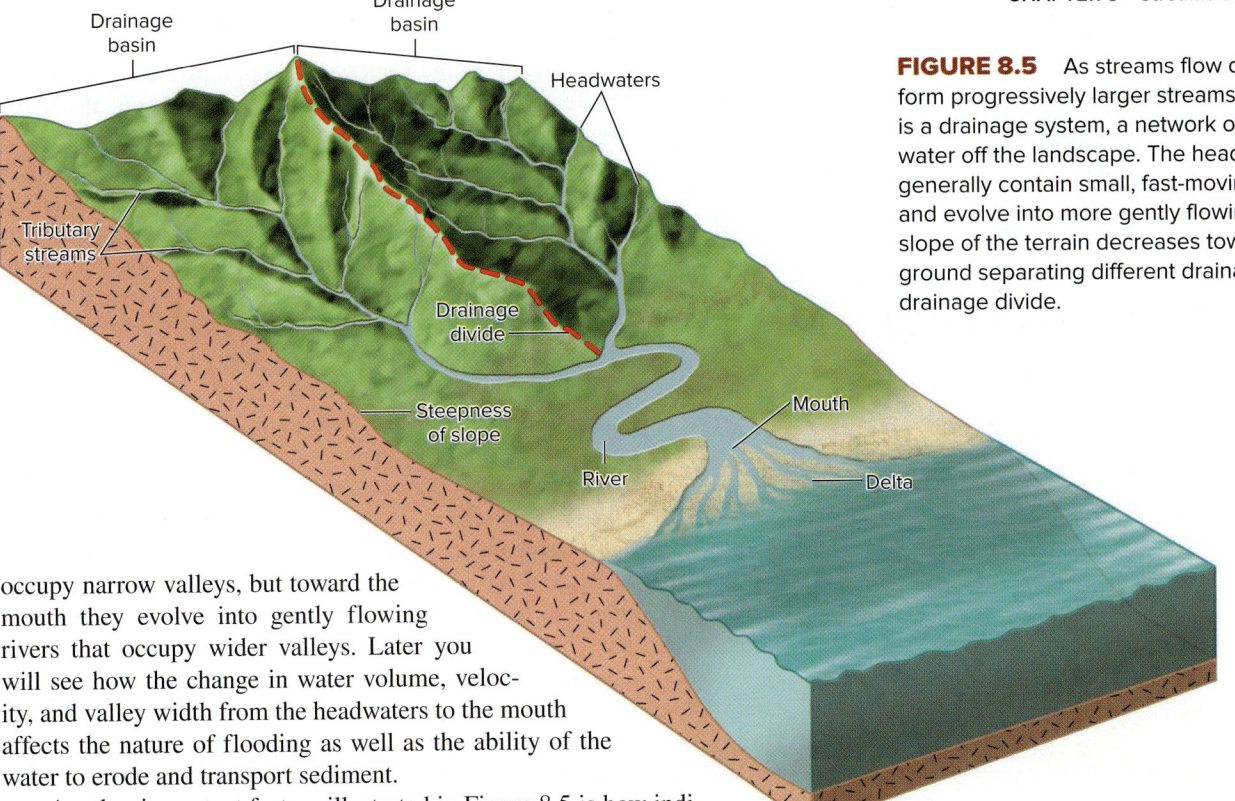

FIGURE 8.5 As streams flow downhill, they merge and form progressively larger streams and rivers. The result is a drainage system, a network of channels that drain water off the landscape. The headwaters of this network generally contain small, fast-moving streams that merge and evolve into more gently flowing rivers as the overall slope of the terrain decreases toward the mouth. The high ground separating different drainage systems is called a drainage divide.

occupy narrow valleys, but toward the mouth they evolve into gently flowing rivers that occupy wider valleys. Later you will see how the change in water volume, velocity, and valley width from the headwaters to the mouth affects the nature of flooding as well as the ability of the water to erode and transport sediment.

Another important feature illustrated in Figure 8.5 is how individual drainage systems are separated from one another by a topographic line along crests in the landscape called a **drainage divide.** From the example in Figure 8.6, notice how rain or snow that falls on different sides of a drainage divide will eventually enter different drainage networks. A large-scale example is the Continental Divide of North America (Figure 8.7), which acts as the boundary between surface waters that flow into the Pacific Ocean from those making their way into the Atlantic and Arctic oceans. Note the additional drainage divides that separate the waters flowing into different parts of the Atlantic and Arctic oceans. Another interesting feature of this map is the drainage divide that encloses the Great Basin in the western United States. This area is somewhat unusual in that surface waters here are unable to drain to an ocean, which means the water can exit only through evaporation.

Drainage divides are useful for mapping what hydrologists call a **drainage basin** or *watershed*, which represents the land area that collects water for an individual stream or river. For example, Figure 8.8 shows the drainage basin for the Mississippi River—note how part of the basin is bounded by the Continental Divide. Rainfall or snowmelt anywhere within the Mississippi basin will first move into one of the river's numerous tributaries. From there the water will make its way to the main channel of the Mississippi River, and eventually discharge into the Gulf of Mexico at the river's mouth. Figure 8.8 also illustrates how the drainage basin of the Mississippi can be subdivided into smaller basins based on the drainage divides that exist between the river's major tributaries. This process of subdividing a drainage basin into progressively smaller basins can continue down to the level of individual tributaries.

The concept of a drainage basin is also useful in that it helps explain how rainfall can cause flooding in one basin, but yet have no impact in adjoining basins. For example, suppose heavy rains were restricted to the Ohio River basin (Figure 8.8), causing rivers there to overflow their banks. Such an event would raise water levels downstream on the lower Mississippi, but would not

FIGURE 8.6 The red line in this photo traces a drainage divide, which coincides with the trail that hikers are following along a ridge crest.
© Jim Reichard

FIGURE 8.7 Map showing the major drainage divides in North America. These divides serve as the boundaries between surface waters flowing into separate stream networks. The Continental Divide separates waters flowing into the Pacific from those making their way to the Atlantic and Arctic oceans. The Great Basin is an example of an area where surface streams do not drain into an ocean, but rather water exits only through evaporation.

affect streams in the Upper Mississippi, Missouri, Arkansas, and Tennessee basins. The drainage basin concept also helps us understand how rivers that drain large tracts of land tend to carry large amounts of water. This relationship, however, between discharge and drainage basin size is complicated by the role of climate. If two basins are of equal size, then the one located in a more humid region will naturally have greater discharge.

To illustrate the relationship between discharge, basin size, and climate, let us consider the data listed in Table 8.1. Here we see the 10 largest rivers in the world ranked based on the discharge at their mouths. Notice how the Mississippi and Congo drainage basins are comparable in size, but the Congo discharges 2.5 times more water than the Mississippi. This is due to the fact that much of the Congo basin lies in a wet tropical climate, whereas a considerable portion of the Mississippi basin is semiarid. Also interesting is how the discharge of the Amazon dwarfs that of all the other major rivers. The difference here is because the Amazon has such a large drainage basin, and it also happens to be located in a tropical climate. However, if we consider just sediment load, the Ganges River far outranks even the Amazon. The Ganges transports such a large amount of sediment because it drains the Himalayas, where weathering and erosion generate tremendous volumes of sediment.

Stream Erosion, Transport, and Deposition

In Chapters 3 and 4 you learned that tectonic forces raise the land surface above sea level, at which point weathering processes begin breaking down solid rock into particles called *sediment*. This loose material then makes its way downslope toward stream channels by both mass wasting (Chapter 7) and overland flow. Once sediment reaches a channel, it is transported downstream and eventually deposited at the mouth of a river or in low-lying areas within the drainage basin (stream sediment is often called *alluvium*). In addition to transporting water and sediment, streams play a key role in the process called *erosion*, in which earth materials are removed from a given area. Although wind and ice also erode and transport sediment, our focus in this section will be on the work of running water.

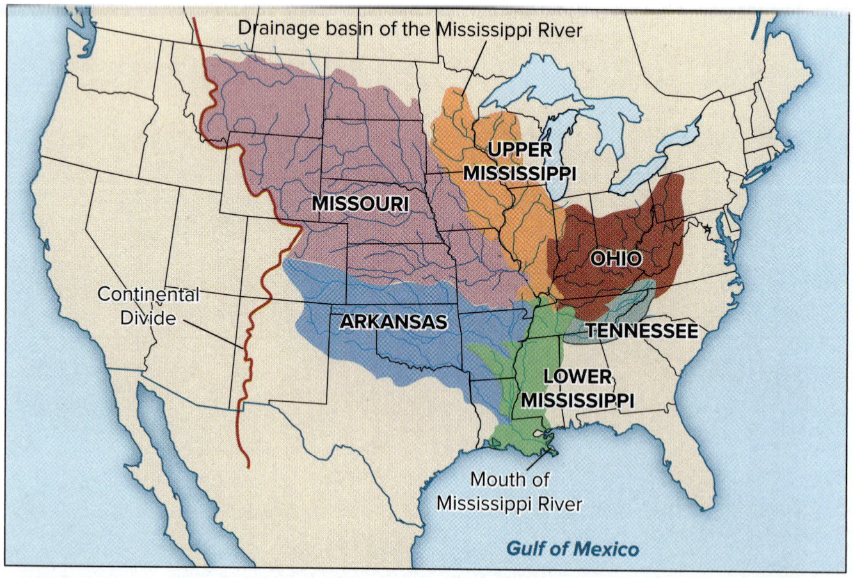

Stream Erosion

One of the key factors in a stream's ability to erode the landscape is the velocity of the water. For example, in Figure 8.9A one can see that when water enters a meander bend, it is forced to slow down on the inner part of the bend, but speeds up on the outer part. This velocity increase along the outer bank greatly enhances the water's ability to erode into what is called the *cutbank*, forming an unstable overhang. Eventually the overhang will fall into the stream by mass wasting, at which point the material will be incorporated into the sediment already being transported by the stream. Along the inner bank where velocity decreases, sediment tends to accumulate and form a deposit known as a *point bar*. The combination of erosion on the outer bank and

FIGURE 8.8 The Mississippi drainage basin represents the land area that contributes water to the Mississippi River, which ultimately discharges into the Gulf of Mexico. The Mississippi basin can also be subdivided into smaller basins, each of which collects water for the river's major tributaries. These smaller basins can be subdivided even further based on their own network of tributaries.

TABLE 8.1   Discharge, drainage basin size, and sediment load of the world's major rivers. Note that the rivers here are ranked based on their discharge.

Rank	River	Continent	Average Discharge at Mouth (ft³/sec)	Size of Drainage Basin (miles²)	Annual Sediment Load (tons)
1	Amazon	S. America	7,750,000	2,670,000	900,000,000
2	Congo	Africa	1,630,000	1,470,000	43,000,000
3	Ganges	Asia	1,570,000	668,000	1,670,000,000
4	Yangtze	Asia	1,110,000	695,000	478,000,000
5	Orinoco	S. America	1,020,000	386,000	210,000,000
6	Paraná	S. America	812,000	1,150,000	92,000,000
7	Yenisey	Asia	681,000	996,000	13,000,000
8	Mississippi	N. America	649,000	1,240,000	350,000,000
9	Lena	Asia	593,000	961,000	12,000,000
10	Mekong	Asia	569,000	313,000	160,000,000

Source: Data from Gleick 1993, Milliman and Mead, 1983, and Zeid and Biswas, 1990.

deposition on the inner bank results in the lateral migration of the stream channels over time. As shown in Figure 8.9B, stream migration and mass wasting work together to form progressively wider river valleys.

In addition to causing lateral erosion, running water also cuts down into the channel or *bed* of a stream. In some cases streams erode down through solid rock, forming deep canyons (Figure 8.10A). It is important to realize that downcutting by streams is not performed by the water itself, but rather by sediment that physically scrapes or wears away rock in a process called *abrasion*. To illustrate the abrasive power of stream sediment, imagine first rubbing your skin with a piece of notebook paper, and then switching to coarse sandpaper. Clearly, the grit within the sandpaper would dramatically increase the amount of abrasion on your skin. In a similar manner, sediment moving along a streambed gives a stream the abrasive power to cut through solid rock. Evidence of stream abrasion in solid rock can be seen in *potholes* that become exposed when water levels are low (Figure 8.10B). Potholes form during periods of high stream discharge when the water column develops a vortex or swirling motion called an *eddy current*.

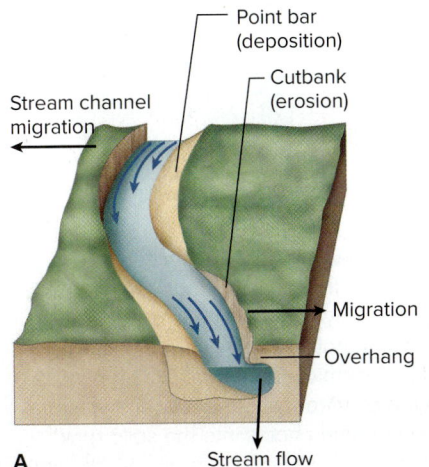

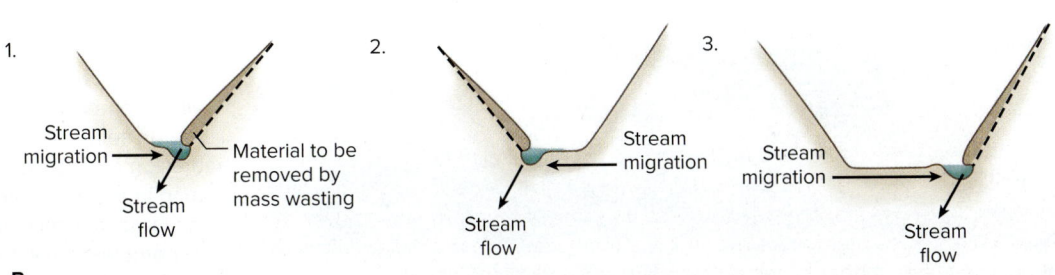

FIGURE 8.9   As flowing water moves through a bend in the channel (A), water velocity increases along the outer bank. This causes erosion and undercutting of the bank, forming an unstable overhang that eventually collapses. The lateral migration of stream channels combined with mass wasting (B) produces progressively wider valleys over time.

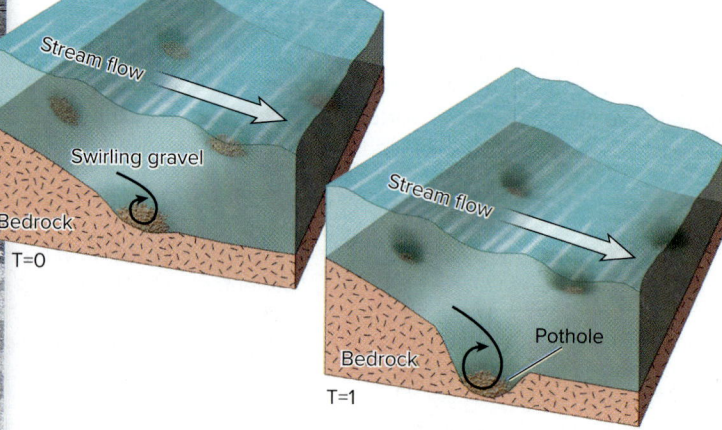

Sediment
filled potholes

Stream flow

Swirling gravel

Bedrock

T=0

Stream flow

Bedrock

Pothole

T=1

FIGURE 8.10 The abrasive power of sediment within the Yellowstone River (A) has enabled the river to cut downward through solid rock, forming the canyon shown here. Sediment abrasion often takes place within potholes similar to those (B) found in a granite riverbed in Yosemite, California. During high discharge, the swirling motion of water causes sediment within the potholes to rotate, slowly grinding holes into the solid rock.
(a–b) © Jim Reichard

A

B

These currents cause sediment on the streambed to slowly rotate, grinding progressively deeper holes into the solid rock over time.

An important factor in a stream's ability to erode is its water velocity. The velocity of a particular stream segment is controlled by the steepness of the channel, called **stream gradient**—also referred to as *grade*. Water velocity naturally decreases whenever the stream gradient decreases, which, in turn, reduces the amount of sediment abrasion. When a river empties into a lake or ocean and the gradient becomes zero, the water stops flowing and erosion ceases. Geologists use the concept of **base level** to describe the lowest level to which a stream can erode. As illustrated in Figure 8.11, sea level is often referred to as *ultimate base level* because the oceans represent the end or low point of most rivers (exceptions include isolated areas below sea level, such as the Dead Sea and Death Valley). Notice that as a river flows toward the ocean, it may encounter a series of *temporary base levels*, which form when its ability to cut downward is reduced by a resistant rock body, lake, or inland sea. In general, the gradient of a river becomes progressively less steep as the channel gets closer to a particular base level. This results in less downcutting and a tendency for a river to meander more, enhancing its ability to undercut its banks and produce a broader valley.

Stream Transport and Deposition

The ability of running water to transport and deposit sediment depends on both the water's velocity and the types of particles being transported. Because velocity varies with stream discharge and gradient, rivers and streams tend to transport and deposit sediment in an alternating manner. To better understand how this

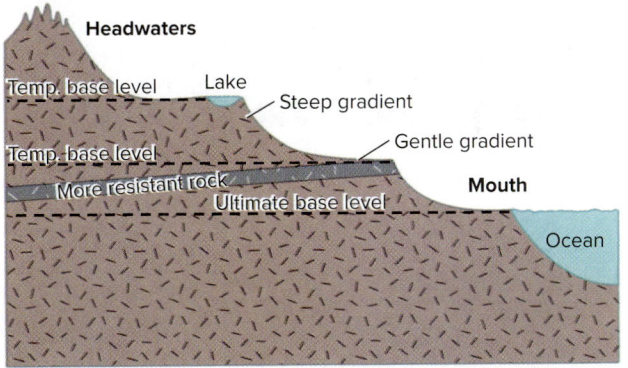

FIGURE 8.11 Base level represents the lowest level to which a stream can cut downward. Sea level serves as the ultimate base level, whereas lakes and resistant rock bodies can create temporary base levels. The photo shows the Grand Tetons, which have been uplifted far above base level and have given tributaries in the headwaters the erosive power to cut down through solid rock. These tributaries eventually merge with the Snake River (foreground), which has a much gentler gradient because it is approaching a temporary base level.
© The McGraw-Hill Companies, Inc./John A. Karachewski, photographer

works, we need to first discuss how solid particles and dissolved ions make up what is called a stream's *load*. The term *suspended load* describes the fraction of solid particles that are held in suspension and moving at the same velocity as the water—suspended load is what makes streams appear muddy. The remaining fraction is called the *bed load*, which consists of sediment particles that roll, bounce, or remain stationary on the streambed. The key point is that when overland flow causes stream discharge and velocity to increase, sediment particles are selectively removed from the bed load and become part of the suspended load. Later when discharge and velocity decrease, particles begin to fall out of suspension and return to the streambed. Note that anytime sediment is in motion, individual particles will undergo abrasion, causing them to become smaller and more rounded the farther they travel.

The process whereby sediment particles move back and forth between the bed and suspended loads depends not just on water velocity, but also on the sediment itself. In Figure 8.12 you can see that when stream discharge and velocity begin to increase, the first particles to be lifted from the bed load are the smallest, least dense, and most angular. What remains on the bed are particles that are too large, dense, and rounded to be carried in suspension at that particular velocity. Conversely, when discharge and velocity eventually decrease, the first particles to fall from suspension and return to the bed are the largest, most dense, and roundest. This process whereby water separates sediment grains based on their size, shape, and density is called **hydraulic sorting.** Over time, the combination of hydraulic sorting and abrasion produces sediment that is progressively finer, more rounded, and more uniform in size as one moves toward the mouth of a drainage basin.

An important aspect of sediment transport is that a considerable amount of the sediment load within a basin remains stationary for extended periods of time, moving only during periodic increases in discharge and velocity. As indicated in Figure 8.12, sediment generally does not move directly from the headwaters to the mouth of a basin, but in a series of steps, each consisting of a period of transport followed by deposition. During this process hydraulic sorting creates deposits where more and more of the grains are similar in terms of their size, shape, and density. Recall from Chapter 3 that the mineral quartz is resistant to chemical weathering, whereas many other common silicate minerals are transformed into various clay minerals. Consequently, sediment that has been transported great distances is usually dominated by tiny clay particles and rounded grains of quartz. The combination of hydraulic sorting and chemical weathering eventually produces relatively pure deposits of sand and clay, two very important natural resources used in society (Chapter 12). Hydraulic sorting also concentrates high-density particles such as gold, platinum, and titanium minerals, thereby creating valuable ore deposits.

Although describing the various types of stream deposits is beyond the scope of this text, we will briefly examine some of the more common deposits. Perhaps the most familiar are mound-shaped channel deposits called *bars* (Figure 8.13). Individual bars consist of sorted material ranging in size from boulders to coarse gravel to fine sand. Boulder deposits are generally found near the headwaters of a drainage system, whereas sand-sized deposits are more common toward the mouth. Crescent-shaped bars, called *point bars*, develop on the inside of meander bends where water velocity decreases. Another common deposit is a *delta* (Figure 8.14), which forms where a river enters a lake or ocean and splits into smaller

FIGURE 8.12 The selective way sediment grains rise and fall from suspension results in progressively finer, more rounded, and better-sorted sediment as a river moves toward its mouth. When stream discharge and velocity increase (A), the smallest, least dense, and most angular particles are the first to rise off the bed and be carried in suspension. When velocity begins to decrease (B), the first particles to return to the bed load are the largest, most dense, and roundest.

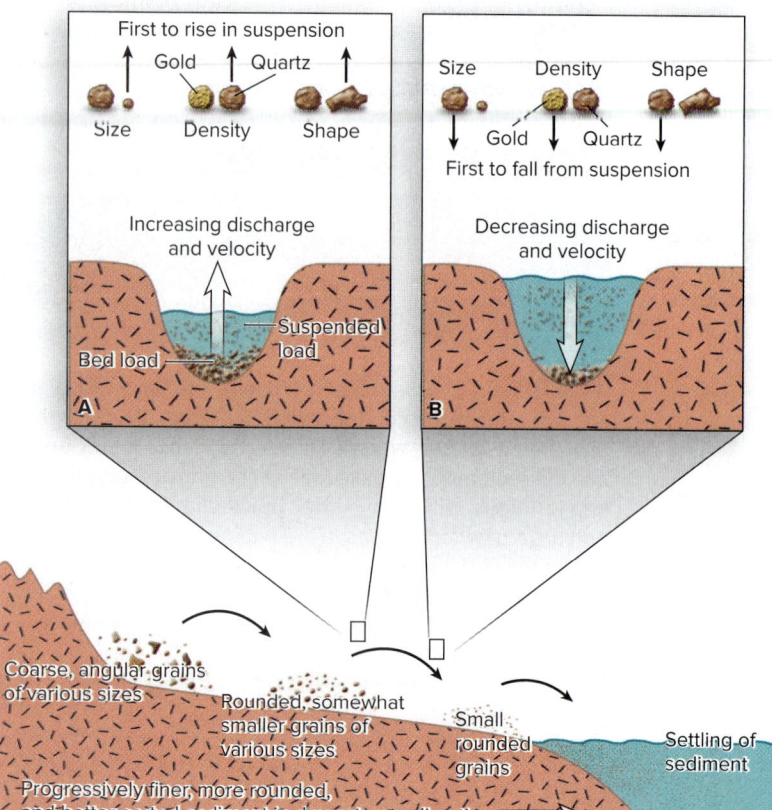

A

B   Point bar

channels and begins depositing sediment as the water velocity decreases. Over time, the weight from this nearly continuous influx of sediment can cause the seafloor to sink and result in *land subsidence* (Chapters 7 and 9). Interestingly, subsidence is what allows deltas to become thicker and grow seaward, thereby creating new land area. A good example is the Mississippi Delta (Figure 8.14), where the deposition of vast amounts of sediment has led to subsidence and, until recently, a greatly expanded shoreline.

Also located at the mouth of rivers are *alluvial fans*, which are large fan-shaped deposits that form where steep mountain streams empty out onto valley floors (Figure 8.15). Here the stream is no longer confined by the valley walls, and the lack of confining barriers combined with the abrupt decrease in gradient and water velocity greatly reduce the stream's ability to transport sediment. The result is a channel choked with sediment called a *braided stream*, which migrates back and forth across the entrance to the valley, creating a characteristic fan-shaped deposit. As described in Chapters 6 and 7, humans have historically located settlements on alluvial fans because of the availability of water and the relatively flat land. However, the potential for large floods, mudflows, and debris slides makes alluvial fans hazardous places to live.

FIGURE 8.13 Hydraulic sorting leads to channel deposits called bars. Bars located near the headwaters (A) are often composed of boulders and coarse gravels, whereas sand-sized material is more common in downstream areas (B) where the gradient is less steep. Point bars are crescent-shaped deposits found on the inside of meander bends. Hydraulic sorting also creates valuable sand and gravel deposits as well as concentrations of gold and other high-density minerals.
(a–b) © The McGraw-Hill Companies, Inc./John A. Karachewski, photographer

FIGURE 8.14 Deltas (A) form at the mouths of rivers where velocity abruptly decreases upon entering a water body, forcing the stream to deposit its sediment. Satellite image (B) showing the Mississippi River as it deposits sediment in the Gulf of Mexico. Light brown colors in the offshore parts of the image represent sand and silt-sized particles falling out of suspension in the delta; green and blue correspond to very fine-grained clay particles that settle out farther offshore.
(b) Courtesy of Liam Gumley, Space Science and Engineering Center, University of Wisconsin-Madison and the MODIS science team

A

Distributaries

Land

River

Delta surface

Accumulated sediment

B

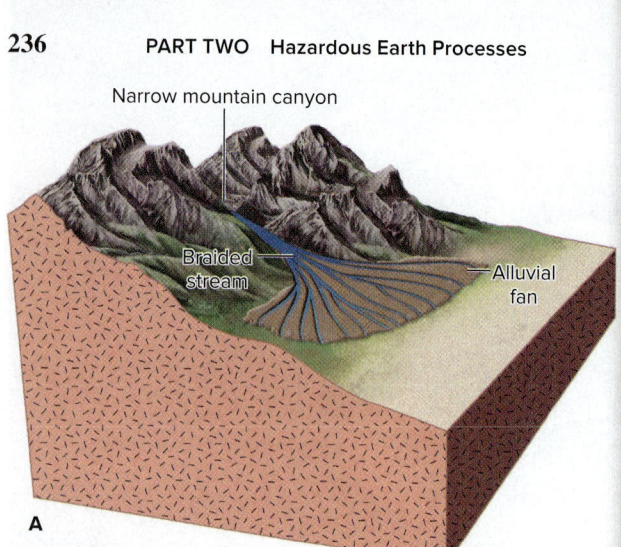

Narrow mountain canyon

Braided stream

Alluvial fan

A

Alluvial fans

B

FIGURE 8.15   Alluvial fans (A) develop where steep mountain streams empty out onto valley floors. The abrupt changes in gradient and velocity cause the stream to become choked with sediment. Over time the stream migrates back and forth across the entrance to the valley, creating a fan-shaped deposit. Aerial view (B) shows alluvial fans in Death Valley, California.
(b) © Doug Sherman/Geofile

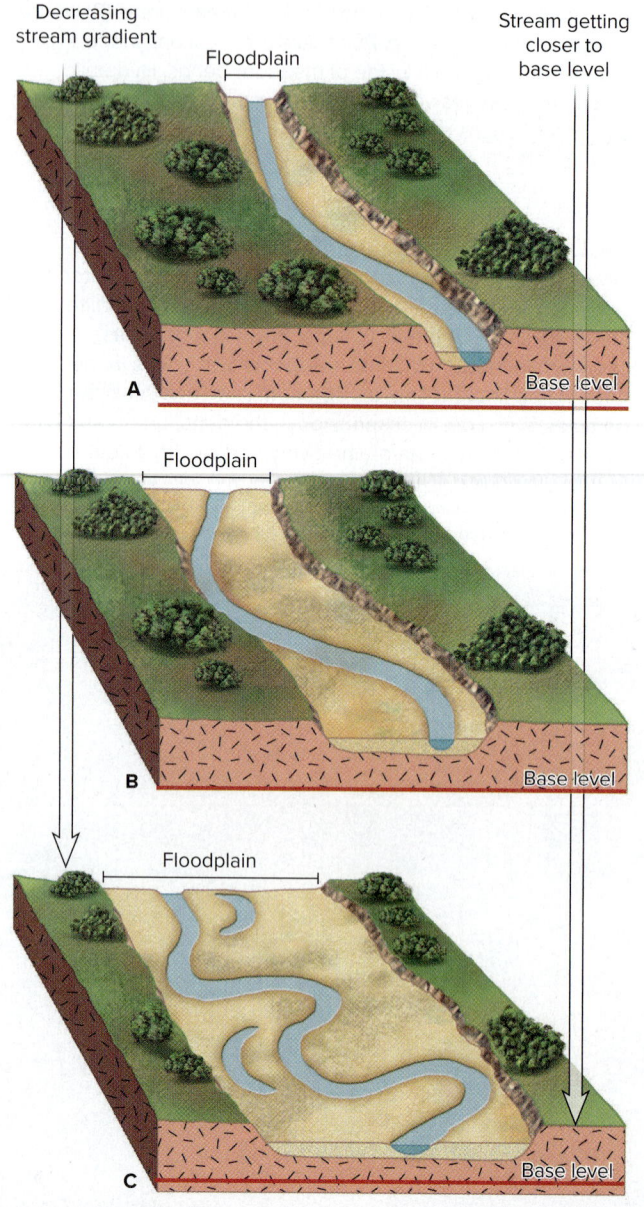

Decreasing stream gradient

Floodplain

Stream getting closer to base level

A   Base level

Floodplain

B   Base level

Floodplain

C   Base level

In addition to solids, streams also transport considerable amounts of dissolved ions (charged atoms), often referred to as the *dissolved load*. The dissolved load originates from the weathering of minerals (Chapter 3), which releases individual ions whose electrical charge allows them to become attached to water molecules. These dissolved ions, also called *salts*, are then carried away with the water as it flows down the drainage system. Ultimately the dissolved load is deposited in an ocean or inland body of water. Here the ions stay behind as water molecules evaporate. Note that dissolved ions are invisible to the human eye because they are on an atomic scale. Therefore, what makes streams, lakes, or oceans appear cloudy is the suspended sediment, not the dissolved ions.

River Valleys and Floodplains

Recall that stream gradient, discharge, erosion, and deposition all vary from the headwaters to the mouth of a drainage basin. Now we want to take a closer look at how valleys change as rivers cut down and approach base level. This is important because the valley shape affects the way flooding occurs. As shown in Figure 8.16, river channels tend to meander more as they approach base level and their gradient becomes less steep. Erosion increases along the outside of the meander bends, producing wider valleys. At the same time, deposition on the inner banks is helping to build a flat plain on the valley floor called a **natural floodplain.** When a river overflows its banks, the first area to be inundated is this flat portion of the valley, hence the name *floodplain*. Floodplains are an integral part of many of Earth's ecosystems, but their primary role with respect to flooding is to periodically store large volumes of water moving through a drainage basin. A floodplain then is not separate from a river, but rather an integral part of a drainage system. Later in this chapter we will examine some of the serious consequences that occur when society attempts to reduce flooding by disconnecting rivers from their floodplains.

Recall that the sediment load of streams approaching base level is typically dominated by sand, silt, and clay-sized particles; overland flow periodically adds decaying organic matter (vegetation) as well. When discharge and velocity increase

FIGURE 8.16   When streams cut down toward base level and their gradient decreases, they tend to meander more, cutting wider valleys with broader floodplains.

during the initial stages of a flood, some of the bed load is naturally picked up and carried in suspension. When a river overflows its channel, there is a sudden decrease in velocity as it begins to spread out onto the floodplain. This forces the largest particles, typically sand, to immediately fall out of suspension and be deposited along the edge of the bank where the velocity change occurs. As illustrated in Figure 8.17, this creates a pair of low ridges called **natural levees** that run parallel to the banks. The remaining particles of finer silt, clay, and organic matter are carried out onto the floodplain where they slowly fall out of suspension and form a blanket-like deposit. When water levels eventually fall, the levees act as a barrier, preventing a portion of the floodwaters from returning to the channel. Draining of the floodplain is further hampered by the fact that it is underlain by fine-grained material that has low permeability. These poorly drained areas of the floodplain are referred to as **back swamps** (Figure 8.17), and they can remain wet long after a flood.

When rivers migrate laterally across their valleys, they also continue to cut downward toward base level. As illustrated in Figure 8.18, this slow lateral movement causes the older deposits in front of the migrating channel to be eroded away while new deposits are laid down behind it. The combination of lateral migration and continued downcutting creates new floodplains in a stair-step fashion, with each being at a progressively lower elevation. The old floodplains left high and dry as a river migrates are called *stream terraces*. Particularly well-defined terraces often develop in response to relatively rapid changes in base level resulting from tectonic uplift or lowering of sea level. Because stream terraces lie above the active flood plain, they are less likely to be inundated during a flood. Consequently, humans have historically made use of terraces as safe locations for building settlements and developing agriculture.

FIGURE 8.17   Natural levees are typically found along river channels that have well-defined floodplains. When a river overflows its banks, the abrupt decrease in velocity causes the largest grains, commonly sand, to fall from suspension and be deposited along the banks. Finer particles of silt, clay, and organic matter are carried with the floodwaters and deposited on the floodplain, forming back swamps where floodwaters are unable to return to the main channel.

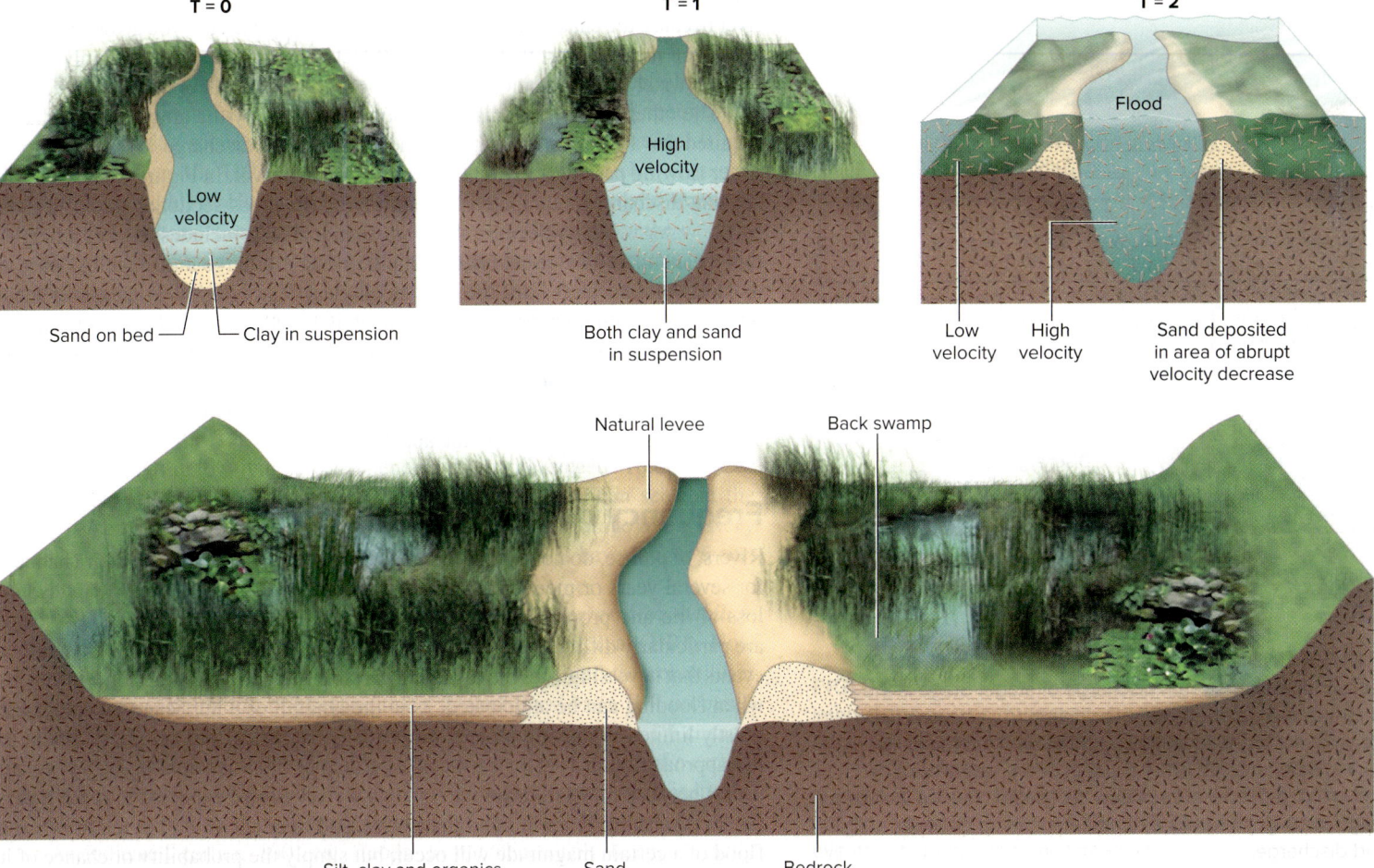

FIGURE 8.30 Urban settings commonly contain large areas of impermeable surfaces where little to no infiltration takes place, generating large volumes of overland flow that rapidly enter the drainage network. The result is more frequent flooding, higher flood crests, and shorter lag time between a rain event and peak discharge.

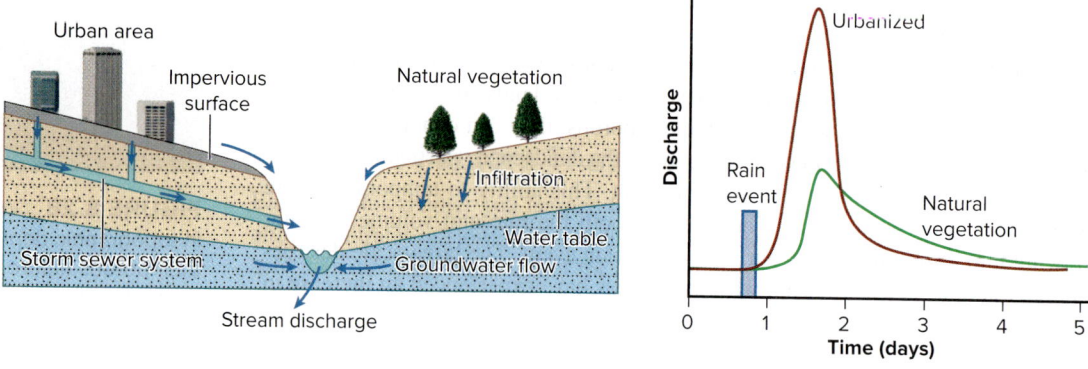

losses by managing human activities that tend to exacerbate flooding. In this section we will explore some of the more common flood mitigation techniques.

Dams

When a dam is built across a river, it gives engineers the ability to control the river's discharge. As illustrated in Figure 8.31, when engineers regulate the discharge of a river, they are able to raise or lower the pool or *reservoir* of water behind the dam. Reservoirs are valuable to society in that they protect against floods, serve as important sources of freshwater and electrical power, and offer a variety of recreational opportunities. In general, engineers allow water to accumulate in a reservoir during wet periods when streamflows are high, thereby creating a reserve or stockpile of water. During dry periods when the demand for water exceeds the river's discharge, the pool is then lowered in order to meet the demand. In the case of a dam that is used to generate hydroelectric power, engineers raise and lower the pool on a daily basis. Regardless of whether the reservoir height is manipulated daily or seasonally, engineers must always ensure that enough water is being released to meet the needs of downstream ecosystems.

The storage capacity of a reservoir can also be manipulated for the purpose of preventing flooding downstream of the dam. Here engineers must maintain a sufficient amount of reserve or emergency storage behind the dam (Figure 8.31). This additional storage allows periodic surges of water from upstream areas to be safely contained within the reservoir. Although a sufficient amount of emergency storage is normally maintained, unexpectedly heavy or prolonged rains sometimes cause a reservoir to reach its maximum capacity. When this occurs, engineers may be forced to release water at such a high rate that it causes downstream flooding, which, ironically, is what the dam is supposed to prevent. The alternative is to risk letting the pool rise to the point that the dam itself is weakened. The dam's structural integrity can also be threatened should water be allowed to flow over the top in an uncontrolled manner. In either case, the structural failure of a dam can generate a truly catastrophic flood, as in the massive flood described earlier that struck Johnstown, Pennsylvania, in 1889 (Figure 8.23).

In addition to the potential for structural failure and devastating floods, dams are also a concern because of their environmental impact on downstream ecosystems. When a river is dammed, the suspended and bed loads are forced to be deposited within the reservoir. The water temperature also becomes much cooler due to the depth of the reservoir. Therefore, what is released from the reservoir is cool, sediment-free water. This can be highly disruptive to aquatic ecosystems that naturally evolved in the presence of warmer water and a certain amount of sediment moving downstream. Moreover, because the volume of water that is

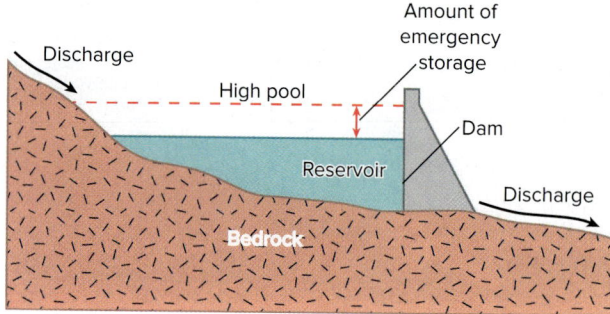

FIGURE 8.31 Dams prevent downstream flooding by intercepting and storing stream discharge in their reservoirs. By carefully regulating the release of water, engineers can maintain a reservoir such that it provides a sufficient amount of emergency storage and enough water to meet the demand for freshwater and electricity.

happening. Determining the
scientists and engineers to q
Naturally the lowest areas wi

Flood frequency is deter
a river for as many years as pc
of flooding, the maximum di
example dataset for the Tar Ri
such a dataset, a **recurrence**
which represents the frequenc
to recur. For example, a discha
average, 100 years should pas
because recurrence intervals a
more than 100 years to repeat,

The recurrence interval i
record is calculated using the

Recurre

where N is the number of val
discharge maximum. Using th
(Table 8.2), one can see that
case makes $N = 110$. To deter
highest to lowest, which make
($M = 1$) and the 1981 event of
intervals for these two dischar

70,600 ft³/sec
(1

3,340 ft³/sec: 1
(1

What this means is that a
itself, on average, every 111 y
year. The remaining recurren
111 years. However, the 1999
ated with the landfall of Hurr
results of this analysis. Theref
tions and plot the remaining d
the graph shown in Figure 8.2(
mated recurrence interval of o
Clearly, this was a major flood

Figure 8.20 also illustra
numerous, and repeat more f
terms of flooding, the Tar Ri
and begin inundating low-lyin
(flood stage is 19 feet above
10,000 ft³/sec minimum for p
one and a half years. Progress
rence intervals.

Another useful way of m
which is simply the inverse or
in Figure 8.20 that the recurre
along the top axis. From this, o
ity of taking place in any giver
Recurrence intervals therefore
size, whereas percent probabili
tant to keep in mind that these
ages. As indicated earlier, 10

during the initial stages of a flood, some of the bed load is naturally picked up and carried in suspension. When a river overflows its channel, there is a sudden decrease in velocity as it begins to spread out onto the floodplain. This forces the largest particles, typically sand, to immediately fall out of suspension and be deposited along the edge of the bank where the velocity change occurs. As illustrated in Figure 8.17, this creates a pair of low ridges called **natural levees** that run parallel to the banks. The remaining particles of finer silt, clay, and organic matter are carried out onto the floodplain where they slowly fall out of suspension and form a blanket-like deposit. When water levels eventually fall, the levees act as a barrier, preventing a portion of the floodwaters from returning to the channel. Draining of the floodplain is further hampered by the fact that it is underlain by fine-grained material that has low permeability. These poorly drained areas of the floodplain are referred to as **back swamps** (Figure 8.17), and they can remain wet long after a flood.

When rivers migrate laterally across their valleys, they also continue to cut downward toward base level. As illustrated in Figure 8.18, this slow lateral movement causes the older deposits in front of the migrating channel to be eroded away while new deposits are laid down behind it. The combination of lateral migration and continued downcutting creates new floodplains in a stair-step fashion, with each being at a progressively lower elevation. The old floodplains left high and dry as a river migrates are called *stream terraces*. Particularly well-defined terraces often develop in response to relatively rapid changes in base level resulting from tectonic uplift or lowering of sea level. Because stream terraces lie above the active flood plain, they are less likely to be inundated during a flood. Consequently, humans have historically made use of terraces as safe locations for building settlements and developing agriculture.

FIGURE 8.17 Natural levees are typically found along river channels that have well-defined floodplains. When a river overflows its banks, the abrupt decrease in velocity causes the largest grains, commonly sand, to fall from suspension and be deposited along the banks. Finer particles of silt, clay, and organic matter are carried with the floodwaters and deposited on the floodplain, forming back swamps where floodwaters are unable to return to the main channel.

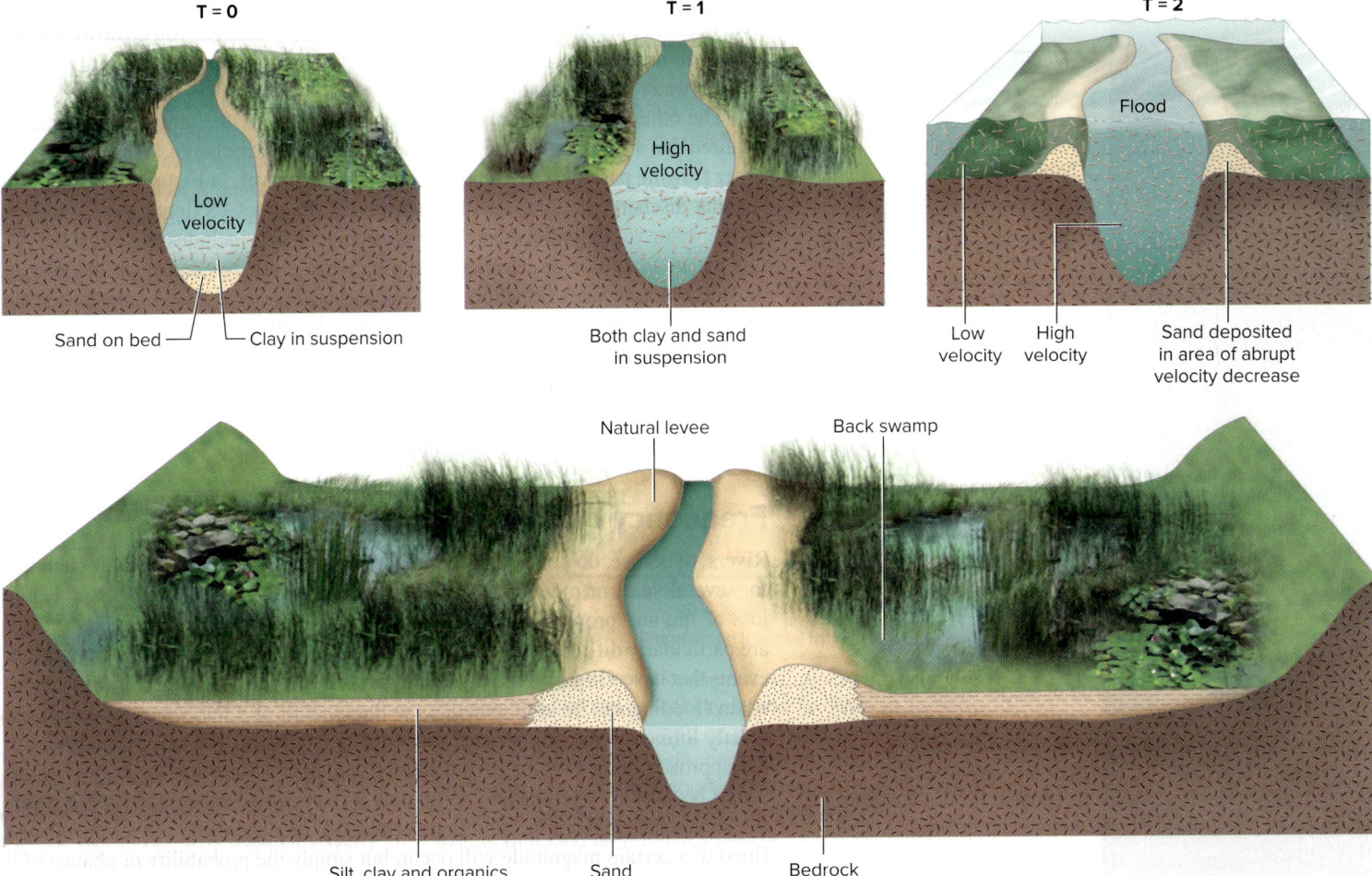

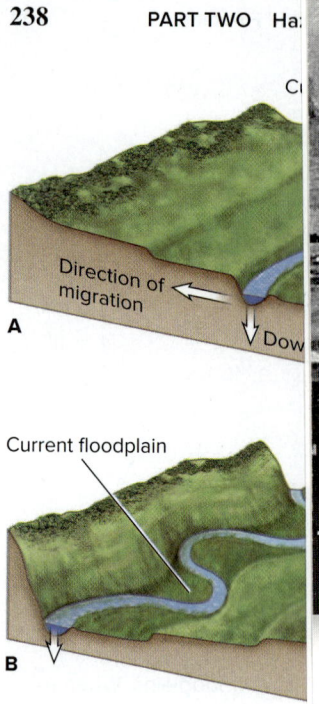

A

Direction of migration ←

↓ Dow

Current floodplain

B

↓

FIGURE 8.18   Terraces are
high and dry as the stream cha
laterally and cuts downward ac
terraces are flat and less pron
found them ideal sites for settl

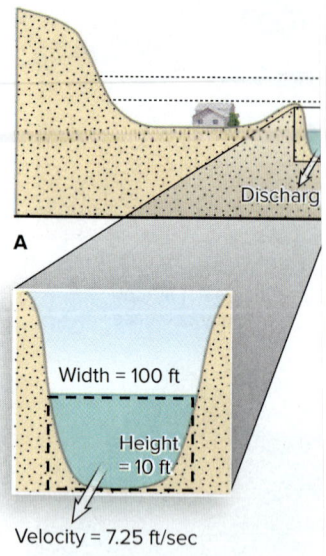

Discharg

A

Width = 100 ft

Height = 10 ft

Velocity = 7.25 ft/sec

B

FIGURE 8.19   The severity c
measuring either stream discha
height of the water above some
stage refers to the level where
banks. Discharge (B) is the proc
and cross-sectional area of flow
stage or height, there is a direct
and discharge.

FIGURE 8.30   Urban settings
commonly contain large areas of
impermeable surfaces where little to
no infiltration takes place, generating
large volumes of overland flow that
rapidly enter the drainage network.
The result is more frequent flooding,
higher flood crests, and shorter lag
time between a rain event and peak
discharge.

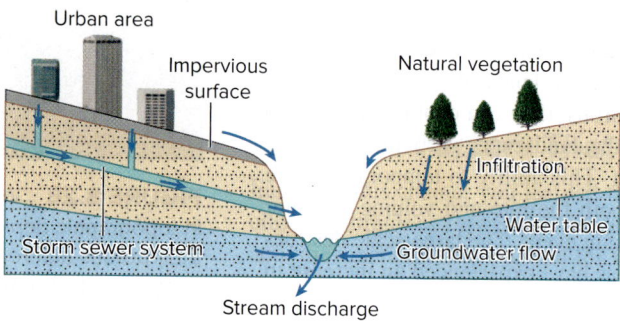

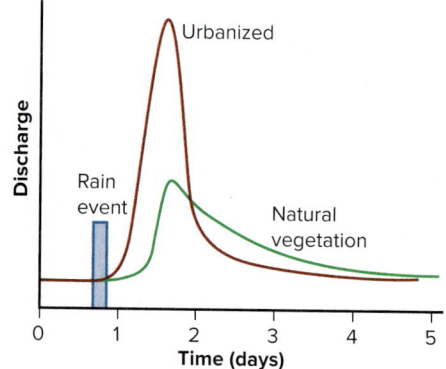

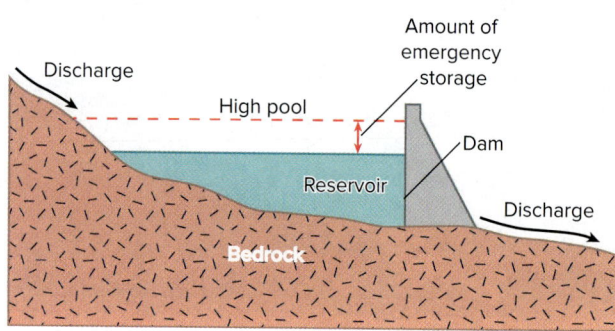

FIGURE 8.31   Dams prevent downstream flooding by
intercepting and storing stream discharge in their reservoirs.
By carefully regulating the release of water, engineers can
maintain a reservoir such that it provides a sufficient amount
of emergency storage and enough water to meet the
demand for freshwater and electricity.

losses by managing human activities that tend to exacerbate flooding. In this section we will explore some of the more common flood mitigation techniques.

Dams

When a dam is built across a river, it gives engineers the ability to control the
river's discharge. As illustrated in Figure 8.31, when engineers regulate the
discharge of a river, they are able to raise or lower the pool or *reservoir* of
water behind the dam. Reservoirs are valuable to society in that they protect
against floods, serve as important sources of freshwater and electrical power,
and offer a variety of recreational opportunities. In general, engineers allow
water to accumulate in a reservoir during wet periods when streamflows are
high, thereby creating a reserve or stockpile of water. During dry periods when
the demand for water exceeds the river's discharge, the pool is then lowered in
order to meet the demand. In the case of a dam that is used to generate hydroelectric power, engineers raise and lower the pool on a daily basis. Regardless
of whether the reservoir height is manipulated daily or seasonally, engineers
must always ensure that enough water is being released to meet the needs of
downstream ecosystems.

The storage capacity of a reservoir can also be manipulated for the purpose of preventing flooding downstream of the dam. Here engineers must
maintain a sufficient amount of reserve or emergency storage behind the dam
(Figure 8.31). This additional storage allows periodic surges of water from
upstream areas to be safely contained within the reservoir. Although a sufficient amount of emergency storage is normally maintained, unexpectedly heavy
or prolonged rains sometimes cause a reservoir to reach its maximum capacity.
When this occurs, engineers may be forced to release water at such a high rate
that it causes downstream flooding, which, ironically, is what the dam is supposed to prevent. The alternative is to risk letting the pool rise to the point that
the dam itself is weakened. The dam's structural integrity can also be threatened should water be allowed to flow over the top in an uncontrolled manner.
In either case, the structural failure of a dam can generate a truly catastrophic
flood, as in the massive flood described earlier that struck Johnstown, Pennsylvania, in 1889 (Figure 8.23).

In addition to the potential for structural failure and devastating floods, dams
are also a concern because of their environmental impact on downstream ecosystems. When a river is dammed, the suspended and bed loads are forced to be
deposited within the reservoir. The water temperature also becomes much cooler
due to the depth of the reservoir. Therefore, what is released from the reservoir
is cool, sediment-free water. This can be highly disruptive to aquatic ecosystems
that naturally evolved in the presence of warmer water and a certain amount of
sediment moving downstream. Moreover, because the volume of water that is

happening. Determining the probability of floods of different magnitudes allows scientists and engineers to quantify the flood risk for areas adjacent to a river. Naturally the lowest areas will have the highest level of risk.

Flood frequency is determined by first acquiring historical discharge data for a river for as many years as possible. Because the goal is to analyze the probability of flooding, the maximum discharge for each year in the record is tabulated. An example dataset for the Tar River in North Carolina is listed in Table 8.2. Based on such a dataset, a **recurrence interval** is calculated for each maximum discharge, which represents the frequency at which that particular discharge can be expected to recur. For example, a discharge with a 100-year recurrence interval means that on average, 100 years should pass before that same discharge occurs again. Note that because recurrence intervals are statistical in nature, a 100-year flood could take more than 100 years to repeat, but could just as easily recur in less than 100 years.

The recurrence interval for each maximum yearly discharge in a historical record is calculated using the following relationship:

$$\text{Recurrence Interval (RI)} = (N + 1)/M$$

where N is the number of values in the record, and M is the rank of a particular discharge maximum. Using the discharge record of the Tar River as an example (Table 8.2), one can see that the record covers a 110-year period, which in this case makes $N = 110$. To determine rank, the discharge values are then sorted from highest to lowest, which makes the 1999 event of 70,600 ft^3/sec the highest rank ($M = 1$) and the 1981 event of 3,340 ft^3/sec the lowest ($M = 110$). The recurrence intervals for these two discharges are calculated as follows:

$$70{,}600 \text{ ft}^3/\text{sec Recurrence Interval} = (N + 1)/M =$$
$$(110 + 1)/1 = 111 \text{ years}$$

$$3{,}340 \text{ ft}^3/\text{sec: Recurrence Interval} = (N + 1)/M =$$
$$(110 + 1)/110 = 1.0 \text{ year}$$

What this means is that a maximum discharge of 70,600 ft^3/sec should repeat itself, on average, every 111 years, and one of 3,340 ft^3/sec should occur every year. The remaining recurrence intervals for this record range between 1.0 and 111 years. However, the 1999 Tar River flood (70,600 ft^3/sec), which was associated with the landfall of Hurricane Floyd, was so large that it tends to skew the results of this analysis. Therefore, if we remove the 1999 event from the calculations and plot the remaining discharge values versus recurrence intervals, we get the graph shown in Figure 8.20. From this we see that the 1999 flood has an estimated recurrence interval of over 500 years, perhaps even as high as 1,000 years. Clearly, this was a major flood of historic proportions.

Figure 8.20 also illustrates the fact that low-discharge events are more numerous, and repeat more frequently, compared to high-discharge events. In terms of flooding, the Tar River at this particular site will overflow its banks and begin inundating low-lying areas whenever discharge exceeds 10,000 ft^3/sec (flood stage is 19 feet above sea level). From the graph we can see that the 10,000 ft^3/sec minimum for producing a flood recurs on average about every one and a half years. Progressively larger floods, of course, have longer recurrence intervals.

Another useful way of measuring flood frequency is *percent probability*, which is simply the inverse or reciprocal of the recurrence interval (1/RI). Note in Figure 8.20 that the recurrence intervals have been converted to probabilities along the top axis. From this, one can see that a 10-year flood has a 10% probability of taking place in any given year, whereas a 100-year flood has a 1% chance. Recurrence intervals therefore tell us how *often* we can expect floods of a certain size, whereas percent probabilities indicate their *chance* of occurring. It is important to keep in mind that these statistical measures are based on long-term averages. As indicated earlier, 100-year floods do not necessarily happen exactly

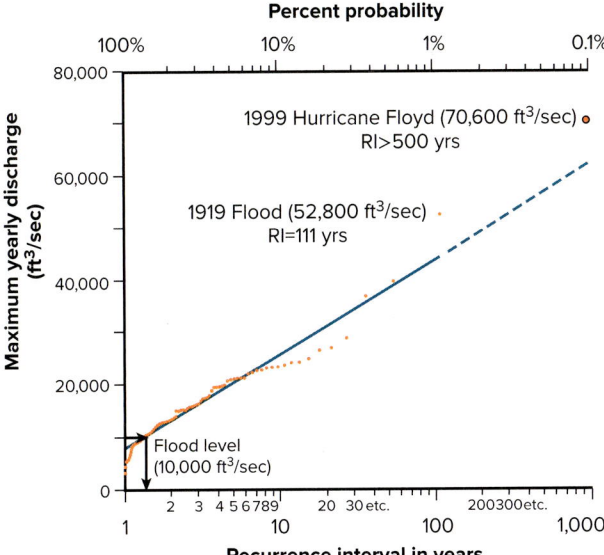

FIGURE 8.20   Plot showing discharge versus recurrence interval and percent probability for the Tar River at Tarboro, North Carolina (graph based on the discharge record in Table 8.2). Note that low-discharge events are not only more numerous, but also have shorter recurrence intervals. Major floods such as the ones in 1919 and 1999 are exceptional events with long recurrence intervals.

FIGURE 8.30 Urban settings commonly contain large areas of impermeable surfaces where little to no infiltration takes place, generating large volumes of overland flow that rapidly enter the drainage network. The result is more frequent flooding, higher flood crests, and shorter lag time between a rain event and peak discharge.

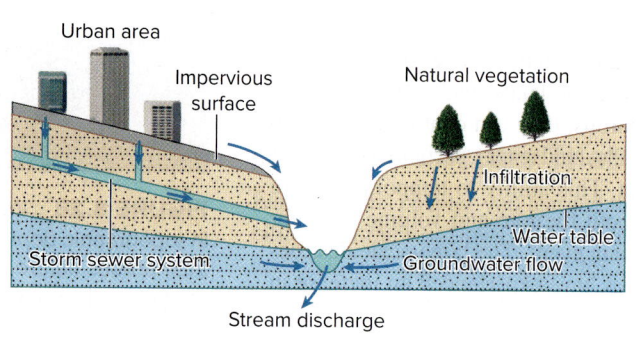

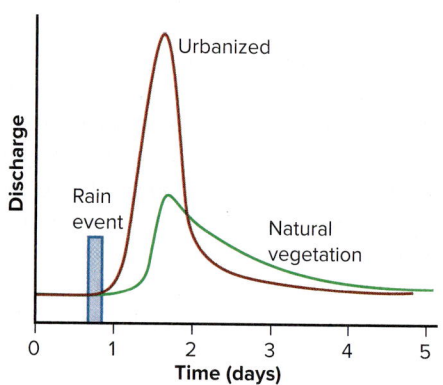

losses by managing human activities that tend to exacerbate flooding. In this section we will explore some of the more common flood mitigation techniques.

Dams

When a dam is built across a river, it gives engineers the ability to control the river's discharge. As illustrated in Figure 8.31, when engineers regulate the discharge of a river, they are able to raise or lower the pool or *reservoir* of water behind the dam. Reservoirs are valuable to society in that they protect against floods, serve as important sources of freshwater and electrical power, and offer a variety of recreational opportunities. In general, engineers allow water to accumulate in a reservoir during wet periods when streamflows are high, thereby creating a reserve or stockpile of water. During dry periods when the demand for water exceeds the river's discharge, the pool is then lowered in order to meet the demand. In the case of a dam that is used to generate hydro-electric power, engineers raise and lower the pool on a daily basis. Regardless of whether the reservoir height is manipulated daily or seasonally, engineers must always ensure that enough water is being released to meet the needs of downstream ecosystems.

The storage capacity of a reservoir can also be manipulated for the purpose of preventing flooding downstream of the dam. Here engineers must maintain a sufficient amount of reserve or emergency storage behind the dam (Figure 8.31). This additional storage allows periodic surges of water from upstream areas to be safely contained within the reservoir. Although a sufficient amount of emergency storage is normally maintained, unexpectedly heavy or prolonged rains sometimes cause a reservoir to reach its maximum capacity. When this occurs, engineers may be forced to release water at such a high rate that it causes downstream flooding, which, ironically, is what the dam is supposed to prevent. The alternative is to risk letting the pool rise to the point that the dam itself is weakened. The dam's structural integrity can also be threatened should water be allowed to flow over the top in an uncontrolled manner. In either case, the structural failure of a dam can generate a truly catastrophic flood, as in the massive flood described earlier that struck Johnstown, Pennsylvania, in 1889 (Figure 8.23).

In addition to the potential for structural failure and devastating floods, dams are also a concern because of their environmental impact on downstream ecosystems. When a river is dammed, the suspended and bed loads are forced to be deposited within the reservoir. The water temperature also becomes much cooler due to the depth of the reservoir. Therefore, what is released from the reservoir is cool, sediment-free water. This can be highly disruptive to aquatic ecosystems that naturally evolved in the presence of warmer water and a certain amount of sediment moving downstream. Moreover, because the volume of water that is

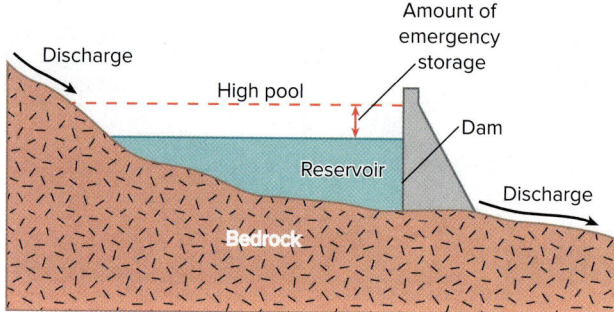

FIGURE 8.31 Dams prevent downstream flooding by intercepting and storing stream discharge in their reservoirs. By carefully regulating the release of water, engineers can maintain a reservoir such that it provides a sufficient amount of emergency storage and enough water to meet the demand for freshwater and electricity.

happening. Determining the probability of floods of different magnitudes allows scientists and engineers to quantify the flood risk for areas adjacent to a river. Naturally the lowest areas will have the highest level of risk.

Flood frequency is determined by first acquiring historical discharge data for a river for as many years as possible. Because the goal is to analyze the probability of flooding, the maximum discharge for each year in the record is tabulated. An example dataset for the Tar River in North Carolina is listed in Table 8.2. Based on such a dataset, a **recurrence interval** is calculated for each maximum discharge, which represents the frequency at which that particular discharge can be expected to recur. For example, a discharge with a 100-year recurrence interval means that on average, 100 years should pass before that same discharge occurs again. Note that because recurrence intervals are statistical in nature, a 100-year flood could take more than 100 years to repeat, but could just as easily recur in less than 100 years.

The recurrence interval for each maximum yearly discharge in a historical record is calculated using the following relationship:

$$\text{Recurrence Interval (RI)} = (N + 1)/M$$

where N is the number of values in the record, and M is the rank of a particular discharge maximum. Using the discharge record of the Tar River as an example (Table 8.2), one can see that the record covers a 110-year period, which in this case makes $N = 110$. To determine rank, the discharge values are then sorted from highest to lowest, which makes the 1999 event of 70,600 ft^3/sec the highest rank ($M = 1$) and the 1981 event of 3,340 ft^3/sec the lowest ($M = 110$). The recurrence intervals for these two discharges are calculated as follows:

$$70{,}600 \text{ ft}^3/\text{sec Recurrence Interval} = (N + 1)/M =$$
$$(110 + 1)/1 = 111 \text{ years}$$

$$3{,}340 \text{ ft}^3/\text{sec: Recurrence Interval} = (N + 1)/M =$$
$$(110 + 1)/110 = 1.0 \text{ year}$$

What this means is that a maximum discharge of 70,600 ft^3/sec should repeat itself, on average, every 111 years, and one of 3,340 ft^3/sec should occur every year. The remaining recurrence intervals for this record range between 1.0 and 111 years. However, the 1999 Tar River flood (70,600 ft^3/sec), which was associated with the landfall of Hurricane Floyd, was so large that it tends to skew the results of this analysis. Therefore, if we remove the 1999 event from the calculations and plot the remaining discharge values versus recurrence intervals, we get the graph shown in Figure 8.20. From this we see that the 1999 flood has an estimated recurrence interval of over 500 years, perhaps even as high as 1,000 years. Clearly, this was a major flood of historic proportions.

Figure 8.20 also illustrates the fact that low-discharge events are more numerous, and repeat more frequently, compared to high-discharge events. In terms of flooding, the Tar River at this particular site will overflow its banks and begin inundating low-lying areas whenever discharge exceeds 10,000 ft^3/sec (flood stage is 19 feet above sea level). From the graph we can see that the 10,000 ft^3/sec minimum for producing a flood recurs on average about every one and a half years. Progressively larger floods, of course, have longer recurrence intervals.

Another useful way of measuring flood frequency is *percent probability*, which is simply the inverse or reciprocal of the recurrence interval (1/RI). Note in Figure 8.20 that the recurrence intervals have been converted to probabilities along the top axis. From this, one can see that a 10-year flood has a 10% probability of taking place in any given year, whereas a 100-year flood has a 1% chance. Recurrence intervals therefore tell us how *often* we can expect floods of a certain size, whereas percent probabilities indicate their *chance* of occurring. It is important to keep in mind that these statistical measures are based on long-term averages. As indicated earlier, 100-year floods do not necessarily happen exactly

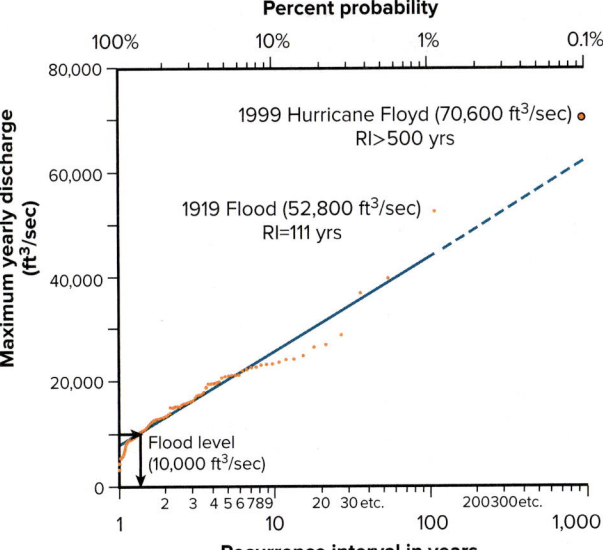

FIGURE 8.20   Plot showing discharge versus recurrence interval and percent probability for the Tar River at Tarboro, North Carolina (graph based on the discharge record in Table 8.2). Note that low-discharge events are not only more numerous, but also have shorter recurrence intervals. Major floods such as the ones in 1919 and 1999 are exceptional events with long recurrence intervals.

TABLE 8.2   Maximum yearly discharge values over a 110-year period for the Tar River at Tarboro, North Carolina. Flood stage at this particular gauging station near the coast is 19.0 ft above sea level, which corresponds to a discharge of 10,000 ft3/sec. Years in which the river did not flood are highlighted in orange; flood years are in green and major flood years are in blue.

Year	Stage	Maximum Yearly Discharge	Year	Stage	Maximum Yearly Discharge	Year	Stage	Maximum Yearly Discharge
1906	23.5	16,600	1943	19.0	10,800	1980	17.6	8,910
1907	21.2	13,400	1944	21.6	13,800	1981	8.9	3,340
1908	29.4	27,300	1945	28.1	24,600	1982	16.0	7,790
1909	19.7	11,500	1946	21.0	13,200	1983	23.7	16,400
1910	27.3	23,100	1947	14.1	6,570	1984	26.4	21,300
1911	17.4	9,210	1948	25.4	19,800	1985	20.8	11,900
1912	17.7	9,480	1949	22.5	15,300	1986	23.4	15,900
1913	21.2	13,400	1950	14.4	6,990	1987	28.4	25,200
1914	19.2	11,000	1951	13.4	6,250	1988	13.2	5,870
1915	19.1	10,900	1952	24.2	17,600	1989	24.2	17,200
1916	16.9	8,770	1953	18.1	9,950	1990	21.7	13,200
1917	23.3	16,200	1954	27.4	23,600	1991	16.4	8,090
1918	21.0	13,100	1955	23.5	16,600	1992	24.5	17,800
1919	34.0	52,800	1956	20.9	13,000	1993	25.7	19,900
1920	19.1	10,900	1957	22.8	15,500	1994	24.5	17,700
1921	17.2	9,030	1958	29.2	26,900	1995	21.6	13,100
1922	26.4	21,400	1959	21.7	14,000	1996	26.6	21,600
1923	22.8	15,500	1960	22.8	15,500	1997	21.2	12,500
1924	20.7	12,700	1961	20.9	13,000	1998	27.6	23,700
1925	33.5	39,800	1962	20.3	12,200	1999	41.5	70,600
1926	19.2	11,000	1963	18.1	9,850	2000	24.5	19,200
1927	17.8	9,570	1964	20.8	12,800	2001	22.5	16,000
1928	30.2	29,200	1965	25.6	20,000	2002	19.0	9,970
1929	25.5	19,800	1966	22.7	15,300	2003	26.3	21,000
1930	27.8	24,000	1967	17.6	8,950	2004	20.3	11,400
1931	17.7	9,480	1968	21.2	12,500	2005	15.4	7,410
1932	20.2	12,100	1969	19.4	10,300	2006	28.0	24,500
1933	16.0	8,050	1970	17.8	9,020	2007	23.0	15,300
1934	22.1	15,900	1971	18.8	9,820	2008	15.3	7,290
1935	27.4	23,500	1972	23.1	15,500	2009	18.6	9,670
1936	25.5	20,200	1973	24.7	18,200	2010	23.0	15,300
1937	26.2	21,500	1974	14.8	6,930	2011	18.9	9,940
1938	21.3	13,500	1975	27.1	22,600	2012	11.0	4,580
1939	27.0	23,000	1976	21.7	13,200	2013	18.5	9,300
1940	31.8	37,200	1977	17.5	8,880	2014	19.4	10,200
1941	16.7	8,460	1978	26.4	21,200	2015	21.9	13,500
1942	14.8	7,310	1979	27.0	22,400			

Source: Data from the U.S. Geological Survey

100 years apart, but rather they occur *on average* every 100 years. This means that it is entirely possible to have two 100-year floods in back-to-back years, two in a single year, or two separated by 200 years or more.

It should be apparent that recurrence intervals can only be computed for streams for which historical discharge records have been compiled. Moreover, the reliability of these values depends on the number of years in a given record. For example, a 100-year record is more likely to contain several major flood events than a 10-year record, thereby producing more reliable recurrence estimates. A longer record also helps smooth out any short-term variations in climate. You can gain an appreciation for this by examining the data in Table 8.2. Notice that from 1968 to 1982 there were 8 floods in a 15-year period, whereas during the next 15 years (1983 to 1997) there were 13 floods, a 63% increase.

Natural Factors That Affect Flooding

Recall that most floods are related to heavy rainfall events, where large volumes of overland flow enter a drainage network. However, flooding can also be caused by a number of different mechanisms. For example, coastal areas are commonly inundated by the surge of seawater that pushes ashore as a hurricane makes landfall (Chapter 9). During volcanic eruptions glacial ice caps may undergo rapid melting, producing catastrophic floods called mudflows (Chapter 6). In some regions rapid snowmelt in the spring frequently leads to downstream flooding, whereas the breakup of river ice and subsequent formation of ice dams cause upstream areas to become inundated (Figure 8.21). Because most floods are caused by rain and snow, we will focus our attention in this section on the factors that affect precipitation-induced floods.

Nature of Precipitation Events

There are many different types of rainfall events, from light, steady rains that may last for days to heavy, torrential rains lasting anywhere from a few minutes to several hours. The potential for flooding naturally increases as the *intensity* and *duration* of rainfall increase. Although regional climate determines annual precipitation and its distribution throughout the year, daily weather conditions are most important with respect to flooding because they govern the rate (intensity) and duration of individual precipitation events. For example, floods are often associated with the intense rainfall from thunderstorms. These storms are common from spring to late summer when warm air masses are more likely to collide with cool dry air. On the other hand, less intense rains can also produce large volumes of overland flow, provided the rains fall over an extended period of time. In many regions, floods related to these moderately intense rains can occur throughout the year. With respect to snow, its ability to cause flooding largely depends on how much accumulates and how rapidly it melts. A good example of a snow-induced flood is the great flood of 2009 along the Red River in Minnesota, North Dakota, and Canada (Figure 8.21). This flood began when spring rains coincided with the sudden melting of large amounts of snow that had accumulated over the winter.

Another important factor associated with rainfall is the *size* of the area over which the rain falls. An isolated thunderstorm for example may generate intense rains, but flooding is generally more localized because such storms are relatively small and tend to keep moving. On the other hand, large regional storms, like those that move inland from the ocean (e.g., tropical storms and northeasters, also known as *nor'easters*), can drop tremendous amounts of precipitation over large areas, thereby producing widespread flooding.

Ground Conditions

The ability of the land surface to absorb water, referred to as **infiltration capacity,** plays a critical role in flooding since water that infiltrates is generally removed from overland flow. Therefore, drainage basins that have large areas with high infiltration capacities are less prone to flooding. The actual rate at which water can infiltrate is determined by the slope of the land surface, type of ground material, and moisture content of the material. Clearly as slopes become progressively steeper, more and more rainwater will move as overland flow, leaving a smaller fraction that can infiltrate. With respect to the material itself, gravel- and sand-rich soils have much higher infiltration capacities than soils rich in clay. Also, the infiltration rates of soils are highest when they are dry, and progressively decrease as the pores take on more water. Once soils reach saturation, infiltration

FIGURE 8.21   The great 2009 Red River flood in the United States and Canada was caused by a combination of spring rains and the rapid melting of snow that had accumulated over the winter. Note how ice jams can block the flow of water in the main channel, making flooding worse.
Carrie Olheiser, The National Operational Hydrologic Remote Sensing Center (NOHRSC)/NOAA

will continue at a constant rate provided the land surface is above the water table; otherwise, infiltration will cease completely. Note that some materials have almost no ability to absorb water, which forces nearly all of the rainwater to move as overland flow. Examples of such materials include unfractured igneous rock, frozen soils, and surfaces covered with asphalt and concrete (e.g., roads and parking lots).

Vegetation Cover

Vegetation is another critical factor because it helps reduce the volume of overland flow, and thus reduces flooding. Vegetation intercepts and stores a fraction of the rain, thereby preventing it from reaching the land surface and moving as overland flow. In some heavily forested areas as much as 30% of the annual rainfall is captured and never reaches the ground; instead, it evaporates and returns to the atmosphere. For the fraction of water that does reach the land surface, vegetation restricts its ability to move downslope, thereby giving it more time to infiltrate. Very dense grasses that blanket or carpet the landscape are particularly effective at increasing infiltration and reducing overland flow. As could be expected, rates of overland flow are significantly higher in arid climates because of the sparser vegetation. In the last section of this chapter we will examine how human modifications to the landscape (removal of vegetation and creating impervious surfaces) lead to significant increases in overland flow.

Types of Floods

In terms of rainfall and overland flow, the potential for flooding depends on the ability of the precipitation to exceed the infiltration rate, the size of the area over which the rain falls, and the vegetation density. For example, under even the most ideal circumstances an isolated thunderstorm in a large drainage basin like the Mississippi is not going to generate enough overland flow to cause flooding downstream on a major river channel. This same storm, however, may be perfectly capable of causing small tributaries to quickly overflow their banks. In this section we will take a closer look at how flooding hazards vary depending on the size of a channel and its location within a drainage basin.

Flash Floods

Earlier in this chapter you learned that the headwaters of drainage basins are dominated by numerous tributaries and small rivers. Because these channels are relatively small, even isolated thunderstorms can generate enough overland flow to cause localized flooding. In addition, since small tributaries are where overland flow first enters a drainage network, the lag time between the precipitation event and peak discharge in the channels is rather short. Small streams and rivers therefore tend to rapidly overflow their banks in what are called **flash floods.** Because small channels are more abundant in the upper parts of a basin, flash floods are also referred to as *upstream floods.* Keep in mind that flash flooding can occur in lower areas of a drainage system along small tributaries that flow into large rivers.

Although flash floods generally affect only localized areas, they are particularly dangerous due to the rapid way in which they develop. What may normally be a pleasant, gently-flowing stream can quickly turn into a raging torrent, leaving little time for people to escape. Adding to the hazard is the fact that the water level and velocity can be far greater compared to normal flow conditions. Recall that headwater areas are typically far above base level and have relatively narrow valleys where floodplain development is minimal or nonexistent. The general lack of a floodplain essentially means that a stream or small river has a very limited area in which to store excess water during a flood. The narrow valley then forces floodwaters to reach greater heights. In addition to the height, the velocity of the water is very high because the stream gradient in the headwaters is considerably steeper than in downstream channels closer to base level. The result can be a raging torrent of water racing down a narrow valley, ripping up trees and moving boulders the size of cars. For people living in narrow mountain valleys, the only means of escape during a major flash flood is to quickly climb to higher ground.

While humans may consider flash floods to be unusual events, in terms of geologic time they are a regular occurrence in many headwater streams. A good example is the 1976 flash flood along the Big Thompson River near Rocky Mountain National Park in Colorado. On July 31 the typical late afternoon thunderstorms began developing along the rugged Front Range of the Rockies (Figure 8.22). But instead of the normal winds, which move the storms eastward out onto the plains, the winds on this day blew in the opposite direction, pushing the thunderstorms westward up against the mountains. One particularly large thunderstorm remained stationary for nearly three hours over the upper reaches of the Big Thompson Valley. Heavy rains began around 6:00 p.m. and continued over the next four hours, dumping as much as a foot of rain. Since the surface of this steep mountain terrain consists mostly of bedrock, very little infiltration took place, forcing nearly all of the rainwater to flow directly into the Big Thompson River. By 8:00 p.m. this normally tranquil mountain stream was transformed into a raging monster, cresting 20–30 feet (6–9 m) above normal stage and reaching speeds of up to 50 miles per hour (80 km/hr). In a matter of two hours the Big Thompson flood killed 145 residents and vacationers, destroyed 418 homes and 152 businesses, all of which were washed down the canyon. Before and after photographs in Figure 8.22 attest to the tremendous power of this flash flood.

In contrast to the thunderstorm-driven event over the Big Thompson River, lower-intensity but longer-duration rainfall conditions persisted up and down the Front Range in September of 2013, generating flash floods in multiple canyons that killed eight people and destroyed 1,500 homes. Because there are large uncertainties in determining the recurrence interval of large floods like the 1976 and 2013 events, hydrologists estimate that such major flash floods along the Front Range recur on the order of every few hundred years or even as infrequently as every 1,000 years.

Deadly flash floods also occur in some desert regions where afternoon thunderstorms are primarily restricted to the headwater areas in nearby mountain ranges. The combination of heavy rainfall with sparse vegetation and steep terrain create flash floods that sweep down through normally dry channels on the valley floor. For example, nearly every year in the western United States people are trapped in dry streambeds by flash floods that develop from distant thunderstorms. The rushing water can rise so rapidly that the victims do not have enough time to scramble to higher ground.

In addition to thunderstorm activity and persistent rains, the sudden failure of a dam can produce a flash flood. Note that dams can either be constructed by humans or composed of natural debris such as logs and sediment or broken-up ice. The worst flood disaster in U.S. history occurred in 1889 when heavy rains contributed to the failure of a poorly maintained dam, located 14 miles upstream of Johnstown, Pennsylvania (Figure 8.23). The sudden failure of the dam sent a surge of water and debris, nearly 40 feet (12 m) high, roaring down the valley, killing 2,209 residents.

Downstream Floods

As indicated earlier, it takes a much greater volume of water to force a large river to overflow its banks than it does for a small stream located near the headwaters of a drainage basin. A river in the lower portion of a basin is also more likely to have a wider valley and more expansive floodplain. Therefore, a **downstream flood** can be defined as one where a river in the lower part of a drainage basin leaves its channel, flowing out onto its floodplain and inundating large areas of the valley floor. Clearly, the volume of water required to cover the floodplain of

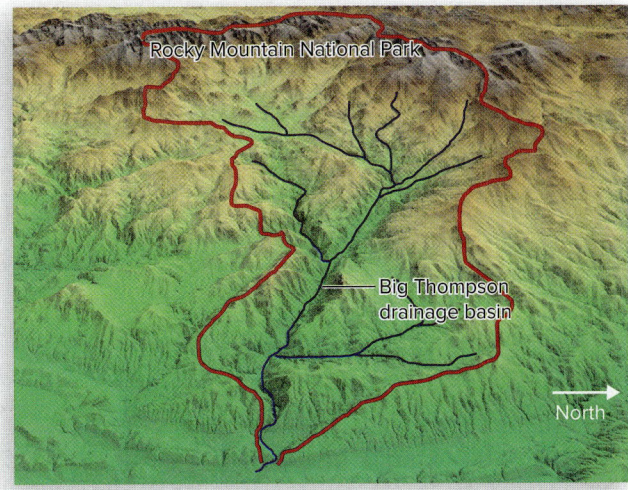

A

B

FIGURE 8.22 The 1976 flash flood along the Big Thompson River, in Colorado, developed when a thunderstorm remained stationary, dumping a foot of rain over the upper reaches of the river's drainage basin (A) in mountainous terrain. Before and after photos (B) showing boulders that were transported during the flood. These boulders are a testament to the tremendous power and velocity of the floodwaters.

(a) National Geophysical Data Center, NESDIS, NOAA, Big Thompson Drainage from USGS NED, J. Varner CIRES (b) (both) USGS

FIGURE 8.23 Photo showing the aftereffects of a failed dam in 1889 that unleashed a catastrophic flash flood, killing 2,209 people in the valley leading into Johnstown, Pennsylvania. As with most river valleys, human development was concentrated here because of the flat terrain, fertile soil, and abundant water resources.
National Park Service/Johnstown Are Heritage Assoc.

a large river cannot be generated locally from a single isolated storm. Most downstream floods are caused by regional accumulations of water in the more elevated parts of the drainage basin. What typically happens is that rain falls for an extended period of time, from days to perhaps weeks, over large areas of the upper basin—in some cases melting snow adds to the volume of water. Discharge from swollen tributaries will then combine, creating ever larger volumes of water as the river flows downstream. If the channel is not capable of carrying this combined flow, it will overflow its banks and begin storing the excess water in its floodplain.

In addition to the volume of water involved, downstream floods differ from flash floods in the amount of lag time between the rain event and peak discharge, and in the length of time the river remains above flood stage. As illustrated in Figure 8.24, a major rain event in the headwaters will result in water entering the network of small tributaries. From there it will collect in progressively larger channels, traveling as a crest or pulse that moves downstream through the basin. Also note in the hydrographs how the lag time between the rain event and peak discharge becomes greater as the flood crest moves downstream. Here we see that flash floods can form in the upper basin where water levels rise and fall relatively quickly. This stands in contrast to downstream floods where the combined flows of the tributaries cause water levels to rise more slowly as the river's floodplain becomes inundated. The hydrographs also show how the river stays above flood stage in the downstream areas for longer periods of time before eventually returning to its channel. Flash floods are naturally more hazardous than downstream floods because they occur so quickly, which means people have less time to evacuate. On the other hand, downstream floods typically cause greater property damage because they inundate much larger areas and can keep buildings under water for days or even weeks (Figure 8.25).

A good example of a downstream flood is the 1993 flood on the Mississippi River (Figure 8.26), which ranks as one of the largest natural disasters in U.S. history. This flood had its origins in the winter and spring when persistent rains kept the ground nearly saturated, thereby reducing its infiltration capacity. By June, streams across the Upper Midwest were already filled to capacity, and the heaviest rains were yet to come. During the summer of 1993, rainfall in the Upper Midwest was 200–350% greater than normal. One of the hardest hit areas was Iowa, where 48 inches (122 cm) of rain fell from April through August. The problem was related to how summer weather patterns cause cool, dry air from the north to collide with warm, moist air coming up from the Gulf of Mexico, producing a belt of thunderstorms that move across the region. In the summer of 1993, however, this frontal system remained fixed throughout June and July, causing wave after wave of thunderstorms to dump rain over the same swath of ground for weeks on end.

In 2008 fixed weather patterns again brought historic flooding to the Upper Mississippi basin (Figure 8.25). Particularly hard hit were parts of Iowa, Wisconsin, Illinois, and Indiana, which received over 12 inches (30 cm) of rain in early June.

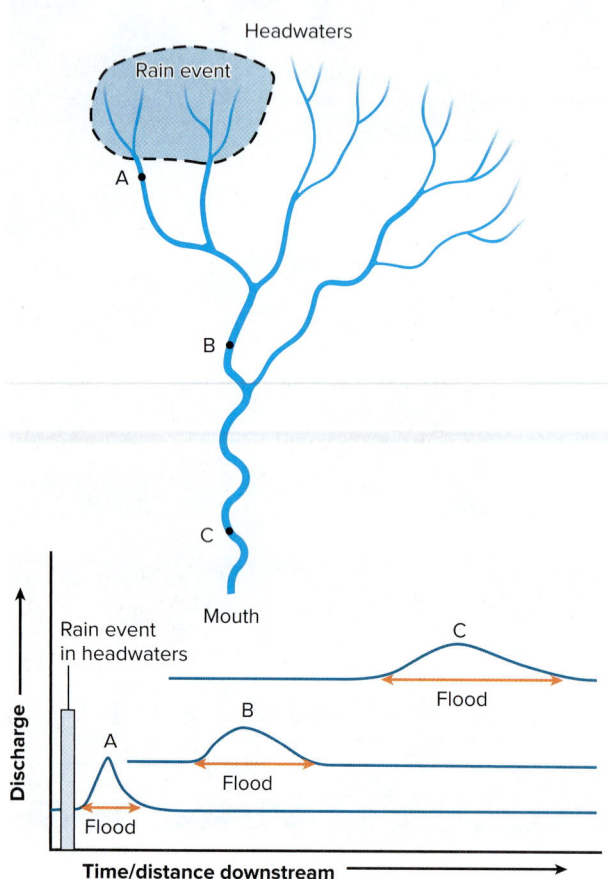

FIGURE 8.24 Idealized hydrographs illustrating how a river will respond to heavy rainfall in the headwaters of a basin. As the flood crest moves downstream, there is a progressive increase in lag time between the rain event peak and discharge; the river downstream also rises more slowly and remains above flood stage for a longer period of time. Hydrograph (A) is characteristic of a flash flood, and (C) of a downstream flood, with (B) representing a transitional phase.

FIGURE 8.25   Photo showing the effects of downstream flooding in Cedar Rapids, Iowa. This event was associated with historic flooding across the Upper Midwest in 2008, leaving some areas inundated for as long as two weeks.
Don Becker, U.S. Geological Survey

Three years later, in 2011, historic flooding occurred yet again, with rains focusing this time on the Upper Mississippi and Ohio basins. During a 14-day period from late April through early May, rainfall in some areas totaled more than 20 inches (51 cm). The floodwaters eventually moved downstream, causing major flooding along the Lower Mississippi. Because the historic floods of 1993, 2008, and 2011 occurred in less than 20 years, but had recurrence intervals on the order of hundreds of years, some climatologists believe that these large floods are taking place more frequently because the global atmosphere is now warmer and has more water vapor that falls as precipitation in thunderstorms and hurricanes (Chapter 16).

Human Activity and Flooding

One of the unintended consequences of human modifications to the natural environment is that the risk of flooding has increased. We have cleared considerable expanses of land for agriculture, built great cities, and constructed vast transportation networks for moving people and material. While these achievements have allowed us to prosper, they are not without consequences. In recent years scientists have discovered that our extensive modifications to the landscape have altered the hydrology of streams such that the frequency and severity of flooding has increased. Ironically, some of our techniques for reducing flooding in one location have led to increased flooding in other areas. Protecting ourselves from floods has also encouraged development in places where the risk would otherwise be deemed too high. The 2005 disaster in New Orleans (see Case Study 8.1) provides a sober lesson about how flood controls allow development to flourish, but then produce catastrophic losses should the controls suddenly fail.

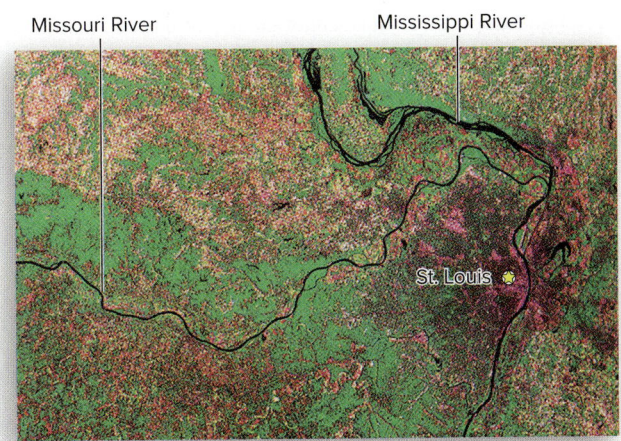

FIGURE 8.26   During the summer of 1993, weather patterns in the Upper Midwest caused wave after wave of storms to drop rain over the same area. The result was regional flooding as well as downstream flooding on the Mississippi River. Shown here are satellite images illustrating how this historic flood inundated floodplains above the confluence of the Mississippi and Missouri Rivers.
(both) NASA

Levees and the Disastrous 2005 Flood in New Orleans

The tragic flood that struck New Orleans in 2005 in the aftermath of Hurricane Katrina was more of an engineering failure related to flood control than a true natural disaster. Interestingly, this flood had its origins back in 1718 when the French built a settlement on a natural levee along a bend in the Mississippi River (Figure B8.1). This location gave the French access to a vital transportation corridor into the continental interior, and the levee provided dry ground in an area otherwise surrounded by cypress backswamps. Although the natural levee remained dry most of the year, early settlers found themselves in danger of being swept away by periodic floods. In order to minimize the hazard, French engineers built crude artificial levees. As New Orleans expanded, its protective levees grew in both height and length.

Until the early 1900s the city of New Orleans was pretty much restricted to the high ground along the natural levees. To enable the city to expand, engineers began draining the backswamps that lay between the river and Lake Pontchartrain to the north. Occupying this land required placing levees along Lake Pontchartrain to hold back high water levels associated with hurricanes. Because the cypress swamps were near sea level, draining the swamps could only be accomplished by lowering the water table and pumping the water into elevated canals (Figure B8.2). From here the canals could carry the water away from the city. However, once the thick, organic-rich soils in the backswamps dried out, they underwent considerable compaction, causing the land surface to subside. This created a bowl-shaped depression between the river and Lake Pontchartrain, parts of which eventually sunk below sea level (Figure B8.3).

Because much of New Orleans was now located in this human-made depression, it became clear that a levee failure could lead to a catastrophic flood in which the depression would fill with water. It was

FIGURE B8.2 Photo showing one of the many canals within New Orleans used to help drain the original cypress swamps. Water is pumped into the canals from the adjacent low areas that now consist of residential housing tracts. The canals are lined with floodwalls in order to keep water from flowing out of the them during hurricanes and flooding the reclaimed low areas.
US Army Corps of Engineers

absolutely critical then that the levees be made high enough to hold back potential floodwaters from not just the Mississippi River and Lake Pontchartrain, but from the canals as well. In the 1960s the U.S. Army Corps of Engineers embarked on a major program of raising and strengthening the levees. This effort included installing concrete floodwalls along the canals to help prevent the levees from being overtopped by water forced up into the canals during a hurricane. It was this intricate system of levees, pumps, and canals that made it possible for the backswamps to be developed, which ultimately allowed New Orleans to grow into a city of nearly 1.4 million people.

Although the engineering efforts appeared to have nature under control, the natural system was starting to respond in ways no one had anticipated. In the 1970s geologists began finding evidence that

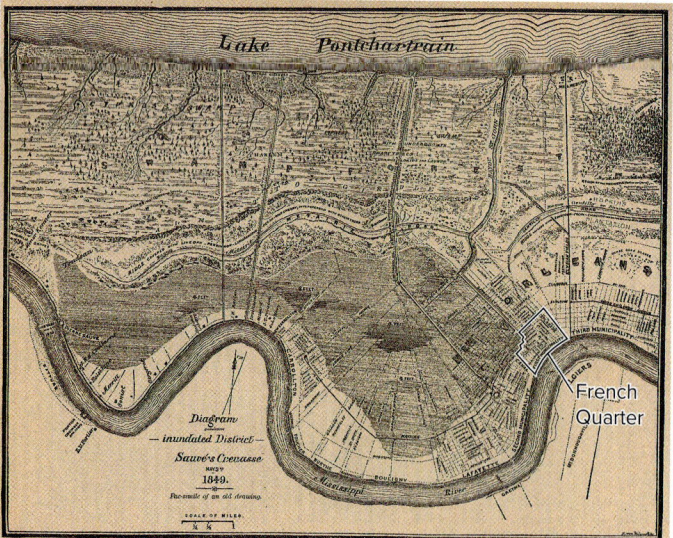

FIGURE B8.1 Map from 1849 showing the growing city of New Orleans and the cypress swamps that once existed north of the city. Note the position of the original French settlement (French Quarter) on the highest part of the levee, and how the swamps once drained to the north into Lake Pontchartrain.
Courtesy of the University Libraries, The University of Texas at Austin

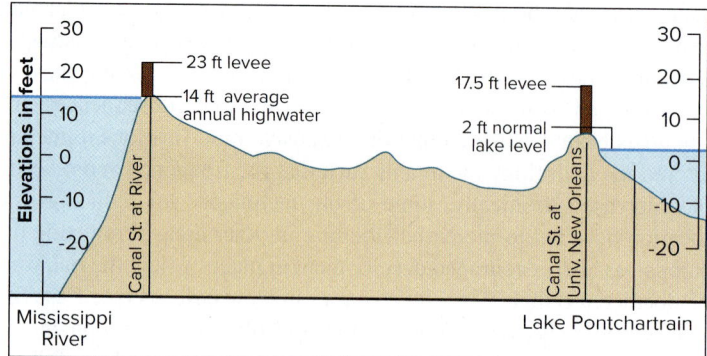

FIGURE B8.3 Draining of the backswamps and subsequent compaction of the organic-rich soils caused the land to subside beneath New Orleans. This subsidence has created a bowl-shaped depression, such that many parts of the city now sit below sea level. Tall levees are all that keep periodic high water levels in the Mississippi River and Lake Pontchartrain from flooding the city.

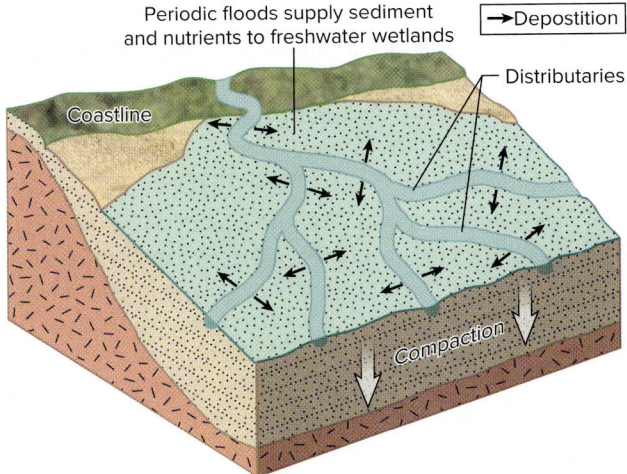

Periodic floods supply sediment
and nutrients to freshwater wetlands

Depostition →

Distributaries

Coastline

Compaction

FIGURE B8.4 Fine-grained sediment within river deltas naturally undergoes compaction due to the overburden pressure from the delta's immense weight. The land surface and extensive wetlands remain above sea level since sediment accumulation naturally keeps pace with compaction. In the case of the Mississippi Delta, artificial levees along the main channel have cut the delta off from its supply of sediment and freshwater, causing the entire delta to begin sinking below sea level.

the entire Mississippi Delta was slowly sinking below sea level. As indicated in Figure B8.4, the immense weight of a delta naturally causes the sediment to compact. Under normal geologic conditions the supply of new sediment being brought into a delta keeps pace with the rate of compaction. Therefore, the land surface does not sink (subside) below sea level. This delicate balance was disrupted in the Mississippi Delta when the artificial levees cut the delta off from its supply of sediment and freshwater. Compounding the problem is the fact that sea level continues to rise (Chapters 9 and 16). The combined effect of land subsidence and sea-level rise has caused some parts of New Orleans to be more than 17 feet (5 m) below sea level.

By lining the Mississippi River with artificial levees and draining its backswamps to allow for greater development, society inadvertently set New Orleans up for a flood of epic proportions. When Hurricane Katrina slammed ashore just east of New Orleans in 2005, the city was spared from a direct hit, but the storm pushed water up into the canal system just as scientists had predicted. Despite the fact the water levels within the canals remained about 4 feet (1.2 m) below the tops of the floodwalls—well within their design limits—the concrete panels failed at nearly 50 locations around the city. Investigators later determined that the failed panels had been improperly anchored in their foundations. During the storm the weakened panels simply fell over due to the weight of water within the canals. This allowed water to rush into the bowl-shaped depression of New Orleans. From Figure B8.5 one can see that while nearly 80% of the city was flooded, certain sections remained dry because some of the canals dividing up the city did not fail. Note that the French Quarter remained dry because the original settlement was located up on the natural levee, which was the highest ground available.

An important lesson from this disaster is that while levees do provide protection from floods, they also disrupt natural systems and produce consequences that society finds undesirable. Levees also encourage development in low-lying areas that otherwise would not be developed. A single event or failure can then result in a major disaster. Because it is only a matter of time before a major hurricane makes a direct strike on New Orleans, questions were raised after Katrina as to whether it is a good use of taxpayer dollars to try to upgrade the levee system. This debate was particularly important since the next levee failure could result in losses far greater than those from Katrina. In light of the fact New Orleans continues to sink while sea level rises, some suggested that the money would be better spent relocating the city farther inland on higher ground. In the end, the federal government approved $14.5 billion to improve and upgrade the city's network of levees, floodwalls, and pumps. The new system is designed to prevent flooding for 100-year storm events and to lessen the risk for even larger storms. Only time will tell whether or not this was a wise investment.

Not flooded   Flooded

Super
Dome

French
Quarter

A

Breached levee

B

FIGURE B8.5 Photos of New Orleans taken on August 29, 2005, the day Katrina made landfall. Satellite image (A) showing how only certain sections of the city were flooded by failed levees. Aerial view (B) showing flooded neighborhoods and a breached canal levee in background.
(a) NASA, (b) Jocelyn Augustino/FEMA

In this section we will explore how human modifications to the landscape have increased the risk of flooding in modern times. We will also examine some of the techniques used to reduce the impact of flooding.

Land-Use Factors That Affect Flooding

Earlier in this chapter you learned that when water accumulates on the surface it either infiltrates or flows over the landscape to the nearest tributary. Many human activities increase the potential for flooding simply by decreasing the amount of infiltration, which increases overland flow. Other activities contribute to flooding simply by bringing about increased erosion. This additional erosion results in the filling of stream channels with excess sediment, making it more likely that streams will overflow their banks during high discharge events. Flooding is also exacerbated by the presence of human infrastructure, particularly roads, that restricts the free flow of surface water. The following sections will describe some of the specific ways in which human activity has increased the frequency and severity of flooding.

Removal of Natural Vegetation

Humans have historically cut down forests in order to obtain lumber and create open spaces for growing crops and building settlements. Likewise, vast expanses of natural grassland have been eliminated for the purpose of creating farmland. One of the consequences of the widespread removal of forests and grasslands is the increased ability of water to flow downslope, which translates into more overland flow and less infiltration. Note that when agricultural crops replace natural vegetation, overland flow still increases since the density of the crops is typically far less than that of the replaced vegetation. The effect of agriculture on flooding is most pronounced when fields are fallow or contain young, immature crops.

Removing natural vegetation from the landscape also leaves soil more exposed to the effects of falling raindrops and overland flow. During impact, raindrops dislodge soil particles, which are then carried into tributary channels by overland flow. This carrying of excessive sediment off the landscape and into drainage systems is called **sediment pollution.** Over time sediment pollution can cause channels to become filled with sediment, thereby reducing their capacity to carry water (Figure 8.27). This filling thus increases the frequency and severity of flooding as streams overflow their banks more easily. Sediment pollution not only exacerbates flooding, it also destroys the natural ecology of streams. Many types of aquatic species are unable to survive when sediment fills the channel and alters the original stream habitat. Also, fish that depend on relatively clear water to see their prey commonly do not survive when visibility is reduced by the higher levels of suspended sediment.

Destruction of Wetlands

As illustrated in Figure 8.28, wetlands (swamps) are commonly found in topographic depressions and adjacent to river channels, in which case they are called *riparian wetlands*. In many parts of the world wetlands have historically been viewed as wastelands because they were not suitable as building sites and represented obstacles for roads and rail lines. Wetlands were also a source of mosquito-borne diseases. It is not surprising then that the practice of draining wetlands became quite common, particularly since they contained organic-rich soils that could be converted into fertile agricultural land. All that was required to drain a wetland was to lower the water table by digging a network of ditches. Once the

FIGURE 8.27 Sediment pollution (A) occurs when natural vegetation is removed from the landscape, which leads to increased overland flow and erosion that fills stream channels with excess sediment. This deposition of sediment reduces the capacity of channels to carry water, making it more likely streams will overflow their banks and cause flooding. Photo (B) showing sediment pollution taking place in a stream. Note that the sediment is being carried off the agricultural field by overland flow.
(b) Paul Ankcorn USGS

surface dried out, people could take uninhabitable "wasteland" and transform it into productive farmland.

Aside from the obvious loss of wildlife habitat and damage to natural ecosystems, the destruction of wetlands has reduced the landscape's ability to store water. Because wetlands are generally quite porous, they can capture and absorb significant quantities of water moving over the landscape, particularly during periods when the organic matter has dried out. This ability to temporarily store water means that wetlands also help reduce both the rate and volume of water flowing into a stream channel during a heavy rain event. The large-scale destruction of wetlands is a key reason why some areas have experienced an increase in the frequency and severity of flooding.

Construction Activity

Most construction activity involves removing natural vegetation and regrading the land surface. This activity exacerbates flooding because it increases overland flow and causes stream channels to fill with sediment (Figure 8.29). In many instances the structures that we build tend to restrict the free flow of water moving through small channels. Most problematic are roadways because of the number of sites where these linear features cross over small channels. Bridges are usually constructed over relatively large streams, whereas concrete or steel pipes called *culverts* are used for small streams that flow intermittently. The problem is that the amount of water that can flow through a culvert is limited by the diameter of the pipe. During large flood events, culverts are often not able to handle the large volume of flow, causing water to back up such that upstream areas become flooded. This problem can be severe in highly developed urban areas with large numbers of culverts.

Urbanization

Of all the changes humans make to the landscape perhaps none has had a bigger impact on flooding than urbanization. Significant portions of urban areas in developed countries are typically covered with impermeable surfaces, chiefly roads and parking lots composed of concrete and asphalt. Water then cannot infiltrate over much of the landscape, but rather is forced to move as overland flow. The roofs of buildings are also impervious, and because urban areas contain large numbers of buildings, they too add to the volume of overland flow. This excess overland flow will make its way to the nearest stream via drainage ditches and storm sewers. As illustrated in Figure 8.30, when urbanization replaces natural vegetation cover with impermeable surfaces, the additional overland flow means that nearby streams will reach flood stage more frequently. Also important is that it takes far less time for overland flow to reach a stream channel than it does for infiltrating water to move through the groundwater system. Consequently, not only does urbanization increase the heights of floods, it also leads to more flash flooding since it decreases the lag times between precipitation events and peak discharge.

Ways to Reduce the Impact of Floods

Because flooding has historically been a serious problem, humans have developed a variety of ways to lessen the impact of floods—a process known as *flood mitigation*. Some mitigation techniques require building structures designed to control the flow of water in a channel, whereas others attempt to decrease the rate of overland flow. Laws have also been passed that are designed to minimize flood

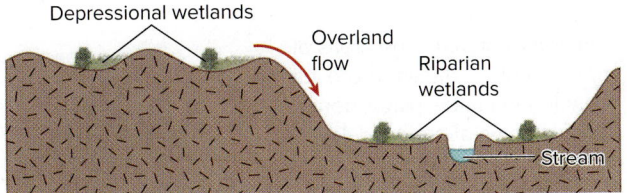

FIGURE 8.28   Wetlands are common along rivers and in topographic depressions in upland areas. Because of their porous nature, wetlands have a great capacity to capture and store water moving across the landscape. The destruction of wetlands has resulted in an increase in flooding as these areas no longer store water.

FIGURE 8.29   Photo showing how excessive amounts of sediment have moved off a construction site and into a nearby drainage ditch. Note how the culvert is being clogged with sediment.
© Jim Reichard

FIGURE 8.30 Urban settings commonly contain large areas of impermeable surfaces where little to no infiltration takes place, generating large volumes of overland flow that rapidly enter the drainage network. The result is more frequent flooding, higher flood crests, and shorter lag time between a rain event and peak discharge.

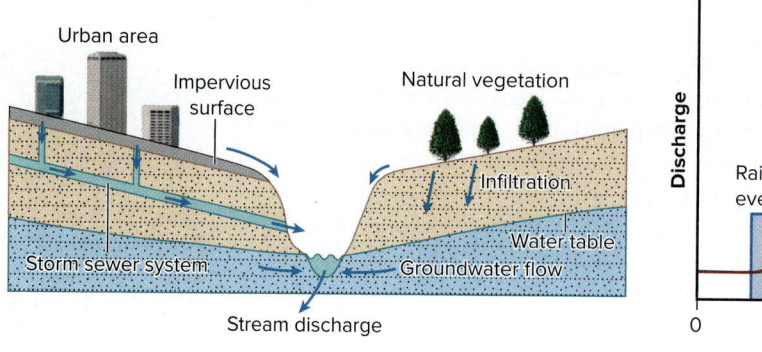

losses by managing human activities that tend to exacerbate flooding. In this section we will explore some of the more common flood mitigation techniques.

Dams

When a dam is built across a river, it gives engineers the ability to control the river's discharge. As illustrated in Figure 8.31, when engineers regulate the discharge of a river, they are able to raise or lower the pool or *reservoir* of water behind the dam. Reservoirs are valuable to society in that they protect against floods, serve as important sources of freshwater and electrical power, and offer a variety of recreational opportunities. In general, engineers allow water to accumulate in a reservoir during wet periods when streamflows are high, thereby creating a reserve or stockpile of water. During dry periods when the demand for water exceeds the river's discharge, the pool is then lowered in order to meet the demand. In the case of a dam that is used to generate hydroelectric power, engineers raise and lower the pool on a daily basis. Regardless of whether the reservoir height is manipulated daily or seasonally, engineers must always ensure that enough water is being released to meet the needs of downstream ecosystems.

The storage capacity of a reservoir can also be manipulated for the purpose of preventing flooding downstream of the dam. Here engineers must maintain a sufficient amount of reserve or emergency storage behind the dam (Figure 8.31). This additional storage allows periodic surges of water from upstream areas to be safely contained within the reservoir. Although a sufficient amount of emergency storage is normally maintained, unexpectedly heavy or prolonged rains sometimes cause a reservoir to reach its maximum capacity. When this occurs, engineers may be forced to release water at such a high rate that it causes downstream flooding, which, ironically, is what the dam is supposed to prevent. The alternative is to risk letting the pool rise to the point that the dam itself is weakened. The dam's structural integrity can also be threatened should water be allowed to flow over the top in an uncontrolled manner. In either case, the structural failure of a dam can generate a truly catastrophic flood, as in the massive flood described earlier that struck Johnstown, Pennsylvania, in 1889 (Figure 8.23).

In addition to the potential for structural failure and devastating floods, dams are also a concern because of their environmental impact on downstream ecosystems. When a river is dammed, the suspended and bed loads are forced to be deposited within the reservoir. The water temperature also becomes much cooler due to the depth of the reservoir. Therefore, what is released from the reservoir is cool, sediment-free water. This can be highly disruptive to aquatic ecosystems that naturally evolved in the presence of warmer water and a certain amount of sediment moving downstream. Moreover, because the volume of water that is

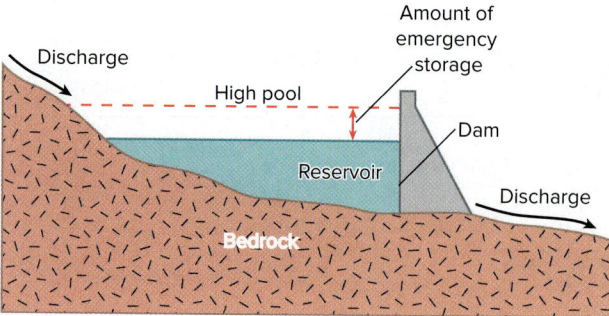

FIGURE 8.31 Dams prevent downstream flooding by intercepting and storing stream discharge in their reservoirs. By carefully regulating the release of water, engineers can maintain a reservoir such that it provides a sufficient amount of emergency storage and enough water to meet the demand for freshwater and electricity.

released from a dam is normally kept within a fairly narrow range, downstream areas no longer experience a natural range of high and low streamflows. Because many ecosystem functions depend on large variations in streamflow, the highly regulated discharge from dams can have dire consequences for the aquatic food chain. Perhaps the most well-known environmental impact of dams is how they block the migration of certain fish species.

Artificial Levees

Earlier in this chapter we discussed how *natural levees* form when a river leaves its channel and deposits sediment, forming a pair of sandy ridges running parallel to the banks. **Artificial levees** are those built by humans for the purpose of keeping a river from overflowing its banks and inundating its floodplain. Most artificial levees are constructed of earthen materials, whereas large concrete panels called *floodwalls* are sometimes used in urban areas. From Figure 8.32 one can see that the increased height of artificial levees reduces the probability that a river will flow out onto its floodplain. While artificial levees are quite effective in reducing the frequency of flooding, they disrupt the natural drainage system by disconnecting a river from its floodplain. This often leads to undesirable consequences for society.

The basic problem is that a channel lined with artificial levees will hold far less water than what the floodplain is capable of storing. As shown in Figure 8.32, when water levels rise, the river must stay within the artificial levees as opposed to flowing out into its floodplain. This means that the water which normally would have been stored in the floodplain is forced to begin backing up in the channel. In essence then, artificial levees create bottlenecks that restrict the flow of a river, resulting in more frequent and severe flooding in areas upstream. Another undesirable consequence is that the river now tends to deposit sediment in its channel rather than in the floodplain. Since the additional sediment takes up space within the channel, it increases the likelihood that floodwaters will reach the top of the levees. The degree of flood protection provided by the levees therefore will slowly diminish over time.

Because levees cause rivers to reach greater heights, upstream property owners are faced with the need for greater flood protection. This creates the incentive for upstream communities to build their own levees. As more of the river becomes confined by levees, floodwaters are pushed to even greater heights. This vicious cycle not only causes more frequent and severe flooding in upstream areas that lack flood protection, the higher water levels also threaten the structural integrity of the levees themselves. As illustrated in Figure 8.33, when water levels are high, the levees must bear the additional weight of the water in the channel. Thus, any weakness within a levee may lead to a collapse and catastrophic breech, allowing floodwaters to inundate areas that were once protected by the levee. Common types of weaknesses include cracks in concrete

FIGURE 8.32 Artificial levees (A) reduce the frequency of flooding by raising the height a river must rise before overflowing its banks. Most artificial levees consist of a ridge of earthen material constructed parallel to a river bank. Concrete levees called floodwalls (B) are sometimes used along urban corridors. Note how levees also create bottlenecks that make floods worse in upstream areas.

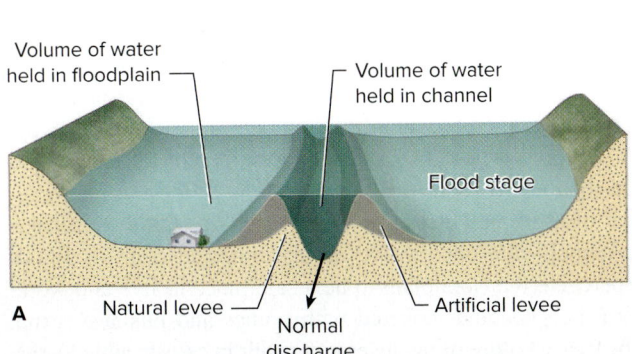

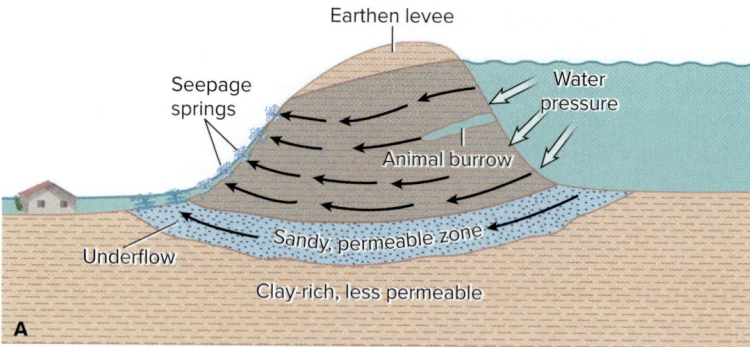

FIGURE 8.33 Earthen levees generally fail or collapse when floodwaters (A) exert additional water pressure and weight on the levee. This added pressure can allow water to flow through zones of weakness within the levee and through underlying permeable zones, causing erosion that weakens the structure and its underlying support. Photo (B) shows floodwaters pouring through a broken levee and flowing onto the floodplain.
Jocelyn Augustino/FEMA

floodwalls and settling of foundation support due to compaction, whereas earthen levees often have problems with burrowing animals creating tunnels. Earthen levees have the additional problem of becoming weak when they remain saturated during extended periods of high water levels.

Actual levee failures are commonly triggered when water flows under or through the structure such that some of the material is removed by erosion. This can occur when weakness zones develop within a levee, or when the structure is built across some permeable zone on the original land surface. Notice in Figure 8.33 how flood conditions create additional water pressure that is exerted on the sides and base of the levee. This additional pressure forces water to preferentially flow along weak zones within the levee itself, and through permeable zones underneath the structure. If the flow or leakage becomes great enough to cause erosion, the structure can weaken to the point where it fails or collapses, thereby allowing flood waters to inundate areas behind the levee.

The sudden failure of artificial levees brings us to perhaps the most serious problem with levees, namely that once a river is disconnected from its floodplain, people tend to perceive the floodplain to be a safe place to live and work. Because the floodplain then becomes covered with homes and businesses that would otherwise not be there, failure of the levee can result in catastrophic losses.

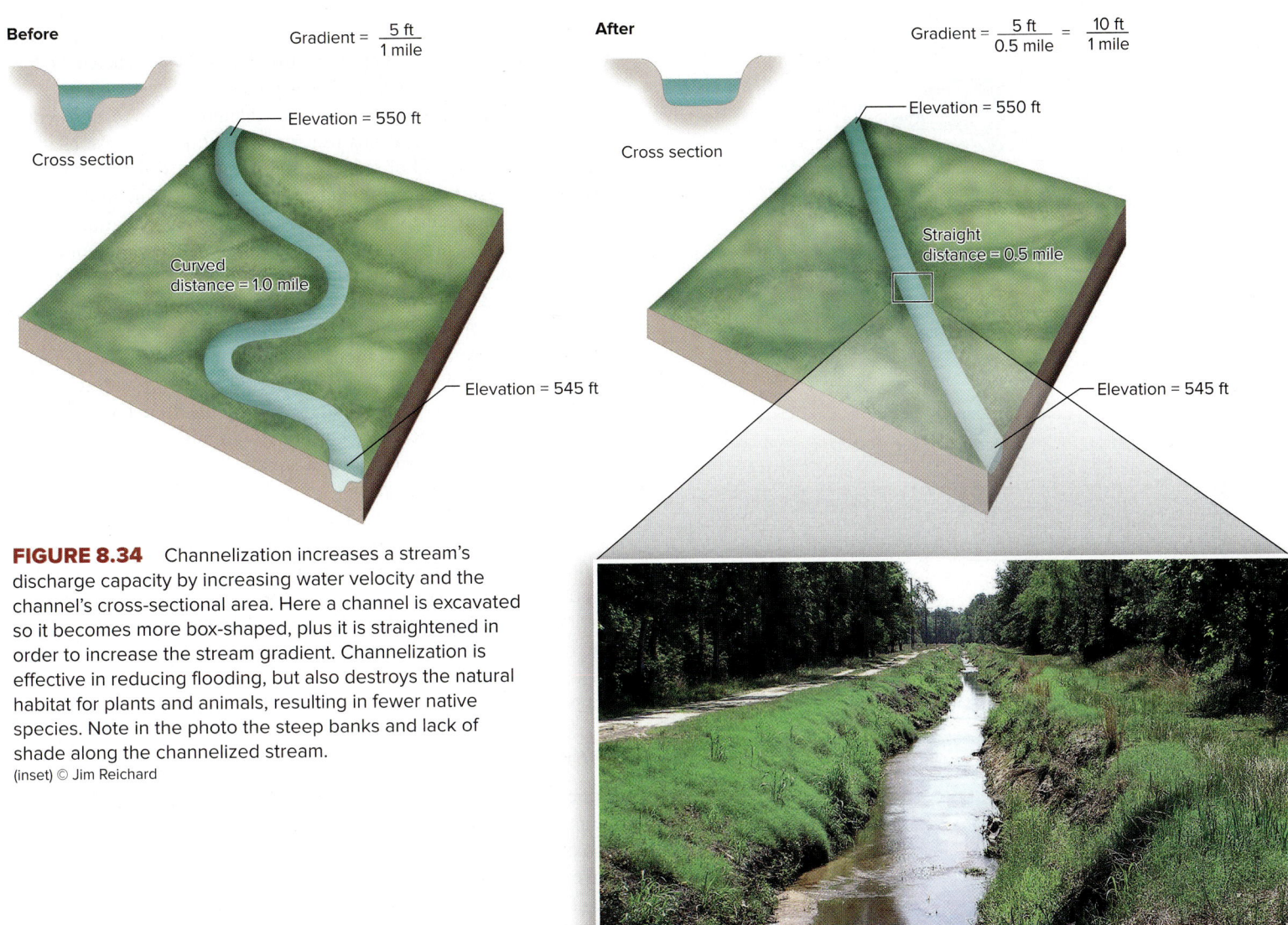

FIGURE 8.34 Channelization increases a stream's discharge capacity by increasing water velocity and the channel's cross-sectional area. Here a channel is excavated so it becomes more box-shaped, plus it is straightened in order to increase the stream gradient. Channelization is effective in reducing flooding, but also destroys the natural habitat for plants and animals, resulting in fewer native species. Note in the photo the steep banks and lack of shade along the channelized stream.
(inset) © Jim Reichard

There is no better example of this so-called *levee effect* than the tragic flooding of New Orleans in 2005, when the levees failed as a result of Hurricane Katrina (Case Study 8.1).

Channelization

Another common flood control technique is **channelization,** which involves straightening and deepening a stream channel in order to increase its *discharge capacity*. When a section of a stream is allowed to carry more water, this reduces the probability that the stream will overflow its banks. Channelization makes use of the basic principle that discharge is a function of water velocity and cross-sectional area of the channel. Anything that increases either of these factors will increase a stream's ability to carry water, thereby reducing the potential for flooding. As illustrated in Figure 8.34, a stream's cross-sectional area of flow is enlarged by excavating the channel such that it becomes deeper and more box-shaped. During this process the channel is also straightened by removing the original bends and curves, allowing water to travel a shorter distance while experiencing the same elevation drop. This results in an increase in stream gradient and water velocity. Water velocity can further be increased by creating a smoother

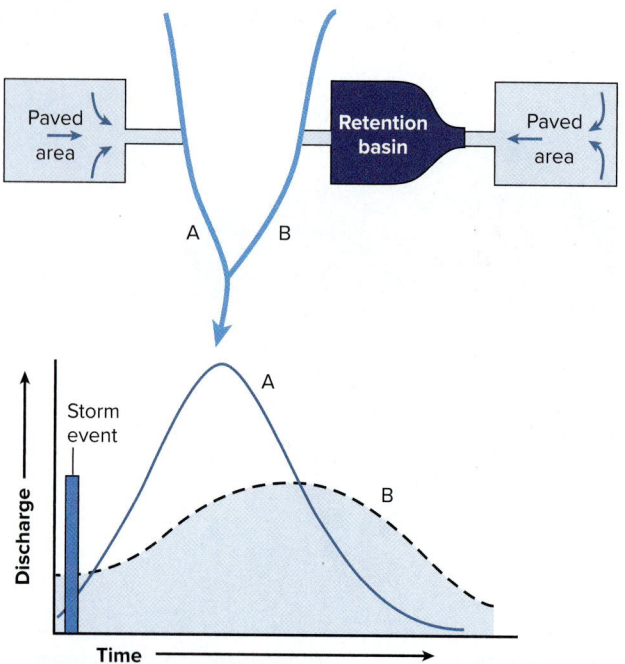

FIGURE 8.35 Retention basins store excess overland flow captured from paved surfaces during storm events, then slowly release the water into the natural drainage system. The hydrograph illustrates how the slow release from a retention basin reduces peak discharge, thereby decreasing the potential for flooding.
Robert Criss Washington Univ St. Louis

streambed so there is less drag or resistance on the flowing water. This can be accomplished by lining the channel with concrete and by removing obstructions such as downed trees and large rocks.

From Figure 8.34 one can see that channelization results in dramatic physical modifications to a stream. Although these changes have the desired effect of reducing the frequency of floods, the natural stream system also responds in ways humans find undesirable. One of the more serious consequences of channelization is that flooding actually becomes worse downstream. The problem is that in nonchannelized sections, the discharge capacity of the stream remains the same. When higher flow volumes move downstream from the channelized segments, the unmodified areas are commonly overwhelmed. Another problem is that the increased water velocity in the channelized area causes the stream to begin cutting downward, leaving steeper banks that are more prone to mass wasting.

Perhaps less noticeable, but equally significant is how channelization affects the ecology of a stream. Natural streams provide native plant and animal species with different water depths, velocities, and amounts of shade. Channelization takes this diverse, natural ecosystem and transforms it into an artificial environment with a single habitat that has a relatively swift current, uniform depth, smooth bottom, and banks generally free of overhanging vegetation. Particularly harmful is how the reduced shade and more uniform water depth leads to higher water temperatures. Ultimately, many native fish and plants are driven out, whereas the new environment is more favorable to alien, and often undesirable, species.

Retention Basins

Earlier in this section we discussed how excessive overland flow from urban areas leads to an increase in both the frequency and severity of floods. The basic problem is that urbanization causes more water to make its way into the drainage network over shorter periods of time. An effective engineering solution is to temporarily store some of this excess water in a series of depressions called **retention basins,** which are constructed within the network of tributary channels as shown in Figure 8.35. During a storm event, overland flow is routed into a retention basin whose only outlet is a relatively small culvert that restricts the outflow of water. A retention basin is similar to a small-scale dam in that it temporarily stores excess water that was generated upstream, then releases it at a controlled rate. The slow release of water effectively reduces peak discharge in downstream areas (Figure 8.35), thereby decreasing the potential for flooding. Note that retention basins are commonly located adjacent to parking lots and within residential subdivisions in order to capture overland flow from paved surfaces.

Erosion Controls

Because filling stream channels with excessive sediment causes streams to overflow their banks more easily, techniques that help prevent sediment pollution also help reduce flooding. Unwanted sediment primarily comes from agricultural fields and construction sites where soils are exposed and readily washed off the landscape during overland flow. There are two basic approaches to reducing sediment pollution. One involves employing practices designed to keep soil particles in place so as to minimize their ability to move downslope. The other approach is to use some type of physical barrier to trap sediment before it can enter the drainage network. In agricultural areas where soils are an irreplaceable resource, farmers quite naturally prefer to use erosion controls that help keep the soil in place. A good example is the practice of using a crop as a cover that protects soil from falling raindrops.

The use of physical barriers to trap sediment is often required because erosion controls by themselves are not entirely effective at keeping exposed sediment in place. *Stream buffers* are a type of barrier in which vegetated strips line the banks of stream channels, trapping sediment before it can enter the drainage network. Stream buffers are normally required in areas where the land has been cleared by logging or agriculture. Another common barrier system employs temporary **silt fences,** which are made of a synthetic fabric fine enough to trap sediment, but yet allowing some water to pass. As shown in Figure 8.36, silt fences are placed downslope of construction activity in order to keep exposed sediment from leaving the site. Although these fences are generally effective, the problem is that they can be completely overwhelmed in places by overland flow, thereby allowing sediment to pour into nearby streams. The last type of barrier system involves the use of *silt basins*, which are ponds constructed for the purpose of trapping sediment that makes its way into a drainage system. However, heavy equipment must periodically be brought in to dig up the accumulated sediment and haul it away—see Chapter 10 for a detailed discussion of erosion-control practices.

FIGURE 8.36   Temporary silt fences are used in an attempt to keep exposed sediment from leaving construction sites. Although generally effective, silt fences can be overwhelmed in areas where overland flow accumulates and begins to follow a channel.
© Skip Metheny

Wetlands Restoration

By building artificial levees and disconnecting rivers from their floodplains, humans have drastically reduced the ability of riparian wetlands to store floodwaters. The same holds true for depressional wetlands located in upland areas, most of which have been drained for agricultural and urban development. Scientists now understand that the loss of wetlands has not only had serious ecological consequences, but has also contributed to the occurrence of more frequent and severe flooding. Because of the negative impacts, various groups in the United States are actively working to preserve existing wetlands and restore as many as possible back to their native state. Much of this effort involves reconnecting lowlands to their natural water supply by filling in drainage ditches and canals and removing levees. In some cases new wetlands are constructed from scratch in order to counteract the effects of increased overland flow.

Flood Proofing

As noted earlier in this chapter, there are a number of reasons why people build and live on floodplains. In many cases they are well aware that their property will periodically be flooded, but choose to live along a river for its beauty, recreational value, or business opportunities. Here people may feel that the benefits simply outweigh the consequences of an occasional flood. Others may consider the risk to be small, and therefore take the chance that a serious flood will not happen in their lifetime, or anytime soon. For example, a property owner may decide that

FIGURE 8.37   Elevating homes is one of the oldest flood-proofing techniques and is still commonly used today.
© Jim Reichard.

investing in a structure with a 50-year life expectancy is worth the risk posed by a flood with a 100-year recurrence interval. Depending on what they feel is an acceptable level of risk, property owners may choose to "flood proof" their buildings or property. As shown in Figure 8.37, a time-honored flood-proofing technique is to raise the building above the expected flood level. Another option is to surround one's property with a permanent levee. For those who do not plan ahead, there is always the possibility of constructing an emergency levee using sandbags. Depending on the particular flood, however, it may not be feasible to build a temporary levee high enough or fast enough. Note that such hastily built and makeshift levees may not be strong enough to hold back the floodwaters.

Floodplain Management

In 1968 the U.S. Congress passed the National Flood Insurance Act and created the National Flood Insurance Program (NFIP), providing federally subsidized flood insurance to property owners. This federal program enables people and businesses to obtain flood insurance that would otherwise be difficult and expensive to obtain. To be eligible, a person has to live in one of the over 23,000 local communities currently participating in NFIP. Here communities must first perform a hydrologic study that analyzes a stream's flood characteristics, the most important being the height of the projected 100-year flood. By comparing the elevation of the land surface to the projected flood height, a flood map is generated. The map is then used to determine insurance rates for individual land parcels based on the relative flood risk. Communities participating in NFIP are also required to restrict development in those areas lying below the 100-year flood level. As illustrated in Figure 8.38, the area within the 100-year flood zone is called the *regulatory floodplain*, which consists of two parts: the *flood fringe* and *floodway*. The floodway is the most critical as this is where floodwaters are the deepest and fastest.

Although NFIP is voluntary, communities that participate are required to manage or restrict development within the regulatory floodplain. This requirement results in what some call *floodplain zoning* as it works similarly to local zoning ordinances, where commercial, industrial, and residential buildings are restricted to certain areas or zones within a community. In the case of floodplain zoning, regulations prohibit the building of new structures within the floodway (Figure 8.38). The idea is to keep the floodway free of buildings and other obstructions that would impede the flow of water, and therefore raise the flood height. Many communities have found that floodways make excellent sites for parks and golf courses because they involve few permanent buildings. With respect to the flood fringe, new buildings are allowed, but must be built at or above the 100-year flood stage, which means they must be elevated. The end result is that NFIP provides both affordable flood insurance to communities, and encourages sensible floodplain management strategies that actually reduce flood losses and insurance claims.

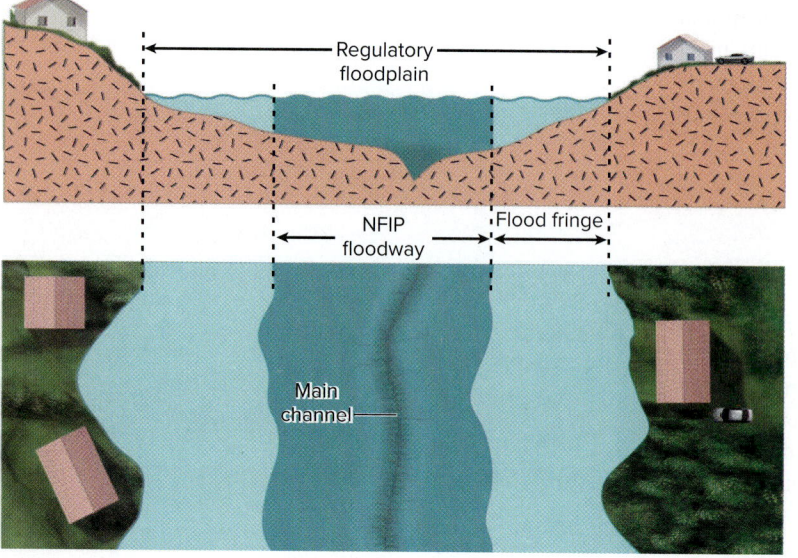

FIGURE 8.38   Floodplain zoning involves identifying areas adjacent to a stream that will be inundated in a 100-year flood. This regulatory floodplain is divided into the flood fringe and floodway, and regulations then restrict the type of development allowed in each of these two zones.

Education

As with other geologic hazards, educating the public about flooding is a very cost-effective means of reducing the number of fatalities and property damage. Even though the vast majority of people are aware that rivers flood, many are ignorant of the fact that their property may lie in an area that periodically becomes inundated. Of course, there are those who are aware of the hazard, but decide to take the chance that a flood will not happen anytime soon. Education then serves to help make people aware of the flood risk, and for those willing to gamble, shows them ways in which they can reduce their risk. For example, in Figure 8.39 you can see that someone *knowingly* built homes directly in the middle of a floodplain, but took steps to reduce the risk by raising the height of the land. While such a strategy might be effective for a typical flood, there is certainly no guarantee it will work for larger, less frequent events. Note that during a flood these people will be living on an island, so in order to access their homes they must either have a boat or wait for the water to recede.

Perhaps the most important issue citizens need to understand is the hazard of driving a vehicle through floodwaters. Most flood-related fatalities occur in flash floods, and of these, approximately 50% are vehicle related. This is due in part to the way flash floods develop, whereby people quickly find themselves trapped in their vehicles. There are also cases where people purposely drive through floodwaters. In either case drivers and their passengers can suddenly find themselves in a life-threatening situation. The problem is that high water decreases the weight of a vehicle such that it begins to float. When this occurs, the vehicle can easily be swept away with the current. As illustrated in Figure 8.40, only 2 feet (0.6 m) of water is needed to float a typical car, and a large SUV will float in as little as 2 to 3 feet. Once a vehicle is swept away, the situation quickly becomes more dangerous, because unlike a boat, a vehicle tends to roll over when it floats. Another problem with driving in floodwaters is that the road itself becomes difficult to see. Therefore, you are more likely to drive off the road into even deeper water or into an area where the roadbed has been washed away. In addition to driving hazards, downed electrical power lines pose an electrocution hazard for those who venture out into floodwaters. The safest thing to do then is simply avoid going into floodwaters if at all possible.

FIGURE 8.39   Being knowledgeable about floods and floodplains can help people avoid placing themselves in high-risk areas. This knowledge can also be used to take steps to reduce the risk for those who purposely build in flood-prone areas, such as the owners of the homes shown here.
The National Operational Hydrologic Remote Sensing Center (NOHRSC), NOAA

FIGURE 8.40   Most people are not aware of how little water it takes for their vehicle to begin floating, and thus may place themselves at risk of being swept away with the current. Approximately half of all flash-flood fatalities are vehicle related.

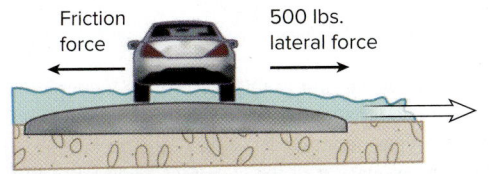

Friction force   500 lbs. lateral force

Water 1 foot deep: Extremely dangerous

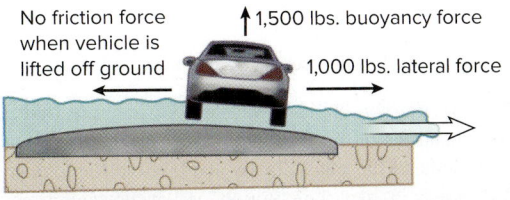

No friction force when vehicle is lifted off ground   ↑1,500 lbs. buoyancy force   1,000 lbs. lateral force

Water 2 feet deep: Fatal
Vehicle begins to float when water reaches its chassis, which allows the lateral forces to push it off the road.

Muddy water hides washout: Fatal
Washed-out roadway can be hidden by muddy water allowing a vehicle to drop into unexpected deep water.

SUMMARY POINTS

1. The hydrologic cycle describes the continuous movement of water within the Earth system. Water falling on the land will either infiltrate, evaporate, be absorbed by plants and animals, or move as overland flow downslope into streams. The basic function of streams and rivers is to transport both water and sediment off the landscape.

2. Stream discharge (volume flowing in a channel over a given time interval) generally increases as channels merge from the headwaters to the mouth. Stream discharge comes from overland flow (rainfall and melting snow) and baseflow (contributions from the groundwater system).

3. A drainage system is a network of channels that merge to form progressively larger streams, and eventually rivers. A drainage basin is the land area that collects water for a given stream network. The upper portion of a basin is referred to as the headwaters; the lowest portion is called the mouth.

4. Streams not only transport water, but play an important role in eroding the landscape down to base level and transporting the resulting sediment to low-lying areas. The abrasive effect of a stream's sediment load enables it to cut down through solid rock.

5. Sediment within a stream channel is transported and deposited in response to changes in velocity. As velocity increases, streams are able to transport progressively larger and heavier particles; when velocity decreases, the largest and heaviest particles are deposited first. This process results in sediment being sorted based on size, shape, and density.

6. As streams approach base level, the following generally occur: (a) sediment size decreases, (b) gradient and velocity decrease, (c) channels meander more, (d) the valley becomes wider, and (e) floodplains, natural levees, and back swamps become more pronounced.

7. Floods originate when heavy rain or melting snow causes water to accumulate on the land surface faster than it can infiltrate. This excess water moves as overland flow into the drainage network, where discharge may increase to the point that a river overflows its banks and inundates normally dry areas.

8. The severity of floods is quantified in terms of either discharge or stage. Recurrence interval or percent probability are used to measure the frequency of a given discharge event. Large floods occur less frequently than small events.

9. Natural factors that affect flooding are: (a) intensity and duration of precipitation events, (b) size of the area over which precipitation falls, (c) ground conditions that affect infiltration, (d) speed at which snow melts, and (e) amount and type of vegetation cover.

10. Flash floods generally occur in the upper parts of basins and are characterized by a large increase in discharge over a relatively short time period. Downstream floods occur lower in the basin and develop more slowly, but involve large volumes of water moving out onto a floodplain.

11. Flood frequency and severity have increased in areas where human activities generate large volumes of overland flow and excess sedimentation. Significant activities include: (a) removal of natural vegetation, (b) destruction of wetlands, (c) blockage of small tributaries, and (d) urbanization.

12. The impact of floods can be reduced by dams, artificial levees, and channelization. Such flood controls, however, tend to encourage development in flood-prone areas and may create more problems upstream or downstream. Thus implementing such controls generates the potential for much greater property losses in the event of a large flood.

13. Flooding can also be lessened by reducing the amount of overland flow and sediment coming into a drainage system. Common techniques include stream buffers, restoration of wetlands, and the construction of retention basins, silt fences, and silt basins. Other important tools for mitigating flood damage include education, elevating structures, and floodplain management.

KEY WORDS

APPLICATIONS

Student Activity   Take a household bucket, go outside, and find some loose dirt of varying grain size. Throw a couple of handfuls of dirt into the bucket and then find a garden hose and add about 4–5 inches of water. Scoop up some of the wet sediment and take a close look at the different types of grains. Now shake the bucket vigorously in a swirling motion, and at the same time, carefully dump out part of the sediment. Grab a handful of the remaining sediment, and examine it closely. Describe the types of grains that have tended to remain in the bucket compared to those that were lost. What is the scientific name of this process you are using to separate the sediment grains? Name a commercial application where this process might be useful.

Critical Thinking Questions

1. There is a common misconception that bare soil, as in a farm field, will allow more rainwater to infiltrate compared to a field covered with natural grasses. Thus, the farm field should generate less overland flow. Why is this incorrect?

2. Hydraulic sorting is the process whereby streams separate sediment particles based on their size, shape, and density. If you were to look for tiny pieces of gold in stream sediment, would you more likely find them in an area of fine sand or coarse gravel? Why?

3. Humans construct artificial levees in order to reduce flooding. How then can these levees actually increase the frequency and severity of flooding?

Your Environment: YOU Decide

© Doug Sherman/Geofile

In 1968 Congress passed the National Flood Insurance Act and created the National Flood Insurance Program (NFIP), which provides federally subsidized flood insurance to property owners. This allows people and businesses to obtain private flood insurance that would normally be difficult and expensive to obtain. However, the property owners must live in a community where the local government manages and regulates development within designated floodplains. The idea is to manage and reduce the risk so that flood losses are minimized. Do you think it is a good idea to regulate development in known floodplains in order to reduce the cost of flooding to society, or should the NFIP insurance option no longer be available so that people are free to build anywhere they like? Explain.

Chapter 9

Coastal Hazards

LEARNING OUTCOMES

After reading this chapter, you should be able to:

▶ Explain the fundamental differences between ocean tides, currents, and waves.

▶ Describe how waves form, travel through open water, and change when they approach shore and enter shallow water.

▶ Discuss how longshore currents develop, and explain the longshore movement of sediment.

▶ Characterize the basic way hurricanes form and the three major hazards they pose to humans.

▶ Explain why the number of hurricane-related deaths has decreased in recent years but property losses have increased.

▶ Characterize how tsunamis form, and explain why they pose such a hazard to people and property.

▶ Identify the basic cause of shoreline retreat, and discuss how humans are making the problem worse.

▶ Describe the different types of coastal engineering techniques, their purposes, and their undesirable consequences.

Aerial view along the North Carolina coast illustrating that when humans build expensive structures too close to the water's edge, their investments are at risk of falling into the sea as the shoreline naturally retreats. Notice how large sandbags have been used in a desperate attempt to stop shoreline retreat.

Introduction

© StockTrek/Getty Images

Our interest in this chapter is the *coastal environment*, which refers to the unique setting where the terrestrial (land) and marine environments meet. Coastlines, often called *shorelines*, are a critical component of the biosphere because a large number of species are connected in some way or another to habitats found only at this interface between the land and sea. In fact, it was along shorelines that marine creatures first ventured onto land, ultimately evolving into the amphibians, reptiles, and mammals that colonized Earth's landmasses. In addition to playing a pivotal role in the evolution of life, coastal environments are immensely important to modern humans, from the food provided by fisheries to the harbors where port cities serve as vital centers of commerce and trade.

Like other environments where people live, coastal zones have natural processes that can pose a hazard to human life and property. For example, each year tropical storms make landfall along shorelines around the world, causing major damage and untold human suffering. Moreover, because the coastal environment functions as a system, human modifications to shorelines commonly have undesirable consequences. Perhaps most troublesome is how humans routinely build engineering structures designed to control coastal erosion. These structures interrupt the natural movement of sand along a shoreline, which ironically causes unwanted erosion and deposition in places along the coast. Today rapid population growth in coastal regions (Figure 9.1) is magnifying the problem of coastal erosion as well as the hazards associated with powerful storms. In the United States, for example, 39% of the population now lives in the narrow fringe of coastal counties (including the Great Lakes region), which represents less than 10% of the nation's land area, excluding Alaska.

In this chapter we will explore some of the natural processes and hazards along coastlines, focusing on how humans interact with this unique environment.

FIGURE 9.1 Plot (A) showing how population density in the United States is much higher in coastal counties, and continues to increase. Satellite image (B) showing high-density development near Hanauma Bay on the Hawaiian island of Oahu. Note how development is concentrated on low-lying terrain closest to the shore and in valleys leading to the sea. These areas make better construction sites compared to the surrounding rugged terrain. Also note the extinct cinder cones, one of which has been breached, forming a small bay.
(B) NASA

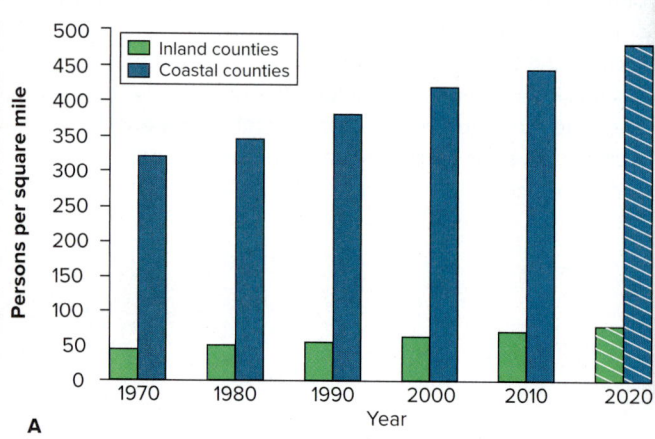

A

Cinder cone

B

Although our emphasis will be on marine coastlines, keep in mind that much of our discussion also applies to shorelines along inland seas and freshwater lakes, such as the Great Lakes in North America.

Shoreline Characteristics

As with most physical features on our planet, shorelines have different characteristics due to the dynamic nature of the Earth system. Some shorelines are relatively straight and have broad beaches, while others are more irregular and rugged and often have beaches restricted to coves. Still others have the appearance of a flooded network of stream channels (e.g., Chesapeake Bay). These different characteristics are important in environmental geology because they influence the way humans interact with each coastal environment. In this section we will examine how shoreline characteristics are related to two key geologic processes: plate tectonics and changes in sea level.

Recall from Chapter 4 that mountain ranges form along convergent and transform plate boundaries. In areas where these tectonic forces deform and uplift the land, we can refer to the shoreline as a *leading-edge shoreline*. Tectonically active shorelines are usually rugged and irregular, with beaches commonly restricted to coves and inlets. This type of shoreline occurs along parts of the Pacific coast of the United States. In contrast, a *trailing-edge shoreline* is one with little to no tectonic activity, commonly resulting in a relatively straight coastline with flat-lying terrain, such as that along the U.S. Gulf and Atlantic coasts. Figure 9.2 illustrates some of the fundamental differences

FIGURE 9.2 Tectonically active continental margins typically have steep terrain that produces irregular shorelines where beaches are commonly restricted to coves. On continental margins where tectonic activity is minimal, shorelines generally have broad, straight beaches and low-lying terrain that extends far inland.
Rear Admiral Harley D. Nygren, NOAA Corps (ret.)/NOAA
Ralph F. Kresge/NOAA

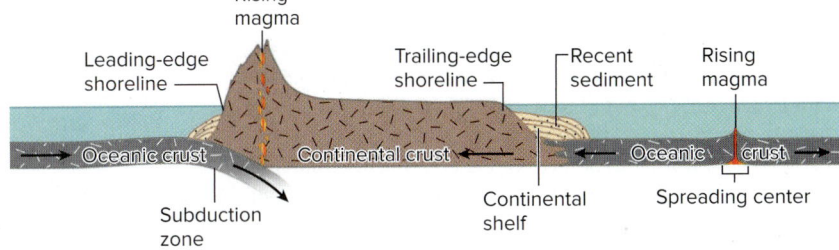

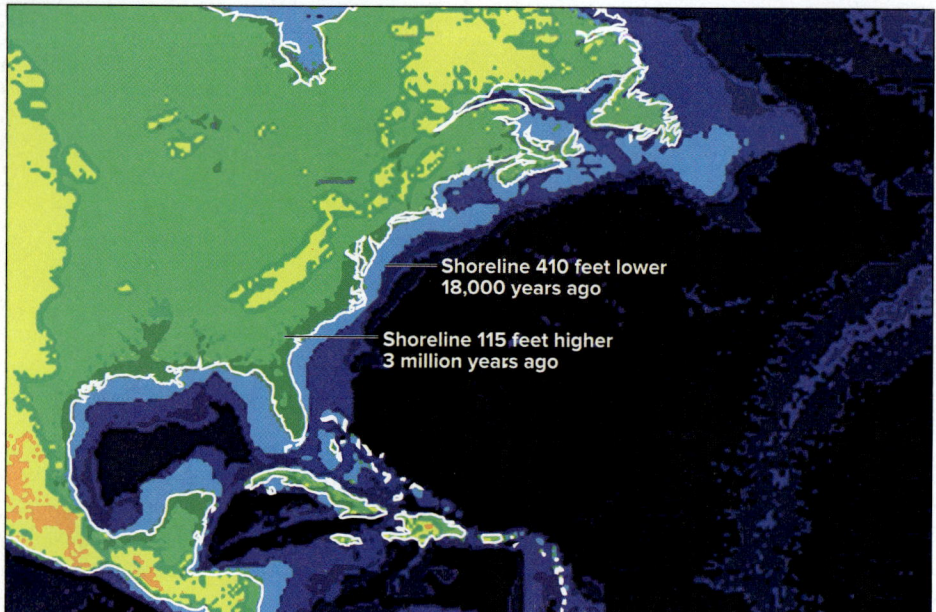

Shoreline 410 feet lower
18,000 years ago

Shoreline 115 feet higher
3 million years ago

FIGURE 9.3   Climate change and the transfer of water between the oceans and glacial ice over the past 3 million years have led to large fluctuations in global sea level and dramatic changes in the positions of shorelines. Sea level worldwide had been rising at a rate of 0.6 feet (0.2 m) per century since 1900, but recent measurements indicate that the rate has increased to nearly 1.0 feet (0.3 m) per century due to global warming. Sea-level rise could accelerate more dramatically should the warming destabilize the ice sheets on Greenland and Antarctica.

between leading- and trailing-edge shorelines. From this figure one can see that mass wasting hazards would be more prevalent along tectonically active shorelines, whereas the low-lying terrain of passive coastlines is conducive to high-density development. While trailing-edge shorelines allow large numbers of people to enjoy living near the sea, the low-lying nature of the terrain can put both people and their associated structures at risk during major storms.

Another important process that affects the nature of shorelines is the relative movement of the shoreline either seaward or landward. For example, in tectonically active areas the land commonly rises, which causes the shoreline to shift seaward. Normally such changes take place slowly over geologic time, and hence pose few problems for people living at the edge of the sea. Shorelines sometimes shift rather rapidly, however, thereby creating serious issues for developed areas, particularly if the shift is landward such that the sea starts to drown low-lying areas. One way a landward shift occurs is when human activity disrupts natural processes such that the land surface begins to sink or subside (Chapter 7). A good example is how the construction of levees have starved the Mississippi Delta of its natural sediment supply (Chapter 8), causing large sections of the Louisiana coast, including New Orleans, to sink below sea level. Another example is the drowning of Venice, Italy, where the withdrawal of groundwater has led to land subsidence (Chapter 10).

Shorelines also shift in response to worldwide changes in sea level that occur when Earth's global climate alternates between cool, glacial periods and warm periods called interglacials (Chapter 16). As the climate cools, huge volumes of water are removed from the oceans and stored on land as glacial ice, lowering global sea level and causing shorelines worldwide to shift seaward—note that in areas experiencing land subsidence, shorelines could still shift landward. Over long periods of time the climate eventually warms and the water begins returning to the sea, raising sea level and causing shorelines to move inland—here shorelines could still shift seaward in areas of tectonic uplift. As illustrated in Figure 9.3, climatic changes over the past 3 million years have resulted in sea level being as much as 115 feet (35 m) higher and 410 feet (125 m) lower than today. The key point here is that sea-level changes are not unusual, both globally and locally. Throughout this chapter you will see how even moderate rates of global sea-level rise are exacerbating the problems of shoreline erosion and the hazards associated with large storms. Finally, since more people live along trailing-edge, low-lying coastlines with wide beaches, we will focus on processes and hazards that are more prevalent on sandy as opposed to rocky shorelines.

Coastal Processes

In this section we will briefly examine several natural processes that play an important role in shaping coastal environments, including tides, currents, waves, and erosion and deposition of sediment.

Tides

If you have ever spent time by the ocean, you no doubt have witnessed the landward and seaward movement of the shoreline that occurs as the tide comes in and goes out. This rhythmic movement of the shoreline is caused by the spinning motion of the Earth, combined with the gravitational interaction between the Earth, Moon, and Sun. Because of the way the Earth, Moon, and Sun move within the solar plane, this complex interaction creates a net outward force along Earth's equator. As illustrated in Figure 9.4, this net or *tidal* force causes the solid Earth and the oceans to bulge outward along the equator. Notice how a given point on the solid surface of the Earth literally moves in and out of the two ocean bulges as the planet rotates about its axis. This creates the periodic rise and fall of sea level known as **ocean tides.** Because of the timing of Earth's rotation and the Moon's orbit, most shorelines experience two high tides and two low tides each day (approximately 12 hours and 25 minutes passes between two high tides). Note that tides also occur on large lakes, but the rise and fall are relatively minor compared to those of the oceans.

Scientists use the term *tidal range* to describe the difference in sea level between high and low tides. From Figure 9.4 one can see that tidal range is influenced by the relative positions of the Moon and Sun with respect to the Earth. Note how the Moon's influence on the bulge is much larger than that of the Sun. Despite the Moon being much smaller than the Sun, the fact that it lies so close to Earth means its gravitational influence on the oceans is much greater. Although ocean tides are dominated by the Moon, the Sun can either enhance or reduce the tides, depending on the position of the Moon. For example, when the Moon and Sun periodically line up such that their gravitational effects reinforce one another, the tidal range is maximized in what is called a *spring tide.* Conversely, when the gravitational pull of the Moon and Sun are at right angles, their tidal effects tend to cancel one another, producing a small tidal range known as a *neap tide.* In addition to being subject to orbital influences, tidal range varies depending on latitude, water depth, shape of the shoreline, and the presence of large storms. Tides are very important in the life cycle of coastal ecosystems, the generation of coastal currents, and the shaping of the shoreline itself. In Chapter 14 we will examine how humans are beginning to harness the tides in order to produce electricity that does not release carbon dioxide, and thus does not contribute to global warming.

Currents

Similar to flowing water in a stream, **ocean currents** involve the physical movement of water molecules from one location to another. Ocean currents are driven by various forms of energy, and like all things in motion, currents flow from an area of high energy to one of lower energy. For example, when Earth's rotation brings a high tide into a coastal area, the tide generates mechanical energy that forces water to funnel up into inlets and river channels, creating strong localized currents called *tidal currents.* When the tide goes out, the situation is reversed and the tidal currents flow back out toward the sea. In contrast, out in open water near the surface of the sea, large-scale *surface currents* form that are driven mainly by winds blowing consistently in the same direction. The effect here is similar to a person gently blowing across a flat pan of water, causing the surface of the water to begin flowing in the same direction as the moving air. Surface currents are also influenced by Earth's rotation and changes in atmospheric pressure.

In addition to wind-driven currents, there are large-scale *density currents* that form in response to differences in ocean temperature and salinity (i.e., amount of dissolved ions). Because cooler and/or more saline water is relatively dense, it tends to sink and flow toward areas where the water is less dense. Oceanographers use the term *ocean conveyor* to refer to the density-driven currents that circulate water, both vertically and horizontally, in a convective manner between tropical and polar regions. Density and wind-driven currents are important components of Earth's global climate system (Chapter 16) since they transfer vast amounts of heat energy from the tropics toward higher latitudes, providing heat to the cooler parts of the globe. These currents also transport nutrients throughout the oceans, making them vital to marine ecosystems.

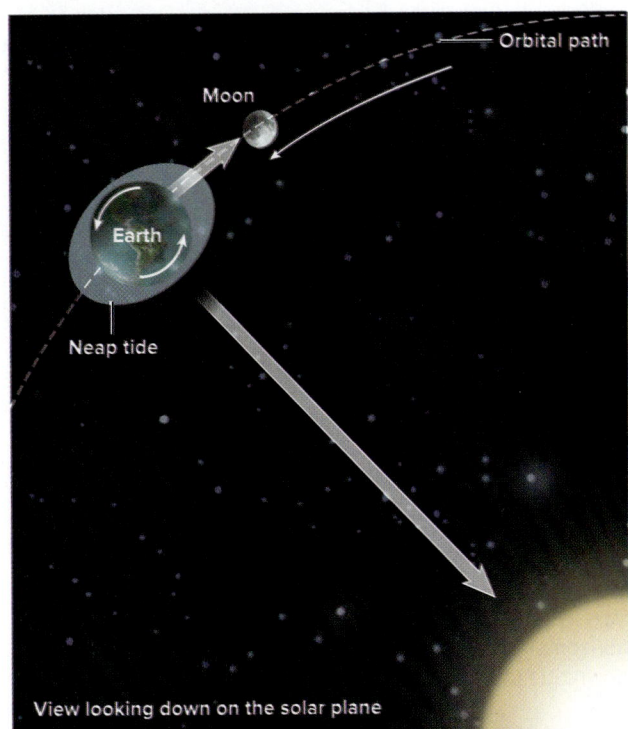

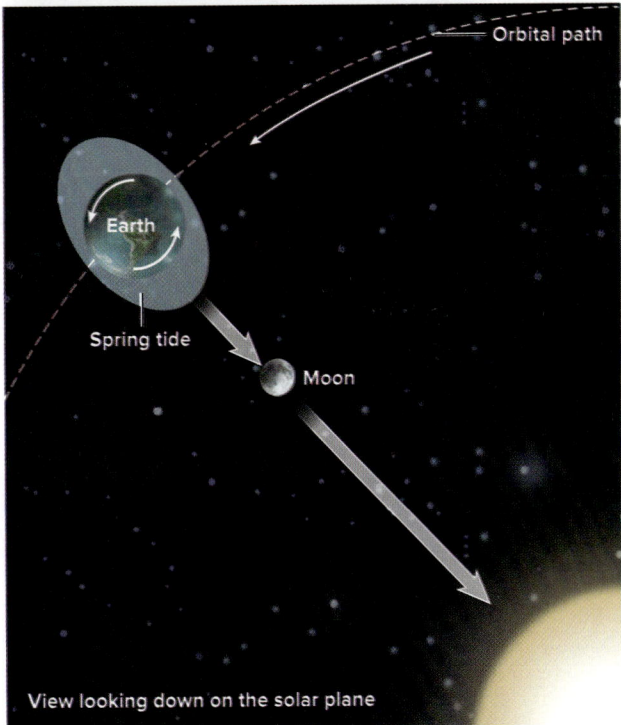

FIGURE 9.4 Earth's oceans bulge outward because of forces created by the planet's spinning motion and gravitational interaction with the Moon and Sun. Ocean tides form as the Earth rotates so that points on its surface move with respect to the bulges within the oceans. Note that the Moon has a greater tidal influence because it is much closer to the Earth than the Sun. The maximum tides, called spring tides, occur when the Moon and Sun align such that their gravitational effects reinforce each other.

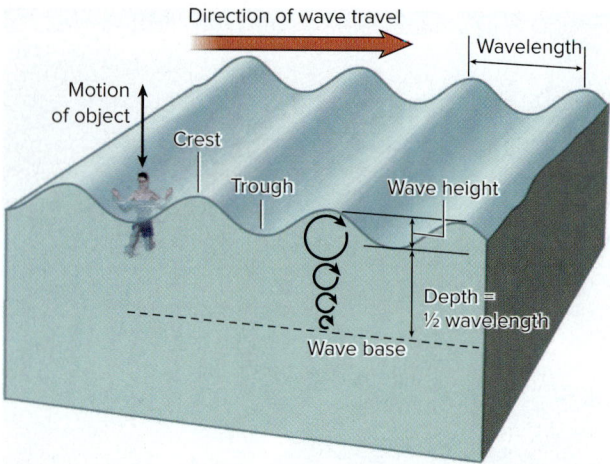

FIGURE 9.5   As wave energy travels horizontally through water, water molecules move in circular paths that get progressively smaller with depth. The level at which all movement stops, called wave base, gets deeper with increasing wave energy. Floating objects do not move horizontally with a passing wave, but rather bob up and down due to the motion of the water molecules.

Waves

Recall from Chapter 5 that during an earthquake, stored energy is released at a point and then travels outward through solid rock in the form of vibrational (seismic) waves. In contrast, **water waves** in the open ocean transport energy such that water molecules move or vibrate in a circular manner as illustrated in Figure 9.5. Water waves not only vibrate differently than seismic waves, but the frictional resistance of water is far less than that of rock. This lower resistance means that deep water waves lose less energy as they travel outward from their energy source; hence they can continue traveling until shore is reached. Note that water waves transport energy in a horizontal manner, but the circular motion of water molecules causes physical objects to move in a vertical manner. This is why a wave moving through open water will not carry a boat or swimmer along with it, but rather cause any floating object to simply bob up and down. Keep in mind that currents are different from waves in that currents transport both water and physical objects, whereas only energy moves in the direction of a traveling wave.

Like all waves, water waves can be characterized by the distance between successive crests (wavelength), the difference in height between the crests and troughs (wave height), and the amount of energy the waves contain. Notice in Figure 9.5 how the circular paths in the water column get progressively smaller with depth, eventually dying out at what is referred to as **wave base.** Because the circular movement of water extends to greater depths as waves become more energetic, wave base can be used as a measure of wave energy. It turns out that wave base in a series of waves is equal to about one-half of their wavelength (i.e., distance between successive crests). Thus, if wavelength is measured at 100 feet, for example, then wave base would be about 50 feet. In the next section you will learn how this relationship between energy and wave base is important when waves approach a shoreline.

Although waves can form when earthquakes, volcanic eruptions, landslides, or asteroid impacts transfer energy to a body of water, ordinary waves are generated by the transfer of energy from wind (i.e., flowing air). Naturally as wind speed increases, more energy is transferred to the water, producing waves with a deeper wave base. The energy of wind-generated waves is also affected by the duration of the wind and the size of the contact area between the wind and water, called *fetch*. For example, a wind that blows steadily for 20 hours will generate higher-energy waves compared to the same wind that blows for only 2 hours. Likewise, a 20-square-mile lake will accumulate much more energy than one whose area is 2 square miles. In general, wave energy is the highest, and wave base the deepest, when high-velocity winds blow for a long period of time over extensive reaches of open water. Such conditions typically occur during tropical storms and hurricanes.

Wave Refraction and Longshore Currents

When waves travel through deep water, they experience little frictional resistance, which helps them maintain energy and continue traveling until they reach a shoreline. But as deep-water waves enter shallow coastal waters they eventually begin to drag on the seafloor and lose energy, a process referred to as *shoaling*. This interaction with the seafloor also affects the shape and velocity of the waves themselves. As shown in Figure 9.6, when wave base comes into contact with the seafloor and the wave begins to slow down as water molecules encounter greater frictional resistance, the circular water motion becomes progressively more elliptical. This results in a net forward motion since the bottom of the wave slows down at a faster rate than the top. The decrease in velocity associated with shoaling leads to the deposition of sandbars and also forces the waves to pile up such that they become taller, whereas their wavelength decreases. In addition, the uneven braking action within the water column slows the front part of the waves more than the rear, causing the waves to become progressively less symmetrical (Figure 9.6). Because the waves continue to rise and become less symmetrical, at some point they literally fall over on themselves, producing what are called *breaking waves*. Once the waves begin to break, water is then pushed up onto the beach, after which it flows back downslope and into the next breaking wave—the *surf zone* refers to the area where the waves break.

Prior to reaching shallow waters, wind-generated waves travel as a series of parallel crests and troughs. As illustrated in Figure 9.7, when a particular wave crest moves into shallower water, the end closest to shore will start dragging on the seafloor while the rest of the wave remains in deep water. As the wave continues toward shore, the result is a progressive decrease in velocity along the length of the wave, forcing it to bend in a process called **wave refraction.** This process is similar to how lines in a marching band begin to turn like spokes in a wheel when band members in a given line walk more slowly the closer they are to the pivot point. Wave refraction is important because as the waves bend, water from the breakers is pushed up onto the beach at an angle (Figure 9.7). But when the water flows back down the slope of the beach face, it follows a path that is generally perpendicular to shore. This action of the water forces individual sand grains to follow a zigzagging path in the surf zone, ultimately transporting the grains parallel to shore in a process called *beach drift*—sometimes referred to as *longshore drift*. Wave refraction also causes water to build up in the surf zone, creating a current called a **longshore current** that flows parallel to shore (Figure 9.7). This longshore current is the reason that swimmers in the surf zone find themselves slowly drifting away from the point where they left the beach.

Because of beach drift, the sediment within the surf zone is always moving in the direction of the longshore current, provided the wave energy is high enough for transport to occur. A key point here is that as wave energy changes, so too does the width of the surf zone. Clearly, the larger the incoming waves, the wider the surf zone and the greater the volume of sand actively being transported. As could be expected, the surf zone is at its widest and sediment transport is at a maximum during storms when wind and wave energy are at their highest. Keep in mind that during transport some sections of the shoreline will undergo a net a loss of sediment, defined as *erosion* (Chapter 3), whereas other sections will experience *deposition*, or a net gain of sediment.

Shoreline Evolution

In the previous section you learned that energy from wind is transferred to bodies of water and then travels as water waves. When the waves strike a shore, this wave energy is transferred to solid land where it drives both erosional and depositional processes. Similar to how weathering and stream processes (Chapter 3 and 8) slowly wear down the landscape in one area, then deposit the material in another, erosion and deposition from breaking waves cause coastlines to evolve over time. The interaction between waves and a landmass can cause the shoreline to slowly move landward, a process referred to as **shoreline retreat.** Landward migration of a shoreline can also occur when there is a rise in global sea level, or when the land itself becomes lower due to subsidence (Chapter 7). Likewise, shorelines can migrate seaward, a process called *shoreline progradation*, due to a lowering of sea level or tectonic uplift.

To help illustrate how shorelines evolve and retreat landward, consider the example of an irregular coastline shown in Figure 9.8. When waves crash into the exposed rocks, the impact generates tremendous pressure, forcing water into tiny cracks and crevices. This repetitive hydraulic action slowly breaks the rocks apart

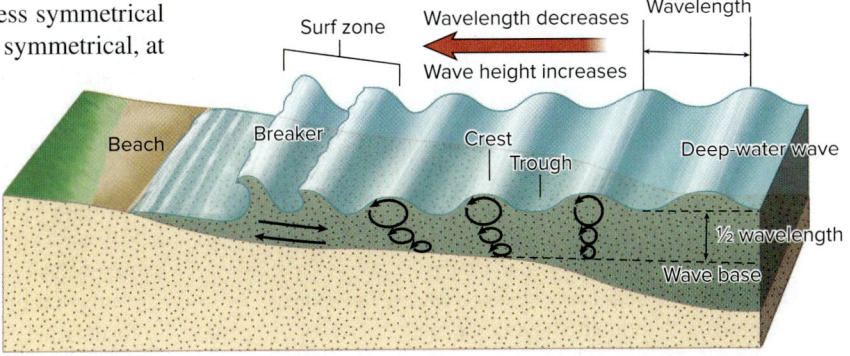

FIGURE 9.6   As waves enter shallow water, wave base will eventually meet the seafloor, creating friction that causes the waves to slow down. This, in turn, causes the wavelength to decrease as the waves grow in height and become less symmetrical. Eventually the waves become so asymmetric that they fall over on themselves and form breaking waves.

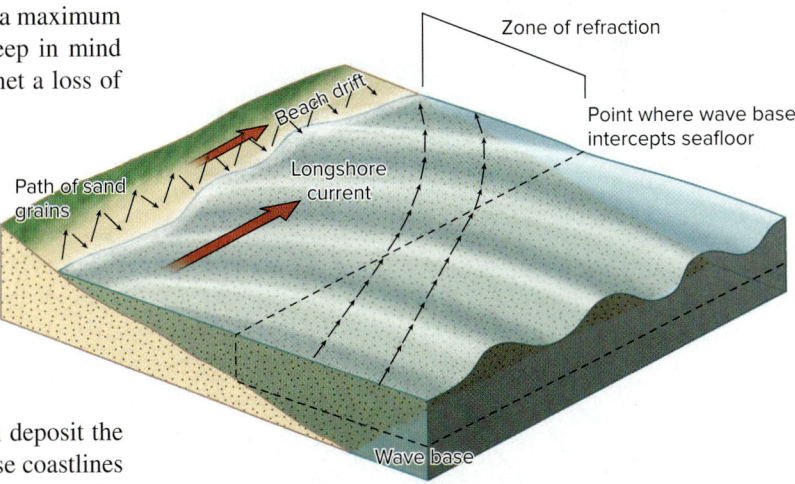

FIGURE 9.7   As a wave approaches land, the end closest to shore encounters the seafloor first, forcing it to slow down while the other end travels at its original velocity. This velocity difference causes the wave to bend or refract toward shore. Breaking waves push water up the beach, creating a zigzagging path as the water flows back into the surf zone. This process is important as it causes sediment to drift parallel to shore. Wave refraction also forces water to flow parallel to shore in what is known as a longshore current.

and forms a notch within the cliff face. As the notch deepens, the overhanging cliff becomes less stable, eventually causing the slope to fail by mass wasting (Chapter 7), at which point the cliff face retreats landward. In some instances the hydraulic pressure from crashing waves will slowly bore a hole through a cliff face, forming a *sea arch*.

Notice in Figure 9.8 that an irregular shoreline has recessed areas called *coves* and points of land called *headlands*. Headlands are important in coastal geology because they are where waves first make contact with land. This interaction causes waves to refract around both sides of the headlands, generating longshore currents that transport eroded sediment into adjoining coves where it is deposited to form beaches. Beaches form in coves because of the way opposing longshore currents lose energy as they eventually collide. Unless there is tectonic uplift, the continuous nature of headland erosion and deposition in coves will cause the shoreline to straighten over time. In Figure 9.9 one can see that as headlands are slowly eliminated, isolated beaches eventually begin to merge and the shoreline becomes less irregular. As the shoreline is being straightened, inland areas are slowly worn down by weathering and erosion such that the land surface gets closer to sea level (i.e., base level). Over geologic time therefore, a once irregular and tectonically active shoreline will develop the characteristics of a trailing-edge shoreline, namely, low-lying terrain with long stretches of continuous beach. Note that during this evolution the shoreline is constantly shifting, which is why structures built along a coastline are often at risk of falling into the sea.

Finally, we need to briefly discuss how shoreline evolution affects longshore currents. In Figure 9.9 one can see that as headlands are eliminated and beaches begin to merge, the small sets of opposing longshore currents combine and begin to flow in the direction of the prevailing wind. Although trailing-edge shorelines are relatively straight, curves still remain and produce broad headlands where erosion processes dominate. Likewise, there are recessed areas where longshore currents tend to deposit sediment. This means that even along trailing-edge shorelines there are places undergoing shoreline retreat (erosion-dominated) and those that are actually growing seaward due to deposition. Later in this chapter you will see that identifying areas dominated by either erosion or deposition is important with respect to human efforts to control shoreline retreat.

FIGURE 9.8 Headlands are places where waves first make contact with land and have the greatest amount of energy; hence, erosion is high at these locations. As the waves refract around both sides of the headlands, eroded material is transported into coves via longshore currents and deposited, forming isolated beaches.

(left): © Michael J Walsh; (right): States of Alderney Photo Library/Ilona Soane-Sands, photographer

Headland (high energy)

Cove (low energy)

Barrier Islands

Many coastlines have elongate sediment deposits called **barrier islands,** which parallel the shore and are separated from the mainland by open water, lagoons, tidal mudflats, or saltwater marshes (Figure 9.10). In the United States barrier islands are common along much of the Atlantic and Gulf coasts; worldwide they are found along about 15% of all coastlines. As their name suggests, barrier islands serve as a protective barrier that helps shield the mainland from powerful ocean storms. Despite being the first point of contact for storms, these islands have become highly prized for residential and recreational development due to their wide beaches composed of well-sorted sand. However, as we will explore in the next section, since barrier islands are basically narrow ribbons of sand only a few feet above sea level, living there presents a grave risk during major storm events.

Scientists have proposed several different hypotheses as to how barrier islands form, but the exact origin is still somewhat uncertain. What we do know is that barrier islands result from the complex interaction between waves, sea-level change, and sediment supply. For example, one way these islands are thought to have formed is by incoming waves causing shoaling and the accumulation of sand offshore, forming shallow sand bars. Eventually enough sand accumulates so that the bar stays above the high tide line, thereby creating a true island. If the sediment supply is sufficient, the island can grow in height as wind piles the sand up into dunes, which are then stabilized by vegetative cover. In general, most barrier islands are no higher than 20 feet (6 m) above sea level, but their length and width vary depending on the sediment supply and relative influence of tides and waves. Along coastlines where the effect of tides is greater than that of waves, barrier islands tend to be short and stubby. Here the constant ebb and flow of the strong tides in the inlets between individual islands commonly produces what are called *ebb-tidal* and *flood-tidal deltas* (Figure 9.10). In contrast, when wave energy increases and the effect of longshore currents becomes more dominant, the result is longer and more slender islands. Finally, geologists believe that the formation of large barrier islands requires a steady sand supply and a fairly stable shoreline, meaning that the rate of change in sea level or land elevation needs to have been relatively slow.

Shoreline retreat along tectonically inactive coastlines can create serious problems for people, particularly on barrier islands. A significant factor here is how the continued rise in sea level since the last ice age is forcing the islands to retreat landward. Because the rate of sea-level rise over the past 7,000 years has been very gradual, barrier islands have been able to migrate fast enough to keep from being flooded. As illustrated in Figure 9.11, actual landward movement takes place during major storm events when higher sea level combined with high surf and wind erodes sediment from the ocean side of the island. The sediment is then deposited on the back side. In general, barrier islands migrate in *two* directions: parallel to the coast due to longshore currents and landward as sediment is transported over the islands. A key point to remember is that barrier islands, like most coastal features, remain fairly stationary, but then move in pulses during periodic storm events when wave energy is at its highest. In the next section you will see that when humans build directly on the coast, expensive structures often fall into the sea as the island migrates landward and/or down-drift (parallel to shore).

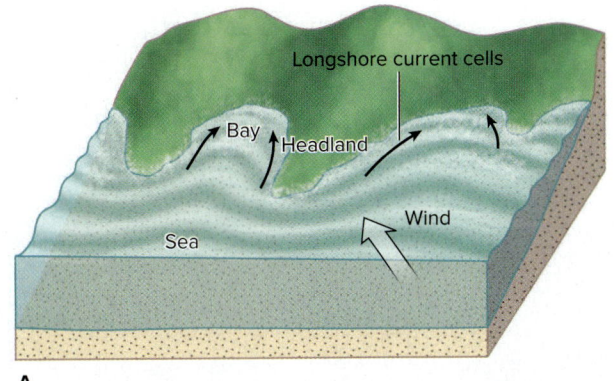

A

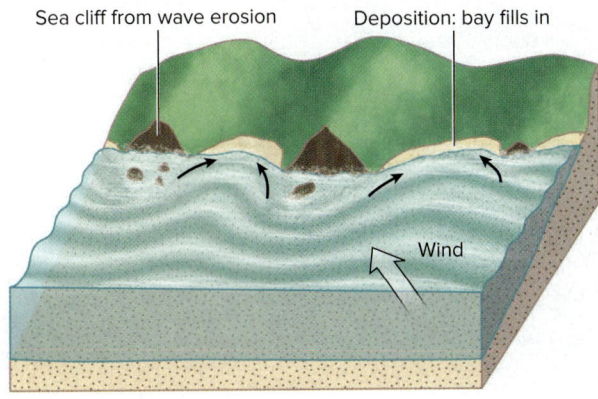

B

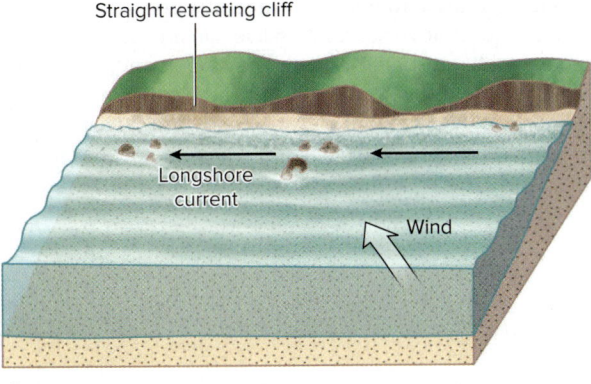

C

FIGURE 9.9 Once tectonic activity ceases, irregular coastlines slowly evolve into trailing-edge shorelines with more low-lying terrain and broad, straight beaches. Initially waves break on headlands, forming longshore current cells that transport eroded material into coves. As the headlands become smaller, the beaches and longshore cells eventually merge, forming relatively straight sections of beach where sediment is transported parallel to shore.

Coastal Hazards and Mitigation

Similar to other geologic processes discussed in this text (e.g., earthquakes, volcanoes, streams, mass wasting), there are several coastal processes that can become hazardous to people and their property. In this section you will learn that

Hutchinson Island, Florida

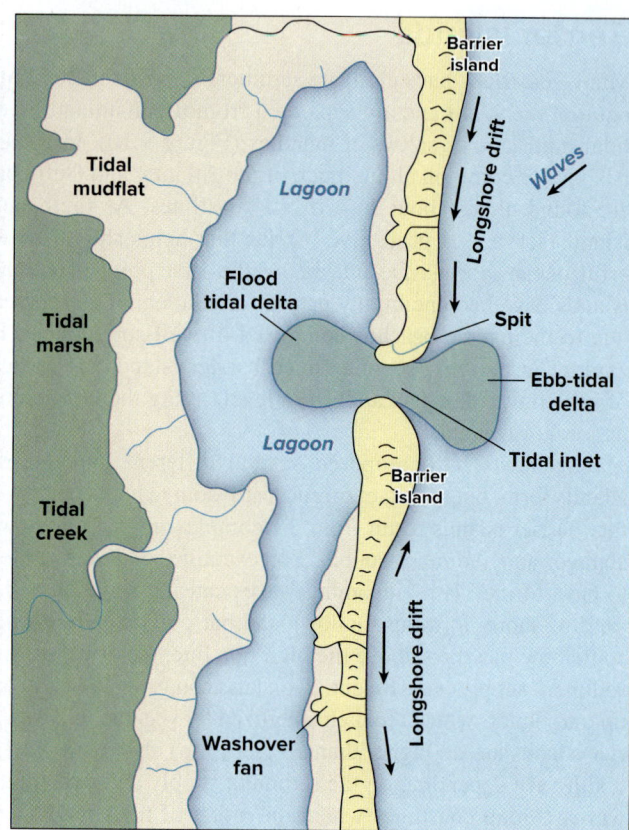

FIGURE 9.10   Barrier islands are elongated sediment deposits separated from the mainland by open water or wetlands. Tides move sand within inlets in an oscillating manner, creating submerged ebb-tidal and flood-tidal deltas. The islands themselves are highly prized locations for development because of their wide sandy beaches, but their low elevation makes them vulnerable to being overwashed during major storms.
USGS

most coastal hazards are related to storm events because this is when both wind and wave energy are at their highest. We will also explore some of the steps that can be taken to mitigate or reduce these hazards, which is becoming increasingly more important as development continues to expand in coastal zones. We will begin by examining tropical cyclones, commonly known as hurricanes in the Western Hemisphere.

Hurricanes and Ocean Storms

Imagine being on a tropical island on a sunny day with only a gentle wind blowing in off the ocean. You then notice the waves getting higher and the surf zone becoming wider and moving up onto the beach beyond the normal high-tide line.

FIGURE 9.11
Shoreline retreat on barrier islands primarily occurs during storms when sea level increases and sediment is more easily transported over the island by wind and waves, allowing the islands to essentially roll over on themselves.

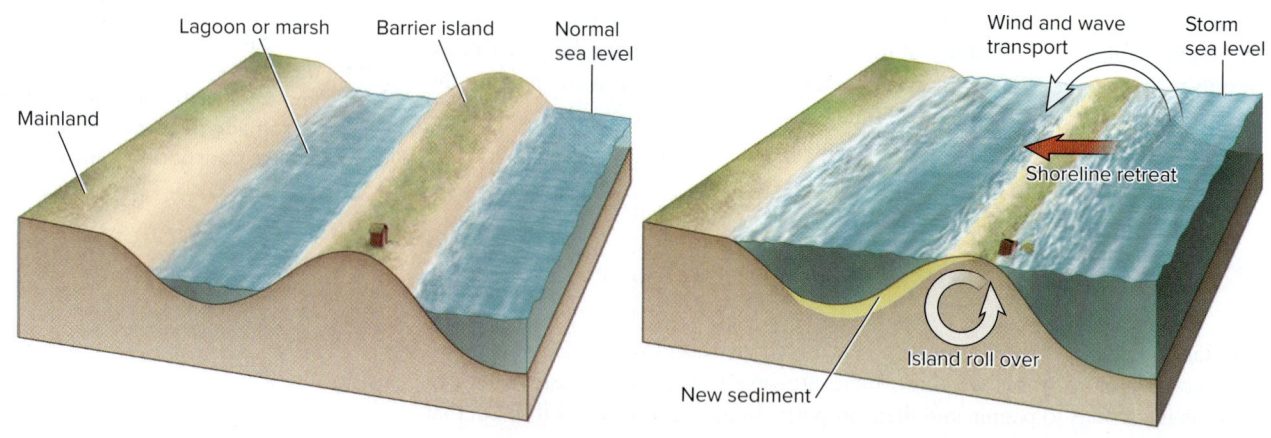

Later the wind begins to pick up and banks of clouds roll in. Intense rain soon follows, along with 150-mile-per-hour winds that begin wreaking havoc on nearly every human structure on the island. Making matters even worse, the sea soon rises nearly 20 feet, on top of which are large breaking waves that sweep most of the remaining structures off the island. This horrific experience is not over in a minute or two as in the case of an earthquake or tornado, but may last several hours. Should you be lucky enough to survive, it is safe to assume that you would never want to repeat the experience. This scenario is not from some fictional movie, but is what people often have to face in coastal areas where powerful tropical storms make landfall.

Scientists use the term *tropical cyclones* to refer to large, rotating low-pressure storm systems that originate in tropical oceans. Exceptionally strong tropical storms are called *hurricanes, typhoons*, or simply *cyclones* depending on where they form in the tropics (Figure 9.12). For the remainder of this chapter we will use **hurricane** to refer to a powerful tropical storm. Of historical interest, the word *hurricane* originates from the Mayan god of wind and storm called "Hurakan," which the Carib Indians later modified to "Hurican" to describe their god of evil.

Hurricanes develop over the warm tropical parts of oceans where low-pressure disturbances periodically become amplified into gigantic, rotating storms that contain high winds and intense precipitation. While evaporation occurs within the low-pressure disturbance and warm humid air begins to rise, large quantities of heat energy and water are removed from the ocean (i.e., a transfer of *latent heat*). The rising air mass eventually cools to the point where the water vapor condenses, and in the process releases the energy it had withdrawn earlier from the ocean and transfers it to the atmosphere. As the warm air rises and begins to condense, it is replaced by descending air that is relatively cool and dry. This convective movement of heat and water produces heavy rains and thunderstorms. As shown in Figure 9.13, a hurricane forms when the convective movement combines with Earth's spinning motion to produce a rotating system centered about an area of low pressure called the *eye*. In a hurricane this intense convection generates high winds and thunderstorm activity in the wall of clouds surrounding the eye, called the *eyewall*, and within the spiral bands that rotate around the eye. This rotating type of motion is called cyclonic, hence the scientific name *cyclone*.

The energy that drives a hurricane is extracted from warm seawater (latent heat) through the evaporation process. A hurricane's spinning motion and internal convection are important because they enable the storm to draw even greater amounts of energy from the ocean. Once in motion, this giant heat engine is able

FIGURE 9.12 Cyclones, hurricanes, and typhoons (A) are different terms used to describe large, rotating storm systems that originate over warm tropical waters. These storms generally follow curved paths toward higher latitudes and can produce winds in excess of 150 miles per hour and dump torrential amounts of rain, wreaking havoc on coastal areas. Satellite image (B) showing Hurricane Katrina prior to making landfall in Louisiana and Mississippi in 2005.
(b) NOAA and Cooperative Institute for Meteorological Satellite Studies/University of Wisconsin-Madison

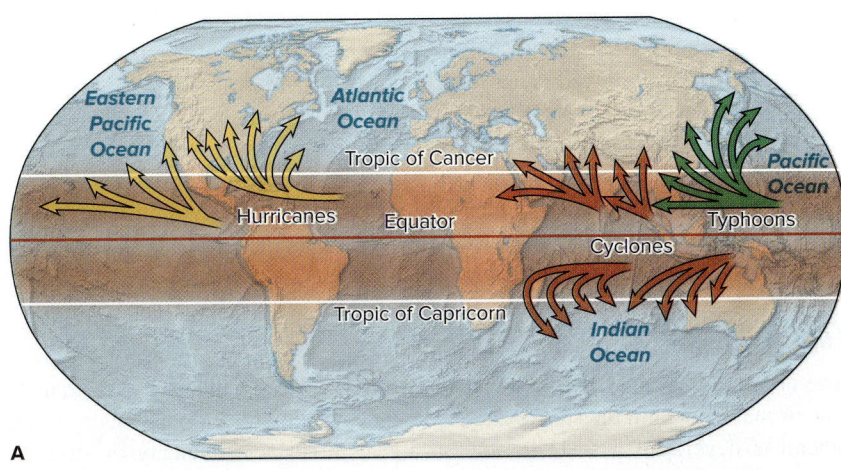

A

B

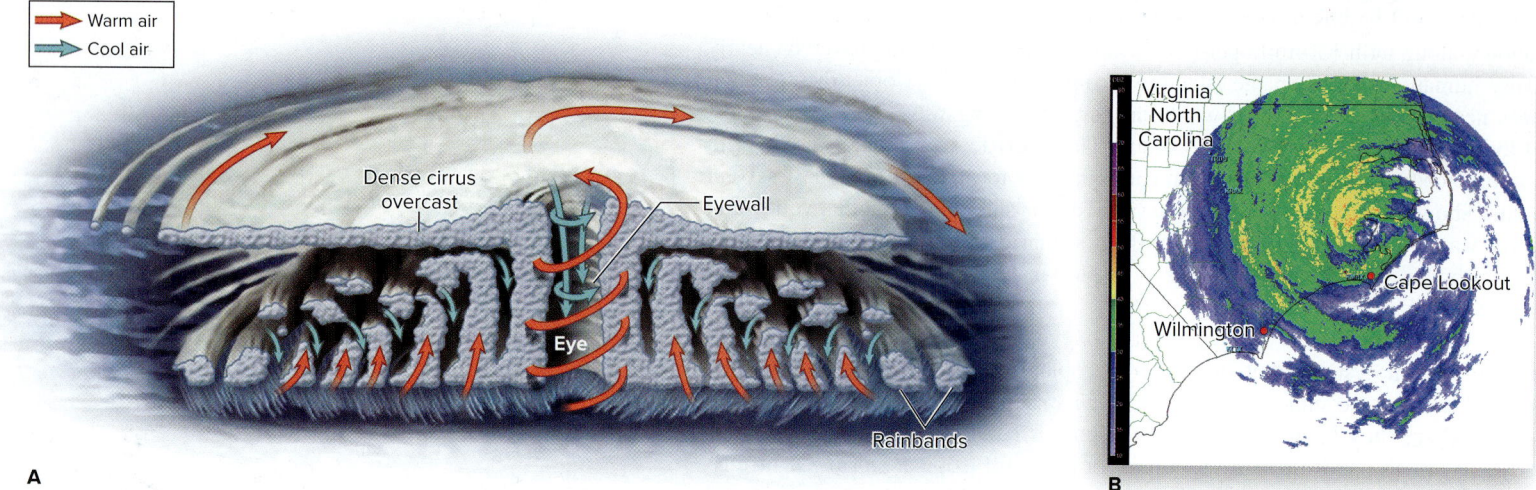

FIGURE 9.13   Hurricanes (A) form around low-pressure disturbances as evaporation removes heat energy and water from tropical waters. The resulting convection combined with Earth's spinning motion produces a rotating storm system with an area of low pressure in the center, or eye. Intense winds, rains, and wave action cause major damage to coastal areas. Radar image (B) of Hurricane Irene in 2011, showing spiral bands of heavy precipitation rotating around the eyewall.
(b) NOAA

to sustain itself and can further intensify given favorable atmospheric conditions and water temperatures. Hurricanes typically begin to weaken (i.e., lose energy) when they move into cooler waters, or encounter upper-level winds that disrupt their internal circulation. They also weaken by coming ashore, at which point they are shut off from their basic source of energy, namely the evaporation of seawater. Clearly, a hurricane that makes landfall is a very undesirable event for people living along the coastline due to the strong winds and high wave energy. Making the situation even more hazardous is the fact that the winds and low pressure within a hurricane produce a rise in sea level called a *storm surge*. This surge in sea level causes flooding and allows heavy surf to pound areas normally above the high-tide line. A hurricane will also produce heavy rains that commonly lead to river flooding far inland from where the storm makes landfall.

Although hurricanes are very destructive, other types of ocean storms produce similar hazards. Strong winter storms can form at higher latitudes when cold and warm air masses collide along frontal boundaries. These storms commonly develop in the northern Pacific, bringing heavy surf and rain to the Pacific Northwest coast of North America. On the east coast, storms called *northeasters* (or nor'easters) are mid-latitude cyclones that form when cold arctic air collides with warm humid air associated with the ocean current known as the *Gulf Stream*. The cyclonic winds from these winter storms blow onshore from the northeast, generating intense surf and coastal erosion and bringing heavy rain and snow to inland areas. In 2012, a late-season hurricane named Sandy moved up the U.S. coast and merged with a cold front just prior to making landfall. This produced an extremely powerful and destructive mid-latitude cyclone (nor'easter) that was one of the most costly storms in U.S. history, causing an estimated $67 billion in damage and leaving over 5 million people without electricity.

The following discussion on coastal hazards will concentrate on high winds, storm surge, and inland flooding. Although we will focus on the effects of hurricanes, much of the discussion will also apply to cool-weather, mid-latitude storms.

High Winds

The unusually strong winds associated with hurricanes result from the circulating air masses within the storm. Because wind speed increases with energy level, scientists developed a scale for measuring a hurricane's intensity or strength that is based on the storm's maximum sustained winds. This scale, listed in Table 9.1, is called the *Saffir-Simpson scale*. Note how different hurricanes are ranked based on their sustained winds, with the lowest category having winds of at least 74 miles per hour—anything less is called a *tropical storm* or *depression*. Although category 1 hurricanes typically cause only minimal damage, their winds are still powerful enough to overturn tractor-trailers (cool-weather ocean storms rarely produce category 1 force winds). At the other end of the scale are category 5 hurricanes, whose sustained winds of over 155 miles per hour are capable of catastrophic damage. Notice in the table that wind speed increases as the air pressure in the eye decreases. The most powerful hurricane on record is the 1979 Pacific typhoon named Tip, whose central pressure was measured at 870 millibars. Based on this pressure, scientists estimate that Tip's sustained winds were an incredible 190 miles per hour!

Hurricanes are clearly capable of producing devastating winds, where the level of destruction naturally depends on building construction and wind speed. For example, mobile homes typically experience the most damage in any hurricane. Winds from a moderate category 2 hurricane can damage even well-built homes by stripping shingles from roofs, causing rain damage to the interior, or toppling trees that fall onto the structure. At still higher wind speeds it is common for windows to be blown out and poorly anchored roofs to be lifted off buildings. At the 131–155-mile-per-hour range of a category 4 storm, roof and structural damage becomes pervasive over wide areas as shown in Figure 9.14A. In category 5 storms where maximum sustained winds exceed 155 miles per hour, structural damage is normally so complete that it is classified as catastrophic (Table 9.1).

Finally, we need to examine the actual way in which wind damages buildings and threatens human life. By picking up loose debris, hurricane-force winds usually contain airborne projectiles that not only pose a mortal threat to people left exposed to the wind, but can completely destroy otherwise intact buildings. For example, Figure 9.14B illustrates the penetrating power of lumber traveling in the 145-mile-per-hour winds of a category 4 hurricane. Serious structural damage can occur even under less extreme conditions when flying debris penetrates a building's windows, thereby allowing high winds to enter the structure. As illustrated in Figure 9.15, the combination of wind flowing through and over a building causes a difference in air pressure. This pressure differential generates an upward force on the roof, which is identical to the vertical lift created by an airplane wing. Therefore, once high winds begin flowing through a building, this lifting force can literally pull the roof off the structure. One of the key reasons for covering windows with plywood or metal sheeting is to prevent wind from entering the structure, thereby keeping both the roof on and the rain out.

TABLE 9.1 The Saffir-Simpson scale is used to rank hurricanes based on their wind speed. Note how wind speed and storm surge increase as air pressure decreases within the eye of a hurricane.

Type	Category	Level of Damage	Maximum Sustained Wind Speed (miles per hour)	Central Pressure (millibars)	Storm Surge (feet)
Tropical depression	TD		<39		
Tropical storm	TS		39–73		
Hurricane	1	Minimal	74–95	>980	4–5
Hurricane	2	Moderate	96–110	965–979	6–8
Hurricane	3	Extensive	111–130	945–964	9–12
Hurricane	4	Extreme	131–155	920–944	13–18
Hurricane	5	Catastrophic	>155	<920	>18

Introduction

© StockTrek/Getty Images

FIGURE 10.1   Photo showing water carrying valuable topsoil off a farm field in Tennessee after a heavy rain. Agricultural activity commonly leads to increased erosion and a net loss of soil because row crops offer far less protection against falling raindrops and flowing water compared to natural vegetation. If left unchecked, soil loss will ultimately lead to a reduction in worldwide food production.
Tim McCabe, USDA Natural Resources Conservation Service.

Soils are a unique part of the Earth system in that they are where the atmosphere, hydrosphere, terrestrial biosphere, and geosphere (solid earth) all interact with one another. For example, consider how atmospheric gases and precipitation cause weathering of rocks and minerals, generating loose sediment that blankets the landscape (Chapter 3). Plants and organisms then thrive on the water and nutrients that exist within the sediment, ultimately producing a life-sustaining body of natural material called *soil*. Moreover, some of the water that reaches the land surface and infiltrates through the soil zone eventually replenishes the groundwater system, whereas some of the remaining water flows over the soil as overland flow (Chapter 8). The types of soil and plants covering the landscape, therefore, play an important role in determining the portion of water within the hydrologic cycle that infiltrates versus the portion that flows over the landscape. Because of their influence on the biosphere and hydrosphere, soils are fundamental to the Earth system and life as we know it.

Although soils may not capture people's attention the way other environmental features do (e.g., earthquake hazards and energy resources), soils are actually far more important because they are critical to our human existence. Were it not for soils, there would be no land plants, and, in turn, no food to eat except for what comes from the sea. Soils thus form the basis for nearly our entire food supply. Because of this fundamental connection, people throughout history have been keenly aware of the difference between fertile and poor soils. In the past most people had to grow their own food in order to survive. In fact, after each fall harvest it was common for people to worry whether they had stored enough food to make it through the winter. Much has changed in today's modern societies where food is grown and stored on a scale that would have been unimaginable just a hundred years ago. This change has also resulted in a dramatic decrease in the number of people living on farms in developed nations, causing many of us to lose sight of the connection between soils and our food supply.

that is based on the storm's maximum sustained winds. This scale, listed in Table 9.1, is called the *Saffir-Simpson scale*. Note how different hurricanes are ranked based on their sustained winds, with the lowest category having winds of at least 74 miles per hour—anything less is called a *tropical storm* or *depression*. Although category 1 hurricanes typically cause only minimal damage, their winds are still powerful enough to overturn tractor-trailers (cool-weather ocean storms rarely produce category 1 force winds). At the other end of the scale are category 5 hurricanes, whose sustained winds of over 155 miles per hour are capable of catastrophic damage. Notice in the table that wind speed increases as the air pressure in the eye decreases. The most powerful hurricane on record is the 1979 Pacific typhoon named Tip, whose central pressure was measured at 870 millibars. Based on this pressure, scientists estimate that Tip's sustained winds were an incredible 190 miles per hour!

Hurricanes are clearly capable of producing devastating winds, where the level of destruction naturally depends on building construction and wind speed. For example, mobile homes typically experience the most damage in any hurricane. Winds from a moderate category 2 hurricane can damage even well-built homes by stripping shingles from roofs, causing rain damage to the interior, or toppling trees that fall onto the structure. At still higher wind speeds it is common for windows to be blown out and poorly anchored roofs to be lifted off buildings. At the 131–155-mile-per-hour range of a category 4 storm, roof and structural damage becomes pervasive over wide areas as shown in Figure 9.14A. In category 5 storms where maximum sustained winds exceed 155 miles per hour, structural damage is normally so complete that it is classified as catastrophic (Table 9.1).

Finally, we need to examine the actual way in which wind damages buildings and threatens human life. By picking up loose debris, hurricane-force winds usually contain airborne projectiles that not only pose a mortal threat to people left exposed to the wind, but can completely destroy otherwise intact buildings. For example, Figure 9.14B illustrates the penetrating power of lumber traveling in the 145-mile-per-hour winds of a category 4 hurricane. Serious structural damage can occur even under less extreme conditions when flying debris penetrates a building's windows, thereby allowing high winds to enter the structure. As illustrated in Figure 9.15, the combination of wind flowing through and over a building causes a difference in air pressure. This pressure differential generates an upward force on the roof, which is identical to the vertical lift created by an airplane wing. Therefore, once high winds begin flowing through a building, this lifting force can literally pull the roof off the structure. One of the key reasons for covering windows with plywood or metal sheeting is to prevent wind from entering the structure, thereby keeping both the roof on and the rain out.

TABLE 9.1 The Saffir-Simpson scale is used to rank hurricanes based on their wind speed. Note how wind speed and storm surge increase as air pressure decreases within the eye of a hurricane.

Type	Category	Level of Damage	Maximum Sustained Wind Speed (miles per hour)	Central Pressure (millibars)	Storm Surge (feet)
Tropical depression	TD		<39		
Tropical storm	TS		39–73		
Hurricane	1	Minimal	74–95	>980	4–5
Hurricane	2	Moderate	96–110	965–979	6–8
Hurricane	3	Extensive	111–130	945–964	9–12
Hurricane	4	Extreme	131–155	920–944	13–18
Hurricane	5	Catastrophic	>155	<920	>18

A

B

FIGURE 9.14   This neighborhood (A) near Miami, Florida, experienced extreme damage from the 145-mile-per-hour winds produced by Hurricane Andrew in 1992. A piece of lumber (B) driven through a tree during Andrew demonstrates the destructive power of airborne debris.
(a–b) NOAA

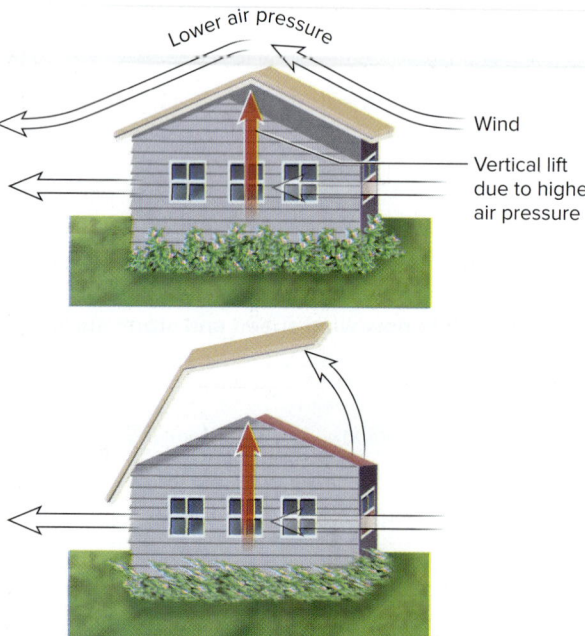

FIGURE 9.15   In addition to being damaged by airborne debris, buildings can be destroyed when a hurricane's high winds blow over and through a structure, which increases the amount of vertical lift on the roof such that it is removed.

Storm Surge

Another serious hazard associated with hurricanes is **storm surge,** which is the rapid rise in sea level that accompanies a storm as it makes landfall and inundates areas above the normal high-tide line. The surge of water moves inland at about the same speed as the hurricane and can be as high as 30 feet (9 m) above normal sea level and flood as much as 100 miles (160 km) of shoreline. This can be particularly devastating along low-lying coastlines since the water is able to move inland considerable distances.

As illustrated in Figure 9.16, storm surge occurs in part due to the abnormally low air pressure found associated with the rotating eye of a hurricane. Because air pressure represents the weight of the atmosphere pushing downward, the area of low pressure beneath a hurricane exerts less weight on the ocean surface, allowing it to rise. The result is a mound-shaped dome of water centered under the eye that literally moves or surges onto land as the storm makes landfall. An even larger rise in sea level takes place ahead of a hurricane due to its forward motion and intense winds. Because hurricanes rotate counterclockwise in the Northern Hemisphere, the combined effects of the storm's forward motion and rotational winds generate the highest storm surge on the northeastern side of the storm's path or track (Figure 9.16). In contrast, the surge height is much less on the southwestern side of the track since the winds are blowing offshore.

In addition to affecting the height of the storm surge, wind is also directly responsible for the heavy surf that accompanies the surge as it moves onto land. It is the addition of large breaking waves that makes a storm surge particularly dangerous and destructive. In fact, more people die from storm surges than any other hurricane hazard. For example, from Figure 9.17A one can see how buildings constructed above the high tide line are not only inundated by storm surge, but are exposed to tremendous forces associated with large breaking waves. Most buildings that take the full impact of such waves are simply demolished, leaving their occupants with little chance for survival. Note in the photos in Figure 9.17B the extensive beach erosion and number of homes that were destroyed by the storm surge from Hurricane Sandy, a strong category 1 storm that struck the New Jersey coast in 2012.

When it comes to a major hurricane, one of the worst places to ride out the storm is on a barrier island. This is due in part to the fact that these islands are where a hurricane first makes landfall; hence they are exposed to the strongest winds. Moreover, since barrier islands are essentially ribbons of sand only a few

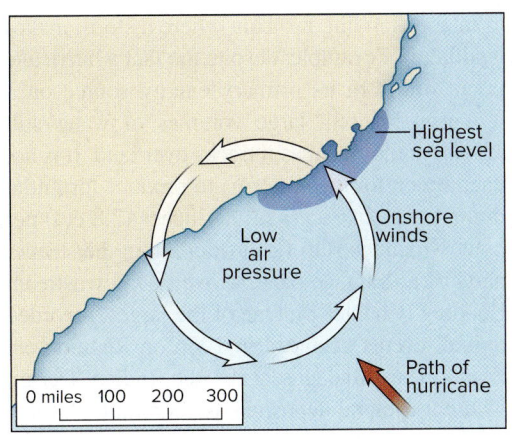
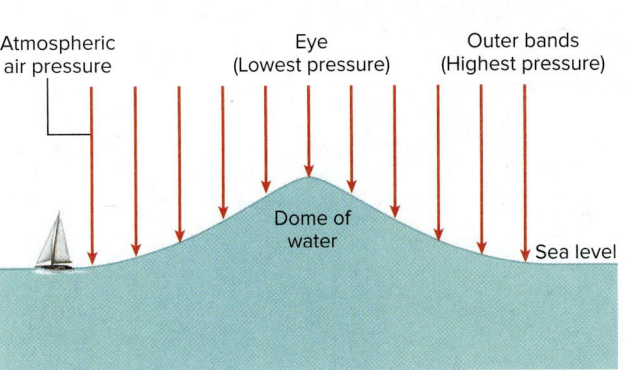

FIGURE 9.16 Storm surge forms in part because of the decrease in air pressure toward the eye of a hurricane. This allows the sea surface to rise, creating a dome of water that follows the storm inland. Even higher storm surge is generated on the northeastern side of the eye due to the storm's counterclockwise rotation and intense winds.

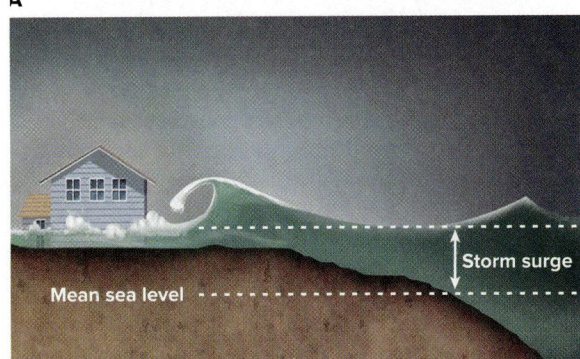

feet above sea level, the most serious threat is that the island will be completely overwashed by the storm surge. Riding out the surge and its breaking waves, even in an elevated building, is a high-risk and often fatal gamble. For those who decide not to evacuate, there is no escape once the leading edge of the storm surge floods the road to the mainland. Prior to modern communications and satellites, the only warning people had that a major hurricane was approaching was the large waves that commonly arrived ahead of the storm. Since many of the people living on barrier islands back then were not familiar with this warning sign, they never had the chance to flee before a hurricane arrived. They simply had no choice but to go to the top floor of a building or climb a sturdy tree, then hope for the best. Such was the case in 1900 for the 35,000 residents of Galveston, Texas, when the barrier island they were living on was completely overwashed during a category 4 hurricane. The storm surge progressively destroyed block after block of the city, creating a pile of debris 30 feet (9 m) high (Figure 9.18) that helped save the remaining buildings from the breaking waves. In the end, an estimated 6,000–10,000 people died, making this the worst natural disaster in U.S. history.

A more recent example of a deadly storm surge is the one that occurred in 2008 when Cyclone Nargis made landfall on the densely populated Irrawaddy Delta in Myanmar (Burma). At landfall this cyclone had sustained winds of 130 miles per hour (209 km/hr) and produced a storm surge that covered large areas of the low-lying delta. Unfortunately, the government failed to evacuate residents prior to the storm and even inhibited the relief effort of various international organizations in the aftermath of the storm. The result was a large death toll, estimated at over 100,000 people. This tragedy is a grim reminder of the risks associated with living in low-lying coastal areas as well as of the importance of evacuating people to safety prior to a storm and providing adequate relief assistance afterward.

Inland Flooding

Hurricanes and ocean storms naturally remove vast amounts of water vapor from the oceans via evaporation; much of this water is eventually returned to the Earth's surface in the form of heavy rain or snow. This intense precipitation commonly leads

FIGURE 9.17 Storm surge (A) not only inundates areas normally above high tide, but also brings breaking waves that demolish structures. Photo (B) of Mantoloking, New Jersey, showing the effects of storm surge and waves associated with Hurricane Sandy in 2012. Arrows mark the same house that appears in both images. Notice the destroyed houses and roads and extensive beach erosion. Also note the large volume of sand that was deposited on the back side of the island.
(b) (both) USGS

FIGURE 9.18 In 1900 a storm surge from a category 4 hurricane swept over Galveston Island, Texas, killing an estimated 6,000–10,000 people in a city of 35,000 residents. Photo showing the pile of debris that formed as breaking waves progressively destroyed city block after city block. Open ocean is to the right in the photo.
© AP Photo

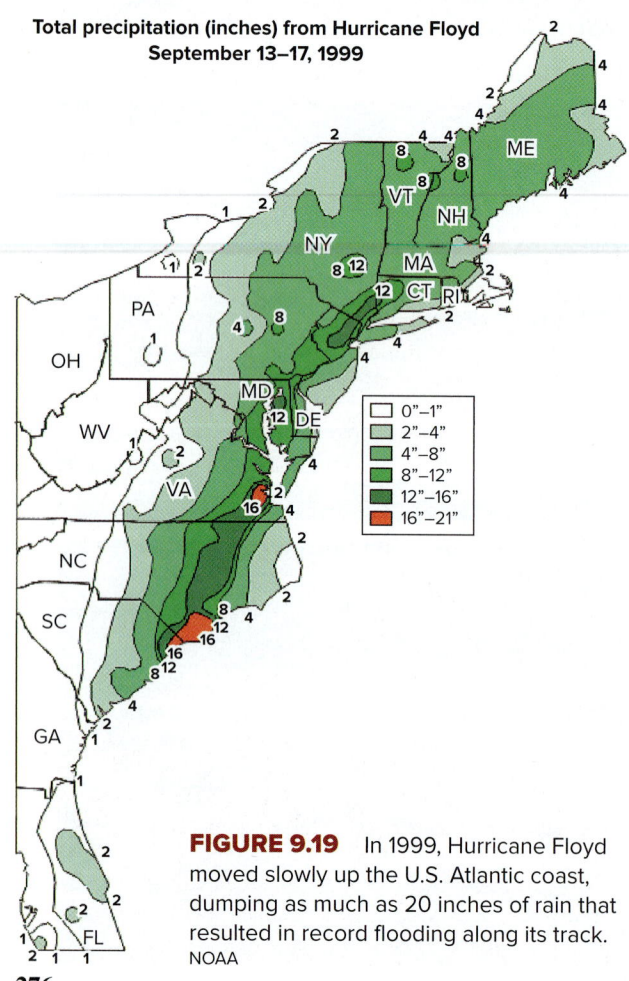

FIGURE 9.19 In 1999, Hurricane Floyd moved slowly up the U.S. Atlantic coast, dumping as much as 20 inches of rain that resulted in record flooding along its track.
NOAA

to inland flooding far from where a storm makes landfall. For example, despite the fact a hurricane is cut off from its primary energy source once it makes landfall, large volumes of water still remain in the storm as it tracks over land. It is not uncommon for these storms to produce torrential rainfall rates in excess of 1.5 inches (3.8 cm) per hour, which leads to flash flooding in the steeper parts of a drainage basin as well as downstream flooding (Chapter 8). One of the largest recorded rainfall events occurred in 1966 on Réunion, an island near Madagascar in the Indian Ocean, where a cyclone deposited an incredible 3.8 feet (114 cm) of rain in just 12 hours.

In addition to the precipitation rate, another key factor affecting the degree of flooding is a storm's forward speed as it moves over land. Slow-moving storms—with speeds less than 10 miles per hour (16 km/hr)—are particularly dangerous since more rain will fall on a given area than with a storm that passes more quickly. Consider Hurricane Floyd in 1999, for example, which moved slowly up the U.S. Atlantic coast and deposited as much as 20 inches (51 cm) of rain along its track as shown in Figure 9.19. This resulted in record flooding, the worst of which was along the Tar River in North Carolina as described in Chapter 8. Notice how heavy rainfall from Floyd was not limited to the immediate area where the eye of the storm made landfall, but extended far inland along the storm's track. In some cases remnants of a hurricane will merge over land with other weather systems, causing even more intense rainfall.

Rainfall intensity also increases when an ocean storm moves inland and encounters rugged or mountainous terrain, forcing humid air masses within the storm to gain elevation. The rapid elevation gain results in a faster cooling rate, which, in turn, increases condensation and precipitation rates. This topographic effect is a major factor in the severe flooding that commonly takes place when hurricanes move inland over the rugged coastlines along parts of Central America and the Gulf of Mexico. Similarly, the Appalachian Mountains in the United States help intensify precipitation rates from ocean storms moving inland from either the Atlantic or Gulf of Mexico. In addition to flooding hazards, the heavy precipitation associated with ocean storms moving inland commonly results in major agricultural losses, in terms of both crops and livestock. In mountainous and hilly terrain, the heavy rainfall often triggers numerous mass wasting events (Chapter 7).

Mitigating Storm Hazards

Similar to other hazards discussed in this text, ocean storms have existed throughout most of Earth's history, and thus are part of our natural environment. Although these powerful storms present many hazards, they have also played a key role in the evolution of coastal ecosystems, which ironically are what help draw people to live along the sea and put them in harm's way. Because people have found it both desirable and beneficial to live close to the ocean, societies have learned how to *mitigate* or minimize the hazards associated with ocean storms. In this section we will explore some of the more common ways of reducing the risk of coastal hazards, particularly with respect to more dangerous hurricanes.

Perhaps the oldest mitigation strategy goes back to ancient cultures, which avoided locating large settlements directly on the coasts where hurricanes frequently made landfall. If one considers how coastal development has been booming in the United States in recent years, it is apparent this strategy is no longer being employed. This modern trend can be explained in part because of the availability of flood insurance that has reduced the financial risk, and because of pro-development policies at the state and local levels. Another major factor has been the fact that relatively few major hurricanes struck the U.S. Atlantic and Gulf Coasts during the post–World War II construction boom. For example, from Figure 9.20A one can see that in a given year the percent probability of a major hurricane (category 3–5) making landfall is actually quite low for most sections of the Gulf and Atlantic shoreline. This means that in many coastal areas individuals could live their entire lives without experiencing a major hurricane. The low probability of a major hurricane tends to make people complacent. In terms of geologic time, however, it is all but inevitable that each section of the coast shown in Figure 9.20A will sooner or later experience a major strike. Note also that because the strike probabilities are statistical in nature, it is possible for a given area to experience more than one major strike in a single year. Such was the case in 2004 when the paths of three major hurricanes (Charley, Frances, and Jeanne) crossed the same area of South Florida within a six-week period (Figure 9.20B). A fourth hurricane (Ivan) was present in the Gulf of Mexico during this same period, but made landfall in Alabama rather than Florida.

In addition to human complacency, another contributing factor in the modern boom in coastal development was the creation of early warning systems. These systems originated in the early 1900s when ocean-going ships began using radio technology for reporting weather conditions back to land-based stations. After World War II, the U.S. Air Force started flying aircraft into hurricanes to record atmospheric data and to track the storms' positions. Today, weather satellites continuously track storm locations, which when combined with aircraft data allow scientists to use computer models to predict the paths of hurricanes with impressive accuracy. For example, Figure 9.21 shows the computer-predicted paths for Hurricane Katrina on two different dates in 2005 as the storm approached land. Note how the forecasted position of the storm takes the shape of a cone that grows outward from the eye. The cone shape reflects the fact that the uncertainty in the storm's future position increases with increasing distance from the eye. As in the case of Katrina, modern forecasting models can predict, out to 48 hours in advance, where a hurricane will make landfall with a fairly high degree of accuracy.

At some point before a hurricane makes landfall, emergency managers must make the decision to order an evacuation and get people to move to a safe location. The goal, of course, is to minimize the loss of life, but managers are also under pressure not to evacuate too early, in part because of disruptions to the local economy. However, as coastal population continues to grow, progressively more time is needed to evacuate, forcing emergency officials to make earlier decisions as to when to evacuate. Because of the need for more evacuation time, officials must begin the evacuation process when a hurricane is relatively far offshore, which is also when forecasters are less certain of the storm's path (Figure 9.21). This uncertainty, in turn, forces emergency officials to evacuate even longer stretches of coastline, most of which will experience little impact from the storm. The result is that many people will have evacuated unnecessarily, making them

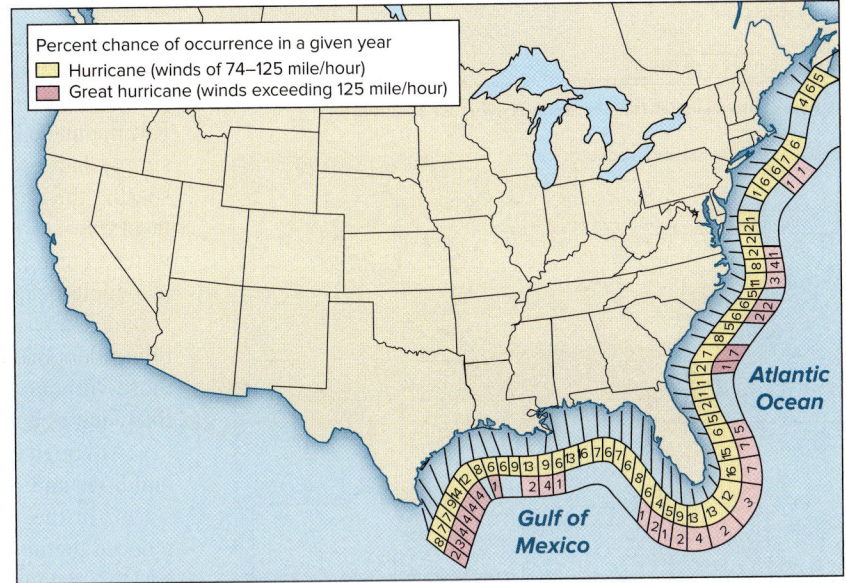

A

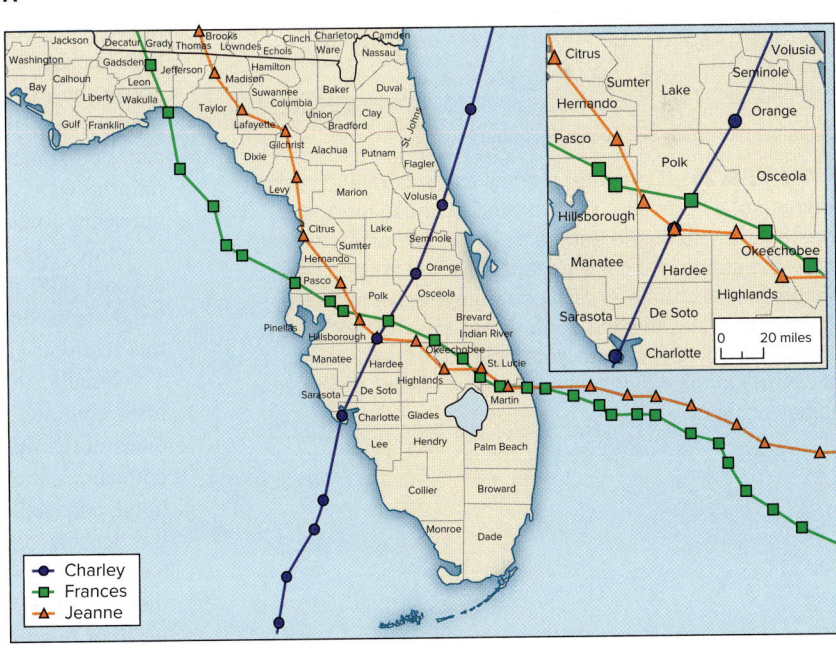

B

FIGURE 9.20 (A) Yellow areas show the percent probability of a moderate hurricane (category 1–2) striking sections of the U.S. coast in a given year. Red shows the chance of a major strike (category 3–5). (B) Because strike probabilities are statistical, multiple strikes are possible in a single year, as was the case in South Florida in 2004 when three hurricanes struck the same region.

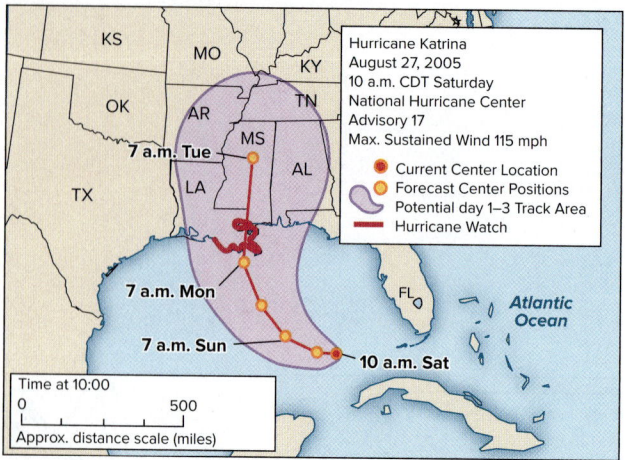

48 hours before landfall (10 a.m., Saturday, August 27, 2005)

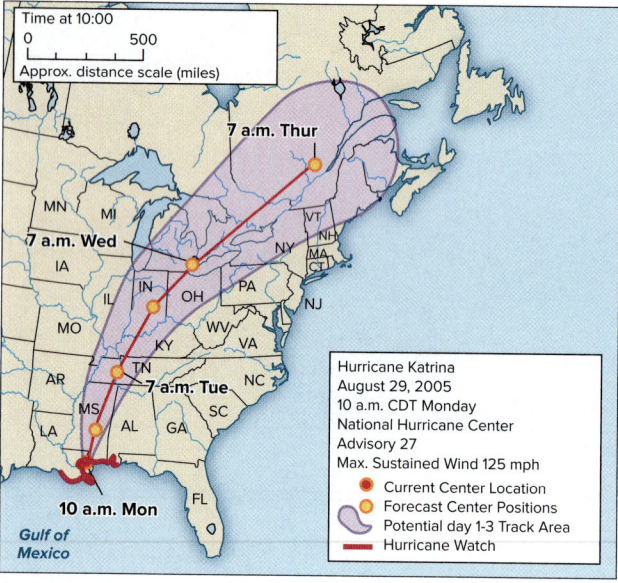

Landfall (10 a.m., Monday, August 29, 2005)

FIGURE 9.21 Computer models can accurately predict the path of a hurricane, as illustrated by these three-day forecasts for Hurricane Katrina in 2005. The projected path takes the shape of a cone because the storm's position becomes less certain as distance from the eye increases. The center line within the cone represents the most likely position at any given time. Note how the 48-hour forecast of where Katrina would make landfall was very close to the actual location.

less likely to evacuate in a future storm and putting them at higher risk of being injured or killed. Should officials decide to wait to order an evacuation until the storm's path is more certain, then they run the risk of not having enough time to carry out the evacuation. This, too, could lead to an unnecessary loss of life.

A good example of the evacuation dilemma emergency officials must face occurred in 2005 when Hurricane Katrina was far offshore in the Gulf of Mexico. As Katrina grew into an extremely dangerous category 5 storm, forecast models showed it was on track to make a direct strike on New Orleans. This was the nightmare scenario scientists had feared for years (see Case Study 9.1). Emergency planners had previously estimated that it would take 72 hours (3 days) to evacuate the 1.3 million residents in the New Orleans metropolitan area. The long evacuation period was deemed necessary in part because the city has only three escape routes, and because an estimated 100,000 residents who relied on public transportation had no means of leaving on their own. Despite the existing plans, city officials did not order a mandatory evacuation until 24 hours before Katrina made landfall. Although most residents were able to flee to safety, many simply had no way to leave. Fortunately, Katrina tracked slightly to the east, sparing the city the worst of the storm surge and high winds. Despite the more favorable storm track, numerous levees within New Orleans broke, sending water pouring into the city. With much of the city flooded, tens of thousands of people became trapped in the Superdome or stranded on the roofs of their homes, many waiting to be rescued for days in the oppressive heat. Had evacuation orders been issued earlier, plans could have been implemented to evacuate those without transportation, and the human tragedy following Katrina would likely have been far less.

Although early warning systems give people the opportunity to flee to safety, buildings and other types of property, of course, must stay behind and face the storm. In the United States, for example, early warning systems have greatly reduced the number of hurricane-related fatalities. However, the increased safety has also encouraged greater population growth in coastal zones, ultimately resulting in increased property losses. To help reduce property losses, engineers have developed improved construction techniques that minimize the amount of structural damage from hurricanes. As illustrated in Figure 9.22, elevating a building above the expected storm surge allows wave energy to pass beneath the structure rather than smashing it off its foundation. Note that wave energy can still cause erosion around the foundation supports such that the building may collapse into the water. Another key design element is to strengthen the structure against hurricane-force winds. Here metal straps are used to secure the roof to the main structure, which in turn is anchored to the foundation. Also, windows should be covered with plywood or metal sheeting prior to the storm. This prevents high winds from entering a building and lifting off the roof as described earlier in this section.

Finally, there is growing concern among scientists that as ocean temperatures continue to rise due to global warming (Chapter 16), hurricanes will produce more rainfall and category 4–5 storms will occur more frequently. Insurance companies and emergency managers are particularly concerned about this likelihood, especially after the devastation caused by Hurricane Sandy in 2012 along the densely populated northeastern U.S. coast. Historical data show that there has been an overall increase in the number of tropical storms and hurricanes in the Atlantic over the last few decades (Figure 9.23). However, it is not yet clear whether this trend is due to global warming or part of natural oscillations in the climate system. Based on current trends in greenhouse gas emissions, climate models are projecting a large increase worldwide in the number of powerful category 4–5 hurricanes over the 21st century, whereas the overall number of named storms will decrease or stay the same. Here the modeling studies indicate that global warming will affect upper-level winds so as to inhibit the formation of these complex storms. Unfortunately, warmer ocean temperatures will make it

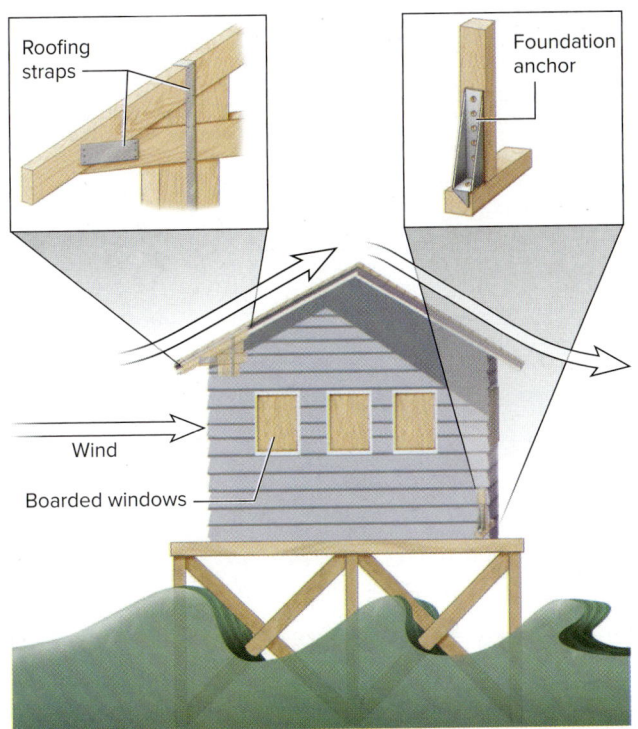

FIGURE 9.22 Structural damage from hurricanes can be greatly reduced by elevating a building above the storm surge so that wave energy can freely pass underneath. Boarding up windows and strapping the roof and frame help keep the roof from being lifted off the structure. The building can be strengthened further by anchoring the frame to the underlying structure.
Dave Gatley/FEMA

more likely that the storms that do form will become more powerful. The projections that hurricanes will produce greater amounts of rainfall are based on the fact that as the atmosphere warms, it holds more water vapor that ends up falling as precipitation.

Tsunamis

In Chapter 5 we discussed *tsunamis,* unusually high-energy waves that form not due to wind, but by the transfer of energy from earthquakes, volcanic eruptions, or landslides to a body of water (*tidal wave* is sometimes used to describe a tsunami, but is a poor choice because tidal forces are not involved). Due to the large amount of energy contained in tsunamis, the circular motion of water molecules extends to a much deeper level, a level we previously defined as *wave base* (Figure 9.5). Recall that the depth of wave base is about half the distance between wave crests (i.e., wavelength). Because tsunamis traveling through deep ocean waters have exceptionally long wavelengths, typically from 6 to 300 miles (10 to 500 km), their wave base is rather deep. Moreover, the height of these waves in deep water is quite small (less than a meter), but they travel at speeds of over 500 miles per hour (800 km/hr). Interestingly, despite the great speed of a tsunami, their small amplitude allows them to pass unnoticed by ships operating in deep waters.

Although harmless out in the deep ocean, tsunamis turn deadly as they approach shore and their wave base starts to encounter the seafloor. This interaction with the seafloor causes the fast-moving tsunamis to quickly decelerate, at which point their enormous energy is translated into progressively taller waves. These large waves eventually break and push water far above the normal surf zone, a process referred to as *run-up*. Depending on the amount of wave energy and shape of the coastline, run-up can be as high as 100 feet (30 m) or more. Because people typically build permanent structures along a

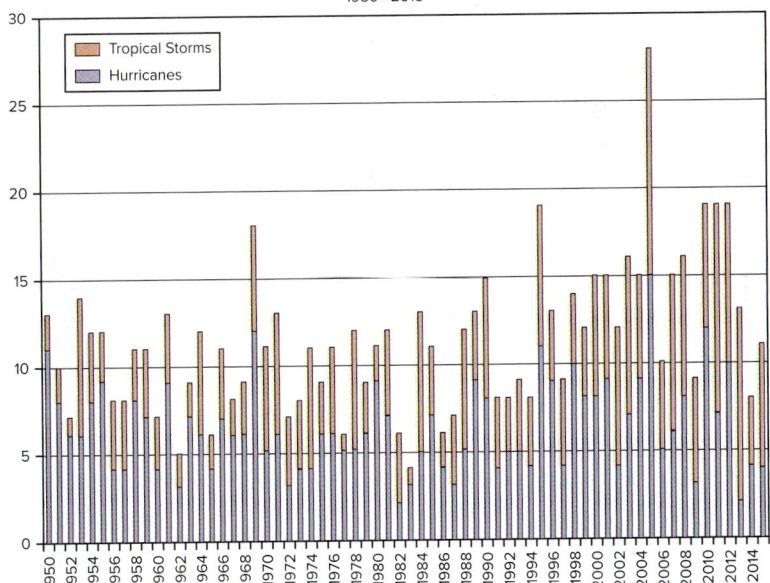

FIGURE 9.23 Histogram showing the frequency of hurricanes (blue) and tropical storms (red) in the Atlantic since 1950. There has been an increase in the number of storms, but it is not yet clear if this trend is due to global warming or natural oscillations. Model projections indicate that hurricanes may not become more frequent, but will produce more rainfall and be more likely to develop into powerful category 4–5 storms. Insurance companies and emergency managers are concerned that hurricane activity may be entering a more active and dangerous phase.

New Orleans and the Next Hurricane Katrina

When Hurricane Katrina approached the Louisiana coastline in 2005, it strengthened into a category 5 storm and was projected to make a direct strike on New Orleans (Figure B9.1). Katrina's 170-mile-per-hour (274 km/hr) winds were capable of causing catastrophic damage and creating a storm surge that would inundate the entire city. This was the nightmare scenario that scientists and emergency managers had been warning of for years, yet it seemed to take government officials by surprise, both before and after the storm. Partly because local officials delayed the implementation of a mandatory evacuation plan, tens of thousands of citizens without their own transportation had no means of escape. Fortunately, Katrina tracked slightly to the east and weakened into a category 3–4 hurricane just before making landfall, sparing the city the worst of the 28-foot (8.5-m) storm surge and high winds. Although Katrina was still a major disaster for New Orleans, it was by no means the nightmare scenario many had predicted. Since the Federal Emergency Management Agency (FEMA) had long anticipated the worst case, many people found it difficult to understand why the agency's response in the aftermath of the storm was so inadequate.

The disaster in New Orleans leads us to two important questions: Why did Hurricane Katrina have such a devastating impact on the city despite the fact it tracked to the east and weakened before making landfall? Also, is it practical to protect the city from the worst-case scenario, namely a direct strike by a category 5 hurricane? The answer to the first question is that the Katrina disaster in New Orleans was basically a flood caused by an engineering failure of the U.S. Army Corps of Engineers levee system (Chapter 8). Here poorly installed concrete levee panels simply fell over as the levees filled with water from the storm surge. Moreover, the Mississippi River Gulf Outlet, built by the Corps in the 1950s and 1960s to provide ships a shortcut into New Orleans, acted as a conduit that brought the storm surge directly into the city (Figure B9.2). Experts estimated that this shipping canal raised the storm surge within the levee system an additional 3 feet (0.9 m). Despite the increase, the water level remained below what the levee system was designed to handle. This means that the city's flood protection system had failed for a category 3–4 storm, the worst of which passed to the east of the city.

The Katrina disaster in New Orleans was due in large part to the historical use of artificial levees to provide flood control and to allow the urban area to expand (Chapter 8). Although the levees constructed along the lower Mississippi River have been effective in controlling floods, they have also cut the delta off from its natural supply of sediment and freshwater. This in turn has caused much of the delta to experience land subsidence as sediment within the delta continues to compact and freshwater marshes die from saltwater contamination.

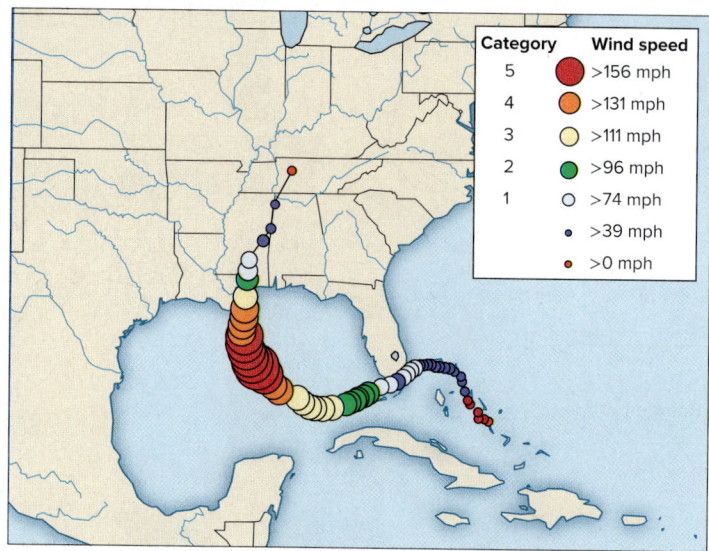

FIGURE B9.1 Map showing the intensity of 2005 Hurricane Katrina in terms of wind speed. Note how the storm developed into a category 5 hurricane, but then weakened into a category 3–4 just before making landfall.

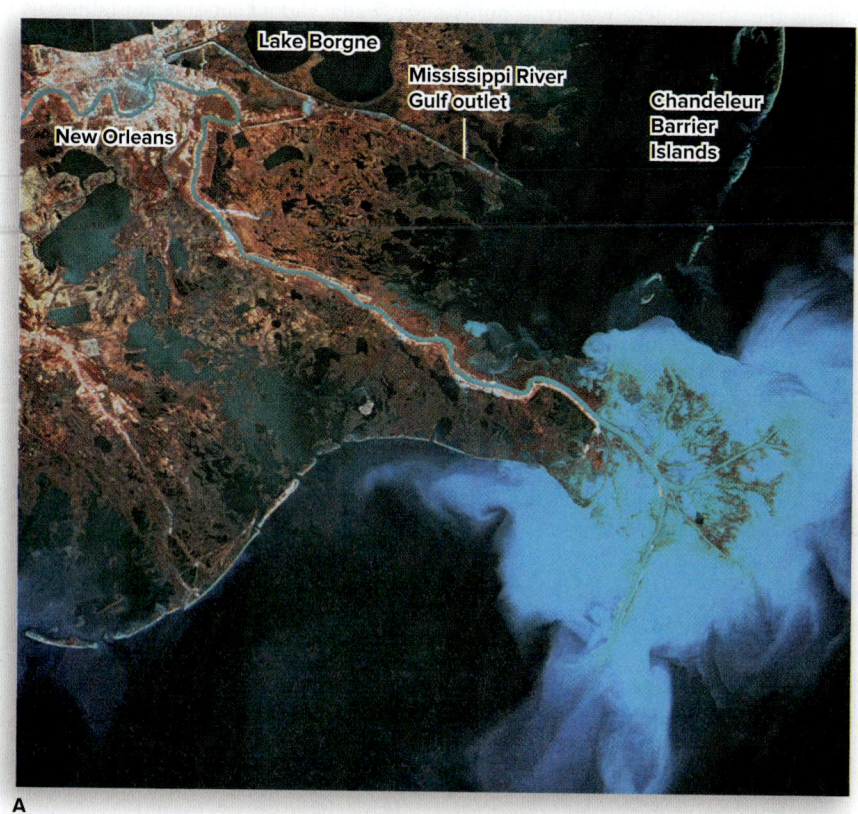

A

New Orleans, therefore, keeps sinking at the same time as global sea level is rising at an accelerated rate. To make matters even worse, the subsidence and decreased sediment supply are resulting in severe erosion of the barrier islands that ring the delta (Figure B9.2). Historically these islands have helped shield both New Orleans and the surrounding wetlands from hurricanes and wave action.

In response to the Hurricane Katrina disaster, the federal government funded a $14.5 billion program to strengthen and improve New Orleans's flood protection system. The improved system, which consists of a network of levees, floodwalls, and pumps, is designed to prevent flooding from a storm surge with a 100-year recurrence interval and significantly lessen the risk for a 500-year event. Because storm surge height is influenced by factors other than wind speed, the U.S. Army Corps of Engineers based their risk assessment on recurrence intervals rather than hurricane categories (i.e., wind speed).

One of the key components of the new flood protection system is a $1.1 billion, concrete and steel barrier, 25 feet high and nearly 2 miles long, located about 12 miles to the east of New Orleans near Lake Borgne (Figure B9.2A). It was here that Katrina's storm surge entered the canal network, causing the levees to collapse and widespread flooding within the city. In 2012 the improved flood protection system had its first real test when Hurricane Isaac pushed about a 15-foot storm surge toward the recently completed barrier near Lake Borgne. Although Isaac's storm surge was smaller than Katrina's, the new barrier proved effective in keeping the surge from entering the city.

This leads us to the question of whether it is practical to protect New Orleans from a direct strike by a category 5 hurricane (i.e., 500-year event). Although the new system is designed to significantly reduce the impact of a 500-year storm, damage could still be quite extensive. Moreover, considering that the surrounding delta continues to sink and is losing its protective marshes and barrier islands while global sea-level rise accelerates, safeguarding the city from a major hurricane will become increasingly difficult over the long term. Building higher and stronger flood control measures will never erase the fact that most of the New Orleans lies below sea level and will always be at serious risk of flooding.

July 17, 2001

August 31, 2005

B

FIGURE B9.2 The Mississippi Delta (A) is subsiding because human levees have cut off its supply of sediment and freshwater, plus various canals are bringing saltwater into its marshes. The lack of sediment and subsidence is causing the protective ring of barrier islands to undergo significant erosion as shown by the photos (B) taken before and after Hurricane Katrina.
(a) © InterNetwork Media/Getty Images; (b) (both): USGS

shoreline, the run-up from tsunami waves presents a serious hazard. As illustrated in Figure 9.24, buildings and other structures have little chance of withstanding the tremendous forces involved, particularly when they lie in the surf zone of the breaking waves. To make matters worse, development is usually concentrated in protected bays, which are particularly dangerous locations since the shape of the shoreline acts to funnel the waves into a smaller area, thereby increasing run-up.

Tsunamis are deadly in part because these high-energy events occur infrequently, which makes it more likely that humans will become unaware of or complacent regarding the hazard. Because several generations or more may pass between major tsunamis, people tend to build towns and cities along highly vulnerable strips of low-lying terrain next to the sea (Figure 9.24A). For example, consider the tsunami associated with the 1883 eruption of the Indonesian volcano called Krakatau (Chapter 6). This eruption was so violent that the 2,600-foot (792-m) mountain was blown skyward, leaving in its place a hole in the seafloor over 1,000 feet (300 m) deep. Some of the energy from this colossal explosion was transferred to the surrounding water, generating a series of tsunami waves that swept throughout the region, killing over 36,000 people in low-lying coastal towns and villages. These people had lived next to the sea for generations and probably never experienced a tsunami of this magnitude.

The 1883 Krakatau tsunami was a major catastrophe, but an even greater tragedy took place 120 years later when another giant tsunami formed, but this time the cause was an earthquake. On December 26, 2004, a magnitude 9.1 earthquake took place in the subduction zone located off the Indonesian coast of Sumatra (Figure 9.25). This massive quake caused the seafloor to suddenly lurch upward nearly 50 feet (15 meters) along 550 miles (950 km) of the subduction zone. This vertical shift of the seafloor displaced an immense volume of water, creating a series of waves that silently began to travel outward in all directions. Within minutes the tsunami crashed on the adjacent coast; then over the next several hours it swept over low-lying coastal communities around the Indian Ocean, killing an estimated 228,000 people and leaving another 2.2 million homeless. This disaster highlighted the grim fact that tsunamis can travel great distances and still pose a serious threat. Consider that of the 228,000 victims, approximately 60,000 lived in coastal communities so far from the epicenter that the people never even knew an earthquake had occurred. On the other hand, people living close enough to feel the ground shake had little time to escape, and they had to face much higher waves.

FIGURE 9.24 Aerial view (A) of Indonesia's coastline where towns and villages once stood, but were obliterated by the 2004 tsunami. Development along this rugged coastline was concentrated on small strips of level ground adjacent to the sea. Notice in the photo how the shape of the shoreline would have helped funnel the waves, thereby increasing the wave height. Photo at ground level (B) illustrating how the powerful waves ripped buildings off their foundations, leaving only the foundations themselves and steel-reinforcing rods that were once embedded in concrete walls.
(a) U.S. Navy photo by Photographer's Mate 3rd Class Tyler J. Clements;
(b) USGS

A

B

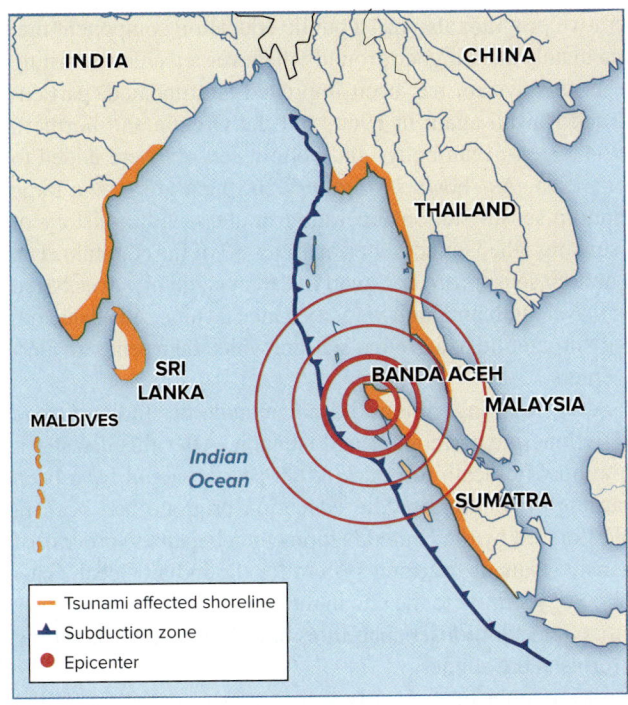

A

Before

After

B

Those hit hardest by the 2004 Indian Ocean tsunami were people living in cities and villages along the coast of Indonesia's Aceh province, a mere 60 miles (97 km) from the earthquake's epicenter (Figure 9.25). Here the waves came crashing ashore in less than 20 minutes and reached run-up heights of 100 feet (30 m). As can be seen from the photos in Figures 9.24 and 9.25, nearly every building and structure in the low-lying areas of Aceh province was ripped off its foundation and crushed by the powerful waves. In the provincial capital of Banda Aceh, approximately one-third of the 320,000 residents were killed.

Many scientists thought that the 2004 Indian Ocean disaster might be a once in a lifetime event, but then a few years later a similar disaster unfolded along the Pacific coast of northern Japan. In 2011, a magnitude 9.0 subduction zone earthquake generated a massive tsunami that devastated coastal communities, killing 21,000 people and crippling the Fukushima nuclear power plant (Chapter 14). Terrible as it was, the death toll would have been far worse were it not for Japan's high level of preparedness (see Case Study 5.1).

Mitigating Tsunami Hazards

One of the important lessons from the Indonesia and Japan disasters was that while humans cannot prevent a tsunami, taking steps to mitigate or minimize the hazard can save large numbers of people from near certain death. After a tsunami killed 170 people on the Hawaiian Islands in 1946, the U.S. government began the first coordinated effort at developing an early warning system. This system, which today includes the cooperation of numerous Pacific nations, utilizes a network of seismograph stations for detecting subduction zone earthquakes that have a potential for generating a tsunami (Chapter 5). Should such a quake be detected, an electronic warning is sent to various coastal centers, which then alert residents via emergency sirens and a public address system (in Japan the system now includes the ability to send automatic alerts via cell phone and text

FIGURE 9.25 In 2004 a magnitude 9.1 earthquake off the Indonesian coast (A) generated a tsunami that swept across the Indian Ocean, killing an estimated 225,000 people. Before and after photos (B) of the city of Banda Aceh, which was the closest to the epicenter, provide a dramatic testament to the devastating power of the waves.
(b) (both): © Digital Globe/Getty Images

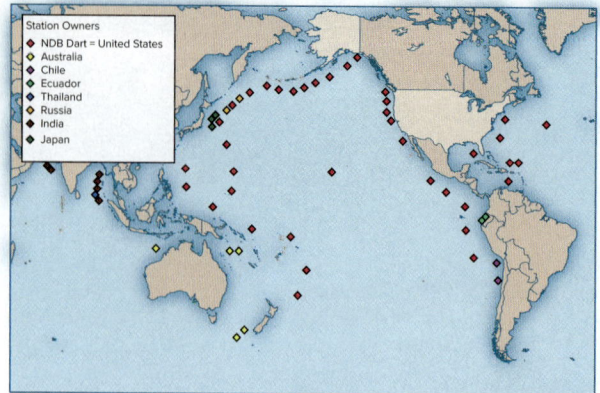

A

B

FIGURE 9.26 Map (A) showing the location of buoy and bottom sensor stations that are part of the tsunami early warning systems in the Pacific and Indian oceans. Photo (B) showing one of the stations operated by the United States. Each station consists of a buoy connected to a bottom sensor; the stations are designed to detect passing tsunami waves and then relay an alert to land-based centers via satellite.

(a–b) NOAA

messages). This cooperative program also has a public education component that teaches citizens to immediately seek higher ground whenever a tsunami warning is issued. Over the years this system has been improved and updated, particularly after tsunamis struck Hawaii again in 1960, and the Oregon and northern California coasts in 1964. Deep-ocean buoys and bottom sensors were added to the system beginning in 1995. As shown in Figure 9.26, there are now a large number of buoy and bottom sensor stations operating in areas with a history of generating damaging tsunamis (the United States operates 39 of the stations). The specialized buoys and bottom sensors are designed to detect tsunami waves based on their unusually long wavelength and high velocity. Once a tsunami is detected, a message is transmitted via satellite to a ground station, which then sends an alert to the various coastal centers.

Unfortunately, this early warning system did not include the Indian Ocean at the time of the 2004 Indonesian earthquake and tsunami. After this disaster it became obvious that the horrific death toll of 228,000 people would have been avoided if an early warning system and public education program had been in place. An international effort led by the United Nations then began the process of developing a comprehensive tsunami-warning system for the Indian Ocean. This system, which uses technology similar to the one in the Pacific, first became operational in 2006. Since that time additional capabilities have been added, including warnings for tropical storms and cyclones.

Finally, in addition to a sophisticated early warning system and education program, Japan has developed various engineering controls designed to mitigate the effects of tsunamis as well as the storm surge associated with tropical storms. The engineering controls include large gates that can be closed to block the flow of water into rivers and harbors, and walls, called *breakwaters*, designed to keep large waves from breaking farther up onto the shore. While Japan's early warning system is credited with greatly reducing the death toll from the 2011 tsunami, many of the engineering structures were simply overwhelmed by the sheer size of the tsunami. Most notable was the failure of the breakwaters at the Fukushima nuclear power plant. Engineers had designed the breakwaters for a worst-case scenario of a tsunami 18.7 feet (5.7 m) high. However, the 2011 tsunami was estimated to be 46 feet (14 m) at the plant, which easily overtopped the breakwaters and flooded the plant. For more details on the Fukushima disaster, see Chapter 14.

Rip Currents

A serious risk for people on beaches is getting caught in a strong current that flows away from shore, called a **rip current,** sometimes inappropriately referred to as a *rip tide*. Recall from earlier in this chapter that when waves break onto shore, the resulting backwash flows down the beach toward the sea. Once at the water's edge this backwash will generally flow parallel to shore until it can escape through a break in an underwater sand bar. As illustrated in Figure 9.27, the water can then funnel through this break and create a narrow, but powerful current that flows toward deeper waters where it eventually spreads out and dissipates. Rip currents are particularly strong, hence most dangerous, when the surf becomes higher as this creates greater volumes of backwash that must exit the beach via rip currents. It is estimated that in the United States alone, over 100 people drown each year by getting caught in rip currents. In fact, officials believe the actual number is far higher, because many deaths are listed simply as drowning and are not reported as being caused by rip currents. Whatever the true number, even 100 deaths a year is much greater than the number of people killed by sharks in the United States, which averages less than one per year. Although the media focus a great deal of attention on shark attacks, swimmers clearly face a much higher risk from rip currents.

Rip currents are dangerous because of the way they carry swimmers into deeper water. Even strong swimmers often drown after becoming exhausted in a futile attempt to swim back to shore against these powerful currents. Rip currents can even take the lives of nonswimmers who are simply wading in shallow water, but get knocked down by a wave and then carried into deep water by the current. Should you ever find yourself being swept out to sea in a rip current, the best approach is to stay calm and swim parallel to shore. Once you get beyond the narrow zone where the current is operating, it then is easy to swim back to shore. For people who cannot swim, it is best to stay out of the water during periods of heavy surf. As shown in the photo in Figure 9.27, the location of a rip current can often be recognized by the way the surf is disrupted and by the absence of foam floating on top of the water.

Note that rip currents are sometimes confused with *undertows*, which are outward-flowing currents that move downward along the sea floor as opposed to along the surface of the water. Undertows are more common along beaches with a relatively steep slope and where underwater sandbars are lacking. Here the back-wash is able to flow along the seafloor rather than being forced to flow through a breach in a sandbar as with a rip current. Because the current in an undertow is less concentrated or focused, undertows are far less dangerous to swimmers than rip currents.

Shoreline Retreat

Earlier in this chapter we defined *shoreline retreat* as the landward migration of the shoreline that occurs whenever global sea level rises or the land is lowered due to subsidence. However, shoreline retreat also depends on the supply of sediment along a coastline as well as the amount of wave energy. Recall that when waves strike a shoreline, wave energy is transferred to the land where it drives both erosional and depositional pro-cesses. In areas where there is a net loss of material, the shoreline will migrate landward. During storm events wave energy is at its highest, which means erosion and deposition rates are also at a maximum. Storms therefore can produce dramatic changes along a shoreline, sometimes literally overnight. In some instances large volumes of sediment are deposited in places that society finds undesirable. For example, sediment deposition in shipping chan-nels reduces their depth below that required (draft) for heavily loaded vessels to navigate safely. The sediment must ultimately be removed via expensive dredging operations. Sediment is also frequently deposited on land during major storms, covering roads and other valuable property and adding to the cleanup costs.

Although sediment deposition can interfere with human activity, shoreline erosion and retreat generally cause far more serious and costly problems. Perhaps the biggest problem occurs when people place buildings too close to the water's edge, put-ting their investments at risk of falling into the sea as the shore-line retreats. Clearly, the level of risk is directly related to the rate of shoreline retreat, which itself is largely controlled by the fre-quency and intensity of storms. In the following sections we will look at the factors that control the rate of shoreline retreat as well as the engineering controls commonly used to mitigate the problem. We will also examine how human modifications to the coastal environment often cause coastal systems to respond in ways society finds undesirable.

Increased Frequency of Major Storms

Because large ocean storms are relatively rare, coastal erosion rates are generally low on a day-to-day basis. This means that a given stretch of a shoreline may remain relatively stationary

FIGURE 9.27   Rip currents (A) form when backwash from the surf zone funnels through a break in underwater sand bars. Photo (B) showing a rip current flowing back out to sea through the surf zone in the Monterey Bay area of California. Note that the rip current can be recognized by how it disrupts breaking waves within the surf zone.
(b) Wendy Carey, Delaware Sea Grant

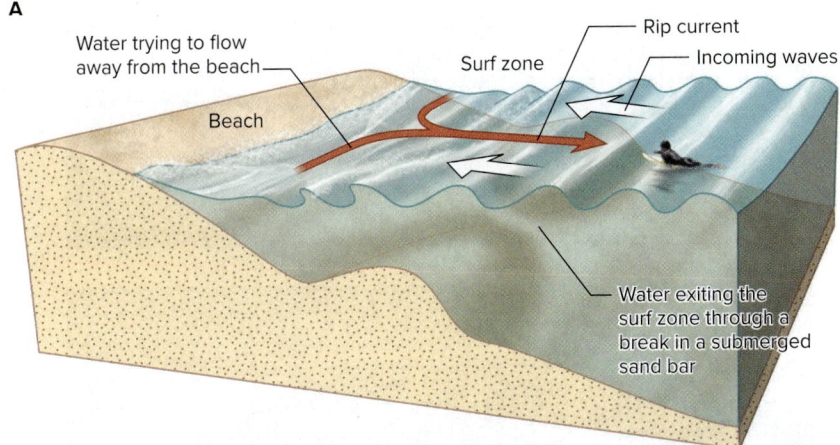

2001: prior to Ivan

2004: after Ivan

2005: after Katrina

FIGURE 9.28   Photo sequence of Dauphin Island near Mobile Bay, Alabama, showing how shoreline retreat occurs in pulses during major storm events. As the island retreats, homes become closer to the surf zone. Note in the bottom photo the missing homes and the oil rig that came ashore during the storm.
(all) USGS

for several years, perhaps even decades, before retreating suddenly in a major storm. It is during the stable periods that humans tend to become complacent, building structures that may soon end up in the surf zone. The sequence of photos in Figure 9.28 illustrates how shoreline retreat occurs in pulses, causing homes to move closer to the surf zone after each major storm.

Because shoreline retreat is directly related to storm activity, any increase in the frequency or intensity of storms will cause the overall rate of retreat to accelerate. Storm frequency and intensity are functions of ocean temperature and climatic patterns, both of which can change over time scales of decades or more in a complex and cyclic manner. As noted earlier in this chapter, there is growing concern among scientists and insurance companies that global warming (Chapter 16) is leading to an increase in the number of powerful category 4 or 5 hurricanes. Should this trend continue as climate models predict, then shoreline retreat is likely to accelerate, putting buildings at even higher risk of falling into the sea.

Effects of Sea-Level Rise

Global sea level rose rapidly after the end of last glacial period about 18,000 years ago; then around 6,000 years ago it began to slow and has remained fairly gradual ever since. Today, there is growing evidence that sea-level rise is starting to accelerate due to global warming. This is a serious concern because it would exacerbate the problems associated with shoreline erosion and retreat, and also make shorelines more vulnerable to storm surge during major storms. Consider that from 1900 to 2000 sea level increased a total of about 0.6 feet (0.2 m) worldwide, but by 2100 climate experts project that sea level will rise an additional 0.9 to 3.2 feet (0.28 to 0.98 m). While this may not seem like much, even a foot of rise would cause inundation problems for low-lying areas already close to sea level. For example, since the slope of the land along much of the U.S. Atlantic coastal plain averages around 3 feet/mile (0.57 m/km), a 1-foot sea-level rise would inundate 1,760 feet (540 m) of land extending back from the coastline. There is also a distinct possibility of even higher rates of sea-level rise. Today many climatologists are concerned over growing evidence that large portions of the ice sheets covering Greenland and Antarctica are becoming unstable. For example, collapse of the Greenland ice sheet alone could raise sea level an estimated 23 feet (7 m) over the course of several centuries. Should this occur, many coastal cities would simply have to be abandoned—see Chapter 16 for more details on climate change and sea-level rise.

Although no one is certain as to the exact amount of future global sea-level rise, what is certain is that the problems associated with shoreline erosion and retreat will worsen. The basic problem is that as sea level continues to increase, it becomes easier for storm waves to reach the top of the active beach. Along more rugged shorelines this means that waves are able to pound away at the sea cliff on a more frequent basis. The slope is then undercut more rapidly such that mass wasting puts human structures at greater risk of falling into the sea. A similar process takes place along low-lying coastlines as higher sea level makes it easier for waves to reach the upper part of the active beach and nearby buildings.

In some areas around the world the problems associated with global sea-level rise and shoreline retreat are compounded by the fact that the land is also sinking (Chapters 7 and 11). Perhaps nowhere in the United States is this more of an issue than in coastal Louisiana, where an average of 34 square miles (88 km²) of land has been lost to the sea each year for the past 50 years. The basic problem is that artificial levees lining the lower Mississippi River have cut the delta off from its natural supply of freshwater, nutrients, and sediment, plus a vast network of canals has allowed salt water to move into the delta (Chapter 8). Because sediment deposition is no longer able to keep pace with natural compaction within the delta, the land surface is now sinking below sea level. The subsidence problem is exacerbated as the freshwater marshes continue to die due to the increased presence of salt water. The combination of low-lying terrain, land subsidence, and accelerated sea-level rise gives coastal Louisiana the unfortunate distinction of having one of the fastest rates of shoreline retreat in the world. Figure 9.29 illustrates the dramatic loss of land expected for this area with a relative sea-level change of 3 feet (0.9 m) over the next century. Clearly, shoreline retreat in coastal Louisiana not only means that an entire culture and way of life will be lost, it also means that New Orleans will be at far greater risk from hurricanes and storm surge (Case Study 9.1).

Disruptions in Sediment Supply

Another key factor in the rate of shoreline retreat is the amount of sediment moving with the longshore current. Should anything disrupt the supply of sediment, beaches naturally become narrower. When this occurs, storm waves will reach the top of the active beach more frequently, accelerating shoreline retreat and putting human structures at greater risk of falling into the sea. One of the ways beaches become starved of sand is due to dams, which prevent rivers from transporting sediment to the sea, leaving less material to be distributed onto beaches by longshore currents. The degree to which dams choke off the sediment supply to a coastline depends upon their number and location within a drainage system. More dams clearly mean less sediment reaching a coastline, whereas a single dam located near the mouth of a river can prevent nearly the entire sediment load from reaching the sea. Because rivers in the more developed parts of the world are heavily dammed, coastlines there are generally starved of sand and are retreating at an accelerated rate.

In some cases the sediment that does make it to a river's mouth is prevented from entering the longshore current system. Consider how the levees along the lower Mississippi River (Chapter 8) have kept the river from migrating across the delta, causing much of its sediment to be deposited in deep water far beyond any longshore currents. This has resulted in a reduced supply of sediment along the Louisiana coast, and as noted in Case Study 9.1, is contributing to the severe erosion problem on barrier islands that once formed a protective ring around the delta.

Perhaps the most common way longshore currents are cut off from their sediment supply is through *dredging*, a process where sediment is removed from the bottom of a river, harbor, or coastal inlet. Dredging is normally done to create a deep channel so that large ships can gain access to port facilities. Due largely to cost considerations, dredged material has historically been used to create artificial islands or to fill wetlands adjacent to the channel rather than moving it downdrift along the coastline where it can enter the longshore current system. A good example is the deep shipping channel in Figure 9.30A that leads into the Savannah River and port of Savannah, Georgia. This shipping channel has a long history of dredging that has allowed progressively larger ships to

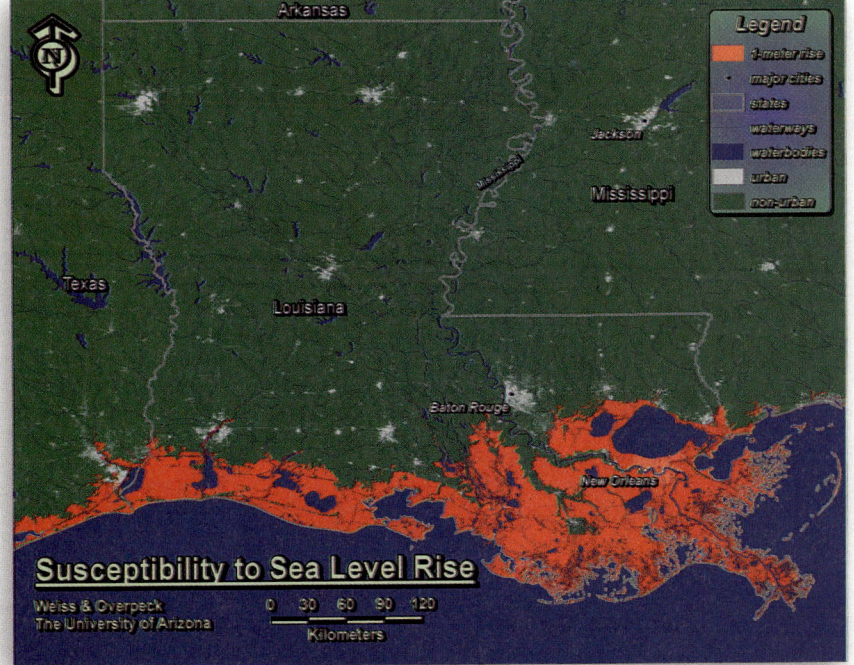

FIGURE 9.29   Map showing the amount of shoreline retreat in coastal Louisiana expected from a 3-foot (0.9-m) relative rise in sea level. This change is due to land subsidence within the delta and accelerated sea-level rise, both of which are directly related to human activity.
Weiss and Overpeck, The University of Arizona

enter the port, which is one of the oldest in the United States. While dredging has proved valuable in allowing both the port and Savannah's economy to expand, it has unfortunately reduced the amount of sediment entering the coastal environment. Moreover, because of the way the channel extends offshore, dredging has also interrupted the southward movement of sediment in the longshore current system. Not surprisingly, the barrier islands immediately south of the shipping channel (Figure 9.30B) are experiencing serious erosion problems since they are being starved of sand.

Reducing the Adverse Effects of Shoreline Processes

When humans place buildings and varied infrastructure directly on the edge of a coast, erosion and shoreline retreat eventually force the decision to either let these investments fall into the sea or take steps to stop the retreat. In other situations sediment may accumulate in places where we find it disruptive, like shipping channels, which require dredging. Other times we simply want a quiet area protected from waves in order to dock boats. Because society faces different types of issues in coastal environments, humans have developed various engineering solutions to help mitigate the problems. In this section we will explore some of the basic engineering techniques and the problems they are designed to address. We will also examine how the techniques themselves further disrupt natural coastal processes. Here you will learn that in our attempt to solve one problem, we often create yet another set of problems, all of which can be rather costly.

Seawalls When shoreline retreat threatens valuable real estate or buildings, one solution is to install a **seawall,** which is a physical barrier made of concrete, steel, wood, or large rocks built against the shore (Figure 9.31)—seawalls are also called *bulkheads* and *revetments*. Because of the way seawalls physically prevent waves from impacting the landward side of the shoreline, this technique is often referred to as *hard stabilization* or "armoring a shoreline." Although seawalls are

FIGURE 9.30 Map (A) showing the shipping channel leading into the port of Savannah, Georgia. Dredging of the channel has reduced the volume of river sediment entering the coastal environment, and has stopped the southward movement of sediment in the longshore current system. The nearby barrier island, Tybee Island (B), is consequently being starved of sand and experiencing serious erosion problems. (b) Olsen Associates, Inc.

effective in reducing shoreline retreat, there is a rather undesirable side effect in that the beach is eventually lost. What happens is that under natural conditions, storm waves are able to remove sand from the upper part of the beach, which then gets deposited just offshore. Later when wave energy is low, the sand is brought back and the beach is rebuilt. However once a seawall is installed, reflection of wave energy off the seawall prevents sediment from accumulating. Instead, the energy ends up transporting some or all of the sand back offshore. As shown in Figure 9.31B, the beach then gets smaller over time until it can only be found at low tide; at high tide the water is directly up against the sea wall. Once the beach is gone, wave action can begin to undermine the seawall, ultimately causing the structure to collapse. The irony of building seawalls in order to save buildings is that we end up losing the recreational use of the beach, which is the very thing most people come to the coast to enjoy.

Groins   In areas where shoreline retreat is a problem because of beach starvation, an alternative to building seawalls is to widen the beach by trapping sand that is moving with the longshore current. This method involves installing a barrier called a **groin,** which is made of large rocks, wood, or steel sheets and is built seaward as shown in Figure 9.32—a group of groins is called a *groin field*. Because a groin interrupts the longshore current and corresponding movement of sand, the beach becomes wider on the up-drift (up-current) side of the structure. This is desirable, because as the beach becomes wider the rate of shoreline retreat will decrease. Eventually the beach grows to the point where sand is able to go around the groin and continue moving down-drift with the longshore current. This means that once a groin fills, the overall movement of sand is about what it was prior to the structure being installed. However, the problem is that while the groin is filling, the down-drift areas will experience beach starvation and increased erosion and retreat (Figure 9.32). Long-term erosion problems can also occur if a groin is built too long, in which case the sand at the tip of the structure moves out into deeper water and is lost from the longshore current system. While groins are effective in doing what they are designed for, namely reducing shoreline retreat, they have a rather undesirable side effect of commonly causing greater retreat in down-drift areas.

Jetties   The sand that moves with longshore currents must eventually cross various inlets located along a coastline. The degree to which these inlets are kept open is due in part to the flushing action of tidal currents and rivers emptying into the sea. Throughout history humans have found inlets valuable in that they provide access to areas where ships can anchor and remain protected from the waves of the open ocean. As described earlier in this section, engineers learned to dredge sediment from inlets in order to create deeper shipping channels so

FIGURE 9.31   Seawalls (A) are physical barriers designed to keep waves from impacting areas landward of the shoreline. Over time, waves reflecting off the seawall will cause the beach to be redeposited offshore. As shown in the photos of Jekyll Island, Georgia (B), this movement of sediment ultimately results in the beach being present only during low tide.
(b) (both): © Jim Reichard

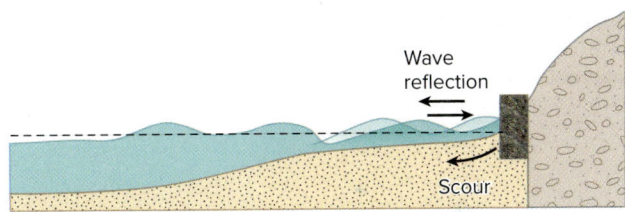

A

B

Low tide

High tide

larger ships could come into port. Keeping navigation channels open requires periodic and costly dredging operations. Barriers made of large rocks, concrete, or steel walls called **jetties** are often installed at the mouth of an inlet (Figure 9.33) to keep sediment from clogging channels, thereby reducing dredging costs. Note that jetties normally come in pairs because changing winds sometimes cause longshore currents to temporarily change direction.

Although jetties are used for navigation purposes rather than erosion control, their impact is similar to that of groins. For example, because jetties interrupt the longshore current, beaches get wider on the up-drift side of the structures and become starved in the down-drift direction (Figure 9.33). In addition, because of the need to prevent sand from flowing around the structures and clogging the channel, jetties are generally long—some over a mile—so that the sand is forced out into deep water. This means that most of the sand is lost from the longshore current system, resulting in beaches becoming starved down-drift. Although jetties are effective in keeping shipping channels open, an undesirable side effect is that they create severe erosion and shoreline retreat in down-drift areas.

Breakwaters Because some coastlines have relatively few natural harbors, large linear structures called **breakwaters** are placed offshore to keep waves from breaking onto land, thereby creating a protected area as shown in Figure 9.34. Breakwaters have traditionally been made of large rocks placed on the seafloor to create a quiet area for mooring boats. These structures have been employed in other situations to help reduce coastal erosion by reflecting incoming waves, which allows the beach to grow. Regardless of the application, the low-energy environment created by a breakwater causes the energy within the longshore current to abruptly decrease. This results in sand accumulating behind the breakwater, which for a marina is undesirable because it eliminates the space needed for docking boats. Moreover, because breakwaters disrupt the longshore current, they also increase shoreline retreat in down-drift areas since the beaches there become starved of sand.

To help remedy the erosional and depositional problems associated with traditional rock breakwaters, floating systems are now being deployed in some areas. Because floating breakwaters can be set at different water depths, engineers can adjust the system and reduce the impact on the longshore current, thereby minimizing the problems associated with erosion and unwanted deposition. Floating breakwaters have another advantage in that they give engineers greater flexibility in designing wave-protected areas.

FIGURE 9.32 Groins are built perpendicular to shore in order to trap sand moving with the longshore current. As the beach widens on the up-drift side of a groin, shoreline retreat is reduced. Note however that beaches down-drift become starved, causing shoreline retreat to accelerate there.
© Doug Sherman/Geofile

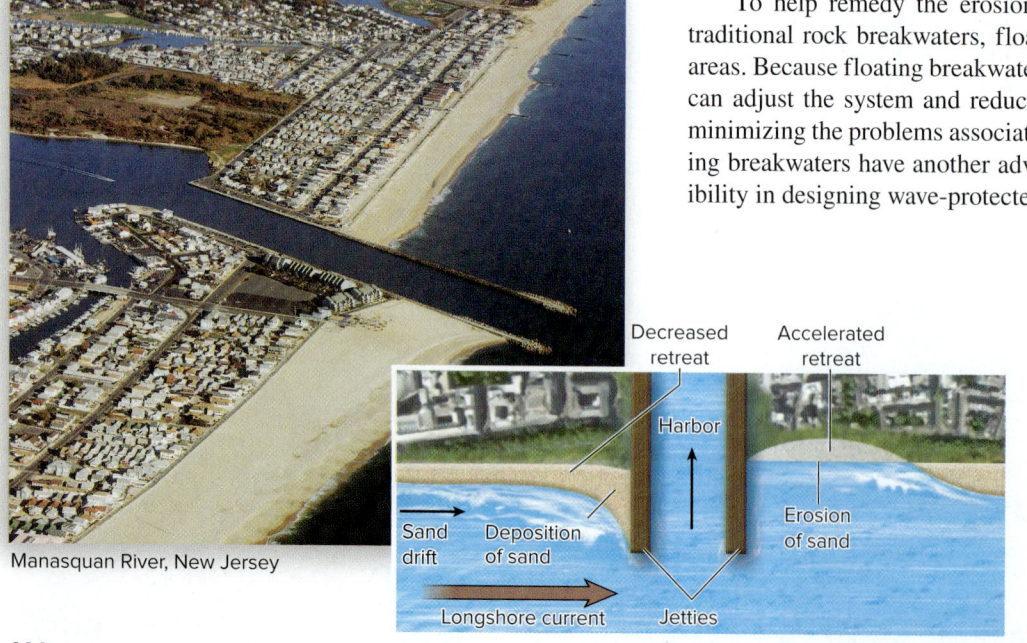

Manasquan River, New Jersey

FIGURE 9.33 Jetties normally come in pairs and are placed at the mouths of inlets to help keep the longshore movement of sand from clogging navigational channels. This unfortunately also causes down-drift areas to experience rapid shoreline retreat as beaches become starved of sand.
© Wallace C. Smith Jr.

Beach Nourishment   In many cases the only real solution to beach starvation is to manually add sand to the beach in a process called **beach nourishment.** The most cost-effective way of nourishing a beach is usually by pumping sand up onto the beach from offshore sand deposits (Figure 9.35)—in some situations trucks are used to transport sand from land deposits. Widening the beach not only reduces erosion, but also enhances the recreational use of the beach, which typically means more tourists and tourist dollars. Another desirable attribute of beach nourishment is that it does not disrupt the longshore current, and therefore does not contribute to down-drift erosion as do groins. The problem, however, is that adding sand to the beach does not address the underlying cause of why the beach is being starved in the first place. This means that the new material being added will eventually be lost, making beach nourishment not just an expensive option, but one that must be repeated indefinitely. Another drawback is that the sand supply for nourishment projects typically comes from offshore sand deposits, whose mud content is too high or whose texture is different from that of natural beaches. For example, offshore sand commonly contains abundant shell fragments that are relatively coarse, which makes walking barefoot or lying on the beach somewhat uncomfortable.

Because it is expensive, beach nourishment is not always an economically feasible option. The cost effectiveness depends on how frequently a beach needs to be renourished, the value of the property involved, and the amount of tourist revenue gained from the recreational use of the beach. In cases where the local economy is highly dependent on tourists, the cost-benefit analysis usually comes out in favor of beach nourishment. For example, in the 1950s and 1960s Miami Beach in Florida developed into a major tourist destination, but then declined in popularity due in part to the slow loss of its beach. To help reverse the decline, $64 million was spent from 1976 to 1981 on a major nourishment project, which is largely credited with bringing tourists back and revitalizing the area. Beach attendance reportedly increased from 8 million in 1978 to 21 million in 1983. In 2001 it was estimated that on an annual basis visitors were spending $4.4 billion, making the original $64 million a very good investment.

Natural Retreat   Because erosion controls and beach nourishment projects are expensive, they typically are not cost effective in areas with a small economic base and high erosion rates. This leaves one last option, and that is to let the shoreline retreat naturally. Here existing structures will either be allowed to fall into the sea or be relocated farther inland, while new buildings will have to be set back a certain distance from the shoreline. This type of management strategy has been used sparingly in the United States, and is found mainly on barrier islands designated as state or federal parks. Consider for example the Cape Hatteras Lighthouse on the Outer Banks of North Carolina shown in Figure 9.36. The present lighthouse there was built in 1870 and located 1,500 feet (460 m) from the shoreline. But by the 1930s, the structure was at the edge of the active beach because of shoreline retreat. A series of erosion control efforts held the retreat in check, but then in the 1990s the decision was made to move the lighthouse to safer ground. In 1999 this historic landmark was moved 2,870 feet (875 m) to its present position. Although this move was successful, it is not feasible for most buildings to be moved on heavily developed coastlines simply because of the lack of vacant land.

For undeveloped coastlines, particularly those on barrier islands, some people propose restricting development of overnight accommodations and other facilities, forcing most visitors to return to the mainland in the evening. In addition to saving money on expensive erosion control and beach nourishment projects, the combination of natural retreat and limited development allows people to enjoy the beach and adjacent wooded areas in a more natural setting. Clearly this is not a desirable option for those who like the convenience of beachfront

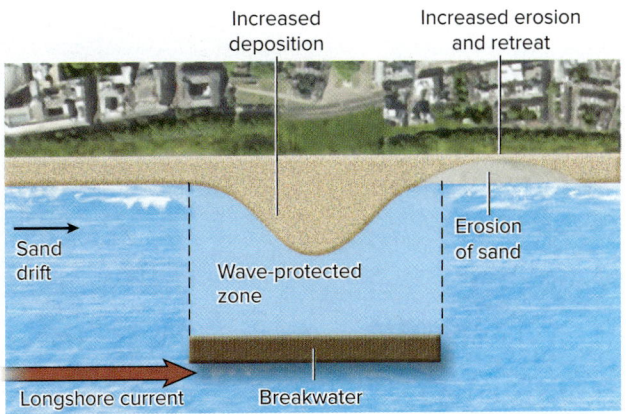

FIGURE 9.34   A breakwater is a barrier placed just offshore and used to keep waves from directly impacting the shoreline. Although effective in reducing erosion and providing a quiet area for mooring boats, a breakwater also disrupts the longshore current and creates unwanted deposition behind the structure and increased erosion in down-drift areas.

FIGURE 9.35   Beach nourishment involves removing sand from offshore deposits and spreading it on an eroding beach. Although expensive, this is often the only solution for bringing back a recreational beach in areas of chronic erosion.
Olsen Associates, Inc.

South Amelia Island, Florida

FIGURE 9.36 Aerial view showing the Cape Hatteras Lighthouse on the Outer Banks of North Carolina shortly after it was moved in 1999. The historic lighthouse was moved inland to a safer location as shoreline retreat had progressed to where the structure was at the edge of the active beach. Note the groin that had previously been installed to widen the beach in front of where the lighthouse once stood.
Photo by Elizabeth Pendleton/Woods Hole Science Center, USGS

hotels or condominiums, nor would it be popular among investors who wish to continue developing our coastlines. However, as government budgets become tighter, property owners can no longer rely on receiving federal tax dollars for erosion control and beach nourishment projects. In many areas such projects are now being funded by local property and user taxes. As sea level continues to rise, society will ultimately have to decide where to spend its limited financial resources on stopping shoreline retreat, and leave the rest to retreat naturally.

SUMMARY POINTS

1. Shorelines are unique in that they are where Earth's two most fundamental environments meet: the terrestrial (land) and marine (ocean), forming a desirable habitat for humans. Because population growth is much higher in coastal zones compared to inland areas, more people are exposed to coastal erosion and hazards such as hurricanes and tsunamis.

2. Ocean tides are caused by the gravitational pull of the Moon and Sun on Earth's oceans, whereas ocean currents move in response to winds, differences in water density, and wave action along coastlines.

3. Wave energy travels horizontally and causes water molecules to move vertically in circular paths. As a wave enters shallow water, the moving water molecules begin to drag on the bottom, causing the wave to decelerate. This causes the wave to increase in height and to become more asymmetric, eventually falling over to form a breaking wave.

4. When waves crash onto shore at an angle, water is pushed parallel to shore in a longshore current. Grains of sand in the surf zone move in a zigzagging pattern in the direction of the longshore current in the process of beach drift.

5. Irregular shorelines with more isolated beaches are commonly found in tectonically active areas and in places where sea level is rising relative to the land surface. Erosion and deposition from wave action slowly causes shorelines to evolve into ones with longer and wider beaches. At the same time, weathering and erosion of the landscape tend to produce more low-lying terrain.

6. Hurricanes are a serious coastal hazard as they generate powerful winds, storm surge, and heavy rains. Satellite early warning systems have greatly reduced the number of fatalities, but increased coastal development has caused property losses to escalate.

7. Tsunamis most commonly form during subduction zone earthquakes as water is displaced by movement of the seafloor. When a tsunami reaches shallow water, the tremendous wave energy translates into tall waves that break far beyond the normal surf zone, causing death and destruction along developed coastlines.

8. Rip currents pose a serious risk to swimmers as water from the surf funnels back out to sea through breaks in shallow sand bars. People are unable to swim back to shore against the strong currents, but can get out of the current by swimming parallel to shore.

9. The interaction between waves and a landmass can cause a shoreline to naturally retreat landward. The slow migration of a shoreline can also occur when there is a global rise in sea level or when the land itself becomes lower due to subsidence. Accelerated shoreline retreat is occurring in many areas due to human activity, increasing the hazards associated with ocean storms.

10. Humans attempt to protect their property and reduce shoreline retreat through engineering techniques such as seawalls, groins, and beach nourishment. Jetties are used to keep navigational channels free of sediment, and breakwaters provide quiet areas by keeping waves from impacting on the shoreline. Some of these techniques result in beach starvation and accelerated retreat in down-drift areas.

KEY WORDS

APPLICATIONS

Student Activity

Find a shop vacuum cleaner that can be used for both wet and dry cleanups. Next, take a standard 9-by-13-inch cake pan, place it on a level surface, and fill it with water to about a half-inch from the top. Now turn on the wet/dry shop vacuum and, holding the nozzle about a foot above the center of the pan, slowly lower it toward the surface of the water until you see the water surface start to rise in the center of the pan. Explain the basic principle involved, and describe how the observed effect relates to storm surge in a hurricane.

Next, change the vacuum setup so air blows out the nozzle—most models let you do this. Turn it on and, holding the nozzle horizontally about 2 feet from the pan, blow air across the surface of the water. What happens to the water surface this time? How is this effect similar to storm surge in a hurricane? Which setup produced the largest storm surge?

Critical Thinking Questions

1. Have you ever been on boat sitting on open water when the wake from a passing boat causes yours to move up and down, and at the same time roll sideways? What exactly is it about wave energy passing through water that makes your boat roll sideways?

2. Beach drift occurs when surf pushes sand grains up the beach at an angle, and then, as the water flows back to the sea, the grains are carried straight down the beach. What happens to this process when wave energy increases dramatically during a storm? How does this increase affect the beach?

3. Suppose you head to the beach to go swimming, but find that the wave energy is particularly high and generating heavy surf. What should you look for in the surf zone so that you can avoid getting caught in a dangerous rip current?

Your Environment: YOU Decide

Natural shoreline retreat poses a serious threat to commercial and domestic structures in coastal areas, such as beachfront resorts, condominiums, and private homes. Due to the importance of tourism to the economy of many coastal regions, state and federal agencies often spend tax dollars on various engineering projects designed to reduce shoreline erosion and retreat. These projects include building groins and seawalls and engaging in beach nourishment. Ultimately, these engineering projects are government subsidies that encourage the construction of expensive resorts and vacation homes. In the past, funding came from general tax revenues, but today more is coming from user taxes that target people who patronize the businesses that benefit from the shoreline engineering projects.

Discuss whether everyone in society should share the costs of government projects to reduce shoreline erosion (i.e., a general tax) or the costs should be covered by a user tax. Would it be even better not to have commercial buildings and homes built directly on the shoreline in the first place, but instead require them to be built farther back? Explain why.

Program for the Study of Developed Shorelines, Western Carolina University.

Chapter **10**

The soils that cover the landscape are a critical component of the Earth system and have made it possible for the land plants and animals that we see. In terms of society, soils are essential as they provide the basis for most of the world's food supply. However, human use of the landscape has upset the natural balance between soil formation and erosion, leading to the steady loss of soils worldwide. The loss of such a critical resource will present challenges as humans attempt to increase food production to meet the growing population.

Soil Resources

LEARNING OUTCOMES

After reading this chapter, you should be able to:

▶ Describe how soils are derived from rocks, and explain why soils are composed primarily of quartz and clay mineral particles.

▶ Characterize the process by which soil horizons develop, and describe why the number of horizons typically increases as the soil evolves.

▶ Discuss how soil color can be used to indicate the presence of organics and the drainage characteristics of a soil.

▶ Identify the five soil-forming factors, and describe how they control the type of soils that develop.

▶ Explain how quartz and clay mineral particles are different and how this affects soil properties.

▶ Discuss how the weathering of silicate rocks leads to aluminum-rich minerals.

▶ Characterize the relationship between soil erosion, soil loss, and sediment pollution.

▶ Explain why soil loss is a problem for humans, and list some of the ways it can be reduced.

▶ Describe why salinization and hardpans are problems for agriculture.

Introduction

© StockTrek/Getty Images

FIGURE 10.1 Photo showing water carrying valuable topsoil off a farm field in Tennessee after a heavy rain. Agricultural activity commonly leads to increased erosion and a net loss of soil because row crops offer far less protection against falling raindrops and flowing water compared to natural vegetation. If left unchecked, soil loss will ultimately lead to a reduction in worldwide food production.
Tim McCabe, USDA Natural Resources Conservation Service.

Soils are a unique part of the Earth system in that they are where the atmosphere, hydrosphere, terrestrial biosphere, and geosphere (solid earth) all interact with one another. For example, consider how atmospheric gases and precipitation cause weathering of rocks and minerals, generating loose sediment that blankets the landscape (Chapter 3). Plants and organisms then thrive on the water and nutrients that exist within the sediment, ultimately producing a life-sustaining body of natural material called *soil*. Moreover, some of the water that reaches the land surface and infiltrates through the soil zone eventually replenishes the groundwater system, whereas some of the remaining water flows over the soil as overland flow (Chapter 8). The types of soil and plants covering the landscape, therefore, play an important role in determining the portion of water within the hydrologic cycle that infiltrates versus the portion that flows over the landscape. Because of their influence on the biosphere and hydrosphere, soils are fundamental to the Earth system and life as we know it.

Although soils may not capture people's attention the way other environmental features do (e.g., earthquake hazards and energy resources), soils are actually far more important because they are critical to our human existence. Were it not for soils, there would be no land plants, and, in turn, no food to eat except for what comes from the sea. Soils thus form the basis for nearly our entire food supply. Because of this fundamental connection, people throughout history have been keenly aware of the difference between fertile and poor soils. In the past most people had to grow their own food in order to survive. In fact, after each fall harvest it was common for people to worry whether they had stored enough food to make it through the winter. Much has changed in today's modern societies where food is grown and stored on a scale that would have been unimaginable just a hundred years ago. This change has also resulted in a dramatic decrease in the number of people living on farms in developed nations, causing many of us to lose sight of the connection between soils and our food supply.

Like other parts of the Earth system, human survival is clearly linked to the natural processes that form soil as well as those that cause soil erosion. The basic problem we now face is that humans have increased soil erosion because of the way our activity disturbs the natural vegetation covering the landscape. We have therefore upset the natural balance between soil formation and erosion, which is leading to a net loss of soil (Figure 10.1). When Earth's human population was small, there was little concern when rich topsoils were lost by erosion or depleted of nutrients by repeatedly growing the same crop. When soils became lost or depleted, people simply moved on and cleared new land. As population grew, it became increasingly more difficult to find new land, forcing people to learn how to grow crops on the same soil year after year. This continuous use of the land, combined with the practice of exposing (i.e., baring) soils so crops can be planted in rows, has led to severe soil loss in many areas. Although farmers are now adopting practices designed to reduce soil erosion, it remains a serious problem in agricultural regions around the world. For society the continued loss of valuable topsoil means that food production cannot keep increasing forever. Ultimately then, soil loss and food production are on a collision course with our exponentially expanding population (Chapter 1).

In addition to serving as a natural resource that is the basis for our food supply, soils have important uses that are based on the physical and chemical properties of the minerals they contain. For example, some soils are composed of minerals that are ideal for making bricks to construct homes. Others contain minerals that serve as important raw materials, such as those used for making aluminum, a metal that has numerous applications in modern society. Some soils contain so much organic matter that they have historically been used as an energy source for heating and cooking. Soils are also important to people because they help determine the rate at which fluids are allowed to infiltrate into the subsurface. This ability to transmit fluids affects flooding (Chapter 8) and has implications for such things as civil engineering projects and the disposal of human and animal wastes. Soils are even used by geologists to determine the frequency of earthquakes, volcanic eruptions, floods, and other sudden events. For this, geologists use radiometric dating techniques (Chapter 1) to determine the age of the buried organic matter.

We will begin this chapter by examining how soils form, and then take a brief look at some of their properties that are important to society. Because soils are intimately tied to food production, we will pay particular attention to soil erosion and the ways in which this problem can be minimized.

Formation of Soils

The land surface in most places is covered with weathered rock and mineral fragments, some of which may be quite coarse and others so fine that the individual grains can only be seen with a powerful microscope. In areas where solid rock or *bedrock* is exposed (Figure 10.2), one can often see how this fragmental debris covers the Earth's surface, similar to how snow blankets the landscape. Although the everyday term for this loose material is simply *dirt*, scientists and engineers use different terms depending on the field of study. For example, geologists refer to fragmented material as *regolith*, but call it *sediment* if it has been transported by wind, water, or ice. Engineers use the term *soil* rather broadly to describe any type of broken-up material that lies above bedrock. On the other hand, soil scientists define **soil** more narrowly as a layer of weathered mineral and/or organic material that is capable of supporting plant life. Because our focus in this chapter is primarily on soils as they relate to plant growth and food production, we will use the scientific definition of *soil* as the loose material on Earth's surface that is capable of supporting plant life. Likewise, we will use *regolith* to describe weathered material that has not been transported and *sediment* for material that

Soil zone   Bedrock

FIGURE 10.2   Photo showing a soil that developed from the breakdown of the underlying rock into individual particle grains. Notice how the soil covers the landscape as a thin blanket of loose weathered material, which geologists refer to as regolith.

Howard Woodward, Plant Science Department, South Dakota State University has been transported by wind, water, or ice. We will begin by examining how weathering processes break rocks down into smaller particles. For a thorough discussion on weathering, refer to Chapter 3.

Weathering

The origin of soils ultimately begins when rocks physically disintegrate and chemically decompose in a process known as *weathering*. Recall from Chapter 3 that the rate of weathering is generally the highest at Earth's surface because this environment is relatively hostile compared to the environment in which many types of rocks form. Rocks therefore tend to break down at the surface because they are now exposed to liquid water, atmospheric gases, biologic agents, and relatively large fluctuations in temperature related to solar heating. Geologists use the term *physical weathering* to refer to those processes that cause rocks to disintegrate into smaller particles by some mechanical means. For example, solid rock is commonly broken down mechanically due to the force water exerts as it freezes and expands within pore spaces and fractures. Plants are also effective at breaking rocks into smaller fragments as their roots force their way into existing fractures. In addition to arising due to tectonic activity, fractures can form near the surface as erosion reduces the amount of confining pressure on underlying rocks (Chapter 4). Even wildfires and solar heating can produce fractures when large temperature changes cause rock to expand. Finally, climate is an important factor in physical weathering because of the way many mechanical processes are affected by temperature and the presence of water.

An important consequence of the physical weathering of rock into smaller particles is that it causes a significant increase in surface area, exposing more of the rock to chemical reactions—similar to how grinding salt into finer grains makes it dissolve more rapidly. Geologists use the term *chemical weathering* to refer to the process whereby individual mineral grains within a rock decompose due to chemical reactions (Chapter 3). Climate also plays an important role in chemical weathering because many of the chemical reactions involving rock-forming minerals are affected by temperature and the presence of water. Another important factor in determining weathering rates is the types of minerals a particular rock contains. For example, the mineral calcite ($CaCO_3$), which makes up limestone and marble, is highly susceptible to chemical weathering since it slowly dissolves when exposed to naturally acidic water near Earth's surface. As calcite-rich rocks undergo dissolution, calcium (Ca^{2+}) and carbonate (CO_3^{2-}) ions get carried away with the water, but the insoluble impurities within the rock are left behind to form a layer of soil.

Because of their sheer abundance, the most important rock-forming minerals are those rich in aluminum (Al) and silicon (Si), commonly called *aluminosilicates* or simply *silicates* (Chapter 3). Recall how the silicate mineral named *quartz* (SiO_2) is a major constituent in granitic rocks. Moreover, because quartz is highly resistant to chemical weathering, it is a dominant constituent in many types of sediment and sedimentary rocks. Other common silicate minerals, however, are rather susceptible to chemical weathering. Of particular interest are feldspar and ferromagnesian silicate minerals that make up the bulk of the rocks in Earth's crust. When acidic water near the surface comes into contact with these minerals, they react chemically and are slowly transformed into a variety of extremely fine-grained silicate minerals collectively known as *clay minerals*. As feldspar and ferromagnesian minerals are transformed into clay minerals, ions such as sodium (Na^+), potassium (K^+), magnesium (Mg^{2+}), and calcium (Ca^{2+}) are released and get carried away with the water. Because these ions are slowly lost, the clay minerals that are left behind become enriched in aluminum and silicon. Note that the chemical weathering of both ferromagnesian and iron pyrite (FeS_2) minerals releases iron ions (Fe^{3+}) into the water. However, these

A Stone Mountain, Georgia

B Grand Canyon, Arizona

ions will quickly combine with free oxygen (O_2) to form iron-oxide minerals such as hematite (Fe_2O_3). Soils therefore not only contain residual deposits of clay minerals, but if free oxygen is available, they typically appear bright red or yellow due to the presence of iron-oxide minerals.

FIGURE 10.3   Photo (A) showing a bowl-shaped depression in solid granite that has been filled with soil formed from the weathering of the rock itself. The roots of the tree in (B) have grown into fractures within the rock, extracting moisture and nutrients from soil within the cracks. (a–b) © Jim Reichard

Development of Soil Horizons

Soils can form on any type of fragmental material, such as transported sediment or regolith (untransported). For example, consider the solid mass of granite in Figure 10.3A where weathered fragments have been transported into a small depression. This accumulation of sediment is considered soil because it is supporting plant life, namely grass. Likewise, we can consider the small amount of regolith that exists within the fractured rock in Figure 10.3B as soil since it supports the growth of the tree and smaller plants. In both cases, as vegetation becomes established, organic matter is incorporated into the uppermost portion of the soil. The addition of organic matter combined with continued weathering creates a layer that is compositionally different from the soil below. People generally refer to this organic rich, uppermost soil layer as *topsoil*. Scientists, on the other hand, use the term **soil horizons** when referring to the different soil layers that have developed due to continued weathering and the downward transportation of material by infiltrating water. In this section we will explore how various soil horizons develop over time.

Since all fragmental materials are ultimately derived from the weathering of rock, a useful place to start our discussion on soil horizons is with what are called *residual soils,* those that form on the original regolith that overlies solid rock. For example, consider the cross-sectional or *profile* view of the block of granite shown in Figure 10.4. Note that soil scientists refer to unweathered rock that is within a few meters from the surface as the *R horizon*. Since the granite in this example is composed mainly of quartz and feldspar minerals, once it is exposed to physical and chemical weathering, the feldspar minerals will start to transform into clay minerals. Because clay minerals are rather soft and weak, the uppermost part of the granite will literally begin to crumble, producing a layer of regolith consisting of mostly quartz, clay, and broken rock fragments (step 2 in Figure 10.4). Plants soon start to establish themselves, and through the decay of

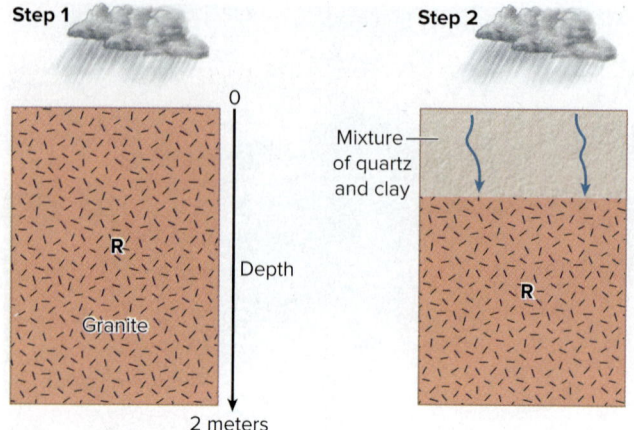

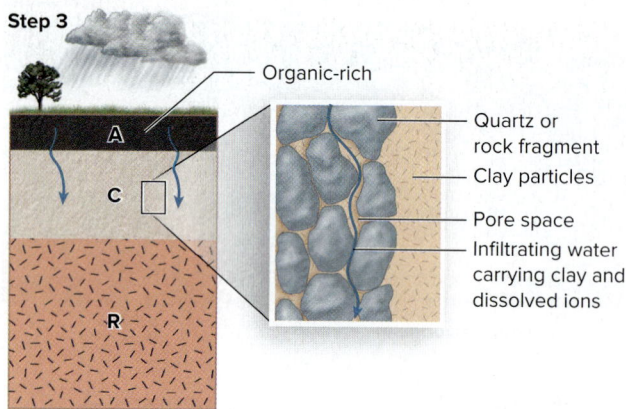

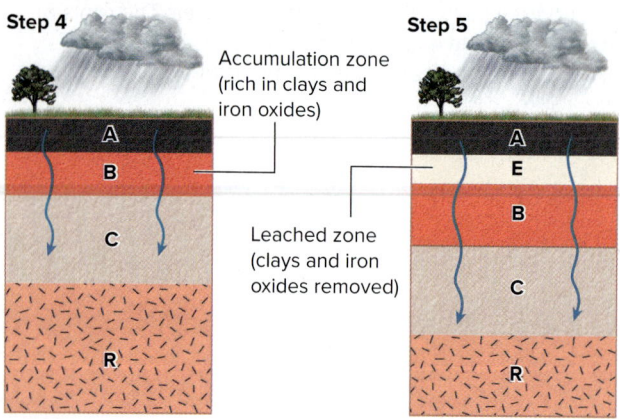

FIGURE 10.4   A time sequence illustrating the order in which soil horizons will develop when granite bedrock becomes exposed to weathering processes on Earth's surface. Notice how clay minerals, dissolved iron, and other elements are carried downward with infiltrating water and then accumulate in the B horizon.

their leaves, stems, and roots, organic matter becomes incorporated into the uppermost portion of the soil zone, resulting in two distinct soil horizons (step 3). Soil scientists call the uppermost organic-rich zone the *A horizon* (i.e., topsoil), and the remaining mixture of weathered material the *C horizon*. In very young soils such as this, the primary difference between the A and C horizons is that the A horizon contains significant amounts of organic matter.

In step 4 of our time sequence (Figure 10.4), notice how continued weathering and infiltration of water has lowered the original bedrock surface, producing an older and thicker soil zone. During the time it took for this to occur, a new layer forms called the *B horizon*, which is enriched in minerals such as different types of clay and iron and aluminum oxide minerals. Because clay particles are extremely small, infiltrating water is able to carry these particles downward through the pore spaces that exist between the much larger grains of quartz and rock fragments. Eventually a B horizon forms as the clay particles accumulate between the A and C horizons. While the clay minerals are accumulating, minerals near the surface continue to undergo chemical reactions that release iron and other ions (electrically charged atoms) into the infiltrating water. As the dissolved ions move down through the soil, iron quickly oxidizes and forms iron-oxide minerals. These iron minerals, as well as those containing aluminum, tend to accumulate in the B horizon along with the clay particles. Because the B horizon is enriched in clay and other minerals, it is often referred to as the *zone of accumulation*.

The final step in our sequence (step 5) shows the development of a layer called the *E horizon*, also known as the *zone of leaching*, where clay and other minerals have been flushed from the upper soil zone by the infiltrating water. Note that E horizons are less common and typically found in older, well-drained soils that have been heavily leached by infiltrating water. They can also be found in conifer forests where pine needles make the infiltrating water acidic. Finally, the uppermost soil layer in low-lying and poorly drained areas with lush vegetation is often exceptionally rich in organic matter and referred to as the *O horizon* (not shown in Figure 10.4). Because the ground in these conditions is typically saturated with water, oxygen levels within the soil zone are at a minimum. This lack of oxygen slows down the decay of organic matter to the point where it can accumulate faster than it decays, allowing for the development of an O horizon.

Soil Color, Texture, and Structure

In addition to horizons, other important soil characteristics include color, texture, and structure. These characteristics can best be seen by digging a trench to obtain a side view called a *soil profile*. For example, from the two soil profiles in Figure 10.5, you can see that the horizons can be distinguished in part by their color. Soil colors are the result of different types of pigments (i.e., coloring substances). For example, organic matter gives soil a blackish to brownish appearance and iron-oxide minerals generate yellowish to reddish colors. Note that it takes only a small amount of pigment to give soil a color—similar to how very little pigment is needed to turn white paint into colored paint.

Because A and O horizons form at the surface, they contain decaying plant matter, which gives these horizons a brownish to blackish color. In contrast, less common E horizons appear whitish or blonde because they lack pigmenting materials. This lack of color in E horizons results from organic matter and various oxide minerals being leached and transported downward into the soil profile by infiltrating water. B horizons, however, exhibit a range of colors because they represent the zone where various oxide minerals accumulate within the soil profile. The specific color of B horizons varies depending on the presence of free oxygen (O_2) within the pore space of soils. For example, in well-drained soils where oxygen is readily available, the dissolved iron in infiltrating water will combine with the oxygen to form iron-oxide minerals, giving the B horizon a yellowish or reddish color. In areas where little oxygen enters the pore spaces because water

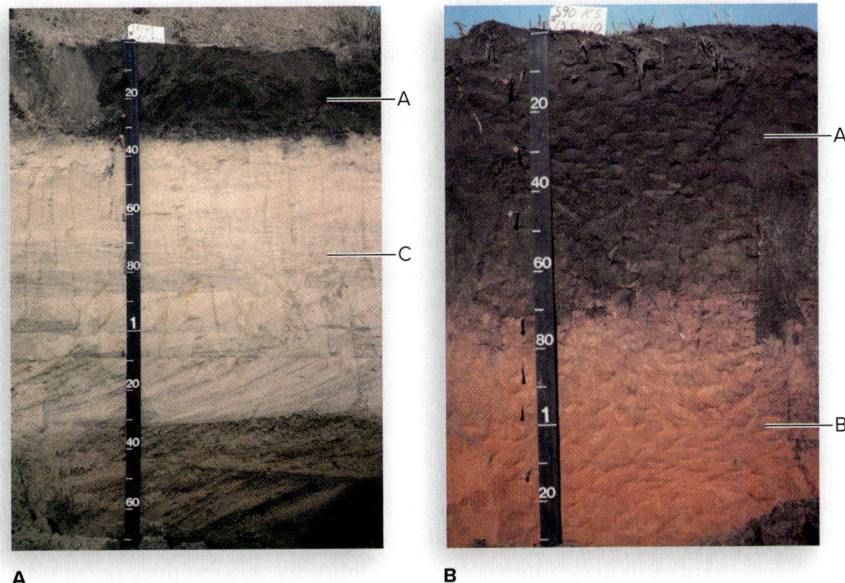

A

B

FIGURE 10.5   Soil horizons commonly have distinct colors due to the presence or absence of pigments. The organic content of the A horizon in (A) gives it a black color, which is in marked contrast to the C horizon that is light colored because it lacks pigments. The older, more developed profile in (B) shows a much thicker A horizon that overlays a B horizon that is reddish in color due to the presence of iron-oxide minerals.
(a–b) Jim Fortner, USDA-NRCS

saturates the soil, iron oxides generally do not form, producing a grayish-colored B horizon. Note that because of the relationship between iron-oxide minerals and availability of free oxygen, the color of B horizons is a useful indicator as to the level of drainage and aeration within a soil.

Soil scientists also classify soils based on *texture*, which refers to the amount of sand, silt, and clay-sized material within a particular horizon (note that geologists define these sizes somewhat differently). As illustrated in Figure 10.6, there are 12 textural classes based on the percentage of each grain size within a particular soil. For example, a soil that is composed of 40% sand grains, 40% silt, and 20% clay would be classed as a *loam soil*. A more sand-rich soil (60% sand, 30% silt, and 10% clay) would be called a *sandy loam*. Also notice in Figure 10.6 the vast difference in particle size, with clay particles being exceedingly small compared to sand grains. Note that although sand and silt-sized grains commonly consist of the mineral quartz, and clay is typically composed of various types of clay minerals, this classification system is based strictly on size, not composition. The distribution of grain sizes within a soil is important since it plays a key role in determining a soil's permeability, ease of tillage, drought resistance, and fertility. These properties, in turn, help determine the type of crops that a particular soil can support; thus they are of great importance in agriculture. For example, sandy soils may have good drainage, but are generally not very fertile and do poorly in droughts because they do not retain water very well. Clay-rich soils on the other hand may be more fertile and drought resistant, but do not drain well and are difficult to work due to the plastic and cohesive nature of clay particles. The best agricultural soils are those classified as loam (Figure 10.6) since they provide the best combination of drainage and fertility.

FIGURE 10.6   Scientists break soils down into 12 textural classes based on the percentages of sand, silt, and clay-sized particles. Texture is important because it helps determine the drainage and fertility characteristics of a soil. Note that soil scientists define the size range for sand, silt, and clay differently than do geologists.

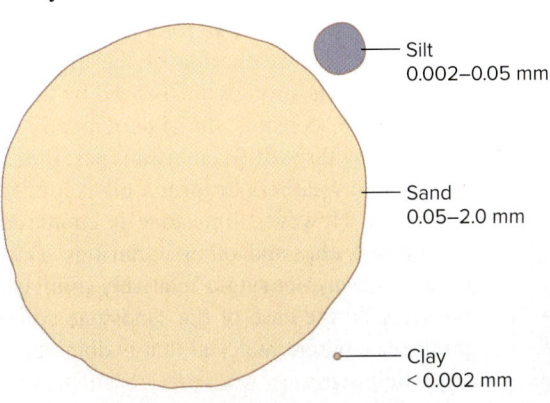

Silt
0.002–0.05 mm

Sand
0.05–2.0 mm

Clay
< 0.002 mm

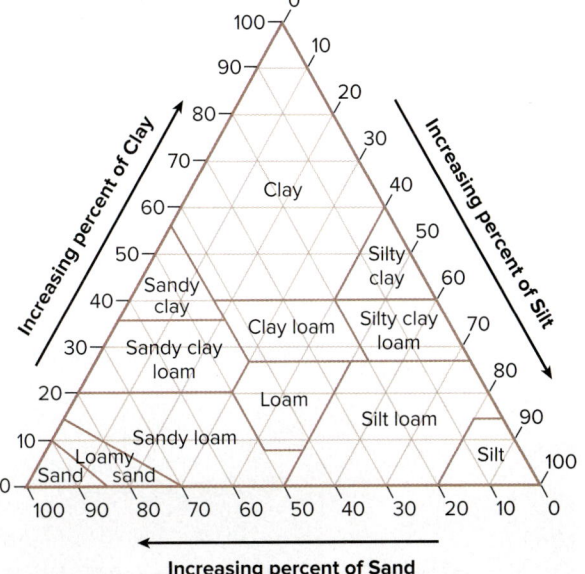

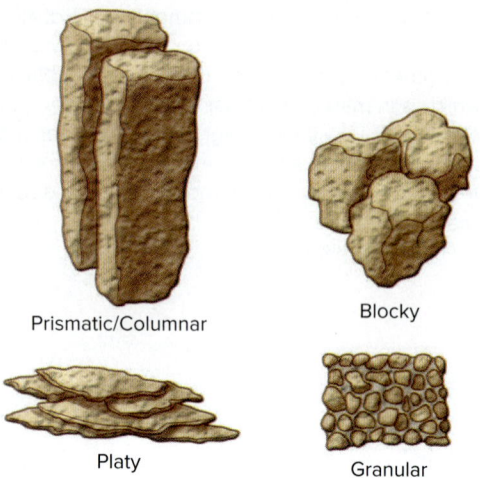

Prismatic/Columnar

Blocky

Platy

Granular

FIGURE 10.7   Illustration showing various shapes of soil peds (aggregates). The size and shape of peds determine a soil's structure and influence root development and infiltration of water.

FIGURE 10.8   Illustration showing how some soils form on parent material that is derived from weathering of the underlying bedrock, whereas other soils form on transported sediment that bears no relationship to the bedrock. Soils that form on river-transported material are commonly quite fertile due to the abundance of organic matter that grows under the moist conditions and is also deposited during periodic floods.

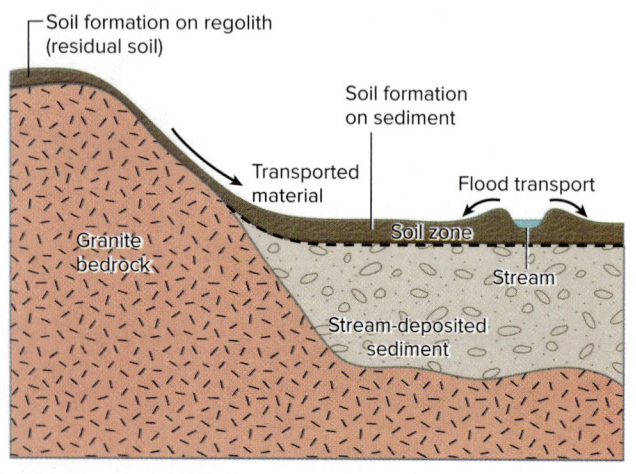

Soil formation on regolith (residual soil)

Soil formation on sediment

Transported material

Flood transport

Granite bedrock

Soil zone

Stream

Stream-deposited sediment

Finally, soils can also be characterized based on *soil structure*, which refers to the way in which soil particles are arranged. When previously undisturbed soil is dug up, it will naturally break up into separate clumps, something soil scientists refer to as *peds* or *aggregates*. As illustrated in Figure 10.7, individual peds of soil can take on different shapes, including clumps that are granular, platy, blocky, and prismatic or columnar. Soil structure is important because the size and shape of the peds greatly influences the ability of water to infiltrate and the ease with which roots can penetrate into a soil. Thus, soil structure is important to farmers as it helps determine the productivity of their soil. Farmers therefore should try to avoid running heavy equipment in their fields when they are wet so as to avoid compacting the soil, altering its structure such that infiltration and root development is inhibited.

Soil-Forming Factors

Earth's various landscapes contain many different types of soils. For example, in some areas the soil will have only A and C horizons, whereas there may be well-developed A-E-B-C horizons in other regions. Soils may also be thick in one area and thin in another. Clearly, there must be different controls or factors that determine why soils have different characteristics. There are actually five soil-forming factors recognized by soil scientists: parent material, organisms, climate, topography, and time. Moreover, these factors are not independent of one another, but rather commonly work in conjunction. For example, climate strongly influences the weathering rates of minerals and the types of plants and animals found in a given area. Soils therefore are commonly thought of as a system comprised of various components all working together, somewhat analogous to the Earth system described in Chapter 1. In this section we will briefly explore the five soil-forming factors and how they affect soil development.

Parent Material

Of all the different types of soils, approximately 99% are derived from the byproducts of weathered rock; the remaining 1% develops from thick accumulations of organic material. Soil scientists define **parent material** as the C horizon, which consists of the original weathering product or organic material from which soil horizons develop. Although bedrock is ultimately the source of weathered material from which soils form, the underlying bedrock in a given area is not necessarily related to the parent material of the soil. As illustrated in Figure 10.8, soils also develop on sediment that has been transported and deposited by streams. Clearly, the parent material in this case bears no relationship to the underlying bedrock. Note that soils that develop on recent river sediment, called *alluvium*, are commonly quite fertile due in part to the abundance of organic matter that is deposited along with the sediment during periodic floods. Soils also commonly develop on sediment that has been deposited by wind, called *loess*, and by glaciers.

In contrast, *residual soils* develop from parent material (C horizon) that forms from the weathering of the underlying bedrock. Here the type of soil that develops is strongly influenced by the mineral composition of the bedrock. For example, consider the situation in Figure 10.9, where the land surface is underlain by three different rock types: limestone, sandstone, and granite. Recall that granite weathers to form a blanket of loose rock, quartz, and clay-mineral fragments. However, limestone is composed primarily of calcite and varying amounts of clay and other impurities. During weathering, the calcite simply dissolves away, leaving a relatively thin layer of insoluble material rich in clay minerals. In the case of the sandstone composed mostly of quartz, weathering produces a parent material that is dominated by quartz sand grains (sand refers to particle size).

Recall that it is only after the parent material or C horizon becomes available that plants will establish themselves. Once plants start to grow, an A horizon will form as organic matter is incorporated into the soil. Here the type of parent material can strongly influence how the soil develops. For example, consider what would happen when rain fell on the three different parent materials shown in Figure 10.9. Clearly the highest infiltration rate would be in the sandy material that developed over the sandstone, whereas the lowest rate would be over the clay-rich material that formed from the limestone. Pine trees and other plants whose roots require well-drained soils would preferentially grow on the sandy material. Likewise, oak and other hardwood trees and plants that prefer poorly drained conditions would tend to grow on the clay-rich material. An even different assemblage of trees and plants would develop on the mixture of quartz and clay minerals blanketing the granite terrain. The important point here is that the differences in vegetation would affect the amount of organic matter that is incorporated into the A horizons. This means that the A horizons in our example would all be of different thicknesses. Also, the high infiltration rate of the sandy material means it would experience the greatest flushing action of water moving downward through the soil zone. This flushing then would increase the rate at which any fine particles and dissolved ions would move down into the B horizon.

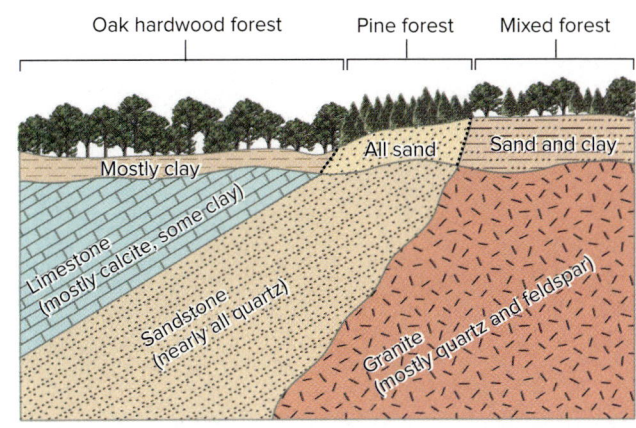

FIGURE 10.9 Because rocks contain assemblages of minerals, the weathering of different rock types can produce parent material with varying proportions of quartz and clay minerals. This variation causes residual soils to differ in their drainage and water storage properties, ultimately leading to the preferential growth of different plant communities.

Organisms

It should be clear that parent material helps determine the plant communities that develop on a given soil. However, soils also contain other living organisms, such as burrowing animals, insects, and microbes that can impact soil development. Soil then not only supports life on the surface, but within the soil itself. This is why some scientists consider soil to be a living system (Figure 10.10). Similar to plants, these other organisms contribute organic matter to soil, help break down minerals, and create passageways that allow oxygen and water to circulate more freely within the soil. Consider, for example, how the familiar earthworm creates burrows and processes organic matter within a garden. Likewise, certain types of insects and animals bring material up to the surface and create mounds, a process which overturns soils and aids in the development of the A horizon.

Climate

Another important factor in soil development is climate, because rainfall and temperature help determine the abundance and diversity of organisms, and also the weathering rates of rocks and minerals. Rich, productive topsoils generally contain abundant organic matter, which, in turn, requires moderate temperatures and adequate amounts of water. Therefore, in areas of extreme temperatures and or limited rainfall, rich soils are uncommon due to a decrease in the number and types of organisms. Also important are how physical and chemical weathering rates increase with temperature and rainfall. This accelerated weathering results in faster rates of soil formation in warm and humid climates since mineral and organic matter tend to break down more quickly. However, note that exceptionally warm and humid climates generally have rather poor soils due to more extreme rates of leaching and chemical decomposition. In areas where the parent material remains permanently frozen or dry, soil development simply stops until there is a change in climate that produces more favorable conditions.

Climatic zones also vary in terms of rainfall patterns. Recall from Chapter 8 that flooding in some regions is related to weather patterns that tend to produce sporadic but intense rainfall events. Because the rainfall is infrequent, vegetation

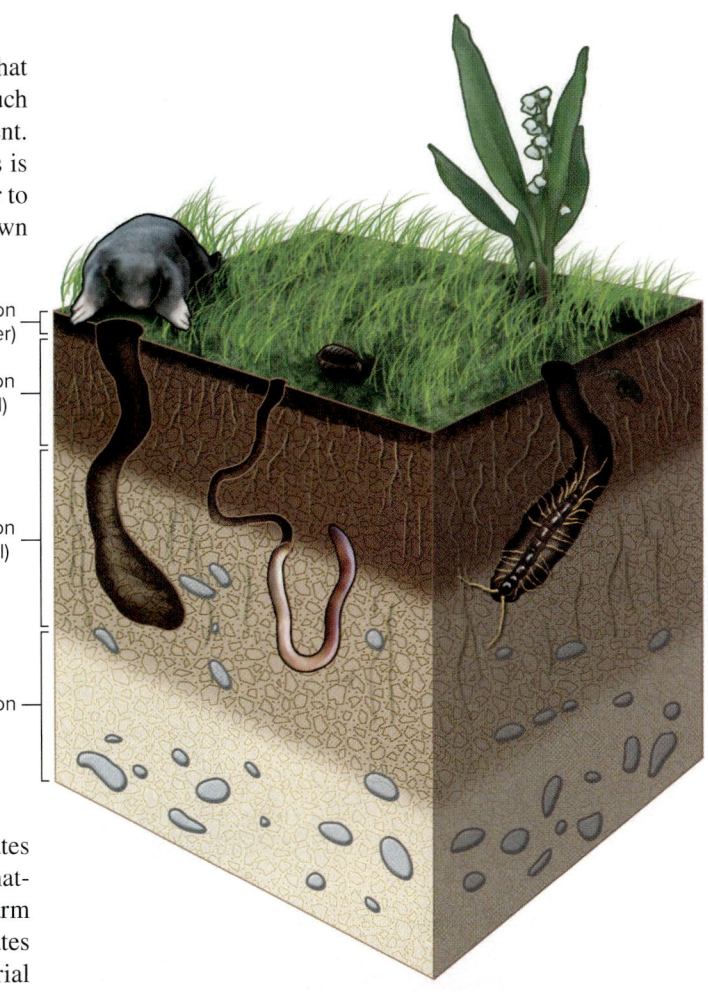

O horizon (leaf litter)

A horizon (topsoil)

B horizon (subsoil)

C horizon

FIGURE 10.10 Soil can be thought of as a living system, supporting life both on the surface and in the subsurface. Organisms aid in soil development by adding organic matter, overturning the soil, and providing passageways for air and water.

in these regions is more sparse, which leads to higher erosion rates during the occasional intense rains. The higher erosion rates naturally make it more difficult for thick A horizons to develop because the exposed topsoil is easily washed away.

Topography

Earth's landscape is highly variable, ranging from flat plains to rugged mountains. *Topography* simply refers to the shape of the land surface, including the amount of slope and vertical relief (elevation difference between high and low points). In terms of soil formation, topography is important because it helps control infiltration, erosion, and chemical decomposition rates as well as the types of organisms that inhabit the landscape. To help understand these relationships, scientists often examine a given location in terms of its *position* on a slope, its *steepness*, and the orientation of the slope toward the sun—called *aspect*. With respect to position, we can see from Figure 10.11 that the water table is relatively deep on the upper portions of a slope, and shallow in the flat-lying areas at the bottom of the slope. Water therefore drains more readily through the soil zone in topographically high areas, whereas the drainage is rather poor in low areas. This results in greater leaching of the soils in the upland areas where the infiltration rate is high. The flushing action of the water in these areas also transports clay minerals and dissolved ions deeper into the soil profile.

The varying depth of the water table throughout the terrain also plays an important role in determining the amount of organic matter that accumulates in the topsoil. Notice in Figure 10.11 that vegetation is more lush in low-lying areas due to the greater availability of water. Soils here also tend to be saturated, which helps keep free oxygen (O_2) from entering the pore spaces within the soil. Because of the limited free oxygen, the decay of organic matter by oxygen-dependent or *aerobic* bacteria is greatly reduced. This slower decay rate combined with more lush vegetation helps produce thick, organically rich soils in topographically low areas. Such areas make for highly productive agricultural lands.

From Figure 10.11 we can also see that soils tend to be thinner on steeper slopes. As a slope becomes steeper, a greater fraction of the precipitation will move directly down the slope, which means less water can infiltrate. This reduced infiltration produces greater erosion and lower rates of chemical weathering within the soil profile, both of which lead to thinner soils. Finally, we need to consider the orientation of a slope toward the sun. In the Northern Hemisphere, south-facing slopes receive more direct sunlight; hence they are naturally warmer and retain less moisture compared to slopes facing north. The result can be significant differences in vegetation cover and chemical decomposition rates, both of which influence the types of soils that develop.

FIGURE 10.11 Soils on topographically high areas generally contain less organic matter because of better drainage and higher rates of decomposition. Soils in low areas commonly have more organic material due to more lush vegetation and poorer drainage, which tends to preserve organic matter. On steeper portions of a slope, soils are thinner due to the higher rates of erosion.

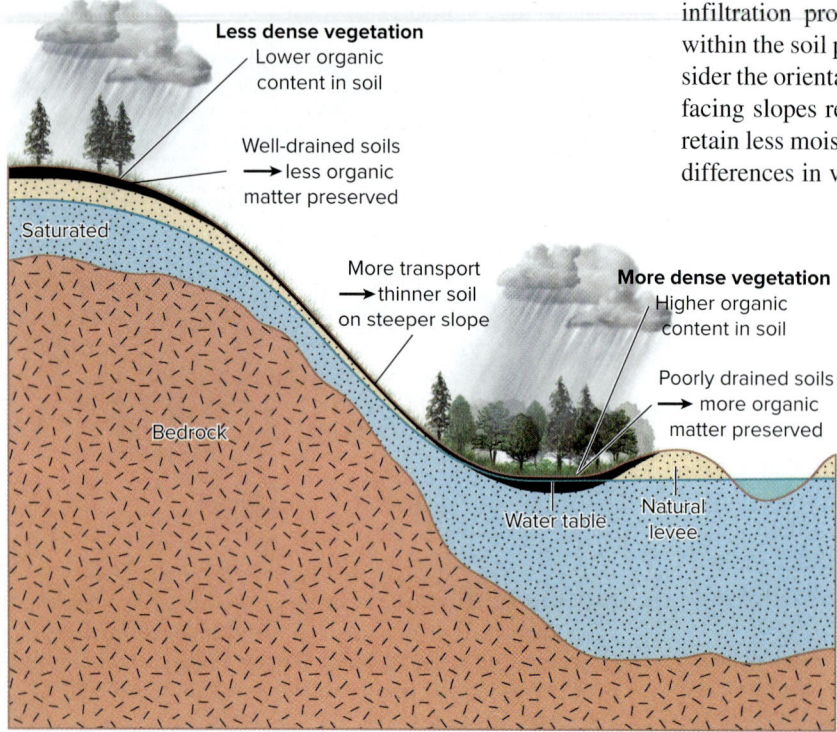

Time

Because the various physical and chemical processes involved in soil formation operate slowly, time is a significant factor in horizon development. As indicated earlier in this section, these processes can speed up or slow down depending on the manner in which the other soil-forming factors (parent material, plants and animals, climate, and topography) interact with one another. For example, soil horizons develop much more rapidly in warm, humid climates and where slopes are gentle and covered with lush vegetation. To provide some perspective of the time involved, under suitable conditions a simple soil sequence of just A and C horizons may take a hundred years or less to form. On the other hand, several hundred years are normally needed for

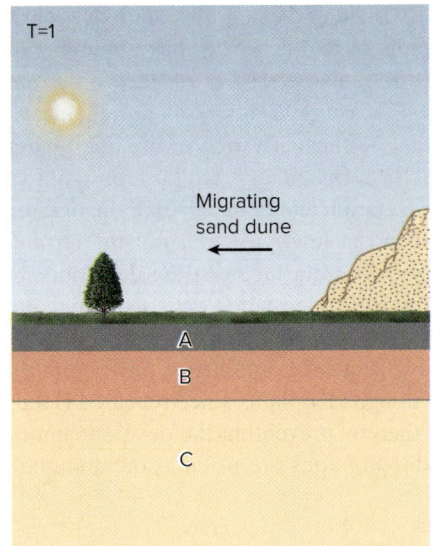

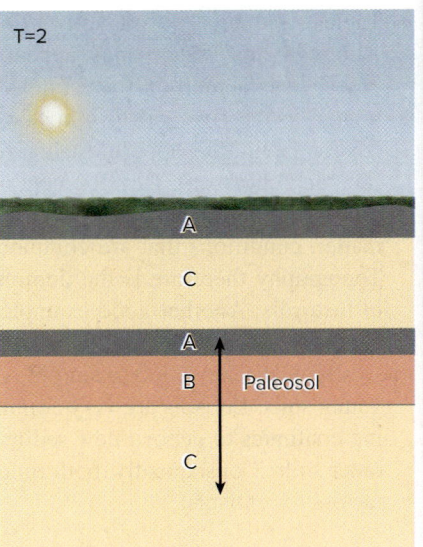

A

B

FIGURE 10.12 Illustration (A) showing how a paleosol forms when new sediment is deposited over an existing soil sequence, creating an important time marker that can be dated by carbon-14 radiometric dating. Photo (B) shows a paleosol in Finland that formed when wind-blown sand was deposited over the existing landscape.
(b) © Jukka Käyhkö, University of Turku

the development of an A-B-C sequence. More deeply weathered soils require about 5,000 to 10,000 years to form, whereas intensely weathered tropical soils that are highly enriched in aluminum need approximately 100,000 years.

At any point in the development of a soil sequence, certain geologic events, such as floods, volcanic eruptions, and migrating sand dunes, can quickly bury the soil with new sediment, generating what is referred to as a **paleosol** (Figure 10.12). Once the old soil is buried, a new sequence of horizons will begin to form, essentially resetting the clock on soil formation back to zero. Since paleosols represent distinct time events, they are quite useful in many types of geologic investigations. Moreover, because the burial helps preserve organic matter contained within the A horizon, scientists are able to use carbon-14 and radiometric dating techniques (Chapter 1) to accurately date the burial event. Although the upper limit of carbon-14 dating is only about 50,000 years, paleosols are useful in environmental studies to help determine the age of relatively recent hazardous events, such as floods, earthquakes, and volcanic eruptions. By dating events that occurred prior to written history, scientists can more accurately determine the recurrence intervals of certain types of hazardous processes.

Classification of Soils

Because of the complex way in which the five soil-forming factors interact with each other, a wide variety of soil types are found on different landscapes around the world. Both scientists and engineers have tried to make sense of the different soil types by grouping or classifying them based on common characteristics. For example, some soils have A horizons that are exceptionally rich in organic matter, whereas others have thick and highly weathered B horizons. Because scientists and engineers define soils differently and tend to be interested in different characteristics, they have developed separate soil classification systems. In this section we will briefly explore the two systems.

Soil Science Classification

The classification system used by soil scientists, called *soil taxonomy*, is based on the characteristics of the horizons found in a particular soil as well as the soil's temperature and moisture regime. This system, listed in

Table 10.1, breaks soils down into 12 different categories called *orders*. Although this system may appear complicated due to the large number of orders, it is actually rather simple since the characteristics of each order are related to the five soil-forming factors. For example, histosols are characterized as having an uppermost horizon that is unusually rich in organic matter (i.e., an O horizon). Development of such an organically rich horizon requires saturated conditions that promote lush vegetation and help preserve the organic matter, conditions that are commonly found in low-lying areas of the terrain. Topography therefore is the dominant soil-forming factor in the development of histosols. Another good example is the soil order called entisols, which are characterized by a general lack of subsurface horizons. This means that entisols typically have only a simple A and C sequence, and hence must be fairly young soils. Entisols are very common along floodplains since repeated flooding continues to deposit new sediment, thereby preventing the development of older soils. Consequently, both time and topography are listed as the dominant factors for entisols.

TABLE 10.1   Simplified version of the soil classification system used by soil scientists. Soils are broken down into 12 major categories called orders, which are based on the characteristics of different horizons found within a soil. Although the dominant soil-forming factor (or factors) is listed for each soil order, all five factors are involved in the development of any soil.

Order	Simplified Description	Dominant Soil-Forming Factor(s)
Alfisols	Soils that are not strongly leached and have a subsurface horizon of clay accumulation. Common in forested areas where the climate is humid to subhumid.	Climate and living organisms
Andisols	Soils that form in volcanic ash and contain aluminum-rich silicates that actively bind with organic compounds.	Parent material
Aridisols	Soils that form in dry climates, are low in organic matter, and often have subsurface horizons with salt accumulations.	Climate
Entisols	Young soils lacking subsurface horizons because the parent material recently accumulated or because of constant erosion. Common on floodplains and steep mountain terrain.	Time and topography
Gelisols	Weakly weathered soils that contain permafrost in the profile. Common in higher latitudes.	Climate
Histosols	Soils with a thick organic-rich O horizon that contains very little mineral matter (e.g., quartz and clay). Common in poorly drained areas.	Topography
Inceptisols	Soils that have weakly developed subsurface horizons because either they are young or the climate does not promote rapid weathering.	Time and climate
Mollisols	Soils that are not strongly leached and have an organic-rich A horizon. Common in grasslands where the climate is semiarid to subhumid.	Climate and organisms
Oxisols	Very old, extremely leached and weathered soils with a subsurface accumulation of iron and aluminum oxides. Common in humid tropical climates.	Climate and time
Spodosols	Soils that have a well-developed B horizon rich in iron and aluminum oxides. Form in cold, moist climates under pine vegetation and sandy parent material.	Parent material, organisms, and climate
Ultisols	Strongly leached soils (but not as strongly leached as oxisols) with subsurface accumulation of clay. Common in humid tropical and subtropical climates.	Climate, time, and organisms
Vertisols	Soils that develop deep, wide cracks when dry due to the presence of swelling clays.	Parent material

Source: Modified after Brevik, *Journal of Geoscience Education*, v. 50, n. 5, 2002.

The relationship between soil orders (Table 10.1) and the five soil-forming factors is nicely illustrated by the map shown in Figure 10.13. Perhaps the most obvious relationship here is the one between climate and the extensive belt of gelisols found in the high-latitude regions of North America. Because gelisols are characterized by having permafrost within the soil profile, temperature is clearly the most dominant factor in the development of these soils. Another good example is the relationship between deserts and aridisols, which are notable in that they have a low organic content. The major control here is how the lack of rainfall greatly limits vegetation growth, hence the low organic content of aridisols. Finally, note how ultisols cover almost the entire southeastern portion of the United States, where the landscape is geologically old and the climate is quite warm and humid. This fits nicely with the fact that ultisols are highly leached and have thick B horizons, characteristics that require considerable amounts of both time and rainfall.

Engineering Classification

It is important to emphasize here that scientists and engineers use the term *soil* quite differently. To scientists, soil is considered to be the narrow zone of fragmental material near the surface where physical and chemical processes have created soil horizons. In contrast, engineers simply view soil as any type of fragmental earth material (i.e., nonbedrock), which is what geologists refer to as sediment (transported) or regolith (untransported). Although the focus in this chapter is on the scientific view of soils, the engineering classification is also presented for the benefit of those students who are also learning about soils from an engineering perspective.

Engineers are clearly more interested in the physical properties of fragmental earth materials (sediment and regolith) than in knowing how and where these materials form in nature. Consequently, the system developed by engineers, called the *Unified Soil Classification System*, is based on physical properties. A simplified version of this system is presented in Table 10.2. Note how soils are classified based primarily on the proportion of gravel, sand, silt, and clay-sized particles. Also notice how the soil names are quite descriptive. Take *clayey sand*, for example, which indicates a material that consists mostly of sand-sized grains, but yet contains a significant amount of clay particles. On the other hand, the modifier *well-graded* means that the soil contains a diverse range of particle sizes, whereas *poorly graded* implies more uniform-sized particles (note that the geologic term *sorting*, used to describe the distribution of grain sizes, has the opposite meaning of the engineering term *grading*).

In addition to being descriptive, this classification system is practical for engineering purposes since many other soil properties are directly related to a material's grain-size distribution. Therefore, once a soil has been properly classified, an engineer can get a pretty good idea of its general

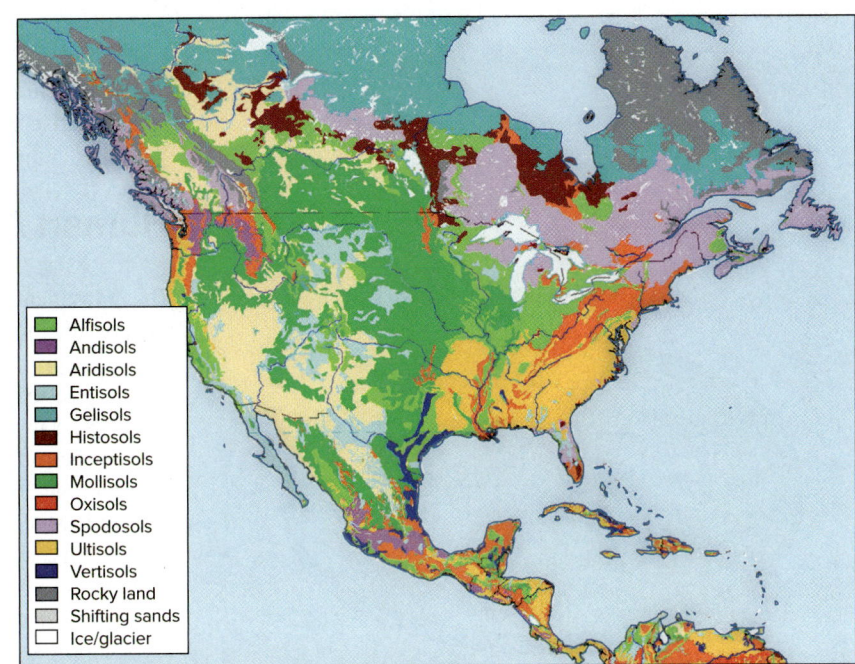

FIGURE 10.13 Map showing the distribution of soil orders in North America. Many of the patterns shown here are related to variations in climate and geologic history, both of which strongly influence soil formation.

Legend:
- Alfisols
- Andisols
- Aridisols
- Entisols
- Gelisols
- Histosols
- Inceptisols
- Mollisols
- Oxisols
- Spodosols
- Ultisols
- Vertisols
- Rocky land
- Shifting sands
- Ice/glacier

TABLE 10.2 Simplified version of the Unified Soil Classification System used in engineering. In this system soils are actually sediment or regolith and grouped primarily according to the proportion of different grain sizes (gravel, sand, silt, and clay).

Major Divisions	Subdivisions	Soil Name	Symbol
Coarse-Grained Soils (>50% of grains visible with naked eye)	Gravels	Well graded gravel	GW
		Poorly graded gravel	GP
		Silty gravel	GM
		Clayey gravel	GC
	Sands	Well graded sand	SW
		Poorly graded sand	SP
		Silty sand	SM
		Clayey sand	SC
Fine-Grained Soils (<50% of grains visible with naked eye)	Silt	Plastic silt	ML
		Nonplastic silt	MH
		Organic silt	OL
	Clay	Low-plastic clay	CL
		High-plastic clay	CH
		Organic clay	OH
Organic Soils	Organic Matter	Peat	PT

properties simply from its name. For example, one can predict that a poorly graded (i.e., well-sorted) sand would be more permeable than a clayey sand. In the next section we will take a closer look at how soil properties, as viewed by both scientists and engineers, are important in terms of human activity.

Human Activity and Soils

The introduction to this chapter pointed out that soils are critical to humans because they form the basis of our food supply and provide minerals that are raw materials for producing such things as aluminum metal. However, the importance of soils goes beyond just food and aluminum. For example, nearly every construction project involves excavation, or digging into loose material (regolith or sediment). The properties of a particular material will determine how hard it will be to dig a hole, as well as the hole's tendency to collapse and the speed at which it fills with water. Soil properties are also an important consideration for building foundations, which must be designed so that the weight of a building can safely be supported by the underlying soil. One of the first tasks for an engineer in any large construction project is to determine key soil properties at the site. Measuring soil properties is also important at contaminated sites, where it is necessary to know the rate and direction in which pollutants will migrate (Chapter 15). In this section we will examine some of the key soil properties and consider how they affect the ways in which people interact with soils. To avoid confusion over the different definitions of soil used in science and engineering, we will equate soil properties with regolith properties, which is the general term that describes any type of fragmental earth material.

Soil Properties

Recall that Earth's crust is largely composed of silicate minerals, many of which are transformed into various clay minerals during chemical weathering. An important exception is the mineral quartz. Quartz is not only very abundant in granitic rocks, it is also highly resistant to chemical weathering, which explains why most sand and silt-sized particles are composed of quartz. Therefore, because of the weathering characteristics of silicate rocks, soils tend to be chiefly composed of quartz and clay mineral particles along with lesser amounts of organic material.

Also recall that clay particles are exceedingly small compared to sand and silt-sized grains of quartz. In addition to size, quartz and clay minerals are quite different in terms of their crystalline structure (i.e., internal arrangement of atoms). Because the atoms in clay minerals are arranged in complex sheetlike structures, individual clay particles are plate-shaped and have a negative electrical charges on their outer surfaces. The surfaces of quartz grains on the other hand are electrically neutral. Later in this section you will see how the particle size and electrical properties of quartz and clay minerals play a major role in many soil properties.

Porosity

As illustrated in Figure 10.14, sediment naturally contains void space or *pore space* between the solid particles. Geologists use the term **porosity** to describe the fraction of sediment or rock that consists of void space. In terms of total volume, soils average about 45% mineral matter and 5% organics; hence, porosity averages around 50%. Since water can infiltrate a soil and fill the pore spaces, this means that about 50% of a soil's total volume can be taken up by water. Note that when the pores become completely filled with water, the soil is said to be *saturated*, but when the pores contain both air and water, it is considered

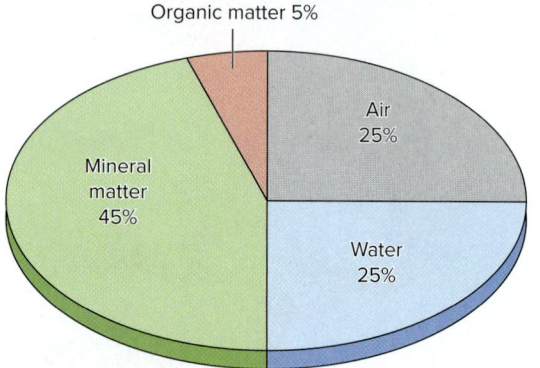

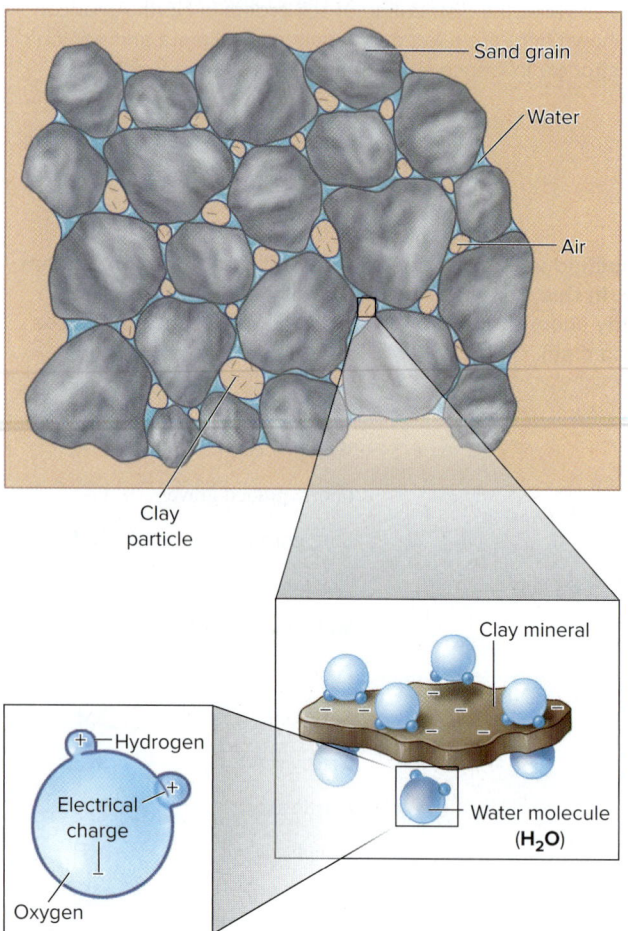

FIGURE 10.14   Soils consist of about 45% mineral matter and 5% organics, with the remaining 50% being void space that is filled with air and water. Within the pores, dipolar water molecules are strongly attracted to the extremely small clay-mineral particles, whose crystal structure results in negative charges on their outer surfaces.

unsaturated—rarely are soils completely dry. One of the reasons porosity is an important property is because it indicates the amount of water that is potentially available for plants, particularly agricultural crops.

Soil Moisture and Drought Resistance

Some soils naturally lose water from their pore space more easily than others, making certain crops more susceptible to being lost or damaged during a drought. The *drought resistance* of soils is controlled by their mineral composition and the dipolar nature of the water molecule (H_2O). As illustrated in the inset for Figure 10.14, the orientation of hydrogen and oxygen atoms in the water molecule causes one side of the molecule to be slightly positive and the other side slightly negative. Although water is electrically neutral overall, this uneven distribution of electrical charge results in individual water molecules being attracted to one another—scientists use the term *cohesive* force to describe this attraction between similar molecules. Water molecules are also attracted to other types of molecules, such as those of a solid surface, in which case the attraction is referred to as an *adhesive* force. For example, as shown in Figure 10.15, a hot shower will cause water vapor to build up to the point that water droplets begin falling from the ceiling. Initially a thin film of water forms on the ceiling as water molecules attach themselves through adhesive forces. Additional molecules then become attached to this film of water by cohesive forces. As the number of layers of water molecules continues to grow, gravity pulls the layers downward, creating a water droplet. Eventually gravity overcomes the cohesive forces, at which point the droplet falls, leaving some water molecules firmly attached to the ceiling by adhesive forces.

The simple example of water droplets falling from a ceiling is a useful analogy to how water moves through pore spaces within a soil. In the case of soils, the different sizes and electrical properties of quartz and clay-mineral grains play a critical role. For example, imagine we have two soils that are completely saturated, with one composed mostly of clay minerals and the other mostly quartz. Because the surfaces of the clay particles are electrically charged, whereas those of quartz grains are not, the adhesive forces between water and the clay particles are much stronger. Moreover, clay particles are extremely small compared to quartz sand grains, which provides significantly more surface area for water to be held by adhesive forces. Within the large pores of the sand-rich soil, much of the water is held in place by cohesive forces between individual water molecules since the adhesive forces between water and the quartz grains are not particularly strong. In contrast, water within the tiny pores of the clay-rich soil is held there by very strong adhesive forces.

Now imagine that we allow our two saturated soils to suddenly drain due to the downward force of gravity. Because the forces holding water in the soil composed mostly of quartz sand are rather weak, much of the water will drain quite easily, making this soil susceptible to drought. On the other hand, very little if any water will drain from the clay-rich soil due to the strong attraction between water and clay particles. This is what makes a clay-rich soil rather susceptible to drought. Here plants commonly begin to wilt when they are unable to pull water from the clay particles, even though considerable amounts of water remain. One would think that an ideal soil in term of drought resistance would be one with an even mixture of sand and clay. It turns out that the most drought-resistant agricultural lands are soils with a high percentage of silt and moderate amounts of sand and clay. These soils, referred to as *loam*, are ideal because they retain more water for plants than a pure sand soil, but yet give up their water more freely than a soil composed of all clay.

Permeability

The concept of water being held by cohesive and adhesive forces within a soil is also useful for understanding the property called **permeability,** which is the ability of a fluid to flow through interconnected pore space. Soils rich in clay

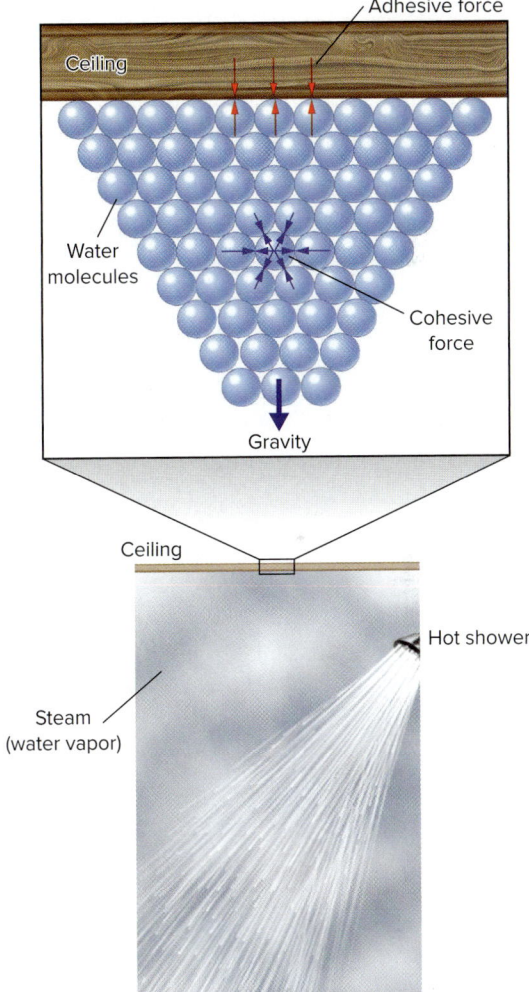

FIGURE 10.15   Illustration showing how a water droplet is composed of individual water molecules that are attracted to one another by cohesive forces. The droplet is attached to the ceiling by adhesive forces that exist between the solid surface and the water molecules. Additional water molecules will cause the droplet to grow in size until gravity overcomes the cohesive forces, at which point the droplet will fall. These same forces operate in soils and control the ability of water to flow through the pore spaces.

minerals have low permeability since the strong adhesive forces within the soil greatly inhibit the flow of water. In contrast, the permeability of sand-rich soils is high because much of the water is held by relatively weak cohesive forces, and thus can flow much more freely. Anything then that increases the fraction of water being held by the weaker cohesive forces will also cause permeability to increase. Consequently, the most permeable soils are those with large pores (e.g., coarse sands) and with grains of uniform size so that smaller grains do not plug the pore spaces and restrict the flow. Although similar forces are involved, drought resistance refers to the *amount* of water in the soil that is available to plants, whereas permeability refers to the ability of water to move through the soil.

In terms of human interactions with the landscape, permeability is important because it determines how fast soils are able to drain. For example, suppose a heavy rain saturates the upper soil horizons in a farmer's field. If the water does not drain quickly enough so that air can get back into the soil, certain types of crops may drown. Clay-rich soils are more prone to this type of crop damage because they drain more slowly. Excessive water can also lead to pest and disease problems. In addition, permeability affects the proportion of rainwater that is forced to flow across the landscape (overland flow), which is a major factor in the flooding characteristic of a drainage basin (Chapter 8). Likewise, permeability plays a role in many mass wasting events (Chapter 7) where infiltrating water decreases the stability of slopes by reducing the frictional forces within the subsurface.

Plasticity

Related to the discussion on plastic deformation in Chapter 4, **plasticity** is the ability of a material to deform without breaking when a force is applied—similar to how modeling clay can be shaped into various forms, but not break. In terms of soil, plasticity generally increases with clay content depending on the specific type of clay minerals present. On the other hand, plasticity always decreases as sand content increases, with pure sand having essentially no plasticity. Another important factor affecting plasticity is water. When some fine-grained soils take on too much water, they can start to flow like a liquid. On the other hand, if allowed to dry completely, many clay-rich soils will behave more like solid rock, in which case plasticity approaches zero (similar to how modeling clay becomes hard when it dries). Water therefore has the ability to drastically change the plasticity of soils, which, in turn, can have profound implications in engineering. For example, the weight of a large structure, such as a tall building, dam, or bridge, exerts a tremendous vertical force on the subsurface. If the plasticity of subsurface materials increases because of additional water, they could begin to flow, thereby threatening a structure's integrity. Therefore, when designing foundations for structures, engineers take great interest in soil properties and moisture conditions around the site.

Strength and Sensitivity

The **strength** of soil refers to its ability to resist being deformed or, in other words, how well its particles stick together. For example, if you were to dig a trench, the walls would be less likely to collapse in a strong soil compared to a weak soil. Note that the strength of a soil depends on the cohesive and frictional forces that exist between the grains. Of interest here is the fact that some soils experience a loss of strength when they are suddenly disturbed. Engineers therefore use the term **sensitivity** when referring to how easily a soil will lose its strength when disturbed. For example, if a saturated sandy soil is shaken during an earthquake, the individual sand grains can easily lose contact with one other. When this occurs, the grains become suspended in the water, causing the material to behave as a liquid, a process called *liquefaction* (Chapter 5). During the time the soil is liquefied, it has virtually no strength, which allows heavy objects to sink into the subsurface. Note that although water binds tightly to most clay minerals, certain types of clay minerals known as *quick clays* are highly sensitive to vibrations and will readily undergo liquefaction.

Compressibility

We can define **compressibility** as the ability of a material to compact and reduce its volume when placed under a force or load. In general, the compressibility and volume reduction of soil increase as the clay-mineral content increases. As shown in Figure 10.16A, when a load is placed on a soil containing only quartz grains, the grains will reorient themselves so that the pore space is at a minimum. Once this occurs, the sand will be nearly incompressible since there is no further room for movement, plus the quartz grains themselves are strong and incompressible. In contrast, when a heavy load is placed on a clay-rich soil where the plate-shaped particles are more randomly oriented (Figure 10.16B), the particles will rearrange themselves in a parallel manner. This process is facilitated by water in the soil that exerts pressure within the pore space, which tends to keep the particles apart, making it easier for them to rearrange and become more tightly packed. Compaction in clay-rich soils therefore can be quite significant, resulting in considerable volume reduction that can lead to settling problems at the surface. Perhaps the most famous example is the Leaning Tower of Pisa in Italy (Figure 10.16C). Variations in the clay content of the underlying soils, combined with the weight of the tower, led to different amounts of compaction, which caused the structure to lean.

Clearly, compressibility and compaction of soils is important to engineers when designing the foundation supports for a heavy structure. Compressibility is also of great concern to farmers, particularly since compaction is essentially irreversible. Agricultural fields can undergo compaction due to the use of heavy farm equipment, which permanently reduces the ability of soils to drain and makes it more difficult for roots to penetrate the soil. As could be expected, compaction is more of a problem in wet, clay-rich soils. To help minimize compaction, farm vehicles are equipped with extra wide tires, and farmers generally try to avoid running heavy equipment when their fields are overly wet.

Shrink-Swell Potential

Like many types of materials, soils swell or expand when they take on water, then shrink as they dry out—similar to how a sponge expands and contracts. Clay-rich soils generally have the greatest capacity to shrink and swell due to the small particle size and the fact that water molecules have such a strong attraction to clay minerals. As noted earlier in this section, clay minerals have a sheetlike structure dominated by aluminum, silicon, and oxygen atoms. Because the atomic structure of these sheets varies among the different types of clay minerals, it turns out that each clay mineral has a different ability to store water between its sheets. The term **swelling clays** or **expanding clays** refers to those clay minerals capable of

FIGURE 10.16   When a heavy load is applied to a soil, the individual grains will attempt to rearrange into a more tightly packed configuration. Sandy soils (A) have low compressibility because the reduction in volume that can occur when rounded grains are rearranged is relatively small. Clay-rich soils (B) are highly compressible because the random orientation of small clay particles allows for a significant reduction in volume. The uneven settling of the Leaning Tower of Pisa (C) was caused by differences in compaction that were related to variations in the clay content of the soils.
(c) © Goodshoot/Fotosearch

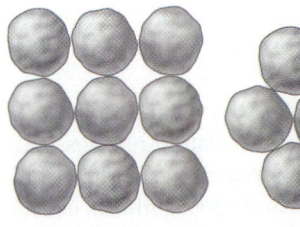
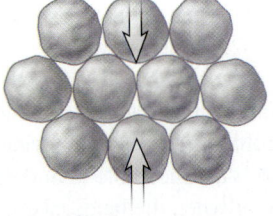
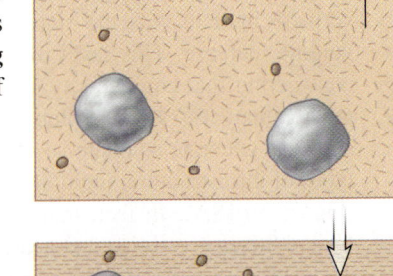
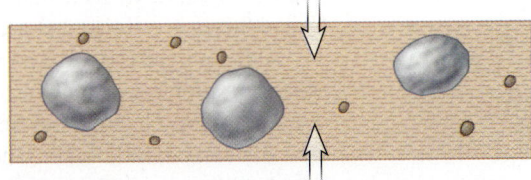

A   **B**   **C**

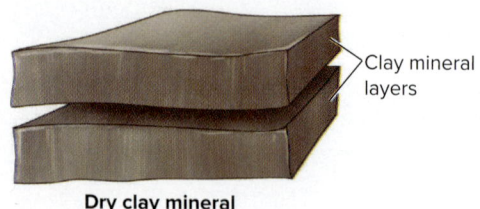

Dry clay mineral

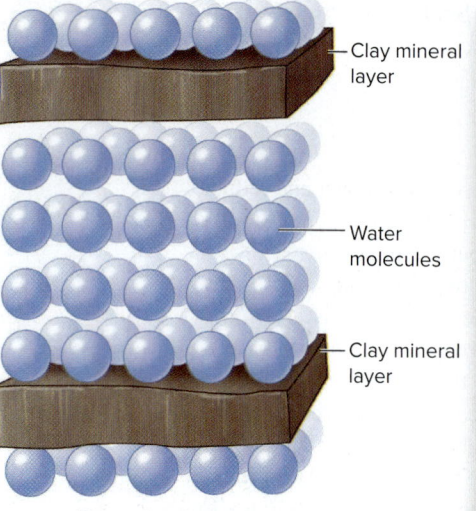

Expansion due to absorption of water

Clay mineral layers

Clay mineral layer

Water molecules

Clay mineral layer

Camera

FIGURE 10.17   The number of water molecules that can be held within the sheetlike structure of clay minerals varies among different types of clays. Expanding clays have a great capacity to take on water, causing soils called vertisols to increase in volume. When vertisols are allowed to dry, they shrink considerably, creating cracks, like those in the dried-out lake bed shown in this photo—note the yellow camera for scale.
© Paul McDaniel, University of Idaho

incorporating large numbers of water molecules within their structure, thereby producing significant volume changes, as illustrated in Figure 10.17. Note that the presence of expanding clays is a characteristic feature of soils known as vertisols (refer to Table 10.1).

Most of the swelling clays found in vertisols belong to a family of clay minerals referred to as *smectites*, notable for their great capacity to take on water and expand. For example, a pure deposit of the smectite mineral called *montmorillonite* can expand up to 15 times its original volume. Because of their ability to absorb water, humans have found a variety of applications for expanding clays. Perhaps the most familiar application is the dried material used in the litter boxes for cats and for cleaning up spills involving oil, antifreeze, and other liquids. Smectite clays are also commonly used as an additive that increases the density of fluids used in the drilling of water and oil wells. After the drilling is complete, expanding clays are then used to create an impermeable seal around the metal or plastic casing of the well.

Ironically, the same properties of expanding clays that people find useful can sometimes cause serious problems for society. For example, the map in Figure 10.18 shows areas in the United States where vertisols contain significant amounts of expanding clays; hence the ground has the capacity to shrink and swell as soil moisture conditions change. Note that the different colors in this map indicate the *shrink-swell potential* of the ground, which is related to both the amount and type of expanding clays present in the soil. It is not uncommon for vertisols in areas with the highest shrink-swell potential to experience a 25–50% volume increase. This level of volume change is highly significant, especially considering that engineers typically require specialized foundations for changes greater than 3%. Large volume changes present problems because expanding soils can generate as much as 20,000 pounds per square foot of force on various types of infrastructure, including sewer and utility lines, highways, and buildings (Figure 10.18A). In fact, damage from expanding and contracting soils in the United States is estimated at nearly $2 billion per year.

Although vertisols can be found across broad areas of the landscape, fewer expansion problems occur in humid climates since soil moisture remains relatively constant. The most severe problems are found in drier climates where soils typically contain little water. The soils are then able to undergo significant expansion during occasional wet periods, or when the landscape is put under irrigation.

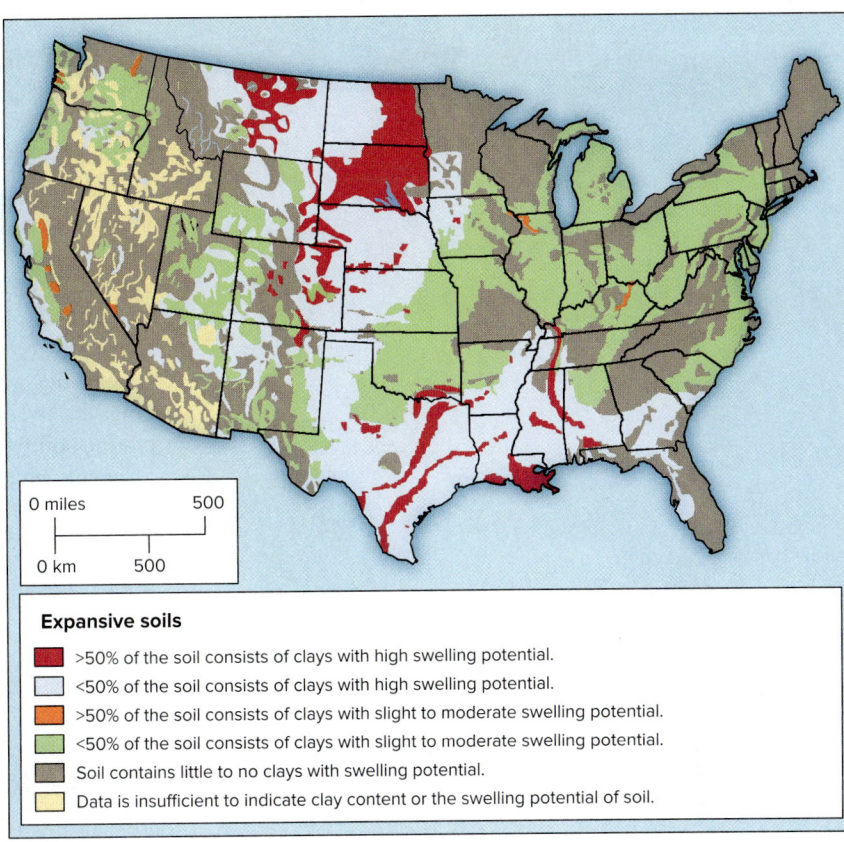

Expansive soils

- >50% of the soil consists of clays with high swelling potential.
- <50% of the soil consists of clays with high swelling potential.
- >50% of the soil consists of clays with slight to moderate swelling potential.
- <50% of the soil consists of clays with slight to moderate swelling potential.
- Soil contains little to no clays with swelling potential.
- Data is insufficient to indicate clay content or the swelling potential of soil.

B

FIGURE 10.18   Soils known as vertisols contain significant amounts of expanding clays, which can cause serious structural damage should the soil go through repeated drying and wetting cycles. Photos (A) illustrate the types of damage that can occur when the underlying soil expands and contracts. Map (B) showing areas in the United States where soils have a high swelling potential. (a) (top): Colorado Geological Survey/Photo by David Noe; (bottom): P. Camp, USDA-NRCS

A

Structural damage associated with expanding soils can largely be avoided if a proper soil investigation is undertaken prior to construction. Should the investigation uncover a potential problem, specialized construction techniques can be used to avoid damage when the soil eventually does expand. Another effective technique is to ensure that any landscaping around buildings is done in such a way that soil moisture changes are kept to a minimum, thereby reducing the potential stress on the structure.

Ion Exchange Capacity

Recall that when minerals undergo chemical weathering, charged atoms and groups of atoms called *ions* are released into the water. Some of the ions, such as potassium (K^+) and calcium (Ca^{2+}), are important in soils because they serve as essential nutrients for plants. Other important nutrients include nitrate (NO_3^-) and ammonium (NH_4^+) ions that form not by chemical weathering of minerals, but by the interaction of atmospheric gases and biological processes within the soil. **Ion exchange** is the process whereby dissolved ions attach themselves to soil particles, and are then removed in a selective manner by growing plants and by water moving through the soil zone. This process in which various ions are stored and then removed from soil particles is critical to plant growth and crop production.

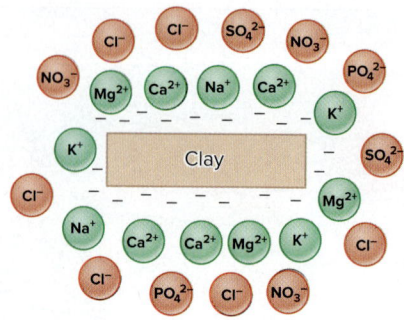

A

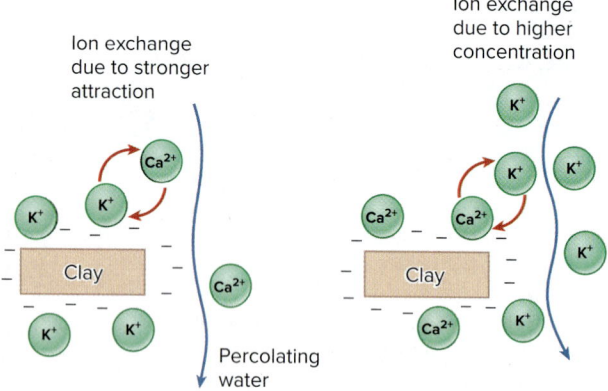

Ion exchange due to stronger attraction

Ion exchange due to higher concentration

Percolating water

B

FIGURE 10.19   Positively charged ions naturally attach themselves in a layer (A) around the negatively charged surfaces of particles of clay minerals and organic matter within a soil. Notice how negative ions surround the positive ions. As percolating water (B) carries ions through the soil, they selectively exchange with the ions already attached to the soil particles. The actual exchanges that occur depend on the attraction and concentration of ions in the percolating water.

As shown in Figure 10.19, dissolved ions will attach themselves in a layered manner around clay-mineral and organic particles in a soil. Notice that when water flows through a soil, there are two basic ways ion exchange can take place between the water and soil particles. The first is when water contains ions that have a stronger attraction to the soil particles than the ions already attached. For example, because of their smaller size and greater charge, calcium ions (Ca^{2+}) have a stronger attraction to soil particles than do potassium ions (K^+). Therefore, when calcium-rich water flows through a soil, potassium ions are removed from soil particles and replaced by calcium ions. Ion exchange can also work in reverse, where a strongly held ion (e.g., calcium) is removed from the soil and replaced by an ion with a weaker attraction. For this to occur, the water must contain a very high concentration of the weaker ion so that the overwhelming numbers of that ion force the more tightly held ions from the soil particles.

Finally, note that scientists use the term *ion exchange capacity* to describe the ability of a particular soil to exchange ions with water. Soils with a high ion exchange capacity naturally contain significant amounts of electrically charged clay and organic particles. In the next section we will examine how ion exchange plays a critical role in soil fertility and agricultural production. In Chapter 15 we will examine how this exchange process is important in the immobilization of contaminants within the soil zone, thereby minimizing the spread of pollution.

Soil as a Resource

Earlier in this chapter it was noted that land plants within the biosphere would not exist were it not for soils. Of particular importance to society are agricultural plants such as wheat, rice, and corn since they make up the bulk of the world's food supply. In this section we will explore the connection between soils and our food supply and then discuss some other important resources we obtain from soils.

Agricultural Food Production

The single most important use of soils in society is for growing food. Soils are commonly referred to as being *fertile* in areas where crops grow particularly well and infertile where crop production is low. In addition to a certain range of water and temperature conditions, different types of plants require a particular set of elements in order to function and grow properly. **Soil fertility** is the term used to define the ability of a soil to supply the elements necessary for plant growth. The most critical elements, often referred to as **essential nutrients,** include nitrogen (N), phosphorus (P), potassium (K), calcium (Ca), magnesium (Mg), and sulfur (S). These elements come from the weathering of minerals and the interaction of atmospheric gases and biological processes within the soil. They are then stored within the soil in the form of electrically charged ions, which plant roots selectively remove depending on the needs of a given plant. For example, because some plants have a relatively high demand for calcium, they will selectively remove calcium ions and leave the other types of ions attached to the soil particles.

The ion exchange capacity of a soil is clearly tied to soil fertility as it helps determine whether essential nutrients in the form of charged ions are available in the soil. As indicated in Figure 10.20, the key to this is the amount of nutrients being produced by weathering and biological activity and the amount of electrically charged clay and organic particles within the soil. Soils that are rich in organic matter and have a moderate amount of clay are naturally fertile as they contain a large number of exchange sites for storing nutrients. However, if there

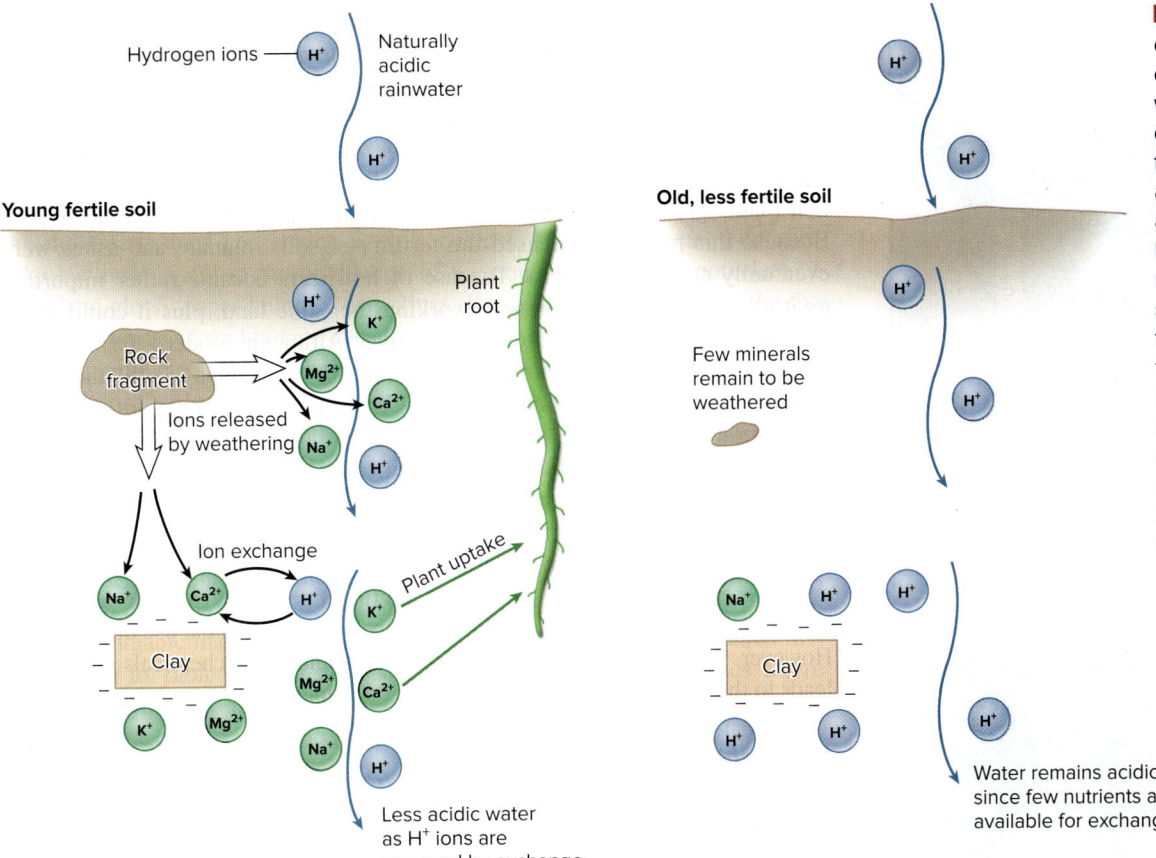

FIGURE 10.20   Young soils commonly have an abundance of rock and mineral fragments, which weather and produce dissolved ions that are essential for plant growth. These nutrients attach themselves to clay minerals and eventually exchange with hydrogen ions in rainwater, making it less acidic. In older soils few fragments remain to be weathered, and thus fewer nutrients are produced. Percolating rainwater then remains acidic since the ion exchange sites are occupied by nonessential elements and hydrogen ions, rendering the soil infertile and acidic.

are few nutrients moving through the soil, then fertility will be low regardless of the number of exchange sites. Older soils are generally not very fertile because chemical weathering has progressed to the point where few minerals remain that can release important ions such as potassium, calcium, and magnesium. When certain nutrients are no longer being released, their continued removal by plants combined with the flushing action of naturally acidic rainwater eventually leaves the exchange sites occupied by nonessential elements and hydrogen (H^+) ions (Figure 10.20). Old soils therefore, such as oxisols and ultisols (refer to Table 10.1), are characterized as being both infertile and acidic (i.e., containing numerous H^+ ions).

Long before scientists learned about ion exchange and soil nutrients, people recognized that there was a relationship between crop production and the amount of organic matter in the upper soil horizon, called topsoil. Because clay minerals alone can provide a sufficient number of ion exchange sites, organic material must play an additional role in soil fertility. Modern scientists have learned that organic matter acts as a food source for bacteria and fungi within the soil, which are needed to break down complex organic materials into simpler substances plants can use directly. Although the amount of organic material in topsoils may be relatively small, it is critical because it helps produce certain essential nutrients that are not available through the chemical weathering of minerals. Of particular importance is how biological activity within soil converts atmospheric nitrogen (N_2) into ammonium (NH_4^+) and nitrate (NO_3^-) ions, which are the essential forms of nitrogen needed by plants. The breakdown of organic matter within a soil also produces nitrogen as well as sulfur and other important nutrients. Consequently, soils simply will not support much plant life without organic matter.

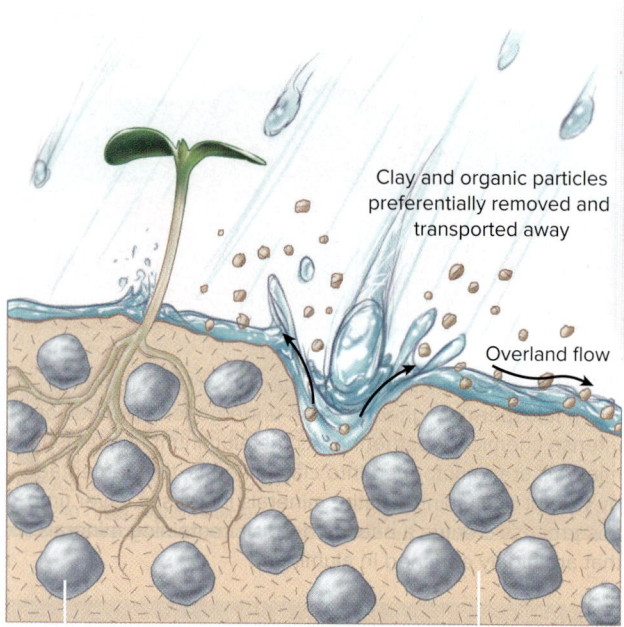

FIGURE 10.24 The impact of raindrops on an unprotected soil creates an explosive effect that preferentially ejects clay and organic particles onto the surface. Soil erosion occurs when the loose particles are transported by wind or water moving downslope as overland flow. Over time this process reduces soil fertility.
USDA

have when they make impact with the soil. In addition to acting as a shield, plants have roots that help hold the soil in place. As illustrated in Figure 10.24, when a raindrop hits bare soil, it results in a small impact crater. The ejected soil particles are then easily picked up by wind or water flowing across the landscape. This movement or transport of soil particles away from their place of origin is called **soil erosion.** Keep in mind that water and wind preferentially transport the smaller and lighter organic and clay-mineral particles within the soil, leaving behind the much larger sand grains. This preferential loss of organics and clay minerals is a serious problem, as these particles are the key to soil fertility.

Soil erosion is actually a natural process that has been taking place for hundreds of millions of years. Long before people began inhabiting the planet, there would always be certain parts of the landscape experiencing severe erosion as fire or disease destroyed the natural vegetation cover. But in many parts of the landscape, soil formation and erosion were in balance such that there was no net loss of soil for long periods of time. Today, however, the natural vegetation cover over much of the landscape has either been removed or disturbed by human activities that include farming, logging, and construction. Our use of the landscape has caused soil erosion to increase to a point that it far exceeds the rate of soil formation in many areas, resulting in a net loss of soil called **soil loss.** From the graph in Figure 10.25 you can see how erosion and soil loss vary over time in response to land-use changes in a particular area. Although soil loss can result from a variety of human and natural causes, our interest in the next section will be on those related to human activity. In particular we will focus on agricultural land use as it leads to the most serious consequence, namely a drop in food production.

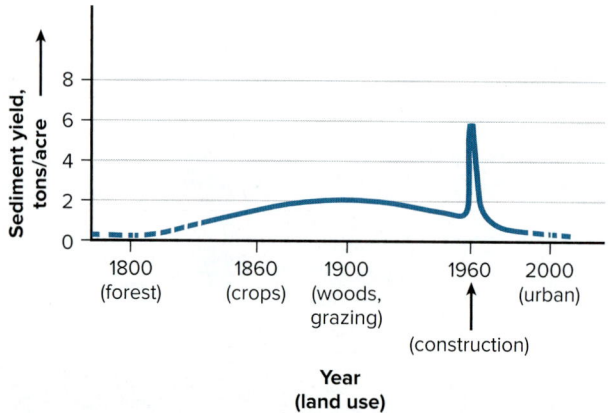

FIGURE 10.25 Graph showing how sediment loss changed over time in response to different land uses at a site in the eastern United States. Note the significant increases in soil loss that accompanied changes to agriculture and a construction boom.

Consequences of Soil Loss

Soil loss presents farmers with a difficult dilemma, namely that the very act of growing crops requires that natural vegetation be removed from the landscape, allowing topsoil (A horizon) to be blown or washed off the fields (Figure 10.26). From the map in Figure 10.27 one can see that soil loss on agricultural land in the United States is not evenly distributed due to variations in the steepness of the terrain and the intensity of agricultural activity. For example, much of the Great Plains today experiences relatively low rates of soil loss simply because agriculture is limited by the dry climate. In these areas much of the land is used for grazing livestock, so soil erosion is relatively low since the cover material generally remains in place. Growing wheat is common here, which also tends to generate lower rates of soil erosion because it provides fairly good soil protection. However, plowing the natural grasslands of the Great Plains to grow wheat had at one time led to severe wind erosion and the collapse of local farm economies. The combination of irrigation and improved soil management practices greatly reduced soil loss in the region (see Case Study 10.1).

In contrast to the Great Plains, the eastern half of the United States experiences more severe soil loss because of the types of crops and the more intense level of agricultural activity. Here corn and soy beans are common row crops, both of which provide far less protection against erosion than does natural

A Southern Iowa

B Cass County, Iowa

FIGURE 10.26 Photos showing the irreplaceable loss of topsoil from agricultural fields. Soil is actively being removed during a rainstorm in (A), whereas (B) shows a small drainage channel where eroded topsoil has accumulated.
(a–b) Lynn Betts, USDA Natural Resources Conservation Service.

vegetation. Note how soil loss is not uniform in states such as Iowa and Illinois (Figure 10.27), indicating that factors other than the level of agricultural activity and types of crops are involved. The primary reason for the variation in these areas is the steepness of the terrain. Agriculture fields located on hilly terrain naturally have higher rates of erosion compared to fields on flat-lying terrain.

Another important consequence of soil loss is *sediment pollution*, which is the choking of drainage systems by excessive sediment moving off the landscape (Figure 10.28). As described in Chapter 8, the process called *overland flow* carries both water and sediment off the landscape and into stream channels. Once in a channel, sediment is transported downstream through drainage networks. The problem is that in areas of extensive soil erosion, the stream network simply cannot transport all the additional sediment, in which case the channels become choked or filled with sediment. This increases the frequency of flooding because there is less room in the channels for water. Sediment pollution also causes problems in rivers used as transportation corridors since the channels must be dredged more often in order to keep the rivers open for shipping. Finally, the natural ecology of streams is often destroyed when channels become choked with sediment. For example, fish such as bass and trout are no longer able to see their prey in the sediment-laden water, plus the areas of coarse sediment where they normally spawn become buried by finer material. Moreover, the aquatic life that once formed the basis of

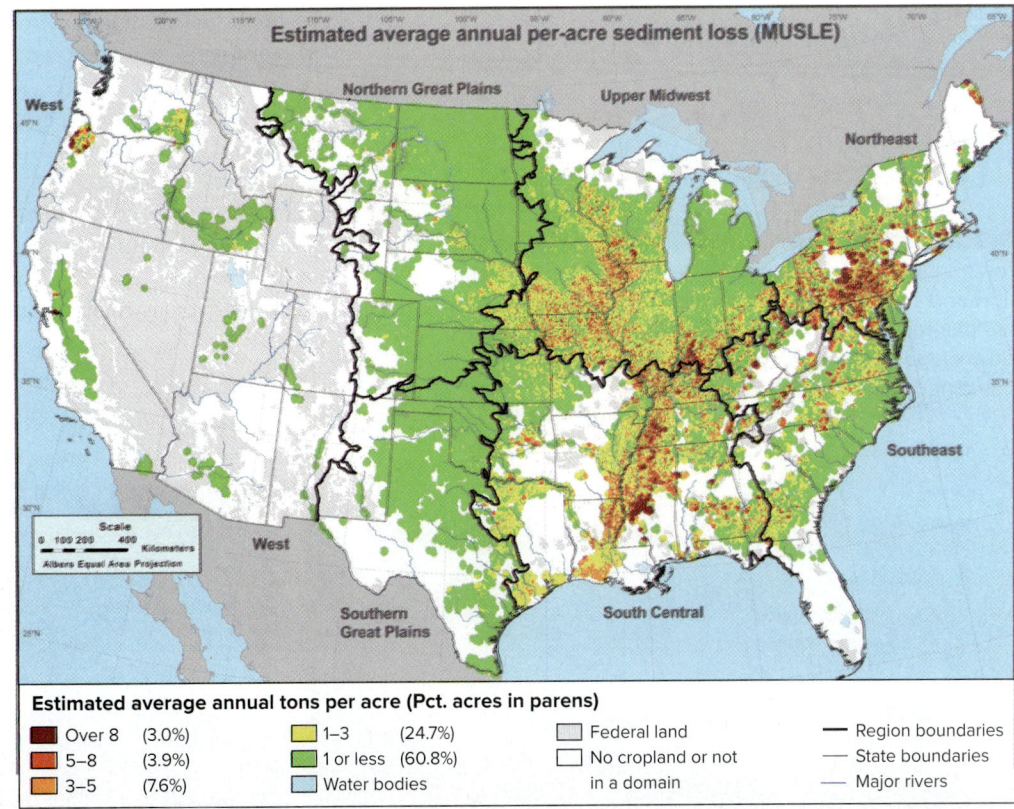

Estimated average annual tons per acre (Pct. acres in parens)

■ Over 8 (3.0%)	■ 1–3 (24.7%)	☐ Federal land	— Region boundaries
■ 5–8 (3.9%)	■ 1 or less (60.8%)	☐ No cropland or not	— State boundaries
■ 3–5 (7.6%)	■ Water bodies	in a domain	— Major rivers

FIGURE 10.27 Map showing the estimated average rate of soil loss per acre on agricultural land in the United States. Variation in soil loss is largely due to the level of agricultural activity and the steepness of the terrain.
USDA

Obion River in Central Tennessee

FIGURE 10.28   Photo showing how soil loss leads to sediment pollution where stream channels become choked with excess sediment. Sediment pollution increases the likelihood of flooding and destroys the natural ecology of streams.
Tim McCabe, USDA Natural Resources Conservation Service

FIGURE 10.29   Contour plowing (A) involves planting rows parallel to the slope of the land, reducing overland flow and soil erosion. In strip cropping (B) different crops are planted in alternate strips to trap sediment moving downslope—note in the photo how strip cropping is combined with contour plowing.
(a–b) Tim McCabe, USDA Natural Resources Conservation Service

A Southwest Iowa

B Iowa-Minnesota border

the stream's food web is also greatly impacted. Sediment pollution therefore creates a rather sterile environment in which only a small fraction of the original species survive.

Mitigating Soil Loss

Because both soil loss and sediment pollution are detrimental to society, various techniques have been developed over the years to minimize the problems. These techniques are generally intended to either keep the soil in place so it does not blow away or move downslope, or trap the sediment before it can enter the drainage network. In many situations these two basic approaches are used in conjunction with one another in order to minimize both soil loss and sediment pollution. In this section we will briefly explore the more common techniques, some of which were developed in response to severe wind erosion on the Great Plains during the 1930s (Case Study 10.1). Although most of the following techniques are associated with agriculture, some also address problems resulting from construction and logging activities.

Contour plowing is an old practice whereby farmers plant rows of crops parallel to the contours of the land surface. Although contour plowing originally came about because people naturally found it was easier to plow parallel to the slope, it also happens to be very effective in reducing soil erosion. From Figure 10.29A one can see how the rows reduce the ability of water to flow directly down the slope, thereby helping to keep soil particles from being carried off the fields. Contour plowing is particularly important in areas of rolling or hilly terrain where cultivated fields have more of a slope. This technique is also important in more arid climates as a means of increasing infiltration and soil moisture.

Strip cropping, where crops are planted in strips within the same field (Figure 10.29B), is often performed in conjunction with contour plowing. Strip cropping originated from the practice of rotating crops on a particular field each year in order to minimize nutrient depletion. Because the foliage and root system of some crops are more effective in reducing soil erosion than others, rotating crops on sloping fields can result in erosion being relatively high one year and low the next. Rather than rotating the entire crop on fields in rolling terrain, farmers today rotate in strips so the crop that is more effective in reducing soil erosion will help trap sediment moving down from the less effective crop. The result is an overall reduction in soil erosion compared to what occurs with crop rotation involving entire fields. Note that in semiarid climates strip cropping is used in conjunction with keeping fields fallow (planting no crops) to help maintain soil moisture (Case Study 10.1).

No-till farming is a technique where farmers leave the remains of the previous crop standing in their fields rather than tilling (plowing) the residue into the soil. In the past, fields were routinely tilled prior to planting to help control weeds and aerate the soil, but the exposed soil led to high erosion rates. From Figure 10.30 one can see that with no-till farming the new crop is simply planted between the rows of the previous year. The old crop residue then acts as a cover material, preventing raindrops from making direct impact with the ground and dislodging soil particles. The crop residue further reduces erosion by helping to hold the water in place, thereby increasing infiltration and minimizing the amount of water moving downslope as overland flow. Although no-till farming is highly effective in reducing both soil loss and sediment pollution, some farmers have been slow to adopt the practice because of the need to purchase new equipment and additional chemicals for controlling weeds. Note that because no-till farming requires additional chemicals, this technique represents a trade-off between two environmental problems, namely soil loss and pollution of streams from the runoff of agricultural chemicals (Chapter 15).

Grassed waterways are naturally low areas within a field where a farmer plants grass rather than crops. These low areas, also called *swales*, are where water from overland flow starts to collect, and therefore represent the very beginning of small channels within a drainage network. When farmers plant rows of crops across these swales, water flowing through the swales due to overland flow will form gullies as shown in Figure 10.31A. Therefore, not only is a portion of the field taken out of crop production, but the gullies are obstacles that farm machinery must go around. This makes it more difficult and time-consuming for farmers to work their fields. Farmers have learned that the small amount of extra production they get from growing crops in drainage swales is simply not worth all the problems it creates. Besides, the production is eventually lost in those areas anyway because of soil loss. Today, many farmers wisely choose to keep swales covered with grass (Figure 10.31B) so that soil erosion and gully development are minimized.

Terracing, an ancient technique used in areas of steep terrain, involves construction of retaining walls (Chapter 7) in order to create flat surfaces for

FIGURE 10.30   Rather than plowing the remains of the previous crop into the soil, in no-till farming the residue is left in place to help shield the soil from the direct impact of raindrops.
© Doug Sherman/Geofile

FIGURE 10.31   Overland flow naturally collects in the low areas or swales of a field and leads to the formation of gullies (A) in the exposed soil, creating obstacles for farm machinery and reducing crop production. Planting grass in the swales rather than crops (B) prevents the formation of gullies.
(a–b) Fred Gasper, USDA Natural Resources Conservation Service

A Ionia County, Michigan

B Missouri

Lessons Learned from the 1930s Dust Bowl

In the late 1800s and early 1900s, agriculture in the United States expanded westward onto the prairies of the Great Plains. Early settlers used the fertile grasslands, with topsoils as much as 6 feet (1.8 m) deep, mostly for grazing cattle. However, a combination of higher grain prices and the availability of mechanized farm equipment encouraged farmers to plow up vast tracts of the prairie to grow wheat. No one realized it at the time, but the bumper crops of wheat that farmers enjoyed from 1918 to 1929 were not just a result of the rich soils, but also because of rainfall being 20% higher than the long-term average. Although the tilled prairie was very productive, its soils were now completely exposed rather than covered with deep-rooted grasses. Because rainfall in this semiarid climate was inconsistent, the rich farmlands were susceptible to severe wind erosion. When a prolonged drought gripped the southern parts of the Great Plains from 1931 to 1939, the exposed soil particles were easily picked up by the strong winds and carried away, leaving behind a 100-million-acre (400,000-km²) ecological and economic dead zone known as the Dust Bowl (Figure B10.1).

During the Dust Bowl, the combination of strong winds and static electricity allowed soil particles to be lifted over 10,000 feet (3,000 m) into the atmosphere. Driven by high-altitude winds, the dust clouds turned into massive storms called *black blizzards*, which rolled across the open plains and turned day into night (Figure B10.2). Some of the dust storms transported soil all the way to the Atlantic Ocean, blocking out the sun in New York, Boston and Washington, D.C. Farming was now impossible in the Dust Bowl and living conditions were unbearable. Families began losing their farms since they could no longer pay their mortgages, triggering the collapse of local economies and the mass migration of over 2.5 million people. The Dust Bowl had become the largest environmental disaster in U.S. history.

To address the disaster, in 1935 the federal government created the Soil Conservation Service (SCS) within the Department of Agriculture—it was later renamed the Natural Resources Conservation Service. Because the Great Plains has a semiarid climate with limited rainfall, scientists with the SCS recommended farming practices that would limit runoff and increase soil moisture to help soils better withstand periods of drought. Specifically, the SCS recommended that farmers employ contour plowing to minimize runoff and maximize infiltration and soil moisture. Since it was now recognized that crop residue protects soil from erosion, it was also recommended that farmers stop using tilling methods that overturn the topsoil and bury the crop residue. The SCS suggested using no-till methods that preserve the protective crop residue, while at the same time eliminating weeds and increasing soil moisture. In addition to the new tilling practices, the SCS recommended that rather than planting only wheat, farmers should grow wheat and sorghum

FIGURE B10.1 Map showing the area known as the Dust Bowl that was affected by severe wind erosion and soil loss during the 1930s.

in alternating strips to minimize erosion and maximize infiltration. In conjunction with strip cropping, farmers learned to systematically allow portions of their fields to lie fallow (dormant) so that soil moisture could be replenished during the summer rains. To help encourage farmers to participate in the new program, the SCS established soil conservation districts at the local level so that individual farmers could have direct input in new farm policies—today there are over 3,000 conservation districts across the United States.

By 1941, the drought ended and the worst of the Dust Bowl crisis was over. Due in large part to the federal government's efforts, local farm economies had stabilized and the mass migration of people slowed considerably. However, when a drought in the 1950s brought the region to the brink of another disaster, Congress decided to begin providing subsidies that encouraged farmers to convert some of their tilled acreage back to grazing livestock. Farmers were also encouraged to increase irrigation from the Ogallala Aquifer (see Case Study 11.2) to protect against the loss of soil moisture and variable rainfall patterns. In the end, efforts by the federal government to address the farming crisis went far beyond the Dust Bowl itself. The soil and water conservation techniques developed during the Dust Bowl have been improved over the years and are now widely adapted by farmers across the United States.

A

FIGURE B10.2 Dust Bowl era photos showing tilling practices (A) that exposed the soil to severe wind erosion and led to the development of large dust storms known as black blizzards (B). The result was a collapse of local farm economies (C) and the mass migration of people from the region.

(a–c) USDA Natural Resources Conservation Service

B

C

FIGURE 10.32 Terracing allows agriculture to take place on steep hillsides that would otherwise be impossible to cultivate due to severe soil erosion. Note how the rice fields shown here follow the contours of the land surface.
© PhotoLink/Getty Images

growing crops (Figure 10.32). This technique has long been used in parts of Asia to grow rice on steep hillsides. Here a relatively high population density combined with limited natural farmland forced people to develop agriculture on steep slopes. Clearly, without terraces, sustaining agriculture on such slopes would be virtually impossible since the soil would be quickly lost. Note that in the case of rice, terraces also enable farmers to flood their fields during the critical planting season.

Stream buffers are strips of grass or forested areas left between cultivated fields and stream channels (Figure 10.33), thereby minimizing the amount of sediment that can make its way into a drainage network. In the case of logging operations, stream buffers are created by leaving uncut strips of forest adjacent to the stream network. In agricultural areas farmers often plant grass along the edges of their fields to form buffers. Both types of stream buffer are effective in reducing the amount of sediment entering a drainage network, which minimizes sediment pollution and its negative impacts on stream ecology and flooding (Chapter 8). Note, however, that because stream buffers do not prevent the downslope movement of soil on a field, this technique does not address the problem of soil loss.

Silt fences are another means of preventing sediment from entering a drainage system and causing sediment pollution. However, because silt fences are not designed to be permanent, they are primarily used around construction sites where land disturbances are temporary. Silt fences are commonly seen along highways and around construction projects (Figure 10.34A) where they are used to keep sediment from filling nearby drainage channels. Because of the 1972 Clean Water Act (Chapter 15), construction projects in the United States are now required to install silt fences to help prevent sediment pollution. Although developers and contractors generally do not like having to incur the costs associated with installing silt fences, the costs are minor compared to what taxpayers would otherwise bear in terms of increased flooding and destruction of stream ecosystems. Silt fences are a good example of how environmental regulations

Story County, Iowa

FIGURE 10.33 Stream buffers reduce sediment pollution by trapping sediment moving off adjacent fields before it can enter the drainage system. Most stream buffers consist of a combination of grass strips and uncut forest along stream channels.
Lynn Betts, USDA Natural Resources Conservation Service

A George L. Smith State Park, Georgia

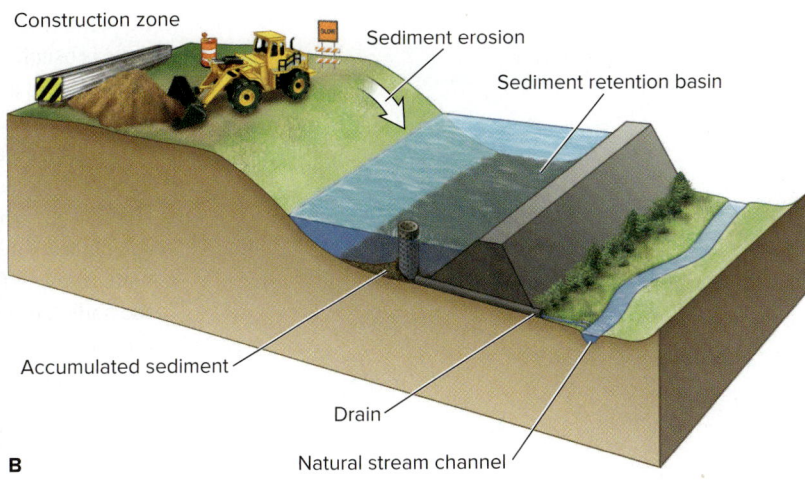

B

FIGURE 10.34   Properly installed silt fences (A) help prevent sediment pollution by keeping sediment at construction sites from entering a drainage network. Retention basins (B) are used to collect or trap sediment before it can enter a stream channel. Farmers also use retention basins to collect valuable topsoil coming off their fields.
© Jim Reichard

are often perceived as being a burden on society, but in reality help protect people and their property from damage caused by someone else's use of the land.

Retention basins are human-made depressions excavated within a drainage system and designed to store excess water and/or sediment (Figure 10.34B). Although most retention basins serve as a flood-control measure (Chapter 8), they are sometimes built with the additional purpose of trapping sediment in order to prevent sediment pollution. Retention basins are commonly located within residential developments and adjacent to large parking lots, both of which have sizable areas where impermeable asphalt and concrete cover the surface. Because the first phase of residential and commercial development projects typically involves installing a drainage system, retention basins are already in place prior to the major construction phase when erosion rates are highest. As the basins accumulate sediment, heavy equipment is periodically used to remove the material and haul it away. Farmers also use retention basins to collect valuable topsoil that gets carried off their fields. The topsoil is then placed back onto the fields.

Slope vegetation cover is a simple and effective way to reduce sediment pollution. Many construction projects involve excavating and regrading the land surface, leaving the soil exposed and highly vulnerable to erosion. Covering the sloped areas with vegetation immediately following the regrading is critical to prevent the loss of valuable topsoil. If the topsoil is lost, then it becomes very difficult for vegetation to take hold, which allows for continued erosion and the eventual development of deep gullies. As described in Chapter 7, there are several techniques used to keep soil in place while permanent vegetation takes hold, including hydroseeding and the use of synthetic meshes and fabrics.

Salinization of Soils

Recall that water within subsurface materials contains dissolved ions due to the chemical weathering of minerals. During the summer months when high evaporation rates cause soils to lose moisture, the dissolved ions in the soil zone can combine and form new minerals, commonly called salts. In more humid climates, where freshwater is flushed through the soil zone fairly regularly, the mineral salts are eventually redissolved and carried downward through the soil profile and into the groundwater system (Chapter 11). In arid climates where the flushing of freshwater is much less frequent, mineral salts can accumulate in the soil zone. However, even in arid climates there is usually enough rainfall to flush the salts out of the uppermost part of the zone. The result is a soil profile consisting of topsoil with relatively few mineral salts that overlies a salt-rich horizon—soil scientists call these aridisols (see Table 10.1).

Deserts have historically been areas with little agricultural activity due to a lack of water. However, modern irrigation practices have allowed agriculture to flourish on certain desert soils that would otherwise be unproductive. Note that agriculture in some arid zones can be highly profitable in part because the climate allows crops to be grown year-round. For example, the combination of fertile soils, climate, and irrigation have made the Great Central Valley in California one of the most productive agricultural regions in the world. Unfortunately, heavy irrigation on poorly drained desert soils can also lead to **salinization,** a process in which the salinity of soil water increases to the point that plant growth is reduced. As illustrated in Figure 10.35, the presence of a horizon with low permeability can prevent the downward movement of water through the soil profile. As water accumulates above this horizon, it dissolves some of the salts, making the soil water highly saline. Eventually, this saline water moves up toward the surface via capillary (i.e., wicking) action into the root zone of the plants, reducing their ability to grow. To lessen the effects of salinization, farmers can install drainage systems in their fields (consisting of perforated pipes) so that water can flush through the soil zone, thereby preventing the buildup of saline water. Another strategy is to closely monitor soil moisture conditions in order to avoid overirrigating the fields and allowing saline water to rise into the root zone.

FIGURE 10.35   Irrigating poorly drained desert soils that naturally contain mineral salts often leads to salinization. Poor drainage allows the irrigated water to accumulate and dissolve the salt minerals. The resulting saline water then moves up into the root zone, reducing crop production.
Ron Nichols, USDA Natural Resources Conservation Service

Soils with Hardpans

Some soil profiles contain what soil scientists refer to as a **hardpan,** which is a layer whose physical characteristics limit the ability of either roots or water to penetrate into the soil (Figure 10.36). Hardpans commonly consist of dense accumulations of clay minerals, or layers where the soil particles have been cemented together by minerals (e.g., calcite and iron oxides) that have precipitated from water within the soil. In some cases a hardpan will develop when repeated use of heavy farm equipment causes compaction of clay-rich soils. As noted earlier, farmers try to avoid creating a hardpan, sometimes called a *plowpan*, by staying out of their fields when they are wet and most susceptible to compaction.

Hardpans are a problem in agricultural areas because they inhibit the movement of water and air in the soil and limit the growth of plant roots. Because a hardpan forces plants to develop shallow root systems, the plants are more susceptible to wilting during dry periods when soil moisture becomes depleted. Moreover, the lack of drainage can cause the upper soil zone to become saturated during wet periods, causing plants to drown due to the lack of oxygen in the soil. To remedy the problems, farmers can sometimes break up a hardpan through deep plowing. In the case of a homeowner wanting to plant a tree or shrub, it is best to dig out a bowl-shaped depression within the hardpan, making it large enough to accommodate the root system when the plant is fully mature (Figure 10.36). It is also important to penetrate the entire hardpan to prevent the depression from filling with water and drowning the tree or shrub. Finally, note that the impermeable nature of hardpans is actually desirable for growing rice since it allows farmers to maintain standing water in their fields, a condition necessary for planting young rice seedlings.

Permafrost

In higher latitudes it is quite common for a soil profile to contain a layer of **permafrost,** which is a subsurface horizon that remains frozen throughout most or all of the year—the frozen zone can extend downward as much as several thousand feet before thawing upon encountering a sufficient amount of Earth's internal heat. As shown earlier in Figure 10.13 and listed in Table 10.1, a permafrost layer is characteristic of soils known as gelisols, which cover a considerable portion of

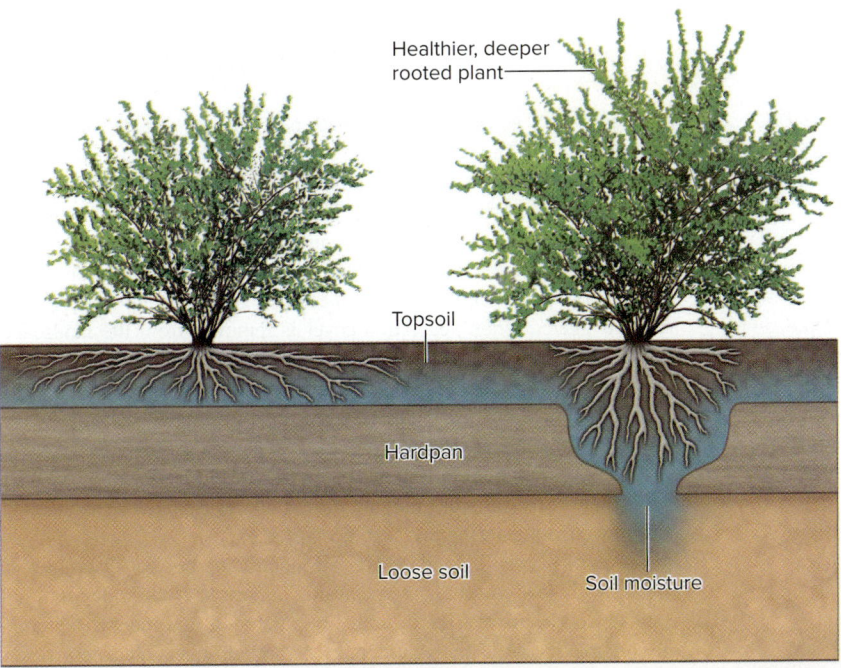

FIGURE 10.36 A hardpan is a soil layer rich in clay or one cemented together by minerals, making it difficult for roots or water to penetrate the soil. Plants are then forced to have shallow root systems, making them more susceptible to wilting during droughts, and drowning during wet periods. When planting trees or shrubs, it is best to dig through the hardpan to provide better drainage and more room for root growth.

FIGURE 10.37 Construction of this highway in Alaska caused a portion of the underlying permafrost to thaw, creating a subsurface void that then collapsed into a sinkhole. Damage caused by human-induced melting of permafrost is common in polar regions.
Joe Moore USDA-NRCS

Earth's land area. Depending largely on latitude, the entire soil zone remains frozen year-round in some regions, but in other areas the uppermost soil layers will thaw during the summer period. Human activity, such as the construction of roads and buildings, can alter the heat balance of the soil zone, thereby causing the permafrost layers to thaw. When this occurs, subsurface materials can begin to flow and undergo compaction, damaging buildings and infrastructure (Figure 10.37). Of much greater concern is the impact from global warming. As the planet continues to warm, scientists expect that permafrost will begin to thaw over vast stretches of the Arctic, releasing great quantities of greenhouse gases. We will discuss this topic in more detail in the chapter on global climate change (Chapter 16).

SUMMARY POINTS

1. To soil scientists, soils are natural mixtures of organic matter and rock and mineral fragments that are capable of supporting plant life; engineers refer to soil as any fragmental earth material that lies above bedrock.

2. Soils originate from the physical and chemical weathering of solid rock, which generates finer particles of rock and minerals. Most soil particles are composed of quartz and various clay minerals.

3. Soil horizons (layers) develop when organic matter is incorporated in the uppermost horizon and infiltrating water carries clay-mineral particles and dissolved ions downward where they accumulate in subsurface layers.

4. The color of soil horizons is an indicator of organic content and drainage characteristics. Organic matter gives topsoil a blackish to brownish color, whereas iron oxides impart a yellowish to reddish color, indicating a well-drained soil.

5. Five soil-forming factors determine the type of soil horizons that will develop for a particular soil: parent material, organisms, climate, topography, and time.

6. Soil properties are largely a function of water content and the relative proportions of clay, sand, and silt-sized particles. The small size and electrical charge of clay-mineral particles play a key role in determining

properties such as drought resistance, permeability, plasticity, strength and sensitivity, compressibility, and shrink-swell potential.

7. Soil fertility depends upon the ion exchange capacity of the soil and the availability of essential nutrients. The decomposition of minerals and organic matter in the soil produces important nutrients. Both clay minerals and organic matter help store these nutrients for later use by plants.

8. Soil loss occurs when soil erosion becomes greater than soil formation and is triggered when vegetation cover is removed from the landscape. The loss of topsoil is a serious problem because it ultimately reduces worldwide food production.

9. Excess soil erosion also causes sediment pollution of streams, which increases flooding and destroys the streams' natural ecosystems.

10. Soil loss and sediment pollution can be minimized by a variety of techniques that keep soil from moving downslope or by preventing soil that does move from entering drainage networks.

11. Irrigation of poorly drained desert soils containing horizons rich in mineral salts can lead to salinization, causing decreased crop production.

12. Hardpan soil layers restrict the ability of plant roots and water to penetrate the soil, making crops more susceptible to both drought and excessive soil moisture.

KEY WORDS

compressibility   311
essential nutrients   314
expanding clays   311
fertilizers   316
hardpan   327
ion exchange   313
paleosol   305

parent material   302
permafrost   327
permeability   309
plasticity   310
porosity   308
salinization   326
sensitivity   310

soil   297
soil erosion   318
soil fertility   314
soil horizons   299
soil loss   318
strength   310
swelling clays   311

APPLICATIONS

Student Activity

Find a spot in a wooded area with lots of trees where you can dig a small hole. Using a shovel, preferably one with a flat blade (not curved and pointed), carefully dig a 2-by-2-foot pit with vertical sides about 2 feet deep. Describe what happens to the layer of leaves or pine needles as you get deeper. How thick is the underlying topsoil? How do you think the overlying organic material actually gets into the topsoil? When it rains, what percentage of the water do you think is going to move as overland flow compared to infiltrating? Now, imagine that the trees are all cut down and the land becomes an open farm field. How much rainwater do you think will now move as overland flow compared to infiltrating? How will the change in land cover affect the topsoil? When finished, be sure to fill the hole back in and cover it with leaves.

Critical Thinking Questions

1. Climate is one of the five soil-forming factors. All other things being equal, what differences would you expect to see in a soil that forms in a very humid climate compared to one that forms in a drier, semiarid climate?
2. How does aluminum ore, called bauxite, become concentrated in tropical soils? Does infiltrating water carry aluminum deeper into soil, or is it simply left behind? Explain.
3. To grow properly, plants need to extract certain elements, called nutrients, from the soil. Nitrogen-based nutrients ultimately come from nitrogen gas (N_2) that is in the atmosphere. However, nutrients such as calcium (Ca) and magnesium (Mg) are derived from the soil itself. Where exactly do these types of nutrients come from, and how do they become available to plants?

Your Environment: YOU Decide

In this chapter you learned that excessive soil erosion leads to soil loss and sediment pollution. Soil loss is a problem because it affects our ability to increase food production, whereas sediment pollution fills stream channels with sediment, which results in more severe and frequent flooding and destroys the natural ecology of streams.

Clearly, everyone in society must ultimately bear the consequences of excessive soil erosion, namely higher food prices, greater insurance premiums for flooding, and loss of recreational fishing. Most farmers are already making a serious effort to reduce soil erosion because it affects their livelihoods. However, other than existing federal regulations, development and construction companies have little incentive to prevent sediment pollution. Explain whether these companies should be required to install temporary silt fences or the regulations should be abolished in order to make their businesses more profitable, and hence more likely to grow.

© Glow Images

Chapter 11

Water Resources

LEARNING OUTCOMES

After reading this chapter, you should be able to:

- ▶ Discuss how freshwater originates and moves within the hydrologic cycle.
- ▶ Explain the difference between off-stream and consumptive uses of water.
- ▶ Describe how a water-supply reservoir is harmful to stream ecosystems.
- ▶ Characterize the benefits and undesirable side effects of off-stream water diversions by aqueducts and pipelines.
- ▶ Describe the difference between unconfined and confined aquifers, and explain what causes groundwater to move through the subsurface.
- ▶ Explain what is meant by leakage in groundwater systems and why it is important.
- ▶ Define *groundwater mining,* and describe its undesirable consequences.
- ▶ Discuss why groundwater is generally better than surface water as a water-supply source.
- ▶ List and describe the various alternative ways humans can increase water supplies.

Freshwater is our planet's most precious resource. Humans need water not only for personal consumption, but also to run industries and to sustain large-scale food production. In places where water is scarce, people commonly make a serious effort to conserve it and use it more efficiently. As human population continues to grow, so too will the demand for this limited resource.

Introduction

© StockTrek/Getty Images

Of all the natural resources, the most important to people are air, water, and food. In terms of human survival, an individual can live only a few minutes without oxygen, one to two weeks without water, and approximately a month without food. Although slightly less critical than oxygen, water is essential for the body to function properly (the average human body is 70% water by weight). Interestingly, Earth's water is directly related to our oxygen and food supply. The basis for our food supply is land plants, whose growth depends on rainfall as well as on the soil that forms by chemical weathering reactions involving water and rocks (Chapter 10). The oxygen we breathe is a by-product of plant photosynthesis. The entire biosphere then, including humans, operates within the Earth system (Chapter 1), in which liquid water is the key that links everything together.

Another key element of the Earth system is the uneven distribution of precipitation around the globe due to global climate patterns. Because water is critical to life as we know it, the abundance and diversity of species within the biosphere closely correspond to the distribution of rainfall. Likewise, human settlement patterns have historically followed the availability of water. Hunters and gatherers eventually formed permanent settlements in areas where rainfall was sufficient to support agriculture. The establishment of settlements led to the development of more complex societies in which specialized workers produced goods and services for other members of society. Rivers were soon being used for both water supply and for transporting goods and people deep into the interiors of continents.

In modern societies water is used for a wide array of activities, from the manufacturing and processing of consumer goods to the production of concrete and vast amounts of electrical power. Individuals use water in their homes to take showers, wash clothes, flush toilets, maintain landscaping, and more. The modern use of water has certainly made our lives more comfortable and pleasant. However, human population growth coupled with more uses for water is pushing the available water supply in many regions past its natural limit. At the same time our supplies are being stretched, climate change (Chapter 16) is threatening to exacerbate the problem by altering global precipitation patterns. Some societies will likely find themselves without adequate supplies of water, severely restricting their future growth and, in some cases, threatening their very existence. Because water is unquestionably our most precious natural resource, we will begin by examining its origin and distribution.

Earth's Hydrologic Cycle

Recall from Chapter 2 that the origin of Earth's water is related to the formation of the solar system. There is strong evidence that Earth experienced a Mars-sized impact early in the solar system's history, which ejected into Earth's orbit debris that eventually formed the Moon. The evidence also suggests that this impact caused the Earth to melt, and the subsequent volcanic activity released large volumes of gas. These volcanic gases combined with countless comet impacts to form a dense atmosphere, rich in water vapor. As the planet cooled, water condensed from the atmosphere and fell as precipitation, resulting in an ocean. Tectonic activity later caused landmasses to rise above sea level, which allowed rainwater to begin flowing across the young landscape and into rivers that carried the water back to the ocean. While water was falling as precipitation, evaporation was simultaneously sending water vapor back into the atmosphere from both the ocean and landmasses, thereby creating a continual and cyclic transfer of water between the ocean, atmosphere, and land.

Today scientists refer to the cyclic movement of water within the Earth system as the **hydrologic cycle.** As illustrated in Figure 11.1, the hydrologic cycle is driven by solar radiation that causes water to evaporate from the oceans and land surface. The water vapor rises into the atmosphere, where it cools and eventually falls to the surface as precipitation (rain, snow, sleet, or hail). Some of the rain and water from melting snow and ice will move downslope and into stream channels and return directly back to the ocean. Some of the precipitation also infiltrates permeable sediment and rocks and then enters the groundwater system, eventually discharging into a stream, lake, or ocean. Recall from Chapter 10 that a fraction of the infiltrating water is removed from the soil zone by evaporation and plant transpiration and then returned to the atmosphere. Also recall from Chapter 8 that streams eventually transport water back to oceans. The emphasis in this chapter is on how humans make use of precipitation that moves through the hydrologic cycle.

An important feature of the hydrologic cycle is the variation in *salinity,* which is the amount of electrically charged atoms called dissolved ions, or salts, in the water (Chapters 3 and 10). Low-salinity water with very few dissolved ions is referred to as **freshwater,** whereas highly saline water is called *salt water.* Freshwater is produced whenever liquid water undergoes evaporation. In this process individual water molecules (H_2O) move into the vapor state, but the dissolved ions remain with the original liquid. As indicated in Figure 11.1, the majority of Earth's freshwater is produced by the evaporation of salt water in the oceans. Notice how some of the freshwater precipitation falls on the landmasses, where it eventually flows across the landscape or infiltrates the subsurface. Here the freshwater chemically reacts with minerals, which release various ions that act to increase salinity.

The production of freshwater within the hydrologic cycle is critical to Earth's terrestrial (land-based) life-forms, including humans. These organisms have evolved in the presence of freshwater over long periods of geologic time and cannot survive on salt water. In fact, every terrestrial organism has a certain limit on (tolerance for) the salinity of the water it can safely consume. In the United States the maximum salinity level for human drinking water is set at 500 milligrams of salt per liter of water (mg/l). Chickens, on the other hand, can tolerate a salinity level up to 3,000 mg/l and cattle up to 7,000 mg/l. For comparison, seawater averages around 35,000 mg/l of dissolved salts; thus it is fatal to any terrestrial organism if consumed in sufficient quantity.

In addition to the terrestrial environment, much of the marine ecosystem is dependent on the discharge of freshwater from rivers into coastal estuaries. Here the freshwater reduces the salinity of the seawater, producing a zone of *brackish water.* Some marine organisms depend on brackish water during certain periods of their life cycle, particularly reproductive and growth phases. In fact, coastal estuaries with their brackish water serve as critical nursery grounds for nearly 70% of all commercial fish and shellfish in the United States. Therefore, not only does the terrestrial biosphere depend on freshwater, but a significant portion of the marine ecosystem does as well.

Where Freshwater Is Found

Earth is often called the "blue planet" because of its abundance of water. However, as can be seen in Figure 11.2, only 2.8% of the planet's total supply consists of freshwater. If we examine the breakdown of this relatively small amount of freshwater, we see that 77.4% is stored as glacial ice and 22.1% as *groundwater* or subsurface water. This leaves about 0.5% as *surface water,* which is the term hydrologists use for the water in rivers, lakes, and soils. To give you a better sense of the water-supply distribution, examine the detailed breakdown of Earth's water resources listed in Table 11.1. Most striking perhaps is how the rivers in the world account for a mere 3 thousandths of a percent of all freshwater. Although rivers and other forms of surface water represent a tiny fraction of all freshwater, surface

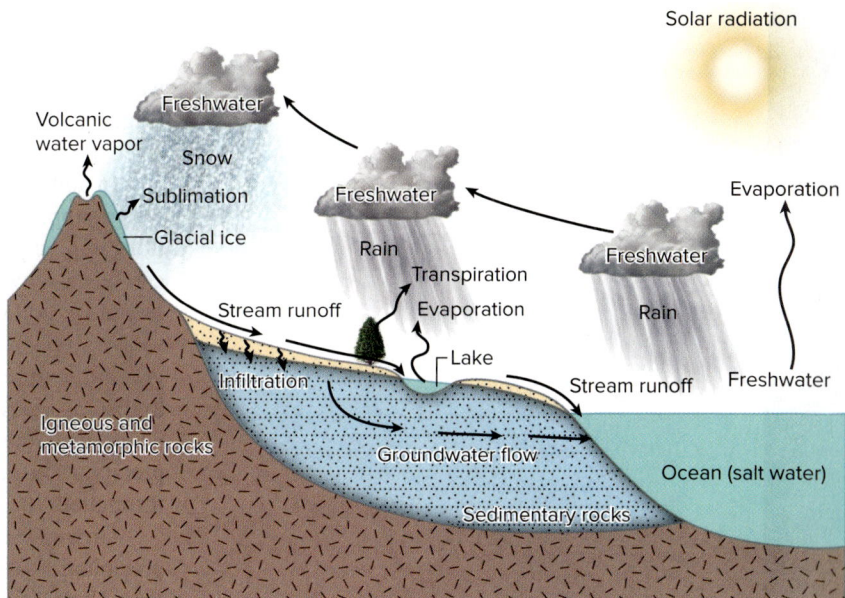

FIGURE 11.1   The hydrologic cycle describes the cyclic movement of water through the Earth system. The cycle is driven by solar energy that causes water to evaporate from the oceans and land surface and allows for plant transpiration. Evaporation of seawater produces large quantities of freshwater, which is vital for humans and the terrestrial biosphere. Humans' primary sources of freshwater are streams, lakes, and groundwater systems. Note that groundwater is found in fractured igneous and metamorphic rocks, but the vast majority occurs in porous sedimentary material.

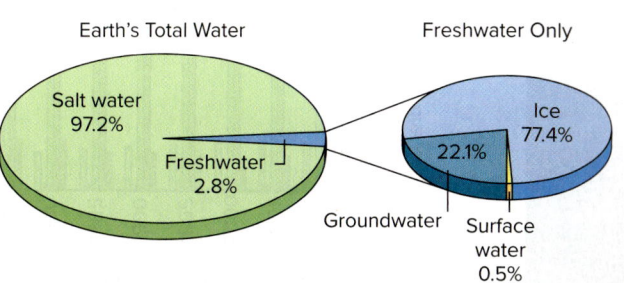

FIGURE 11.2   Breakdown of the water in Earth's hydrosphere.

[left margin partial text]

Source:
http://pr

Precipit
<4
4–8

Figure
from 19
graph p
water is
tions. F
withdra
associa
after 19
more el
lic wate
12% of
the 45%
tion, re
make u
comes f

Sta
esting t
tomary
get the
effect o
tion in
(4,542
water b
good m
For exa
ages ab
consum

Early stage

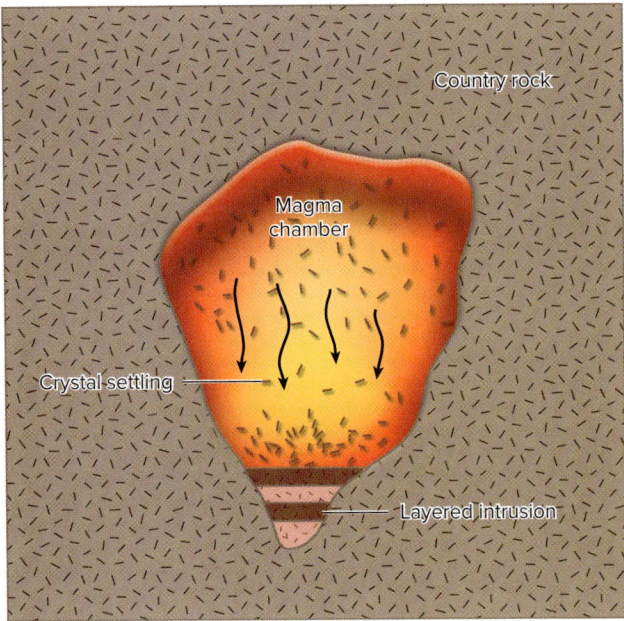

A

B

FIGURE 12.5 Dense minerals that crystallize early in the cooling process can settle to the bottom of a magma chamber and form a layered ore deposit (A). The photo (B) shows layers of chromium-rich minerals that are part of a layered intrusion in Bushveld, South Africa.
(b) © David Waters, Oxford University

Late stage

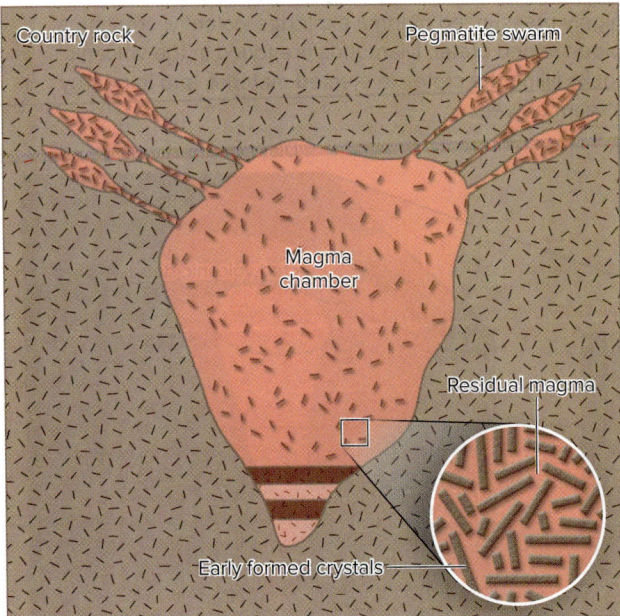

FIGURE 12.6 In the later stages of cooling, a magma chamber consists mostly of mineral crystals, with the remaining magma residing between the crystals. This residual magma is enriched with certain elements that can form valuable mineral deposits if injected into surrounding rocks, forming small intrusive bodies called pegmatites.

called *crystal settling,* whereby dense minerals that form early fall or settle to the bottom of the magma chamber. As shown in Figure 12.5, this process can create layered ore deposits that geologists refer to as **layered intrusions.** Layered ore deposits typically contain metallic minerals that are valuable sources of chromium, titanium, and vanadium.

Igneous ore deposits can also form when magma migrates away from newly formed minerals, as opposed to the minerals settling out of the magma. This process is believed to occur later in the cooling process, when the remaining magma resides between the mineral grains—similar to how water exists in the pore spaces of sedimentary materials. Pressure changes can then force this residual magma to leave the magma chamber and be injected into fractures within the surrounding rock. As illustrated in Figure 12.6, the magma then cools and forms elongated bodies of igneous rocks. These small intrusive bodies can contain valuable ore minerals when the crystallization process leaves the residual magma enriched with certain elements. One type of mineral deposit that forms in this manner is unusually coarse-grained deposits of silica-rich rock called *pegmatites*. These pegmatites are similar to granite in composition, but sometimes contain concentrations of minerals that contain rare elements (e.g., lithium, beryllium, niobium, and tantalum), which are normally found dispersed throughout an igneous rock body. Some pegmatites are mined solely for their exceptionally large grains of quartz and feldspar.

Hydrothermal Deposits

In Chapter 6 you learned that all magmas contain varying amounts of water. When magma slowly cools beneath the surface, some of the water becomes incorporated into the crystalline structures of certain minerals, and some remains with the magma during the final stages of crystallization. Water is also commonly present in the rocks surrounding the magma, which geologists refer to as *country rocks*. Naturally, the intense heat from the magma body raises the temperature of any groundwater present in the country rocks. The water from

Mean Annual Precipitation (1981–2010)

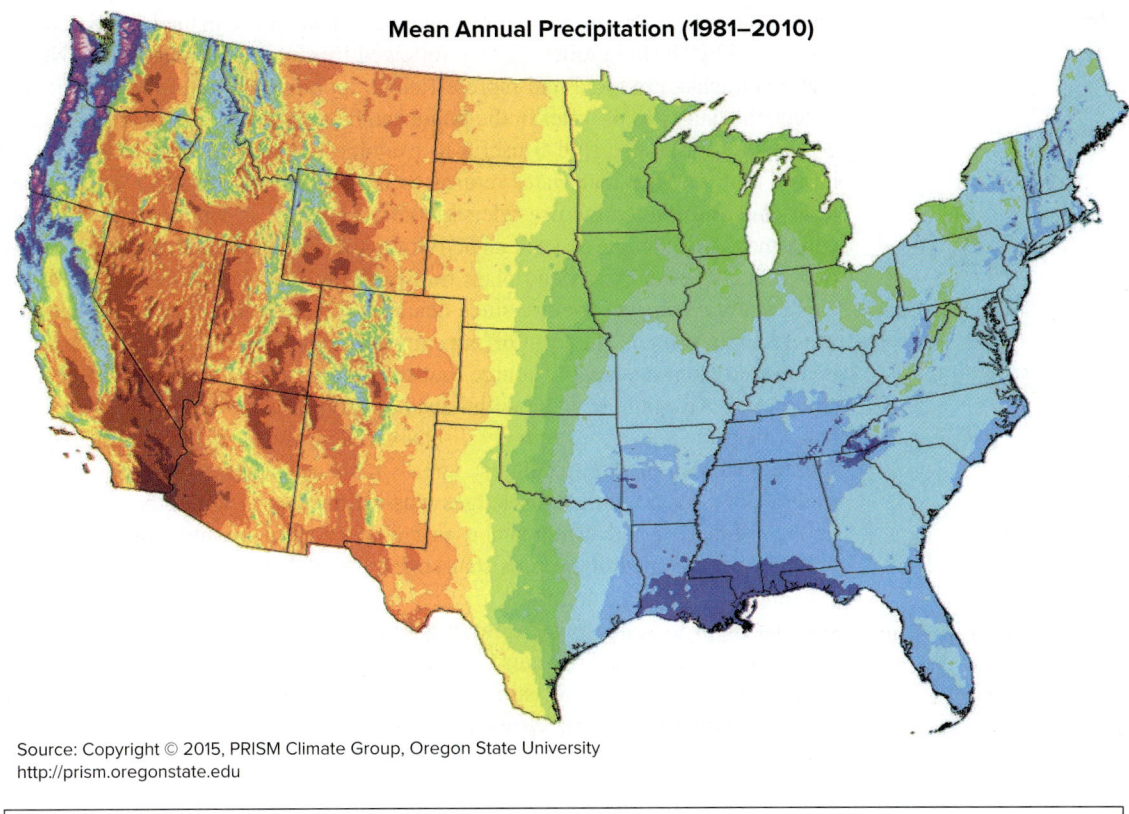

Source: Copyright © 2015, PRISM Climate Group, Oregon State University
http://prism.oregonstate.edu

FIGURE 11.3 Average annual precipitation varies widely across the United States, resulting in significant differences in the availability of freshwater supplies. © 2009, PRISM Climate Group, Oregon State University, http://www.prismclimate.org Map created September 2009

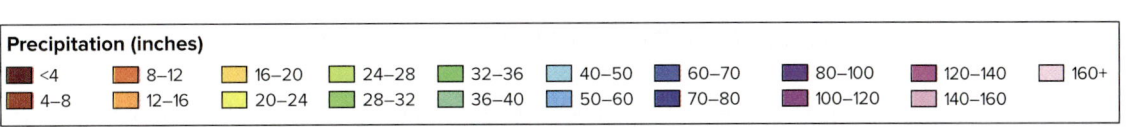

Precipitation (inches)

<4	8–12	16–20	24–28	32–36	40–50	60–70	80–100	120–140	160+
4–8	12–16	20–24	28–32	36–40	50–60	70–80	100–120	140–160	

Figure 11.4, which shows total water usage (fresh and saline) in the United States from 1950 to 2010. Since 86% of the withdrawals in 2010 were of freshwater, this graph provides a good representation of freshwater usage—nearly all of the saline water is used for cooling purposes in electrical production and for mining operations. Perhaps the graph's most obvious feature is the dramatic increase in total withdrawals in the period leading up to 1980, with most of the increase being associated with irrigation and electrical production. Total withdrawals declined after 1980 and then declined again after 2005 due to conservation efforts and the more efficient use of water across nearly all sectors of society. Interestingly, public water supply, often called *municipal* or *city water,* accounted for only 12% of all withdrawals in 2010. This amount is rather small compared to the 45% and 32% that is withdrawn for electrical generation and irrigation, respectively. Note that in the United States surface-water sources make up 75% of the freshwater supply, whereas only 25% of freshwater comes from groundwater.

Statistics on total water withdrawals are useful, but it is also interesting to examine the rate at which a society uses water. Here it is customary to take the withdrawal volume and divide it by the population to get the *per capita* (per person) consumption rate, thereby eliminating the effect of population growth on the data. In terms of per capita consumption in the United States, the average citizen uses about 1,200 gallons (4,542 liters) of water per day. However, because this represents the water being used to support all aspects of society, it does not provide a good measure of the *personal* consumption rate in the nation's homes. For example, the personal consumption rate for a typical American averages about 69 gallons per day (262 liters/day), which is far less than the per capita consumption of 1,200 gallons per day that includes industrial and irrigation use.

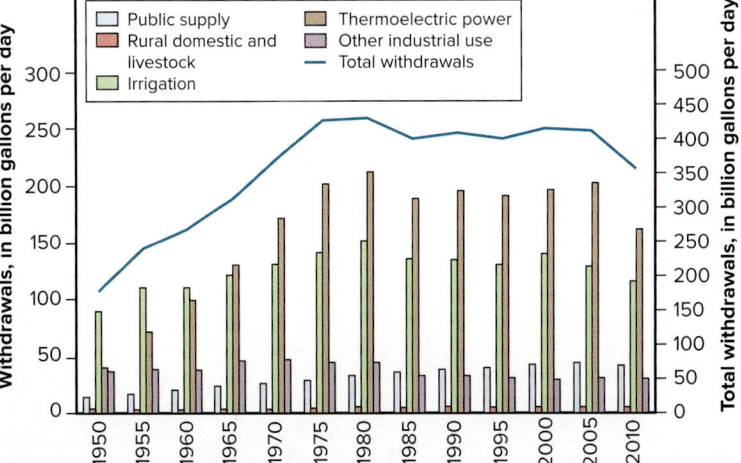

FIGURE 11.4 Total water withdrawals in the United States between 1950 and 2010.

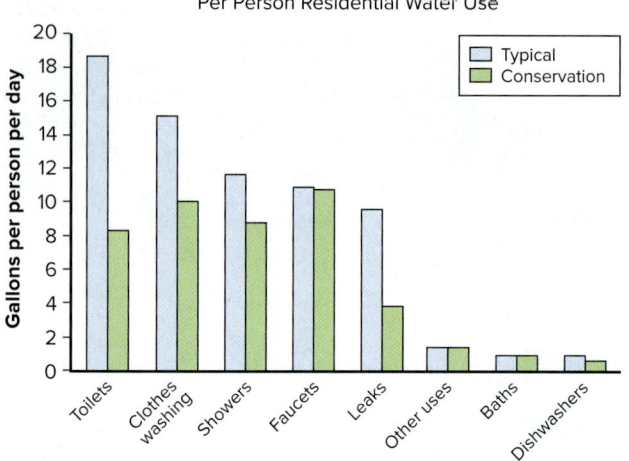

Per Person Residential Water Use

FIGURE 11.5 Breakdown of the rate of personal water usage inside the average U.S. household.

From Figure 11.5 we can see that personal indoor water use in the United States is dominated by flushing toilets (27%), followed by clothes washing (22%). Notice that when leaks are fixed and more efficient appliances are installed, especially toilets, the daily use of water in the home can be reduced by approximately 35%.

Hydrologists also find it useful to examine how much water is lost during its use and whether that water returns to its origin or some other source. The term **consumptive water use** refers to those activities in which water is lost or consumed. An example of such an activity is the irrigation of crops, where water is lost due to evaporation and plant transpiration. Washing dishes, on the other hand, is a *nonconsumptive* use since nearly all of the water can be returned to a water supply source. The term **off-stream use** refers to activities in which water is removed from one supply source, but then returned to a different source after being used. An example is the withdrawal of groundwater for a public water system, where the water is eventually discharged as wastewater into another supply source, such as a stream. In contrast, an *in-stream use* occurs when a city removes water from a river, but then discharges it as wastewater back into the same river.

Consumptive and off-stream uses are of concern because they can cause problems in the natural systems that supply us with freshwater. In the case of a river, substantial water losses may reduce discharge to the point where ecosystems begin to collapse and downstream users are no longer able to withdraw the quantities they need. Because of the nature of subsurface withdrawals, most groundwater is used in an off-stream manner, which almost invariably has a negative impact on groundwater systems.

Traditional Sources of Freshwater

In this section we will explore society's traditional sources of freshwater and consider how increased human population, combined with higher per capita consumption rates, is pushing some natural systems beyond their ability to supply water. This unsustainable use of water is not only harming the very systems that provide us with water, but is beginning to limit the growth and development of population centers around the world. Compounding the problem is the fact that precipitation patterns are changing in response to global warming (Chapter 16). Because of climate change, areas already pushing the limits of their water supply may soon find themselves without adequate freshwater. When faced with a water shortage, government officials are generally left with the following options: (1) increase storage capacity by building reservoirs; (2) transport water from another source via pipelines and aqueducts; (3) institute conservation programs; (4) develop nontraditional sources (e.g., desalinated seawater); or (5) voluntarily limit growth. We will examine these options as we take a closer look at the different sources of freshwater, beginning with rivers and lakes. Note that stream processes were described in considerable detail in Chapter 8; hence, only a brief overview of the subject will be provided here.

Surface Water Resources

Rivers and Streams

Recall that rivers and streams collect freshwater that falls on Earth's landmasses as precipitation, and then transport most of this water to the oceans. The remainder returns to the atmosphere by evaporation and plant transpiration or infiltrates and enters the groundwater system. The term *water table* refers to the depth in the subsurface where porous materials are completely saturated. In more humid climates where rainfall is plentiful, the water table is normally higher than the nearby stream channels, which causes infiltrating water to eventually flow into streams as *groundwater baseflow* (Chapter 8). Consequently, streams in humid areas not only receive water from periodic precipitation events, but also receive a fairly continuous supply of groundwater. The contribution of groundwater is very important because it allows streams to keep flowing during prolonged dry periods. In more arid climates where water tables are relatively deep, streambeds may go dry during the summer. Note that some rivers are able to flow year-round across desert landscapes because much of the water comes from melting snow or from rains in distant parts of the drainage basin. A good example is the Colorado River, which is sustained by snowmelt in the Rocky Mountains as it flows across the desert Southwest region of the United States.

In terms of human water supply, humid landscapes naturally have a greater number of streams capable of providing water on a year-round basis than do arid landscapes. This is why human settlements in arid climates have historically been restricted to narrow corridors along major rivers—such as the Nile flowing through the desert of Egypt and Sudan and the Tigris and Euphrates of Mesopotamia (modern Iraq). In the humid regions of Europe and North America, settlements were less restricted geographically because greater precipitation and higher water tables produced abundant rivers and streams that flow year-round.

Although streams serve as reliable sources of freshwater, human population growth and demand can outstrip the ability of streams to provide water, particularly during low-flow periods in the summer. Rather than limiting growth and development, societies learned how to increase supply by diverting water from distant sources via aqueducts and canals (an off-stream use). Perhaps the most famous example is the elaborate system of aqueducts the Romans built throughout their empire (Figure 11.6A). This aqueduct system not only increased the empire's water supply, but the elevated nature of the system allowed Roman cities to have running water and sewers. A modern water diversion example is the California aqueduct system (Figure 11.6B), where massive amounts of water are transferred to the Los Angeles basin. This large-scale diversion has allowed Los Angeles to grow into a metropolis of over 18 million, which is far beyond what the local water supply could have supported in this arid environment.

Whenever water is removed from a stream or river, the impact on natural ecosystems will depend on the volume that is removed and the quantity and quality of what is returned. In cases of off-stream withdrawals where the water is not returned, such as aqueduct systems and use of water for irrigation, stream discharge can be reduced to the point where entire ecosystems collapse (Case Study 11.1). Reduced flows also mean existing users in the basin from which the water is being diverted are unlikely to be able to expand their supply, thereby limiting future economic growth and development.

With respect to in-stream withdrawals, the impact on the natural environment largely depends on the quality of the water being returned to a river or stream. For example, municipal (public) water systems typically collect the bulk of their wastewater from homes and businesses, then process the water at a sewage treatment plant before discharging it back into a river (Chapter 15). Although this process is effective in removing harmful bacteria, nutrients such as nitrogen are not removed and can negatively impact the water quality of a steam. Another important in-stream use is by electrical power plants, which are usually located along rivers because they require large volumes of water. Here the problem is the discharge of warm water, which causes thermal pollution that is harmful to the natural aquatic life. To minimize thermal pollution, plant operators are now required to recycle water within the plant to lower its temperature before returning it to the river.

Lakes and Reservoirs

Although natural lakes are often excellent sources of freshwater, they are relatively rare because they are generally restricted to geologically young landscapes. For example, the Great Lakes and surrounding smaller lakes all formed after the last glacial ice sheet began its final retreat about 12,000 years ago. As the ice retreated, lakes began forming because the newly exposed landscape did not

A

B

FIGURE 11.6 Roman aqueducts (A) were off-stream diversion projects that allowed areas with inadequate water supplies to flourish. The modern aqueduct system serving the greater Los Angeles, San Diego, and San Francisco areas (B) provides water to millions of residents living in an arid environment.
(a) © Image Source/PictureQuest

Off-Stream Water Usage and the Aral Sea Disaster

Prior to the 1970s the Aral Sea in Kazakhstan and Uzbekistan (formerly part of the Soviet Union) was the fourth largest lake in the world. The lake was located on the arid plains of central Asia and supplied with freshwater by two major rivers, the Amu Darya and Syr Darya. Because the Aral Sea was in a closed basin where water could only exit through evaporation, the salinity of the lake increased over time and its water was considered brackish (i.e., between freshwater and salt water). Although salinity had increased, the lake level remained relatively steady as the rate of evaporation reached a state of natural equilibrium with the volume of freshwater flowing in from the two rivers. For thousands of years, this system supported abundant fisheries, which in turn supported thriving local economies and communities around the shoreline.

Then in the 1960s, the Soviet government embarked on a massive engineering project designed to create an agricultural economy on the arid plains. Large volumes of water from the Amu Darya and Syr Darya rivers were diverted and used to irrigate millions of acres of land to grow cotton and other crops. Although the planners realized that diverting most of the rivers' flow would cause the Aral Sea to recede, they believed that the benefits of the new agricultural economy would outweigh the loss of the fishing industry. It was also thought at the time that letting freshwater flow into the lake only to be lost by evaporation was a waste of valuable freshwater.

By the 1970s the Aral Sea began losing water rapidly as evaporation was now greatly exceeding the relatively small volume of freshwater still flowing into the lake. As shown in Figure B11.1, satellite imagery provided dramatic evidence that the lake was dying. In 2007 the volume of water in the Aral Sea had shrunk from 168 cubic miles (700 km³) to a mere 18 cubic miles (75 km³), whereas the salinity had increased from 14,000 milligrams of salt per liter (mg/l) to over 100,000 mg/l, making the lake more than twice as saline as ocean water. As the shoreline receded, local fishing communities not only found themselves having to travel farther to reach the lake, but the increased salinity was destroying the commercial fisheries. By 1980 the once vibrant fishing industry had collapsed.

Although the Soviet planners knew that the lake would dry up and become hypersaline, what they failed to consider was that some of the irrigation water would carry pesticides and fertilizers off the agricultural fields before making its way to the Aral Sea via the two rivers. As the lake continued to evaporate, the dissolved salts and agricultural chemicals eventually precipitated out and formed a layer of contaminated salts on the exposed lakebed. The Soviets also did not anticipate that the exposed salts and lake sediments would get picked up by the strong winds that periodically blow across the flat plains, creating toxic dust storms. These storms would then deposit the toxic salts on the surrounding landscape, causing decreased crop production and widespread respiratory problems in the human population. For example, the rate of esophageal cancer in the region is now 25 times greater than the worldwide average. Another unintended consequence is that as the Aral Sea dried up, there was a large reduction in the amount of water returning to the atmosphere via evaporation. This altered the regional climate, producing hotter summers and colder winters. Ultimately, this shortened the growing season and reduced crop production. In the end, the Soviets did more than just trade one economy for another, they created an environmental catastrophe that led to a host of economic and human health problems.

While the human modifications to the natural hydrology of the Aral Sea have produced severe environmental consequences, they are not unique. There are many other lakes around the world that have been negatively impacted by major off-stream water usage. Hopefully, policymakers and planners can learn lessons from these water-diversion projects that will help minimize the negative consequences of future projects.

1977	2000	2015

A B C

FIGURE B11.1

Satellite images showing how the Aral Sea was lost over time due to the off-stream use of water from the rivers flowing into the lake. Note that the 1977 image is false color, in which vegetation is shown in red and water appears black. The other two images are in natural color—the lakes appear green due to differences in suspended sediment.
(a) USGS EROS Data Center and NASA; (b–c) NASA'S Earth Observatory

FIGURE 11.7 In addition to creating large reservoirs for supplying freshwater, modern dams are built to control flooding and to generate electricity. Shown here is Lake Powell behind the Glen Canyon Dam on the Colorado River in Arizona.
© Jim Wark/Airphoto

have an established network of streams to carry away the surface water. Like all lakes, those in the Great Lakes region will slowly fill in with sediment as the terrain and drainage system mature. Natural lakes are also common in areas of recent tectonic uplift and where streams have been blocked by debris from mass wasting events (Chapter 7).

Perhaps the most common way of creating a large and reliable source of freshwater is by damming rivers to create artificial lakes called *reservoirs*. As shown in Figure 11.7, a reservoir can stockpile an enormous amount of water simply by capturing part of a river's flow during periods of elevated discharge—the volume of water behind some dams is greater than the river's total annual flow. As a surplus store of water, a reservoir can be used to support both population growth and the irrigation needs of agriculture. In fact, some of the Colorado River water in the reservoir shown in Figure 11.7 is sent to Los Angeles via the aqueduct system described earlier in this section. Note that many modern dams do more than just increase water supplies; they are also used for flood control (Chapter 8) and for generating clean hydroelectric power (Chapter 14). Another important benefit is the recreational use of a reservoir for boating and fishing.

Although there are many benefits associated with dams, there are also negative consequences for both society and the natural environment. When a reservoir fills with water and drowns a river valley, it almost invariably results in the displacement of people, businesses, and agricultural activity. In terms of the natural environment, a reservoir's large surface area means that it will lose considerably more water to evaporation than did the section of the river it has replaced. In addition, when a free-flowing river encounters the still waters of a reservoir, it is forced to deposit its sediment load behind the dam. All reservoirs therefore have a certain life expectancy as they will eventually fill in with sediment. Also note that because reservoirs are relatively deep, the water temperature decreases significantly. The

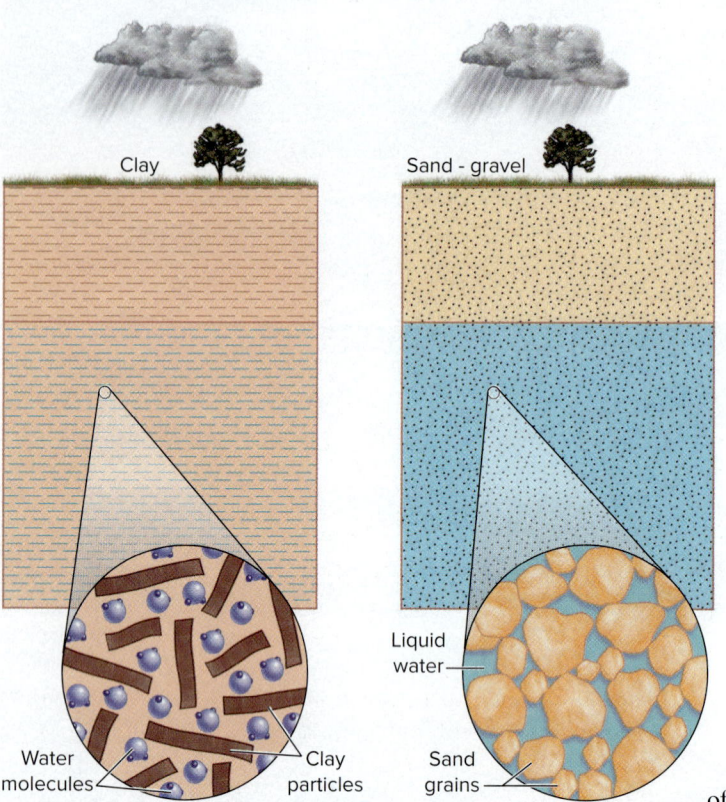

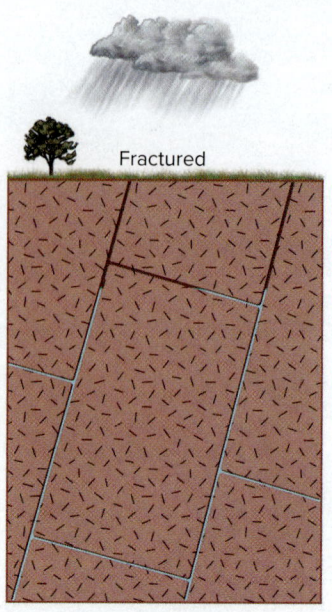

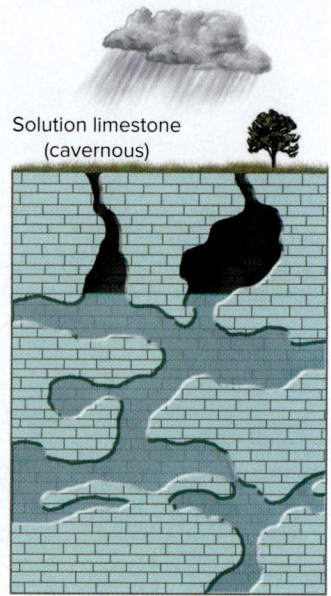

FIGURE 11.8   Porosity determines the volume of groundwater that can be held in subsurface materials. Sedimentary materials composed of clay, sand, or gravel-sized particles normally have high porosity, whereas crystalline rocks have little porosity and usually contain water only in fractures. Water moving through certain types of limestone can create high porosity by dissolving the rock to form passageways and caverns.

water that is released below the dam then is both sediment-free and considerably cooler than the water was in the natural system. This can have devastating impacts on downstream ecosystems that have evolved in the presence of warmer, sediment-laden water. Another problem is that the volume of water being released from a dam is highly regulated and thus downstream areas no longer experience natural variations in streamflow, especially extreme highs and lows. Because many aquatic ecosystems depend on natural variations in flow, the controlled discharge of water from dams can cause serious disruptions in the aquatic food chain, particularly in coastal estuaries where freshwater mixes with seawater.

Groundwater Resources

In addition to surface water, the other major source of freshwater is **groundwater,** which is defined as water that resides within the voids or *pore spaces* of subsurface materials. Three major factors determine whether groundwater is a viable source for a given area: (1) the quantity that exists; (2) the ease with which it can be withdrawn; and (3) the quality of the water. When these three conditions are favorable, groundwater often makes a better source of freshwater than a stream or lake. Keep in mind that the conditions that make groundwater a suitable source vary from user to user. For example, a small supply of groundwater may be sufficient for a single household, but be totally inadequate to meet the irrigation needs of an agricultural user. Likewise, groundwater whose salinity makes it unfit for human consumption might be fine for poultry and cattle.

From Figure 11.8 you can see that the volume of groundwater beneath a given area is directly related to the amount of void space within subsurface materials, a property scientists call *porosity* (Chapter 10). Rocks with a crystalline texture (interlocking mineral grains) have virtually no porosity since there is very little void space between the mineral grains. Consequently, areas underlain by crystalline rocks such as granite normally hold only small amounts of groundwater within fractures. Large volumes of groundwater are generally found in areas with thick sequences of sediment or sedimentary rock. Porosity in these granular materials, which are composed of sand, gravel, or clay-sized particles, can be quite high. Note that porosity is highest in granular material that is unconsolidated, and decreases in material whose grains have undergone compaction and become cemented together, forming sedimentary rock such as sandstone, shale, or limestone. Because limestone contains the soluble mineral calcite, groundwater circulating through fractures can dissolve away this rock and form large passageways and caverns. Areas underlain by such rock, which geologists refer to as *solution limestone* or *karst limestone,* often contain large volumes of water in the open void spaces (Figure 11.8).

Although groundwater volume is an important factor in selecting a water supply, how easily the water can be withdrawn is equally important. Recall from our discussion on soils (Chapter 10) that most sediment is composed of rock and quartz and clay-mineral fragments. Also, clay particles are extremely small and have a strong electrical attraction to dipolar water molecules, which makes it very difficult for water to move through clay-rich materials. Much coarser sand and gravel-sized sediments typically consist of quartz and rock fragments that are electrically neutral. The large pore size and lack of electrical attraction allow water to move rather freely through gravels and sands. The property known as **hydraulic conductivity** describes the ease with which a material transmits a fluid—similar to how electrical conductivity relates to the flow of electricity. Note that hydraulic conductivity takes into consideration the transmission property of the material (its permeability) and the properties of the fluid itself (its density and viscosity). In Chapter 13 you will see how these fluid and material properties affect our ability to extract oil and gas from the subsurface.

Geologists use the term **aquifer** to describe a water-bearing rock or sediment layer that readily *transmits* water, in other words, a material whose hydraulic conductivity is relatively high. Suitable aquifers normally consist of layers of relatively coarse sediment (sand and gravel) or granular sedimentary rocks (sandstone and limestone). Fractured crystalline rocks can also serve as aquifers, but provide only limited amounts of water due to their low porosity. Conversely, materials with low hydraulic conductivity are considered **aquitards;** these include layers of clay-rich sediment, shale, and unfractured crystalline rock, including certain types of dense limestone. Figure 11.9 illustrates how hydraulic conductivity varies greatly among the more common types of rocks and sediment. Note that each tick mark on the horizontal scale represents a tenfold increase in conductivity. From this graph you can see that gravel is about a hundred times more conductive than sand and a billion times more conductive than clay.

Types of Aquifers

Recall from Chapter 3 that sedimentary sequences generally consist of varying amounts of quartz, clay, and carbonate (e.g., calcite) minerals and rock fragments. Because there can be vast differences in the hydraulic conductivity among the layers, sedimentary sequences usually contain multiple aquifers and aquitards. From the illustration in Figure 11.10, one can see that an **unconfined aquifer** is a highly conductive layer that is open to the atmosphere and surface waters. In contrast, a **confined aquifer** is overlain by an aquitard that helps seal it off from the surface environment. Note that aquitards are commonly called *confining layers* because of the way they restrict the vertical movement of water between different aquifers.

There are several important differences between confined and unconfined aquifers that we can examine using the wells shown in Figure 11.10. When a well is drilled into an unconfined aquifer, the hole will eventually reach the **water table,** which is the depth below which all the pore spaces are completely filled with water. The area above the water table is referred to as the *unsaturated zone* because the pores contain both air and water, and the area below it is called the *saturated zone.* Notice how the water table is not necessarily flat, but mimics the topography of the land surface. In topographically high areas, the water table is at a higher elevation, which gives the groundwater greater potential energy. This difference in potential energy

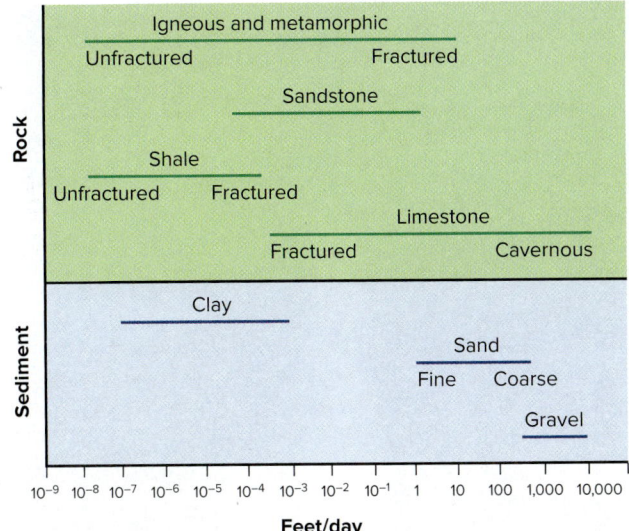

FIGURE 11.9   Graph showing the range of hydraulic conductivity for various geologic materials. Each tick mark on the horizontal scale represents a tenfold change in conductivity. Some materials have a wide range of hydraulic conductivity, and thus can be regarded as an aquifer in some instances and an aquitard in others.

FIGURE 11.10   An unconfined aquifer is open to the surface environment and has a water table that marks the top of the saturated zone. A confined aquifer has an overlying aquitard that limits the vertical movement of water, causing the aquifer to become pressurized. When a well penetrates an aquitard, water rises to the potentiometric surface, which represents the amount of pressure within the confined aquifer.

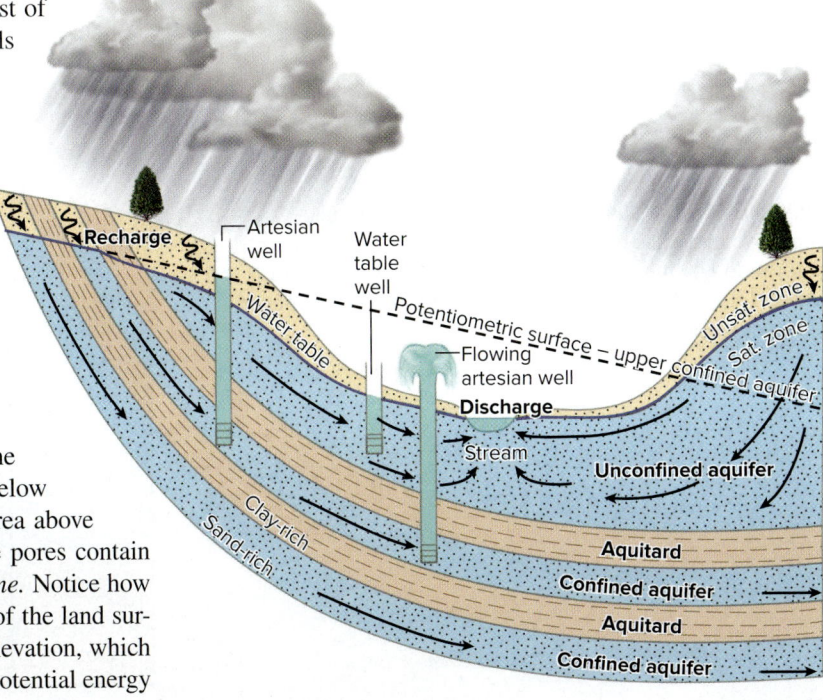

FIGURE 11.11   A free-flowing artesian well occurs when the potentiometric surface of a confined aquifer is higher than the land surface, allowing pressurized water to rise above the surface.
Alfonso Rivera, Geological Survey of Canada

forces water to flow through an unconfined aquifer away from high areas until it eventually discharges into topographically low areas such as lakes, wetlands, and free-flowing streams. Note that the water table naturally rises and falls depending on the amount of infiltrating water that makes its way into the groundwater system; this infiltration process is called *groundwater recharge.*

The confined aquifers in Figure 11.10 behave differently than the unconfined aquifer because the aquitards (confining layers) limit the ability of water to move vertically. This allows water within confined aquifers to become pressurized. Consequently, when a well is drilled into a confined aquifer, the pressurized water will rise up into the well. Hydrologists use the term **potentiometric surface** to refer to the height water will rise in a well, which is a measure of the potential energy within a confined aquifer. Note that groundwater within a confined aquifer flows from areas where the potentiometric surface is high to areas where it is lower. Moreover, the potentiometric surface has no relationship to the water table or the surface topography. Finally, notice in Figure 11.10 that wells in unconfined aquifers are known as *water table wells,* and those in confined aquifers are called *artesian wells.* A free-flowing artesian well is one where water flows freely to the surface because the potentiometric surface is higher than the land surface—an example is shown in Figure 11.11.

Movement of Groundwater

Recall that groundwater always flows from areas of higher potential energy to areas of lower energy. Although the energy that causes groundwater flow can involve changes in density and temperature, the flow in most situations is due to differences in elevation and water pressure. Because both elevation and pressure are affected by gravity, hydrologists measure the potential energy within an aquifer system in terms of **hydraulic head,** which in practical terms is the height of the water table or potentiometric surface. Figure 11.12 shows how a measuring tape is used to record the depth to the water level in a well. To get the hydraulic head (distance) above some datum such as sea level, the depth to water is simply subtracted from the elevation at the top of the well.

Measuring hydraulic head is important because it allows hydrologists to determine the direction in which groundwater is flowing in the subsurface. As shown in Figure 11.12, the **hydraulic gradient** is defined as the slope or steepness of a water table or potentiometric surface, and it is easily calculated from the difference in hydraulic head and distance between any two points. The direction of groundwater

FIGURE 11.12   Illustration showing how hydraulic head and hydraulic gradient in an unconfined aquifer are determined by measuring the depth of the water in two wells. Note that groundwater flow is always in the direction of the hydraulic gradient.

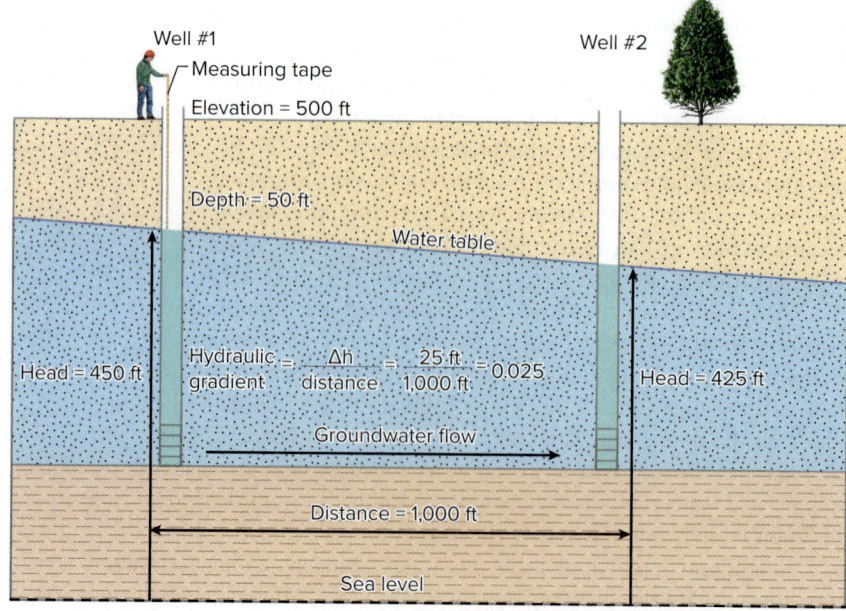

TABLE 11.2   Average groundwater velocities computed from Darcy's law for representative layers of sand and clay under the same hydraulic gradient. Travel time is based on horizontal flow through material of uniform composition.

	Hydraulic Conductivity	Porosity	Hydraulic Gradient	Average Groundwater Velocity	50-mile Travel Time
Sand	10 ft/day	0.25	0.025	1.0 ft/day	723 years
Clay	0.00005 ft/day	0.50	0.025	0.0000005 ft/day	1.4 billion years

flow is always from higher to lower head, and the steepness of the gradient indicates how much potential energy is available for driving the water. Groundwater velocity therefore should be higher in places where the gradient is steeper because more energy is available. However, porosity and hydraulic conductivity also play an important role. Remember that groundwater flows more readily through high-conductivity layers, such as gravel, as opposed to layers rich in clay minerals. On the other hand, higher values of porosity cause groundwater to move more slowly through granular material because of the greater proportion of void space.

In 1856 a French engineer named Henri Darcy experimentally derived a mathematical law that describes the relationship between hydraulic gradient and conductivity and the flow of groundwater through granular material. By including porosity, the following version of *Darcy's law* can be used to compute the average velocity of water flowing through a granular material.

$$\text{Groundwater velocity} = \frac{\text{Hydraulic conductivity}}{\text{Porosity}} \times \text{Hydraulic gradient}$$

Calculating the average groundwater velocity allows hydrologists to predict the amount of time it will take water to travel a given distance through porous material. For example, Table 11.2 shows the calculated velocities for water flowing through average sand and clay layers in response to a typical hydraulic gradient. Although the velocity of 1.0 foot (0.3 m) per day in sand may seem slow, it is actually quite fast in terms of geologic time. Most striking perhaps is just how much faster groundwater moves through sand compared to clay. Note that the vast difference in velocity is mostly due to the difference in hydraulic conductivity between these two materials—the effect of porosity is comparatively minor. We can also see that because of the extremely slow water velocity in clay, travel time is very long.

It should be clear that clay and other types of aquitard materials act as barriers to water trying to flow between different aquifers. As illustrated in Figure 11.13, although aquitards inhibit the flow of water, they almost always allow some

FIGURE 11.13   Water travels between different aquifers by the process of leakage. Downward leakage occurs when the water table is higher than the potentiometric surface of a confined aquifer. When the potentiometric surface is higher, leakage is upward.

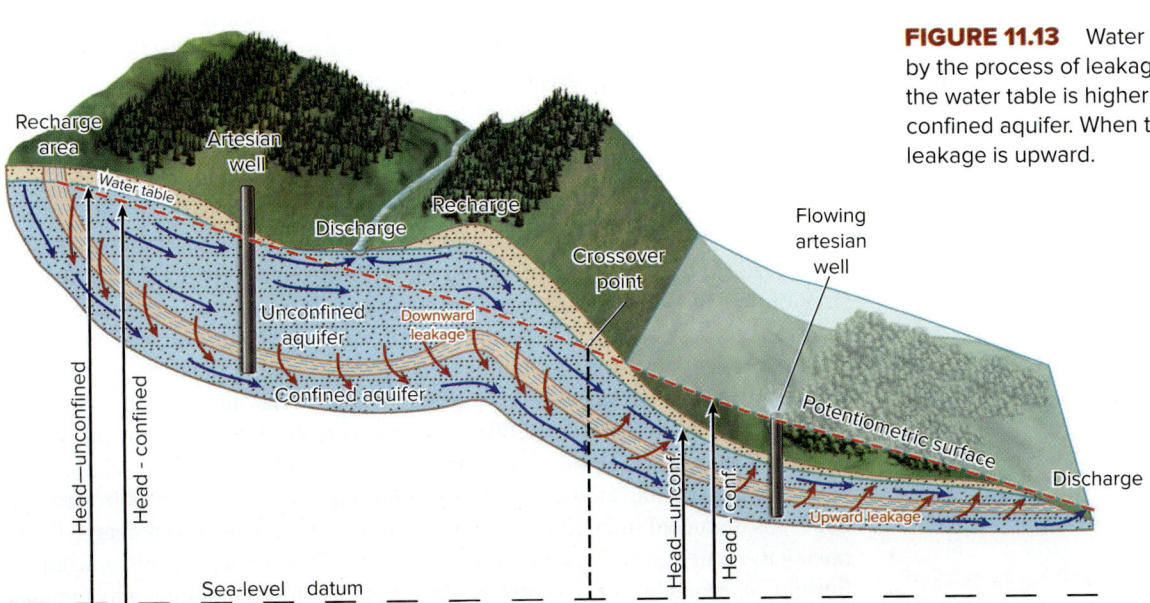

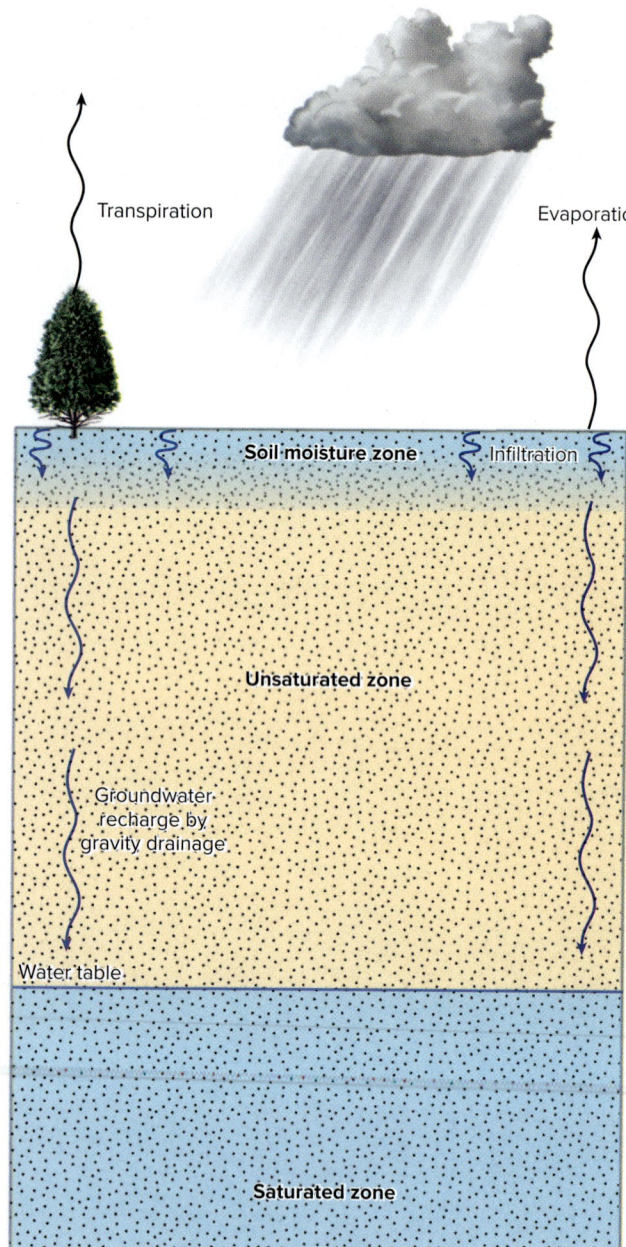

FIGURE 11.14 Groundwater recharge occurs when soil moisture builds to the point where water begins to drain due to gravity. Recharge is less common during hot summer months since most of the water entering the soil zone is quickly returned to the atmosphere by evaporation and transpiration.

water to pass in a process called **leakage.** Note that the actual amount of leakage depends on the aquitard's hydraulic conductivity and the difference in hydraulic head between the two aquifers. Moreover, the direction of leakage can be either upward or downward depending on the hydraulic head within the aquifers. Notice in the figure how the relative positions of the water table and the potentiometric surface switch at the crossover point. Above this point the hydraulic head in the unconfined aquifer is greater, which forces water to leak downward and enter the confined aquifer. Below the crossover point the confined aquifer has the higher head, which means water is flowing toward the surface via upward leakage. Later you will learn how aquitards play a critical role in determining the movement and spread of contaminants in groundwater systems (Chapter 15) and in keeping oil and gas trapped within the subsurface (Chapter 13).

Groundwater Recharge

Recall that *groundwater recharge* refers to the process whereby aquifers are replenished by infiltrating water that makes its way to the water table. As shown in Figure 11.14, this process begins when rain or melted snow infiltrates the soil zone. However, unless the water table is at the surface, infiltrating water can only reach the groundwater system if soil moisture increases to the point where the downward force of gravity can pull the water from the pore spaces. During the hot summer months, evaporation and plant transpiration normally prevent soils from accumulating enough water for gravity drainage to take place. Groundwater recharge typically occurs on a seasonal basis when temperatures are cooler and precipitation is more plentiful. The cooler temperatures reduce evaporation and transpiration rates such that soil moisture can accumulate to the point that water drains by gravity.

Because of climatic controls, groundwater recharge occurs periodically and on a more seasonal basis, whereas the discharge of groundwater into streams, lakes, and wetlands is a fairly continuous process. Another key point is that whenever the recharge rate becomes greater than the discharge rate, water levels within an aquifer will rise. In this sense an aquifer is similar to a leaking tire that requires you to keep adding air. The tire pressure goes up only if you are able to add air faster than it is leaking out. Similarly, if we looked at a natural groundwater system over a period of several years or decades, we would see that the system reaches a state of *dynamic equilibrium* in which the overall rate of recharge equals the discharge rate. In other words, while water levels may go up and down on a daily or seasonal basis, they remain fairly steady over the long term as long as recharge equals discharge. An important point here is that this natural equilibrium can be disrupted by changes in climate, by excessive groundwater withdrawals by humans, or by the loss of natural vegetation cover due to human activity. Vegetation loss tends to reduce infiltration and, ultimately, the contribution of groundwater baseflow to streams. Reduced groundwater baseflow can be of critical importance to stream ecosystems during the summer months (Chapter 8).

The dynamic relationship between recharge and discharge is illustrated in Figure 11.15, which shows the groundwater level in a semiconfined aquifer 75 feet (23 m) below the surface. Notice how the groundwater level consistently rises during the winter months when recharge is greater than discharge, then reaches a peak prior to the hot summer months. Despite some significant rains, water level declines over the summer when recharge is minimal because of the high rates of evaporation and transpiration. Of particular interest is the fact that recharge during the spring of 2001 was able to replace nearly all of the water that was lost in the previous summer, returning the groundwater level close to the same maximum. However, the dynamic equilibrium from 2000 through 2001 was disrupted in 2002 by drought conditions that prevented recharge from raising the water level to the previous maximum. This concept of dynamic equilibrium in groundwater systems is important for understanding why human

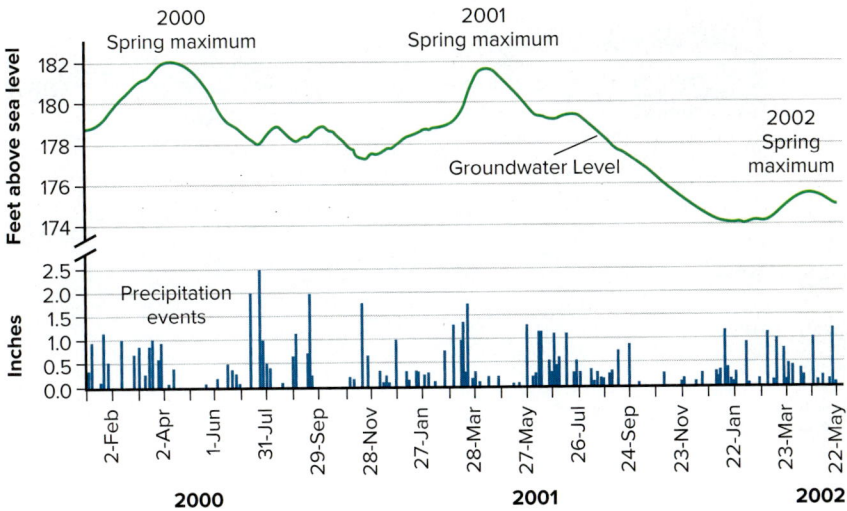

FIGURE 11.15 Graph showing how the groundwater level in a semiconfined aquifer in Georgia rises and falls with the seasons. Between 2000 and 2001 the system was in balance as recharge replaced most of the water lost due to discharge. Drought conditions in 2002 disrupted the system's natural equilibrium, causing the groundwater level to fall. Groundwater levels will return to their normal maximum provided there is a wet year or series of years with enough recharge to make up for the water deficit in the system.

withdrawals of groundwater are unsustainable in some situations. Later in this section we will examine some of the problems that develop from excessive groundwater withdrawals.

Springs

A **spring** is simply a place where groundwater discharges at the surface in a concentrated manner. As shown in Figure 11.16, springs can discharge directly on the land surface or under a body of water, such as a stream, lake, or ocean. Historically springs gave humans unique access to groundwater long before the advent of modern water wells. In desert climates a spring can create a natural *oasis* that is often the only available source of water in a large area. Also interesting is how society has used springs for different purposes, depending on the geology of the springs themselves. For example, springs that discharge from relatively shallow aquifers usually have low salinity because the water is fairly young and has not acquired many dissolved ions. The low salinity combined with the fact that groundwater contains very little sediment and bacteria means that shallow springs make exceptionally good sources of drinking water. These types of springs, though, are more prone to going dry during extended droughts than those that originate from a deeper source.

Springs that are fed by deep aquifers commonly have such high salinity levels that the water is unfit for human consumption. However, high-salinity springs, sometimes referred to as *mineral springs,* have long been a source of valuable salts for cooking and preserving food. In addition, Earth's internal heat causes the temperature of water from springs fed by deep aquifers to be much warmer than that from springs that originate in shallow groundwater systems. Depending on the actual temperature, the term *hot spring* or *warm spring* is often used to describe saline spring water. Hot springs have long been used to fill pools at resort spas, which advertise the mineral-rich water as having therapeutic value. In some volcanic areas the temperature of the groundwater is so high that it discharges as steam or boiling water. Hot groundwater is also used as a source of clean geothermal energy (Chapter 14). Finally, because the chemistry of spring water is often quite different from the surrounding surface water, springs create unique habitats for certain species within an ecosystem.

Water Wells and Drawdown Cones

Other than springs, the only way people could access groundwater before modern drilling equipment existed was through hand-dug wells. Because a person can safely dig a hole only so deep, wells were normally located where the water table was relatively close to the surface. Once the hole was completed, the well was

FIGURE 11.16 Springs occur where groundwater discharges at the surface in a localized area. The geology of the site determines the depth of the groundwater source, which in turn influences the spring's salinity, temperature, and consistency of flow. The spring in (A) forms when water becomes trapped in the unsaturated zone, then flows laterally until discharging along a hillside. In (B) the spring discharges from a solution passageway in limestone. Example (C) illustrates how water from deeper aquifers can flow to the surface along faults or fractures.

(a) © The McGraw-Hill Companies, Inc./John A. Karachewski, photographer; (b–c) © Jim Reichard

FIGURE 11.17   Diagram showing the construction of a modern water well. Note the clay seal and concrete placed around the well casing to prevent surface contaminants from moving down into the well. Photo showing a completed well and pressure tank at the surface.
© Jim Reichard

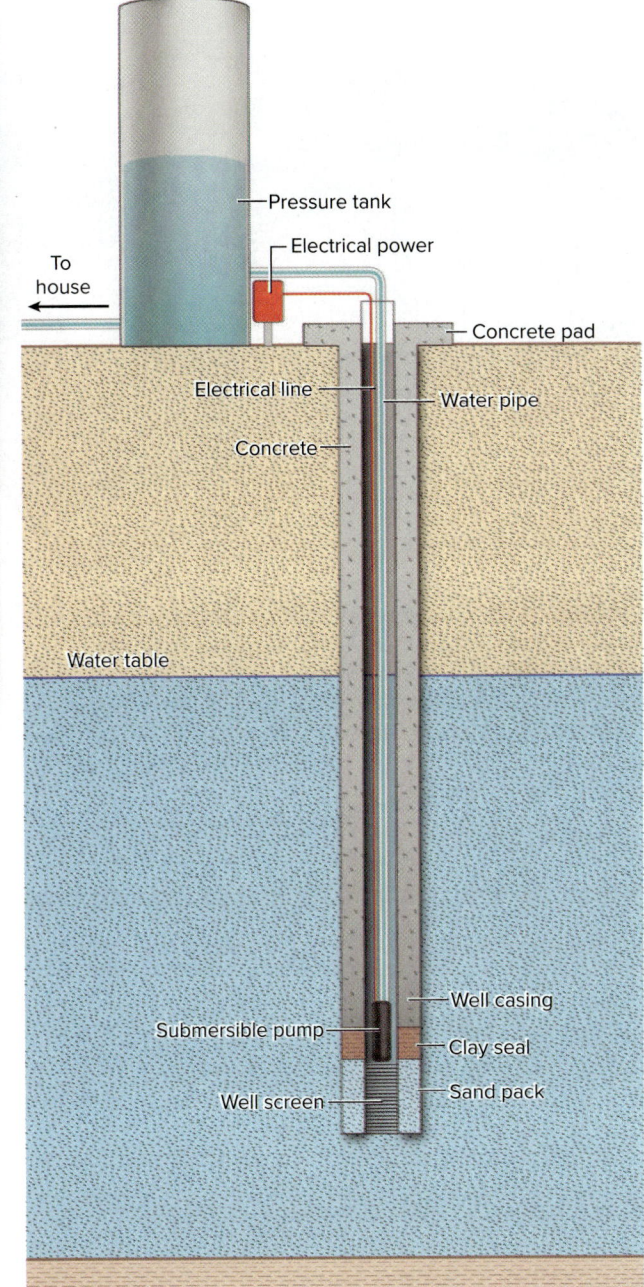

usually lined with stones or bricks to prevent the walls from collapsing. All that was required to extract the groundwater was a bucket and rope. The problem with shallow wells is that they are easily contaminated by chemicals and animal and human wastes. Hand-dug wells pose a special hazard because their diameter is large enough for a person to fall to the bottom. Abandoned wells of this type are particularly dangerous because they are commonly covered with boards, which eventually rot and become obscured by a blanket of leaves.

Modern water wells are usually installed with a truck-mounted drilling rig that uses a rotating drill bit to bore a small-diameter (about 10 inches) hole into the subsurface. After the drilling is finished, a metal or plastic pipe called a *well casing* is lowered into the hole—the bottom section has a screen with fine slits to keep sediment from entering the well. Figure 11.17 shows a cross section of a modern well used for supplying a home or small development. Note how sand is packed around the screened portion of the casing, followed by a clay seal and concrete to prevent surface contaminants from reaching the screen. To bring water to the surface, a submersible pump is placed in the well, then connected to a pressurized storage tank and electrical power source on the surface. Most modern wells are installed at depths ranging from 50 to 1,000 feet (15–300 m), which means they are less likely to become contaminated compared to shallow hand-dug wells. Keep in mind that confined aquifers normally provide the safest groundwater supply because the aquitard helps block the downward movement of contaminants.

When the pump is turned on and begins to push water to the surface, the water level (hydraulic head) within the well is lowered. The decreased hydraulic head causes groundwater to flow into the well in a radial or 360-degree manner. As shown in Figure 11.18, this pumping or "sucking" action creates a cone-shaped depression called a **cone of depression** in the water table or potentiometric surface of the surrounding aquifer. Depending on the amount of pumping, a cone of

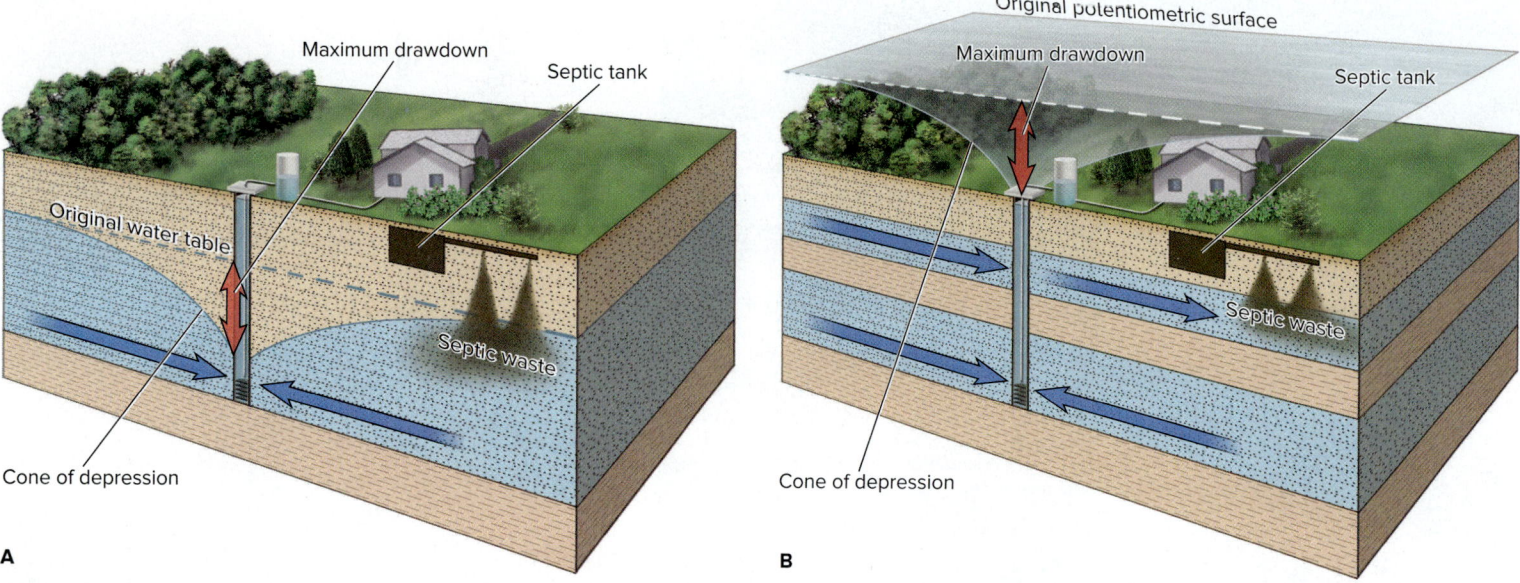

FIGURE 11.18 A pumping well in an unconfined aquifer (A) draws in water and creates a cone of depression in the water table. A well in a confined aquifer (B) creates a cone in the potentiometric surface, which in this case lies above the land surface. Such a drawdown cone can pull contaminants into a well and reduce the flow of water to nearby rivers or springs.

depression can actually reverse the horizontal and vertical flow of groundwater around the well. This means that a well can pull in contaminated water that would have otherwise flowed away from the well.

Impacts of Groundwater Withdrawals

Society has enjoyed many benefits from the use of modern water wells to extract freshwater from the subsurface. Perhaps the most significant benefit is that groundwater has allowed population and agricultural production to blossom in areas where surface water is naturally scarce. Consider, for example, how naturally pure groundwater has made it possible for individual families to live in rural settings, far from any river and water filtration plant. Farmers are also now able to use groundwater to help ensure consistent harvests, particularly in areas facing a high risk of crop failure due to scarce and unreliable rainfall. Although the availability of groundwater has been a tremendous asset to society, this resource is being used in an unsustainable manner in many regions, which is giving rise to a host of serious issues. The underlying problem is that groundwater levels are falling because the withdrawal rates are much faster than the rates at which aquifers are being replenished by natural recharge. Since subsurface water levels are not something people can readily see and appreciate, overuse commonly continues until serious problems begin to develop, such as wells starting to go dry. We will examine some of the problems associated with the overuse of groundwater in the following sections.

Groundwater Mining Hydrologists use the term **groundwater mining** to describe situations where water levels in an aquifer get progressively lower because recharge is unable to replace all the water being removed from the system by pumping. When this occurs, groundwater is no longer considered to be a renewable resource, which makes its withdrawal similar to mining in that what is being removed is not replaced. Groundwater mining is a serious problem because it depletes the water resource, potentially leaving future users without a water supply. Arid regions are particularly susceptible because recharge rates are naturally low, and in some cases nonexistent. In fact, the aquifers beneath many arid regions have experienced only minor recharge since the climate became more arid after the last ice age, which ended about 12,000 years ago. These aquifers have essentially been acting as storage reservoirs that are not being refilled. Here water levels are rapidly falling and will eventually reach the point where water quality declines and or the cost of pumping becomes so great that the aquifer is no longer useable.

A good example of groundwater mining is occurring in Tucson, Arizona. This city of over a half-million people (Figure 11.19) lies in a desert environment where streams are small and flow intermittently, making groundwater the only reliable supply. The problem in Tucson is that the groundwater recharge rate is very low, whereas the withdrawal rate is quite high. Officials estimate that approximately 10% of the city's groundwater supply has been mined since 1945, which was when the city began experiencing explosive growth. While a considerable volume of groundwater remains in storage, the stark reality is that once this supply is gone, it is gone forever. Because the city's very existence is at stake, Tucson officials have wisely adopted various strategies designed to extend the water supply farther into the future.

Groundwater mining is also a serious problem in regard to society's ability to maintain its current rate of food production. In many regions around the world farmers have been able to substantially increase crop production by using groundwater irrigation to supplement natural rainfall—in some arid environments irrigation provides the only water the crops receive. As water levels around the world continue to fall because of groundwater mining, society is faced with the prospect of decreased food production at the same time as population keeps expanding (Case Study 11.2).

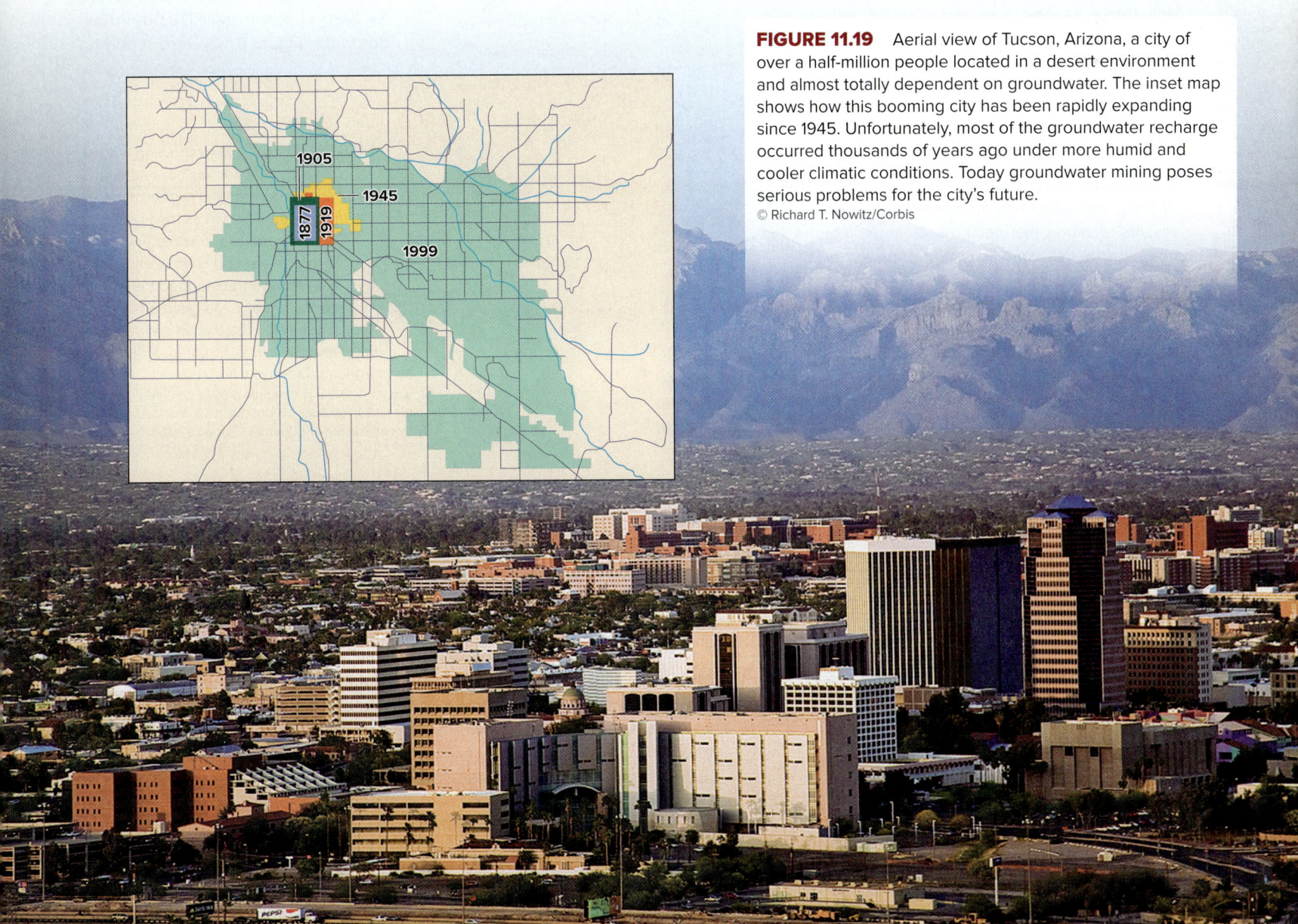

FIGURE 11.19   Aerial view of Tucson, Arizona, a city of over a half-million people located in a desert environment and almost totally dependent on groundwater. The inset map shows how this booming city has been rapidly expanding since 1945. Unfortunately, most of the groundwater recharge occurred thousands of years ago under more humid and cooler climatic conditions. Today groundwater mining poses serious problems for the city's future.
© Richard T. Nowitz/Corbis

Groundwater Mining in the Breadbasket of the United States

The High Plains region in the central United States (Figure B11.2) is well known for its extensive grasslands that supported roaming herds of buffalo and for the Native Americans that used to inhabit the area. Beneath the Plains lies a thick sequence of sedimentary rocks containing vast deposits of sand and gravel known as the Ogallala Aquifer. This enormous aquifer system has allowed farmers to transform the High Plains into one of the world's major grain-producing regions, often referred to as the breadbasket of North America. Unfortunately, groundwater mining is threatening the continued high rates of agricultural production.

As with all groundwater mining, the basic problem in the High Plains is that groundwater recharge is less than the rate of withdrawal by humans. Rainfall across this semiarid region ranges from less than 16 inches (41 cm) per year near the Rocky Mountains and West Texas, to 28 inches (71 cm) in central Kansas. Because evaporation and transpiration rates nearly equal the amount of annual precipitation, soil moisture rarely reaches the point where groundwater recharge can occur. Similar to other aquifers in the American West, much of the groundwater being removed today represents recharge that occurred during cooler and more humid conditions that prevailed at the end of the last ice age, 12,000 years ago.

Prior to the 1880s settlers found the sparse rainfall of this semiarid region suitable mostly for grazing cattle. However, the climate pattern was one of alternating periods of unusually wet years followed by several dry years. Farmers soon discovered that they could grow bumper crops of wheat during consecutive wet years. Eventually they plowed large sections of the Plains to grow

wheat, only to experience plummeting production during dry years. A major drought in the 1930s, combined with tilling practices ill-suited for the semiarid climate, led to severe wind erosion and the collapse of local farm economies (see Case Study 10.1)

In response to the crisis in the 1930s, the federal government recommended a number of new farming practices, including irrigation to maintain soil moisture. Irrigation was made possible by the advent of modern drilling techniques and electric pumps. Toward the end of the 1930s farmers began ensuring bountiful harvests on the High Plains by extracting groundwater from the Ogallala Aquifer. Grain became so plentiful that it was used to feed cattle; soon the High Plains accounted for 40% of all grain-fed cattle in the United States. By 1980 approximately 170,000 irrigation wells (Figure B11.3) were supplying

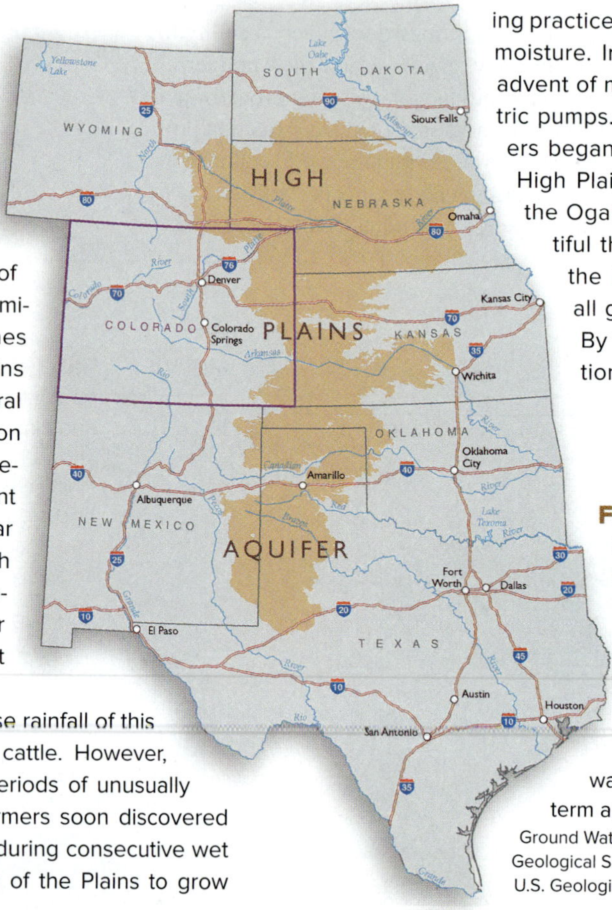

FIGURE B11.2 Map showing where the Ogallala Aquifer lies beneath the surface in the semiarid region of the United States known as the High Plains. Groundwater withdrawals from this vast and complex aquifer system have transformed the region into the nation's top grain producer, but have also resulted in dramatic water-level declines, threatening long-term agricultural production.

Ground Water Atlas of Colorado (2003): Colorado Geological Survey Special Publication 53, modified after U.S. Geological Survey Open File Report 99-267

Increased Well Costs In areas where water levels continue to fall due to groundwater mining, well owners will eventually be forced to incur the cost of lowering the pumps inside their wells. Should water levels drop below the well's intake screen, then the well itself must be deepened, which is considerably more expensive than lowering just the pump. Another problem is that falling water levels result in increased operating costs as the pumps require more electricity to lift the water greater distances to the surface. Although water will probably always be withdrawn for drinking purposes regardless of the electrical costs, the same does not apply for irrigation. At some point farmers will find that irrigation is no longer profitable,

A

FIGURE B11.3 Groundwater irrigation uses a center-pivot system (A) where a large-capacity well, located in the center, supplies water to a wheeled piping system that travels in a circle. In arid and semiarid climates, little if any of the applied water recharges deep aquifers due to the high evaporation and transpiration rates. False-color satellite image (B) showing part of Kansas reveals the large number of irrigation systems—each circle represents a center-pivot system. Such extensive groundwater withdrawals for irrigation in arid and semiarid climates are simply not sustainable.
(a)(left): © David Frazier/Corbis; (right): © Getty Images; (b) NASA

B Garden City, Kansas

water to 13 million acres of cropland, whereas only 2 million acres were irrigated in 1949.

Although irrigation has transformed the High Plains into a highly productive agricultural region, groundwater levels have fallen as much as 200 feet. Similar to other areas where extensive groundwater irrigation occurs in arid or semiarid climates, high rates of evaporation and transpiration allow very little if any of the applied water to return to the deep aquifers. In the case of the Ogallala Aquifer, the unsustainable use of groundwater means it is highly unlikely that farmers will be able to maintain today's high rate of agricultural production. Ultimately, grain production can be expected to fall, followed by higher food prices. Farmers will either have to depend solely on rainfall again or revert to grazing cattle.

forcing them to rely on natural rainfall. This in turn will result in lower crop production, or land being removed from production altogether (Case Study 11.1).

Reduced Stream and Spring Flow In regions with humid climates, large groundwater withdrawals are more likely to be sustainable because of greater recharge. In such an area a cone of depression will typically expand until it captures enough recharge to equalize what is being withdrawn by a well or series of wells. The cone will then stabilize and water levels will stop falling because the amount of recharge equals the discharge. Although the system reaches

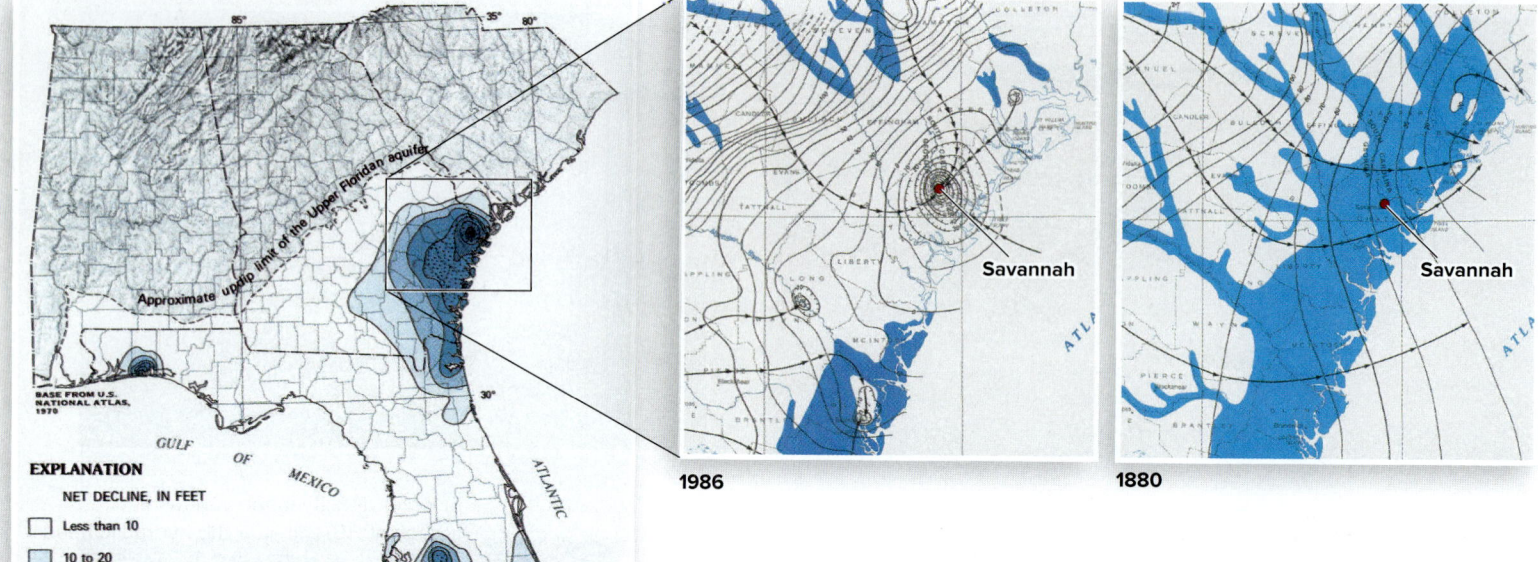

1986   1880

FIGURE 11.20 Large pumping withdrawals from a confined aquifer along the Georgia coast have resulted in a massive cone of depression. Contours show the potentiometric surface before and after major pumping began. Note how the areas of upward artesian flow (in blue) have been dramatically reduced.
(left): Bush, P. W., and R. H. Johnson (1988), U.S. Geological Survey. Professional Paper 1403-C.; *(middle-right):* Krause, R. E., and R. B. Randolph (1989), U.S. Geological Survey Professional Paper 1403-D

equilibrium, the hydraulic head of the aquifer is now lower than before the pumping began. This means that the amount of groundwater discharging to streams and springs will decrease. Such reductions in discharge can prove detrimental to many aquatic species, especially during dry periods when streams are sustained solely by groundwater discharge. Reduced flows also have the potential to impact important commercial and recreational fisheries. In some cases groundwater withdrawals can cause springs and streams to go completely dry, particularly in more arid climates.

Coastal Georgia provides a good example of an area where heavy pumping has reduced groundwater discharge to the surface environment. Over 60 years of industrial and municipal withdrawals from a confined aquifer have resulted in a massive cone of depression located near the city of Savannah, Georgia (Figure 11.20). Although the cone has stabilized, the potentiometric surface is now as much as 135 feet (41 m) lower than the prepumping level. In areas where the potentiometric surface was above the land surface (blue areas in insets of Figure 11.20), the hydraulic gradient was upward, which caused water from the confined aquifer to leak through fractures in the aquitard and discharge into the surface environment. This upward flow of artesian groundwater has now largely been shut off because of the drawdown cone. Researchers are currently studying how this reduction in groundwater flow may have impacted the ecology of the coastal region.

Salt Water Intrusion   Large groundwater withdrawals along a coastline can reverse the hydraulic gradient and cause **saltwater intrusion,** in which saline water migrates into the freshwater portions of an aquifer. From the example in Figure 11.21 you can see how pumping reverses the natural flow of groundwater around a well, drawing seawater into the coastal aquifer. This situation is very undesirable as the intruding seawater can raise the salinity of the aquifer to the point where it becomes unusable. Because a pumping well draws in water from all directions, a coastal aquifer can also become contaminated when salt water moves vertically along faults and fractures that cut across aquitards. Note that even in areas located far from any ocean, the upward movement of saline water from a deep aquifer can cause saltwater contamination of freshwater aquifers. Regardless of how it occurs, saltwater intrusion is a serious problem because it permanently degrades the water quality of an aquifer.

Land Subsidence   As described in more detail in Chapter 7, water exerts pressure inside the pore spaces of saturated materials. This pore pressure in turn helps support the weight pressing downward from the overlying column of rock

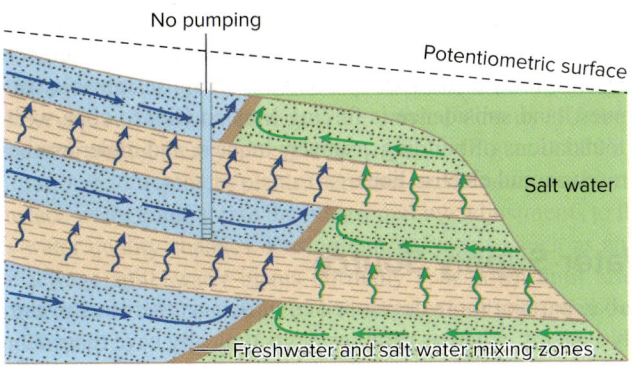

No pumping

Potentiometric surface

Salt water

Freshwater and salt water mixing zones

A

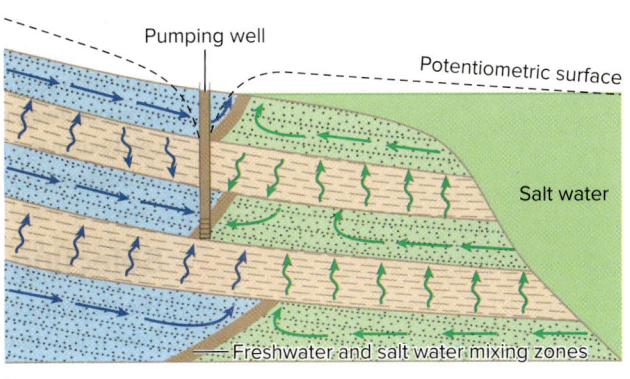

Pumping well

Potentiometric surface

Salt water

Freshwater and salt water mixing zones

B

Freshwater
Salt water
Brackish water

FIGURE 11.21 Under natural conditions (A), coastal aquifers contain both freshwater and salt water, which flow toward a mixing zone, the position of which depends on the hydraulic head and water density in the various aquifers. Large pumping withdrawals (B) will alter the hydraulic head within the system, causing the position of the mixing zone to move toward the well and allowing saltwater to contaminate the water supply.

or sediment. However, when water is pumped from the subsurface, this withdrawal decreases the pore pressure, which forces aquifers and aquitards to support more of the overlying weight. These materials then undergo compaction and experience a reduction in volume, which causes the land surface to slowly sink or subside as shown in Figure 11.22A. Because clay minerals are much more compressible than quartz or rock fragments, subsurface layers rich in clay will experience considerable volume reductions when pore pressure is reduced. Therefore, compaction of the aquifers themselves is relatively minor compared to that of aquitards. Problems associated with land subsidence are typically more severe in groundwater systems with thick, clay-rich aquitards. Note that once compaction occurs, it is largely irreversible, which means that the resulting subsidence becomes permanent.

One of the best examples of land subsidence is found in Mexico City, a metropolis of over 21 million people where nearly 70% of the water supply comes from a thick sequence of sand and clay layers. Because the withdrawal rate in this arid environment is greater than the rate of recharge, groundwater levels have been falling by about 5 feet per year. The corresponding loss of pore

FIGURE 11.22 Illustration (A) showing how land subsidence occurs when pumping in a confined aquifer creates a cone of depression, causing leakage and a reduction in pore pressure within the system. Most of the subsidence is due to compaction of highly compressible, clay-rich aquitards. Photo (B) shows the casing from a well in Mexico City that became exposed when heavy pumping withdrawals across the city caused the land surface to subside.
(b) © Alan V. Morgan, Earth and Environmental Sciences, University of Waterloo

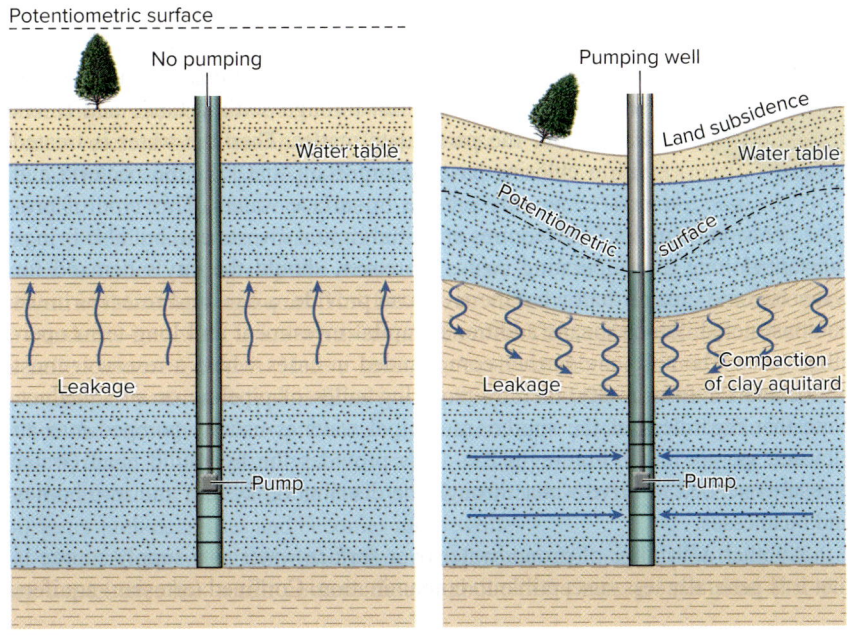

Potentiometric surface

No pumping

Water table

Leakage

Pump

Pumping well

Land subsidence

Water table

Potentiometric surface

Leakage

Compaction of clay aquitard

Pump

A

B

seawater, thereby creating a disposal issue. When discharged directly into the marine environment, the extreme salinity of the wastewater can be harmful to sensitive ecosystems. One successful solution is to blend it with freshwater being discharged from a nearby sewage treatment plant. In addition to the wastewater issue, desalination requires significant amounts of energy, making it considerably more expensive than extracting and processing conventional water resources. It currently costs about 75 cents to desalinate 250 gallons (950 l) of seawater, whereas a traditional supply system can produce the same volume of freshwater for around 15 cents.

Because of the high energy costs, desalination is primarily being used only in those areas with acute water-supply problems and by nations with the necessary financial resources. For example, desalination is common in some of the Persian Gulf nations where freshwater is extremely limited, but oil wealth enables those countries to pay the higher costs. In fact, the world's first large-scale desalination plant was built in Kuwait in 1965. According to the International Desalination Association, as of 2013, 150 countries around the world desalinate a combined total of 21 billion gallons (80 billion liters) of water per day. To put this number in perspective, the world's largest plant, which is in Saudi Arabia, desalinates 264 million gallons (1 billion liters) per day. For comparison, New York City's municipal system supplies 1.2 billion gallons (4.5 billion liters) of freshwater from surface reservoirs to over 9 million residents each day.

Chronic supply problems related to population growth and limited water resources have prompted other nations to begin desalination projects. For example, Israel completed a large reverse osmosis plant (Figure 11.24) in 2005, and it is now producing 87 million gallons (330 million liters) of freshwater per day. Israel subsequently added another major reverse osmosis plant with double the capacity. Together these two desalination plants meet about 15% of the country's total freshwater demand. In the United States the first large-scale desalination plant was built in Tampa Bay, Florida, where population growth had outstripped the natural water supply. There are also major desalination plants in California, Texas, North Carolina, and Arizona.

Desalination is likely to become a more attractive option in the future as demand for freshwater continues to grow. Moreover, technological improvements are making desalination more economical and efficient. For example, some plants are reducing costs by using excess heat from the process to generate electricity. Reverse osmosis is perhaps the most promising method of desalination since it can also be made more cost-effective by utilizing water of moderate salinity. In some areas brackish water from an estuary or deep aquifer is being used instead of seawater. However, even with increased efficiency, desalination is unlikely ever to be a cost-effective means of producing water for agricultural irrigation.

FIGURE 11.24 Aerial view showing Israel's reverse osmosis plant at Ashkelon, which is one of the largest in the world. Plants such as this produce highly saline wastewater, which can disrupt marine ecosystems if discharged directly into the ocean.
© IDE Technologies

Reclaimed or Recycled Wastewater

Municipal Wastewater Recycling

In addition to filtration and treatment plants for drinking water, municipal governments in the United States also operate sewage treatment facilities that break down organic matter in the wastewater before discharging it into the environment (Chapter 15). In areas with limited water supplies, municipalities are discovering that the supplies can be extended by using reclaimed wastewater rather than highly treated drinking water for certain applications. Moreover, it is far cheaper to expand water supplies in this manner than with the traditional approach of building new reservoirs or installing additional wells. Wastewater today is increasingly being viewed as a valuable resource as opposed to a financial liability.

Some sewage treatment plants in the United States are employing ultraviolet (UV) disinfection techniques capable of treating wastewater to the point where it meets drinking-water standards. Despite the fact that reclaimed water can be made safe to drink, the idea of drinking water that had once been flushed down toilets is unappealing to most people. Because of public perception and the potential for operator error that could result in the contamination of drinking-water supplies, reclaimed wastewater is generally not placed directly back into the supply system. Reclaimed water is instead being used for a variety of non-drinking applications, such as irrigating golf courses and parks, and for various industrial operations. The use of reclaimed water is currently being limited by the capital costs municipalities must incur to install distribution lines from the treatment plant out to potential users. Consequently, lines carrying reclaimed water are usually first extended out to the largest users, such as golf courses and large industrial plants.

Due to high population growth and limited water supply, Florida has been a leader in developing a comprehensive program to promote nondrinking uses of reclaimed wastewater. Some counties in California with similar issues have taken the next step and are putting treated wastewater back into the drinking-water system. In 1990 Los Angeles County officials announced a plan to use reclaimed water to recharge water-supply aquifers. The proposal, dubbed "toilet-to-tap" by critics, met stiff public opposition that eventually forced officials to shelve the project in 2001. When nearby Orange County officials announced a similar plan, they did so in conjunction with an extensive public education program. This effort was successful and resulted in a $481 million plant that is now processing 100 million gallons (379 million liters) of water per day, which is equivalent to the freshwater production at some of the world's largest reverse osmosis plants Based on the success of the public education approach, similar projects are now underway in San Francisco and San Diego. Drought-stricken towns in western Texas are also starting to adopt the toilet-to-tap approach, although on a much smaller scale.

Industrial and Domestic Recycling

Similar to municipal governments, industrial plants and homeowners have found it cost-effective to recycle their wastewater, particularly in areas where municipal water rates are high. For example, when water was relatively cheap and plentiful, factories would often use water in certain applications and then promptly discharge it as waste. Today, there are financial and political incentives for businesses to reuse water whenever possible. A good example is the reuse of so-called *noncontact cooling water,* which is water that is used only for cooling and does not come into direct contact with the heat source. Other than being at a higher temperature, the quality of noncontact cooling water is unchanged. Industrial plants therefore have found it economical to store the water on site so it can cool and be used again.

Reclaimed wastewater can also be used in homes in order to reduce demand on the drinking-water supply. Here the plumbing system is typically modified so that all the wastewater in the home except that from the toilets is collected. Because this water has never been in contact with human waste, it is called *graywater,* which distinguishes it from sewage. Graywater is mostly used to irrigate outside plants and gardens, but can also be used in any number of applications that do not require drinking water.

Aquifer Storage and Recovery

A management technique known as **aquifer storage and recovery (ASR),** sometimes referred to as *water banking,* involves storing surplus surface water in aquifers and then removing (i.e., recovering) it for use at a later time. Aquifers that are well suited for this are generally composed of granular materials that have high porosity and are very permeable. In this technique surface water is either injected into the aquifer using wells or spread out on the surface in specially constructed recharge beds or basins and allowed to infiltrate. The water is then removed by pumping wells during periods of high demand. In ASR an aquifer functions like a surface reservoir in that it simply stores water for later use.

Although ASR and surface reservoirs function in a similar manner, there are important differences that make storing water in aquifers inherently risky. Some ASR systems use treated water obtained from rivers during seasonal periods of elevated discharge. Careful management is required since withdrawing too much water can damage critical stream ecosystem functions that depend on periods of high discharge. Other ASR systems use treated effluent from sewage treatment plants (Chapter 15). Although the source water in both cases is carefully treated, there is the potential that contaminated water could inadvertently be introduced into the aquifer. Because of the porous nature of most aquifers, it is extremely difficult, if not impossible, to remove the contaminants. ASR therefore could permanently render water from an aquifer unfit for human consumption. Another potential problem stems from the fact that surface water has a much higher dissolved oxygen content than does most groundwater. Injecting such oxygen-rich water into an aquifer can result in undesirable chemical reactions (e.g., release of arsenic from the aquifer material) and/or bacterial growth.

Despite the potential risks, ASR is being used in many parts of the world, including the United States. By injecting water into an aquifer, ASR not only produces a useful stockpile of water but helps to rebuild water pressure (hydraulic head) in the subsurface. This, in turn, can help remediate the effects of saltwater intrusion and land subsidence problems associated with excessive groundwater withdrawals.

Rainwater Harvesting

Collecting and storing rainwater is an ancient practice commonly referred to as *rainwater harvesting.* A typical rainwater harvesting system for a home or business (Figure 11.25) consists of a set of gutters to collect rainwater that runs off a roof and a piping system for carrying it to a storage tank called a *cistern.* The building's plumbing system is then modified to make use of the cistern water. Some homeowners choose to have the cistern provide water to the washing machine, toilets, and outside faucets, leaving only the dishwasher, shower, and inside faucets plumbed to the primary drinking-water supply. Others may use cistern water strictly for irrigating a garden and outside plants. Some businesses harvest rainwater from both roofs and parking lots and then store the water in ponds for irrigation purposes.

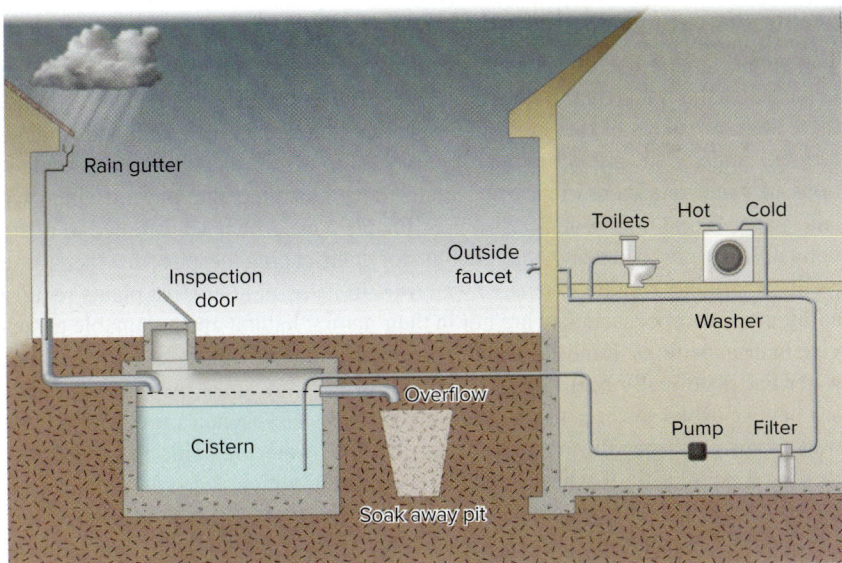

St. Petersburg, Florida

FIGURE 11.25   Rainwater harvesting systems involve collecting rainwater from a roof, then storing it either above or below ground in a tank called a cistern. With modifications to a building's plumbing system, cistern water can be used in place of drinking water for certain applications.
© Rainwater Services, LLC

People are beginning to return to this age-old practice because of the need to reduce the demand on their primary water supply. For example, many rural homes must rely on an aquifer that is either thin or composed of fractured rocks, and therefore not capable of supplying large volumes of water. This means they may not have enough water for such things as washing multiple loads of clothes, irrigating a garden, or even frequent toilet flushing. In other cases homeowners may simply want to reduce their use of municipal water because of its high cost or because of restrictions on outdoor water use.

Conservation

Use of alternative water supplies essentially involves adding new or additional freshwater to a region's overall supply. Water conservation is different in that it is about using an existing supply more efficiently so that some of the water becomes available for other uses. Although conservation does not increase the total supply, it has the same net effect. An analogy is a person who saves money by reducing the amount spent on nonessential items, such as going to the movies and eating in restaurants. The savings are then used to purchase more essential items, like gasoline for getting to work. Similar to being fiscally conservative, water conservation means you have freed up some of your water supply for more essential uses.

All water conservation efforts fall into one of two categories. The first consists of *engineering practices* in which plumbing systems, appliances, and operational procedures are modified so less water is used during some process or application. An example is a more efficient dishwasher. The other category encompasses *behavioral practices* in which people change their water-use habits. Here an example is spending less time in the shower. Although conservation is an effective means of increasing a water supply, it often faces resistance because it requires people to change their personal habits or spend money to purchase more efficient hardware (toilets, shower heads, etc.). This resistance is usually overcome when: (a) the price of water becomes sufficiently high; or (b) there is a shortage and users have no other means of increasing supply. Next, we will explore some of the more common conservation techniques, with the most desirable generally being those that save the greatest volume of water with the least amount of hardship and capital investment.

A Phoenix, Arizona

B Las Vegas, Nevada

FIGURE 11.26 Large reductions in home water use can be realized through landscaping changes. Xeriscaping involves using native plants (A) rather than nonnative vegetation (B) that requires extensive irrigation.
(a) © Desert Crest, LLC; (b) USDA

Domestic and Commercial Users

The single largest use of water by residents and commercial businesses in the United States is for irrigating lawns, shrubs, flowers, and gardens. Interestingly, there are more acres of turf grass in the United States than the combined acreage farmers dedicate towards growing corn, wheat, and fruit trees. Substantial reductions in water use therefore can be realized by changing the way in which we landscape around homes and businesses. People have grown accustomed to using nonnative shrubs, flowers, and grasses for landscaping despite the fact these plants commonly require extensive watering. In general, nonnative plants require additional water because they are not in their natural habitat and are unable to survive in drier settings. Landscape irrigation has not been much of an issue because water has historically been cheap and plentiful in many areas. As water supplies are being pushed to their limits, many local and state agencies are now raising water rates and imposing restrictions on outdoor watering. This has led to the increased popularity of what is called **xeriscaping,** which is a type of landscaping using native plants that require little to no irrigation (Figure 11.26). People are discovering that native vegetation not only saves them money by dramatically reducing their water use, but can also be aesthetically pleasing.

With respect to indoor water usage, the highest-volume activity is toilet flushing. It is estimated that the average American uses nearly 7,000 gallons (26,000 l) of water a year just to flush toilets. Traditional toilets use between 3.5 and 5 gallons (13–19 l) per flush and account for approximately 27% of indoor domestic use. In order to conserve water, so-called *low-flow toilets,* which use only 1.6 gallons (6 l) per flush, have been required in new U.S. homes since 1994. The potential savings from low-flow toilets is significant considering the volume of water being used and the fact that the new toilets use 2 to 3 times less water than traditional models. Another high-volume activity is showering, which accounts for 12% of indoor water use. Because older shower heads usually release far more water than necessary, replacing them with highly efficient low-flow shower heads is a simple and inexpensive way of conserving water.

In addition to physical changes, there are many behavioral practices people can alter to reduce water usage. For example, running a dishwasher only when it is full can save 10 to 20 gallons (38–75 l) per day per household. Another simple method is to turn off the water while shaving or brushing your teeth—some people even choose to take shorter showers and turn the water off while soaping. Similarly, when washing dishes by hand, water can be saved by not allowing the faucet to run continuously—this has the additional benefit of saving energy to make hot water. When laundering, one can adjust the water level in the washing machine so that it matches the size of the load being washed. Behavioral changes are important outside of the home as well. Examples include turning the hose off while washing a car and using a broom instead of a hose to sweep off driveways and sidewalks.

Although the individual savings from any single appliance or personal habit may seem small, added together they can reduce a household's consumption rate by 30% or more. Moreover, if every household in a city conserved at this same rate, the municipal supply system would have 30% more water for additional users. You will see in later chapters how the collective power of conservation can play an important role in our mineral and energy supplies.

Municipal Supply Systems

In order to reduce the overall demand for water, municipal supply systems across the United States are developing educational programs designed to encourage water conservation. Because low-flow toilets and shower heads are relatively inexpensive and make a significant impact on consumption rates, many municipalities have instituted retrofit programs that offer homeowners rebates for installing more efficient units. In some cases owners are reimbursed for the entire cost of the retrofit. Local governments have discovered that increasing water supplies through conservation programs is often less costly than physically expanding their supply system, which requires a large capital investment.

Education is helpful, but water utilities commonly find it difficult to get people to conserve when the price of water is low. Considering that municipal water averages only 30 cents for 250 gallons (950 l), there is little financial incentive for people to conserve. To help discourage excessive water use, many municipalities are changing from a flat water rate to a tiered rate in which users pay progressively higher rates as the volume they use increases. Another effective conservation measure is repairing leaks in municipal supply systems. Leaks develop over time as water mains deteriorate with age or develop cracks when the ground undergoes compaction and settles. Reducing the amount of water being lost through leaks means that additional water can be made available for new users. Municipalities that repair leaks also save the processing and treatment costs they were paying for water that is lost into the ground.

Agriculture

Recall that agricultural irrigation is one of largest uses of freshwater. Moreover, modern irrigation techniques are one of the primary reasons agricultural productivity has increased in recent years, allowing the world to feed an ever-growing population (see Case Study 11.1). As water supplies continue to become stretched, improving irrigation efficiency will become more critical, particularly in areas of where groundwater mining is underway. For example, depending on temperature and wind conditions, as much as 30% of the water leaving traditional irrigation systems (Figure 11.27A) is caught by the wind and evaporates before ever reaching the ground. Because of poor efficiency and the sheer volume of water involved, minimizing irrigation losses can save enormous amounts of water.

One of the simplest and least expensive ways of improving irrigation efficiency is for farmers to employ techniques that cause soils to retain more moisture—the more moisture that is retained, the less the fields need to be irrigated. Recall from Chapter 10 that no-till farming is the technique in which the previous season's crop is left standing so as to reduce soil erosion. By keeping the soil covered, this crop residue also reduces the evaporative loss of water from soils, producing the same effect as placing mulch around home flower beds. Furrowing and contour plowing are erosion control practices that decrease overland flow, but have the additional benefit of increasing infiltration and soil moisture. Another inexpensive technique is to gather data on soil moisture, soil properties, and climatic conditions so that irrigation schedules can be fine-tuned for specific crop types. The idea here is to avoid irrigating fields when moisture conditions are high and the crops are unable to benefit from the additional water—similar to the situation where watering your lawn right after a heavy rain does nothing but waste water.

Finally, irrigation efficiency can be increased by making physical modifications to an existing irrigation system or by installing an entirely new system. Either approach results in a farmer using less water to achieve the desired soil moisture for a particular crop. A good example is shown in Figure 11.27B, where a spray-irrigation system has been modified so that water is directed downward to reduce evaporative losses. Under certain conditions a farmer may find it cost-effective to install a *drip-irrigation system,* where a piping system applies water to the base of the plants one drop at a time (Figure 11.28).

FIGURE 11.27   In older irrigation systems (A) as much as 30% of the water evaporates and never reaches the ground. Systems can be retrofitted (B) so that water is directed toward the ground, which greatly reduces evaporative losses.
(a) © Don Tremain/Getty Images; (b) © Clark Dunbar/Corbis

FIGURE 11.28   Drip-irrigation systems, like this one in a vineyard in New Mexico, are the most efficient as water is applied only to the root zone of each plant. However, irrigating in this manner is labor-intensive and involves considerable material costs.
Jeff Vanuga, Natural Resources Conservation Service, US Dept of Agriculture

Drip-irrigation systems are the most efficient in terms of water usage, but are also quite labor-intensive. Although physical improvements in irrigation systems typically involve considerable capital investments, farmers can eventually recover the cost through increased crop yields made possible by the additional water.

SUMMARY POINTS

1. Earth's abundance of liquid water is what makes human life possible and our planet so unique within the solar system.

2. The hydrologic cycle creates freshwater through evaporation. Freshwater eventually falls on landmasses, where it flows into rivers and lakes and infiltrates the subsurface. Humans tap into this freshwater as it moves through the surface and subsurface portions of the hydrologic cycle.

3. Consumptive water use refers to human activities in which water is lost or consumed and cannot be used again. Off-stream use occurs when water is removed from a supply source, but discharged elsewhere after being used.

4. Off-stream diversion of surface water via aqueducts provides increased water supplies for those receiving the water, but can adversely affect water supplies, ecosystems, fisheries, and other activities in the basin from which the water is being removed.

5. Dams are valuable for stockpiling water, but also discharge cooler water and trap sediment, which can damage downstream ecosystems and fisheries.

6. Aquifers are earth materials that easily transmit water. Granular materials like gravels and sands are most desirable because they can transmit larger volumes of water. Fractured rocks and solution limestones also serve as aquifers. Clay-rich materials and unfractured crystalline rocks do not transmit much water, and hence are called aquitards.

7. Unconfined aquifers are open to the atmosphere, whereas confined aquifers are overlain by an aquitard and their water is pressurized. The water level in an unconfined aquifer is defined by the water table, whereas the potentiometric surface represents the pressure in a confined aquifer.

8. Hydraulic head is a measure of water's potential energy within an aquifer and is related to the elevation of the water table or potentiometric surface. Groundwater always flows toward areas of lower hydraulic head, and thus it can move both horizontally and vertically in the subsurface.

9. Groundwater mining occurs when the withdrawal rate from an aquifer is greater than the natural recharge, causing the water level to fall progressively. When this occurs, water is no longer a renewable resource. In addition, withdrawals of groundwater can lead to saltwater intrusion and land subsidence.

10. With respect to use as a water supply, groundwater is generally more desirable than surface water because of its higher purity, which results in lower infrastructure and treatment costs.

11. Alternative ways of obtaining freshwater are needed in areas where traditional supplies are no longer able to meet demand or where withdrawals are creating environmental problems.

12. Desalination and reclaiming of wastewater are alternative means of increasing existing supplies, whereas various conservation techniques make more water available by using existing supplies more efficiently.

KEY WORDS

aquifer   341
aquifer storage and recovery (ASR)   358
aquitards   341
cone of depression   347
confined aquifer   341
consumptive water use   336
desalination   355
freshwater   333

groundwater   340
groundwater mining   348
hydraulic conductivity   341
hydraulic gradient   342
hydraulic head   342
hydrologic cycle   332
leakage   344
off-stream use   336

potentiometric surface   342
saltwater intrusion   352
spring   345
unconfined aquifer   341
water table   341
xeriscaping   360

APPLICATIONS

Student Activity   Although consumptive and off-stream uses of water are different, in both cases the water does not return to the original source. Make a list of the different ways in which you use water around your home and categorize each use as being either "consumptive" or "nonconsumptive" and either "off-stream" or "in-stream." To complete this activity you will need to do an Internet search or call your local water department to determine the source of your water, such as a river, lake, or aquifer—if you are using your own well, then it is an aquifer. Likewise, find out whether the wastewater plant in your community is discharging into a river, lake, or ocean—if you are on a septic system, then you are most likely discharging into an aquifer. Under which category does most of your water usage fall? If you were to try to conserve water, which category would be affected the most?

Critical Thinking Questions

1. Dams can create large reservoirs for greatly expanding a region's water supply. If a dam releases water and always maintains the flow of its river, then how can it have a negative impact on downstream ecosystems?

2. Most water-supply aquifers are composed of granular materials, such as sand and gravel. In some areas though, municipalities withdraw water from limestone (karst) aquifers, in which groundwater moves through open passageways. How might this type of aquifer be more susceptible to contamination than a sand and gravel aquifer?

3. When a municipal water system needs to expand its supply because of growing demand, why is it often preferable to accomplish this through conservation efforts as opposed to building a new reservoir or drilling additional wells?

Your Environment: YOU Decide

Tim McCabe, USDA Natural Resources Conservation Service

Freshwater supplies are clearly being pushed to their limits in many areas, causing a variety of problems, including damage to ecosystems, groundwater mining, saltwater intrusion, and land subsidence. Water conservation is an effective way of extending our existing water supplies without further damaging the environment. However, it is difficult to get people to conserve, given the fact that in most municipalities water is still very cheap, averaging about 15 cents per 250 gallons. Some local governments have imposed tiered rate structures on water users (i.e., people pay progressively higher rates the more they use) so as to discourage excessive consumption. Explain whether you think encouraging people to conserve through a tiered rate structure is a good idea, or should conservation be purely voluntary?

Chapter 12

Mineral and Rock Resources

LEARNING OUTCOMES

After reading this chapter, you should be able to:

▶ Discuss the degree to which people in industrialized societies rely on rock and mineral resources.

▶ Explain why certain minerals are used in specific applications.

▶ Characterize the difference between a mineral reserve and mineral resource, and explain what is meant by an economic deposit.

▶ Explain the key geologic processes that concentrate minerals into economic deposits in the basic rock types: igneous, sedimentary, and metamorphic.

▶ Describe two mining-related factors that determine whether a deposit will be economical to mine.

▶ Characterize the basic steps in ore processing.

▶ Explain why mineral resources are unevenly distributed on Earth's continents and why some minerals are called strategic minerals.

▶ Discuss the two primary ways in which humans can meet future demand for finite mineral resources.

▶ Explain how sulfide minerals are related to acid mine drainage and acid rain.

A spinning bucket excavator removing large quantities of rock from a surface mine. Modern society requires tremendous quantities of mineral and rock resources, which must be extracted from the Earth, creating a host of environmental issues. As population continues to grow, the demand for these resources will increase, presenting additional problems as some types are already in short supply.

Introduction

© StockTrek/Getty Images

Nearly all of us who live in modern societies enjoy a lifestyle that was unimaginable a mere 100 years ago. Today there are highly engineered automobiles and jet airliners that can take us anywhere we want to go in climate-controlled comfort, plus we have the convenience of smart phones. We are also able to live and work in comfortable buildings and enjoy a fantastic array of foods and consumer goods. This modern lifestyle is entirely dependent on two key resources: energy and minerals. Minerals are needed for everything we build, from pencils to farm equipment to the very factories that make all these things. Energy resources such as oil, gas, and coal provide the power that keeps everything running, particularly the lighting, heating, and cooling systems in our buildings, as well as our vast transportation network. Moreover, energy is what makes it possible for us to mine the minerals from which most things are built. Although water resources are essential for life itself (Chapter 11), our modern way of life simply would not exist without Earth's abundant, but limited, mineral and energy resources. Our focus in this chapter will be on the human use and extraction of minerals, whereas Chapters 13 and 14 will be devoted entirely to energy resources.

One of the more interesting facts about mineral resources is the amount of minerals required to support the modern lifestyle we enjoy. Table 12.1 lists the amounts of different mineral resources used by the average American citizen on a yearly basis. Keep in mind that the per capita consumption rate of almost 16,000 pounds of stone, sand, and gravel, for example, does not mean you are using this amount each year at your house; rather it represents your portion of the total amount of this resource being used in constructing the nation's highways, buildings, factories, and more. We can therefore think of each person as having a *mineral-resource footprint* on the planet, representing the amount of minerals needed to support his or lifestyle—this is similar to the concept of an ecological footprint (Chapter 1) and a carbon footprint (Chapter 16). Clearly, a person living in a modern industrialized nation has a larger mineral footprint than a person in a developing country.

TABLE 12.1　**Average yearly per capita consumption rates of various mineral resources in the United States.**

	Mineral Resource	U.S. Yearly Per Capita Consumption	Percentage of All Mineral Resources
	Sand, gravel, and crushed stone	15,892 lb (7,215 kg)	87%
Nonmetals	Cement	616 lb (280 kg)	3.4%
	Salt	422 lb (192 kg)	2.3%
	Phosphate rock	212 lb (96 kg)	1.2%
	Clays	145 lb (66 kg)	0.8%
	Iron	330 lb (150 kg)	1.8%
	Aluminum	66 lb (30 kg)	0.36%
	Copper	13 lb (5.9 kg)	0.071%
Metals	Lead	11 lb (5.0 kg)	0.060%
	Zinc	7 lb (3.2 kg)	0.038%
	Manganese	6 lb (2.7 kg)	0.033%
	All other minerals	592 lb (269 kg)	3.2%
	Total	**18,312 lb (8,314 kg)**	**100%**

Source: Minerals Education Coalition, 2015.

Another interesting aspect of Earth's mineral resources is that, unlike water, people rarely see minerals in their original state; thus few of us make the connection between the goods and services we use and the minerals involved. Moreover, people generally do not appreciate the effort that goes into finding, extracting, and processing minerals. For example, we all use glass every day. But do you know what glass is made of or what it takes to make it? Glass is actually made by melting quartz, which is one of the dominant minerals found in rocks and sediment near Earth's surface. Every time you drink from a glass, look through a window, or enjoy the comfort of a well-insulated building, you are making use of quartz grains just like those you see on a beach or in a river. To help remind people that their everyday lives are dependent on minerals, mining geologists often say "if it can't be grown, it must be mined."

Although people commonly lose sight of the fact that minerals have made our modern lifestyles possible, they are generally aware that the mining and processing of minerals have historically led to environmental problems. Despite the fact that recent environmental regulations have greatly reduced the negative impacts of mining, people often oppose the opening of any new mines near their homes. This attitude is based on the concept of NIMBY, or "not in my back yard." While it is understandable why people would not want a mine next to their house, it is important to realize that we all enjoy the benefits of minerals, hence creating the demand to which mining companies are responding. Therefore, each of us is partly responsible for any negative consequences associated with the mining and processing of minerals. This also means that as citizens we all need to do our part to help ensure that these precious resources are used in a wise and efficient manner, and that they are extracted as safely as possible.

In this chapter we will explore the ways in which geologic processes concentrate minerals into deposits that are economical for people to mine. We will also examine the basic mining and processing techniques used to turn these resources into goods and services, along with the associated environmental problems. We begin by looking at how society uses minerals and why certain minerals are used in specific applications.

Minerals and People

Recall from Chapter 3 that a *mineral* is a naturally occurring inorganic solid whose individual atoms are arranged in an orderly manner (i.e., in a crystalline structure). Some minerals, such as diamond (C), are composed of only one type of atom, but most contain different types of atoms, as in the case of the copper-rich mineral called bornite (Cu_5FeS_4). Also recall that a *rock* is simply an assemblage of one or more minerals. We can think of a **mineral resource** as any rock, mineral, or element that has some physical or chemical property humans find useful. Mineral resources can range from rocks that contain scattered grains of a specific mineral to a rock body composed almost entirely of a valuable mineral, such as the layers of common table salt or halite (NaCl) shown in Figure 12.1. Mineral resources also include sorted river sands and gravels and crushed rocks used as fill material in construction projects. Certain types of rocks are even cut into slabs and used for decorative countertop and flooring materials or as cemetery stones. Notice in Table 12.1 that geologists typically subdivide mineral resources into metals and nonmetals, and that rocks are categorized as nonmetals.

Certain minerals have physical or chemical properties (Chapter 3) that have practical applications. A good example is the mineral *diamond,* which is useful because of its hardness, or the physical ability to resist scratching. From Figure 12.2 you can see that diamond is over four times harder than the next hardest mineral (corundum), making diamond by far the hardest known substance. People have learned to make use of diamond's extraordinary hardness by placing tiny industrial-grade diamonds on the edge of rotating saw blades or drill bits, thereby creating tools that will literally cut through anything. Metallic copper is another good example of a mineral that is valued for its specific properties. Copper metal is widely used for wiring because it easily conducts electricity. Steel, which is made from iron, is also a good electrical conductor and is considerably cheaper than

FIGURE 12.1   Mineral resources include a variety of different rocks and minerals. This underground mine is in a layered sedimentary deposit composed almost entirely of the mineral halite. Finely ground halite is used as common table salt, whereas coarsely ground halite (inset), called rock salt, is used for de-icing roadways in the winter.
Ohio Department of Natural Resources; (inset): © Salt Institute

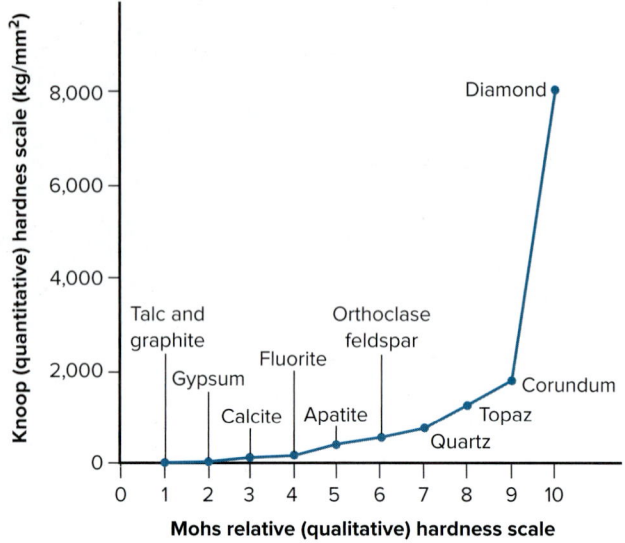

A

B

FIGURE 12.2   Graph (A) comparing the quantitative and qualitative hardness of common minerals illustrates the vast difference in hardness between diamond and the next hardest mineral, corundum. Diamonds are well known as gemstones, but their greatest use is in cutting tools (B), where their extraordinary hardness allows humans to cut through any type of material.
(b) U.S. Navy photo by Journalist 1st Class Mark H. Overstreet (RELEASED)

copper. Why then is copper used for electrical wiring and not steel? The answer is that copper has an additional physical property of being *malleable,* meaning it is able to deform without breaking. Malleability is critical in that it allows copper wiring to be wrapped around a shaft or to make tight bends in a machine, something that would cause steel wiring to break or crack, stopping the flow of electricity.

Table 12.2 lists some of the more commonly used metallic and nonmetallic mineral resources, to give you a better appreciation for the connection between mineral properties and their use in society. A good example is the use of the mineral quartz as the raw material for making glass. Recall that a natural form of glass,

TABLE 12.2   Applications and properties of selected metallic and nonmetallic mineral resources.

Metallic and Nonmetallic Mineral Resources	Important Applications	Key Physical and Chemical Properties
Gold (Au)	Electronics, jewelry, currency, bullion	Electrical conductor, noncorrosive, malleable
Silver (Ag)	Electronics, jewelry, photographic films	Electrical conductor, malleable
Copper (Cu)	Electrical wiring, plumbing, coins, alloys	Electrical conductor, malleable
Lead (Pb)	Batteries, solder, bullets, weights	High density, soft, low melting point
Zinc (Zn)	Rust-proofing steel, paint, alloys, coins	Corrosion resistant
Iron (Fe)	Iron, steel, yellow to brown pigments	High strength
Aluminum (Al)	Aluminum metal, chemicals	Lightweight, high strength, corrosion resistant
Titanium (Ti)	White pigment, metal for aircraft, ships, and human joint replacements	Lightweight, high strength
Graphite (C)	Dry lubricant, graphite compounds, pencil leads	Extremely soft
Diamond (C)	Cutting tools, gemstones	Extremely hard
Quartz (SiO_2)	Glass, sand for mortar and cement, watch crystals	Transparent, hard, chemically resistant
Calcite ($CaCO_3$) (limestone)	Main ingredient of Portland cement, concrete, agricultural lime	Chemically reactive
Gypsum ($CaSO_4\,2H_2O$)	Sheetrock (drywall), plaster of Paris	Low density
Kaolinite clay	Paper filler or coating, filler and extender in paint, rubber, plastics, cosmetics, and medicine, ceramics	Extremely soft, white color, absorbent

obsidian, forms when silica-rich (SiO_2) magma cools so quickly that the atoms within the magma are unable to arrange themselves into a crystalline structure and form minerals. People long ago learned how to produce their own glass by melting pure quartz sand, which is composed entirely of silica (SiO_2). When quartz is melted, it forms a viscous liquid that can be molded into practically any shape, which can then be preserved by allowing the molten glass to cool and harden. By mixing small amounts of other minerals with quartz sand, people learned to make glass with slightly different properties.

Humans have found a large number of applications for glass because of its ability to take on any shape and its set of physical properties, which are similar to those of the mineral quartz. In addition to making containers for storing liquids and solids, glass is molded into small fibers we call *fiberglass*. Because a blanket of fiberglass contains considerable air space, it inhibits the transfer of heat, making fiberglass an ideal insulating material for buildings and plumbing systems. Another key property of quartz-based glass is that it is highly *transparent* (i.e., readily transmits light). Consequently, modern societies use large amounts of quartz for making thin sheets of crystal-clear window glass. The fact that quartz-based glass is also quite hard gives it a distinct advantage in certain applications over other transparent materials that are relatively soft. For example, plexiglass (made from a plastic) is rather soft and is never used for automotive windshields since it would become so heavily scratched that it would be difficult to see through. Quartz-based glass is also chemically inert, which means that glass products do not break down when exposed to most chemicals. Keep in mind that quartz is just one of the minerals that we use in society. Finally, note that we will refer back to many of the minerals and properties listed in Table 12.2 throughout this chapter.

Economic Mineral Deposits

In Chapter 3 you learned that there are over 4,000 minerals found on our planet, but only a dozen or so make up the bulk of Earth's crust and mantle. Even more surprising is that 98.3% of the crust by weight is composed of just eight elements (oxygen, silicon, aluminum, iron, calcium, sodium, potassium, and magnesium). All the remaining elements and their compounds therefore are actually quite rare in the rocks and minerals we find near Earth's surface. Fortunately, most of these less abundant elements are not evenly dispersed in trace amounts throughout the crust, but rather are often found concentrated in mineral deposits that result from geologic processes. Because it takes energy to extract minerals from the earth, the concentration of a particular deposit is a major factor in determining whether or not it is economical to mine. A more concentrated deposit means there is less waste material to remove, which in turn makes it more profitable since less energy is required to extract the resource. Geologists use the term **enrichment factor** to describe the degree to which a mineral resource is concentrated above its average concentration in the crust. A deposit may be deemed economical to mine if it reaches a certain concentration factor. Table 12.3 lists the average crustal composition of several important metals and the amount of enrichment typically required to create an economical deposit. Notice how gold can be profitably mined at a much lower concentration than aluminum or iron. The variation in enrichment factors reflects the value of these resources to society and the energy required to extract them.

Since society has developed applications for a relatively small number of Earth's 4,000 or so minerals, we

TABLE 12.3 The average crustal concentration of some commonly used metals and the enrichment factor needed to create an economical ore deposit.

Metallic Element	Average Percent Concentration in Earth's Crust	Percent Concentration Needed for Economical Mining	Approximate Enrichment Factor (Economic Concentration/Crustal Concentration)
Aluminum (Al)	8	35	4
Iron (Fe)	5	20–69	4–14
Titanium (Ti)	0.57	32–60	56–105
Copper (Cu)	0.0063	0.4–0.8	80–160
Lead (Pb)	0.0015	4	2,500
Gold (Au)	0.0000004	0.001	2,500

Source: Data from U.S. Geological Survey Professional Paper 820, 1973.

need some means of describing those which are economically valuable. Although the term *ore* is often associated with metals, for our purposes we can define *ore minerals* as those which contain an element or compound that has some value to society. Likewise, an **ore deposit** will be defined as a body of rock or sediment whose concentration of some mineral is sufficiently high that the deposit is economically feasible to extract. Note that the terms *low-grade* and *high-grade* refer to the enrichment level of ore deposits. Later in this chapter we will examine how geologic processes concentrate crustal elements and form ore deposits.

Resources and Reserves

Before mining can begin, it is obvious that an economical mineral deposit must first be located. In the past high-grade deposits were often discovered on the surface because they looked noticeably different from the surrounding rocks. Deposits that remain today are generally more difficult to locate because they are either buried or contain fine-grained minerals that are dispersed throughout a rock body. In many cases mineral deposits are simply not economical to mine. The volume of a deposit may be too small, or its mineral concentration might be too low to justify the cost of mining and processing. Keep in mind that changes in technology or in supply and demand can alter the economics of mineral extraction and processing. Therefore, deposits that are currently considered uneconomical to mine may become profitable in the future.

In order to quantify the value of their mineral resources, mining companies and governments typically classify their holdings based on economic status. For example, the U.S. Geological Survey uses the term *mineral resource* to describe a deposit that is *feasible* to mine using existing technology. This definition simply means the technical ability to extract the mineral exists; it says nothing about the profitability of extracting it. A **mineral reserve,** on the other hand, is a deposit that is economical to extract under current conditions. As shown in Figure 12.3, this classification makes it possible for geologists to quantify what is referred to as *total mineral reserves,* which represents all the known deposits currently economical to mine. The remaining known deposits are then combined with some estimate of deposits yet to be found, providing what is called the *total mineral resource.* This concept is important because it allows economists to estimate how much of a particular resource is available to the global marketplace at any given time. The ratio of reserves to resources, of course, will vary as supply and demand affect the market price for a particular resource. Should demand decrease or supply increase, prices will fall and many marginally economical mining operations will shut down, causing the total reserves to decrease. The opposite will occur when demand goes up or supply goes down. Often the driving force behind these changes is improved technologies that allow lower-grade deposits to be mined.

Geology of Mineral Resources

In this section we will explore some of the more common geologic processes that result in crustal elements becoming concentrated, thereby forming enriched mineral deposits. These enrichment processes operating within the Earth system can be broken down into igneous, metamorphic, and sedimentary processes (includes weathering).

Igneous Processes

Diamond Pipes

Diamond deposits are unique in that they are found associated with an unusual type of igneous rock that forms from magmas originating at depths of 75 to 125 miles

	Identified deposits	Undiscovered deposits	
		Known mining areas	Unknown mining areas
Economic	Reserves	Hypothetical resources	Speculative resources
Marginally economic	Marginal reserves	Hypothetical resources	Speculative resources
Subeconomic	Subeconomic resources		

FIGURE 12.3 Mineral resources that have been identified and are profitable to mine are called reserves. Those believed to exist in areas of known deposits but yet to be discovered are considered hypothetical resources; all other mineral resources are speculative.

(120–200 km) in the upper mantle. As indicated in Figure 12.4, diamonds form in certain places in the upper mantle where temperature and pressure conditions allow carbon atoms to arrange themselves into the crystalline structure (Chapter 3) of the mineral we call diamond—closer to the surface carbon atoms develop the atomic structure of graphite. Should a magma that forms at depth rise through this diamond-bearing zone, it can then carry diamond crystals toward the surface by forcing its way through fractures. Because the magma is highly pressurized, it decompresses suddenly and violently (Chapter 6) when encountering a weak zone in rocks near the surface. The result is a small, but very explosive volcanic eruption that produces a carrot-shaped crater known as a diamond pipe or *kimberlite pipe* (named after the town of Kimberly in South Africa). The pipes themselves are filled with magnesium-rich volcanic rocks that typically are quite different from the surrounding rocks.

Diamonds are relatively rare because the formation of kimberlite pipes is geologically somewhat uncommon, plus not all pipes contain diamonds. Interestingly, the volcanic eruptions that form kimberlite pipes create circular deposits of volcanic debris that may contain diamonds (Figure 12.4). It is believed that the first diamonds ever found were in volcanic debris lying on the surface. Diamonds were later discovered by gold prospectors in stream gravels that had been carried away from the debris by running water. Eventually people recognized that diamonds were associated with these circular volcanic deposits. This association led to the discovery of new deposits and the mining of diamond-bearing rocks from within the kimberlite pipes themselves. Note that kimberlite pipes tend to occur in swarms or clusters, which is why diamond mines are usually found within small geographic areas, often called *mining districts.*

Intrusive Deposits

Another important class of mineral deposits is associated with large intrusive bodies of igneous rocks. Recall that magma forms when the combination of temperature and pressure within the Earth is such that the minerals within a particular rock body begin to reach their melting points—the presence of water is also a factor in the melting process. As the minerals melt, the chemical bonds between atoms break, releasing charged atoms called *ions* (e.g., K^+, Ca^{2+}, Cl^-, SiO_4^{4-}) that make up the magma. When the magma eventually begins to cool, the ions are incorporated into new minerals in a sequential manner according to the specific crystallization temperatures of the different minerals. Because the minerals crystallize at different temperatures, the magma chamber contains both molten material and solid mineral grains during the early phases of the cooling process.

Since certain types of minerals are crystallizing from the magma before others, this creates the potential for the development of an economic mineral deposit, provided that the minerals in question have some value to society. What needs to occur is for the early-forming minerals to somehow become separated from the remaining magma before the cooling is complete. One way this can happen is through a process

FIGURE 12.4 Diamonds form in the upper mantle under conditions of high temperature and pressure, and are then carried to the surface by magma. Near the surface the highly pressurized magma explodes violently, creating a pipe-shaped crater. Diamonds are found in the volcanic rocks filling the crater, in ejected material, and in nearby stream gravels.

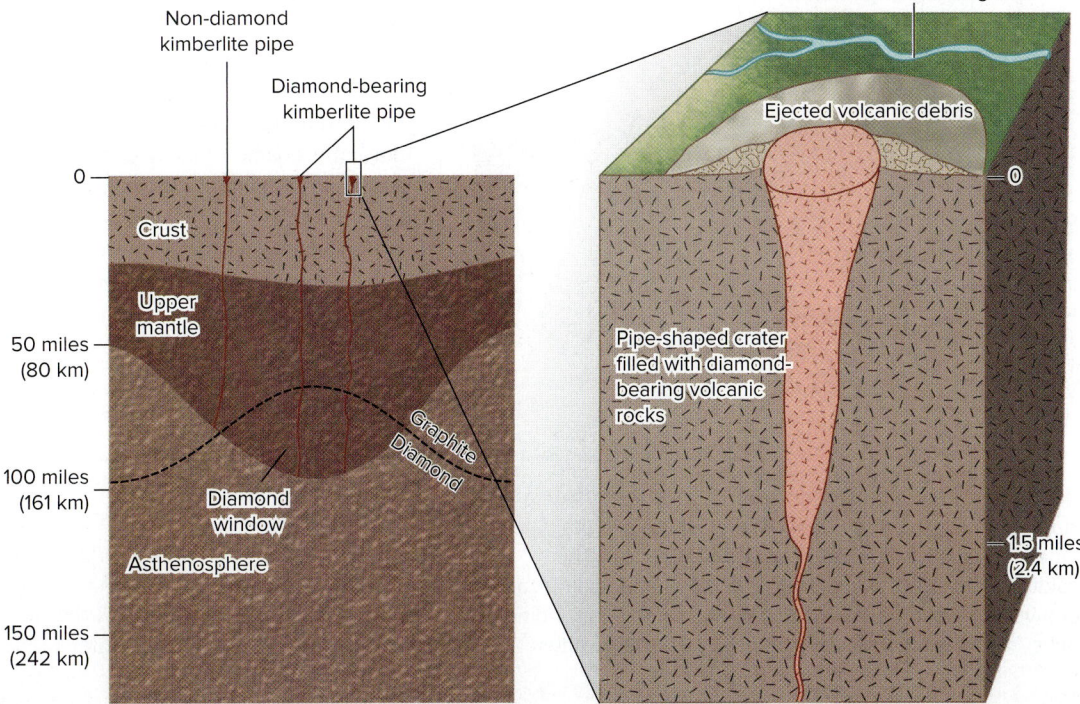

Early stage

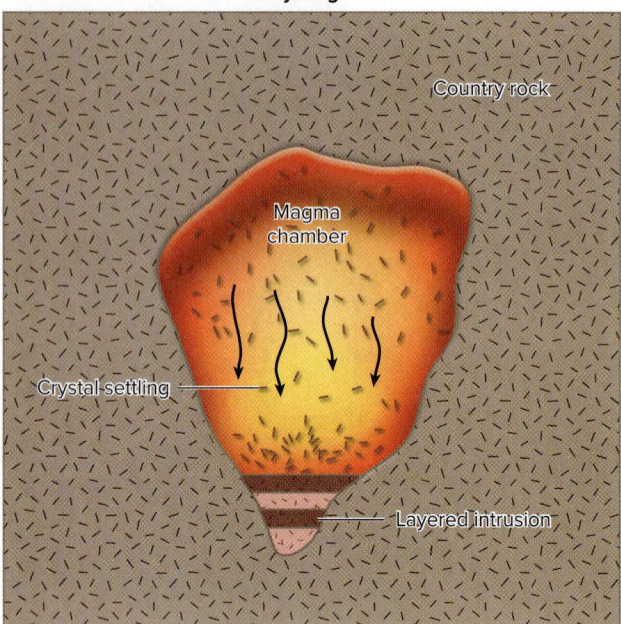

A

B

FIGURE 12.5 Dense minerals that crystallize early in the cooling process can settle to the bottom of a magma chamber and form a layered ore deposit (A). The photo (B) shows layers of chromium-rich minerals that are part of a layered intrusion in Bushveld, South Africa.
(b) © David Waters, Oxford University

Late stage

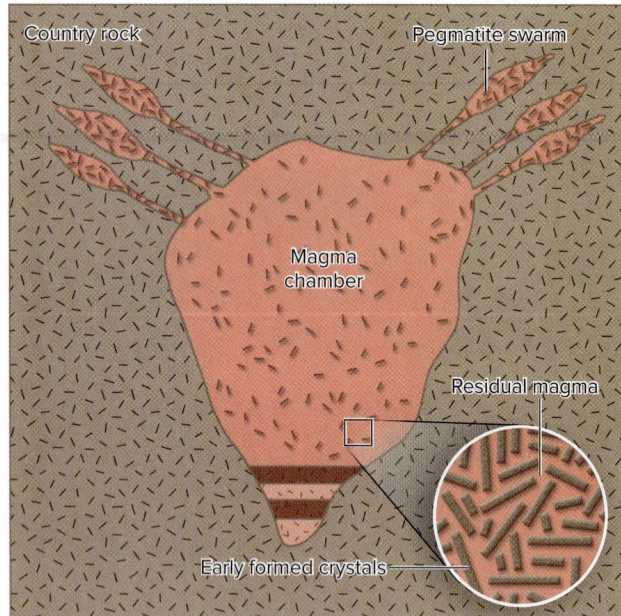

FIGURE 12.6 In the later stages of cooling, a magma chamber consists mostly of mineral crystals, with the remaining magma residing between the crystals. This residual magma is enriched with certain elements that can form valuable mineral deposits if injected into surrounding rocks, forming small intrusive bodies called pegmatites.

called *crystal settling,* whereby dense minerals that form early fall or settle to the bottom of the magma chamber. As shown in Figure 12.5, this process can create layered ore deposits that geologists refer to as **layered intrusions.** Layered ore deposits typically contain metallic minerals that are valuable sources of chromium, titanium, and vanadium.

Igneous ore deposits can also form when magma migrates away from newly formed minerals, as opposed to the minerals settling out of the magma. This process is believed to occur later in the cooling process, when the remaining magma resides between the mineral grains—similar to how water exists in the pore spaces of sedimentary materials. Pressure changes can then force this residual magma to leave the magma chamber and be injected into fractures within the surrounding rock. As illustrated in Figure 12.6, the magma then cools and forms elongated bodies of igneous rocks. These small intrusive bodies can contain valuable ore minerals when the crystallization process leaves the residual magma enriched with certain elements. One type of mineral deposit that forms in this manner is unusually coarse-grained deposits of silica-rich rock called *pegmatites.* These pegmatites are similar to granite in composition, but sometimes contain concentrations of minerals that contain rare elements (e.g., lithium, beryllium, niobium, and tantalum), which are normally found dispersed throughout an igneous rock body. Some pegmatites are mined solely for their exceptionally large grains of quartz and feldspar.

Hydrothermal Deposits

In Chapter 6 you learned that all magmas contain varying amounts of water. When magma slowly cools beneath the surface, some of the water becomes incorporated into the crystalline structures of certain minerals, and some remains with the magma during the final stages of crystallization. Water is also commonly present in the rocks surrounding the magma, which geologists refer to as *country rocks.* Naturally, the intense heat from the magma body raises the temperature of any groundwater present in the country rocks. The water from

FIGURE 12.7 Vein and disseminated deposits result when hot, mineral-rich fluids chemically react with minerals in an igneous intrusion and surrounding rocks, and then transfer elements within a zone around the igneous intrusion. The photo shows a vein deposit containing valuable tungsten and tin minerals in a Portuguese mine.
© Fernando Tornos, Instituto Geológico y Minero de España

these various sources produces hot, mineral-rich fluids that transport ions and chemically react with rocks in a zone around an igneous intrusion. Minerals that crystallize from these highly enriched fluids form what are referred to as **hydrothermal deposits.** Because sulfur (S) is a highly mobile element common in groundwater and magmatic water, hydrothermal deposits typically contain valuable minerals in which the sulfide ion (S^{2-}) is bonded to metals such as copper (Cu), lead (Pb), and zinc (Zn). Gold (Au) and silver (Ag) ores are also common in hydrothermal deposits.

Hydrothermal fluids normally deposit minerals in concentrated masses or in a dispersed manner throughout a rock body as shown in Figure 12.7. For example, *vein deposits* occur when ore minerals crystallize from the hot fluids and fill fractures and small fissures within rocks. Vein deposits are commonly considered to be high-grade deposits because the ore minerals are highly concentrated. In addition to filling fractures, hydrothermal fluids also move through the country rocks in a diffuse manner, resulting in low-grade deposits where the ore minerals are widely dispersed in what are known as **disseminated deposits.** Note in Figure 12.7 that disseminated deposits can be found in both the intrusion and the surrounding country rocks. Some of the world's largest surface mines exist for the removal of massive bodies of rock where the disseminated copper minerals are so small that they cannot be seen with the naked eye.

Another type of hydrothermal deposit is associated with divergent plate boundaries where active volcanism occurs along mid-oceanic ridges. Recall from Chapter 4 that in areas of active seafloor spreading basaltic magmas create new oceanic crust along extensive ridge systems on the ocean floor.

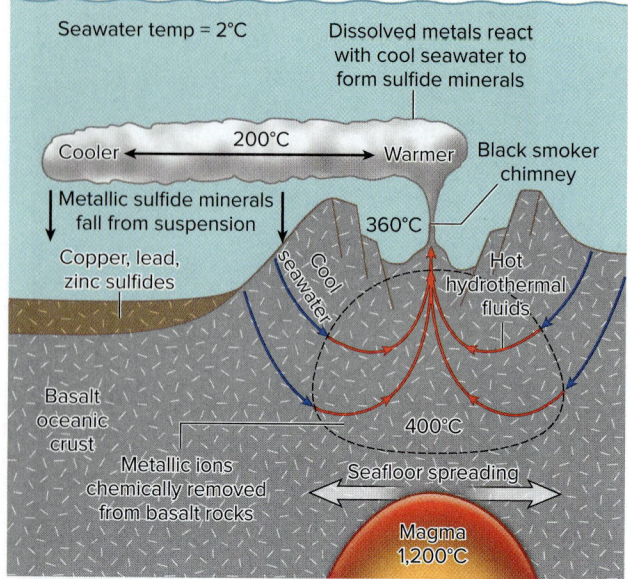

B

FIGURE 12.8 Massive sulfide deposits (A) form when hydrothermal fluids discharge from mid-oceanic ridges and then mix with cool seawater. Here metallic ions bond with sulfur, forming sulfide minerals that eventually accumulate on the seafloor. Note how heat convection pulls cold seawater into the ridge, where it reacts chemically with basalt to form hydrothermal fluids. Photo (B) shows sulfide minerals precipitating as hydrothermal fluids escape from vents on the seafloor.
NOAA

As illustrated in Figure 12.8, heat from shallow magma bodies causes hydrothermal fluids to flow upward through fractured rocks within a ridge; the fluids are replaced by cold seawater flowing downward along the flanks of the ridge. This convective motion is believed to cause strong chemical reactions between seawater and the basaltic rocks, generating hydrothermal fluids rich in copper, lead, and zinc. As the hot fluids discharge along the crest of the ridge, the dissolved metals react with cold ocean water to form sulfide minerals, producing a thick mineral deposit called a **massive sulfide deposit.** Notice in Figure 12.8 how these sulfide deposits form where hydrothermal fluids discharge from *black smoker chimneys* on the seafloor.

Massive sulfide deposits have long been important sources of copper, lead, and zinc. These deposits are quite ancient and much geologic evidence suggests that some found on land today actually originated along oceanic spreading centers. It is also believed that some sulfide deposits are now exposed on land because of simple tectonic uplift, whereas others may have been welded onto an existing landmass when an oceanic plate descended into a subduction zone. A good example is the island of Cyprus in the Mediterranean, where a massive sulfide deposit has been an important source of copper for nearly 4,000 years. In fact, the word *copper* itself is derived from the island's original Greek name, *Kupros.* Cyprus was one of the ancient sites where people made copper implements from pure copper metal found at the surface. After exhausting the supply of pure metal, people continued to obtain copper by placing copper-sulfide minerals into large fires. The intense heat would cause the chemical bonds within the minerals to break, resulting in molten copper draining into a pit at the base of the fire.

Metamorphic Processes

When rocks undergo metamorphism (Chapters 3 and 4), they often experience important physical and chemical changes that can lead to the development of valuable mineral deposits. However, because igneous activity is commonly associated with metamorphism, it is sometimes difficult to classify an ore deposit as being strictly metamorphic or igneous in origin, particularly when hydrothermal fluids are involved. In this section we will briefly explore the role metamorphic processes play in generating economic mineral deposits.

Regional Metamorphism

Recall from Chapter 3 that *regional metamorphism* takes place when rocks become deeply buried or involved in a mountain-building episode, exposing them to intense levels of heat and pressure. During this process many types of minerals are transformed into more stable minerals. Also, platy minerals are reoriented such that the mineral grains become parallel to one another (i.e., foliation). A good example of this process is the transformation by regional metamorphism of clay-rich sedimentary rock (shale) into slate (Figure 12.9A). The transformation of shale into slate creates a much harder and stronger rock that easily breaks along foliation planes, producing thin sheets of slate that are useful for roofing shingles and flooring tiles (Figure 12.9B). Today slate shingles have largely been replaced by less durable, but less expensive asphalt shingles. Another common use of slate is for making high-quality pool tables—those made of plywood tend to take on water and become warped.

Regional metamorphism also produces *marble,* a rock whose color and attractive crystalline texture makes it highly desirable for use as ornamental building stones, flooring, statues, and so forth. Marble forms from beds of limestone rock during regional or contact metamorphism. Here elevated temperatures cause calcite ($CaCO_3$) grains to recrystallize, creating a new rock with a much coarser texture. Although still composed of calcite, marble is usually white or pinkish in color as the impurities in the original limestone tend to migrate into other rock layers. Unfortunately, marble monuments and building stones, particularly those in Europe, are experiencing serious decay due to acid rain (Chapter 15).

Yet another important mineral resource related to regional metamorphism is the collection of fibrous silicate minerals known as *asbestos.* These minerals have widespread applications in industry and in construction materials due to their fibrous nature and ability to resist extreme levels of heat and chemical breakdown. Tragically, the very same properties that made asbestos so versatile also resulted in fatal lung and stomach diseases among workers who breathed in airborne asbestos fibers. Asbestos not only resulted in the loss of lives, but led to a costly and controversial effort to reduce the health risks by removing asbestos-containing materials from buildings (Case Study 12.1).

Contact Metamorphism

Contact metamorphism occurs when rising magma comes into contact with rocks, thereby subjecting the rocks to higher temperatures, but not necessarily higher pressures (Chapter 3). Also commonly present during contact metamorphism are hydrothermal fluids, which react chemically with many of the minerals within the surrounding country rocks. In essence then, rocks that were once in a relatively cool and stable environment suddenly find themselves being baked and infiltrated by hot, corrosive fluids. The result is what geologists refer to as an *alteration zone* or *halo* that surrounds the intrusion, as shown in Figure 12.10. Note that the width of the alteration zone largely depends on the susceptibility of the minerals in the country rocks to the heat and hydrothermal fluids. Also note that vein and disseminated ore deposits can occur in some rock types within the alteration zone. Complex chemical reactions between the mineral-rich hydrothermal fluids and surrounding rocks play an important role in determining where the ore minerals precipitate. Limestone rocks typically have deep alteration zones composed of marble, as the mineral calcite ($CaCO_3$) easily recrystallizes into much coarser grains.

Sedimentary Processes

As is the case with igneous and metamorphic rocks, the various processes that form sedimentary rocks tend to concentrate certain types of minerals, many of which society has found useful. In Chapter 3 we discussed the two basic types of sedimentary rocks: *clastic rocks* composed of fragmental material and *chemical rocks* formed by the chemical precipitation of minerals from water. In this section we will examine some of the minerals extracted from these two basic groups of sedimentary rocks and their applications in society. Although coal forms by sedimentary processes, it is used as an energy resource and will therefore be discussed in Chapter 13 on conventional energy resources.

A Slate outcrop in Antarctica

B

FIGURE 12.9   Slate (A) forms when shale undergoes regional metamorphism. Because slate breaks into thin sheets that are hard and durable, it has long been used for roofing shingles (B) and flooring tiles.
(a) USGS; (b) © Jim Reichard

White marble

Black basalt

B

FIGURE 12.10 Magmatic heat and hydrothermal fluids commonly create a zone of altered rocks (A) surrounding an intrusion in which ore minerals are deposited. Some rock types are more reactive than others, and thus have wider alteration zones and accumulate different types of minerals. Photo (B) showing a marble alteration zone surrounding a basaltic intrusion into limestone beds in Glacier Park, Montana.
© Jim Reichard

Labels in figure A: Ore minerals, Alteration halo, Vein deposit, Disseminated deposit, Magma chamber, Granite

A

Sand, Gravel, and Clay

Recall from Chapter 8 that *hydraulic sorting* refers to the separation of sediment grains by flowing water based on the size, shape, and density of the individual grains. This sorting occurs whenever the velocity of a fluid (water or wind) increases such that sediment grains are selectively picked up and then deposited in a sequential manner as the velocity decreases. This natural enrichment process is important because it provides humans with valuable deposits of relatively pure gravel, sand, and clay. Depending on the source material, hydraulic sorting by wind or water (in streams and along beaches) can produce sand deposits composed of nearly 100% quartz (SiO_2) grains. Humans have long sought such deposits of quartz sand as the raw material for making glass.

The hydraulic sorting that occurs when fine particles settle out of the water column in lakes and ponds often results in sediment composed almost entirely of clay minerals (Chapters 3 and 10). In addition to sand, clay is one of the most basic types of mineral resources. For example, because of its low permeability, clay is used to line ponds and earthen dams so that water can be stored for irrigation or power generation. Historically one of the greatest uses of clay has been in making clay pots and bricks. Here moist clay is molded into various shapes and then dried in the sun to drive off the water, causing the material to harden. When these molded materials are placed in a kiln, the temperature can be raised to the point where clay minerals are transformed into stronger and harder minerals, thereby producing more durable clay implements; this process is analogous to how shale is transformed into slate during metamorphism.

In terms of weight, stream-deposited sands and gravels are one of the most widely used mineral resources in modern societies. As indicated in Table 12.1, the combination of sand, gravel, and crushed stone represents 87% of all the mineral resources used in the United States. Huge quantities of sand and gravel are mined for use in construction projects and for making building materials. Some of the more common applications of sand and gravel include as *aggregate* for strengthening mortar and concrete and as fill material that forms the base for roadways and building foundations.

Placers

When rocks containing valuable minerals are exposed to weathering, they naturally begin to break down. In some cases the ore minerals are resistant to chemical weathering and are eventually incorporated into the sediment load of nearby streams, as illustrated in Figure 12.11. Here flowing water will cause any dense and chemically resistant minerals to be sorted hydraulically from the rest of the sediment load, forming a concentrated mineral deposit called a **placer deposit.** Gold-bearing stream placers are perhaps the most well known, but placers can also contain valuable platinum, tin, and titanium minerals as well as diamonds. Historically, some prospectors were able to locate rich hydrothermal vein deposits, which they called the *mother lode,* simply by walking upstream and testing the sediment for traces of gold (Figure 12.11). Note that rich placer deposits are also found in stream terraces (Chapter 8), which are older stream deposits located on ground that is higher than the sediment currently being deposited.

Because most streams ultimately transport their sediment load to the ocean or large lakes, wave and wind action along shorelines will also hydraulically sort minerals that have survived the journey, forming what are referred to as *beach placers.* In fact, the major source of titanium comes from the mining of beach placers, where several types of titanium minerals are found because of their resistance to chemical weathering—on active beaches these titanium minerals often appear as thin layers of black sand. In southern Africa, beach placers are mined for gold and high-quality diamonds. Keep in mind that most of the beach placers currently being mined are located inland and formed during periods of higher sea levels.

Residual Weathering Products

In Chapter 10 you learned how soils develop from the physical and chemical weathering of the minerals that make up rocks. Minerals that are resistant to chemical weathering, such as quartz, will remain in newly formed soils and the less resistant minerals will break down into secondary minerals, often called *weathering products.* This chemical transformation of minerals into secondary weathering products results in the release of ions (charged atoms) that get carried downward with infiltrating water. The secondary minerals within the soil column therefore become enriched in those elements that remain, which ultimately may represent minerals that have value to society. Likewise, the ions that are carried away with the water may recombine and form yet another economical mineral deposit.

In the case of rocks such as granite and basalt, recall that many of the aluminum- and silicate-rich minerals (i.e., *aluminosilicates*) chemically break down into various types of clay minerals that are enriched in aluminum (Chapters 3 and 10). As this process continues, the resulting clay minerals become progressively more enriched in aluminum. Given a long period of time and a climate with considerable rainfall, the weathering of silicate rocks can lead to a deposit of highly enriched aluminum minerals known as **bauxite.** It is through the mining and processing of bauxite ores that we obtain pure aluminum metal, which of course has found numerous applications in modern societies. Aluminum is particularly useful in applications where both weight and strength are an issue, such as in airplanes and fuel-efficient cars. Note that soil deposits rich in iron oxide minerals, referred to as *laterite,* can also form from the intense weathering of silicate rocks. Laterite deposits have historically been used for making bricks that are relatively strong and easy to cut.

The weathering of low-grade sulfide deposits also produces highly enriched metallic ore deposits. For example,

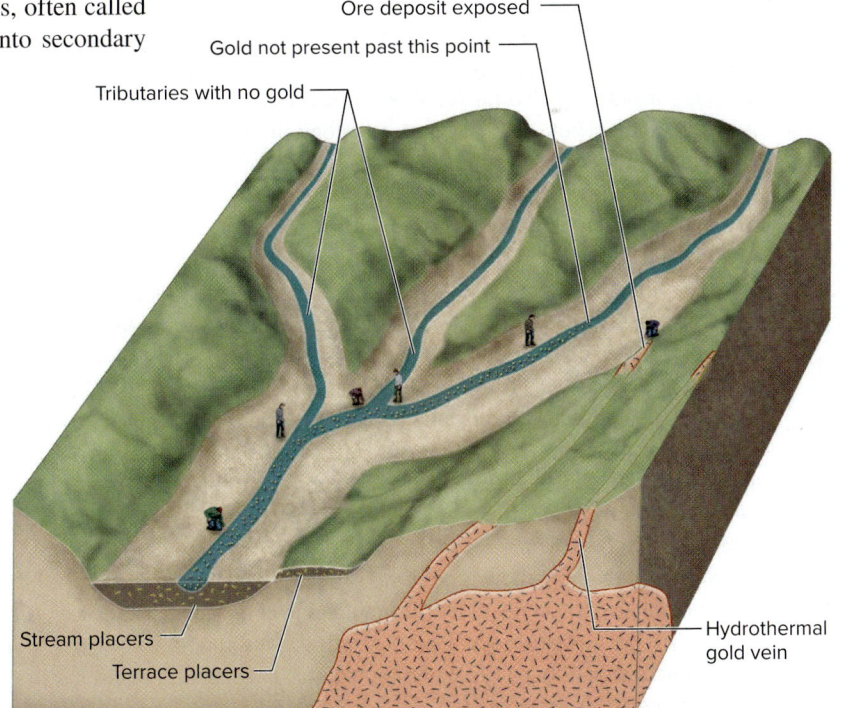

FIGURE 12.11 Stream placers form when weathering liberates valuable minerals from a primary ore deposit; erosion then carries the minerals to a stream where they become concentrated through hydraulic sorting. By following traces of gold in the sediment upstream through the drainage system, prospectors could sometimes locate the primary deposit, referred to as the mother lode.

Asbestos: A Miracle Fiber Turned Deadly

Asbestos is the general name applied to a family of fibrous minerals that have an extraordinary ability to resist heat and chemical breakdown (Figure B12.1). The earliest documented use of this unique set of minerals goes back nearly 5,000 years; in fact, the term *asbestos* originated from a Greek word meaning "inextinguishable." Both the Greeks and the Romans were known to weave asbestos fibers into fireproof cloth. Roman restaurants were said to have used asbestos tablecloths, which were cleaned by throwing them into a hot fire! It was not until the Industrial Revolution, however, that the use of asbestos became widespread. The ability of the fibrous minerals to withstand extreme levels of heat made asbestos ideal as an insulating material for steam boilers, kilns, and ovens and for making fireproof clothing (Figure B12.2). Later applications included thermal insulation for hot water pipes, fireproofing of structural steel in buildings, automotive brake pads and gaskets, and as a binder for holding together ceiling and flooring tiles. By the 1970s these so-called miracle fibers were being used in thousands of products. Even in homes asbestos could be found in products ranging from glue to toaster ovens to hair dryers.

Unfortunately, modern society ignored earlier warnings that these miracle fibers can create serious human health problems, something that was known as far back as Roman times. It was not until the 1960s that medical studies proved that workers who had been exposed to airborne asbestos in the 1930s and 1940s developed high rates of fatal lung and stomach diseases linked exclusively to certain types of asbestos. For example, in one study of shipyard workers who had installed asbestos insulation, over 50% were found to have contracted *asbestosis*—a progressive scarring of lung tissue. Also shocking was the fact that over 10% of the workers' spouses had contracted the same disease, presumably from washing contaminated work clothes.

Medical research showed that the properties that made asbestos so versatile, particularly its fibrous nature and chemical resistance, also made it quite deadly to humans. The needle-like shape of certain types of asbestos (Figure B12.1) and the chemical resistance made it nearly impossible for the human lung to expel individual fibers or to break them down chemically. Researchers found that both asbestosis and *mesothelioma,* a cancer in the lining of the lungs, were caused exclusively by asbestos. Lung and stomach cancers were also linked

FIGURE B12.1   Photograph taken with an electron microscope showing needle-like fibers of the asbestos mineral called anthophyllite. Asbestos fibers can easily become lodged in lung and stomach tissue, but cannot be broken down chemically, leading to scarring of the lungs and fatal lung and stomach cancers.
USGS

to asbestos, but here the link was statistical in nature because these cancers had other causes, such as cigarette smoking. Another important aspect of asbestos-related diseases is that the time period between exposure and the onset of disease, called *latency period,* is rather long, ranging from 10 to 40 years. This delayed recognition by the public that a medical crisis was slowly building. Between 1940 and 1979 approximately 27 million workers were exposed to airborne asbestos. Based on insurance claims, 23–25% of the workers exposed to high levels of asbestos are estimated to have died of asbestosis, 7–10% of mesothelioma, and 20–25% of lung cancer.

Although definitive medical studies did not take place until the 1960s, U.S. companies and government agencies had documented the risks of asbestos to U.S. workers as early as the 1930s. It was not until 1971, however, that the U.S. Occupational Safety and Health Administration (OSHA) began to regulate exposure levels of asbestos in the workplace. The use of asbestos greatly declined after 1989 when the U.S. Environmental Protection Agency (EPA) banned its use in most commercial products. The combination of environmental regulations with the threat of liability lawsuits quickly led to the widespread and controversial removal of asbestos-containing materials from schools and commercial buildings (Figure B12.3). This costly effort was seen by many as not only a poor use of scarce funds, but also an activity that could create an unnecessary health risk. At issue was the fact that the asbestos was firmly contained within many types of building materials, and therefore presented very little risk of becoming an airborne hazard. Removing these materials could actually create a greater risk than keeping them in place. Moreover, it was discovered that the most serious health threats were primarily related to just two types of asbestos minerals (amosite and crocidolite), which lowered the risk even further for many types of materials that were being removed.

The modern use of asbestos presents an interesting example of how a highly useful mineral was abruptly dropped by society due to its threat to human health. While liability concerns and panic helped drive the wholesale removal of asbestos from buildings—an action that was not necessarily justified in terms of the health hazards—the fact remains that certain forms of asbestos can be deadly. Due to the environmental regulations passed in the 1970s and 1980s, about

FIGURE B12.2 Photo showing World War II British firefighters training in asbestos suits.
© Hulton-Deutsch/Corbis

the only people at risk from asbestos in the workplace today are maintenance and construction workers who periodically disturb materials containing asbestos. The key to minimizing even these lower levels of exposure has been for businesses to survey their buildings and identify those materials that contain asbestos, and then train their staff to avoid disturbing these materials. Should maintenance or construction work require that the asbestos be removed or disturbed, workers must wear protective respirators and take steps to ensure that asbestos fibers are contained and therefore not released into the building (Figure B12.3).

FIGURE B12.3 The removal of asbestos from buildings is costly and requires workers and building occupants to be protected from airborne fibers. Photo shows asbestos soundproofing insulation being removed from the ceiling of a building.
© Alain LE BOT/Gamma-Rapho via Getty Images

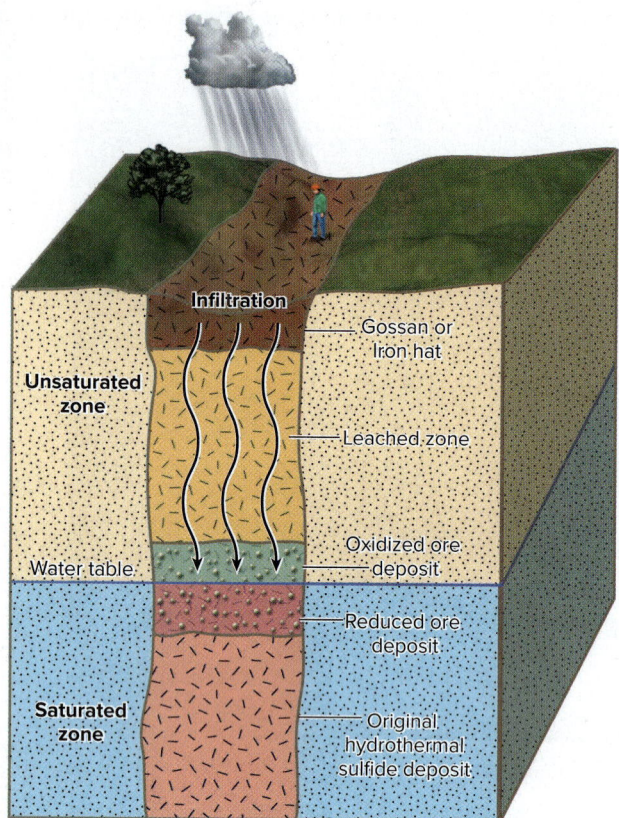

Infiltration

Unsaturated zone

Gossan or Iron hat

Leached zone

Oxidized ore deposit

Water table

Reduced ore deposit

Saturated zone

Original hydrothermal sulfide deposit

FIGURE 12.12   The weathering of hydrothermal veins causes sulfide minerals to break down, allowing water to carry metallic ions downward, where they recombine to form more stable minerals. This commonly results in two enriched ore deposits, one in the oxidized zone above the water table and the other in the reduced zone below, where oxygen levels are low. A low-grade deposit of residual iron minerals lies at the surface, which helps prospectors locate the underlying enriched zones.

Figure 12.12 shows a hydrothermal vein that contains disseminated metallic sulfide minerals that have undergone extensive weathering at the surface. Above the water table the vein lies in the unsaturated zone, where it is exposed to both free oxygen (O_2) and infiltrating water. In this zone the original sulfide minerals, such as pyrite (FeS_2), are chemically unstable and will oxidize relatively quickly. When this oxidation occurs, many of the metallic ions (e.g., iron, copper, lead, and zinc) will be stripped or *leached* from the sulfide minerals and transported downward with the infiltrating water. Some of the transported metals will then form new minerals by combining with more stable negative ions such as oxygen (O^{2-}), carbonate (CO_3^{2-}), or sulfate (SO_4^{2-}). The result is an enriched deposit of oxide, carbonate, and sulfate minerals lying just above the water table.

Other metallic ions that are leached from the primary sulfide minerals, particularly copper (Cu^{2+}), remain soluble and reach the water table, where oxygen levels are commonly lower and reducing conditions prevail. Once the metallic ions are in this reducing environment, they can precipitate back into sulfide minerals, such as bornite (Cu_5FeS_4). Thus the weathering of the original hydrothermal vein can result in two separate high-grade metallic ore deposits: one lying above the water table and consisting of oxide, carbonate, and sulfate minerals, and the other one located below the water table and composed of sulfide minerals. Note that the *gossan* or *iron hat* in Figure 12.12 is a surface deposit consisting of iron oxide minerals and quartz, which are the final weathering products of the sulfide minerals in the original hydrothermal vein. The iron oxide minerals in the iron hat give it a strong yellowish-brown color, a feature that prospectors can use for locating the much more profitable ore deposits lying below.

Banded Iron Formation

The ability of humans to mine and process large quantities of iron ore was one of the key factors that made the Industrial Revolution possible. Steel made from iron is still used for making the basic machinery that runs our factories, which, in turn, make nearly everything we use. The trucks, trains, and ships of our transportation system are all made from iron as is the farm equipment that produces our food. Iron-based steel also provides the strength necessary for supporting bridges and large buildings.

A sizable portion of the world's iron reserves (Figure 12.13) consists of alternating layers of quartz and iron oxide minerals, particularly hematite (Fe_2O_3) and magnetite (Fe_3O_4). Geologists generally agree that these deposits represent chemical sedimentary rocks that formed between 2.6 and 1.8 billion years ago as iron began to precipitate out of shallow seas. During this period Earth's plant life was quite limited, which means very little free oxygen (O_2) was being produced by photosynthesis. Therefore, the iron that was being released from the weathering of iron-rich silicate rocks remained in solution because of the lack of atmospheric oxygen. Later, as photosynthetic algae began to proliferate, the concentration of O_2 in the atmosphere and oceans increased, causing dissolved iron to precipitate and form iron oxide minerals. It is thought that these minerals then fell out of suspension in shallow seas and accumulated in layers. Although the ancient iron deposits could have formed in other ways, this mechanism nicely explains the sheer volume of the deposits within the geologic rock record and the fact that their age coincides with the proliferation of photosynthetic algae and free oxygen.

Evaporites

Another economically important class of sedimentary minerals are those that form when water bodies undergo evaporation. As water molecules vaporize and enter the gas state, dissolved ions or salts will stay behind in the water body, raising its salinity. Depending on the rate of evaporation versus the inflow of any new water, the remaining water, called *brine*, may become so concentrated that mineral grains begin to precipitate from the dissolved salts. These mineral or salt grains then fall to the bottom of the lake or seabed and form layers of chemical sedimentary rock known as **evaporites.** Should a water body evaporate entirely, then all

Red chert layers

Gray hematite layers

A

of its dissolved salts will precipitate and form evaporite minerals—you can try this yourself by taking a glass of salt water and letting it evaporate in the sun.

There are two basic types of evaporite deposits: marine and freshwater (Figure 12.14). In the case of marine evaporites, the minerals themselves reflect the chemical composition of seawater, which is dominated by chloride and sodium ions. Most economical marine deposits consist of minerals such as halite (NaCl), gypsum ($CaSO_4·2H_2O$), potassium chloride (KCl), and calcium chloride ($CaCl_2$). These minerals are important because they serve as raw materials for the chemical industry and are required in the processing and preservation of foods. Perhaps the most familiar mineral is halite since it is used for common table salt and for keeping highways free of ice during the winter. Because evaporite beds have extremely low permeability, they also serve as confining layers (Chapter 11) within the subsurface that are nearly impermeable. These types of confining layers have played a key role in the accumulation of oil and gas in various sedimentary basins around the world, including those in Texas and Louisiana (Figure 12.14A).

Marine evaporites typically form when tectonic activity causes a shallow body of seawater to become restricted, thereby limiting its connection with the open ocean. The restricted circulation allows evaporation to raise the salinity to the point where minerals begin to precipitate. Keep in mind that the different evaporite minerals do not all precipitate at the same place and time, but in a sequential manner, because salinity will vary throughout a basin. For example, gypsum will precipitate first in areas of lower salinity, followed by halite where the salinity is higher. This sequential pattern of precipitation results in geographically separate and relatively pure deposits of different salts. Achieving thick accumulations requires steady salinity levels and a basin that is both actively subsiding and receiving a steady supply of ions to replace those being removed by the deposition

B

FIGURE 12.13 Banded iron deposits (A) are believed to have formed between 2.6 and 1.8 billion years ago when free oxygen became abundant in the atmosphere. Photo showing gray bands of the iron mineral hematite separated by alternating layers of red, fine-grained chert (SiO_2)—note that the layers have been deformed since they were deposited. Photo (B) shows a large open-pit iron mine in the Upper Peninsula of Michigan.

(a) © Jim Reichard; (b) © Jim Wark/Airphoto

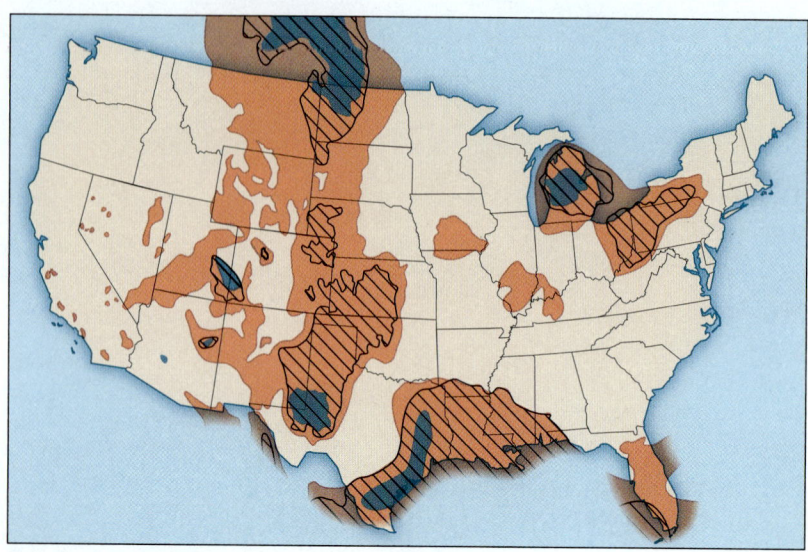

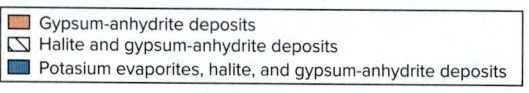
- Gypsum-anhydrite deposits
- Halite and gypsum-anhydrite deposits
- Potasium evaporites, halite, and gypsum-anhydrite deposits

A

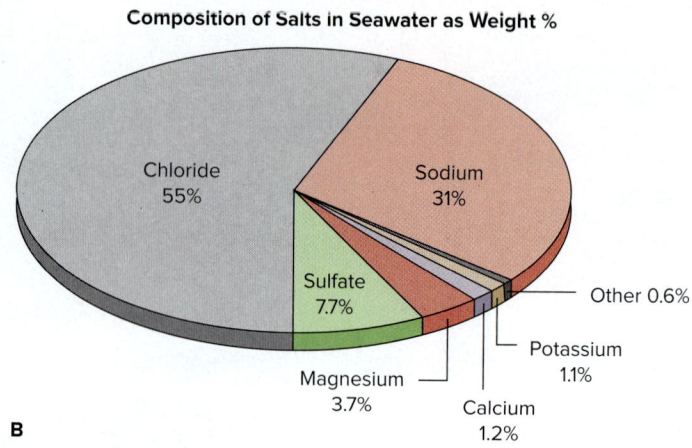

Composition of Salts in Seawater as Weight %

Chloride 55%
Sodium 31%
Sulfate 7.7%
Other 0.6%
Potassium 1.1%
Magnesium 3.7%
Calcium 1.2%

B

FIGURE 12.14 Map (A) showing both marine and freshwater evaporite deposits in the United States. Marine deposits consist chiefly of salts based on chloride and sulfate ions, which are the dominant negatively charged ions in seawater (B).

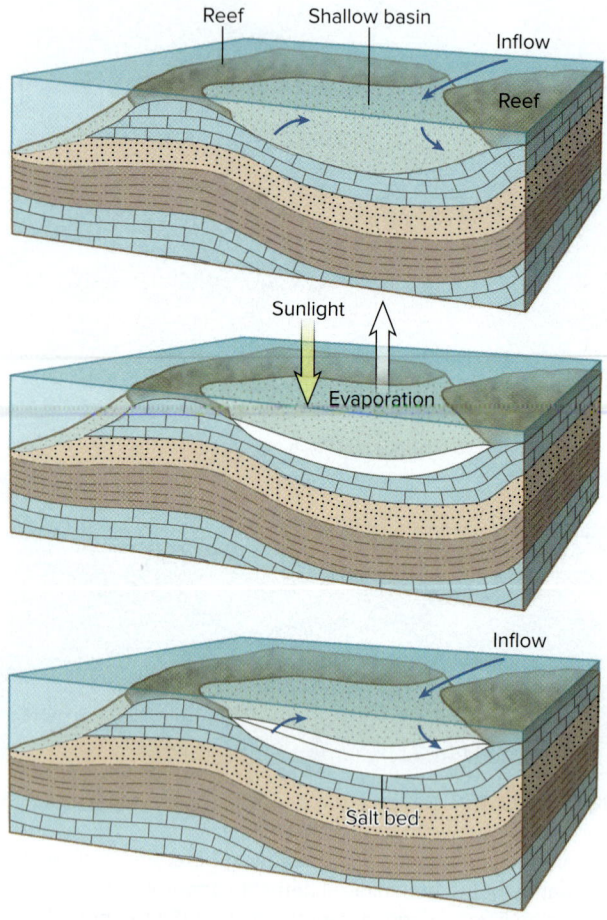

FIGURE 12.15 Evaporite deposits form when surface-water bodies undergo evaporation, increasing salinity to the point where dissolved salts precipitate and fall to the bottom to form layers of salt. Around 400 million years ago a restricted inland sea over much of the present state of Michigan underwent subsidence and intense evaporation, resulting in thick evaporite and reef limestone deposits that are currently being mined.

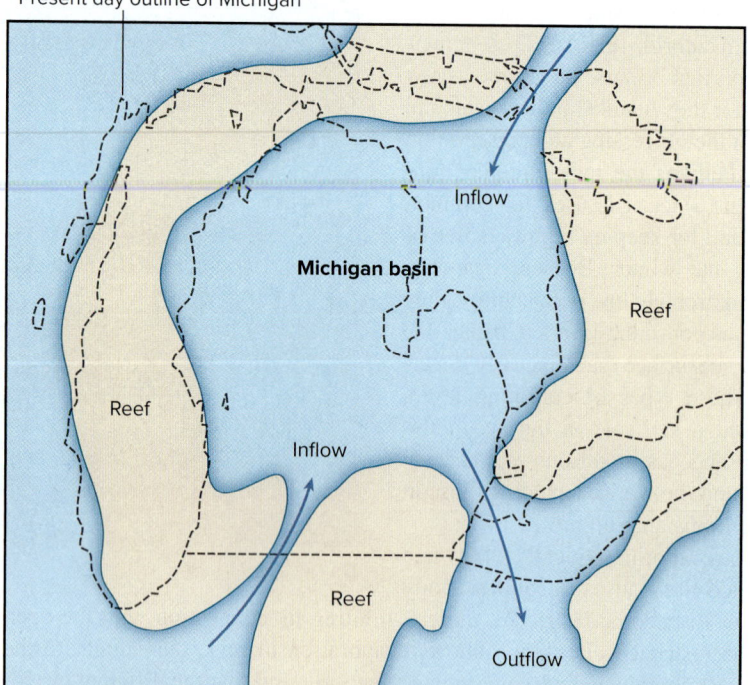

of salt. Note that thick evaporite deposits are relatively rare due to the need for maintaining steady salinity levels over long periods of geologic time.

The area known as the Michigan basin is an excellent example of a location where steady geologic conditions allowed for the deposition of thick evaporite deposits. As illustrated in Figure 12.15, a shallow inland sea covered much of present-day North America during the Silurian period around 400 million years ago. Here a large barrier reef system surrounded what is now the state of Michigan, creating a shallow and highly restricted marine environment. Intense evaporation along with a subsiding seafloor allowed deposits of halite to reach thicknesses of nearly 1,600 feet (490 m) in the middle of the basin. As shown in Figure 12.15, gypsum and reef limestone (calcite) deposits surround the halite in a concentric manner, forming an extensive and important evaporite sequence that is actively being mined.

Because the chemical composition of freshwater is considerably different from that of seawater, lakes produce certain types of evaporite minerals not commonly found in marine deposits, such as those based on the borate (BO_3^{3-}) and nitrate (NO_3^-) ions. Of particular interest are borate minerals because they serve as the source of boron used in making glass and ceramics as well as lightweight metallic alloys. Examples of freshwater evaporite deposits can be found in tectonic valleys in the western United States. As recently as 13,000 years ago, at the end of the last ice age, many of these valleys, which formed due to extension of the crust by tensional forces (Chapter 4), contained deep lakes whose water was relatively fresh. Since then most of the lakes have completely evaporated as the climate has become more arid, leaving evaporite deposits on the now exposed lake beds. One of the few remaining lakes is the now highly saline Great Salt Lake in Utah, shown in Figure 12.16. This lake lies in a topographic depression that receives water from streams, but the only outlet is through evaporation. Because the overall evaporation rates are relatively constant, water levels today in the Great Salt Lake rise and fall primarily in response to yearly changes in stream runoff.

Phosphorites

Similar to the element nitrogen, phosphorus (P) is one of the essential plant nutrients found in soils (Chapter 10). Consequently, economic mineral deposits containing phosphorus are important in the production of modern fertilizers. Phosphorous-bearing minerals, however, are normally widely dispersed within igneous rocks, and are therefore not economical to mine. But when rocks

FIGURE 12.16   Satellite photo of the Great Salt Lake in Utah. This lake (shown in black) represents the remains of a large body of freshwater that once occupied a series of tectonic valleys. White indicates areas where the lake has completely evaporated, leaving vast salt deposits on the now exposed lake bed.
© R. Douglas Ramsey, Utah State Univ., Remote Sensing and GIS Lab; (inset): Bureau of Land Management

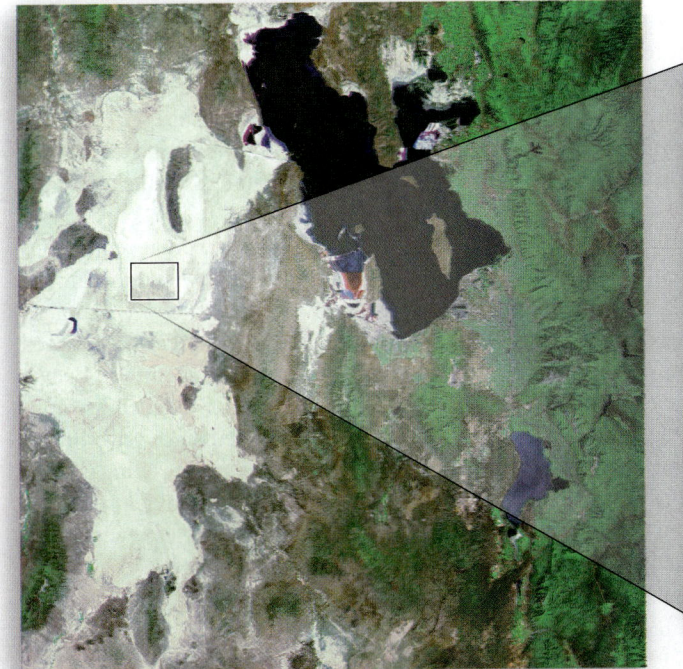

undergo chemical weathering, the phosphorus is released in the form of the phosphate ion (PO_4^{3-}), which is then transported to water bodies where it accumulates. Once the phosphate ions are in aqueous form, animal life is able to extract both those ions and calcium ions to form teeth and bones composed mainly of the mineral apatite ($CaPO_4$). As a result, most limestone rocks that contain the skeletal remains of marine organisms are relatively rich in phosphorus compared to other rock types. In locations where unusually large amounts of skeletal matter accumulate, the material can represent an economic deposit of phosphate that geologists call *phosphorite*.

Rare Earth Elements

Finally, there is an important group of ore minerals that contain metallic elements collectively referred to as *rare earth elements*—see the periodic table located on the last page of the text. Deposits of these minerals are sometimes associated with unusual types of igneous rocks geologists refer to as *alkali* and *carbonatite* rocks. Rare earth minerals are also found in placer deposits, pegmatites, and deep weathering zones of igneous rocks. Today, rare earth elements such as lanthanum, neodymium, and europium are being used in a variety of important applications, including smartphones, televisions, wind turbines, hybrid car batteries, and even night-vision goggles for the military. Despite the name, rare earth elements are actually found in Earth's crustal rocks at much higher concentrations than the major industrial metals of copper and zinc. In fact, even the rarest of the rare earth elements is almost 200 times more abundant in rocks than is gold. The problem is that, unlike gold, copper, and zinc, geologic processes seldom concentrate rare earth elements into economic deposits. Of considerable geopolitical importance is that in 2015, China held 42% of the world's reserves of rare earth elements and accounted for 86% of all mining production. In contrast, the United States held only 1.4% of the world reserves and accounted for just 6.4% of world production.

Mining and Processing of Minerals

The mining and processing of mineral resources have a long history of environmental degradation. The basic problem is that mining by its very nature is destructive to the surface environment. Both ore and waste rock must be removed from the earth, leaving underground passageways or vast open pits at the surface. In some cases entire mountaintops are removed. In addition to the scars on the landscape, mining creates large volumes of noneconomical rock and processed ore called **mine tailings,** which must be placed on the land surface. All this can lead to poisoned water supplies and serious air pollution problems. In addition, miners can be exposed to hazardous dust and fumes and fatal accidents involving tons of rocks.

As in many other industries in the United States, environmental regulations implemented since the 1970s have greatly increased worker safety and minimized the environmental impact of mining. For example, the fatality rate in U.S. mines in 2013 was 12 workers per 100,000, which represents a 54% decline between 1993 and 2013, making mining a much safer occupation than fishing (75 deaths per 100,000) and even farming (23 deaths per 100,000). Despite the fact that mining operations in the United States have dramatically improved since government regulations were implemented, many people still object to new mines being opened near their homes or communities. Part of the problem is that many older mines are now abandoned and continue to cause problems, creating a negative perception of modern mining. People also object to new mines because of concerns over lower property values, noise, dust, heavy truck traffic, and, of course, pollution.

Although many of these concerns about mining are valid, we must keep in mind that mineral resources make it possible for us to enjoy the comforts and conveniences of modern society. Like it or not, we are all very much dependent on minerals and help create the demand for mining. It seems reasonable that, as citizens, we help to ensure that the impact of mining on the environment is kept at a minimum by practicing conservation and supporting effective environmental regulations. In this section we will first examine some basic mining and processing techniques and then discuss their impacts on the environment. A thorough discussion of pollution and environmental regulations can be found in Chapter 15.

Mining Techniques

Prior to the development of modern machinery, the mining of minerals was done using simple hand tools to dig open pits or tunnels into the earth. People would typically begin digging where an ore body was exposed at the surface, then continue down into the subsurface; an example of this can be seen in Figure 12.17. Similar techniques are used today, but on a massive and more efficient scale involving large powerful machines and computerized equipment. Because increased efficiency means greater profitability, mine operators are always seeking the most efficient means possible for extracting a mineral deposit. Improved mining techniques have also made it possible to extract lower-grade deposits that were once considered uneconomic.

Similar to other businesses, mineral extraction has a profitability that is dependent on the market value of a particular mineral resource and the costs associated with getting it out of the ground and to market. One of the major costs involves removing undesirable rock or sediment called **overburden** in order to gain access to a mineral deposit. The problem with overburden is that energy (money) is required to physically remove it from the earth, plus it must be stored somewhere nearby on the land surface. If the amount of overburden becomes too great, then the mining operation will no longer be profitable. Another important factor affecting profitability is the concentration or grade of the mineral deposit itself. A higher-grade deposit allows a mine operator to handle more mine tailings (waste material) and still make a profit. The most desirable situation then is obviously a high-grade deposit with very little overburden. Other important costs that factor into the economic feasibility of mining include costs associated with transportation, labor, and environmental regulations.

For those who work in the mining industry, an unfortunate reality is its long history of so-called boom-and-bust cycles. Fluctuations in the market value of a particular mineral, due to supply and

FIGURE 12.17   Photo shows the trace of a gold-bearing hydrothermal vein that was mined in the late 1800s in Central City, Colorado. Mining began as a surface operation, but later developed into an underground mine as the ore body was followed into the subsurface. Note the piles of mine tailings (waste material) along the trace of the vein.
© Jim Reichard

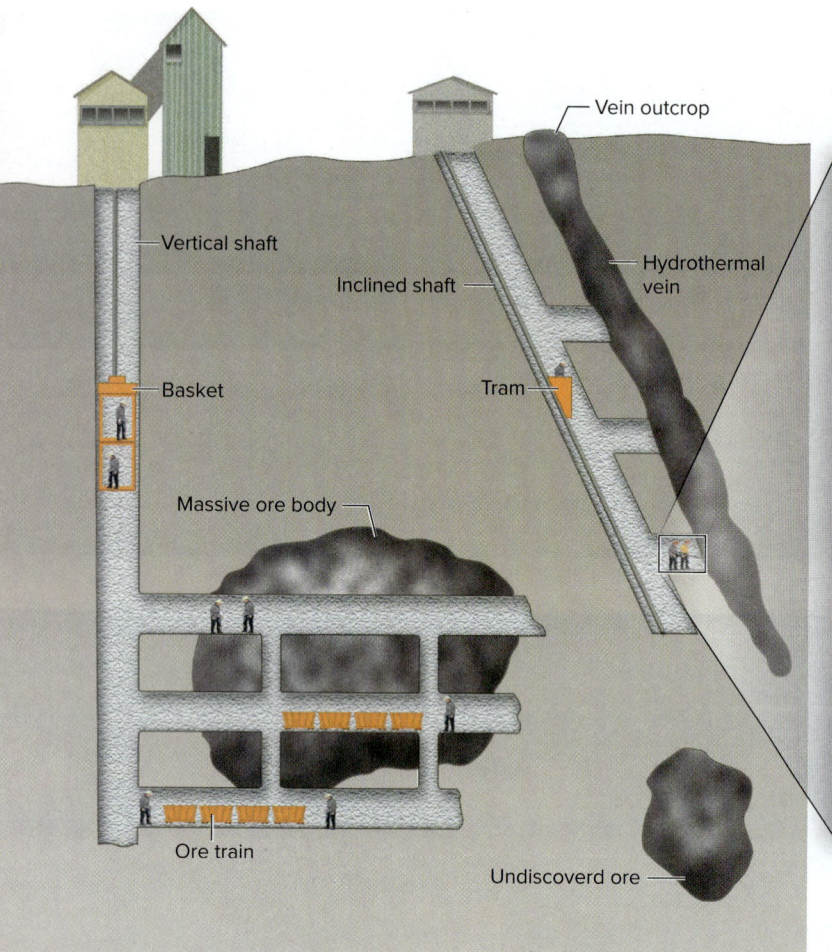

FIGURE 12.20   Underground mining involves blasting shafts and horizontal tunnels to access mineral deposits. Although safety has greatly improved, surface mining is generally preferred over underground mining as it presents fewer hazards and lower operating costs.
(inset) © Digital Vision/Punchstock

is then piled onto dry land using a conveyor. Note that ancient placers are often found above the stream in terrace deposits. During the California gold rush of 1849, miners made extensive use of *hydraulic mining* techniques to mine rich placers located in dry terraces. Mining companies built elevated wooden flumes so that water could be brought down from upstream sites to the locations of terraces. This setup allowed miners to force tremendous amounts of water through relatively small nozzles, generating high-pressure streams of water. When these streams of water were directed onto the terraces, placer material was washed into wooden sluice boxes (Figure 12.21), where the gold was

FIGURE 12.21   Hydraulic mining was developed to facilitate the removal of gold-bearing placers from terraces. An elevated flume was used to collect water from upstream, which was then forced through nozzles to create a high-pressure stream. The sediment on the terraces was washed into sluices where the gold was separated hydraulically. Photo from 1890 showing hydraulic mining in Nevada County, California, in apparent violation of the 1884 ban of the practice.
California Geological Survey Library

then hydraulically separated from the rest of the sediment. Although hydraulic mining was very effective, it was banned in 1884 because it created serious sediment pollution (Chapters 8 and 10) that degraded drinking-water supplies and increased the risk of flooding. Interestingly, California's ban on hydraulic mining represents one of the earliest pieces of environmental regulation in the United States.

Mineral Processing

When mineral resources are removed from the earth, a certain amount of physical and chemical processing is normally required to obtain the desired material. One resource that requires a relatively small amount of processing is the sand, gravel, and crushed stone used as a fill and for strengthening concrete (see Table 12.1). To obtain crushed stone, dynamite is used to blast solid rock into pieces that are small enough to be placed on a truck. The material is then hauled to a large mechanical crushing machine located on-site, which physically crushes the rock into various sizes, from the fine gravel found in parking lots to the extremely coarse material used for drainage along highways. In contrast, river deposits of sand and gravel have the advantage of already being composed of particles of different sizes, which means only sorting is required. Here the sand and gravel is first excavated using earth-moving equipment, and then hauled to a nearby facility where it is sorted using different sized screens. After being sorted, the material is washed. In some cases crushing is required to obtain material of a size that was not present in the natural deposit.

Many mineral resources contain certain ore minerals mixed together with noneconomical minerals and rock. In order to extract the desired ores, the mixture is first crushed and then sieved so it has a uniform grain size. After this process a variety of techniques can be used to physically separate or extract the valuable minerals. One approach is to make use of the fact that any mixture is going to contain minerals with different densities. Flowing water can thus be used to hydraulically separate the different minerals; this method works particularly well for high-density ores such as gold. In contrast, floatation is where relatively light minerals are extracted using a high-density fluid, in which the minerals rise to the surface and literally float. Another means of physical separation makes use of powerful magnets for removing iron ores from the waste material.

After physically separating a desirable mineral, it is sometimes necessary for the ore to undergo chemical processing. In many cases, what society finds useful is not necessarily a mineral itself, but rather some specific element it contains. For example, copper (Cu) is the desired element in the mineral bornite (Cu_5FeS_4), but is chemically bonded to iron (Fe) and sulfur (S). The oldest-known means of chemically breaking down minerals is the process known as **smelting,** in which metals are obtained by heating certain minerals to the point where their chemical bonds are broken—smelting was used in ancient times to extract copper metal from sulfide minerals. In a similar manner, limestone rock is heated in order to chemically transform calcite ($CaCO_3$) into *lime* (CaO). Note that there are many applications for lime because of its acid-neutralizing ability, but perhaps its most significant use is as the raw material for making cement products.

Another common chemical processing technique is **leaching,** in which a solution is allowed to permeate through crushed ore and the resulting chemical reactions liberate the desired element(s). At this point the mineral resource consists of the ions that are dissolved in the solution, normally referred to as *leachate*. The leachate is then collected and piped to a processing plant where the ions are chemically removed. A good example of leaching is the use of cyanide to dissolve metals such as gold and silver from crushed ore. Although cyanide solutions have been used for over 100 years, not until the 1980s did mine operators begin applying the method to what is now called *cyanide heap-leaching*. As illustrated

FIGURE 12.22 Cross-sectional view of a cyanide heap-leaching system for extracting heavy metals from crushed ore. Applied to a pile of ore, a cyanide solution percolates and chemically dissolves metallic ions from the ore minerals. Leachate collected from the bottom of the pile is sent to a processing plant where the metals are chemically removed. A synthetic liner is used to keep cyanide from escaping into the environment.
(top): Courtesy of Barrick Gold Corporation; (bottom left): © Doug Sherman/Geofile; (bottom right): Southern Peru Earthquake Geo-Engineering, Extreme Event Reconnaissance (GEER) team

in Figure 12.22, heap-leaching involves spraying a cyanide solution on top of a large pile or heap of crushed ore. The solution percolates down through the pile and chemically dissolves metallic ore minerals. Buried at the bottom of the heap is a system of pipes that collect the resulting leachate and drain it into a pond; both the pond and heap are lined with a heavy synthetic material designed to prevent leachate from escaping. From there the leachate is sent to a nearby processing plant where the valuable metals are removed, after which the cyanide solution is returned back to the heap and used again.

Cyanide heap-leaching has proven to be very effective in dissolving fine-grained ore minerals that are difficult to process by other means. Moreover, this technological advance has made it possible to mine low-grade deposits that were once considered uneconomical. This process has also made it feasible to recover valuable metals from abandoned mine tailings. Although cyanide heap-leaching represents a significant improvement in our ability to obtain certain metals, it is also rather controversial because of its potential threat to the environment.

Distribution and Supply of Mineral Resources

Humans clearly expend a great deal of energy extracting and processing mineral resources. As could be expected, the different types of mineral deposits are not found just anywhere, but in very specific geologic environments where natural processes concentrate certain minerals. For example, the map in Figure 12.23 shows the location of disseminated copper deposits that are associated with igneous intrusions. Note that many of the deposits along the Pacific coasts of North and South America correspond to convergent plate boundaries and the igneous activity of the Ring of Fire described in Chapters 4 and 6. Other deposits on this map represent very old intrusions that formed in the distant geologic past when the configuration of plate boundaries was considerably different from today. The key point here is that mineral deposits are closely tied to the rock cycle and Earth's tectonic history. Therefore, the search for new mineral resources by modern mining companies requires trained geologists who understand both the natural processes that concentrate minerals and the tectonic history of a particular region.

Because mineral deposits are not evenly distributed around the globe, it is highly unlikely that an individual country will have all the different minerals required to run a modern industrial society. Large countries such as the United States, Russia, and China have a greater proportion of the minerals they need simply because they are geographically large and have a wider variety of geologic terrains within their borders. Even so, these large countries are not entirely self-sufficient and have to import certain mineral resources in order to

FIGURE 12.23 Disseminated copper deposits are unevenly distributed around the world because they are associated with igneous intrusions and tied to plate tectonics. Note how some of these deposits closely correspond to the convergent plate boundaries along the Pacific coasts of North and South America.

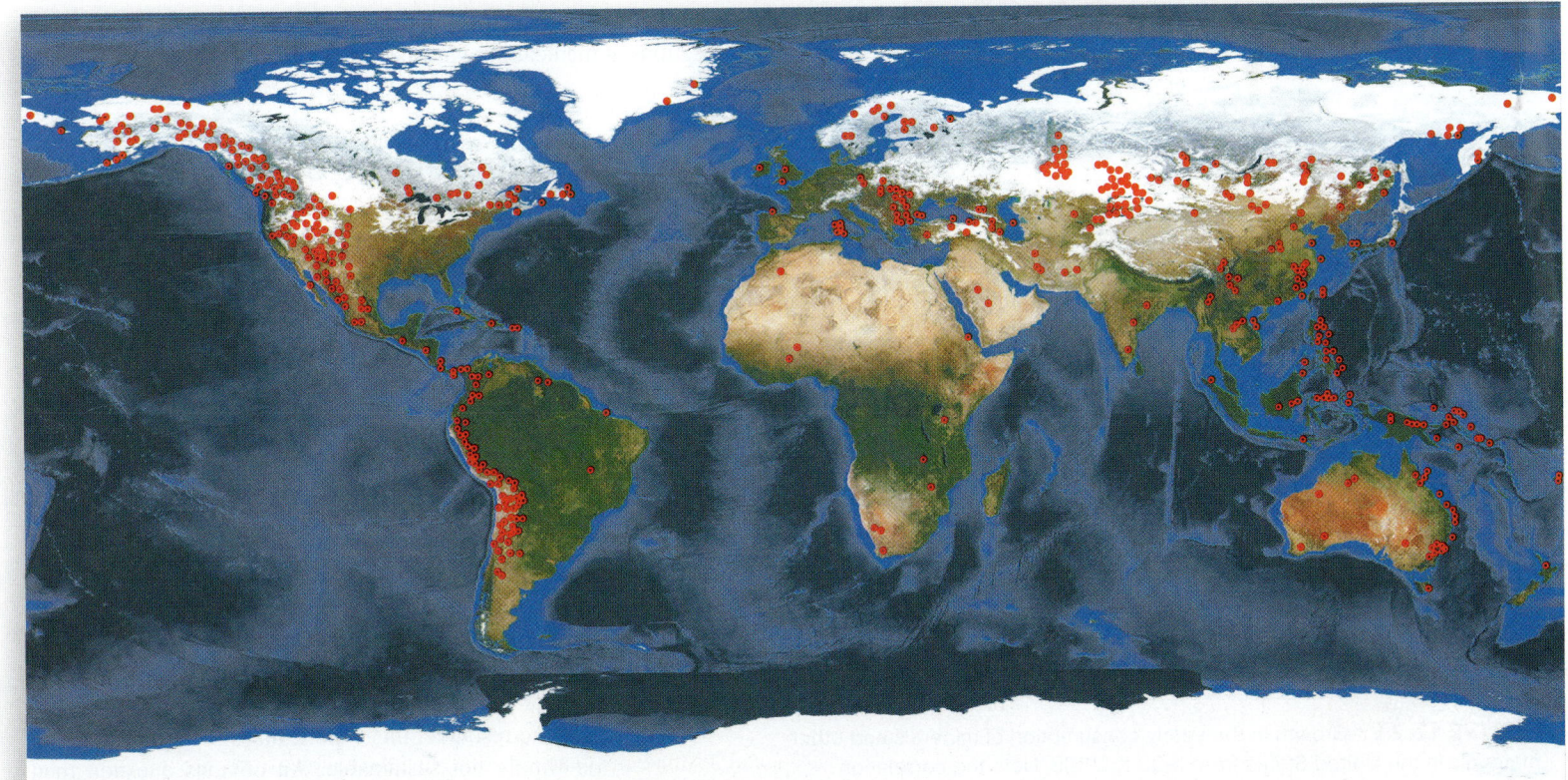

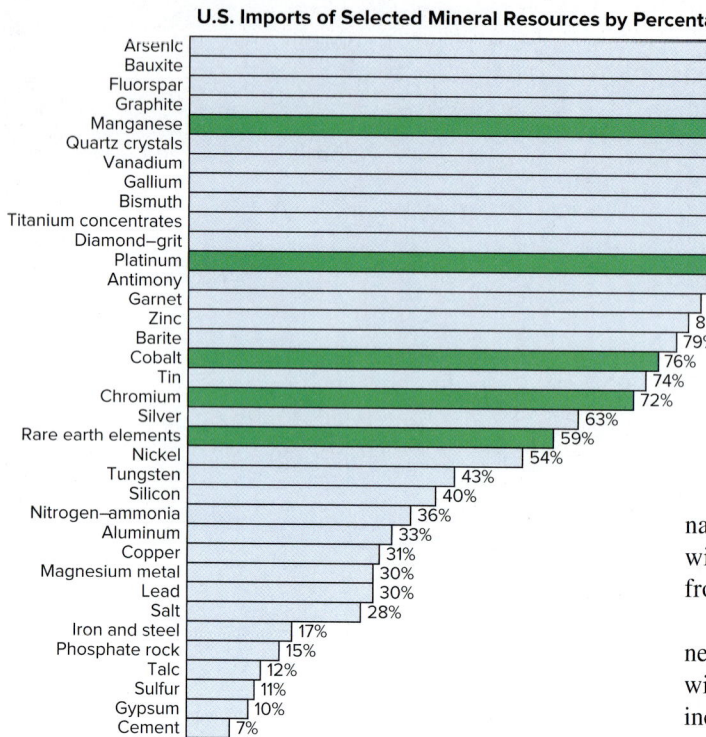

U.S. Imports of Selected Mineral Resources by Percentage - 2015

Mineral	Percentage
Arsenic	100%
Bauxite	100%
Fluorspar	100%
Graphite	100%
Manganese	100%
Quartz crystals	100%
Vanadium	100%
Gallium	99%
Bismuth	94%
Titanium concentrates	91%
Diamond–grit	86%
Platinum	85%
Antimony	84%
Garnet	83%
Zinc	81%
Barite	79%
Cobalt	76%
Tin	74%
Chromium	72%
Silver	63%
Rare earth elements	59%
Nickel	54%
Tungsten	43%
Silicon	40%
Nitrogen–ammonia	36%
Aluminum	33%
Copper	31%
Magnesium metal	30%
Lead	30%
Salt	28%
Iron and steel	17%
Phosphate rock	15%
Talc	12%
Sulfur	11%
Gypsum	10%
Cement	7%

FIGURE 12.24 List of selected mineral resources and the percentage of each that the United States imports (green denotes strategic minerals). Some minerals are imported simply because it is less expensive than mining existing U.S. deposits.

Source: Data from U.S. Geological Survey Mineral Commodity Summaries 2015.

meet their needs. For example, Figure 12.24 lists percentages of selected mineral resources being imported into the United States. The percentages are somewhat misleading, as many of the mineral resources are imported not because the United States has no deposits of its own, but rather because importing is less expensive than domestic mining. The lower cost is generally due to a combination of lower-grade deposits and higher domestic mining costs associated with better wages and environmental regulations designed to protect citizens from pollution.

Since modern societies must rely on imports to meet part of their mineral needs, there is a certain level of risk that the supply of a particularly vital mineral will be disrupted. The risk obviously increases as the percentage being imported increases and when the supply largely comes from politically unfriendly or unstable countries. This brings up the concept of **strategic minerals,** which are those that are considered critical by a given country but that must be imported in significant quantities. For the United States strategic minerals include chromium, cobalt, manganese, and platinum because they are vital to civilian and defense industries. For example, the U.S. military depends on high-performance jet engines with operating temperatures that are considerably higher than those of conventional engines. Such engines are made possible by special alloys containing chromium and cobalt, which are largely imported (Figure 12.24). Just one of these high-performance engines requires 1,600 pounds of chromium and 1,000 pounds of cobalt. Finally, note that, besides their military importance, cobalt alloys have important applications in medical implants because of their low rejection rates by the human body.

Meeting Future Demand for Minerals

As with all natural resources, the demand for minerals continues to increase due to population growth and the fact that more countries are becoming industrialized and developing consumer societies. For example, the graph in Figure 12.25 illustrates the dramatic rise in the consumption of metallic and nonmetallic minerals in the United States as the growing nation built a large industrial base and consumer society. The sharp declines associated with the Great Depression in the 1930s and the oil-related recessions of the 1970s and 1980s demonstrate the close relationship between economic activity and mineral demand. China and other Asian nations that are currently undergoing rapid industrialization are following consumption patterns similar to those of the United States. Since minerals are a finite resource, this exponential rise in consumption is simply not sustainable. An obvious question then is, how long will the world's mineral reserves last?

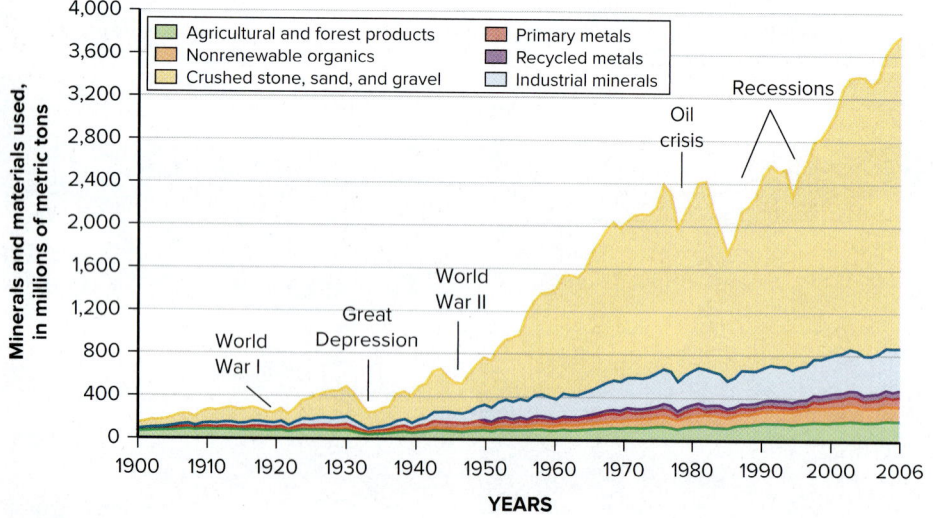

FIGURE 12.25 Growth in the yearly consumption of minerals and other materials in the United States from 1900 to 2006. Note the correlation between economic activity and resource consumption.

To help answer this question, Table 12.4 shows current worldwide production rates and total reserves for some of the more common mineral resources. Notice how the life expectancy of different mineral reserves varies, ranging from as little as 16 years for lead and tin and 19 years for gold to over 300 years for phosphate rock. Keep in mind that these estimates are based on current production rates and reserve estimates, which can change along with economic conditions and technological improvements.

For years scientists have warned about the prospect of important mineral reserves being depleted due to the exponential growth of demand. However, depletion has not occurred because total reserves have kept increasing. One reason for this is that geologists have improved their ability to locate new reserves through more sophisticated exploration techniques. Another factor is that technological advances in mining and processing have allowed many low-grade deposits to become economical to extract, thereby causing more mineral resources to be classified as reserves (see Figure 12.3). This does not mean that all mineral extraction can go on indefinitely. History shows that it is becoming harder to locate new deposits, and those that are found are usually of lower grade. Perhaps an even bigger issue is that mineral extraction and processing typically require large amounts of energy. With rapidly rising energy prices, particularly for diesel fuel, less profitable mining operations will likely begin to shut down, thereby reducing total reserves. The remaining reserves will not necessarily be depleted any faster since higher energy costs will eventually lead to an economic slowdown, followed by a reduction in demand.

It should be mentioned here that due to the critical nature of strategic minerals, manufacturers and governments sometimes create their own stockpiles. Although stockpiling is not a long-term solution, it does provide a means of minimizing the impact caused by a short-term disruption in supply. In fact, ever since World War I the U.S. government has routinely stockpiled strategic minerals for possible emergency use.

Recycling and Reuse

At some point humans are not going to be able to afford to keep on expending greater amounts of energy to mine ever deeper and lower-grade mineral deposits. This means we can expect that some mineral resources will become increasingly scarce, which in turn will drive up their market price. As prices rise, there naturally will be a greater incentive for recycling and for effort to develop substitute materials. From Table 12.5 one can see that the recycling rates for metals in the United States average around 40%, so there is still considerable room for improvement. Recycling is actually quite desirable due to the high cost of mining and processing of metallic ores. For example, obtaining metallic aluminum from recycled scrap requires only 5–8% of the energy that is required to produce new metal from bauxite ore. Although aluminum and other scrap metal is easy to recycle, many people unfortunately still

TABLE 12.4 World mineral production and projected life of estimated reserves. Reserves represent mineral deposits that are economical to extract under current conditions.

Mineral	Production (thousands of metric tons)	Reserves (thousands of metric tons)	Estimated Life of Reserves (years)
Iron ore	3,320,000	190,000,000	57
Aluminum ore (bauxite)	274,000	28,000,000	102
Phosphate rock	223,000	69,000,000	309
Gypsum	258,000	n/a	
Chromium	27,000	>480,000	>18
Copper	18,700	720,000	39
Manganese	18,000	620,000	34
Zinc	13,400	200,000	15
Titanium concentrates	6,090	790,000	130
Lead	4,710	89,000	19
Nickel	2,530	79,000	31
Tin	294	4,800	16
Cobalt	124	7,100	57
Silver	27	570	21
Gold	3.0	56	19
Platinum group	0.39	66	171

Source: Data from U.S. Geological Survey Mineral Commodity Summaries, 2016.

TABLE 12.5 Metal recycling rates in the United States. Although manganese is a strategic metal used in steel, it is not economical to recover and is not listed.

Metal	Supply (thousands of metric tons)	Recycling of Old and New Scrap (thousands of metric tons)	Percentage Recycled
Iron and steel	117,000	58,600	50%
Aluminum	7,010	3,640	52%
Copper	2,450	804	33%
Lead	1,660	1,130	68%
Zinc	1,140	248	22%
Chromium	558	157	28%
Nickel	199	89	45%
Tin	46	13	27%

Source: Data from U.S. Geological Survey Minerals Yearbook, 2014.

throw recyclable metals in the trash. A quick look in most garbage cans will likely reveal significant numbers of aluminum cans.

In addition to creating energy savings and a reduction in environmental impacts, recycling provides society with a nearly inexhaustible supply of certain resources. Of course not all materials are so easily recycled; in fact, some are not recyclable at all. Similar to our uses of water (Chapter 11), certain mineral applications are *consumptive* in nature, meaning that the mineral resource is dispersed into the environment or incorporated into some material from which it is difficult to recover. For example, once salt is used on a highway to melt ice, it becomes dispersed, making it impractical to try to recover the dissolved ions. In the case of manganese, which is used to harden steel, once it is alloyed with iron, it becomes too difficult to extract it from steel scrap. This brings up the concept of *reuse* as opposed to *recycling*. Although manganese in steel may not be recyclable, a steel beam salvaged from a building can indeed be reused for some other purpose. Yet another option is for society to *reduce* consumption of a particular mineral resource through more efficient use. Take for instance the consumptive use of aluminum foil in the home. By minimizing the amount of foil used in covering or wrapping food, one could make a small but noticeable reduction in personal use. Note that public education programs aimed at improving conservation efforts commonly use the concept of *3 Rs,* which stand for *reduce, reuse,* and *recycle*— see Chapter 15 for more details.

Mineral usage can also be reduced when manufacturers develop substitute materials for certain applications. A good example of substitute materials is graphite-based compounds, whose strength and light weight make them ideal replacements for metals in applications ranging from tennis rackets to spacecraft.

Environmental Impacts and Mitigation

As emphasized throughout this chapter, mineral resources are absolutely critical to maintaining our modern societies, and mining is a necessary part of our use of these resources. Although mining activity disrupts a rather small fraction of Earth's total land area, a legacy of poisoned water supplies and devastated ecosystems demonstrates that the impact of mining can be severe. Much of the controversy over mining in the United States is related to the **General Mining Act** passed by Congress in 1872—better known as the 1872 Mining Law. This act governs the mining of precious metals (excluding oil, gas, coal, and nonmetallic ores) on nearly 270 million acres of public lands, mostly in the western United States and Alaska. Under this law a prospector who found a mineral deposit would be able to purchase the land for no more than $5 per acre, and then extract the resource without having to pay any royalties to the government. The intent at the time the law was passed was to provide an incentive to those willing to develop the newly acquired lands out West. Some people take issue with the fact that even today, the 1872 Mining Law allows mining to take precedence over all other land-use activities, outside of national parks. In addition, mining companies still have to pay only $5 an acre to purchase public lands that often yield hundreds of millions of dollars' worth of minerals.

The 1872 Mining Law certainly served its purpose in that it encouraged the mining of precious minerals that were needed as the United States underwent rapid industrialization. Unfortunately, environmental controls on mining or other industrial activity were basically nonexistent until the passage of the Clean Air Act in 1970 and the Clean Water Act in 1972 (Chapter 15). The result was the uncontrolled release of highly toxic metals and other compounds into the environment, leaving streams and rivers throughout the western United States severely impacted even to this day. The pollution from modern mines and processing plants has been minimized due to the regulations enacted since the 1970s and the use of improved mining techniques. Problems and accidents still occur, but the overall amount of pollution from today's mining operations is far less than in the past.

Twin Buttes, Arizona

Bingham Canyon, Utah

FIGURE 12.26   The volume of mine tailings can be staggering, particularly from the open-pit mining of low-grade copper deposits. For scale, note the size of buildings in these photos.
(a) © Jim Wark/Airphoto; (b) © Earthworks

Despite the reduction in pollution, the controversy over mining has not ended in the United States. There are some who view environmental regulations as a costly and unnecessary burden, causing jobs to shift overseas to countries where lax regulations result in lower operating costs. In a global market system, U.S. mines often simply cannot compete. Others see regulations as a way of forcing companies to incorporate the cost of controlling pollution into their operating expenses, as opposed to passing the cleanup and medical costs on to the tax-payers. Moreover, many point to a lack of enforcement and persistent pollution problems as evidence that the regulations designed to protect people and the environment are falling short of their goals.

The issue of pollution versus jobs is complex and there are no easy answers. We will discuss pollution and environmental regulations in more detail in Chapter 15, but for now our focus will be on some of the specific problems, both past and present, which are associated with the mining and processing of minerals. We will also examine some of the mitigation and prevention techniques used to minimize impacts.

Heavy Metals and Acid Drainage

The mining and processing of nearly any mineral resource produce waste material (tailings) whose mineral concentration is too low to justify further processing. Depending on the type of deposit, the volume of tailings can be significantly larger than the mine excavation because the broken-up rock occupies more space than it did underground. In the case of open-pit mining of low-grade ores, the vast majority of the material that is removed is waste, creating tailing piles whose dimensions are truly staggering (Figure 12.26). Although the permanent storage of mine tailings creates what some view as an eyesore, a more serious problem occurs when melting snow or rainwater percolates through the material and chemically reacts with the remaining ores and other minerals, thereby producing leachate. Tailings from metallic ore deposits are of particular concern as they normally contain sulfide minerals that weather easily, creating leachate that is rich in metallic ions such as lead, copper, zinc, arsenic, cobalt, nickel, selenium, and cadmium. This toxic brew of metals eventually drains from the bottom of the tailing pile, where if left unchecked, it will enter both surface streams and groundwater systems.

A Silverton, Colorado

B Jerome, Arizona

FIGURE 12.27   Leachate draining from underground mines and from beneath tailings is commonly highly acidic and laden with heavy metals due to the chemical interaction of water and sulfide minerals. Dissolved iron quickly precipitates to form iron oxides, giving impacted streambeds a characteristic reddish or yellowish color (A), whereas a bluish or greenish color from copper precipitates (B) is less common.
(a) Philip L. Verplanck, USGS; (b) © Jim Wark/Airphoto

The sulfide minerals commonly present in mine tailings not only release heavy metals as they break down chemically, they also release sulfide ions (S^{2-}) that react with oxygen and water to produce sulfuric acid (H_2SO_4). The result is leachate that is extremely acidic and contains highly toxic metals. This same type of leachate also forms in the workings of underground mines where sulfide minerals are exposed to oxygen and water, which can then drain into streams lying below the level of the mine. The term **acid mine drainage** is used to describe acidic leachate, normally rich in heavy metals, that drains from either tailings or underground mines. As shown in Figure 12.27, when acid mine drainage encounters the more oxidizing conditions in a stream, some of the heavy metals will form a precipitate that covers rocks and sediment in the streambed. Because dissolved iron is present in most leachate and easily oxidizes, streams affected by acid mine drainage typically have a characteristic reddish or yellowish color from iron-oxide precipitates. A bluish or greenish color due to copper precipitates is much less common.

In addition to the toxic effect of dissolved metals in acid mine drainage, a serious problem is that the leachate can greatly increase the acidity (i.e., lower the pH) of surface waters. In some areas the pH of a stream can fall from around 6 to 3, which corresponds to a thousand-fold increase in acidity. The combined effects of highly acidic conditions and heavy metals can devastate the ecology of a stream, producing sections completely devoid of life for miles downstream. Even less severe cases of acid mine drainage can disrupt an ecosystem by creating conditions that make it impossible for aquatic species at the base of the food web (e.g., insects) to survive. The loss of those species sets off a chain reaction that ends in the death of numerous species throughout the food web.

Most of the problems with acid mine drainage in the United States are associated with abandoned mines that operated prior to the environmental regulations of the 1970s. Current laws governing mine tailings are complex due to the different ways the waste is classified, which in turn has led to numerous exemptions. In general, most modern mines attempt to minimize the effects of acid drainage by limiting the amount of leachate that is generated, and then collecting and treating what does form. For example, the volume of leachate can be reduced by covering the top of the tailings pile with material that has low permeability, thereby reducing infiltration. Whatever leachate that does form is then collected and run through a treatment process that lowers the acidity and removes most of the heavy metals.

Different techniques are used to treat acid mine drainage, but perhaps the most common involve what are known as *constructed wetlands*. As illustrated in Figure 12.28, the first step often involves having the acidic water flow through a bed of crushed limestone, which lowers the acidity as the water dissolves the calcite ($CaCO_3$) making up the limestone. After this initial treatment the water is routed into a constructed wetland in which high levels of dissolved oxygen are maintained, causing iron and manganese to precipitate and form hydroxide minerals (e.g., $FeO(OH)$), thereby removing these metals from the system. Next the water moves into an oxygen-poor wetland where bacteria convert sulfate (SO_4^{2-}) into sulfide (S^{2-}) ions, causing the heavy metals (e.g., copper, zinc, cadmium) to precipitate as sulfide minerals. Note that special plants are grown in the wetland to help remove the heavy metals.

The problem of acid mine drainage gained national media attention in 2015 when 3 million gallons of contaminated water burst from the abandoned King Gold Mine near Silverton, Colorado, and flowed into a nearby creek and into the Animas River (Figure 12.29). This event highlighted the pre-1970s lack of state and federal laws that required mines to be reclaimed once mining was complete. Mine owners could just walk away, leaving behind large tailing piles and mountainsides riddled with open passageways. In Colorado the lack of regulation resulted in approximately 23,000 abandoned mines, most of which are concentrated in the mountainous parts of the state. Like many of the abandoned workings, the King Gold Mine was leaking heavy metals and acidic water into a nearby creek. Even worse, a natural dam had formed near the entrance of the mine, causing the acid mine water to accumulate in the abandoned passageways. Then on August 5, 2015, the dam burst while workers with the U.S. Environmental Protection Agency (EPA) were attempting to install a drain that would slowly capture and treat the leaking wastewater. While some blamed the EPA for the spill because of the risks involved in their remediation effort, others pointed out that it was only a matter of time before the dam burst anyway. It was also noted that the basic problem stemmed from mine owners being able to simply abandon their mines. Moreover, local resistance prevented the EPA from doing a major cleanup prior to the accident out of fear the project would hurt the regional tourist industry.

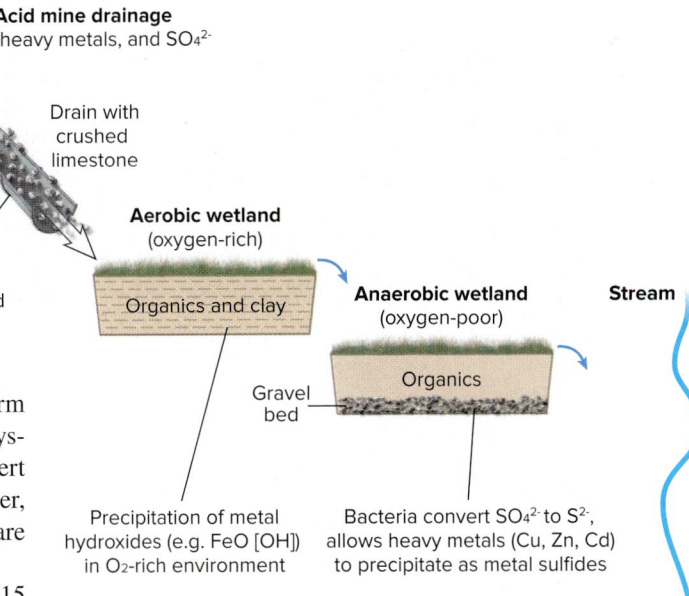

Acid mine drainage
Iron, heavy metals, and SO_4^{2-}

Drain with crushed limestone

Reduces acidity and raises pH

Aerobic wetland (oxygen-rich)

Organics and clay

Gravel bed

Anaerobic wetland (oxygen-poor)

Organics

Stream

Precipitation of metal hydroxides (e.g. FeO [OH]) in O₂-rich environment

Bacteria convert SO_4^{2-} to S^{2-}, allows heavy metals (Cu, Zn, Cd) to precipitate as metal sulfides

FIGURE 12.28 Remediation of acid mine drainage includes neutralizing the acid by letting it first react with crushed limestone, followed by having the leachate flow through a series of constructed wetlands where dissolved oxygen is removed. Here heavy metals are removed via plant uptake and by precipitation under the reducing conditions.
© IntraSearch Inc

FIGURE 12.29 Photo (A) showing the entrance to the King Gold Mine in Colorado where 3 million gallons of acid mine water suddenly burst through a natural dam while workers with the EPA were trying to install a drain. The contaminated water then flowed into the nearby Animas River (B), causing health and safety concerns. After 4 days, the plume had passed the nearby city of Durango and pollutant levels soon returned to normal. Concerns remain about the heavy metals that precipitated out of the water and accumulated on the riverbed.
(a–b) US Environmental Protection Agency

FIGURE 12.30 A cyanide heap-leaching operation at a historic mining area in southern Colorado began developing leaks, contaminating nearby streams that fed into the Alamosa River. The EPA took over the operation, preventing a potentially catastrophic release of cyanide and heavy metals into the surrounding environment.
© IntraSearch Inc.

Map labels:
Summitville and San Luis Valley, Colorado
San Luis Valley
San Juan Mountains
Monte Vista
Rio Grande
Wightman Fork
Monte Vista National Wildlife Refuge
Alamosa National Wildlife Refuge
Summitville Mine
Alamosa
Terrace Reservoir
Alamosa River
Mineralized areas
Capulin
La Jara Creek
0 km   10   20
0 miles   10
Colorado
N
Farmlands irrigated at least in part with Alamosa River water
Wildlife habitat potentially affected by Alamosa River water

Processing of Ores

In addition to leachate from waste material stored in tailings piles, the processing of ore minerals can release a variety of harmful toxins and compounds into the environment. One of the concerns with the cyanide heap-leaching method is that the impermeable liner beneath the pile or holding lagoon might develop a leak, releasing cyanide and heavy metals into surface streams and the ground-water system. Such fears were realized in 2000 when a dam holding mine tailings from a Romanian cyanide operation failed, releasing 260 million gallons (100,000 m^3) of contaminated liquid into a tributary of the Danube River. This contamination caused a massive fish kill and poisoned the drinking water of more than 2 million people. A similar dam failure occurred in 2006 near the town of Miliang, China, resulting in a landslide that killed 17 people and injured 130 and, again, poisoning water supplies.

Cyanide operations in the United States have good safety controls, but leaks and accidents do happen. Critics often point to how a disaster was narrowly averted in 1992 near Summitville, Colorado, when the EPA took over a cyanide heap-leaching operation from a mining company that had declared bankruptcy (Figure 12.30). In addition to cyanide leaks discovered earlier beneath the heap-leaching pad and piping systems, the EPA found that the facility's treatment system was in immediate danger of being overwhelmed by snowmelt. The result could have been a catastrophic release of cyanide and heavy metals into the headwaters of the Alamosa River. The potential disaster was fortunately averted, but with the company bankrupt, the estimated remediation and cleanup cost of $120 million had to be paid from the U.S. government trust fund known as *Superfund* (Chapter 15). Due to the previous leaks, concerns were also raised about the uptake of heavy metals by agricultural crops irrigated with water drawn from the Alamosa River. Later studies revealed that a relatively small portion of the heavy

398

metals found in the irrigated soils was related to the mining operation. Most of the metals are believed to be from the natural weathering of ore-bearing rocks in the region, which were then transported downstream and deposited in the ancient sediment. As in similar areas, the stream here was "naturally polluted" long before humans began mining.

Pollution from mining operations can also involve mercury, a metal that was used extensively in processing gold and silver ores and is now a persistent problem in streams draining historic mining districts throughout the western United States. Mercury has several unique properties, the most notable being that it exists as a liquid metal under the temperatures and pressures found at Earth's surface. Because mercury is so dense, even denser than iron, most earth materials will literally float on liquid mercury. Moreover, mercury will chemically react with gold and silver to form a paste-like alloy called *amalgam,* once commonly used in dental fillings. These properties can be exploited to extract gold and silver by mixing liquid mercury with either placer material or crushed ore. Because gold and silver are denser than mercury, only gold- and silver-bearing particles will sink into the mercury and chemically combine to form amalgam. The mercury can then be removed from the amalgam by heating or by dissolving it in nitric acid. The result is pure gold and silver, and the mercury can be recovered and used again.

Although the mercury-amalgam technique is effective, not all of the mercury is recovered. For example, during the historic gold rushes in the western United States, an estimated 10–30% of the mercury was not recovered and ended up in stream sediment. In California alone, 3–8 million pounds (1.4–3.6 million kg) of mercury are believed to have been lost between the 1860s and early 1900s. The problem is that mercury is highly toxic and persists in the environment, plus it can be easily absorbed through the skin, inhaled, or ingested. Mercury also combines with carbon to form methylmercury, which is ingested by aquatic organisms and can then move up through the food chain and into humans, primarily through the consumption of fish (Chapter 15). Unfortunately, the problems associated with the past use of mercury in the United States are now being repeated throughout South America where recent gold rushes have resulted in the unregulated use of mercury.

In the case of sulfide minerals, ores are typically processed by *smelting,* which is an ancient method where ore is heated to the point that chemical bonds within the minerals are broken, liberating both sulfur and metal ions. Once the sulfur atoms are free, they quickly react with oxygen to form sulfur dioxide (SO_2) gas. The gas then rises into the atmosphere where it reacts with water to form sulfuric acid (H_2SO_4), eventually falling as *acid rain,* devastating land plants and aquatic life in areas downwind of the smelting operations. In addition to sulfur dioxide, some of the metallic ions exit via the smokestack, then fall back to earth as particulate matter, thereby contaminating soils downwind. Acid rain and heavy metal fallout associated with smelting operations have largely been eliminated in the United States as regulations now require that sulfur dioxide and particulate matter be removed before emission gases are released into the atmosphere (Chapter 15).

Collapse and Subsidence

Underground mining naturally creates void spaces in the subsurface, which tend to close in on themselves in response to the tremendous weight of the overlying column of rock. Sometimes the voids close slowly; at other times they collapse quite suddenly. The result at the surface can be large cracks, a collapsed sinkhole, or a gradual lowering of the land surface in a process known as *subsidence.* These surface disruptions can cause serious localized damage to buried utilities and other infrastructure, including homes. Mine-related movement of surface materials is more likely to occur in locations where large workings are relatively close to the surface and in rocks that are rather weak. As shown by the example in Figure 12.31, coal mines are notorious for collapsing, which can lead

FIGURE 12.31   Aerial photo showing numerous pits on the land surface that were caused by the collapse of shallow underground coal mines in Wyoming.
USGS

FIGURE 12.32 Abandoned mines, such as the vertical shaft shown here at a long-abandoned site in New Jersey, pose deadly hazards to humans and animals. Note that the fence was erected by state officials at taxpayers' expense.
© Doug Paddock

to subsidence problems at the surface. More detailed information on collapse and subsidence can be found in Chapter 7, whereas Chapter 13 provides additional information on coal mining.

Abandoned Mine Hazards

Every year people and animals are killed and injured in abandoned mines, mostly by falling into quarries, open pits, or the vertical shafts of underground mines (Figure 12.32). Because mine excavations are permanent features, fences and barricades need to be maintained on a long-term basis in order to keep people from entering, either inadvertently or by intentional trespassing. Unfortunately, there are many historic mines whose owners have long since gone out of business, leaving it up to state agencies (i.e., taxpayers) to erect the barriers. Even for modern mines in the United States, where mining companies are required to restrict access, there is the question of who will maintain such barriers far into the future.

SUMMARY POINTS

1. Nearly every aspect of modern societies depends on Earth's mineral resources. Because mineral extraction is costly and can cause severe pollution, society needs to practice conservation and ensure that minerals are extracted safely.

2. Certain minerals are used in society for specific applications because of their physical and chemical properties. Various geologic processes concentrate these minerals into deposits that can be mined.

3. Enrichment refers to the degree to which mineral deposits are concentrated. Those deposits that are economical to mine become part of the overall mineral reserves.

4. Economic mineral deposits formed by igneous activity include diamond pipes, layered intrusions, pegmatites, and hydrothermal (vein and disseminated) deposits.

5. A variety of valuable mineral deposits result from both contact and regional metamorphism.

6. Important sedimentary deposits include sand, gravel, clay, placers, banded iron formations, bauxite, evaporites, and phosphorites.

7. Mining techniques vary depending on the mineral resource and can be classified as underground or surface (open-pit, strip, and placer). Mining profitability depends on the amount of waste material to be removed and the energy this requires. Other factors include ore grade, mining technique, and labor costs.

8. Ore processing involves crushing and physical or chemical separation of ores from waste materials. Common separation methods include hydraulic sorting, floatation and magnetic separation, smelting, and leaching.

9. Mineral deposits are associated with specific types of geologic environments and are therefore not evenly distributed on Earth's continents. Locating deposits is further complicated because these environments shift over geologic time due to tectonic activity.

10. Geographically large countries generally contain a greater diversity of geologic terrain and mineral resources. Strategic minerals are those that are vital to a nation's economy but whose supply is naturally limited.

11. Minerals are finite resources, so future demand will best be met by recycling and conservation, along with technological improvements that allow lower-grade deposits to be mined economically. Rising energy costs, however, will make mining less economical.

12. Numerous environmental problems arise from mining and processing, particularly acid mine drainage and release of cyanide and mercury. Pollution from modern mine operations has been minimized due to regulations and improved mining techniques.

KEY WORDS

acid mine drainage   396
bauxite   377
disseminated deposits   373
enrichment factor   369
evaporites   380
General Mining Act   394
hydrothermal deposits   373

layered intrusions   372
leaching   389
massive sulfide deposit   374
mineral reserve   370
mineral resource   367
mine tailings   384
open-pit mine   386

ore deposit   370
overburden   385
placer deposit   377
smelting   389
strategic minerals   392
strip mine   386

APPLICATIONS

Student Activity

If you live near the ocean, take home a sample of seawater in a plastic beverage container. Otherwise, get some inexpensive salt from a grocery store and add 4 tablespoons to about 4 cups of water. Then stir until the salt completely dissolves. Now, pour your salt water into a pan and bring it to a slow boil on the stove—eventually you will see salt crystals forming in the water. Do this until nearly all of the water evaporates. Does the salt that precipitates form a solid mass, or does it consist of individual flakes or crystals? Describe how this process is similar to the formation of mineral deposits called evaporites. When seawater evaporates, do you think the minerals precipitate all at the same time, or in a sequence? Explain.

Critical Thinking Questions

1. How can some valuable elements and minerals be fairly abundant in Earth's crust, yet rarely be found in concentrations that are economical to mine?
2. Slate is a metamorphic rock that was once widely used as a roofing material, but has largely been replaced by much less expensive shingles made of asphalt. Although more expensive, why would slate shingles still be a better choice over those made of asphalt?
3. Conservation is important when it comes to any of Earth's resources. How is conserving metals, such as aluminum and copper, any more important than conserving water?

Your Environment: YOU Decide

© Neil Beer/Getty Images

Diamonds come from many places in the world, but most come through official channels and have a set, fixed price. There are also diamonds that come from areas of the world where they are sold for the purpose of financing uprisings and civil wars. These diamonds are commonly known as "conflict" or "blood" diamonds. They end up in regular shops, but do not go through official channels; hence, they are usually less expensive. If you were in the market to buy a diamond or a piece of diamond jewelry, would you buy one that is conflict-free (as some are now certified to be) or go with the cheapest price, not wanting to know its history? Explain your reasoning.

Chapter 13

Conventional Fossil Fuel Resources

LEARNING OUTCOMES

After reading this chapter, you should be able to:

▶ List the basic forms of energy, and describe some of the common transformations between different energy forms.
▶ Describe why petroleum was favored over coal and how it eventually became the dominant resource it is today.
▶ Characterize how organic matter accumulates geologically and is transformed into coal or petroleum.
▶ Discuss the basic process by which petroleum migrates from a source rock and accumulates in a reservoir rock.
▶ Identify the locations of major fossil fuel reserves, and discuss the impact on energy security.
▶ Explain the peak oil theory, and describe how it may affect the world economy.
▶ Discuss why it will take years to scale up production of nonconventional oil resources to make up for declining supplies of conventional oil.
▶ Explain why conservation and efficiency must play key roles in limiting the impact of future shortages.

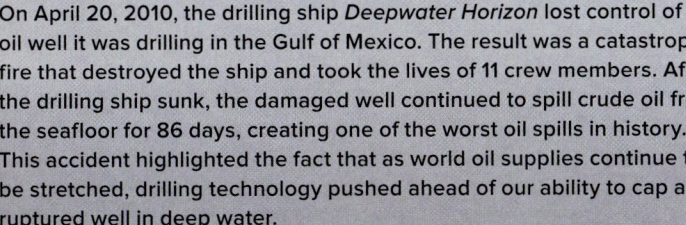

On April 20, 2010, the drilling ship *Deepwater Horizon* lost control of an oil well it was drilling in the Gulf of Mexico. The result was a catastrophic fire that destroyed the ship and took the lives of 11 crew members. After the drilling ship sunk, the damaged well continued to spill crude oil from the seafloor for 86 days, creating one of the worst oil spills in history. This accident highlighted the fact that as world oil supplies continue to be stretched, drilling technology pushed ahead of our ability to cap a ruptured well in deep water.

Introduction

© StockTrek/Getty Images

In previous chapters you learned that the very existence of human civilization depends upon fertile soils and adequate water resources. From Earth's mineral resources we have made steel, concrete, and various materials that form the basis of our cities and factories, which, in turn, produce machines and countless types of products. This has allowed us to grow large quantities of food on highly mechanized farms and build vast transportation networks that move goods and people around the globe. What makes all these activities possible are energy resources. Although there are different forms of energy, about 88% of the energy consumed by humans comes from the burning of coal, oil, and natural gas, fuels that are collectively referred to as **fossil fuels** because they are derived from remains of ancient plants and animals. Although new technology has led to significant increases in supplies of natural gas, and to a lesser extent of oil, these new supplies are much more expensive to bring to market. This has led to price instabilities that threaten to disrupt the global economy. Moreover, the long-term demand for oil still threatens to outstrip supply. Equally troubling is the fact that our massive use of fossil fuels is contributing to the growing problem of global warming.

The world's energy problem is not necessarily that we are running out of oil, but rather that we are running out of *cheap oil*. For more than 125 years humans have extracted the oil that was the easiest to locate and pump from permeable rock layers. The viscosity (resistance to flow) of this oil is relatively low, which makes it easier to extract, plus it produces more of the highly desirable transportation fuels such as gasoline, diesel, and kerosene (jet fuel). Much of what remains in the ground today is more expensive to remove because it is more viscous (heavy oil) or is located in areas that are difficult to access (e.g., offshore and in the Arctic). Other oil exists in rock layers with low permeability and requires expensive technology to extract. Because nearly every aspect of our society depends on this highly versatile liquid, increases in the cost of oil can have a ripple effect through the entire global economy. For example, food prices rise along with the cost of diesel fuel because it is used to operate farm equipment and irrigation pumps as well as to fuel the trucks that transport the food to market. Likewise, higher oil prices make it more expensive to drive to work or school in our gasoline-powered vehicles. We also feel the impact of more expensive oil in the cost of air travel since jet aircraft use kerosene as fuel. In the end, higher oil prices mean that each of us will spend progressively more of our income on the basic necessities of life, leaving less to spend on a house, vacation, or new consumer items.

In order to minimize the risk of higher oil prices, society needs to increase supply by finding new oil deposits, and at the same time decrease demand through serious conservation efforts. However, because of the threat of global warming (Chapter 16), burning even more fossil fuels is not a long-term option. Ultimately we must increase the production of energy from sources that do not contribute to global warming, while also finding ways to reduce the impact of burning fossil fuels. For us to succeed, today's younger generation will need to become knowledgeable about the science behind energy and global warming. After all, it will be young people who will help guide society through the necessary transformations in the coming decades. Keep in mind that although energy use and global warming are interwoven problems that pose serious challenges for society, they also present an enormous number of business and career opportunities.

This chapter will focus on the *conventional fossil fuels* (coal, oil, and natural gas) that have provided the vast majority of the world's energy needs since the Industrial Revolution. In Chapter 14 we will examine the nonconventional, or alternative, energy sources that can help meet the world's future energy needs. For the purpose of this book, *alternative energy sources* are defined as any sources other than the coal, natural gas, and low-viscosity crude oil that currently make up the bulk (86%) of our energy supplies. Using this definition, alternative sources consist of the remaining (14%) sources currently in use, including hydroelectric, nuclear, wind, and solar power and biofuels. Alternative energy sources also include heavy oils and synthetic oils made from tar sands, oil shale, and coal, which presently make up a relatively small fraction of world supply.

Human Use of Energy

Energy is a term we often use when referring to such things as the electric and gas bills for our homes. We may also describe ourselves as being "low on energy," and then go out and buy an energy drink to help stay awake in class. To a scientist or engineer, **energy** is the capacity to perform work or transfer heat, whereas **work** involves moving an object (mass) a certain distance against some force (i.e., work = mass × distance). For example, you do work at the gym when lifting weights against the force of gravity—hence the term "workout." The energy that enables you to do this work comes from the food you eat, which itself is solar energy that has been stored by plants and animals. The key point here is that we are literally surrounded by different forms of energy and, as individuals, are constantly making use of this energy. Because the remaining topics in this textbook require a fundamental understanding of energy, we will take a brief look at the basic forms of energy.

Electrical energy results from the movement of electrons through a material called a conductor, such as a copper wire or a circuit board. In modern times humans have learned to construct electrical generators that produce a steady flow of electrons through wires, which we call *electricity*. Electrical power is one of the most useful forms of energy in society today, running everything from smartphones to air conditioners to high-speed trains.

Chemical energy comes from the energy stored in the chemical bonds between the different atoms making up a compound. All chemical reactions either release heat energy (exothermic) or require heat (endothermic). When we eat, acids and enzymes in the stomach break down the chemical bonds within the food, resulting in the release of heat energy and the formation of new compounds. The heat keeps us warm, whereas the new compounds are absorbed into the blood and used for various body functions, such as muscle movement. Another example of chemical energy comes from the battery in your smart phone. When you turn on the phone, chemical compounds in the battery begin to react, producing electrons that flow through the phone's electrical circuits.

Thermal energy, often called *heat energy*, is produced by the vibrations of individual atoms or molecules. What we call temperature is a measure of the frictional heat produced by these vibrations. For example, a metal bar pulled from a fire is hot because its atoms are vibrating fast and generating a considerable amount of frictional heat. When electrons flow through a wire, the wire becomes warm because there is a certain amount of resistance in the wire that forces the electrons to slow down.

Kinetic energy is the form of energy that results from the velocity of an object (mass) in motion. It is computed by taking half the object's mass and multiplying it by the square of its velocity (kinetic energy = $\frac{1}{2}mv^2$). Heavier and faster objects therefore have greater amounts of kinetic energy. For example, an asteroid a mile in diameter has an enormous amount of kinetic energy, whereas a fist-sized asteroid traveling at the same velocity has comparatively little energy. Conversely, if two asteroids were the same size, the faster one would naturally have more kinetic energy.

Potential energy is simply the energy stored in an object that is not in motion, but is capable of moving. Familiar forms of potential energy are *gravitational*, *elastic*, and *electrical*. For example, a dam creates an elevated pool of water called a reservoir. Because of the water's increased height, its gravitational potential energy has increased. When the water is released and allowed to fall, the stored energy immediately begins converting to kinetic energy, which can be used to spin a generator and make electricity. An example of elastic potential energy is in the use of a bow and arrow. When an archer pulls on a bow, it bends and stores elastic energy. As the bow is released, the energy is quickly transferred to the arrow and transformed into kinetic energy.

Nuclear energy is the energy that binds or holds the nucleus (protons and neutrons) of an atom together—in contrast to chemical energy, which involves the bonds *between* atoms. Nuclear energy is a form of potential energy because it is stored energy, which is released only when the nucleus undergoes one of two types of reactions. *Fission* reactions involve the splitting of the nucleus and *fusion* reactions occur when two separate nuclei combine; both of these types of reactions release tremendous amounts of thermal (heat) energy. Note that fusion is the dominant reaction in the Sun, whereas modern nuclear power plants rely on fission.

Radiant energy is the general term for electromagnetic energy that travels in the form of waves (Chapter 2). Radiant energy is associated with the different wavelengths in the electromagnetic spectrum (radio, microwave, infrared, visible light, ultraviolet, X-ray, and gamma ray). This energy is also commonly called *solar energy* because the various electromagnetic wavelengths are produced by nuclear fusion in stars, similar to what takes place in our Sun.

Energy Conversions

You may have noticed that one form of energy can convert to another. Chemical energy stored in batteries is converted into electricity used to run your cell phone; a stove converts electrical into thermal energy for cooking. In fact, almost everything we do is a series of energy conversions. For example, consider all the energy conversions shown in Figure 13.1 that are related to growing and cooking food. Our food originates from the conversion of nuclear energy to solar radiation, which is then converted into chemical energy by plant photosynthesis. Our digestive system converts this chemical energy into thermal (heat) energy and other chemical compounds, some of which are used to move our muscles. Whenever we use our muscles to walk or lift things we are ultimately converting some of the original food into kinetic energy—often called *mechanical* energy. This kinetic or mechanical energy can then be used to do *work*, such as lifting sacks of groceries or pushing a lawn mower. Prior to the Industrial Revolution and the widespread use of fossil fuels, physical work was mostly performed by people (manual labor) and animals, powered solely by chemical energy derived from plants.

Today's modern societies are powered mainly by the conversion of fossil fuels into different forms of energy we find useful. During this process a certain portion of the energy is converted into forms that we do not need, which means part of what we pay for is wasted. The reason for this is related to the *first law of thermodynamics*, which states that energy cannot be created or destroyed, only transformed from one form to another. Moreover, during every energy transformation there are at least two conversions, one of which results in thermal energy, or simply *heat*. A good example is when you flip a switch and electricity begins to run through a an older incandescent light bulb, causing the filament inside to glow. Here the electrical energy is being converted into visible light, which is just radiant energy in the visible part of the electromagnetic spectrum. However, the bulb also gets hot, telling us that an electrical to thermal conversion is also taking place.

Energy efficiency is a useful measure of how much of the original energy actually goes toward its intended use; the

FIGURE 13.1   Humans constantly make use of energy conversions to fulfill their needs. Consider the number of energy conversions involved in cooking and growing food, which provides chemical energy.

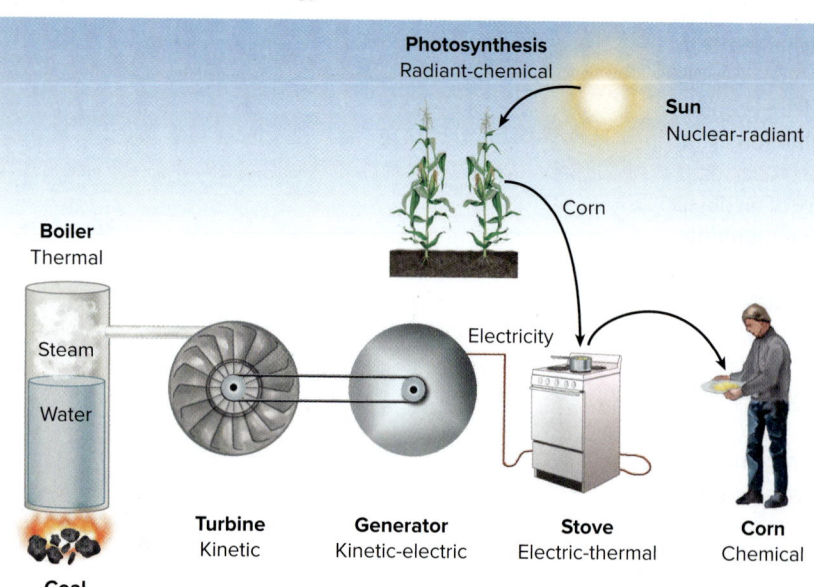

Photosynthesis
Radiant-chemical

Sun
Nuclear-radiant

Corn

Boiler
Thermal

Steam

Water

Electricity

Turbine
Kinetic

Generator
Kinetic-electric

Stove
Electric-thermal

Corn
Chemical

Coal
Chemical

unwanted heat is referred to as *heat loss.* Table 13.1 lists the energy efficiency of some common applications. In the case of an incandescent light bulb, only 5% of the electricity is converted into what you want, namely visible light. In contrast, compact fluorescent bulbs are more efficient, with nearly 20% of the electricity going into producing light, whereas LED lights convert 80% of the electricity into light and are by far the most efficient. Notice that gasoline engines are only 25% efficient, which means that for every dollar you spend on gas, seventy-five cents goes to making heat that is lost through the exhaust pipe and radiator. Because we pay for the energy we use, it is in our own financial interest to try to be as energy efficient as possible.

Renewable versus Nonrenewable Energy

Although there are many different forms of energy available on our planet, solar radiation is the most essential to the biosphere and life as we know it. Through the process of *photosynthesis,* land plants and the ocean's phytoplankton take radiant energy flowing from the Sun and convert it into chemical energy. The result is the growth of cell tissue and the formation of what is called *biomass.* Plants are a critical form of stored chemical energy since they serve as the primary food source for both land and marine animals. In addition, some of the organic matter from the biosphere becomes buried and is then transformed by geological processes into the more concentrated forms of chemical energy we call coal, oil, and natural gas. These fossil fuels are often referred to as "stored sunlight" since they formed from the conversion of solar energy into biomass.

Ever since the Industrial Revolution began over 200 years ago, society has been relying on the Earth's supply of highly concentrated fossil fuels, which have accumulated over the past several hundred million years. Similar to mineral resources, fossil fuels are still forming, but they are considered finite or *nonrenewable* because we are consuming them at a much faster rate than they are forming. Since society's consumption of fossil fuels is not sustainable, we will eventually be forced to turn to alternative supplies of renewable energy such as hydroelectric, wind, and solar power. Our basic problem is that no other source is as concentrated, or *energy dense,* as fossil fuels. Moreover, other energy sources do not provide the type or quantity of liquid fuels currently being used to run our transportation systems. Note that electricity is considered to be a *secondary energy resource* because it has to be produced from some primary source, such as coal or nuclear, wind, or solar power.

Historical Energy Usage

For much of history people depended on sunlight for warmth, but eventually learned how to make wood (biomass) fires for cooking and keeping warm. About 5,000 years ago humans expanded their energy supply beyond wood by domesticating animals, using them to perform work. By the 1800s large numbers of mills had been constructed for harnessing the kinetic energy of wind and water (Figure 13.2). These flowing forms of energy provided mechanical power that could be used to perform a variety of useful work, including grinding grain, cutting lumber, and running textile looms. Mills operating on these renewable sources were later replaced by steam engines as the primary source of mechanical power. Steam engines converted *stored* chemical energy (mainly wood) into the kinetic energy needed to operate machinery. The invention of the steam engine was a major technological advance because factories could now be located almost anywhere; no longer were they restricted to waterways or places where the wind was reliable. Steam engines were also put to use powering ships and trains, a step that revolutionized transportation and commerce.

TABLE 13.1   **Efficiency of some of the most common energy conversions used in modern societies.**

Equipment	Desired Conversion	Efficiency (Percentage Undergoing the Desired Conversion)
Incandescent light bulb	Electrical to radiant	5%
Compact fluorescent light bulb	Electrical to radiant	20%
LED light bulb	Electrical to radiant	80%
Gasoline engine	Chemical to kinetic	25%
Diesel engine	Chemical to kinetic	35%
Electric engine (motor)	Electrical to kinetic	70%
Coal-burning power plant	Chemical to electrical	30%
Home gas furnace	Chemical to thermal	90%
Electrical heating	Electrical to thermal	100%

Source: Data from Alternate Energy Guide and International Panel on Fissile Materials.

FIGURE 13.2   Prior to steam engines, wind and water mills were the dominant means of generating mechanized power. The kinetic energy from wind or falling water was used to turn shafts that provided power to perform tasks such as grinding grain, cutting lumber, and making textiles.
(a) © PhotoLink/Getty Images; (b) © PhotoLink/Photodisc/Getty Images

As indicated in Figure 13.3, the rapid development of steam power during the 1800s greatly increased U.S. consumption of wood as an energy source. By 1875 coal was mined in increasing quantities and soon replaced wood as the dominant fuel source. In addition to being abundant, coal had the advantage of giving off more heat per ton than wood because it is a much more concentrated form of chemical energy (i.e., more energy-dense). Being able to burn coal meant that trains and ships could carry more cargo and less fuel to travel a given distance. Another important development during this period was the switch from whale oil to kerosene as a fuel for lanterns used for indoor lighting. Kerosene was refined from unprocessed or *crude* oil that was collected at seeps where the oil would naturally rise and leak onto the land surface. This switch not only saved whales from near extinction, but the new demand for kerosene led to the first U.S. well being drilled in 1859 in Pennsylvania. Oil was soon being extracted directly from the subsurface in great quantities and refined into a variety of products, including grease, lubricating oils, kerosene, and gasoline.

The consumption of oil started to increase significantly after 1900 as the internal combustion engine was put to use in a variety of applications,

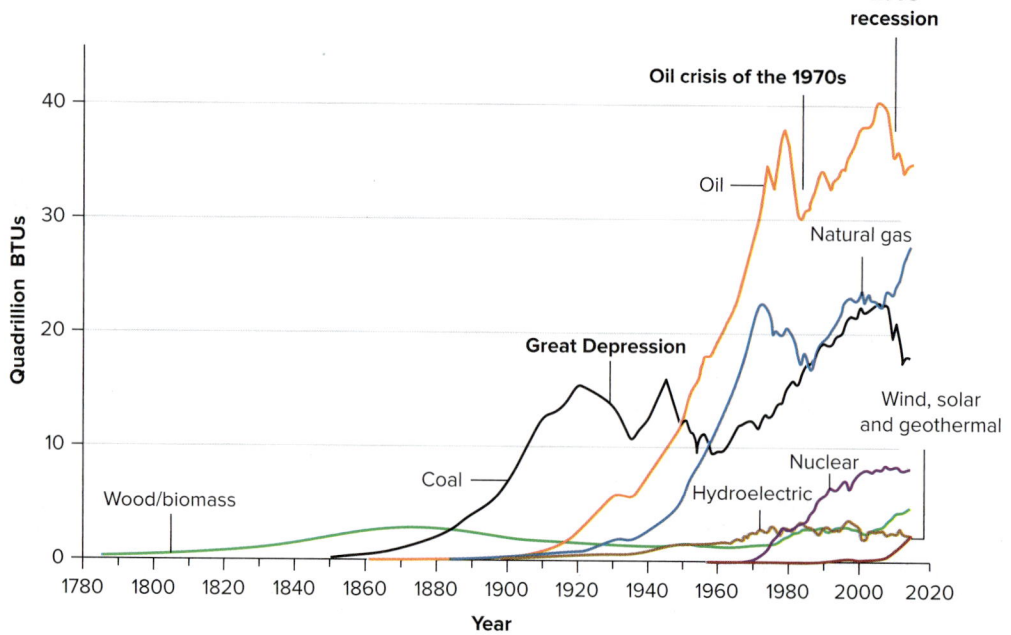

FIGURE 13.3   Plot showing annual U.S. consumption of different energy sources since the Industrial Revolution. Coal usage declined after 1945 as some applications switched to oil and gas, but then increased again after 1960 due to a greater demand for electricity. Note the relatively small contribution of renewable sources compared to the nation's overall energy needs.

particularly cars, trucks, tractors, ships, and airplanes. Increasing use of these engines, in turn, led to the development of large-scale oil exploration (Figure 13.4) and the oil refining industry. As can be seen in Figure 13.3, the dramatic rise in oil consumption was slowed by the Great Depression of the 1930s and later by the oil crisis of the 1970s and the recession of 2008. Oil eventually eclipsed coal as the dominant energy resource because it has higher energy density and its liquid nature makes it more versatile. For example, a relatively small internal combustion engine can produce tremendous amounts of power by burning refined gasoline or diesel fuel. The liquid nature of these fuels makes them easy to transfer and store, and their high energy density means that a vehicle does not have to carry as much weight in fuel. These characteristics gave internal combustion engines a distinct advantage over heavy and dangerous steam engines, whose fuel supply of coal was bulky and more difficult to transport. Coal also lost its place as the primary fuel for heating buildings when it was overtaken by natural gas, which soon followed a consumption trajectory similar to that for oil (Figure 13.3). The advantage of natural gas over coal is that it burns much cleaner (does not produce soot or ash) and is easily transported over land via pipelines. The consumption of coal finally began to increase again after 1960 due to the growing demand for electricity, which is produced in the United States mainly by burning coal. This increased demand for electricity was brought on in part by the widespread use of air conditioning.

Perhaps the most striking aspect of Figure 13.3 is the explosive growth in overall energy consumption that accompanied the Industrial Revolution. The amount of energy that had been harnessed and put to use by burning wood was clearly minuscule compared to what is now being obtained from fossil fuels. By unlocking this enormous reservoir of chemical energy, human civilization has been able to undergo a phenomenal transformation. The world went from largely agrarian societies to those with high-tech service economies, whose citizens enjoy unprecedented comforts and consumer products. Unfortunately, the conventional fossil fuels that made all this possible are finite. Consider that these fuels accumulated over several hundreds of millions of years of geologic time, whereas humans will have consumed the bulk of these irreplaceable resources in about 200 years. This massive use of fossil fuels has also released tremendous quantities of carbon dioxide (CO_2) gas into the atmosphere, which is affecting Earth's global climate system, a subject we will explore in Chapter 16.

The following sections will cover the basic geologic processes that have produced our conventional deposits of coal, oil, and natural gas. We begin with coal, in part because it was the first fossil fuel used on a large scale, and because it forms in a different manner than do oil and gas.

Coal

Geologic studies have shown that the world's coal deposits range in age from 2 million to 350 million years old and that they formed from ancient plants that thrived under swamplike conditions in humid climates. The plant material eventually became buried, at which point geologic processes transformed the material into sedimentary deposits of coal. Although individual beds of coal can be less than an inch thick, economic deposits generally start at a thickness around 6 inches (25 cm), but some reach over 100 feet thick (30 m). It is estimated that about 3 to 10 feet of compacted organic matter is required to produce a foot of coal. However, it takes a somewhat unusual set of geologic conditions for organic matter in a swamp to be transformed into an economic deposit of coal. First, a swamp needs to remain saturated for a considerable period of time. This helps block out free oxygen (O_2), greatly reducing the microbial breakdown of dead plant matter. The abundance of water also promotes lush vegetation growth, which when combined with the slow decay rates leads to thick accumulation of partially decayed plant material. Next, this organic matter must become buried in order for it to be preserved within a sedimentary sequence.

FIGURE 13.4   Oil's liquid nature and high energy density, combined with the advent of the internal combustion engine and well-drilling technology, helped make oil the most desirable energy source. Shown here is the booming Spindletop oil field in Texas in 1903.
Courtesy of Texas Energy Museum, Beaumont, TX

Based on the types of sedimentary rocks found associated with coal, geologists have determined that the environment in which coal originates is similar to today's river deltas located in warm, humid climates (Chapter 8). As illustrated in Figure 13.5, deltas are ideal because they contain expansive back swamps with lush vegetation. Here the stagnant water conditions help produce low levels of dissolved oxygen (O_2), greatly slowing the decay of dead plant material, and allow more organic matter to accumulate. Moreover, large deltas typically experience subsidence as the tremendous weight of sediment presses down on the crust. This progressive sinking allows for even greater amounts of plant matter to accumulate, which then act to compress the previously deposited organic material into a more compact form called *peat*. At some point the river channels within the delta may migrate or sea level will rise, burying the peat with younger sediment. This burial seals the peat deposits off from atmospheric oxygen, increasing the chance they will be preserved.

As the delta continues to receive new sediment and subside, layers of peat are buried even deeper and exposed to progressively higher levels of heat and pressure. Eventually the peat may reach the depth where very low-grade metamorphism transforms it into the combustible sedimentary rock known as **coal.** During this transformation the original plant material undergoes physical and chemical changes that drive off water and other volatile compounds (those that easily form a gas), leaving behind solid coal that is more concentrated in carbon. Because this process is a function of temperature and pressure, the carbon concentration can increase over time if the coal becomes more deeply buried. As illustrated in

FIGURE 13.5   Coal is a sedimentary rock that originates in the back swamps of large river deltas where dead plant material accumulates, eventually compacting into layers of peat. Because large deltas typically undergo subsidence, thick accumulations of peat can develop and become buried by shifting stream channels or rising sea level. If peat becomes deeply buried, the increased heat and pressure can turn it into coal.
© Fred Rich, Georgia Southern University

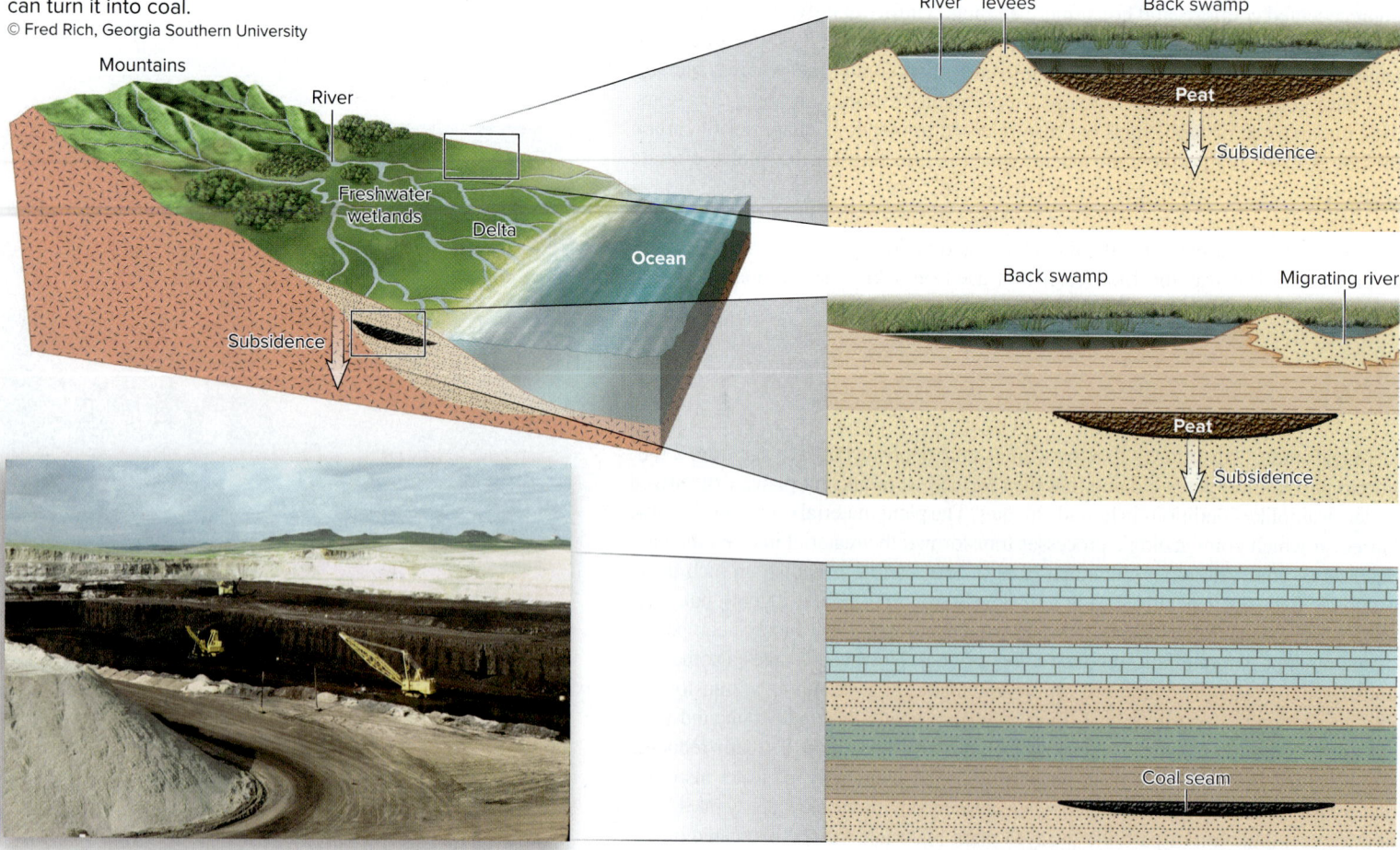

Rawhide mine, Wyoming

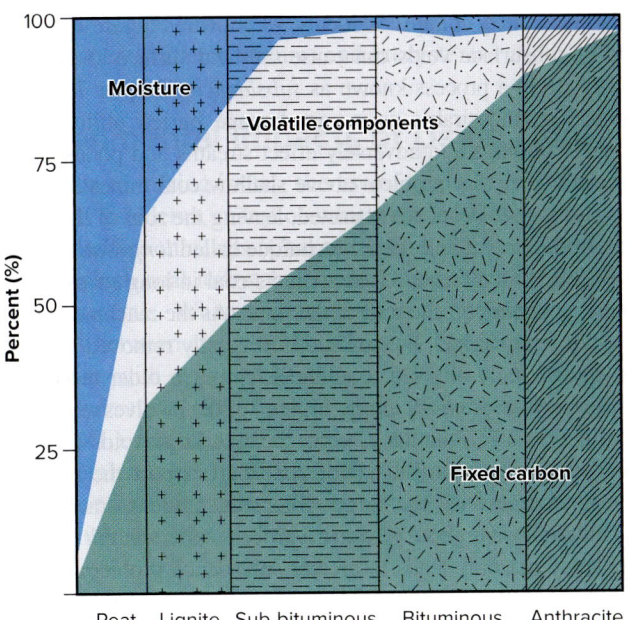

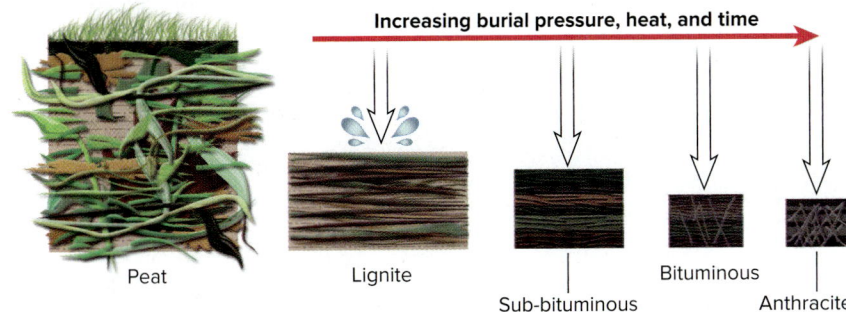

FIGURE 13.6   As peat becomes more deeply buried, the higher temperatures and pressures drive off progressively greater amounts of water and other volatiles, leaving behind a deposit more concentrated in carbon. Higher grades of coal are generally the most desirable as they are more energy-dense and thus release greater amounts of energy.

Figure 13.6, coals are ranked based on the amount of carbon and volatiles they contain. Lignite is the lowest grade, with progressively higher grades called sub-bituminous, bituminous, and anthracite. Notice how the higher-ranked coals are generally more compact, which is a reflection of being buried at greater depths where carbon concentration increases as volatile compounds are driven off.

In general, the most desirable coals are bituminous and anthracite because they are the most energy-dense, and thus release the most energy. Anthracite is relatively rare and is considered to be a metamorphic coal because of the high temperature and pressure at which it forms—beyond anthracite, the coal will metamorphose into the mineral graphite. Figure 13.7 shows the distribution of different coal deposits in the United States. Notice how coals in the eastern United States are generally higher-ranked than those in the West, and that anthracite is restricted to a narrow belt within the Appalachian Mountains. The differences in rank are due to differences in geologic conditions (temperature and pressure) under which the coals formed.

Environmental Impacts of Mining Coal

Most of the bituminous coal in the United States was historically mined in the Appalachian region using underground techniques (Chapter 12). The coal layers there were found exposed along numerous stream valleys that cut through relatively flat-lying sequences of sedimentary rock. Mining typically began where coal seams were exposed at the surface, which the miners then followed horizontally into the subsurface. As mining progressed, vertical shafts were used to access layers of coal buried deep within the hillsides, creating a network of tunnels and shafts. Here the underground miners faced a high risk of rockfalls and cave-ins in the relatively soft sedimentary rocks. Other common hazards included explosions and fires associated with the buildup of methane gas and coal dust. Long-term exposure to coal dust also caused high incidences of lung cancer among miners, more commonly known as *black lung disease*. Today's modern safety regulations and mining techniques have dramatically reduced the number of fatalities and injuries in underground coal mines in the United States. Unfortunately, coal mining remains a very hazardous occupation in many parts of the world as developing nations tend to be slow in adopting similar safety standards.

In addition to the risks posed to miners, underground mining creates void spaces that tend to close in on themselves due to the tremendous weight of the overlying column of rock (overburden pressure). Sometimes the voids close slowly,

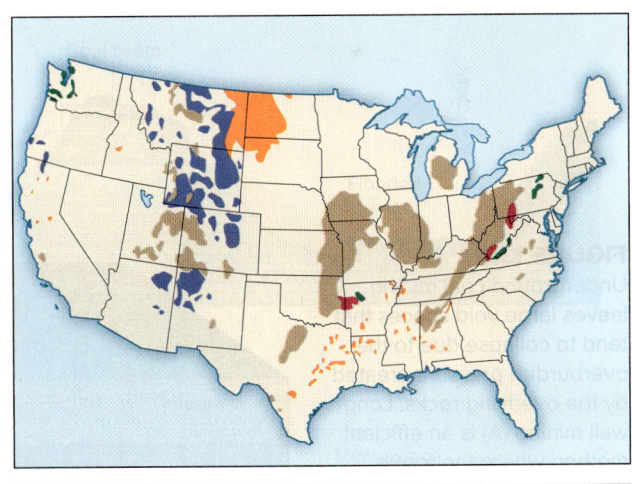

Coal reserves in the United States

■ Anthracite	■ Sub-bituminous coal
■ Low-volatile bituminous coal	■ Lignite
■ Medium- and high-volatile bituminous coal	

FIGURE 13.7   Major coal deposits in the United States. Although western coals generally have lower energy content than eastern coals, they also contain less sulfur, which reduces the acid rain problem associated with smokestack emissions. This has resulted in a mining boom for western coals and a more depressed market for eastern coals.

As illustrated in Figure 13.13, hydrocarbon molecules will form in a sedimentary basin provided there is a suitable source rock that had at one time been buried to a depth within the oil or gas window. Historically, petroleum has been extracted from deposits where hydrocarbon molecules migrate away from a source rock and then accumulate in more permeable rocks. Modern directional drilling techniques, however, now allow petroleum to be removed from the source rock where the permeability is quite low.

When hydrocarbon molecules form within a source rock, their low density causes them to tend to migrate toward the surface through the rock's pore spaces—similar to how gravity allows air bubbles to rise in a swimming pool. The migrating oil and gas molecules will eventually reach a more permeable layer within the sedimentary basin, such as a sandstone or limestone. Recall that these permeable rocks also serve as aquifers because they readily transmit water (Chapter 11). As the hydrocarbons enter such an aquifer, they can be transported laterally with the flowing groundwater. Here the less dense hydrocarbon molecules rise and flow on top of the water within the aquifer. Keep in mind that a considerable amount of petroleum never migrates, but instead remains trapped within the source rock.

In order for migrating oil and gas molecules to form an economic deposit, they must somehow become trapped and accumulate in a sedimentary basin; otherwise, they will disperse and continue to rise toward the surface. For example, in Figure 13.13 one can see that the original sedimentary layers have become inclined and folded by tectonic forces. As groundwater and petroleum move along the permeable layer and encounter the fold, the low-density hydrocarbons become trapped, whereas the water continues to flow. Geologists use the term **petroleum trap** to describe any configuration of subsurface rocks that allows hydrocarbons to accumulate, and the term **petroleum reservoir** for a permeable subsurface rock where oil and gas are being stored. Depending on the history of the source rock and its position within the oil and gas windows, a petroleum reservoir may contain oil or gas, or both. Whenever both are present, the natural gas will lie on top of the oil because it is less dense. In some cases the pressure is such that the gas will be dissolved within the oil itself—similar to how carbon dioxide is dissolved in a soft drink.

One of the more common types of petroleum traps and reservoirs are those where sedimentary rocks are folded into a dome or arch that geologists call an *anticline,* such as the example in Figure 13.13. Notice in the aerial photo how the pattern of layered rocks at the surface reflects the domelike structure that extends into the subsurface. Figure 13.14 provides examples of other common types of traps and reservoirs. Salt domes are particularly common in the Gulf Coast of the United States, where evaporite deposits of halite (Chapter 12) flow upward due to the plastic-like nature of halite. Some traps (e.g., buried limestone reefs and unconformities) have little or no expression at the surface, and are therefore more difficult to locate. Interestingly, many of the early oil fields were associated with domes because geologists could easily identify the telltale pattern of rocks at the surface.

An important feature of all petroleum traps is the presence of **cap rock,** which is a rock layer with low permeability that overlies the trap. A suitable cap rock,

FIGURE 13.13 Petroleum forms when organic-rich source rocks reach the oil and gas windows where higher temperatures transform the organic matter into hydrocarbon molecules. The molecules rise until encountering a permeable reservoir rock, then flow laterally with the groundwater until reaching a trap where the lighter hydrocarbons accumulate. Petroleum that remains in the source rock can also be extracted using directional drilling and hydraulic fracturing. Aerial photo shows the characteristic pattern that a dome trap makes at the surface. Note the individual wells extracting petroleum trapped in the rocks below.
© IntraSearch

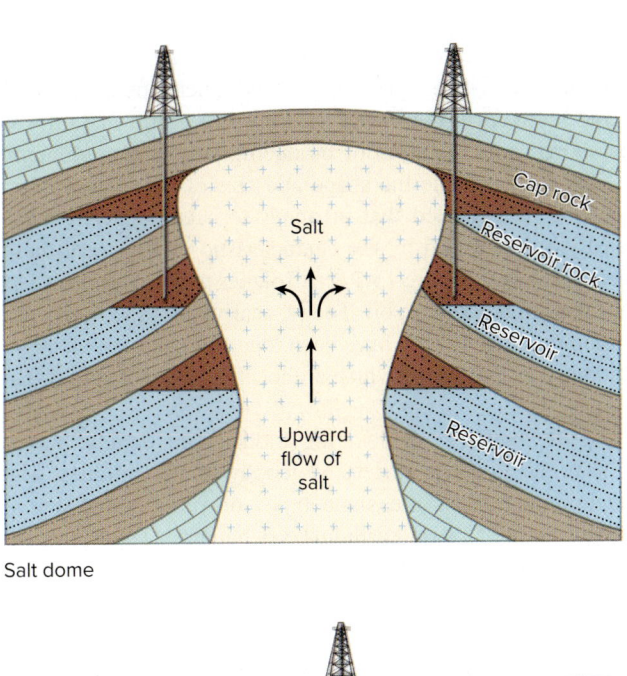

Salt dome

Fault

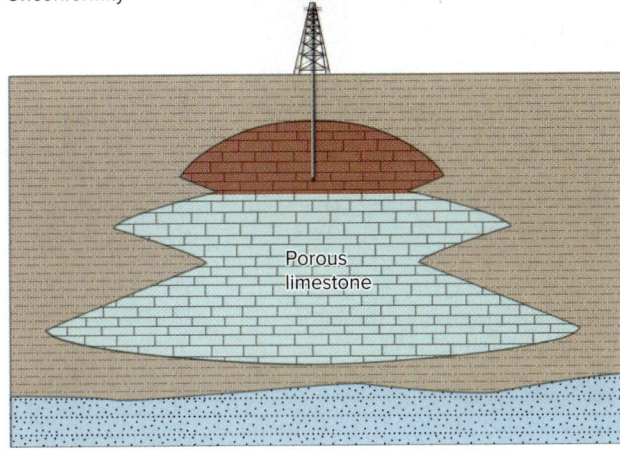

Unconformity

Limestone reef

FIGURE 13.14 In addition to domes or arches formed by anticlines, there are other common types of petroleum traps and reservoirs. All traps must be overlain by low-permeability cap rock in order to keep oil and gas from escaping over time. Conventional reservoir rocks must be fairly porous in order to store significant quantities of petroleum, and also permeable enough to allow the petroleum to be extracted.

typically composed of clay or evaporite minerals, is critical because it limits the ability of oil and gas to escape from the reservoir (Figure 13.14). A good trap with cap rock, however, does not necessarily mean there will be economical amounts of oil or gas. For example, any one of the following circumstances will lead to a reservoir being filled with water as opposed to hydrocarbons:

1. a suitable source rock did not exist;
2. a source rock was present, but never reached the oil window;
3. temperature went above that in the gas window, destroying the hydrocarbons;
4. a leaky cap rock allowed petroleum to escape over time;
5. petroleum formed and migrated before the trap developed.

From a geologic perspective then, traps that contain economical amounts of petroleum are relatively rare. In addition to the petroleum that migrates and becomes trapped in permeable aquifers, oil and gas are now being extracted from organic-rich source rocks and aquifers with very low permeability—natural gas called *coalbed methane* is also extracted from buried coal seams. Later you will see that these so-called *tight oil* and *tight gas* deposits are currently making a significant contribution to the world's energy supply.

Exploration and Production Wells

Similar to water wells (Chapter 11), oil and gas wells are installed by drilling through solid rock using a diamond-tipped cutting tool attached to a rotating steel pipe. This basic technology was used over a hundred years ago for

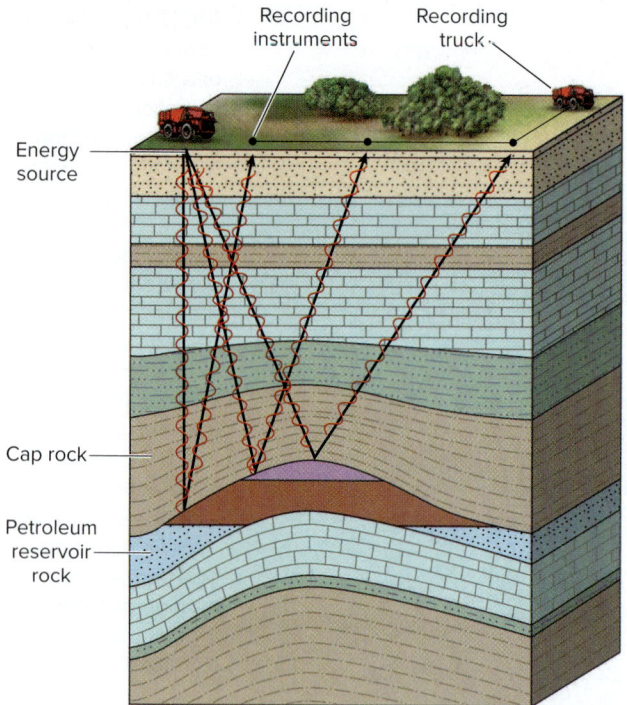

FIGURE 13.15 Seismic exploration techniques use an energy source to generate vibrational waves that are reflected and refracted (not shown) by rock layers of different densities. Based on data obtained by instruments that record the waves that return to the surface, three-dimensional views can be constructed that reveal subsurface rock structures that may contain oil and gas.

FIGURE 13.16 Depending on water depth, large production platforms (A) are towed out to sea and then placed on the seabed where oil reserves have been confirmed by exploration wells. Modern land and offshore operations utilize directional drilling technology to install multiple production wells (B) from a single platform.
(a) © BP Oil

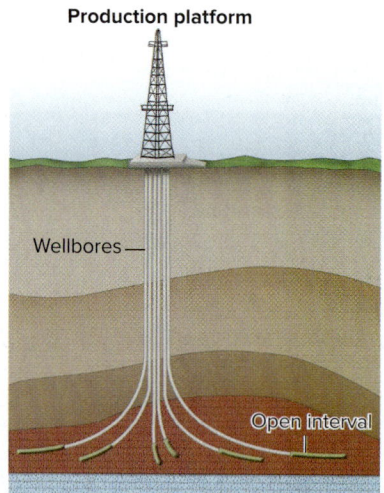

installing the first successful oil wells. By examining the chips of rocks that came to the surface during the drilling, geologists could determine the type of rocks being encountered by the drill bit. When this subsurface information was combined with maps showing the location of producing versus nonproducing wells, it became possible to determine the shape and size of the oil reservoir. This type of mapping quickly revealed that oil tends to accumulate in dome traps, which could easily be identified by the characteristic rock pattern these structures make at the surface (see Figure 13.13). Airplanes were soon put to use in locating new domes that might contain oil. Today, nearly all of the traps with an obvious expression on the land surface have been identified and tested. What remains are mostly hidden traps, many of which are located in remote areas that are hard to access.

To meet the difficult challenge of locating hidden and deeply buried traps, oil companies have developed a variety of geophysical techniques. The most widely used technique utilizes the same types of *seismic waves* that travel through the subsurface during an earthquake (Chapter 5). As shown in Figure 13.15, *seismic exploration* uses an explosive device or a special vibrating truck as an energy source for generating seismic waves. The waves undergo reflection and refraction as they encounter rock layers of different densities; then a series of instruments, called *geophones*, record the waves that return to the surface (a similar approach is used by ships exploring for offshore deposits). From these data geophysicists are able to obtain detailed, three-dimensional views of the subsurface in sedimentary basins both on land and on continental shelves.

Once a seismic survey is completed, teams of geologists and geophysicists determine which traps have the greatest probabilities of containing oil and gas. Despite all the advanced seismic technology, the only sure way to know if a prospective trap is commercially viable is to drill one or more *exploration wells*—over 50% of all exploration wells do not find economical amounts of petroleum. The exploration process also provides the opportunity to perform tests on the actual reservoir rock. Data from these tests allow engineers to estimate how much petroleum is stored in the reservoir and how much can ultimately be extracted or recovered. In the case of oil the recovery rate ranges from 5% to 80%, with the highest rates associated with more permeable reservoir rocks and relatively light, low-viscosity crude oils.

A petroleum reservoir will typically be developed if the exploration data show that there is enough recoverable oil or gas to justify the cost of bringing it to market. Enormous sums of money can be required for installing and operating long-term *production wells*, processing facilities, and pipelines for sending oil or gas to a refinery. A single production well on land can cost $5–15 million, depending on the depth and complexity of drilling. As shown in Figure 13.16, some offshore sites make use of large and very expensive production platforms that are towed out to sea and then allowed to settle on the seafloor. Once in place, multiple production wells are installed using directional drilling techniques in which the drilling bit is steered through the subsurface by technicians on the platform.

New technology has historically been important in expanding oil and gas reserves since it allows us to access deposits that had previously been uneconomical to extract. For example, floating drilling platforms, or drill ships, have opened up new reserves

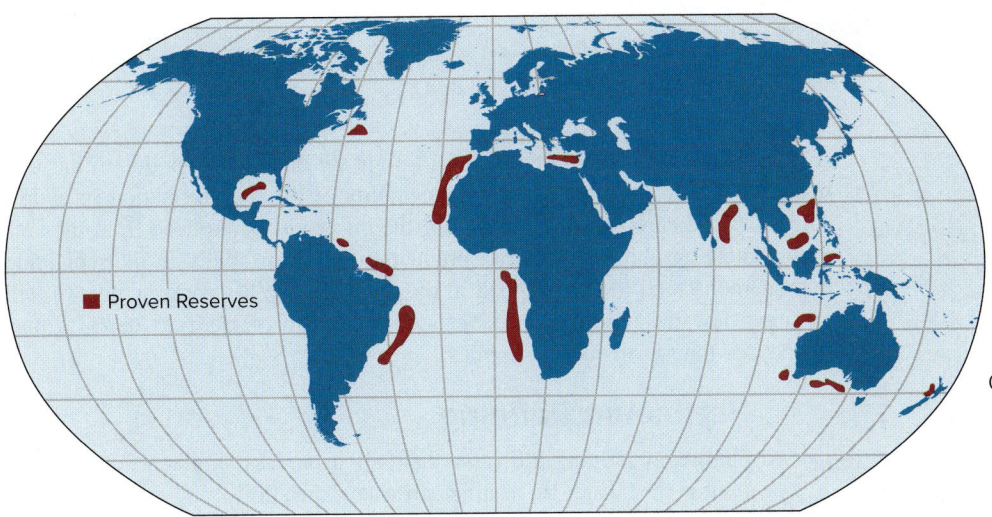

A

FIGURE 13.17   Map (A) showing deep-water oil and gas deposits now being extracted due to the development of floating drill ships (B). These ships can hold their position thousands of feet above the sea floor via global positioning system (GPS) satellites.

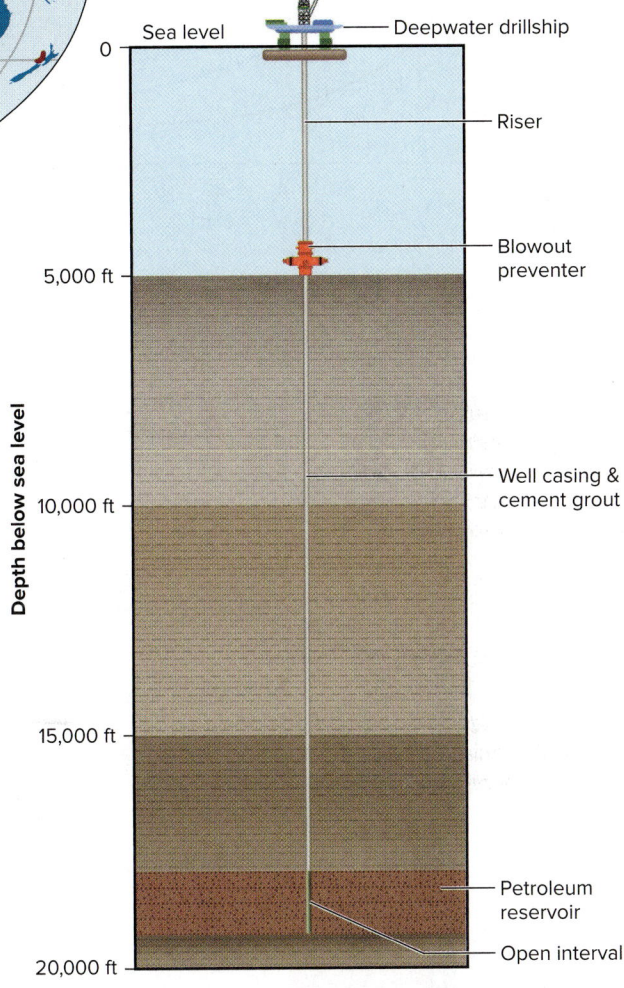

B

in deep-water fields around the world (Figure 13.17). Directional drilling has also been key to the recent development of tight oil and tight gas deposits, where the hydrocarbons are trapped in rocks with very low permeability. Production is achieved by dramatically increasing the permeability of the rocks along the length of the well through a technique known as *hydraulic fracturing,* in which water, chemicals, and fine sand are forced into the rock under high pressure. Once the pressure returns to normal, the sand grains keep the newly formed fractures propped open, thereby allowing the natural gas or oil to flow into the well. Note that hydraulic fracturing has become a controversial technique, a topic we will explore later in this section.

Because all petroleum reservoirs contain a finite amount of oil or gas, it is only a matter of time before the production rate begins to decline. In the case of large oil reservoirs, the initial fluid pressure within the permeable reservoir rock is often high enough that oil will flow freely to the surface—similar to artesian water wells (Chapter 11). As the reservoir pressure eventually declines, pumps must be used to extract the oil, causing production costs go up. To keep the reservoir pressure up, engineers will often inject water or unused gas into the permeable reservoir rock. This technique drives the oil toward the pumping wells and increases the overall recovery rate. Should the reservoir contain heavy, high-viscosity crude oil, recovery rates can be increased by reducing the viscosity of the oil using steam injection systems.

Although injection and other techniques can improve recovery rates, production will ultimately decrease to the point where the cost of removing the oil from the ground and getting it to the market is greater than its market value. The early stages of production are normally quite profitable since oil wells produce mostly crude oil and lesser amounts of water. As time goes on, the wells will produce progressively more water and less oil, making their operation less profitable. Eventually the operating costs will no longer be justified, forcing the owners to stop production despite the fact a certain amount of oil still remains in the reservoir. Of course, if the market price of crude increases, it may be profitable to continue extracting oil for a longer period of time, thereby increasing the overall amount of oil that is recovered. With respect to natural gas reservoirs, production will continue until the pressure declines to point that gas no longer freely flows into the wells.

In the case of tight oil and gas deposits found in low-permeability rocks (organic shales, fine sandstones, coal, etc.), initial production rates are normally quite high, but then decline more rapidly than does the rate for a conventional well

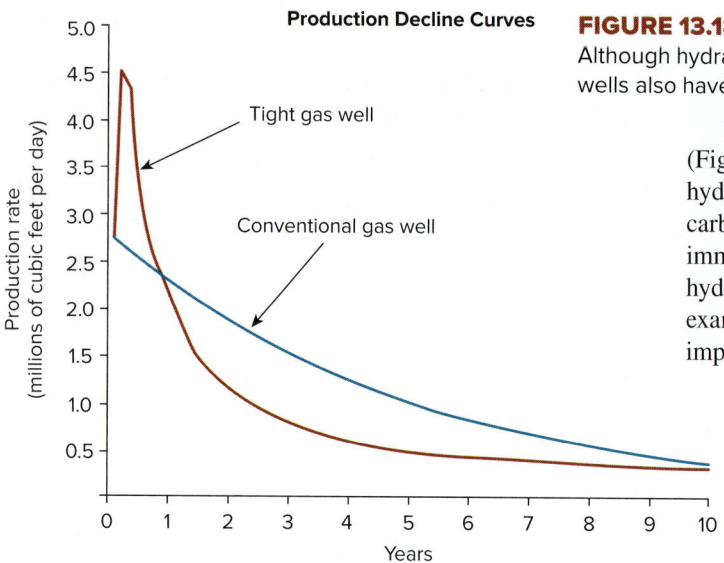

Production Decline Curves

Tight gas well

Conventional gas well

Production rate (millions of cubic feet per day)

Years

FIGURE 13.18   Simulated production decline curves for conventional and tight gas wells. Although hydraulic fracturing of tight wells causes initial production to be quite high, these wells also have higher depletion rates.

(Figure 13.18). The rapid depletion of a tight oil or gas well is due to the way hydraulic fracturing increases the permeability of the rock formation so hydrocarbons can escape into the well. Once the oil or gas is removed from the area immediately around the fractures, production falls off rapidly as the remaining hydrocarbons tend to stay trapped in the low-permeability rocks. Later we will examine how the production characteristics of tight oil and gas deposits may impact our long-term energy supplies.

Petroleum Refining

Petroleum refining utilizes a distillation process similar to the production of freshwater by heating salt water, then collecting the water vapor and letting it cool and condense into a liquid (Chapter 11). Unlike water however, crude oil contains various types of hydrocarbon molecules, each having a different density and thus condensing at a different temperature. As illustrated in Figure 13.19, the refining process begins when

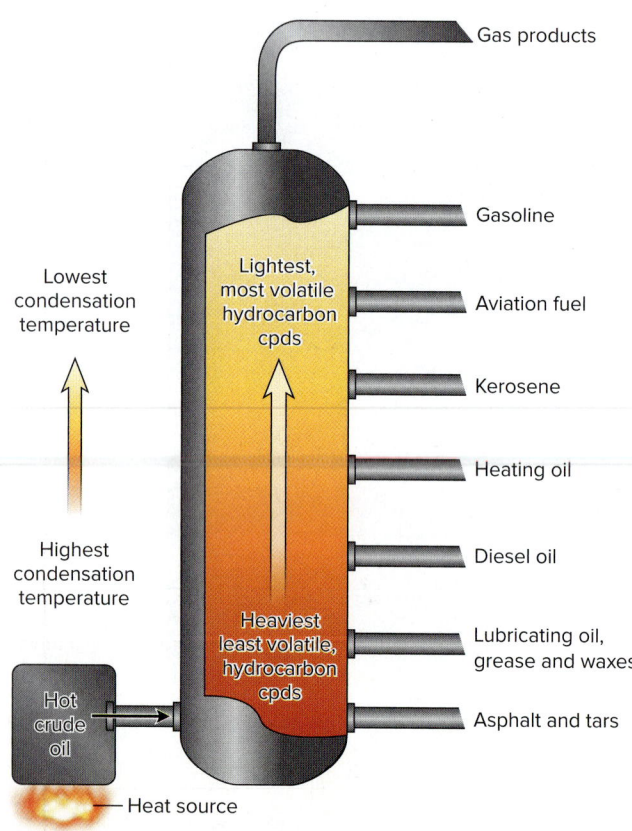

Gas products

Gasoline

Lowest condensation temperature

Lightest, most volatile hydrocarbon cpds

Aviation fuel

Kerosene

Heating oil

Highest condensation temperature

Diesel oil

Heaviest least volatile, hydrocarbon cpds

Lubricating oil, grease and waxes

Hot crude oil

Asphalt and tars

Heat source

A

FIGURE 13.19   Refining of crude oil into various products (A) begins by heating the crude, then sending the vapors into a distillation tower. At the bottom of the tower, where temperatures are highest, the heaviest hydrocarbon molecules are able to condense, at which point they are collected. As the temperature in the tower gets cooler toward the top, progressively lighter and more volatile molecules condense. Shown here are just a few of the more common types of products obtained from crude oil. Aerial view (B) of the British Petroleum refinery in Texas City, Texas. Note the numerous distillation towers and storage tanks.
(b) © BP Oil

B

crude oil is heated to the point where most of the hydrocarbon molecules vaporize into gases and then rise up into a *distillation tower*. Temperatures within the tower are hotter near the bottom and become progressively cooler toward the top. Because the heaviest hydrocarbon molecules (tars, greases, oils, etc.) condense at higher temperatures, they are removed from the lower parts of the tower and then piped to separate holding tanks. Farther up the tower where temperatures are lower, progressively lighter and more volatile molecules are collected and condensed into their respective liquids (e.g., kerosene, aviation fuel, and gasoline).

Some crude oils, called *heavy crudes*, are heavier and more viscous compared to what are known as *light crudes*. The basic difference is that heavy crudes have a higher proportion of long chains of hydrocarbon molecules, making them denser and more viscous. Refiners prefer light crudes since they contain more of the shorter molecules, which means that more gasoline can be obtained from a barrel of this type of oil. Because gasoline is in high demand as a transportation fuel, light crudes command a higher market price than heavy crudes. Refiners are able to increase the amount of highly profitable gasoline obtained by using a process called *cracking*, which breaks long, heavy hydrocarbon chains into shorter and lighter molecules. Light crudes are the most desirable not only because they produce more transportation fuels, but also because they are easier to extract from petroleum reservoirs and have higher recovery rates. Therefore, most of the crude oil that has been extracted for the past 100 years has been relatively light. This is why the bulk of the world's remaining untapped deposits are composed of more viscous and heavy oils, which of course are far more difficult to extract.

Finally, it is important to understand that crude oil not only provides the gasoline and diesel fuels that run our cars and trucks, it also provides the tar and asphalt that pave the highways we drive on. There are also countless consumer products that originate from refined petroleum, forming the basis of what is known as the *petrochemical industry*. Perhaps the most familiar products are those made of nylon and various types of plastic, from you smartphone, water bottle, clothes, and tennis shoes to the interior and exterior of your home and car. Even more important are the agricultural pesticides and nitrogen-based fertilizers made from oil and natural gas. Along with diesel-based farm equipment, these fertilizers and pesticides have been largely responsible for the dramatic increases in food production over the past century. Without petroleum, it simply would not have been possible for Earth's population to reach its current level of over 7 billion people.

Environmental Impacts of Petroleum

Whenever oil and gas are burned (i.e., undergo combustion), a variety of pollutants are released into the atmosphere, many of which are hazardous to human health. Although some of these combustion by-products such as soot can be seen with the naked eye, most are invisible gases. During the combustion process hydrogen and carbon atoms bond with oxygen to form water (H_2O) and carbon dioxide (CO_2). Because combustion is never 100% complete, this process also produces unburned hydrocarbons and carbon monoxide (CO) gas. Unlike carbon dioxide, a carbon monoxide molecule is similar in size to an oxygen molecule (O_2) and can be taken in through the lungs and absorbed in the blood. Carbon monoxide is extremely poisonous and often kills people who fall asleep in cars with the motor running or live in homes with faulty chimneys. Another category of air pollutants includes various nitrogen oxide (NO_x) gases that form when atmospheric nitrogen (N_2) is pulled into the combustion chamber of gasoline and diesel engines (a more thorough discussion of emissions and air pollution can be found in Chapter 15).

In addition to air pollution, society's use of oil has led to an assortment of other environmental problems, particularly oil spills that can damage ecosystems and contaminate water supplies. The drilling of oil wells has historically resulted in spills, especially when a high-pressure zone is encountered and operators lose control of the well. When this occurs, oil or gas will escape under high pressure

Mississippi Delta

■ Approximate location of Deepwater Horizon rig

Oil slick

A

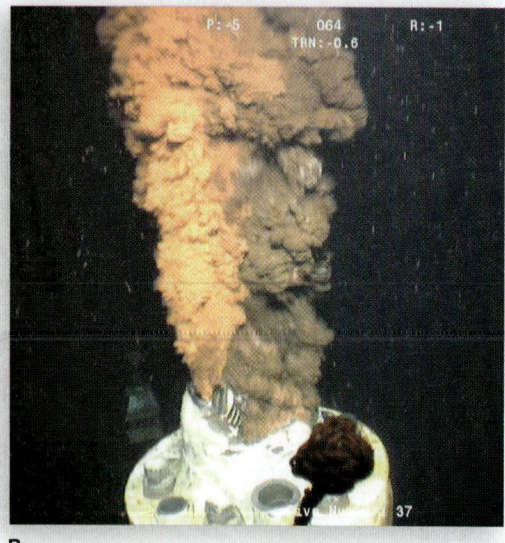

B

FIGURE 13.20   Satellite image (A) showing the oil slick in the Gulf of Mexico approximately a month after the *Deepwater Horizon* lost control of the well it was drilling. The ruptured well on the seafloor (B) continued to discharge large volumes of crude oil for 86 days after the accident before finally being brought under control.
(a) NASA; (b) USGS

in what is known as a *blowout*. A blowout not only creates an oil leak, but can also produce a very intense and dangerous fire that is difficult to extinguish.

A recent example is the 2010 blowout that occurred in the Gulf of Mexico when the drilling ship *Deepwater Horizon* (Figure 13.17) lost control of the well it was drilling in nearly 5,000 feet (1500 m) of water. Due to a series of human errors and equipment failures, a mixture of gas and crude oil shot up the drilling pipe, resulting in an explosion and fire that destroyed the ship and took the lives of 11 crew members (see the chapter-opening photo). After the ship sunk, the ruptured well continued to spill crude oil from the seafloor for 86 days, creating one of the worst oil spills in history (Figure 13.20). The well is estimated to have released 206 million gallons (4.9 million barrels) of crude into the Gulf of Mexico. Much of the oil stayed within the water column due the extreme depth of the well and the use of chemicals to disperse the oil slicks on the surface. While conditions along the coastline appear to have returned to normal, scientists are still gathering data on the expected long-term impacts to the marine ecosystem.

Oil spills can also occur while transporting crude via ships or rail cars to distant refineries. One tragic example is the spill that occurred in 1989 when the supertanker *Exxon Valdez* ran aground in Alaska's Prince William Sound, discharging 11 million gallons (260,00 barrels) of crude oil into the pristine waters (Figure 13.21). Damage to the delicate ecosystem was extensive and subsequent cleanup costs were over $2 billion. Although the beaches today look normal, unweathered crude still lies buried along the shoreline. Recent studies indicate that some species have still not recovered, likely due to persistent low-level pollution.

Finally, one of the lessons from the *Deepwater Horizon* spill in the Gulf of Mexico is that advances in drilling technology can push development ahead of our ability to quickly stop the flow of oil from a ruptured well in deep water. A variety of new and untried methods were used to stop the flow of oil in the *Deepwater Horizon* disaster, but they all failed. Ultimately it took 86 days using existing methods before a new well successfully captured the flow of oil. Today there is concern that the drilling technology that has allowed us to extract tight oil and gas from low-permeability rocks may also lead to serious environmental problems. Of particular concern is that the boom in the use of hydraulic fracturing could lead to contaminated drinking water supplies and that natural gas (methane) may escape from the wells and exacerbate global warming (see Case Study 13.1). While scientists gather data to help answer these questions, policymakers are trying to balance the need for expanding our energy supplies with protecting society from any adverse consequences.

Current Energy Supply and Demand

The world's demand for energy has grown tremendously over the past several hundred years due to the Industrial Revolution and rapid population growth. This demand is presently being met primarily through the burning of fossil fuels as they are relatively inexpensive and highly concentrated sources of energy. The problem is that fossil fuel supplies are finite, which means continued economic growth based on these resources is simply not sustainable. Another problem with using fossil fuels is that large quantities of carbon, which were stored in the subsurface for millions of years, have suddenly been released into the atmosphere in the form of carbon dioxide. This increase in atmospheric carbon dioxide is affecting Earth's climate system and is contributing to the problem of global warming. We will examine climate change in Chapter 16, but for the remainder of this chapter our focus will be on the dilemma society faces with respect to its limited supply of fossil fuels.

FIGURE 13.21   In 1989 an oil tanker ran aground in Prince William Sound in Alaska, polluting beaches and killing wildlife along a vast stretch of Alaska's irregular shoreline. Despite years of cleanup efforts, beaches still contain buried oil that continues to be a source of low-level pollution.
(right): © Exxon Valdez Oil Spill Trustee Council

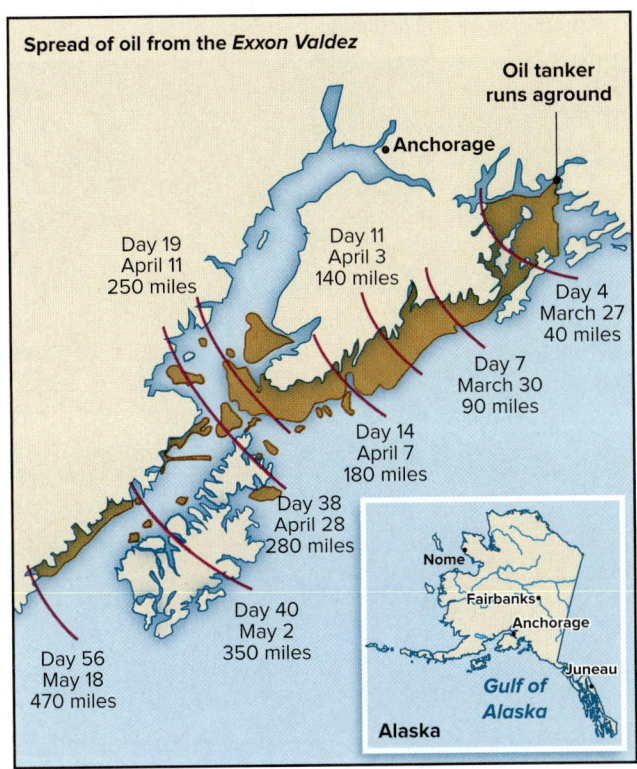

Economic Development and Energy Demand

Recall that the per capita consumption of resources (water, minerals, and energy) depends largely on a nation's level of economic and technological development and on its efforts to improve efficiency and conservation. Developed countries with large populations naturally consume the most energy. This relationship explains the uneven rates of energy consumption by different nations around the world. For example, from Figure 13.22 one can see that China and the United States consume substantially more energy than any of the other major users. The United States, though, with its much smaller population, consumes almost as much energy as China. This means that the per capita consumption of energy in the United States is considerably higher than in China. Although both China and India have very large populations, their per capita energy consumption rates are relatively low because they are still developing.

This brings up a very important point: relative to their populations, more-developed countries consume a disproportionate share of the world's energy resources. Consider that the United States consumes 18% of the world's energy resources with just 4.4% of the world's population. In contrast, China has 19% of the planet's population, but consumes 23% of the energy supply, which is less on a per capita basis than for the United States. The problem is that China's per capita consumption rate has been increasing as the country becomes more industrialized. Consider that China put an average of 25,000 additional cars on the road each day between 2006 and 2010. Today, the total number or cars in China exceeds even that in the United States. Moreover, China nearly tripled its consumption of coal

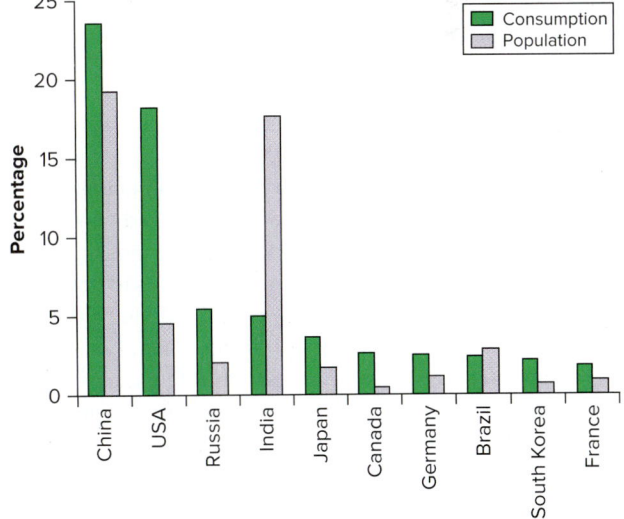

FIGURE 13.22   Plot showing the top 10 energy users in terms of their percentages of consumption of the world energy supply; also shown for each country is its percentage of the world population. Considering population size, the United States and other developed nations use a disproportionate amount of energy compared to developing countries such as China and India.

Controversy Over Hydraulic Fracturing and Tight Oil and Gas Development

Geologists have long known that substantial amounts of oil and natural gas exist in organic-rich shales and other fine-grained rocks with low permeability, which are found in sedimentary basins around the world. In fact, some of the shales are the source rocks for hydrocarbons we find trapped in conventional reservoirs. These permeable reservoirs have served as the world's primary source of oil and natural gas for nearly 150 years. However, new reservoirs are getting harder to find and existing ones are steadily being depleted. This, combined with the world's growing demand for energy, led to higher oil and gas prices over the years. During the 1990s energy companies found it economical to begin large-scale production of tight oil and gas from organic-rich shales and other fine-grained rocks in the United States (Figure B13.1). This production boom has now spread to sedimentary basins around the world.

These tight deposits were not economical to develop in the past because the extremely low permeability of the rocks made it virtually impossible for the hydrocarbons to flow into a traditional well. Petroleum engineers had faced a similar problem in conventional reservoirs with relatively low permeabilities. Their solution was to place explosives within the well and fracture the reservoir rock, thereby increasing its permeability. By the 1940s explosive techniques were being replaced by *hydraulic fracturing,* in which fractures are produced by injecting water into the reservoir rock under high pressure. Hydraulic fracturing, commonly called *fracking,* has been refined over the years and now involves injecting a mixture of water, chemical agents, and sand into hydrocarbon-bearing formations of low permeability. The chemicals are used to produce a more viscous fluid that makes fracturing more efficient—in some cases they are used to make the fluid chemically reactive. The sand grains help keep the fractures from closing when the fluids are pumped back out and the pressure returns to normal.

When energy prices rose in the 1990s, the combination of hydraulic fracturing and directional drilling made it economical to develop petroleum-bearing shales and other fine-grained rocks. Directional drilling had been refined over the years, with improvements in navigating and steering the drilling bits through the subsurface. As illustrated in Figure B13.2, energy companies gained the ability to drill horizontally along the layers of tight rock formations, while hydraulic fracturing allowed hydrocarbons to enter along the length of the well. Keep in mind that for this method to be economical, oil and gas prices must be high enough to offset the additional costs of installing large amounts of steel pipe and hydraulically fracturing the wells. By the early 2000s significant volumes of oil and gas were being produced

FIGURE B13.1 Map showing the locations in North America of organic-rich shales and other fine-grained rocks with the greatest potential for producing oil and natural gas.

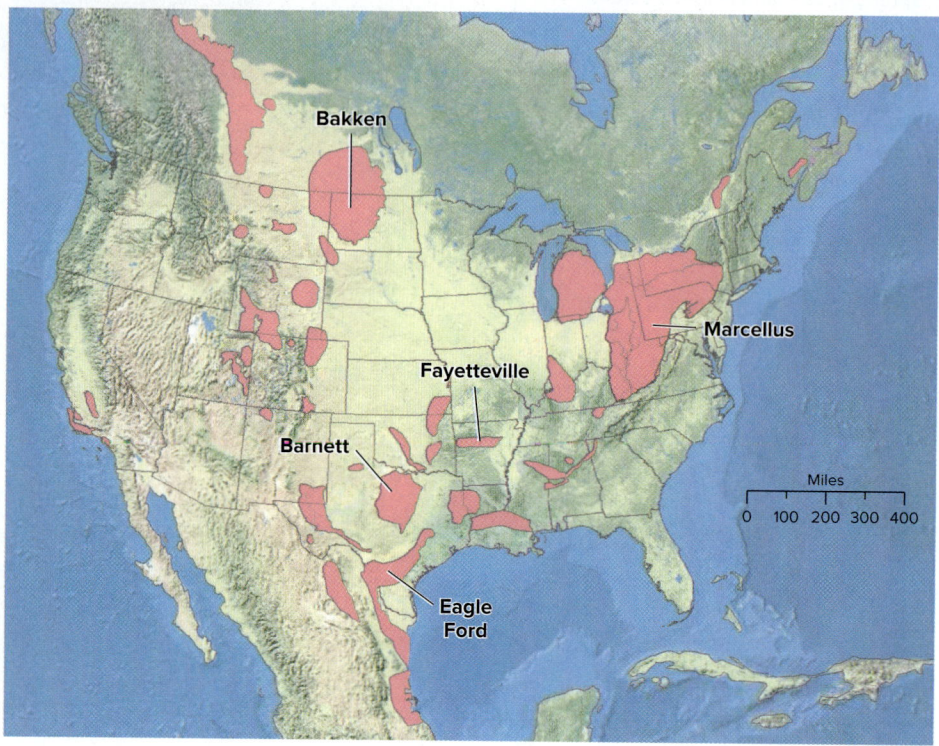

from the Bakken Shale in North Dakota and the Barnett Shale in Texas (Figure B13.1). Considerable attention then began focusing on developing natural gas from the Marcellus Shale in the eastern United States. By 2015, the amount of oil being extracted from the Bakken Shale hit record highs, causing U.S. oil production to actually increase. Production of tight gas also became so great that natural gas prices fell to the point where electric-generating plants started switching from coal to natural gas. Since natural gas generates less carbon dioxide and acid rain, and involves almost no fallout of heavy metals (Chapter 15), many view this as an environmentally positive development.

Despite the benefits of lower energy prices and cleaner electricity, there has been growing concern that the production boom due to hydraulic fracturing is resulting in serious environmental problems. Of particular concern are reports that natural gas and other contaminants are being found in groundwater wells that supply drinking water to nearby residents. However, it is unlikely that the contamination is directly related to hydraulic fracturing itself since the petroleum deposits generally lie several thousand feet below any freshwater aquifer. Most experts believe the contamination is resulting from leakage along improperly sealed well casings as they rise to the surface and pass through a drinking water aquifer (Figure B13.2). Another possibility is that natural gas and fracking fluids could be rising to the surface along old, abandoned oil and gas wells or along natural fractures and faults. In 2015 the EPA released a report stating that while hydraulic fracturing has indeed caused contamination of drinking water, the number of known cases is relatively small compared to the number of hydraulically fractured wells. The EPA did note, however, that the small number of cases could also be due to the limited data that was available for its report.

Other environmental problems are associated with disposing of the 1–5 million gallons (4–19 million liters) of contaminated wastewater that must be pumped from each fracked well before hydrocarbon production can begin. During the fracturing process, the mixture of water and chemical agents spreads into the tight formation and surrounding rock layers. This allows fracking fluids to pick up radioactive elements and heavy metals, which are common in organic-rich shales, along with dissolved salts. When these fluids are pumped back into the well and brought to the surface, radiation and salinity levels will often greatly exceed drinking water standards. This presents the potential for contamination of surface waters should the temporary storage lagoon at the drilling site leak before the wastewater can be safely disposed of.

Environmental regulations in some states have allowed operators to dispose of fracking wastewater at nearby municipal sewage treatment plants, which are not designed to remove radioactive elements or dissolved ions (salts) from the wastewater. This means that these contaminants are discharged directly into nearby rivers, which often serve as the drinking water source for downstream residents. In most states, however, the fracking wastewater is disposed of by deep-well injection (Chapter 15). Here the contaminated wastewater is pumped into saline aquifers that are even deeper than the shales and tight rock formations from which the wastewater was extracted (Figure B13.2).

Due to the large volume of wastewater involved, some of the injection sites are experiencing a rise in earthquake activity. For example, in Oklahoma, where wastewater from both conventional and hydraulically fractured wells is injected into deep aquifers, there has been a 300-fold increase in the number of magnitude 3.0 and larger earthquakes since 2008. Seismologists with the U.S. Geological Survey (USGS) have found that the sudden rise in seismic activity in parts of Oklahoma, Texas, Colorado, New Mexico, and Ohio coincides with the additional volumes of wastewater from oil and gas operations being injected into deep aquifers. The injected wastewater increases the pore pressure in deep rocks, thereby reducing their frictional resistance to the point where they rupture and release their stored seismic energy, causing an earthquake (Chapter 5). Although the majority of the earthquakes have been small, some have been large enough to cause significant damage, such as the magnitude 5.6 quake in 2011 that struck near Praque, Oklahoma, which was followed in 2016 by another magnitude 5.6 quake near Pawnee, Oklahoma. Keep in mind that while the sudden rise in earthquake activity has been linked to wastewater injection, it is very difficult to link a specific earthquake to injection.

In addition to issues associated with fracking wastewater, there are concerns related to the escape of methane and other hydrocarbon gases that are commonly associated with natural gas. Some of the gases, such as benzene, are known to cause cancer and can pose serious health risks to nearby residents. Now that tight gas production is moving into more populated areas of the eastern United States, medical researchers are finding significantly higher rates of human health problems around tight gas operations. This gas leakage presents another potential problem because methane (CH_4), which makes up the bulk of natural gas, is over 20 times more effective than carbon dioxide at trapping heat in the atmosphere. Climate scientists are now worried that the natural gas boom could actually intensify global warming (Chapter 16). The concern may be justified as recent scientific studies have shown that the leakage rate of methane from natural gas operations is much greater than previous industry estimates.

It is quite clear that the recent development of tight oil and gas is making an important contribution to the world's energy supplies and is bringing much needed jobs and tax revenues to many areas. However, as we learned from the *Deepwater Horizon* oil spill in the Gulf of Mexico, the technology that allows us to extract nontraditional sources of petroleum can also bring about new environmental problems. As for the problems associated with tight oil and gas, more scientific information is needed so that policymakers can find a better balance between expanding energy supplies and protecting human health and our environment.

FIGURE B13.2 Illustration showing how directional drilling and hydraulic fracturing are used to extract petroleum from organic-rich shales and other fine-grained rocks. The low-permeability rock is fractured by injecting a mixture of water, chemicals, and fine sand, which then allows hydrocarbons to enter along the length of the well. Contamination of shallow aquifers by fracking fluids and natural gas can occur along improperly sealed well casings or through abandoned oil and gas wells. Wastewater from fracking operations is normally disposed of by deep-well injection.

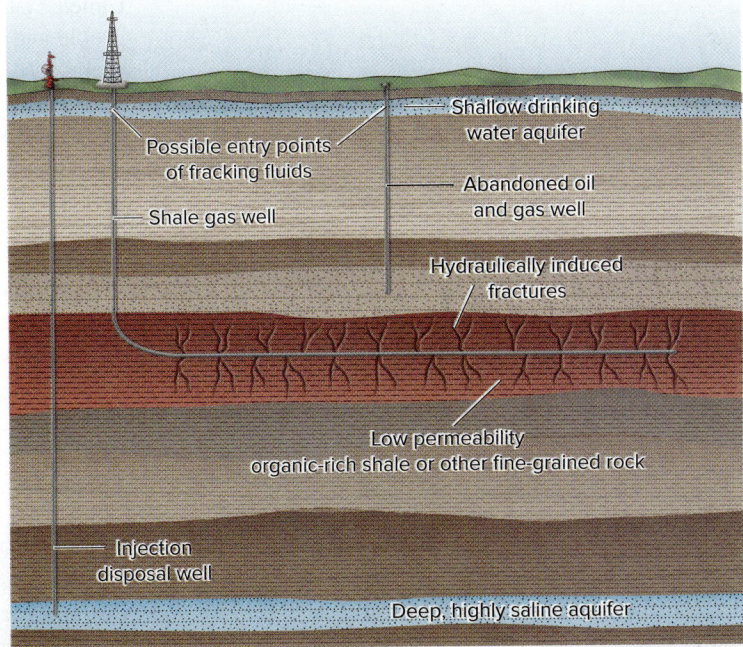

Shallow drinking water aquifer

Possible entry points of fracking fluids

Abandoned oil and gas well

Shale gas well

Hydraulically induced fractures

Low permeability organic-rich shale or other fine-grained rock

Injection disposal well

Deep, highly saline aquifer

FIGURE 13.23

Breakdown of U.S. energy consumption in 2014 by sector (A) and by source (B).

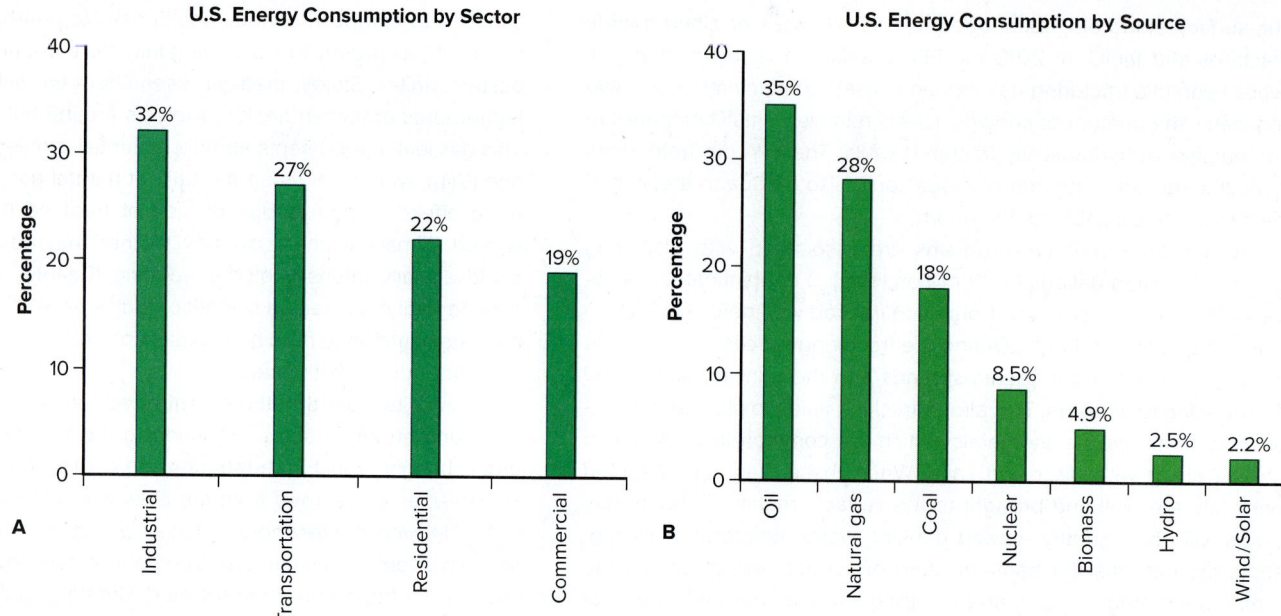

and more than doubled its consumption of oil between 2000 and 2014. India is another country with a large population that is undergoing rapid industrialization. What this means is that as China, India, and other nations strive to develop consumer societies similar to that in the United States, the additional demand on the world's energy resources will be enormous. The question is, will the energy supply be able to meet demand and, if so, for how long?

Types of Energy We Consume

Because a relatively small number of nations consume the vast majority of the world's energy resources, it is important to understand how these nations use energy. Using the data in Figure 13.23A from the United States as an example, we see that energy usage can be broken down into four major sectors: industrial, transportation, residential, and commercial. Notice that 32% of all energy in the United States is consumed by industrial activity, and 27% goes toward moving people and goods throughout the transportation system. From the breakdown of energy use by source (Figure 13.23B), we see that oil accounts for the largest share of the U.S. energy consumption. Interestingly, over 70% of all oil is consumed by the transportation sector.

The use of oil for transportation is a good example of how society converts energy from one form to another to accomplish certain tasks. An internal combustion engine first converts gasoline or diesel fuel (chemical energy) into heat energy. The heat is then transformed into the kinetic energy that propels the car or truck down the highway. Electricity is another very useful form of energy that is widely used throughout the industrial, residential, and commercial sectors of society. As shown in Figure 13.24, most electricity is made by first transforming fossil or nuclear fuels into steam (heat). The steam is then converted into kinetic energy via a turbine that spins a generator, which actually creates the flow of electrons we call electricity. Electricity can also be made by using falling water or wind to spin a turbine and generator—the conversion of sunlight into electricity will be described in Chapter 14. Notice in Figure 13.24 that burning coal is the most common way of making electricity in the United States.

In terms of U.S. energy consumption, the combined use of fossil fuels (oil, gas, and coal) accounts for about 86% of the energy being used across all sectors of the economy (Figure 13.25). China, the world's largest energy consumer, gets 89% of all its energy from fossil fuels, exhibiting a slightly greater dependency than the United States. Some nations, however, are less reliant on fossil

FIGURE 13.24
Electricity in the United States is produced by burning coal or natural gas or using nuclear fuel. These energy sources are transformed into heat and used to produce steam, which then spins a turbine and electrical generator. Electricity is also produced by using water or wind to spin a generator or by using solar panels.

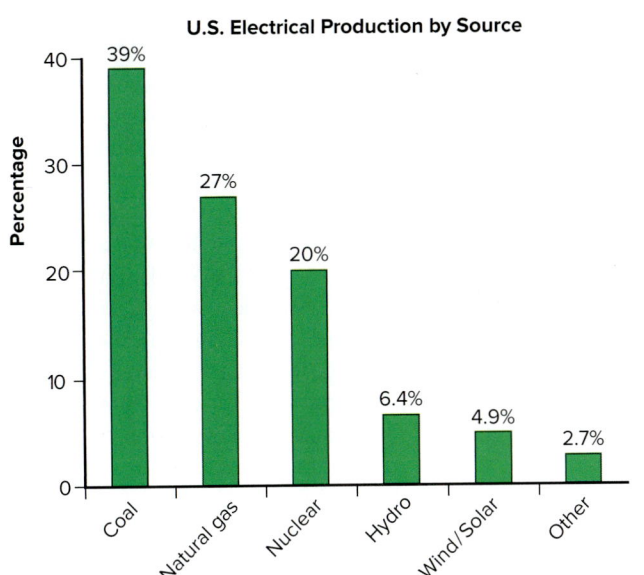

U.S. Electrical Production by Source

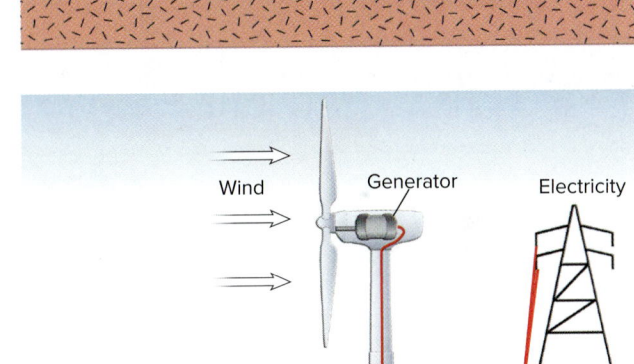

fuels and use considerably more nuclear and hydroelectric (hydro) power and various renewable sources to meet their energy needs—although hydro power is from a renewable source, it is separated here from other renewables such as wind, solar, and geothermal. France, for example, has invested heavily in nuclear power, whereas Sweden has large contributions from both nuclear and hydro and other renewables. Norway, Canada, and Sweden have all taken advantage of their geologic terrain and climate to generate significant amounts of hydro power. Another factor here is that after the global oil crises of the 1970s, many countries embarked on large-scale programs for reducing their reliance on fossil fuels.

Unfortunately, the United States dropped its alternative energy program in the early 1980s when oil prices crashed. When China began undergoing rapid development during this same period, it turned to the least expensive form of energy, namely fossil fuels. The result is that today's top two energy consumers burn tremendous quantities of fossil fuels. Moreover, China's growing demand for oil is helping to drive the continued increase in world oil production. An even more significant problem is that the United States and China are responsible for the bulk of the carbon dioxide being released into the atmosphere, which is altering our climate system and may soon cause habitable zones to shift across the globe (Chapter 16).

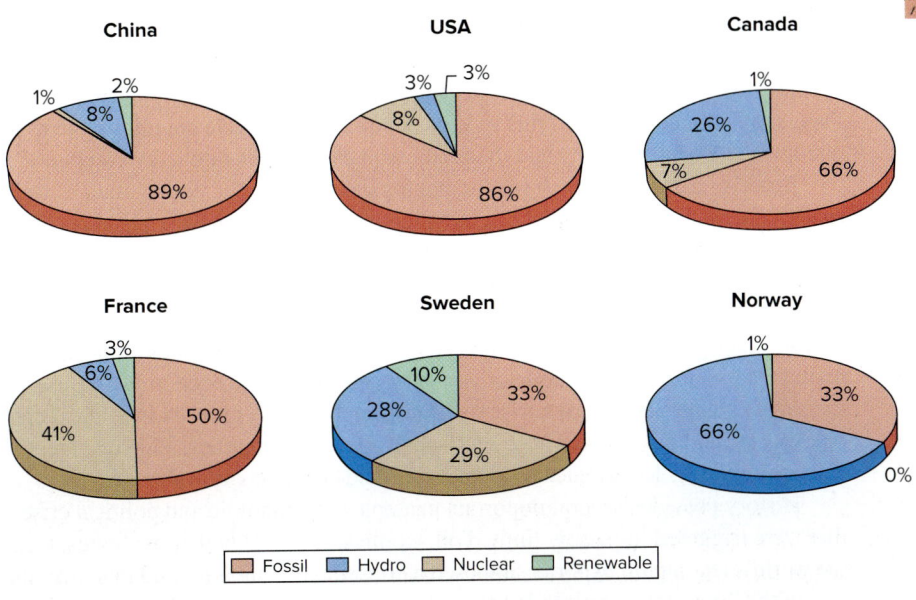

FIGURE 13.25   Countries rely on different combinations of fossil, nuclear, and hydro and other renewable resources to meet their energy needs. China and the United States depend heavily on fossils fuels, whereas Canada, France, and Sweden make greater use of nuclear and hydro and other renewable sources. Norway's geologic terrain allows it to utilize hydro power to a much greater degree than most countries.

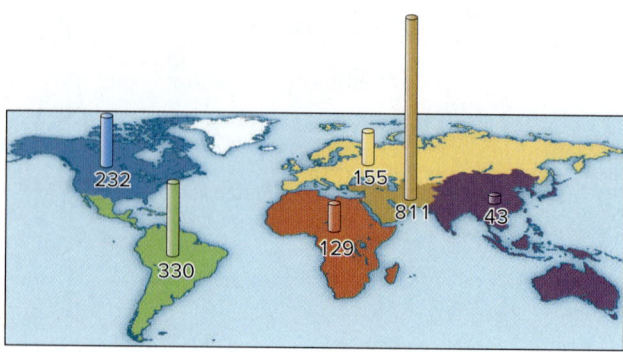

Oil (billions of barrels)

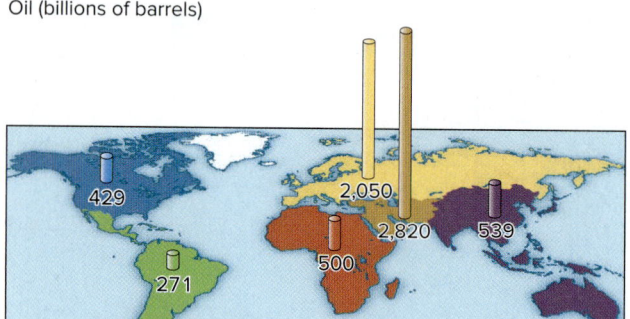

Gas (trillion cubic feet)

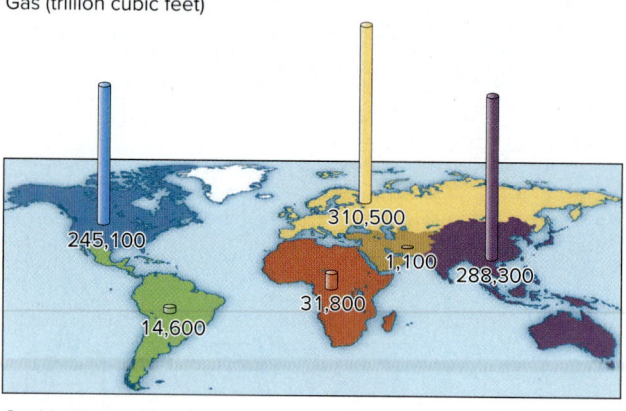

Coal (millions of tons)

■ North America	■ Central and South America	■ Middle East
■ Asia and Oceania	■ Eastern Europe and Russia	■ Africa

FIGURE 13.26   Proven reserves of fossil fuels are not evenly distributed around the globe because economic deposits form only under favorable geologic conditions. Note that the locations of large oil and gas reserves do not necessarily correspond to those of major reserves of coal.

Where Fossil Fuels Are Located

Similar to mineral resources, fossil fuel deposits are not evenly distributed around the globe because they require favorable geologic conditions in order to form. Geographically large countries, such as the United States and Russia, have a higher probability of having sedimentary basins where conditions were just right for petroleum or coal deposits to form. Large geographic size, however, is no guarantee that sizable deposits will be found. Such is the case for Australia, where only minor amounts of oil are found. It just so happens that the Middle Eastern countries around the Persian Gulf are located in an area where ideal geologic conditions created the world's most prolific oil and gas reservoirs. To get an appreciation for how much petroleum this region contains, compare the oil and gas reserves of the Middle East in Figure 13.26 with those of the rest of the world. Just six countries in the region, Saudi Arabia, Iran, Iraq, Kuwait, Qatar, and the United Arab Emirates, hold 47% of the world's proven oil reserves and 42% of natural gas reserves. Although its petroleum reserves are enormous, the Middle East has essentially no coal, whereas the coal reserves of North America, Eastern Europe, and Asia are quite large. The reason for these variations is that petroleum and coal form under different geologic conditions.

Due to the economic dependence of modern societies on fossil fuels, the uneven geographic distribution of fossil fuels plays an important role in global politics and the national security of individual countries. Energy security is a serious concern for developed nations whose own energy reserves are insufficient to meet their needs. For example, the United States was once self-sufficient in terms of oil and was the world's leading exporter. Today, however, the United States has depleted many of its original oil fields and consumes such a large amount of oil that it cannot possibly hope to become self-sufficient again. Japan is in an even more difficult situation as it has virtually no fossil fuel resources of its own, including coal. This situation means that the Persian Gulf nations, with their large petroleum reserves, are able to generate enormous sums of money and wield great political power.

The Future Energy Crisis

In recent years we've seen wide fluctuations in oil and gas prices due to the complex relationship between supply and demand and economic activity. After 2000 energy prices rose steadily as the world economy grew and supplies of oil and gas struggled to keep up with demand. The high prices led producers to begin ramping up production of more expensive tight oil and gas. Demand then suddenly decreased in response to the global recession and financial crisis of 2008. Despite the reduced demand, production of tight oil and gas continued to soar as prices remained high. Energy prices finally began to fall in 2014, eventually reaching levels not seen in nearly a decade and forcing oil and gas producers to put many new projects on hold. This type of boom-and-bust cycle is undesirable as it disrupts the world economy and the production capabilities of energy companies.

Because crude oil is the basis for making gasoline, diesel, and jet fuel, as well as agricultural chemicals and plastics, higher prices for crude have a negative effect that ripples through entire economies. Consumers not only pay more for a gallon of gasoline, but nearly everything they buy, including food, becomes more expensive. As people spend more of their income on basic necessities, they have less to spend on consumer goods and nonessential activities. Ultimately this can lead to an economic recession and the collapse of various sectors within an economy. Because of the severe consequences, countries that must import oil to meet their demand can be expected to compete more aggressively for existing supplies. This heightened competition leads to a greater potential for armed conflicts over oil resources.

History provides several important examples of economic and political crises that were triggered in part by limited oil supplies. In 1941 the United States, then the world's leading oil exporter, imposed a crippling oil embargo on Japan over its war with China. Japan responded five months later by attacking U.S. and British forces throughout the Pacific. This military action allowed Japan, a nation with no petroleum reserves of its own, to seize the oil fields in the Dutch East Indies, thereby eliminating its oil crisis. Germany was in a similar position prior to the war, and had even resorted to producing gasoline and diesel fuel from coal. One of the reasons Hitler made the fateful decision to invade the Soviet Union (Russia) was to capture its oil fields in the Caucasus.

After World War II, the United States remained the world's leading oil exporter until 1970 when domestic production reached its peak. To meet its growing demand, the United States had to begin importing oil. As shown in Figure 13.27, crude oil prices remained quite stable until 1973 when the oil-rich countries of the Middle East imposed an oil embargo and refused to sell to Western nations because of their support for Israel in the 1973 Yom Kippur War. The oil embargo resulted in an immediate shortage and a surge in crude oil prices. This sent shock waves through the Western economies and triggered a global recession. Eventually the embargo collapsed and oil prices stabilized. Then in 1979 yet another oil crisis developed when the Iranian revolution disrupted that nation's considerable oil exports, sending oil markets into a panic and world economies back into recession as oil prices skyrocketed. By 1985 prices collapsed as increased production resulting from the previous round of high prices brought new supplies into the oil market. Additional new production from major fields in Alaska and the North Sea soon helped create an oil glut. This resulted in relatively low prices that fueled an extended period of economic prosperity until 2000, at which point prices began to climb until the global recession of 2008, as noted earlier.

There are several important lessons we can learn from the oil crises of the 1970s and early 1980s. Although the embargo in 1973 created a real oil shortage in Western nations, the world was still capable of producing far more oil than the market demanded. The consequences though were more than just long gas lines; many people lost jobs in the resulting recession. In the second crisis of 1979 there was never an actual shortage since oil production continued to meet demand. Prices were driven upward by the *fear* that unrest in Iran could spread and lead to a significant loss of production. The market then was being driven more by speculation rather than actual supply and demand.

Although predicting the future is always difficult, we know for certain that oil and gas are finite resources. Moreover, we have been extracting petroleum from the least expensive deposits for nearly 200 years. We are now forced to rely on more expensive sources like deep-water deposits in the Gulf of Mexico and tight oil and gas. As population expands, we can expect energy prices to keep fluctuating while continuing on an overall upward trend. Because higher energy prices act as a drag on the world economy, it would be prudent to take steps to minimize the impact. In this section we will examine our crude oil supplies in more detail and some specific steps we can take to help extend our supplies and avoid a future crisis.

Peak Oil Theory

During the early 1900s there was a dramatic increase in the rate at which new oil and gas fields were being discovered in the United States. A boom in overall production followed, but it lagged behind the discoveries since it took years to install the necessary production wells and pipelines. Individual fields showed a steady increase in production and then reached a peak, after which production steadily declined as the reservoir became depleted. This natural depletion did not affect the nation's overall supply since new fields were continually being discovered. However, finding new reservoirs became progressively more difficult, forcing oil companies to develop new technologies for locating hidden traps. Companies also began searching overseas, which eventually led to the discovery of giant oil and gas fields in the Middle East, most of which were found in the 1930s and 1940s.

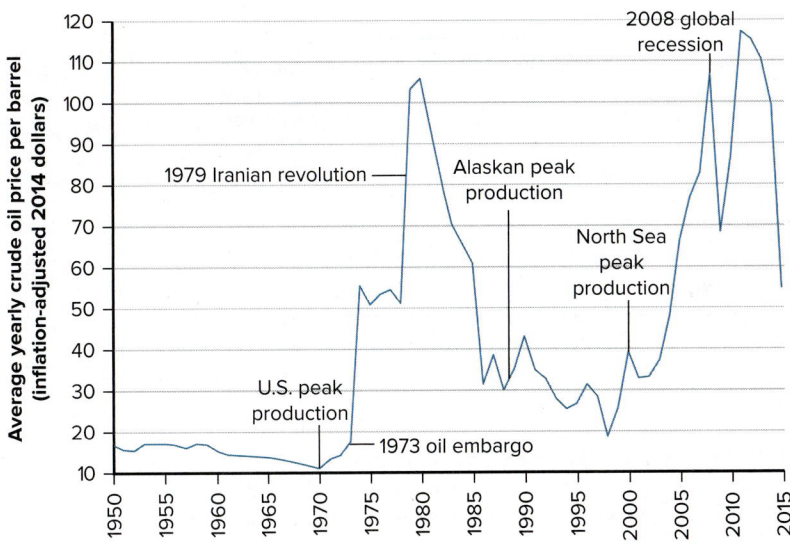

FIGURE 13.27   Crude oil prices have changed dramatically in response to changes in supply and to world events. The major price increases of 1973 and 1979 sent shock waves through the economies of developed nations. Additional production caused prices to fall and eventually stabilize at a lower level until 2000. Prices then rose dramatically until the 2008 recession greatly reduced demand. Recently, additional new supplies and lower economic growth have caused prices to tumble.

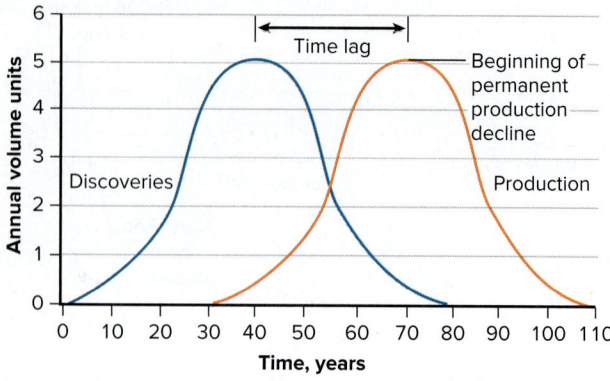

A

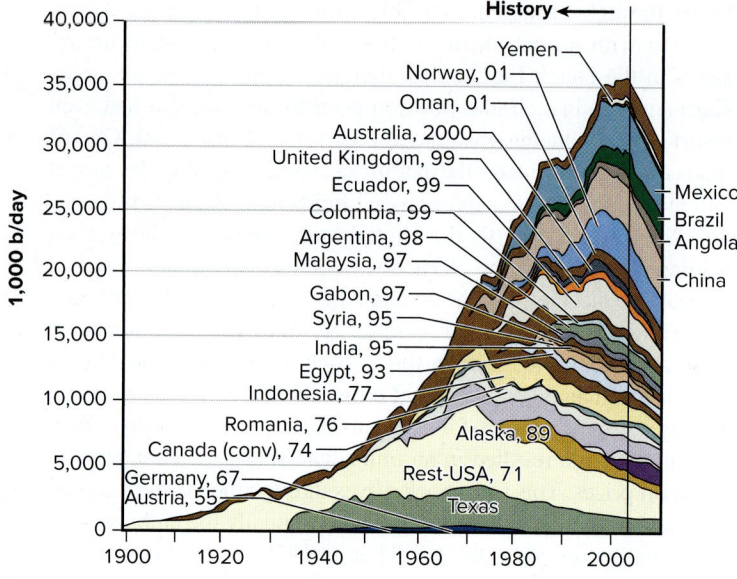

FIGURE 13.28   Hubbert's 1956 model (A) used a statistical approach based on the idea that discoveries of new fields would eventually peak, causing production to peak about 30 years later. Using this statistical model, Hubbert predicted the United States would reach peak production in 1970. A graph from a 2004 study (B) showing that U.S. production peaked around 1970, followed by production peaks in other oil-producing countries.

B

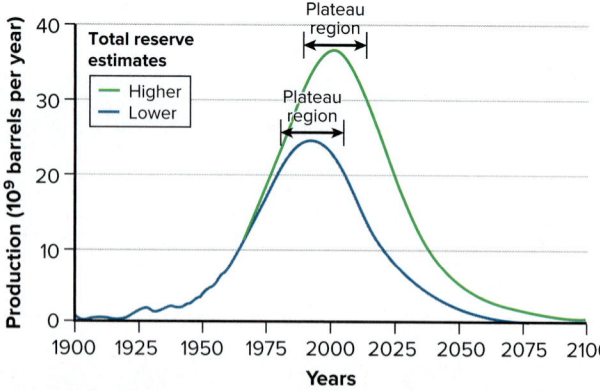

FIGURE 13.29   Based on world production trends and estimated total reserves, Hubbert used his model to project peak world production, with the larger reserve estimate placing the peak at the year 2000. Note how production will reach a plateau before beginning a permanent decline.

In the 1950s a petroleum geologist named M. King Hubbert noticed that the rate at which new oil fields were being discovered in the United States had peaked in the 1930s and then began to decline. Based on the fact that oil is a finite resource, Hubbert developed a model that predicted that exploration and production trends would follow bell-shaped curves as shown in Figure 13.28A. Hubbert's predictive model, often referred to as the **peak oil theory,** was based on a statistical analysis of historical exploration and production data. In 1956 he published his work showing that new discoveries were declining and predicted that U.S. oil production would peak in the early 1970s. Many people at the time found it inconceivable that the world's leading oil producer would reach peak production in the 1970s—that is until 1970, when U.S. production actually peaked, and then began a slow downward decline just as Hubbert's model predicted.

After U.S. production peaked, experts watched as country after country followed similar production declines (Figure 13.28B). The message was now clear: conventional oil supplies were finite and the world as a whole was approaching peak production. In 1969 Hubbert estimated total world supply to be about 2 trillion (2,000 billion) barrels of recoverable oil. Based on this supply estimate, he used his model to create two production curves (Figure 13.29). The more generous of the two curves has world peak production occurring in 2000, which is often referred to as **Hubbert's peak.** In an extensive 2000 study the USGS estimated total recoverable oil reserves to be 3 trillion barrels, which would place world peak production between 2025 and 2040. Many independent experts at the time believed the estimated reserves by the government agency were far too optimistic, and calculated peak production around 2005. While world production from traditional reservoirs appears to be reaching its peak, new supplies from deep-water deposits and tight oil have increased reserves substantially in recent years. It is now estimated that world reserves may be sufficient to meet current production levels for another 50 years.

Natural gas is following discovery and production curves similar to those for oil and will also eventually reach peak production. However, the recent increase in tight gas production is quite large, so supplies should be able to keep up with demand for the foreseeable future. The supply situation for natural gas is also different because its low density makes transporting large quantities of it across oceans using tankers more difficult. The natural gas must first be cooled to a liquid state, called *liquefied natural gas* or LNG, and loaded onto specialized tankers

(Figure 13.30)—unloading also requires specialized shipping terminals. Because of the difficulty and added expense of overseas shipping, natural gas is transported primarily through land-based pipelines and is generally consumed on the same continent on which it is produced. This restriction means that each continent has a unique gas production curve and corresponding date when peak production will occur. Note that the rest of this section will focus on oil because it is critical to the transportation sector of society and the petrochemical industry. In addition, natural gas production from tight formations is such that future supplies are less of an immediate concern.

FIGURE 13.30 Shipping natural gas across oceans is much more difficult than transporting crude oil since gas must be cooled to a liquid state and carried on special tankers.
© BP Oil

Past the Oil Peak

When the world eventually reaches peak oil production, the event will likely occur over a period of time measured in years, creating more of a plateau in the production curve (Figure 13.29) than a sharp peak. This means that the actual date of Hubbert's peak will be known only when production moves beyond the plateau area and the world finds itself permanently on the downward slope of the production curve. Of course, our experience over the past 100 years has been on the rising part of the curve, where production has always been able to exceed demand. This has resulted in relatively inexpensive oil and stable market conditions. Cheap energy allowed industries to expand, whereas steady prices created a stable business environment for companies to grow. Oil experts have predicted that when the world enters the plateau period of global oil production, it will become difficult for supply to satisfy demand, thereby causing higher prices and creating instabilities in the global oil market. This situation will produce dramatic fluctuations in price, in which high prices lead to lower demand, followed by short-lived increases in supply that bring prices down again. The cycle will then repeat as supply is again outstripped by rebounding demand. The volatile nature of the crude oil market since 2008, as described earlier, appears to have followed these predictions.

Many economists, however, do not agree with the peak oil theory. Instead they believe that higher prices will cause market forces to spur more exploration and create advanced technologies, thereby generating sufficient quantities of oil to meet growing demand. For example, assuming that oil prices stabilize at $97 per barrel by 2020, then rise to $141 by 2040, the U.S. Department of Energy estimates that world consumption of liquid fuels will go from 87 million barrels per day in 2010 to 119 million barrels by 2040, a 37% increase. Of the 32 million new barrels per day, about 24 million are projected to be crude oil with the remainder being nonconventional liquid fuels (e.g., biofuels and conversions from coal and natural gas). Based on the historical rate of crude oil production shown in Figure 13.31, the projected 24 million barrel per day increase (green line) appears reasonable. Keep in mind though that for this increase to occur, new production must be added faster than the rate at which older fields are being depleted. If peak oil advocates are correct that production from traditional permeable reservoirs is near its peak, then most of 24 million barrels per day of new crude oil production will have to come from other sources, such as deep-water fields and tight oil deposits. Since world

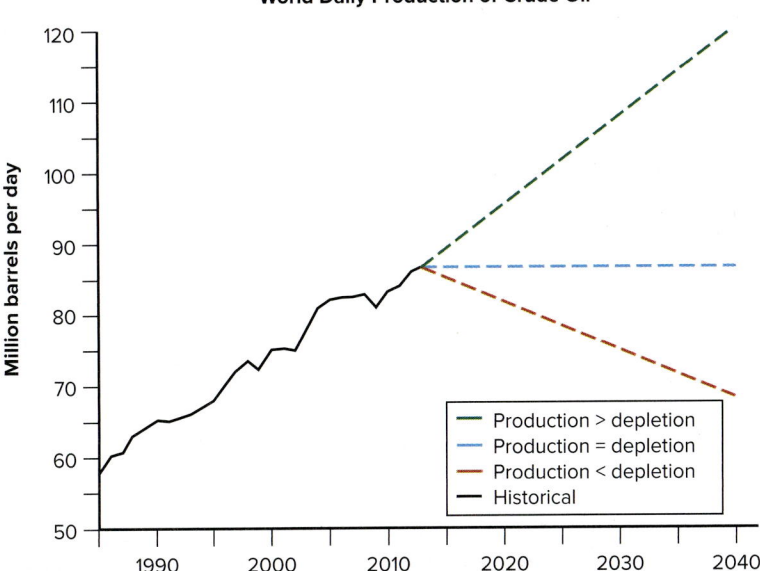

FIGURE 13.31 Graph showing the historic rate of world oil production along with possible future trends. Production rate will increase (green) if new production can be added faster than older fields are being depleted. Flat or no-growth production (blue) occurs if production keeps pace with depletion, but the rate declines (red) if new production cannot keep pace with depletion.

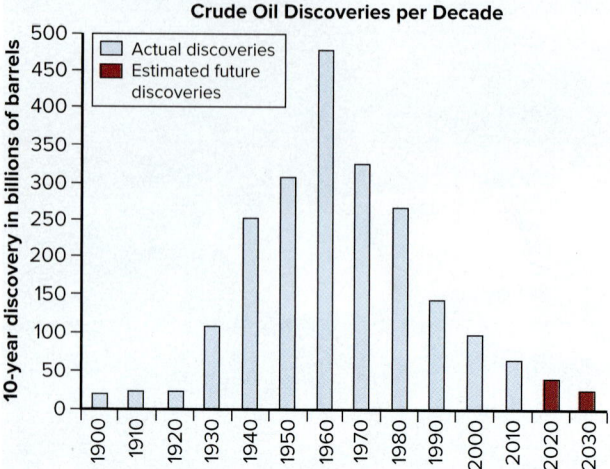

Crude Oil Discoveries per Decade

FIGURE 13.32 Traditional oil reserve discoveries worldwide based on 10-year intervals. Discoveries peaked in the 1960s and have been in decline ever since despite intense exploration efforts.

production of tight oil is projected to reach only 9 million barrels per day by 2040, a considerable amount of the new production will probably have to come from nonconventional sources such as oil (tar) sands (Chapter 14). However, if we are unable to add enough new production to keep pace with both rising demand and depletion of older reservoirs, then the projected increase in supply out to 2040 is unlikely. As indicated by the red line in Figure 13.31, this means that the world will have gone past Hubbard's peak and overall supply will keep decreasing. If we are indeed near Hubbard's peak, then perhaps the best we can hope for is that production can equal depletion for a while so that overall supply remains steady (blue line), buying us time to ramp up production from nonconventional sources.

With respect to conventional crude oil, advocates of the peak oil theory point out that most of the permeable reservoirs containing highly desirable light crude have already been located. In fact, a large portion of the world's oil supply currently comes from giant fields, such as those in the Middle East and Alaska. Most of these giant fields were discovered prior to the 1970s during the major boom in new discoveries illustrated in Figure 13.32. These fields are all slowly being depleted, with many already past peak production and experiencing annual declines in production as large as 10%. This means that a progressively greater number of smaller fields must be found in order to make up for the lost production from the giants. Even more fields must be found to meet the projected increase in the demand for oil. Although there has been much hope that advanced technologies will increase the amount of oil we can extract from a given reservoir, the results so far have been more rapid depletion and only modest increases in overall recovery. It is also important to realize that it requires energy to extract and process oil from a reservoir. Once the amount of energy needed to bring the oil to market equals the energy it can provide, production will stop regardless of how much oil remains in the ground. At this point price becomes irrelevant.

Our best hope for increasing the supply of liquid fuels is to take advantage of the world's vast deposits of *nonconventional oil* (Chapter 14). The problem is that these deposits are much more difficult and expensive to extract as they consist of hydrocarbons that are too viscous to be pumped from the ground, are immature and bound within shale, or exist in ice crystals on the seafloor. Higher prices have already made some of these resources economical to extract using existing technology, but others remain too costly.

One of the concerns with relying on market forces to spur production of non-conventional oil is that when supplies become tight, oil prices tend to skyrocket due to market speculation. These price spikes can have a crippling effect on demand, often referred to as *demand destruction*. For example, high prices force people to consume less gasoline and less of various goods and services that require oil. This drop in consumption creates the potential for an economic recession. Such a scenario occurred in 2008 when oil prices reached nearly $150 per barrel, then fell to a low of around $40. Although the price spike was due in part to market speculation that drove prices beyond normal supply-and-demand relationships, the end result was demand destruction and a significant drop in consumption. This coupled with the global financial crisis of 2008 led to a major recession and a collapse in oil prices. When prices collapse, marginal deposits suddenly become uneconomic since they are extracted only when prices are high enough to offset production costs. This leads to an overall loss of supply known as *supply destruction*. Low prices then spur consumption and economic growth, eventually causing supplies to tighten and prices to climb again. Oil shortages therefore are highly undesirable as they lead to unstable market conditions with repeated cycles of demand and supply destruction that can wreak havoc on the world economy.

Finally, there is always the danger that a sudden and sharp reduction in the supply of crude oil could send prices skyrocketing, plunging the global economy into another deep recession. For example, we could expect a serious oil shock should there be a loss of production from Saudi Arabia, which produces about 13% of the world's crude oil. Such an event is a real possibility given the region's history and previous attacks on the Saudi monarchy. A major oil shock could also occur if Iran made good on its promise, if attacked by Western nations, to shut down oil shipments through the Straits of Hormuz in the Persian Gulf (Figure 13.33).

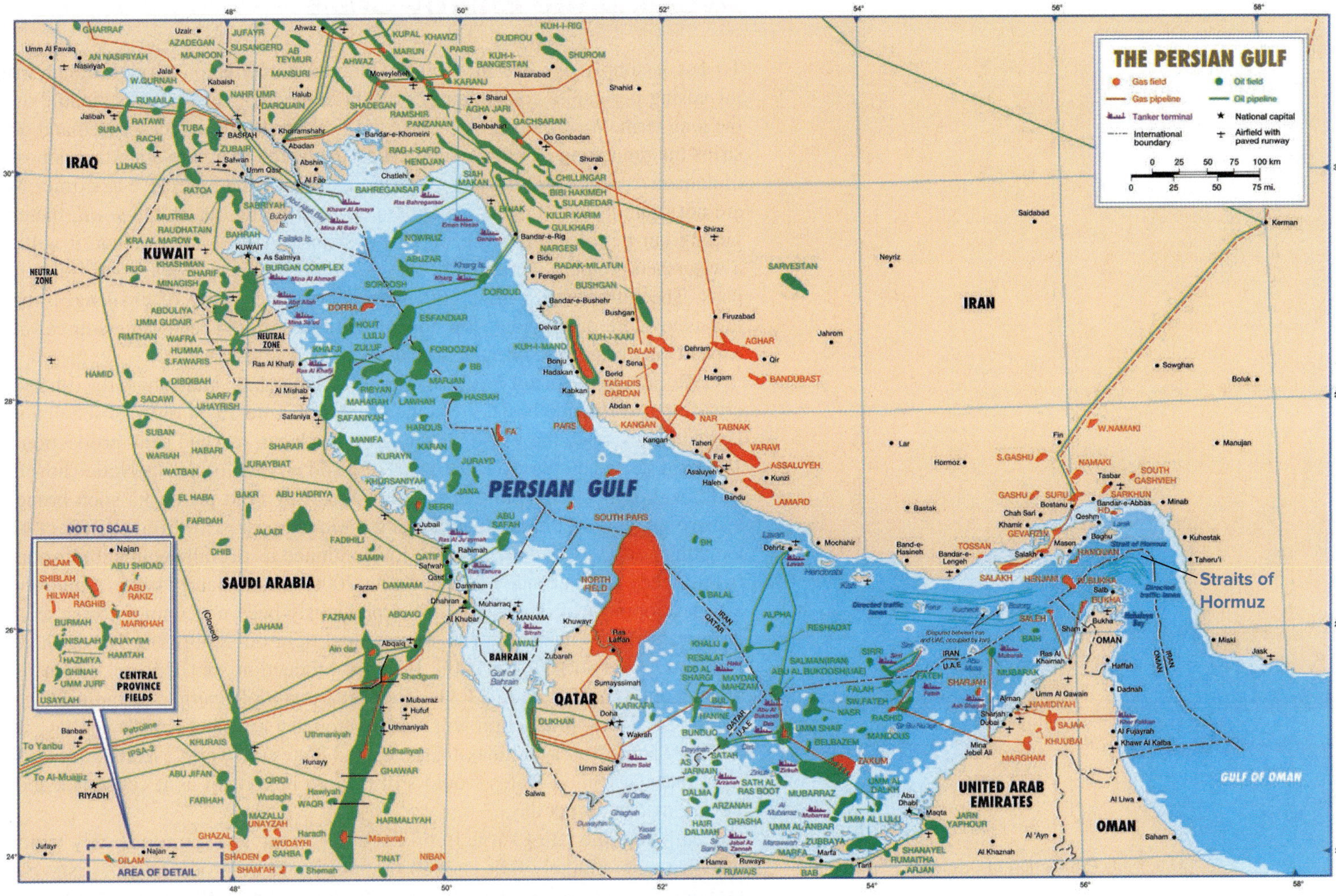

FIGURE 13.33 Nearly 20% of the world's crude oil supply is extracted from the vast Persian Gulf oil fields and loaded onto supertankers. The huge ships must then pass through the Straits of Hormuz, a strategic bottleneck that Iran has threatened to close should there be a military clash with Western nations.

(top): By A. F. Alhajii, reprinted with permission World Oil, September 2007; (bottom): © BP Oil

If this were to happen, 20% of the world's crude oil supply could be immediately removed from the global market, creating a crisis of unprecedented proportions.

Avoiding the Energy Crisis

To put it simply, we are in a bind. Continued population growth along with increasing per capita consumption of light crude oil is just not sustainable over the long term. We will eventually face the situation where increased demand outstrips the remaining supply of light crude. Although there are abundant deposits of nonconventional oil, they are much more difficult and expensive to extract. We are not running out of energy then, but rather running out of *cheap oil*, thereby putting our petroleum-based economic systems at risk. Like it or not, it will be a watershed moment in human history when the world moves beyond Hubbert's peak. The future will bring significant changes and plenty of challenges, forcing us to more quickly develop nonconventional oil and renewable sources and expand our current use of nuclear power.

We can use the history of U.S. oil production to help understand the future oil crisis on a worldwide basis. From the graph in Figure 13.34 one can see that the United States, once the world's leading oil producer, reached peak production in 1970 and started on the downward part of its oil production curve. Notice how the additional production from the giant Prudhoe Bay field in Alaska soon brought a sharp increase in production, creating a second peak. However, the effect was temporary and by 1985 U.S. production continued its overall decline. In response to the steady rise in crude oil prices after 2000, oil companies began exploration and development of deposits that had previously been uneconomic, such as the deep-water fields in the Gulf of Mexico and tight oil deposits in North Dakota (see Case Study 13.1). By 2009, the new production sent U.S. oil production soaring once again (Figure 13.34). The Department of Energy currently projects that by 2025 U.S. production will peak at around 14 million barrels per day, surpassing Saudi Arabia and making the United States the world's leading producer. Note that the United States would still have to import oil as it consumes about 19 million barrels per day.

Although the latest surge in production is impressive, it too is only temporary. It seems unlikely that enough new oil could be produced to satisfy both growing demand and production losses from the older fields currently undergoing depletion. However, some experts believe that the combination of higher prices and new technology will make even more deposits economical to develop; hence production will keep increasing. One problem is that, in the case of tight oil, hydraulic fracturing and directional drilling are expensive and result in wells with high depletion rates. Thus large capital investments would be needed to drill an ever increasing number of wells. Then less capital would be available to invest in other aspects of the economy. At some point the economy would not be able to support both the drain on capital and the higher energy prices required to produce more expensive oil. The United States then would find itself yet again on the downward slope of the peak oil curve. Naturally, the world would face similar constraints in trying to keep increasing its overall supply.

The fact that the world will someday no longer be able to expand its supply of crude oil does not mean that additional exploration will not be needed. Bringing new oil fields online will delay the overall production decline and buy us more time to develop alternative energy sources, thereby lessening the impact of the crisis. Drilling by itself, however, will not avert the crisis. Success will depend on the rapid development of alternative sources combined with serious efforts to improve conservation and efficiency, particularly in the area of transportation.

In this section we will take a closer look at how we can minimize the impact of a global energy crisis. One of the themes is quantity and scale. Because the

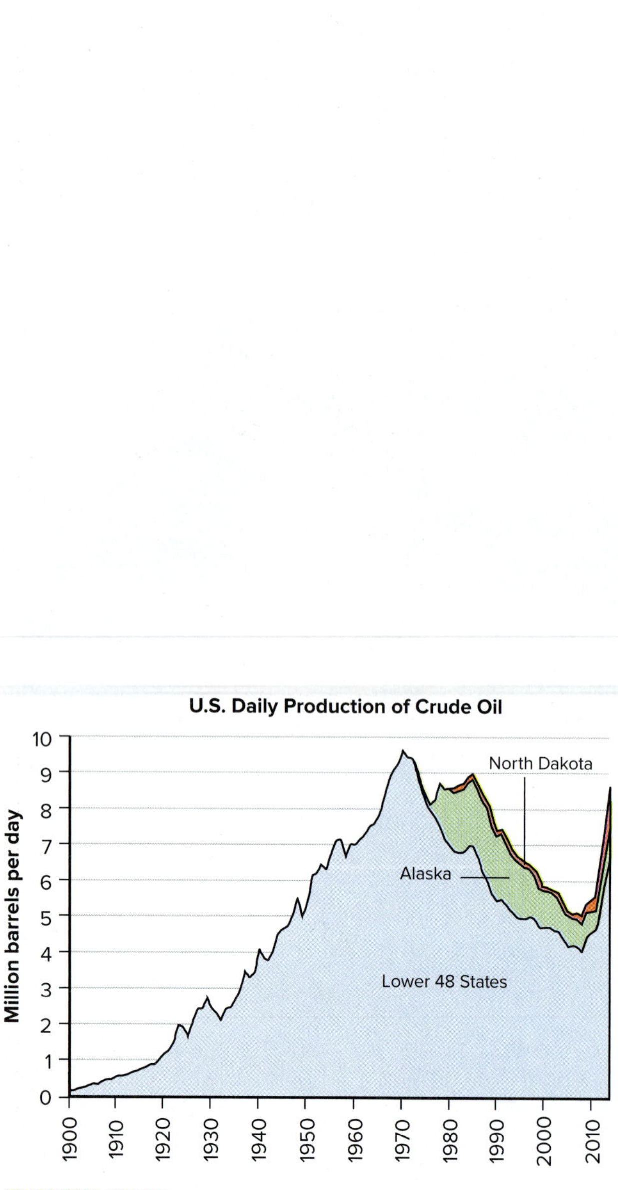

FIGURE 13.34 Graph showing total U.S. crude oil production, which peaked around 1970. Note how the major contribution of oil from Alaska created a second peak and delayed the overall production decline. Significant new production from North Dakota and the lower 48 states has dramatically increased overall production, which is projected to create yet another peak at about 14 million barrels per day in 2018.

world consumes such an enormous amount of oil, it will be necessary to scale up production of alternative energy sources to generate the quantities we need. The other theme is time and money. Large sums of money will have to be invested to bring this energy online and to develop more efficient transportation systems. Also, we will have to work quickly to keep conventional energy prices from rising too rapidly, creating market instabilities that would wreak havoc on the global economy. We will briefly examine how existing technology can be used to quickly scale up production of nonconventional oil and alternative energy sources—see Chapter 14 for a more thorough description.

Replacements for Conventional Oil

To get an appreciation for the size of the problem the world is facing, we can again use the United States as an example because it is responsible for about 20% of the world's total energy consumption. The plot in Figure 13.35 illustrates the tremendous growth in U.S. energy consumption since World War II. This growth reflects the nation's surging economy, increasing population, and growing per capita consumption of energy after the war—consumption today has leveled off due to increased conservation and efficiency and slower economic growth. Of particular interest is how fossil fuels make up the vast majority (86%) of current U.S. energy consumption and how comparatively little energy comes from nuclear and renewables (14%). Moreover, approximately 35% of all energy used in the United States comes from oil. Because the United States consumes such a large proportion of the world's energy, this 35% represents a very significant volume of oil. The energy problem then boils down to one of scale. Developing other energy sources to replace even a fraction of what conventional oil now provides will be a daunting task.

The world faces a dilemma in terms of oil consumption similar to that of the United States, which brings up the question of whether there are existing energy sources that can replace conventional oil. Fortunately, there are significant deposits of heavy crude oil, including oil sands, which can be extracted and refined into fuels and petrochemicals. We also have the technology to make both oil and gas from our abundant coal reserves. We are currently producing crude oil from heavy oil and tar sands, but the amounts are relatively small compared to world demand. Although the production of oil from coal used to be quite common, production today is almost nonexistent largely because it is an inefficient process and generates considerable pollution. Because the technology already exists for producing crude oil from heavy oil, tar sands, and coal, the basic problem becomes one of scaling up production in time to prevent a serious shortage. The difficulty here is that special processing plants must be built, plus the resources themselves are generally more difficult to extract than conventional oil. Scaling up production then will require both time and money.

Another readily available replacement for conventional oil is natural gas. Here special processing plants are not needed because natural gas can be used directly in existing internal combustion engines, with relatively minor modifications. A drawback is the amount of vehicle space taken up by larger fuel tanks. Also, because natural gas has a lower energy density than liquid fuels, even with a larger tank the driving range is much shorter compared to that provided by gasoline or diesel fuel. Another issue is that it will take years to convert a significant portion of our transportation fleet to run on natural gas and build the necessary filling stations. Although supplies are now abundant, natural gas is following a production curve similar to that for oil and will eventually experience depletion. Natural gas therefore cannot serve as a permanent replacement for oil in meeting our transportation needs.

Biofuels, particularly ethanol made from corn, have often been touted as substitute fuels that can lead us to energy independence. The biofuels industry claims that growing corn to make ethanol is cost-effective and that it represents a 25%

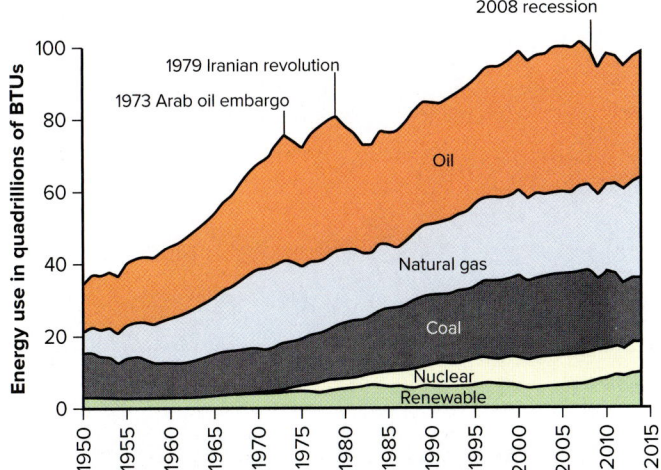

FIGURE 13.35   After the 2008 recession total energy consumption in the United States leveled off due to increased conservation and efficiency and slower economic growth. Oil presently makes up nearly 35% of U.S. energy consumption, representing a large amount of energy that will be difficult to replace.

net gain in energy. However, scientists have estimated that by including the energy required to make fertilizers and pesticides, run irrigation and farm equipment, and operate the distillation process, it takes 30% more energy to produce ethanol from corn than burning the ethanol provides. Scientists also claim that ethanol has been profitable to produce only because of taxpayer-funded subsidies. Now that the U.S. Congress voted to eliminate corn ethanol subsidies, production may actually decline, putting even more pressure on gasoline supplies.

The use of all-electric cars, plug-in hybrids, and vehicles powered by hydrogen fuel cells is often promoted as a way of reducing the demand for oil. However, recall that electricity itself must be made from a primary energy source. The same goes for hydrogen: energy is required to break down water (H_2O) or natural gas (CH_4) molecules to get free hydrogen (H_2). The hydrogen is then used to run a fuel cell that produces electricity for an electric motor. In either case fossil fuels are currently the primary energy source for producing electricity or hydrogen for such vehicles. Because burning fossil fuels releases carbon dioxide and contributes to global warming, we would need to greatly increase electrical production through nuclear, wind, or solar energy for the large-scale use of electric or hydrogen-powered vehicles. While feasible, building the infrastructure for the additional electrical power and production of new vehicles would take years, and thus this is more of a long-term solution.

Although there are several suitable replacements for conventional oil, they will all require time to scale up production. There is an option, however, that would make an immediate and significant impact on oil supplies and buy us precious time for scaling up the production of alternative resources. It is called *conservation*.

Increasing Supply by Reducing Demand

The cause of the world's looming oil crisis is essentially decreasing supply and increasing demand. When supplies become scarce, prices typically spike and trigger cycles of demand and supply destruction. While many people like the idea of a totally free market system, letting the oil market correct itself in this manner is not desirable. In fact, proponents of the peak oil theory have long warned that a permanent and progressive decline in oil production would send the world economy into a major recession, or possibly cause it to collapse. To minimize this possibility, the most prudent approach would be to work both sides of the energy equation: to increase supply while at the same time taking steps to reduce demand.

Reducing demand can easily be accomplished through conservation efforts and increases in efficiency. The great thing about conservation is that it does not necessarily cost anything to implement, and it has a positive and almost immediate effect on both supply and demand. Because the United States consumes about 20% of the world's oil supply, serious conservation efforts in this country can translate into a surplus of oil on the global market, driving prices downward. A good example is the decline in crude oil consumption associated with the 2008 spike in oil prices and onset of an economic recession (Figure 13.35). In the United States alone, oil consumption declined 11% from 2007 to 2009. Although part of the decline was due to reduced economic activity, equally important were the voluntary conservation efforts of citizens when gasoline prices were high. Another example is the dramatic decline in U.S. consumption that occurred in response to the oil crises of the 1970s. Again, the decline was due to an economic recession and conservation efforts, some of which were voluntary and others mandatory.

The twin Middle East oil crises of the 1970s also led to increases in energy efficiency in developed nations around the world. The increased efficiency caused a reduction in oil consumption, with a lasting effect in many countries. In 1975, the U.S. government mandated average efficiency standards for the fleet of cars and light trucks sold in the United States. Manufacturers could sell vehicles of varying efficiencies, but their overall sales had to meet the efficiency standard for a given year. As illustrated in Figure 13.36, the average fuel efficiency went from 16 mpg (miles per gallon) in 1975 to 25 mpg by 1990, a 56% improvement. Unfortunately, during the oil glut from the late 1980s through the 1990s, further efforts at improving efficiency standards failed to pass the U.S. Congress. Lawmakers instead offered tax breaks for low-mileage sport utility vehicles (SUVs), causing fuel efficiency to actually decrease. In contrast, most of the other industrialized nations chose to impose stiff gasoline taxes, on the order of $3 to $4 per gallon, to encourage conservation and the purchasing of more fuel-efficient vehicles. Some of the tax revenue was used to fund research and development of alternate energy resources. Denmark, for example, imported nearly 100% of the energy it consumed from the Middle East in 1973. Today it imports none. The United States, on the other hand, kept gasoline taxes around 40 cents per gallon, encouraging SUV ownership (Figure 13.36A and Figure 13.36B). Finally, in response to the dramatic spike in oil prices leading up the 2008 recession, the United States increased its efficiency standards, resulting in a 16% improvement by 2014. Standards have recently been increased even further, with the fleet average expected to reach 47 mpg by 2025.

Strategy for Living Past Peak Oil

Crude oil is one of our most important resources as it is essential to modern society. For many critical applications, such as jet aircraft fuel, plastics, fertilizers, pesticides, and herbicides, there are no substitutes for crude oil. This presents a problem in that oil is also a finite resource, which means we will eventually find ourselves past peak oil, where demand outstrips supply. Because of the economic consequences of moving past Hubbert's peak, it would be prudent to try and avoid the high oil prices that lead to market volatility and crippling cycles of demand and supply destruction. We should also try to insulate our economies against a major disruption in supply that could result from unrest in the Middle East. A prudent strategy would be to extend our reserves for as long as possible by taking steps that increase supply and reduce demand. Below are some steps that could help the United States and other countries make it past peak oil and develop more sustainable societies that mostly rely on renewable and non-carbon energy sources.

1. **Reduce consumption by increasing efficiency.** Increasing efficiency would result in both decreased demand and increased supply. Measures could include providing financial incentives for purchasing more fuel-efficient vehicles and encouraging development of mass transit, carpooling, and other energy-saving measures.

FIGURE 13.36   (A) After the oil crises of the 1970s the U.S. government mandated efficiency standards for cars and light trucks, resulting in a dramatic 56% increase in overall fuel efficiency by 1990. Efficiency actually fell in the 1990s largely due to Americans switching from cars to SUVs (B). New standards have again caused efficiency to improve.
(b) © Jim Reichard

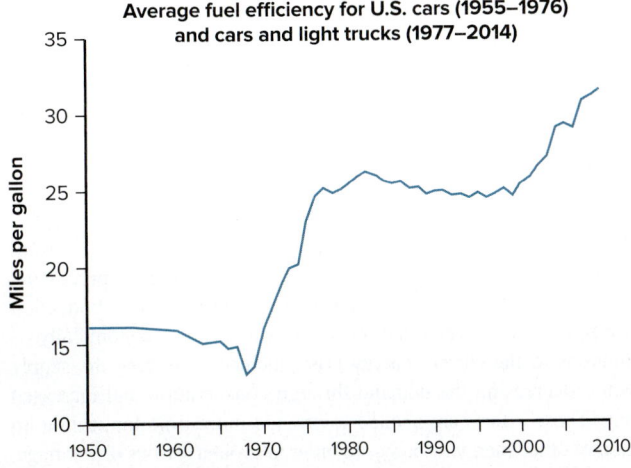

Average fuel efficiency for U.S. cars (1955–1976) and cars and light trucks (1977–2014)

2. **Continue exploration and development of conventional oil deposits.** Exploration for deposits of conventional crude oil would increase supply, whereas new research could lead to technological improvements that result in marginal deposits becoming economical to develop, thereby adding to our reserves.

3. **Increase use of natural gas and electricity in transportation.** This step would reduce demand for oil and create a surplus. Many types of vehicles could quickly be converted to run on natural gas; thus there would be an immediate impact. Rights of way along rail lines could be used to lay power lines for running run freight and passenger trains on electricity.

4. **Begin replacing natural gas used in electrical generation with other energy sources.** The idea here is to use wind, solar, and nuclear power to free up natural gas supplies for meeting the higher demand created by vehicles running on natural gas. Additional power would also be needed for electric trains.

5. **Use biofuels where feasible.** Generating biofuels that do not represent a net energy loss should be encouraged to ease pressure on oil supplies. Ethanol from corn is a net energy loss, so efforts should be directed at more efficient processes.

6. **Focus on using remaining oil to generate aviation fuels and petrochemicals.** The idea here is to direct our limited oil supplies to those areas where suitable replacements do not exist.

It is clear that the United States has placed itself in a difficult position, particularly since China's rapid industrialization is putting major pressure on the world's remaining oil supplies. Fundamentally changing the way our society uses energy to avoid the effects of peak oil production and climate change will be difficult. Fortunately, the United States has world-class research institutions and enormous technical capability that can help it transition from an oil-based economy to one powered mainly by renewable energy resources. In fact, many people believe that the manufacturing and installation of wind towers and solar panels will help form the basis for a so-called *green economy*. Declining oil production and global warming are eventually going to force us to switch to renewable energy resources whether we like it or not. So, why not get started on a more sustainable economy before it is too late?

SUMMARY POINTS

1. Energy is the capacity to do work, and it exists in many different forms (electrical, chemical, thermal, kinetic, nuclear, and solar). Humans make use of energy transformations to fulfill our needs, as in the conversion of fossil fuels into heat, electrical, or kinetic energy.

2. A primary energy resource is a source of energy that humans can use directly, such as fossil fuels or wind—electricity is a secondary energy source because it has to be produced from a primary source.

3. Humans historically relied on wind, water, wood, and then coal to provide mechanical power. Petroleum became the dominant energy source because it has a high energy density and its liquid nature makes it convenient to use and transport.

4. Coal originates when organic-rich deposits in swamps become buried. As temperature and pressure increase, water and volatile compounds are driven off, thereby transforming the organic matter into a carbon-rich combustible rock. Coal and mineral deposits are mined using similar methods.

5. Oil and gas originate from marine plankton and are found in organic-rich sedimentary rocks. Burial of these rocks causes temperature and pressure to reach the point where the organic matter is converted into hydrocarbon molecules, which then migrate upward into permeable rock layers.

6. Geologists locate traps where oil or gas is stored in permeable reservoir rocks. If an economic deposit is found, a production well is installed to extract the hydrocarbons in a manner similar to that used by water wells.

Petroleum is also being extracted from shale source rocks and other fine-grained rocks using hydraulic fracturing and directional drilling.

7. Crude oil is refined into different products by a distillation process in which the oil is heated and resulting vapors are allowed to rise in a cooling tower. The heavier hydrocarbon molecules condense near the bottom of the tower, where temperatures are highest; the lighter, more volatile compounds condense toward the top where it is cooler.

8. Coal and petroleum deposits are widely dispersed around the globe, but nearly 50% of all crude oil reserves are held by the Persian Gulf nations in the Middle East. Most industrialized nations do not have sufficient energy resources to meet their own needs.

9. The peak oil theory holds that production of conventional oil follows a bell-shaped curve that has a peak, after which rising demand is expected to outstrip supply, creating a permanent oil shortage.

10. Current replacements for conventional oil include crude oil produced from coal and deposits of tar sands and heavy oil. However, special processing plants must be built to scale up production from these sources. Production of other types of nonconventional oil are not yet feasible economically.

11. Most solutions to the current energy crisis include increasing the supply of oil while decreasing the demand through conservation and increased efficiency. These efforts are needed to provide more time for scaling up production of other energy resources in order to avoid a serious oil shortage.

KEY WORDS

cap rock   416
coal   410
crude oil   414
energy   405

energy efficiency   406
fossil fuels   404
gas window   415
Hubbert's peak   430

hydrocarbons   414
natural gas   414
oil window   415
peak oil theory   430

petroleum   414
petroleum reservoir   416
petroleum trap   416
work   405

APPLICATIONS

Student Activity

Do an Internet search and find a site that lists the daily market price for crude oil along with the average daily retail price of gasoline in the United States (try using "EIA daily energy prices" in your search). Then track the prices for several weeks. Did the prices behave similarly? Were there any significant news events that may have affected the price of one or both of these commodities?

Critical Thinking Questions

1. Electricity is a widely used form of energy in modern society. Given that the world production of crude oil may eventually be unable to meet demand, why can't we simply use more electricity to meet our energy needs?

2. Coal was the key energy source that made the Industrial Revolution possible. What is it about petroleum that caused society to abandon coal as its primary source of energy?

3. Natural gas supplies have increased significantly in recent years due to development of tight gas deposits. Why can't natural gas be refined into the various petrochemicals used in society?

Your Environment: YOU Decide

U.S. Coast Guard

Crude oil is a global commodity, meaning that its price is set by the world market. The world currently consumes about 89 million barrels per day, whereas U.S. production is about 9 million barrels per day. Even if the United States were to somehow double its production, it could not control the price of oil. Given the fact that the United States cannot set the price of gasoline at the pump, explain whether you think the nation should open up environmentally sensitive areas (e.g., the Florida coast, Arctic National Wildlife Refuge) to drilling or place more emphasis on developing alternative energy resources.

Chapter 14

Alternative Energy Resources

The key to a sustainable future will include making use of available roof space for collecting solar energy, as in this housing development in Lingang, China.

CHAPTER OUTLINE

Introduction

Nonconventional Fossil Fuels
- Synthetic Fuels from Coal
- Heavy Oils and Oil Sands
- Oil Shale
- Gas Hydrates

Carbon-Free and Renewable Fuels
- Biofuels
- Hydroelectric Power
- Nuclear Power
- Solar Power
- Wind Power
- Geothermal Power
- Ocean Thermal Energy Conversion
- Tidal Power

Conservation

Post-Petroleum World
- Transportation Systems
- Generating Electricity in the Future

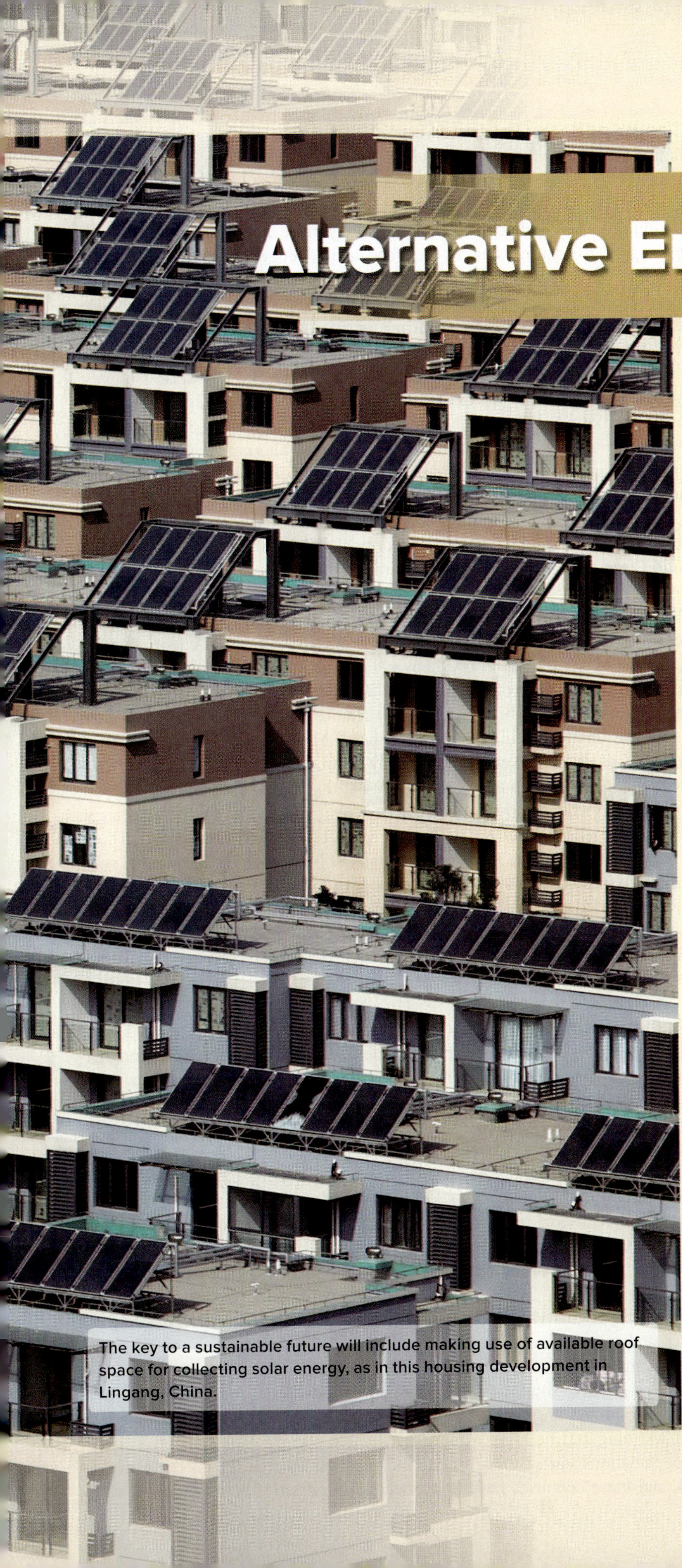

LEARNING OUTCOMES

After reading this chapter, you should be able to:

- Explain the difficulty in replacing existing crude oil supplies with oil produced from coal, heavy oils, and oil sands.
- Discuss why corn-based ethanol represents a net energy loss and how its production affects global food supply and climate change.
- Describe the basic reasons why the nuclear power industry in the United States stopped expanding during the 1990s.
- Characterize the two ways solar radiation can be converted into more useful forms of energy, and explain why solar energy is more cost-effective in some regions than in others.
- Discuss how wind and the tides are used to generate electricity, and describe the advantages and disadvantages of each method.
- Describe the basic operation of a geothermal heat pump, and explain how it differs from using geothermal energy to generate electricity and to operate space heaters.
- Explain how ocean thermal energy systems (OTEC) work and why these systems have geographic limitations.
- Describe the difficulty in moving from fossil fuels to renewable and non-carbon-based sources of energy.

Introduction

© StockTrek/Getty Images

In Chapter 13 you learned that as conventional and tight crude oil production eventually declines, society will have to look toward other energy sources, or else face large swings in oil prices that could cripple the world economy. For our purposes we will define **alternative energy sources** as those other than the coal, natural gas, and light crude oil that currently make up the bulk of the world's energy supply. Alternative sources therefore include renewable energy (e.g., wind and hydroelectric power), nuclear power, and unconventional fossil fuels, such as heavy oils and synthetic fuels made from coal and oil (tar) sands. Although alternative sources make up a significant portion of the energy supply in some nations, the United States and China continue to rely on fossil fuels for nearly 90% of their energy needs. Because these two countries are the world's biggest energy users, consuming 40% of the total supply, their energy usage has a major impact on worldwide supplies. Consequently, despite the success of many industrialized nations at reducing their use of fossil fuels, alternative energy sources still provide only 14 of the world's total energy supply. The continued reliance of the United States and China on fossil fuels not only threatens the global economy, but the release of carbon dioxide from the burning of these fuels is affecting Earth's climate system (Chapter 16). It is critical that the United States and China transition away from fossil fuels and toward alternative energy sources as quickly as possible.

The United States and China continue to rely on petroleum and coal primarily because these are highly concentrated and convenient forms of energy that are comparatively cheap. In contrast, many alternative energy sources are intermittent in nature and have low energy density. Scaling up production so that alternative energy makes a significant impact will require building massive amounts of infrastructure (e.g., windmills and solar panels). Many Americans argue against moving toward alternative energy sources by claiming that it will be too costly for the economy and result in a loss of jobs. What this argument ignores, however, is that the era of cheap and stable oil prices is over and that unstable prices are a serious threat to the economy (Chapter 13). Part of the reason petroleum and coal have remained inexpensive in the United States is because of the substantial subsidies the fossil fuel industry receives in the form of tax breaks. With the exception of corn-based ethanol, U.S. tax incentives for the renewable energy industry have historically been relatively minor and inconsistent, making it difficult for the new industry to develop. Environmentalists argue that if the costs society pays in terms of pollution and climate change caused by fossil fuels are included in the comparison, then renewable energy is actually far cheaper.

The term *green economy* is commonly used today in reference to the inevitable conversion of our fossil fuel-based society to one that is sustainable and that has an energy supply dominated by renewable and carbon-free fuels. Environmentalists envision a green economy where people have jobs building and installing new infrastructure, ranging from solar panels and windmills to an expanded electrical grid to a vast rail system that moves both people and freight. In fact, the green revolution has already begun in Europe, where solar and wind industries are booming. It now appears that China will be a future leader in green technology, in part because of its rapid industrialization and widespread air and water pollution. In fact, Chinese citizens are demanding change because they are literally choking on some of the world's worst air pollution, caused by the country's use of fossil fuels.

The goal of this chapter is to provide an overview of alternative energy sources and to examine their future role in society. Although some of these resources are carbon-based, they will still be needed in the short term to help us get through possible supply shortages of conventional fossil fuels. Others do not contain carbon and are renewable, and therefore will be critical for combating global warming and building a more sustainable society. Many of the world's developed nations started transforming their economies after the oil crises of the 1970s, and these countries have built successful industries based on alternative energy. Meanwhile the United States focused on expanding its supply of conventional oil and coal. For example, alternative energy sources now provide 50% of France's total energy needs, whereas in the United States they provide only 13%. Ultimately, the continued depletion of the world's oil reserves, combined with global warming, will force Americans to make a major shift to alternative energy sources. The United States has come to a fork in the road. One path continues to rely on fossil fuels for as long as possible, leaving the country exposed to supply disruptions and accelerated global warming. The other path leads to energy independence, reduced carbon dioxide emissions, and a more sustainable society.

Nonconventional Fossil Fuels

In this section we will examine **nonconventional fossil fuels,** which we will define as all fossil fuels except for the traditionally used coal, natural gas, and light crude oils. Here we are interested in those fuels that might replace declining production of conventional and tight oil, thereby satisfying the world's energy demand. Some nonconventional fossil fuels have already been used to produce synthetic oil, but at a higher cost than crude oil and in relatively small amounts. During periods of tighter crude oil supplies and higher prices, nonconventional fuels become more competitive and profitable to produce.

Many people have taken note of the fact that nonconventional fossil fuels include deposits that are equivalent to very large reserves of light crude oil, leading to the conclusion that oil will be abundant for the foreseeable future. However, extracting and then converting these deposits into actual crude oil is costly and time-consuming. It will be a challenging task to extract these resources in significant enough amounts to keep up with the world's enormous appetite for oil. Another problem is that nonconventional or synthetic crude requires special processing plants. Proponents of the peak oil theory warn that scaling up production to avoid a serious oil shortage may be difficult simply because of the time required to build the processing plants. We will begin our discussion of nonconventional fossil fuels by taking a look at our abundant coal reserves.

Synthetic Fuels from Coal

Despite the long list of environmental problems associated with coal (air and water pollution, acid rain, mercury fallout, mining hazards, and global warming), it remains the primary fuel for generating electricity. Coal has been highly desirable as a fuel because it has high energy density and is cheap, provided one does not factor in its environmental costs to society. Prior to the widespread availability of petroleum, coal was broken down chemically to produce gas, often referred to as *coal gas* or *town gas.* Coal gas was typically made at a local processing plant, and then piped to homes and businesses where it was used for lighting, cooking, and heating. As shown in Figure 14.1, in the late 1800s gas lanterns made it possible for city streets to become fully lit after sunset for the first time in human history. Later, when automobiles became popular, a similar process was developed for making gasoline and diesel fuel from coal. In fact, during World War II, Germany used coal as its primary source for gasoline and diesel fuel. Note that gas or liquid fuels produced from coal or heavy oil are commonly referred to as **synthetic fuels,** or simply *synfuels.*

With the exception of a single processing plant in South Africa, coal-based synthetic fuels had not been produced on a large scale until recent years because they were not cost-competitive with those derived from conventional petroleum. The basic problem was the additional energy costs required to transform coal into synthetic fuels. However, when energy prices began to rise after 2000, the economics changed, and countries around the world started developing plans for coal-gasification plants. According to the Gasification Technologies Council, an

FIGURE 14.1 Prior to large-scale electrical production, most towns and cities had their own processing plants that converted coal into synthetic gas, which was used for lighting and for heating and cooking. As illustrated by this pre-1890 photo of London, England, gas lighting transformed night into day, allowing city life to thrive after sunset.
© E.O. Hoppe/Corbis

industry trade group, there are currently over 400 coal-gasification plants in operation worldwide, and nearly 250 planned for completion by 2016. For example, China is using synthetic gas derived from coal to help reduce air pollution in its cities. Rather than using local coal-burning power plants to generate electricity, China is building gasification plants near its coal fields. The gas is then piped to cities where it used to generate electricity, thereby dramatically reducing the local emission of harmful gases and particulate matter.

China is also converting some of its significant coal reserves into synthetic oil to help meet its growing demand for transportation fuels, particularly diesel fuel. In 2008, it began operating a new $3.2 billion coal liquefaction plant that produces approximately 1 million tons of synthetic oil per year. Although China is building additional plants with a combined annual capacity of 22 million tons, this output would still represent less than 4% of the 620 million tons of oil that the nation consumed in 2015. Clearly, for coal liquefaction to make a substantial contribution to the world's oil supply, large numbers of liquefaction plants would need to be built.

In addition to the difficulties of scaling up production, there are some other drawbacks to coal-based synthetic fuels. While world coal reserves are large and expected to last about a hundred years at current consumption rates, mass production of synfuels could dramatically shorten the life span of those reserves. Coal therefore cannot be expected to serve as a long-term solution to a decline in crude oil production. Another issue is that the mining of coal would increase dramatically, magnifying the existing environmental problems associated with that activity (Chapter 13). An even more serious concern is that the liquefaction of coal is less efficient and produces considerably more carbon dioxide (CO_2) than does the refining of conventional crude oil. Large-scale coal liquefaction then would exacerbate the current problem of CO_2 emissions and global warming. Synthetic fuels produced from coal are simply not a viable option unless a practical means can be developed for pumping the CO_2 into suitable rock layers for long-term storage in the subsurface, a process called *carbon sequestration* (Chapter 16).

Heavy Oils and Oil Sands

Recall from Chapter 13 that crude oil consists of chains of hydrocarbon molecules whose length and weight varies. Crude oil with a higher proportion of lighter (shorter) molecules and fewer sulfur atoms is referred to as **light crude oil,** and that with heavier molecules and more sulfur is called **heavy crude oil.** Heavy crudes are also referred to as "sour" since they contain more sulfur, whereas light crudes are called "sweet" due to their low sulfur content. This difference is important because heavy crudes produce smaller amounts of valuable transportation fuels (gasoline and jet fuel) when refined, plus the refining costs are higher because of the need to remove the additional sulfur. As a result, light crude commands a higher price at refineries than heavy crude. Heavy crudes are also more viscous, which makes them more difficult and costly to extract and transport. Oil companies therefore prefer to develop deposits of light crude because those are more profitable. Consequently, after more than 100 years of exploration and production, most of the world's remaining oil reserves are on the heavier and more sulfur-rich end of the spectrum. Because of this preferential consumption of lighter and easier-to-produce crude, it is often stated that the world is running out of *cheap* oil.

There comes a point at which crude oil can be classified as nonconventional because it is so heavy and viscous that special techniques are required to extract it from a reservoir rock. For example, despite the additional cost and effort, heavy crude has been produced from several oil fields in California since the early 1900s. At these fields a technique was developed whereby wells are used to inject steam into the reservoir, heating the oil and reducing its viscosity so it can be pumped out. As illustrated in Figure 14.2, one approach is to inject steam into an individual well; then, after waiting for the oil to become thinner, the well is pumped. Pumping continues until the oil again becomes too thick, at which point

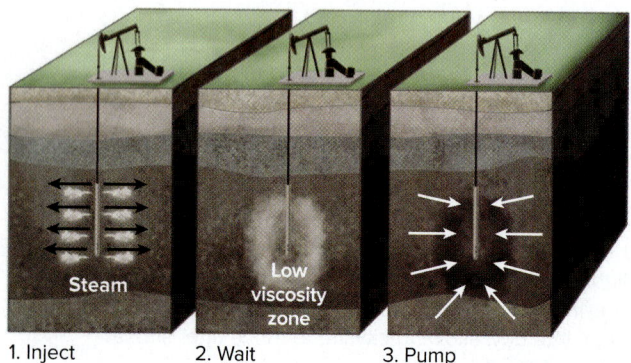

1. Inject   2. Wait   3. Pump

A

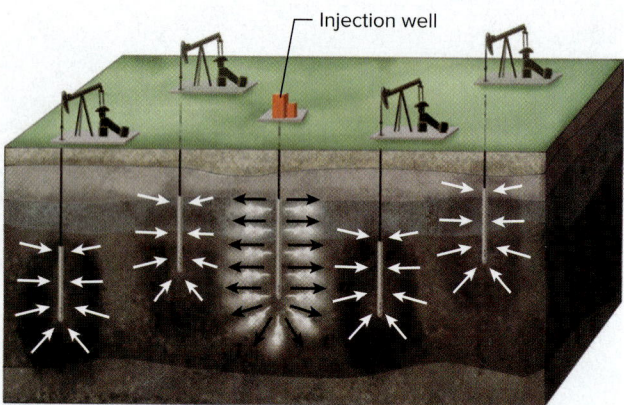

B

FIGURE 14.2   Some heavy oils are too viscous to be extracted directly by pumping. One approach is to inject steam into an individual well (A), thereby lowering the oil's viscosity. The thinner oil is then pumped until the viscosity becomes too great again, at which point the process is repeated. Another method involves a dedicated injection well (B) surrounded by four pumping wells. Here steam is injected continuously while the pumping wells remain in full production.

the process is repeated. A more elaborate and efficient method involves using separate wells for injection and extraction that operate continuously. With either method, once the oil is extracted, it eventually cools down and becomes too thick to transport to a refinery. To make the heavy oil transportable, its viscosity is reduced by mixing it with lighter crude. However, this requires that a supply of lighter crude be available nearby.

A different form of heavy oil is sometimes found near the surface in what are called *tar pits*—a famous example is the La Brea Tar Pits in California, where prehistoric animal remains are preserved. The hydrocarbon substances in tar pits are in a form called *bitumen,* from which the lighter and more volatile hydrocarbon molecules have been lost, leaving behind a highly viscous fluid that is more like asphalt than normal crude oil. Bitumen is also found in sand deposits close to the surface, which geologists call **oil sands** or *tar sands*. These deposits are generally believed to have formed when crude oil from petroleum reservoirs migrated upward and then collected in sand layers near the surface. Here the crude oil was broken down over time by bacteria, leaving behind bitumen. Two of the largest oil sand deposits in the world are located in Canada and Venezuela, and together are estimated to contain enough bitumen to produce the equivalent of several trillion barrels of conventional crude oil. However, only a fraction of this is considered recoverable. For comparison, about 1.3 trillion barrels of conventional crude remain available in the world today.

Presently most of the oil sands production occurs in Alberta, Canada (Figure 14.3). The estimated recoverable reserves of equivalent crude from these deposits is approximately 170 billion barrels (BBL), which ranks third behind the

FIGURE 14.3 Synthetic crude oil is currently being produced from oil sand deposits (A) in Alberta, Canada. The hydrocarbons are in the form of bitumen, a highly viscous substance that is separated from the sand using steam. Photo (B) shows a bitumen sample whose viscosity has been lowered by heating. Approximately 60% of Canadian production involves strip mining (C); the remainder is produced by steam injection and pumping wells.
(B) © Syncrude Canada Ltd; (C) © Photographic Services, Shell International Ltd.

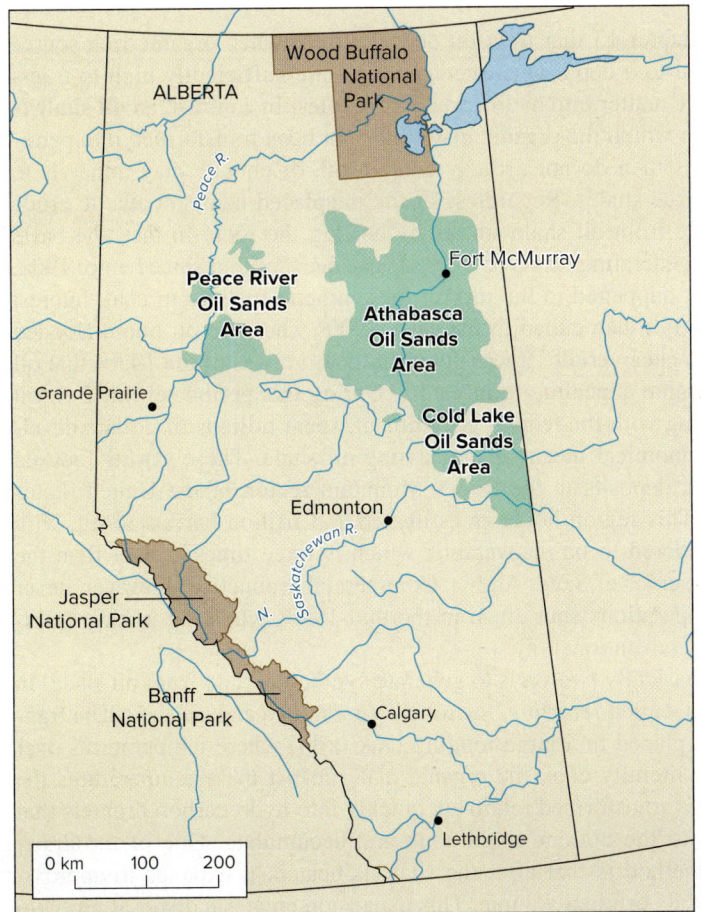

A

B

C Athabasca oil sands, Muskeg River mine

reserves of Saudi Arabia (270 BBL) and Venezuela (220 BBL). The problem is that extracting the highly viscous bitumen is much more costly than extraction of conventional crude. For example, of the oil sand reserves in Canada, 80% can be extracted via wells and steam injection techniques similar to those used for heavy oils. The remaining 20% is relatively close to the surface, making it feasible to remove the material by strip mining (Chapter 12). Here the bitumen-saturated sand is loaded onto trucks and sent to an on-site processing plant where steam is used to separate the bitumen from the sand. Once the bitumen is extracted, natural gas is used in a process to convert it into synthetic crude, which can then be transported via pipelines to refineries.

Canada is currently producing about 2.3 million barrels of synthetic crude per day from its oil sands (55% from wells and 45% by mining), and production is expected to peak at 4 million barrels per day by 2030. Note that approximately 80% of the existing production is exported via pipelines to U.S. refineries. While the Canadians have done an impressive job of scaling up production, the process is energy-intensive and has certain physical limitations. For example, the processing of oil sands requires large volumes of water and natural gas, the supplies of which are not unlimited. Another issue is that the use of oil sands releases more carbon dioxide (CO_2) than conventional crude due to the land disturbances, running the mining operations, and processing of bitumen. The amount of additional CO_2 released has been calculated by various groups, with most estimates falling between 10% and 30%. Regardless of the exact size of the CO_2 footprint, even if Canada's production of synthetic crude oil reaches the projected 4 million barrels per day, it would still represent less than 5% of the approximately 90 million barrels of oil the world consumes each day.

Oil Shale

Recall from Chapter 13 that *tight oil* deposits form when organic-rich source rocks are buried to a depth where temperatures are sufficiently high to transform the organic matter into hydrocarbon molecules. In contrast, an **oil shale** is a source rock in which the organic matter has not been transformed into petroleum. Oil shales then do not actually contain oil, or enough clay minerals to be considered true shales. Regardless of the misplaced name, synthetic crude oil can be made from oil shale simply by heating the rock. In fact, the basic technology for generating this type of crude has been around since before 1900. Similar to what happened to the market for synthetic fuels from coal, interest in oil shale pretty much ended in the early 1900s when the oil boom flooded the market with cheap crude. It was not until the oil crises of the 1970s that oil shale again became appealing to investors. During this period several U.S. oil companies, along with the federal government, spent billions of dollars developing more economical means of processing oil shale. These efforts focused on the oil shale deposits in the Rocky Mountain region of the United States (Figure 14.4). This region holds an estimated 2–3 trillion barrels of oil, with 800 BBL considered to be recoverable, which is three times greater than the conventional reserves of Saudi Arabia. Commercial production, however, never took place as operations shut down in the mid-1980s when the world experienced a glut of conventional oil.

There are basically two ways to generate synthetic crude from oil shale. In the first process, called *retorting,* the rock is first mined and crushed. The fragments are then placed in a large rotating tank (kiln) where temperatures over 900°F (500°C) literally cook the organic material. At these temperatures the organic matter is transformed relatively quickly into hydrocarbon droplets that eventually fall to the bottom of the tank and accumulate. One of the drawbacks of this method is that after the shale is heated, it expands to approximately 120% of its original volume. This expansion creates a disposal problem since the original mining excavation is not large enough to accept the waste.

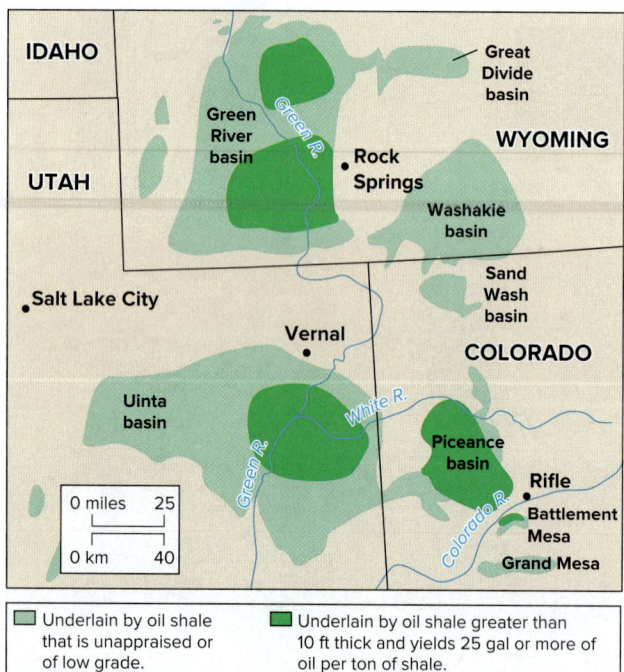

Underlain by oil shale that is unappraised or of low grade.	Underlain by oil shale greater than 10 ft thick and yields 25 gal or more of oil per ton of shale.

FIGURE 14.4 Over 60% of the world's reserves of oil shale are in the Rocky Mountain region of the United States. Past U.S. efforts at commercial oil shale production have failed, but new methods have shown success and may someday prove economical. However, important environmental problems would still have to be overcome, such as the region's limited water supplies and concerns over increased carbon dioxide emissions.

For a large-scale operation the waste material would have to be placed in nearby stream valleys, thereby disrupting the natural hydrology and ecosystems. To avoid this problem, U.S. research efforts during the 1970s focused on *in-situ* (in-place) techniques where the rock is cooked underground. Once the hydrocarbons formed in sufficient quantities, they could then be pumped to the surface via wells.

Although in-situ techniques eliminated the problem of large-scale waste disposal, commercial oil shale production never became viable because of high production costs. These costs are due in part to the large amount of energy needed to keep the temperature elevated during the cooking process. Production costs are also high because of the fact that oil shale produces relatively heavy oil. As with any heavy crude, the refining process requires natural gas to increase the output of highly desirable products such as gasoline and jet fuel. Similar to the problem with oil sands, as long as fossil fuels are used in both the heating and refining processes for oil shale, production will release far more carbon dioxide than is released with conventional oil. Finally, there is the possibility that some of the hydrocarbons generated during the in-situ cooking process will migrate from the site and end up contaminating nearby groundwater supplies.

Despite all the difficulties, the rise in crude oil prices after 2000 led to a renewed interest in oil shale. In 2004 Shell Corporation initiated a pilot program in Colorado based on an improved in-situ method. As illustrated in Figure 14.5, this design uses subsurface electric heaters that allow the oil shale to be cooked at a lower temperature over a longer period of time—on the order of years rather than days or months. The design also involves pumping refrigerants into the subsurface via a series of small wells, creating an ice barrier surrounding the heated section of oil shale. Once the impermeable barrier of ice is in place and the electric heaters are turned on, wells are used to remove the groundwater from within the heated section in order to minimize the contamination of nearby water supplies. Another advantage of Shell's approach is that cooking oil shale at a lower temperature generates lighter crudes, thereby eliminating the need for natural gas in the refining process.

In 2005 Shell Oil reported that their small-scale Colorado tests were successful, and announced plans to continue research and development efforts. However, due to the economics of production and the continued decline in crude oil prices, the company abandoned the project in 2013. Although U.S. oil shale efforts have come to a halt, the technology has been shown to be successful and could be used in the future if crude oil prices rise to a level where production becomes profitable.

FIGURE 14.5   Shell Oil's experimental oil shale project in Colorado used electric heaters to slowly cook the oil shale for several years, producing lighter crude that requires less fossil fuel for refining. Groundwater contamination is minimized by using refrigerants to create an impermeable ice barrier around the heated section of oil shale.
© Photographic Services, Shell International Ltd.

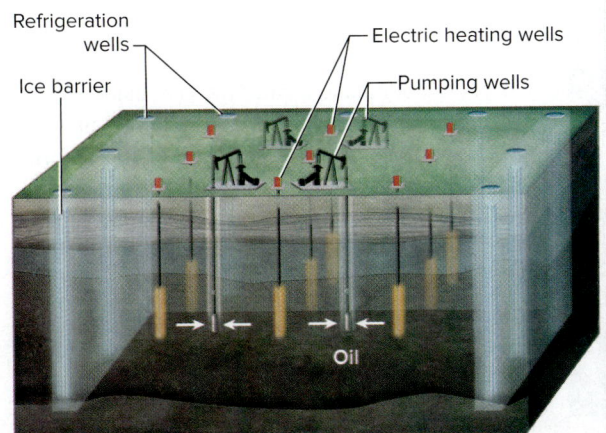

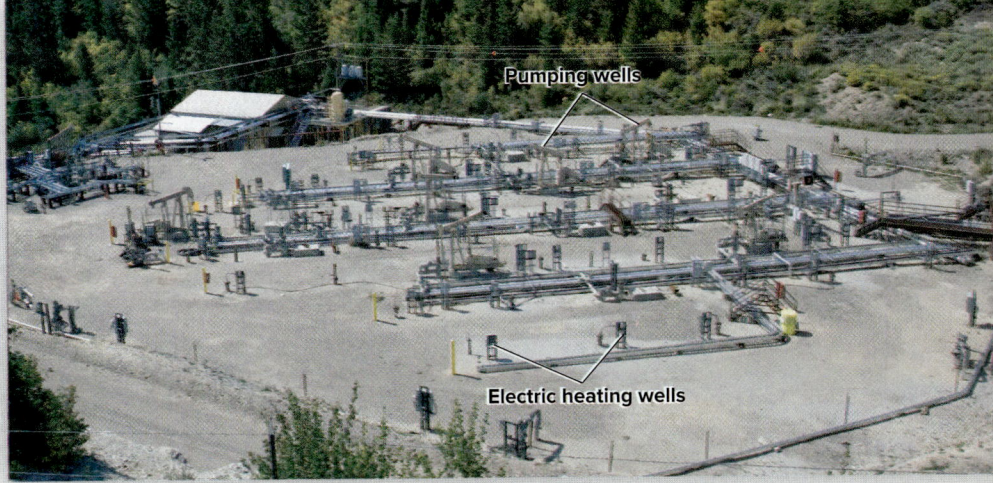

Gas Hydrates

During surveys of the seafloor in the 1970s, researchers discovered deposits of **gas hydrates,** also called *methane hydrates,* icelike substances composed of methane (CH_4) and water molecules bound together. More recent surveys have found thick accumulations in shallow polar waters and in lower latitudes on deeper parts of continental shelves. Deposits are also found buried on land in the permafrost throughout the Arctic. Scientists believe that gas hydrates originate in nutrient-rich waters when the remains of marine organisms fall to the seafloor and accumulate under oxygen-poor conditions. The organic matter slowly breaks down and forms methane gas, which then bonds with water to form gas hydrates. However, this process occurs only under certain temperature and pressure conditions. Should gas hydrates experience a change in pressure or temperature, they can become unstable and begin releasing methane gas (Figure 14.6). In Chapter 16 you will learn that this possibility is a concern since methane is a major greenhouse gas. Because methane (CH_4) is about 20 times more effective than carbon dioxide (CO_2) at trapping heat within the atmosphere, scientists are concerned that as Earth continues to get warmer, gas hydrates will begin releasing large volumes of methane, thereby accelerating global warming.

In addition to being a greenhouse gas, methane makes up the bulk of what we call *natural gas;* hence, gas hydrate deposits represent a potential energy resource. Based on worldwide surveys, scientists estimate that gas hydrates contain nearly twice as much carbon as does the world's current supply of fossil fuels. Of considerable interest is the fact that many of the deposits may be economical to extract due to their thickness and purity. One such deposit, located off the southeastern coast of the United States, is estimated to contain enough natural gas to meet current U.S. demand for nearly 30 years. Because gas hydrates represent a potentially large source of energy, researchers are trying to develop ways of extracting the methane from gas hydrate deposits. Physically removing the material through mining is not practical, so the best approach appears to be one that creates either a pressure decline within a deposit or a temperature increase so that the hydrates begin releasing methane gas. The methane could then be brought to the surface using wells similar to conventional gas wells. In 2008 an international group of scientists completed a series of field and production experiments at a site in Canada, demonstrating that large volumes of methane can indeed be produced by this technique. More recently, in 2013 researchers drilled a test well off the coast of Japan and, using depressurization techniques, succeeded in producing methane gas in a controlled manner.

Although gas hydrates hold much promise, major production is still years in the future due to technical issues and physical dangers. For example, destabilizing the gas hydrates in order to liberate methane creates the risk that methane could be released in an uncontrolled manner. Such a release would pose an explosion and fire hazard to both the workers and the production facility. For offshore deposits, destabilizing the gas hydrates could trigger movements within the seafloor, which could threaten production platforms whose supports rest on the seafloor. Economic deposits also tend to be located in either deep marine environments or harsh arctic climates. Their locations will make it more

FIGURE 14.6   Gas hydrates are icelike substances composed of methane gas and water molecules found in marine sediment and in permafrost on land. When this icelike substance is exposed to either higher temperature or lower pressure, the material breaks down, releasing methane gas. These deposits represent a potential major source of energy.
(both): © MARUM, Bremen University, Germany

difficult and expensive to produce the methane and transport it to distant markets. Assuming that the technical and safety issues will eventually be overcome, the higher costs mean that commercial production will probably not occur until higher energy prices make it economical.

Carbon-Free and Renewable Fuels

With the exception of biofuels, the alternative energy sources in this section do not involve the combustion of organic matter, and thus do not release carbon dioxide. Because of the serious consequences of global warming, these energy sources will be critical in helping us move from a fossil-fuel-based society to one that is largely carbon-free. Many of the energy sources in this section are also considered to be renewable, and therefore their use will help build a more sustainable society. Note that hydrogen fuel cells are not included here because free hydrogen (H_2) is not naturally available in meaningful quantities, which means it is not a primary energy resource. Similarly, as discussed in Chapter 13, electricity is not a primary resource because it has to be made from another energy source. Later you will learn that the long-term solution to our energy needs and global warming will be to transform non-carbon-based energy resources into electricity and hydrogen.

Biofuels

Prior to the widespread use of coal and petroleum, humans had to rely on dried organic matter, such as wood, peat (organic soil), and animal manure (dung), as a fuel for heating and cooking. The term **biofuels** refers to combustible materials derived from modern (nonfossilized) organic matter. Liquid biofuels (e.g., ethanol) had originally been offered as a solution to the oil crisis in the 1980s since they can be used to run transportation systems. Although biofuels have also been promoted as a means of combating global warming, it is strongly debated whether they can be considered *carbon-neutral,* the condition that is achieved when the carbon dioxide that is emitted equals what was removed from the atmosphere by the original plant material. In general, there are currently two types of liquid biofuels: *ethanol,* which is an alcohol made by fermenting plant material (e.g., corn and sugarcane), and *biodiesel,* which consists of fatty acids derived from vegetable oil (e.g., soybeans) and animal fats. We will focus on ethanol because it can be readily used in gasoline engines, and therefore is in greater demand.

Major biofuel production began in the United States during the energy crises of the late 1970s and early 1980s when a portion of the nation's surplus corn crop was turned into ethanol (Figure 14.7). The ethanol was blended with gasoline to create what was known as *gasohol* (10% ethanol and 90% gasoline). Although this effort required government subsidies to be profitable, it was considered reasonable during the oil crisis to use surplus corn to help extend the limited supplies of gasoline. The crisis eventually ended in 1986, but ethanol production and government subsidies continued. Then came the passage of the U.S. Clean Air Act of 1990, which required pollution controls on vehicles and set new air-quality standards. To help meet the new standards, large metropolitan areas began requiring the sale of *oxygenated gasoline* since it produces less carbon monoxide (CO) and ozone (O_3). Because ethanol could be used to make oxygenated gasoline, the sudden demand for cleaner-burning gasoline created a new and important market for ethanol.

As concerns grew in the late 1990s over global warming and rising crude oil prices, industry groups promoted ethanol as a renewable and carbon-neutral alternative to gasoline. The ethanol industry also began pushing to have auto manufacturers make so-called *flex-fuel* vehicles that can run on either

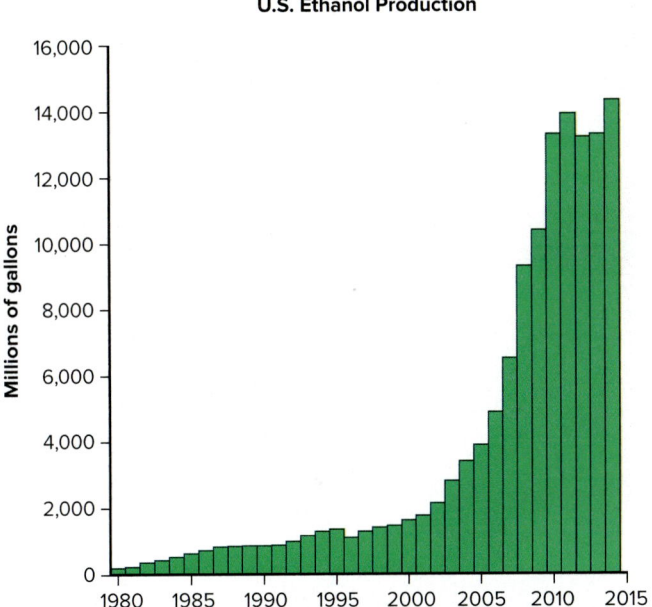

U.S. Ethanol Production

FIGURE 14.7   Production of corn-based ethanol as a transportation fuel began in the United States in 1980 in response to the energy crisis. Demand increased again after 1990, when ethanol started being used as an additive for making clean-burning oxygenated gasoline. After 2000 production further increased in response to higher oil prices.

standard gasoline blends or fuels composed of 85% ethanol, popularly known as *E85*. Proponents often point to the successful flex-fuel program in Brazil, where ethanol supplies nearly half of the country's transportation needs. However, Brazil's ethanol is made from leftover waste material generated when sugarcane plants are processed into raw sugar, whereas the only parts of the corn plant that can be used are the actual ears of corn. Brazil's program therefore is naturally much more cost-effective because it produces two valuable commodities from a single crop. Since sugarcane does not grow well in North America, researchers in the United States are investigating different plants and waste products, with the aim of finding a more efficient means of producing ethanol.

In terms of being carbon-neutral, the problem with corn-based ethanol is the amount of energy needed to produce the fuel itself. Industry groups have claimed that corn ethanol produces 25% more energy than what is used in growing the corn and distilling it into ethanol. When independent scientists include the energy inputs for fertilizers, pesticides, irrigation pumps, and farm equipment, they calculate that corn-based ethanol consumes 30% more energy than it can provide. Critics also claim that large agribusinesses found producing such ethanol to be profitable only because of government subsidies and tax incentives. In addition to not being carbon-neutral, ethanol as a transportation fuel provides 28% less energy compared to normal gasoline. This means that although you may pay the same price for ethanol-blended gasoline at the pump, your gas mileage will be less.

Despite the net energy loss represented by corn-based ethanol, the industry received a boost in 2005 when Congress created the EPA's renewable fuel standard (RFS) program as a means of reducing greenhouse gas emissions and expanding the nation's renewable fuels market. As part of this program, the EPA mandates the annual amount of ethanol that is to be blended into the U.S. gasoline supply. This guaranteed market for ethanol helped cushion the impact in 2011 when Congress ended the ethanol subsidies that had been in place since the 1980s. However, in recent years, greater U.S. fuel efficiency and lower economic growth caused Americans to use less gasoline. Since regular gasoline cannot contain more than 10% ethanol due to its corrosive nature, the reduction in gasoline consumption resulted in ethanol sales falling below the EPA mandates. The ethanol industry then found itself with excess production capacity and less favorable economics. Consequently, industry groups have been pressuring Congress to help expand the market for 15% and 85% ethanol blends.

Finally, there is the issue of how biofuel production affects food supplies and global warming. For example, nearly 40% of the U.S. corn crop in 2014 went toward producing ethanol, which represented less than 10% of the nation's consumption of gasoline. Globally, biofuel production increased four-fold between 2004 and 2014, but it still accounts for less than 5% of transportation fuels. The problem is that to supply even this limited volume of biofuels, a significant amount of farmland is being used that could have gone toward producing food. This sets the stage for a potential food crisis. The danger first became clear in 2008 when a rapid rise in crude oil prices greatly increased the demand for biofuels, which then contributed to a steep rise in world food prices. The higher prices led to food riots in many countries, including Mexico where corn is the main food staple. In addition to its effect on food supply, biofuel production may be contributing to global warming because it is encouraging the conversion of natural forests and grasslands into farmland. A recent scientific study has shown that natural landscapes and soils absorb significantly more carbon dioxide than agricultural lands. Thus, in our quest to produce biofuels that are supposedly carbon-neutral, the creation of new farmland is actually adding carbon dioxide to the atmosphere and exacerbating the problem of global warming. Although biofuels are a renewable form of energy, they do not appear to be a viable solution to our energy needs or global warming.

Hydroelectric Power

In Chapter 13 you learned that humans have a long history of using mechanical energy from falling water, called *hydropower* to perform a variety of work. Today we use falling water primarily to spin a turbine to make electricity, referred to as **hydroelectric power** or **hydroelectricity.** The advantages of hydroelectricity are that it is renewable and does not produce carbon dioxide. Most hydroelectric plants involve damming a river and creating an elevated pool of water (reservoir). Managers can then manipulate the release of water from the reservoir on a daily basis to meet the cyclic demand for electricity. However, by eliminating the natural variations in stream flow, dams not only block the migration of fish, but have a negative impact on entire ecosystems (see Chapters 8 and 11 for details). It is largely because of such environmental impacts that the construction of new dams is commonly met with public resistance. Moreover, hydroelectric dams cannot be built just anywhere, since the reservoir needs to be relatively deep, requiring a somewhat narrow river valley with steep sides (Figure 14.8). Height is important because it gives the falling water more power to move the turbines, thereby making production of hydroelectricity more cost-effective.

Hydroelectricity currently accounts for just 7% of the world's total energy consumption but represents nearly 17% of all electrical production. Even though there remain enough suitable dam sites to double the current level of hydroelectric production, it is unlikely anything near this level will be achieved given the environmental and political obstacles. One interesting solution for generating additional hydroelectric power is to use pumped-storage systems. As shown in Figure 14.9, rather than damming a river, a pumped-storage system uses an elevated reservoir built on a topographic high point. These systems take advantage of the fact that electrical demand peaks during the daytime and reaches a minimum at night. Because generators typically run continuously, this means there is a surplus capacity at night (more electricity is being produced than consumed). Engineers have found it cost-effective to pump water up into a storage reservoir at night using the surplus electricity, then release the water the next day to generate electricity during peak demand. This type of system could also use excess electricity from intermittent sources (solar, wind, and tidal) to pump water in a reservoir. The energy could then be stored and later used to produce electricity during times of peak demand, thereby eliminating a major obstacle to using intermittent energy sources.

FIGURE 14.8   Hydroelectric power plants are most efficient in steeper terrain, which allows for a deeper reservoir. As height increases, so too does the power of the falling water as it moves through the turbines, making production of hydroelectricity more cost-effective. Shown here is Hoover Dam located on the Colorado River in Nevada.
© ThinkStock/Superstock RF

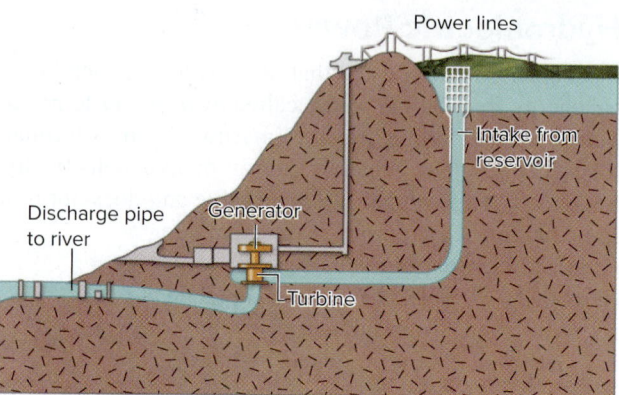

FIGURE 14.9   The Taum Sauk reservoir in Missouri is a pumped-storage system that generates electricity without creating the environmental problems associated with damming rivers. Water is pumped up into the reservoir during the nighttime when demand is low and the electrical grid has surplus capacity. The stored water is then released in the daytime to create electricity to meet peak demand.
© Jim Wark/Airphoto

Nuclear Power

Recall that the nucleus of an atom contains protons and neutrons, and the energy that binds these particles together is called *nuclear energy.* Also, the various combinations of protons and neutrons in atoms of a given element are known as its *isotopes* (the number of protons is fixed, but neutrons can vary). Prior to World War II, physicists determined that the nuclei of certain isotopes were *fissile,* meaning they were susceptible to being split apart. The term **nuclear fission** refers to the process of splitting the nucleus of an atom, resulting in the release of neutrons and the conversion of mass into energy. Research into developing the first atomic bombs focused on making use of the heat energy that would be released from splitting fissile isotopes of the elements uranium and plutonium. One of the technical problems was that uranium's fissile isotope, uranium-235 or U-235 (92 protons and 143 neutrons), makes up only 0.72% of the uranium found in the natural environment—99.28% is nonfissile uranium-238 (U-238). This meant that U-235 atoms had to be separated from the more abundant U-238 atoms. Another option was to produce fissile plutonium, which does not occur naturally, by bombarding U-238 atoms with neutrons in a nuclear reactor. In 1945, American scientists succeeded in initiating the first uncontrolled nuclear chain reaction by splitting just a few pounds of uranium-235 (the first plutonium bomb was developed later that same year). Shortly afterward scientists developed much more powerful atomic bombs, called hydrogen bombs, in which hydrogen nuclei are forced to combine or fuse, a process called **nuclear fusion.**

Fission Reactors

In the years following World War II, physicists designed nuclear reactors for peaceful applications. Here the focus was on harnessing the heat from fission reactions to generate electrical power with steam-driven turbines and generators. In 1954 the first commercial nuclear power plant began generating electricity in the former Soviet Union (now Russia). Today most of the world's nuclear power plants use uranium dioxide pellets whose U-235 concentration has been enriched to somewhere between 3% and 8%. The uranium fuel is then bombarded with neutrons in what is called a *reactor core.* As illustrated in Figure 14.10, the neutrons split U-235 atoms, releasing heat and additional neutrons, which go on to split other U-235 atoms in what is called a *sustained nuclear reaction* (engineers use water or graphite to

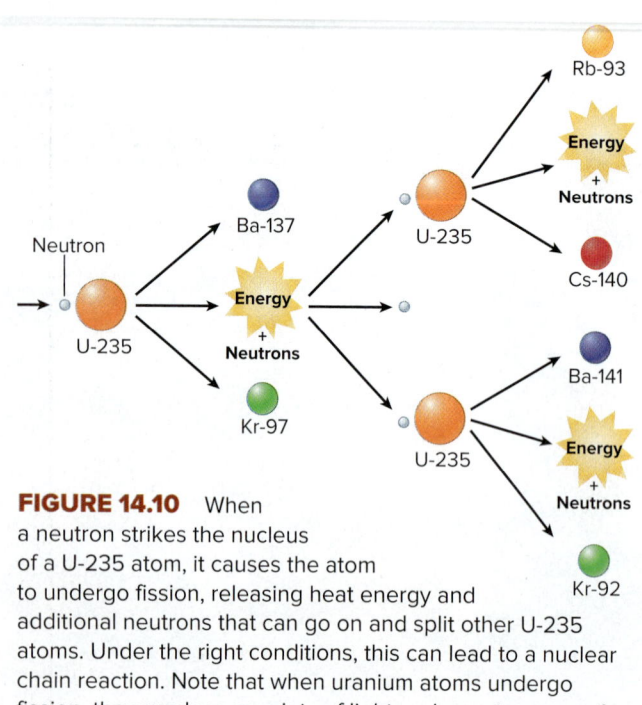

FIGURE 14.10   When a neutron strikes the nucleus of a U-235 atom, it causes the atom to undergo fission, releasing heat energy and additional neutrons that can go on and split other U-235 atoms. Under the right conditions, this can lead to a nuclear chain reaction. Note that when uranium atoms undergo fission, they produce a variety of lighter elements, some of which are radioactive.

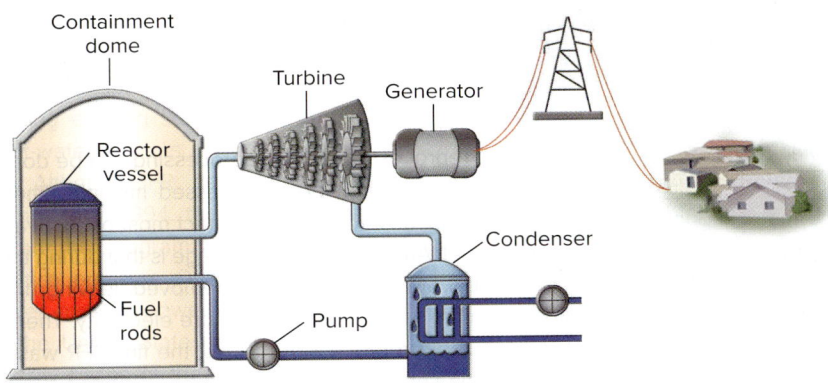

A Boiling water reactor

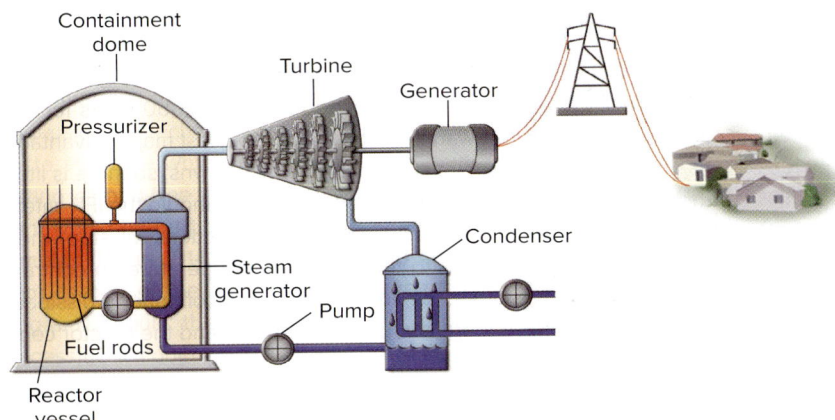

B Pressurized water reactor

FIGURE 14.11 Modern nuclear reactors use water for cooling the reactor core and driving a steam turbine and electrical generator. A boiling water reactor (A) uses the same water for cooling and generating electricity, whereas a pressurized water reactor (B) uses separate loops. Water around the reactor core in both designs makes it possible for the nuclear reactions to be self-sustaining. If a loss of cooling water occurs, the nuclear reactions will stop, but the buildup of heat can cause the fuel rods to fuse together and melt.

slow the velocity of the neutrons so that the reaction is self-sustaining). To keep the reactor core from overheating, circulating water is used to transfer heat away from the core. This hot water is then used to produce steam for driving the electrical generator. Figure 14.11 illustrates the two basic water-cooling designs used in modern reactors. Notice in both designs how the reactor core is housed in a heavily reinforced structure called a *containment building,* whose purpose is to keep radioactive materials from escaping in the event of an accident.

Nuclear power plants have the benefit of being able to generate electricity on a large scale, and at the same time not emit any carbon dioxide, sulfur dioxide, or particulate matter. Nuclear power plants thus do not contribute to global warming or acid rain as do coal-fired plants (Chapters 15 and 16). However, conventional fission reactors generate waste that is highly radioactive, which presents issues of operational safety and long-term waste disposal (see Case Study 14.1). Despite concerns over safety and waste disposal, the construction of new plants boomed during the 1960s and 1970s. Many industrialized nations subsidized the development of nuclear power in an attempt to move away from fossil fuels as the primary means of generating electricity. Then in 1979 a cooling valve malfunctioned at the Three Mile Island nuclear plant in the United States (Figure 14.12), causing one of the plant's reactor cores to overheat

FIGURE 14.12 Aerial view of the Three Mile Island nuclear power plant on the Susquehanna River in Pennsylvania.
© Doug Sherman/Geofile

A

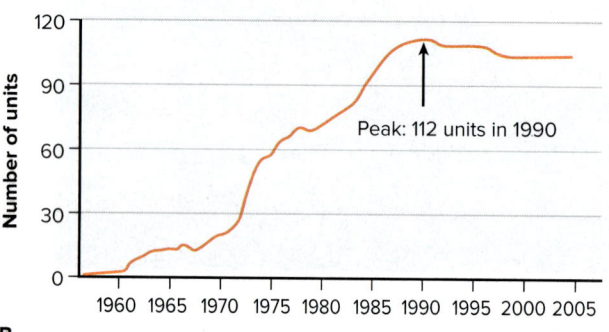

B

FIGURE 14.13   Map (A) showing the location of the 99 nuclear reactors currently operating in the United States. Graph (B) showing how the construction of new nuclear plants came to a halt around 1990 due to high costs and concerns over safety and waste disposal. In 2007, after more than 30 years, utility companies started applying for permits to build new reactors, four of which are currently under construction.

The accidents at Three Mile Island and Chernobyl contributed to a virtual halt in new construction permits for nuclear power plants in the United States. In addition to concerns over reactor safety, there was the issue of there being no long-term strategy for disposing of radioactive waste. Waste disposal is a serious problem because some radioactive isotopes remain dangerous for tens of thousands of years (Chapter 15). To make matters worse, at the time of the accidents utility companies were discovering that operating large nuclear power plants was more expensive than operating plants fired by coal or natural gas. Ultimately, U.S. utility companies canceled plans for constructing additional nuclear power plants. The number of nuclear plants operating in the United States reached a peak of 112 in 1990 (Figure 14.13), falling to 99 by 2015. Some nations, however, continued to expand their nuclear power programs in an effort to become more energy-independent. A striking example is France, which now generates 78% of its electrical needs in nuclear plants, compared to just 20% for the United States.

Concerns over nuclear safety and waste disposal heightened after the 2011 disaster at Japan's Fukushima nuclear power plant. Japan, like the United States, has been storing spent nuclear fuel rods in above-ground pools at its plants for decades. When the tsunami destroyed the emergency power supply at the Fukushima plant, the loss of cooling water caused three of the plant's six nuclear reactors to begin melting down. This resulted in a buildup of hydrogen gas, which then exploded, rupturing the containment buildings and allowing significant amounts of radiation to escape directly into the atmosphere and into the ocean by contaminated groundwater. The earthquake and hydrogen explosions also damaged the buildings that housed the storage pools. There was fear that another earthquake could lead to a complete loss of water from the pools, causing some of the fuel rods to overheat such that additional radiation would escape into the environment. This nuclear disaster will likely bring about improved safety systems and help lead to the permanent storage of high-level nuclear waste.

Despite the safety concerns and higher costs, utility companies in the United States are again constructing new nuclear reactors. Four are currently under construction and applications for 11 more are under review. This renewed interest is partly due to the surge in energy prices after 2000 and the growing likelihood that some type of carbon tax will be imposed on fossil fuels to help combat global warming. Some companies, however, are building natural gas power plants instead because of the recent drop in natural gas prices. A large nuclear power plant can take up to 10 years to build and cost $10 billion, whereas a much smaller natural gas plant with less generating capacity can be built in several years and cost less than $1 billion.

Finally, it will be difficult to address the problem of global warming without making use of nuclear power. One of the key advantages of nuclear power plants is that, aside from the carbon emissions associated with building the plant, generating electricity does not emit any greenhouse gases. Although they are costly to build, nuclear power plants have the additional advantage of being reliable and providing large quantities of electricity to the grid. As we will see in the following sections, wind and solar will be important in our transition to a low-carbon society, but are difficult to incorporate into the existing electrical grid. Thus scaling up the production of non-carbon-based electricity enough to replace conventional coal and natural gas power plants would be difficult using only solar and wind. Moreover, the distributed and intermittent nature of wind and solar inputs to the electrical grid present both technical and physical challenges.

Fusion Reactors

Perhaps the most tantalizing energy source of all is *nuclear fusion* because its basic fuel supply is water and it does not produce long-lived radioactive waste products. Physicists at both private and large government-funded labs are currently testing a variety of experimental reactors of various sizes, most of which

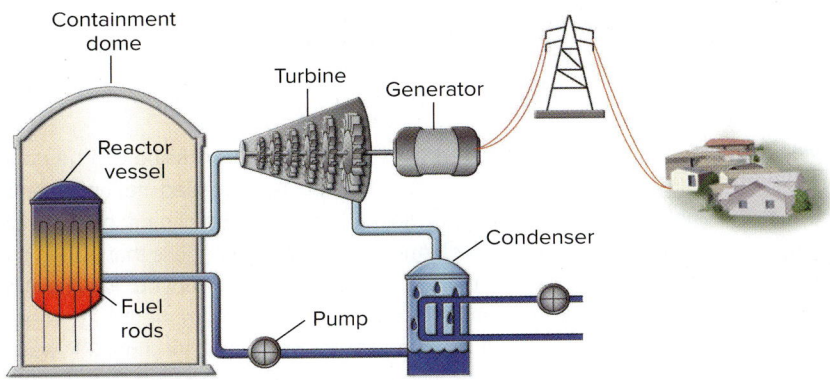

A   Boiling water reactor

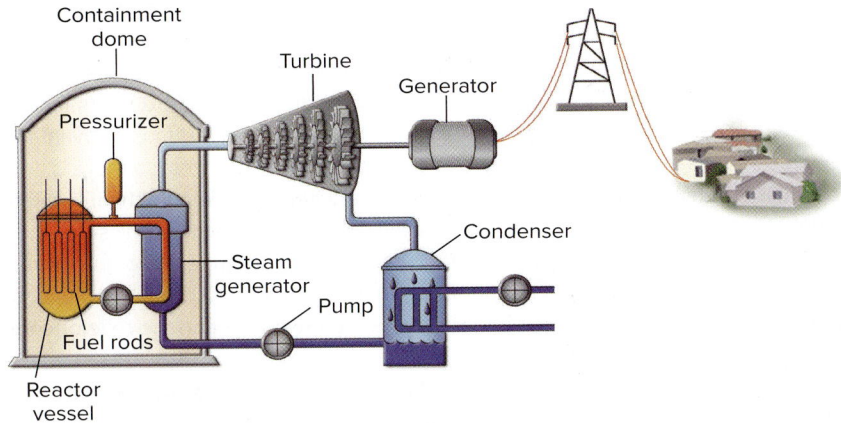

B   Pressurized water reactor

FIGURE 14.11   Modern nuclear reactors use water for cooling the reactor core and driving a steam turbine and electrical generator. A boiling water reactor (A) uses the same water for cooling and generating electricity, whereas a pressurized water reactor (B) uses separate loops. Water around the reactor core in both designs makes it possible for the nuclear reactions to be self-sustaining. If a loss of cooling water occurs, the nuclear reactions will stop, but the buildup of heat can cause the fuel rods to fuse together and melt.

slow the velocity of the neutrons so that the reaction is self-sustaining). To keep the reactor core from overheating, circulating water is used to transfer heat away from the core. This hot water is then used to produce steam for driving the electrical generator. Figure 14.11 illustrates the two basic water-cooling designs used in modern reactors. Notice in both designs how the reactor core is housed in a heavily reinforced structure called a *containment building,* whose purpose is to keep radioactive materials from escaping in the event of an accident.

Nuclear power plants have the benefit of being able to generate electricity on a large scale, and at the same time not emit any carbon dioxide, sulfur dioxide, or particulate matter. Nuclear power plants thus do not contribute to global warming or acid rain as do coal-fired plants (Chapters 15 and 16). However, conventional fission reactors generate waste that is highly radioactive, which presents issues of operational safety and long-term waste disposal (see Case Study 14.1). Despite concerns over safety and waste disposal, the construction of new plants boomed during the 1960s and 1970s. Many industrialized nations subsidized the development of nuclear power in an attempt to move away from fossil fuels as the primary means of generating electricity. Then in 1979 a cooling valve malfunctioned at the Three Mile Island nuclear plant in the United States (Figure 14.12), causing one of the plant's reactor cores to overheat

FIGURE 14.12   Aerial view of the Three Mile Island nuclear power plant on the Susquehanna River in Pennsylvania.
© Doug Sherman/Geofile

Older Technology for the Next Generation of Nuclear Reactors?

After World War II, nuclear fission held the promise of providing society with virtually unlimited amounts of cheap electrical power. Ironically, however, fission was also fueling a nuclear arms race that threatened to destroy all of humanity. During this period, nuclear physicists were experimenting with different types of reactor designs for both civilian and military use. Because of concerns over the scarcity of uranium, they developed what are called *fast breeder reactors,* in which the more abundant nonfissile uranium-238 atoms within the fuel absorb fast neutrons and are transformed into fissile plutonium-239 atoms. Breeder reactors therefore keep creating new fuel at the same time they are generating power. In addition, most breeder reactors are cooled using a liquid metal, such as liquid sodium.

As a result of the exploration boom of the 1950s, however, uranium became fairly abundant by the 1960s, which reduced the need for breeder reactors. Also at this time the U.S. Navy had started using water cooled designs, called *light-water reactors,* to propel its nuclear-powered ships. Therefore, when the U.S. government wanted to quickly ramp up the civilian nuclear industry, efforts focused on light-water designs rather than breeder reactors. As a consequence, the majority of reactors operating in the world today are based on water-cooled designs.

In 1984, as part of an effort to produce safer and more efficient reactors, researchers at the U.S. Argonne National Laboratory began developing a new design based on a fast breeder reactor that had been running successfully since 1964. This new design, called an *integral fast reactor (IFR),* uses a metallic fuel and liquid sodium for cooling the core. One advantage of the IFR design is that the metallic

fuel is much easier to reprocess, plus the reprocessing can be done directly onsite. Since the fuel can be reprocessed multiple times, unlike a light-water reactor, an IFR is able to extract most of the available energy from the fuel. Another major advantage is that during the reprocessing of IFR fuel, waste products are removed, leaving uranium and plutonium in the fuel for producing more electricity. Therefore, at the end of multiple reprocessing cycles, the final IFR waste product contains mostly short-lived isotopes. In contrast, the spent fuel from a light-water reactor contains long-lived isotopes, while at the same time still containing 95% of the energy available in the original fuel. An IFR therefore not only extracts most of the available energy from the fuel, but its waste only needs to be stored and managed for 300–500 years as opposed to over 100,000 years for the spent fuel from a light-water reactor. Finally, the last major advantage is that IFR systems are not susceptible to meltdowns, so there is little risk of radiation escaping into the environment. Should the IFR safety systems suffer a complete failure, the fission reactions will stop when the fuel rods expand due to the increased temperature. In contrast to water-cooled reactors, the liquid sodium will continue to dissipate heat from an IFR, so there can be no meltdown and the reactor components will be undamaged.

In addition to eliminating the risk of a meltdown, being more energy efficient, and producing only short-lived waste, IFR systems can also take the spent fuel from light-water reactors and use it as fuel. Moreover, it is virtually impossible to make nuclear weapons from IFR fuel due to the nature of the material. This means that increased use of IFRs could help address three important goals: generating

and begin fusing together in what is known as a *meltdown.* Although the reactor was permanently destroyed, a complete meltdown was avoided, and only a small amount of radioactive steam managed to escape the containment building. Public attention, however, focused on the prospect that had a complete meltdown occurred, the reactor core would have burned through the floor of the containment building. This would have permanently contaminated the groundwater system, and even worse, created steam explosions, sending large amounts of radioactive materials into the atmosphere.

Then in 1985 a runaway nuclear reaction and fire at the Soviet Union's nuclear plant at Chernobyl sent massive amounts of radioactive gas and dust into the atmosphere. Radioactive particles eventually fell to the ground over many parts of Eastern and Western Europe, with the worst contamination occurring relatively close to the plant in what are now Ukraine, Belarus, and Russia. A total of 50 workers and rescue personnel died after receiving lethal doses of radiation, and 9 children died of thyroid cancer related to the radioactive fallout. In a 2005 United Nations report on the Chernobyl disaster, health experts estimated that 4,000 people will eventually die from radiation-induced cancers. It was also reported that with the exception of the highly contaminated 30-kilometer carbon-free electricity, eliminating the need for long-term storage of nuclear waste, and minimizing the spread of nuclear weapons by consuming existing stockpiles of spent fuel from light-water reactors.

Although IFR development was almost complete in 1994, the U.S. government eliminated the program due to budget cuts and political pressure. One of the reasons was that fewer nuclear power plants were being built in the United States as a result of the lower energy prices of the 1980s and 1990s. Today, IFR technology is gaining attention as a means of reducing the world's stockpile of high-level nuclear waste, which has accumulated over the past 50 years. The British government is evaluating a proposal to build a nuclear power plant using IFR technology to reduce its nuclear waste stockpile while, at the same time, generating carbon-free electricity. The new power plant would be built at Great Britain's Sellafield complex (Figure B14.1), which produced plutonium as part of the nation's nuclear weapons program. Should the Sellafield project prove successful, it could pave the way for IFRs to be used as a means of generating electricity from the world's stockpile of high-level nuclear waste. This is important because to make the transition to a society based on renewable and non-carbon-based energy sources, we will most likely need nuclear power to provide reliable base-load power to our electrical grid.

FIGURE B14.1 View of the sprawling Sellafield nuclear complex in Great Britain where a new IFR electrical power plant is being proposed. Such a plant would help solve the problem of eliminating long-lived nuclear waste while generating carbon-free electricity.
© Steve Allen/Getty Images

(19-mile) exclusion zone around the plant, radiation levels have returned to "acceptable levels" over most of the fallout area.

The Chernobyl accident, which occurred just seven years after Three Mile Island, heightened concern in the United States over the safety of nuclear power plants. The design of nuclear reactors built in Western nations though is fundamentally different from that used at Chernobyl. The Soviet reactor's nuclear fuel was surrounded with graphite, which lowers the velocity of neutrons so that a sustained fission reaction can be achieved (Figure 14.10). Moreover, the Soviet design did not include a containment building. The accident at Chernobyl began when there was a loss of cooling water, but the graphite allowed the fission reactions to continue. The result was a buildup of heat. Eventually the reactor core exploded and the graphite caught fire, sending huge amounts of radioactive material into the atmosphere. In contrast, Western reactors achieve sustained fission by using water to slow the neutrons. This water also happens to be the same water used to cool the reactor. This means that if there is a loss of cooling water, the fission reactions will stop. The reactor may still overheat and begin to melt down, but there is no chance of a runaway reaction and massive graphite-fed fire as at Chernobyl.

A

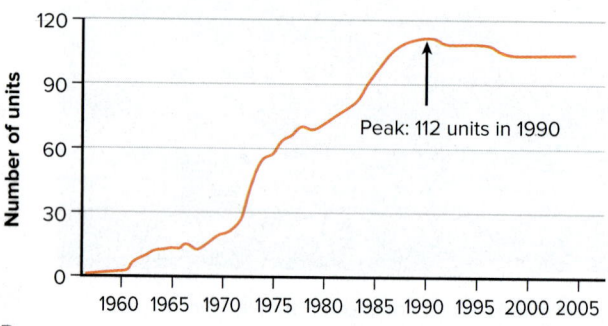

B

FIGURE 14.13   Map (A) showing the location of the 99 nuclear reactors currently operating in the United States. Graph (B) showing how the construction of new nuclear plants came to a halt around 1990 due to high costs and concerns over safety and waste disposal. In 2007, after more than 30 years, utility companies started applying for permits to build new reactors, four of which are currently under construction.

The accidents at Three Mile Island and Chernobyl contributed to a virtual halt in new construction permits for nuclear power plants in the United States. In addition to concerns over reactor safety, there was the issue of there being no long-term strategy for disposing of radioactive waste. Waste disposal is a serious problem because some radioactive isotopes remain dangerous for tens of thousands of years (Chapter 15). To make matters worse, at the time of the accidents utility companies were discovering that operating large nuclear power plants was more expensive than operating plants fired by coal or natural gas. Ultimately, U.S. utility companies canceled plans for constructing additional nuclear power plants. The number of nuclear plants operating in the United States reached a peak of 112 in 1990 (Figure 14.13), falling to 99 by 2015. Some nations, however, continued to expand their nuclear power programs in an effort to become more energy-independent. A striking example is France, which now generates 78% of its electrical needs in nuclear plants, compared to just 20% for the United States.

Concerns over nuclear safety and waste disposal heightened after the 2011 disaster at Japan's Fukushima nuclear power plant. Japan, like the United States, has been storing spent nuclear fuel rods in above-ground pools at its plants for decades. When the tsunami destroyed the emergency power supply at the Fukushima plant, the loss of cooling water caused three of the plant's six nuclear reactors to begin melting down. This resulted in a buildup of hydrogen gas, which then exploded, rupturing the containment buildings and allowing significant amounts of radiation to escape directly into the atmosphere and into the ocean by contaminated groundwater. The earthquake and hydrogen explosions also damaged the buildings that housed the storage pools. There was fear that another earthquake could lead to a complete loss of water from the pools, causing some of the fuel rods to overheat such that additional radiation would escape into the environment. This nuclear disaster will likely bring about improved safety systems and help lead to the permanent storage of high-level nuclear waste.

Despite the safety concerns and higher costs, utility companies in the United States are again constructing new nuclear reactors. Four are currently under construction and applications for 11 more are under review. This renewed interest is partly due to the surge in energy prices after 2000 and the growing likelihood that some type of carbon tax will be imposed on fossil fuels to help combat global warming. Some companies, however, are building natural gas power plants instead because of the recent drop in natural gas prices. A large nuclear power plant can take up to 10 years to build and cost $10 billion, whereas a much smaller natural gas plant with less generating capacity can be built in several years and cost less than $1 billion.

Finally, it will be difficult to address the problem of global warming without making use of nuclear power. One of the key advantages of nuclear power plants is that, aside from the carbon emissions associated with building the plant, generating electricity does not emit any greenhouse gases. Although they are costly to build, nuclear power plants have the additional advantage of being reliable and providing large quantities of electricity to the grid. As we will see in the following sections, wind and solar will be important in our transition to a low-carbon society, but are difficult to incorporate into the existing electrical grid. Thus scaling up the production of non-carbon-based electricity enough to replace conventional coal and natural gas power plants would be difficult using only solar and wind. Moreover, the distributed and intermittent nature of wind and solar inputs to the electrical grid present both technical and physical challenges.

Fusion Reactors

Perhaps the most tantalizing energy source of all is *nuclear fusion* because its basic fuel supply is water and it does not produce long-lived radioactive waste products. Physicists at both private and large government-funded labs are currently testing a variety of experimental reactors of various sizes, most of which produce energy by fusing two different hydrogen isotopes. One isotope, called *deuterium,* can be obtained from seawater, and the other, called *tritium,* can be produced from the element lithium. In order to create a self-sustaining fusion reaction, deuterium and tritium atoms must be brought up to a temperature of at least 100 million °C. The gas molecules are then stripped of their electrons and form what is called *plasma.* The superhot plasma must also be confined under enormous pressure for a long enough period of time (seconds) to allow fusion reactions to begin. Most of the scientific and engineering problems revolve around trying to control the incredibly hot plasma, while simultaneously keeping it confined under terrific pressure. Fusion takes place naturally in the Sun, whose immense gravity provides the confining mechanism. Researchers are developing ways of confining the plasma using either a magnetic field or a series of lasers.

Although physicists have yet to achieve a self-sustaining fusion reaction after 35 years of effort, they have made steady progress and are getting close to achieving the necessary temperature and pressure. Should they succeed, the next step would be to produce a commercially viable fusion power plant. While some researchers are now confident that commercial fusion plants are inevitable, no one knows whether or not their development will occur soon enough to help combat global warming.

Solar Power

It is often said that the Sun is our biggest source of untapped energy. In fact, more solar energy falls on the Earth in an hour than all of humanity uses in an entire year. Since all this solar energy is dispersed over both the land and oceans, the problem, of course, is how to collect it in an economical manner.

Scientists often refer to solar energy as *solar radiation* because it represents electromagnetic radiation that forms from nuclear fusion in the Sun. As described in Chapter 2, electromagnetic radiation is a continuous spectrum of wave energy that travels away from its source. Different portions of the spectrum have been given names based on the radiation's wavelength (e.g., X-ray, ultraviolet, visible, and infrared). When this traveling wave energy strikes the Earth, it is transformed into different forms of energy. For example, plants convert energy from the visible part of the spectrum into cell tissue (biomass), which itself is a form of chemical energy; fossil fuels represent the ancient transformation of sunlight into biomass. We as individuals experience another type of solar energy transformation every time we go out in the sun. As sunlight strikes our skin or clothing, solar radiation is converted into heat energy, making us feel warmer. Sunlight can also be transformed directly into electricity, as in solar-powered calculators. In this section we will focus on two important solar energy transformations: solar to thermal and solar to electrical.

Solar Heating

The process called **solar heating** takes place when solar radiation strikes a solid object and is transformed into thermal or heat energy. People can make use of this free heat energy in one of two ways. The simplest is referred to as *passive* because it does not involve any mechanical apparatus (moving parts). Perhaps the best example is the use of glass windows that allow direct sunlight to enter a building. Solar radiation from the visible part of the spectrum passes through the glass, and is then absorbed by physical objects in the building (walls, flooring, furniture, etc.). As the objects get warmer, they release heat energy, which becomes trapped within the building and raises the indoor temperature (this same process causes the interior of a car to get hot when the windows are up).

As illustrated in Figure 14.14, architects can create passive building designs that take advantage of free solar heating. The basic idea is to maximize the amount of incoming sunlight by orienting the building and banks of windows

Direct sunlight blocked by overhang and tree leaves in the summer

Deciduous trees allow direct sunlight during the winter

FIGURE 14.14 Passive solar homes are designed to let sunlight enter during the wintertime, and then use trees and roof overhangs to block sunlight during the summer months. This passive solar home in Michigan reportedly gets 85% of its heating needs from the sun.

© Michigan Energy OPtions

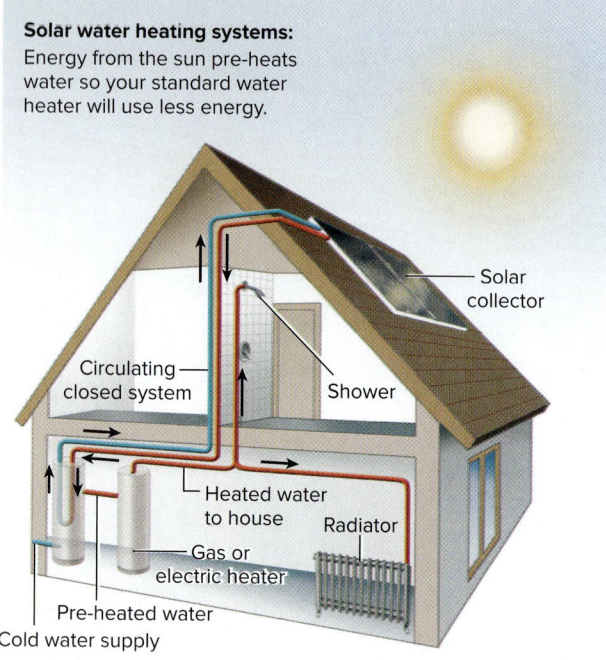

Solar water heating systems:
Energy from the sun pre-heats water so your standard water heater will use less energy.

Solar collector

Circulating closed system

Shower

Heated water to house

Radiator

Gas or electric heater

Pre-heated water

Cold water supply

FIGURE 14.15   Solar collectors mounted on a roof convert solar radiation into heat energy, which is then transferred to a circulating water system. Hot water from a solar system can be used to help reduce the energy load on a conventional hot-water heater, or it can be sent through radiators to help warm the building's airspace.
Solahart/ESTIF

FIGURE 14.16   Aerial view of the Gemasolar concentrated solar power plant located near Seville, Spain. The array of mirrors reflects sunlight onto the tower, which absorbs solar heat to run a steam-driven electrical generator. By using a molten-salt storage system, this plant can produce electricity for up to 15 hours without any sunlight, providing steady power for 25,000 homes.
© Torresol Energy

toward the Sun (south-facing windows are used in the northern hemisphere). In many climatic zones solar heating is desirable in the winter, but is avoided during the summer when the goal is to keep a building cool. One time-proven technique uses deciduous trees and roof overhangs to prevent sunlight from entering the windows in the summer months when the sun angle is high. As wintertime approaches, leaves fall from the trees and the sun angle becomes progressively lower, allowing direct sunlight to stream through the windows. In this way solar heating automatically takes place in the winter, then is blocked during the summer when it is not needed.

The other form of solar heating is called *active* since it involves some type of mechanical system to help collect and distribute the heat. Figure 14.15 shows an active solar heating system in which a solar collector (panel) is mounted on a roof and oriented to maximize the amount of direct sunlight striking it. During the daytime the collector transforms solar radiation into heat energy, which is then transferred to a circulating water system. Many homeowners choose to connect the thermal panel to their hot water supply, thereby reducing the energy demand on the home's gas or electric water heater. The savings can be significant since a conventional hot-water heater accounts for about 15% of the average homeowner's utility bill. Another option is to pipe solar-heated water through radiators, where the heat is transferred to the airspace within the building. Whether the system is active or passive, solar heating is highly desirable because once the energy savings pays for the cost of the system, the heat is essentially free.

Solar heating is also being used on a large scale to generate electricity. A *concentrated solar power plant* utilizes an array of rotating mirrors that track the Sun, focusing sunlight onto a receiver, where the concentrated solar heat is used to power a conventional steam-driven electrical generator. One design focuses sunlight on a central tower (Figure 14.16), while others focus it on a tube running down the length of curved mirrors or on a target in front of parabolic dishes. Modern systems use molten salt to collect and store heat, which is then transferred to water to produce steam. The advantage of molten

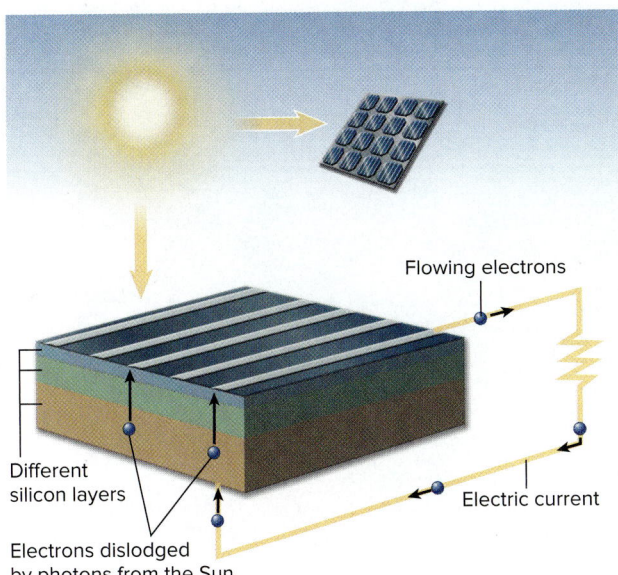

FIGURE 14.17 Photovoltaic panels consist of individual cells composed of silicon alloys that transform solar radiation directly into electricity. Particles of sunlight called photons are absorbed by the uppermost layer, dislodging electrons, which then move in a circulating manner. Electricity is obtained from this circulating flow of electrons. The photo shows both photovoltaic and thermal panels installed on a home.
© Conergy AG

salt is that its temperature can be raised to over 900°F (500°C), and it will then retain enough heat to keep producing electricity long after the Sun goes down. This important feature largely eliminates solar power's disadvantage of being intermittent. The number of concentrated solar power plants is growing rapidly as their cost has become competitive with that of coal and natural gas plants. These new plants are not only highly efficient, but also produce electricity on a large scale without many of the environmental problems associated with fossil fuel and nuclear plants. However, the evaporative loss of cooling water at solar power plants can be a problem for plants located in more arid climates. This problem can be avoided by using dry-cooled systems in areas where water is in short supply.

Photovoltaic Cells

Rather than being collected as heat to drive a power plant, sunlight can be transformed directly into electricity. The key to this transformation lies in the fact that solar radiation consists of tiny particles of energy called *photons*. As illustrated in Figure 14.17, when sunlight strikes a semiconductor, a thin slab of material composed of silicon alloys, some of the photons are absorbed, which dislodges electrons from the alloy. The electrons then start moving in a circular path between the semiconductor's surface and internal layers, creating an electrical current. This device is called a **photovoltaic cell,** sometimes called a *solar cell.* Solar panels consisting of individual photovoltaic cells are commonly placed on roofs and then wired to the electrical system of a building. Note that some home and business owners will install both thermal and photovoltaic panels in order to minimize their energy costs.

One of the problems with photovoltaic cells is that the amount of electricity they produce depends on the intensity of the sunlight hitting the panels. Although photovoltaic systems can generate surplus electricity on bright sunny days, a major drawback has been our limited ability to store electricity. People typically have stored their surplus electricity in banks of interconnected lead-acid batteries, but these batteries are bulky and have a rather limited life span. Rather than using lead batteries, many home and business owners with photovoltaic systems simply switch over to conventional utility power at night, and on cloudy days when electrical production is minimal. Regulations in the United States and other countries often require public utility companies to purchase electricity generated

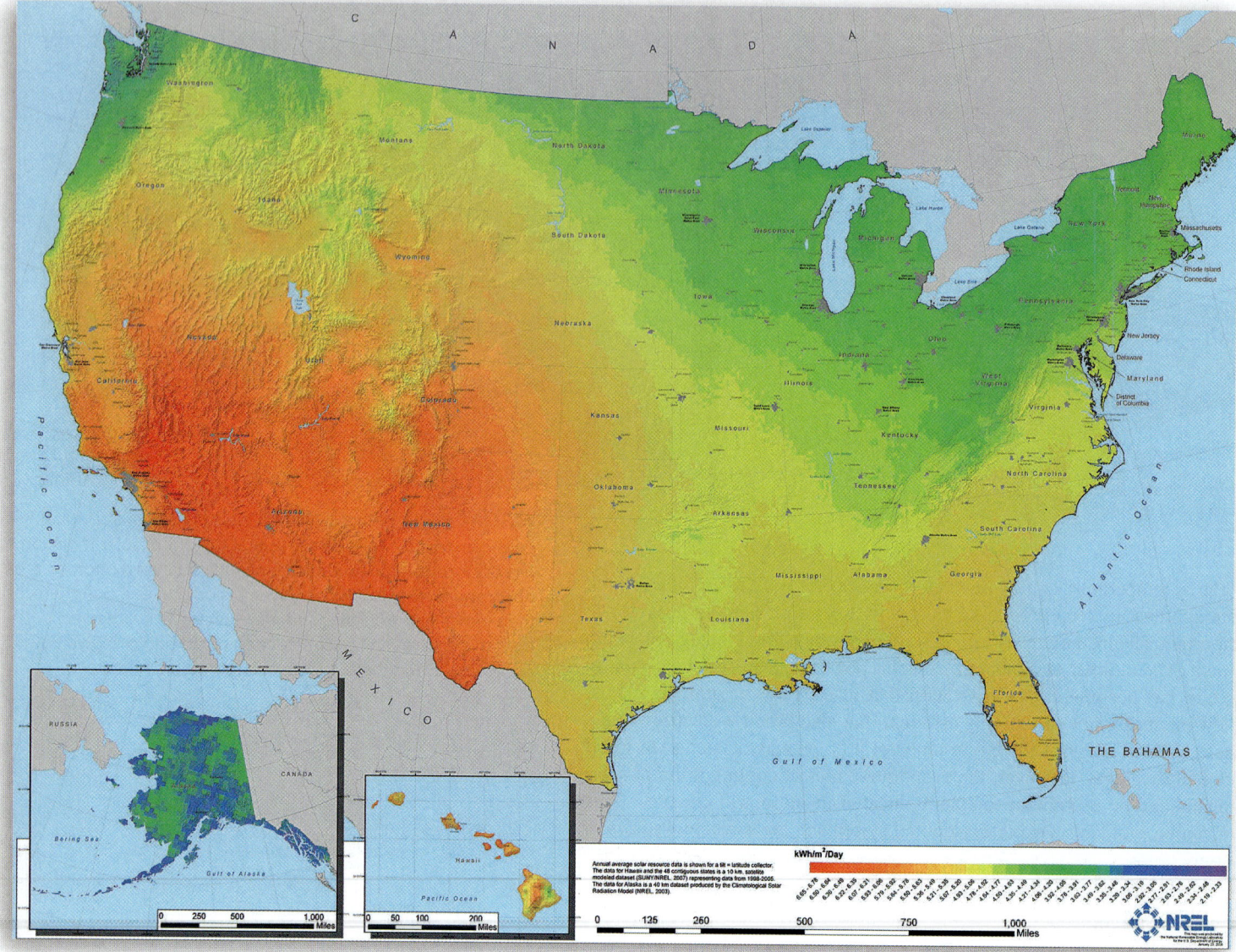

FIGURE 14.18 The primary factor determining the cost-effectiveness of solar panels is average annual sunshine, which is controlled by latitude and weather patterns that determine the amount of cloud cover. From this map of average annual solar radiation, solar power is clearly the most cost-effective in the southwestern portion of the United States.
NREL, National Renewable Energy Lab of the U.S. Dept. of Energy

from small sources in addition to major power plants. Because most homeowners are already connected to the electrical grid, this means they can sell surplus electricity from their photovoltaic system to a public utility company—homeowners are typically given credit on their electric bill. The ability to sell surplus electricity has been important because it helps makes small-scale photovoltaic systems more cost-effective.

Although the price of photovoltaic panels has fallen dramatically over the years, their cost-effectiveness remains an issue because the panels are still relatively expensive. Most people who purchase a photovoltaic system want to save enough on their electric utility bills to pay for their system over some reasonable amount of time. For example, if a system costs $20,000, but saves only $500 a year, it would take 40 years just to recover the initial investment. One of the key factors affecting the *payback period* of a system is the amount of solar radiation the panels receive on an annual basis. Obviously, if the weather patterns in an area result in a large number of cloudy days, then the payback period will be longer. Similarly, the farther north or south of the equator one lives, the lower the sun angle, which reduces the annual amount of radiation the panels can receive. The effects of climate and latitude can be seen in Figure 14.18, which shows the average annual solar radiation across

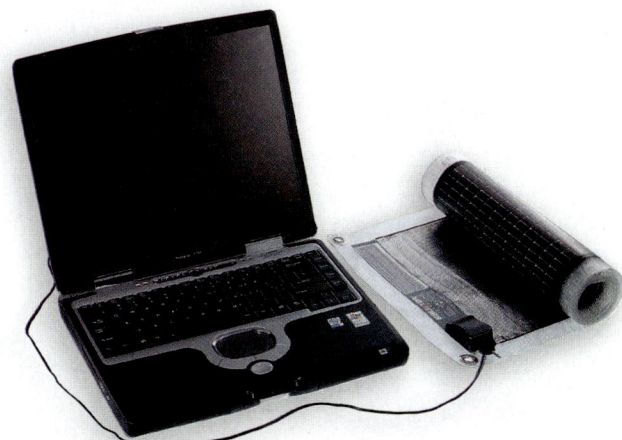

FIGURE 14.19   Thin-film solar cells are flexible and can provide modest amounts of electrical power at remote locations. As the price of thin-film photovoltaic devices continues to decline, this technology may revolutionize the solar energy industry.
(left): © Misty Tosh; (right): Courtesy of Brunton/ Fiskars Outdoor - Americas

the United States. Obviously, photovoltaic systems are the most cost-effective in the southwestern portion of the country due to the lower latitude and fewer cloudy days.

The electric car manufacturer Tesla announced that it would begin selling a new battery system in 2016 called *Powerwall,* which will allow homeowners with photovoltaic panels to store surplus electricity for powering their homes at night. Such a system would be ideal for homeowners who wish to minimize the amount of electricity they purchase from the grid, and for those who find it uneconomical to sell their surplus electricity to utility companies. Tesla plans to sell its Powerwall system, which is supposed to be of sufficient size for most homes, for $3,000. Homeowners with greater power needs will have the option of connecting multiple batteries. While the Powerwall system has the potential to revolutionize the photovoltaic industry, it remains to be seen how well the system actually performs and whether it will be truly cost-effective.

In addition to new battery technology, advances in semiconductor technology have enabled manufacturers to paint thin layers of semiconductor material onto flexible films. The resulting *thin-film solar cells* are 50 times thinner than conventional photovoltaic cells, and designed to be portable (Figure 14.19). Despite being somewhat expensive, thin-film solar cells do offer the convenience of being able to generate modest amounts of electrical power during sunny days anywhere on the globe.

The use of solar power will undoubtedly keep increasing as the price of thermal and photovoltaic panels continues to fall and the need for renewable and non-carbon-based fuels increases. Solar heating systems for homes and businesses are already cost-effective, as are concentrated solar power plants. Although concentrated solar power plants can produce electricity on a large scale, their output is still relatively small compared to that of conventional power plants. For example, the expected production output of the plant shown in Figure 14.16 is 110 gigawatt hours of electricity per year. By comparison, the smallest U.S. nuclear plant generated 4,200 gigawatt hours in 2014. Thus it would take 38 concentrated solar power plants, similar in size to the one in Spain, to equal the output of a *small* nuclear plant. This highlights a problem we face with using all forms of alternative energy for reducing carbon emissions:

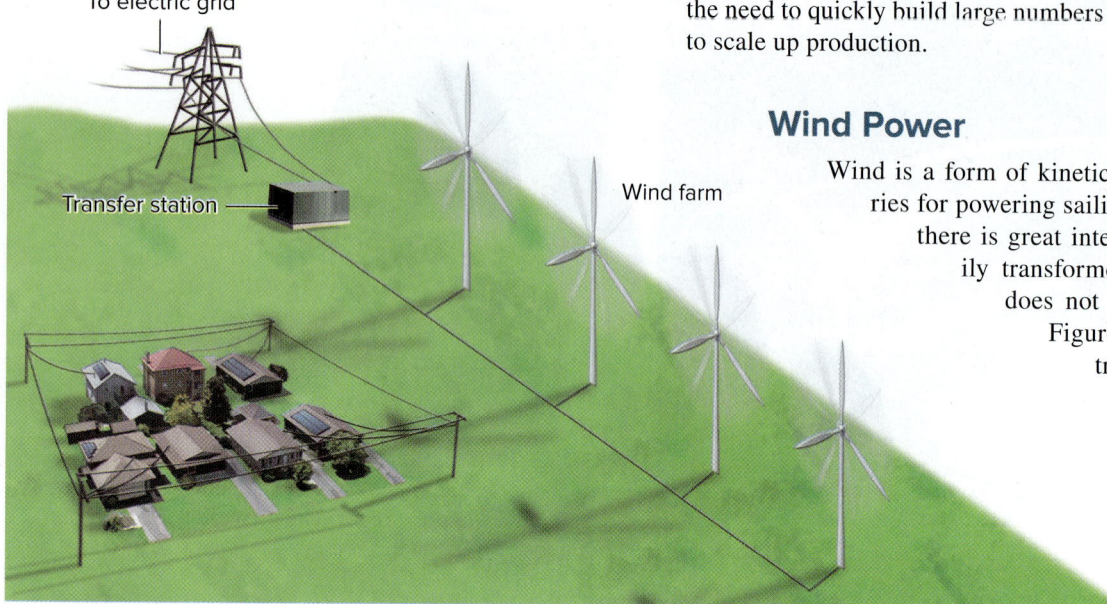

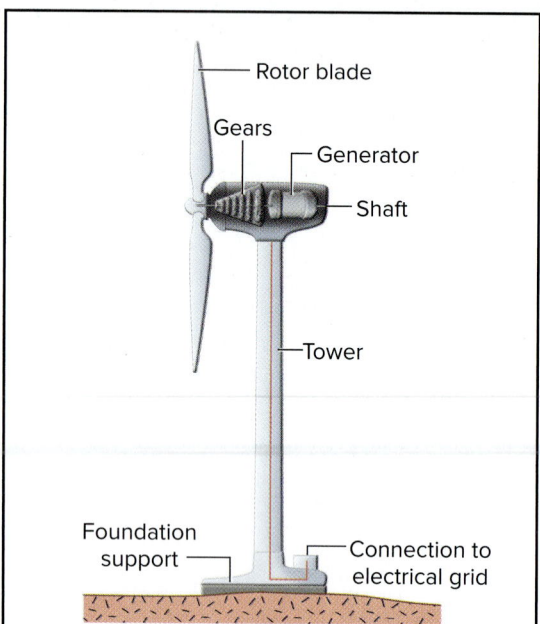

FIGURE 14.20 Wind is a form of kinetic energy that can be transformed into electrical energy when it turns a rotor blade, which then spins an electrical generator. To take advantage of economies of scale, multiple wind towers are usually constructed in what are referred to as wind farms.

the need to quickly build large numbers of plants and individual panels in order to scale up production.

Wind Power

Wind is a form of kinetic energy that has been used for centuries for powering sailing ships and turning windmills. Today there is great interest in wind energy because it is easily transformed into electricity, is renewable, and does not emit greenhouse gases. As shown in Figure 14.20, instead of spinning an electrical generator with hot steam or falling water, the kinetic energy from wind is used to turn a rotor similar to an airplane propeller. The amount of electricity a wind turbine produces depends on how fast and for how long the rotor turns. Clearly, the ideal location is a place where the wind is both strong and steady. Mountainous areas often contain suitable sites because the mountains act as barriers, forcing the wind to pick up speed as it flows over ridges or funnels through gaps or passes (Figure 14.21A). Coastal areas and open plains also make excellent sites because winds tend to be strong and steady. As indicated by the map in Figure 14.21, vast amounts of electricity in the United States could be produced across the Great Plains and Midwest and in most offshore areas, including in the Great Lakes.

Similar to solar energy, the cost-effectiveness of wind-generated electricity depends on both geographic location and the cost of the wind turbines themselves. This means that in most situations, the profits or savings from wind power must be enough to repay the initial investment over some reasonable amount of time. Because wind-generated electricity is now cost-competitive with electricity from coal and natural gas power plants, energy companies are investing heavily in wind projects around the world. In order to maximize their profits, these companies are taking advantage of economies of scale and building multiple wind turbines in what are called **wind farms.** Part of the cost of a wind tower is for installing underground cables and building a transfer station so that the power can be fed into the regional electric grid (see Figure 14.20). By building multiple turbines relatively close together, the cabling and transfer station costs can be shared. In this way the costs go down as the scale or size of a wind farm increases.

Another way of exploiting economies of scale is to make the wind turbines themselves bigger, thereby increasing the amount of electricity an individual turbine can produce. Although the cost of a single tower becomes greater, the increased electrical production more than offsets the additional construction costs. The world's largest wind turbines are now being built by the Danish firm Vestas. In these turbines the rotor assembly has a diameter of 525 feet (160 m) and is mounted to a tower that stands 460 feet (140 m) above the ground. Moreover, a single turbine is capable of producing 8 megawatts of power. Similar to our earlier comparison with the concentrated solar power plant (Figure 14.16), if we assume the best-case scenario, that this turbine can produce power 24 hours day, it would take 60 turbines to produce the same amount of power as the smallest U.S. nuclear plant. Although wind power is currently easier to scale up than solar, we still face the difficulty of quickly installing enough turbines to help make major reductions in greenhouse gas emissions. Small-scale wind turbines

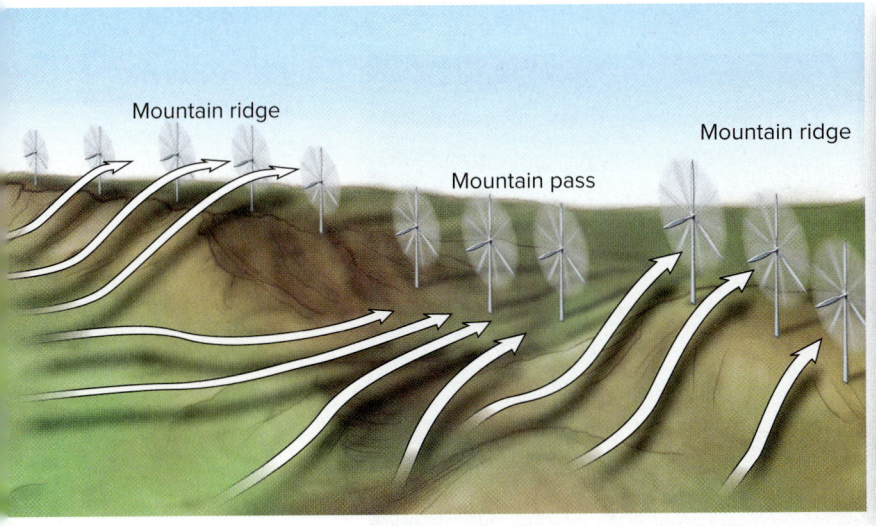

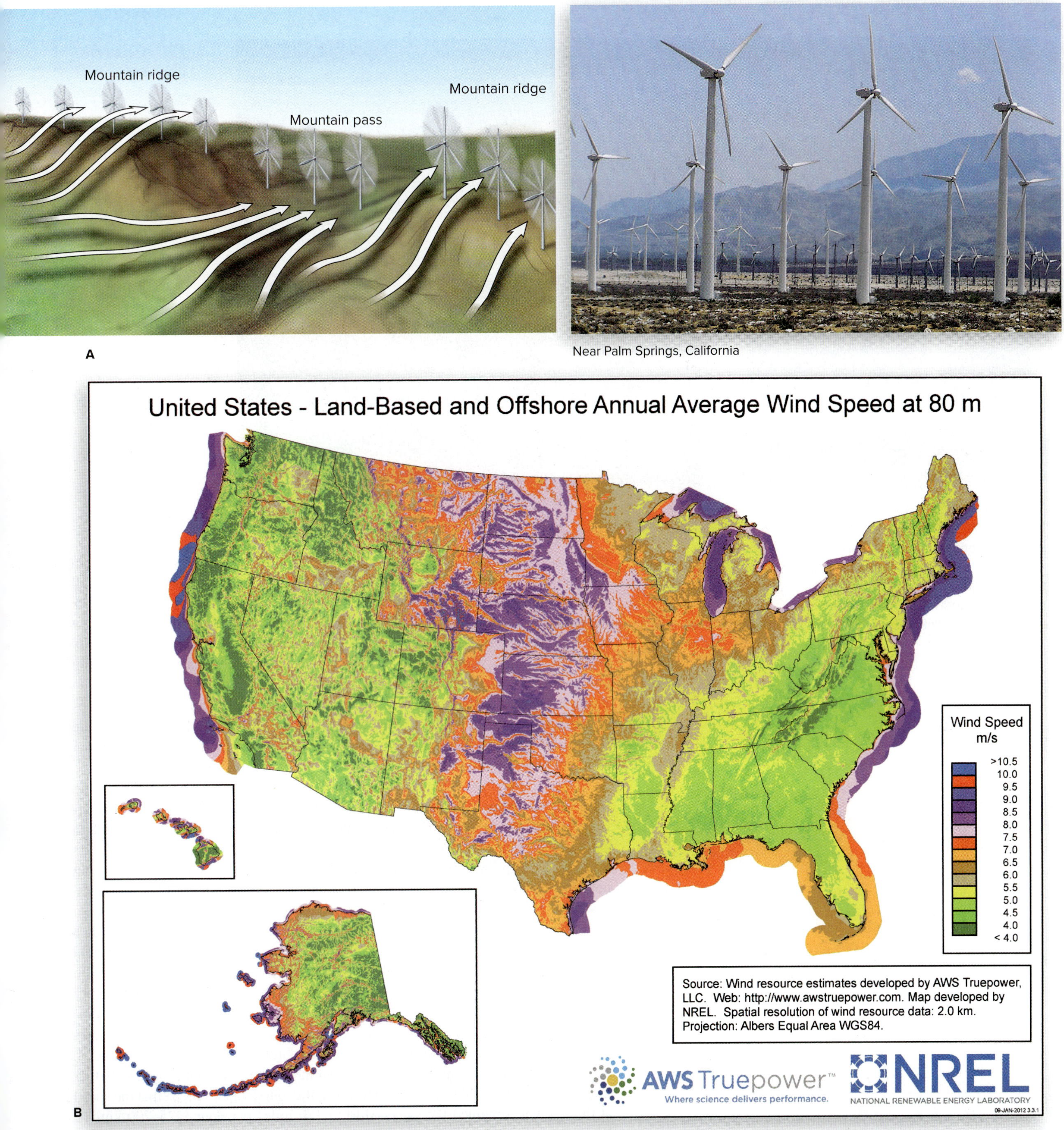

Mountain ridge

Mountain ridge

Mountain pass

A

Near Palm Springs, California

United States - Land-Based and Offshore Annual Average Wind Speed at 80 m

Wind Speed m/s
>10.5
10.0
9.5
9.0
8.5
8.0
7.5
7.0
6.5
6.0
5.5
5.0
4.5
4.0
< 4.0

Source: Wind resource estimates developed by AWS Truepower, LLC. Web: http://www.awstruepower.com. Map developed by NREL. Spatial resolution of wind resource data: 2.0 km. Projection: Albers Equal Area WGS84.

AWS Truepower™
Where science delivers performance.

NREL
NATIONAL RENEWABLE ENERGY LABORATORY
09-JAN-2012 3.3.1

B

FIGURE 14.21 For wind farms to be cost-effective, they need to be located where the winds are both strong and steady for much of the year. Such conditions are common on top of ridges and mountain passes (A), which act to funnel the wind. Other ideal locations include windswept open plains and coastlines. Map of the United States (B) showing the average wind speeds at the height where most wind turbines operate.

(a): © Jim Reichard; (b): U.S. National Renewable Energy Laboratory, U.S. Dept. of Energy

A

B

FIGURE 14.22   Large wind turbines (A) are more cost-effective because the economies of scale allows them to produce more power relative to their size. Residential turbines (B) come in a variety of sizes and can be cost-effective for individual homeowners.
(a) © Construction Photography/Corbis; (b) © Shawn Lessord, Rochester Solar Technologies

are also available and, depending on location, can be cost-effective for individual homeowners (Figure 14.22). As with solar systems, a homeowner can often sell surplus electricity to an electric utility company, thereby reducing the payback period for recovering the initial investment.

Unlike most solar systems, wind turbines can generate electricity on cloudy days and at night. Wind power therefore is generally a better choice than solar in climatic zones dominated by overcast skies. Keep in mind, however, that even in the most ideal locations there are times when the wind slows down or stops altogether. Except for solar systems that store heat, both wind and sunlight are *intermittent* sources of energy, but the times when they cannot be used occur at different times and in different places.

Despite the fact that wind power is clean and renewable, proposed wind farms are often met with opposition from local landowners and environmentalists. In some cases opposition arises because wind farms alter the landscape in ways some people find unattractive. Other times the issue is noise. Residents living in close proximity to wind towers have to endure the constant swooshing noise of the rotating blades. There is also the problem of bird and bat mortality. One of the early wind farms in the United States happened to be located on a mountain pass that serves as a migratory bird route. An estimated 5,000 birds a year were reportedly killed, raising serious concerns. Researchers have also found relatively high rates of bat mortality around nearly every wind facility surveyed in North America. As a result, the wind industry has been working with university and government researchers to better understand the problem and to develop solutions. For example, by avoiding building new wind farms along migratory routes and in areas where birds and bats congregate, fatalities can be greatly reduced. Another factor is that as turbines have grown in size, they are being spaced farther apart and their spin rate has slowed. However, the longer blades mean that the speed at the tips has remained the same—140 to 180 miles per hour (225–290 km/hr). While the wider spacing and slower rotation may have benefited some species, others may be at greater risk due to the expanded size of the area being swept. Recent research has also shown that bats are killed by both direct impact and simply by getting too close to the rotating blades. It turns out that the internal organs of bats are easily damaged when they encounter sudden changes in air pressure that develop around the blades. In addition to locating new wind projects away from bird and bat concentrations, it has been suggested that wind companies curtail operation during high-risk periods based on daily or seasonal flight patterns. The problem with curtailment is that it limits electrical production, thereby reducing a wind company's return on its investment. While research continues, it is worth noting that the number of birds killed by wind turbines is exceedingly low compared to mortality due to other factors, such as moving vehicles, buildings and windows, transmission lines, toxic chemicals and pesticides, and domestic cats.

Geothermal Power

Our planet naturally receives energy from two inexhaustible and renewable sources: the Sun and Earth's interior. **Geothermal energy** refers to the heat contained within the Earth, most of which comes from the decay of radioactive elements. Recall from Chapter 4 that convection cells within the mantle transport heat toward the surface, where it eventually dissipates into space. As indicated by the map in Figure 14.23, the outflow of heat through Earth's crust is not uniform. The higher rates of heat loss generally occur in areas of tectonic and volcanic activity (subduction zones, spreading centers, and hot spots), or where the crust is relatively thin.

Geothermal energy can be extracted from the subsurface in different ways, but perhaps the most common is to utilize the heat contained in water that has come into contact with unusually warm rocks. Areas with the greatest potential for geothermal energy are where the rates of heat flow through Earth's crust are the highest. Volcanic regions, such as Yellowstone National Park (Chapter 6), are ideal since shallow rocks are often hot enough to turn groundwater into steam. Historically, geothermal waters could only be accessed at *natural hot springs,* which are places where unusually warm groundwater is able to make its way to the surface (Chapter 11). Geothermal waters are now extracted using wells drilled directly into aquifers that contain hot water (Figure 14.24). Because wells can extract water that is much closer to the heat source, geothermal wells typically provide water that is warmer than that in natural hot springs. In fact, some geothermal well water is over 900°F (500°C), which is hot enough to melt lead. Figure 14.24 illustrates a situation where hot rocks at depth are not porous, and thus do not contain groundwater. In this case explosives can be used to fracture the rocks deep within a well. Water is then injected into the fractured rocks so that the heat energy can be transferred to the water, which is then removed by a nearby pumping well.

Geothermal Electricity

The water temperature of a geothermal system plays a key role in determining how it can be used as an energy source. For example, if the temperature of the water being extracted is significantly higher than water's boiling point of

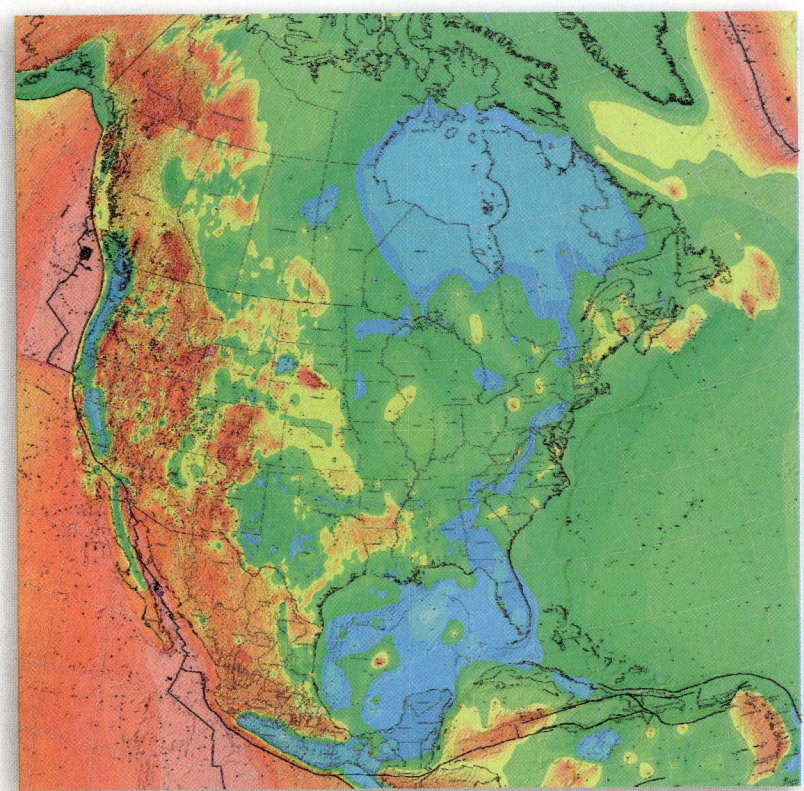

FIGURE 14.23 The rate of heat flow through Earth's crust depends on rock type, crustal thickness, tectonic history, and distribution of heat within the mantle. As shown by the reddish and yellowish colors on the map, crustal rocks are warmer in areas of higher heat flow and can be used as a source of geothermal heat.

© 2004 AAPG, reprinted by permission of the AAPG.

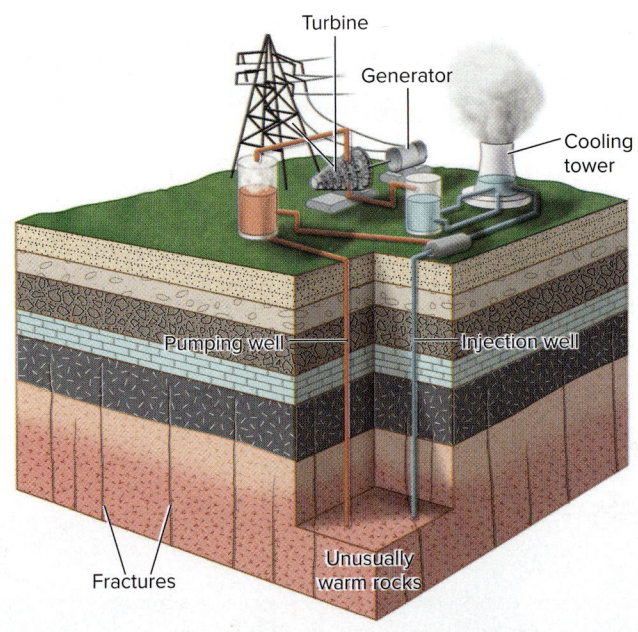

FIGURE 14.24 Geothermal heat can be utilized by extracting water that comes into contact with hot rocks. A series of injection and pumping wells are used to circulate either natural groundwater or surface water pumped down into hot, dry rocks. The water can be used for space heating, or if hot enough, can be turned into steam and used to make electricity.

FIGURE 14.25   Iceland sits atop a hot spot on the mid-Atlantic spreading center (Chapter 4), which allows the nation to meet 45% of its total energy needs using geothermal energy. Photo shows one of Iceland's year-round swimming facilities that is heated by the constant discharge of water from the geothermal power plant in the background.
© Javier Larrea/Age Fotostock

212°F (100°C), the extracted water will quickly turn to steam when exposed to the low-pressure environment at the surface. This steam can be used to turn a turbine and generator to create *geothermal electricity.* Such electricity is highly cost-competitive because free heat from Earth's interior is being used to make steam, as opposed to using coal, natural gas, or nuclear fuel. Another advantage is that geothermal energy does not produce carbon dioxide and is generally a renewable resource.

Several countries are currently producing geothermal electricity in volcanic regions where the rocks are sufficiently hot to produce steam. Many of the sites lie along the convergent boundary of the Pacific plate, often referred to as the Ring of Fire (Chapter 6). In the United States geothermal electricity is presently being produced in several western states where the rocks remain hot from volcanic activity in the recent geologic past. The greatest production is in California, where geothermal power represents 6% of the state's electrical needs. Geothermal electricity is also being produced in Hawaii, from heat associated with a volcanic hot spot.

Geothermal Space Heating

The number of sites around the world where subsurface rocks are hot enough to generate steam is actually quite small. However, there are many places where groundwater temperatures are more than adequate to provide heat for buildings. *Geothermal space heating* involves piping hot water through radiators where the heat energy is then transferred to a building's air space (this method of indoor heating is similar to solar heating, described earlier in this section). Geothermal water can also be used to heat swimming pools. In fact, hot spring resorts often have to mix spring water with water from another source in order to bring the temperature down below 104°F (40°C), which is about the maximum people can tolerate for any reasonable length of time.

Perhaps the best example of where geothermal energy is being utilized on a large scale is Iceland, a country with a population of only 330,000, located just below the Arctic Circle. Iceland is geologically unique in that it lies over a volcanic hot spot and also straddles a major spreading center (Figure 14.25). In fact, Iceland is the only place where the 6,000-mile (10,000 km) long mid-Atlantic spreading center rises above sea level. Due to its abundance of both groundwater and geothermal energy, Iceland is able to heat nearly 90% of its buildings using geothermal space heating. In terms of electricity, geothermal plants produce 25% of the nation's electrical needs, with most of the remainder being generated by hydro power. It is worth noting that Iceland has been testing a pilot program to determine the feasibility of replacing its oil-based transportation sector with one based on hydrogen. The idea is to use the country's excess geothermal capacity to generate the necessary hydrogen fuel. Iceland currently has several hydrogen fueling stations and a fleet of 100 vehicles that are part of the testing program.

Geothermal Heat Pumps

In most climatic zones the air temperature varies considerably from night to day, and also seasonally. However, if you go about 10 feet (3 m) below the surface, the temperature of the rocks or soil no longer corresponds to atmospheric changes. Depending on latitude and elevation, the ground temperature will remain steady somewhere between 45° and 75°F (7–21°C). Note that this is cooler than the average air temperature in the summer, yet warmer than the winter average. A **geothermal heat pump** is a mechanical system that supplies a building with either warm or cool air by making use of the relative difference in temperature between the inside air and the ground. The basic principle involved is similar to how the constant temperature of a cave has a cooling effect on a person who enters during the summer, yet provides warmth during the winter.

As illustrated in Figure 14.26, a geothermal heat pump is a system of interconnected wells or pipes, in which water circulates between the ground and a heat exchanger (metal coils) inside a building. During the summer the heat exchanger receives water from the subsurface that is cooler than the air inside the building. When inside air is blown through the heat exchanger, the water absorbs some of the air's heat. This reduces the temperature of the air, which is then returned back through the building. Meanwhile the water circulates back through the subsurface loops where the heat that had been removed from the building is absorbed by the ground. In the winter the heat transfer works in the opposite direction, namely from the ground to the air. However, this requires an electric-powered compressor in order to extract enough heat to bring the air up to a comfortable temperature. Basically the system removes heat from the subsurface, then transfers it to a compressed gas (*Freon* was used in the past, but was replaced by *Puron* starting in 2010). The gas is then allowed to decompress, at which point the heat that was concentrated is transferred to the inside air. In this mode a geothermal heat pump functions just like your refrigerator, which extracts heat from the food items inside, then transfers the heat to the air outside.

A key point here is that a geothermal heat pump exchanges heat with the subsurface, rather than with the outside air as in a conventional heat pump. A geothermal system ends up using far less electricity to operate because the ground temperature remains constant and is relatively close to the 70–75°F (21–24°C) range of most buildings. In contrast, the outside air temperature can vary widely, ranging from subzero to over 100°F (73°C). This means that a conventional heat pump becomes rather inefficient whenever there is a large difference between outdoor and indoor temperatures. The largest difference occurs on days when the outdoor temperature falls below freezing. A conventional heat pump, therefore, is a poor choice in colder climates since it will consume considerable amounts of electricity while providing minimal heat. Note that geothermal hot-water heaters are also available and are far more efficient that conventional water heaters.

Future of Geothermal Energy

Using geothermal energy is highly desirable because it emits little to no carbon dioxide, plus it can be very cost-effective since it makes use of free heat from the Earth's interior. Unfortunately, geothermal electricity and space heating are geographically limited to areas where the heat flow through the crust is unusually high. Another problem is that rocks are poor heat conductors, which means that some geothermal plants are able to remove heat faster than it can be replaced. In other situations the rate of water being withdrawn is greater than the recharge rate, which not only depletes the resource but can also lead to serious land-subsidence problems (Chapters 7 and 11). In the case of electrical production and space heating, geothermal energy therefore is not always renewable. Fortunately, the problems associated with heat loss and land subsidence can be minimized by injecting the warm wastewater back into the subsurface rather than discharging it on land. Another useful technique is to extract geothermal heat on a cyclic rather than a continuous basis. This allows the thermal energy to build back up between production cycles. Nevertheless, even if mitigation steps are taken, the use of geothermal electricity and space heating will always be restricted to relatively small geographic areas where suitable geothermal reservoirs are found.

In contrast, geothermal heat pumps are viable almost everywhere; therefore they have the potential of making a significant contribution to reducing our use of fossil fuels. According to the EPA, geothermal heat pumps use 44% less electricity than air-based systems, and 72% less than electric heating and standard air

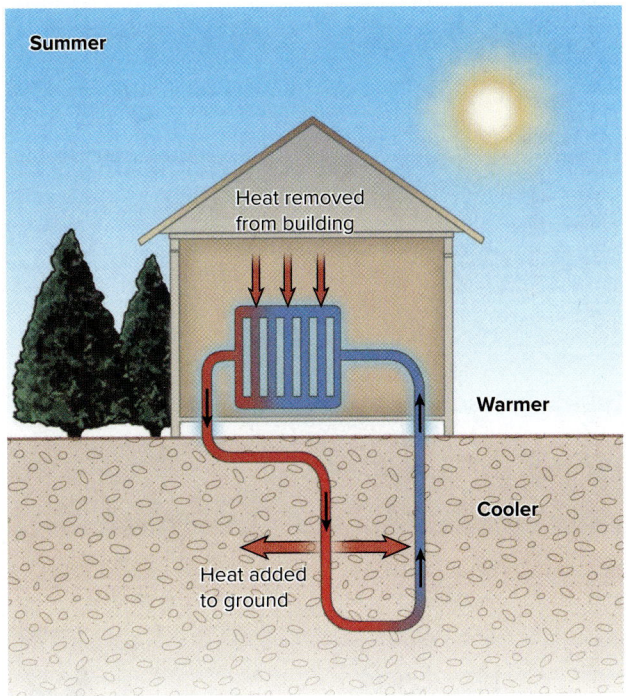

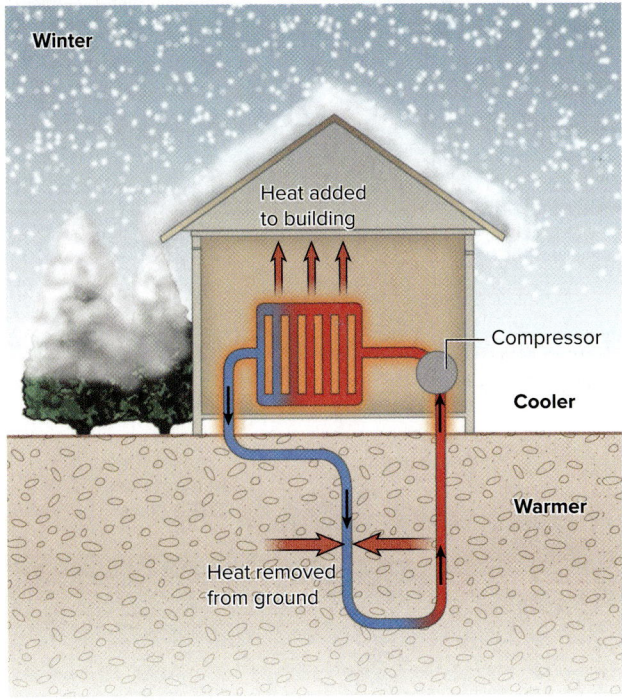

FIGURE 14.26 A geothermal heat pump uses a circulating fluid to exchange heat between the ground and a building. During the summer relatively cool water is piped into the building where it absorbs heat from the air. The warmed water then is piped into the subsurface where the extracted heat is transferred to the ground. In the winter a compressor is used to concentrate heat removed from the ground, which is then released into the building.

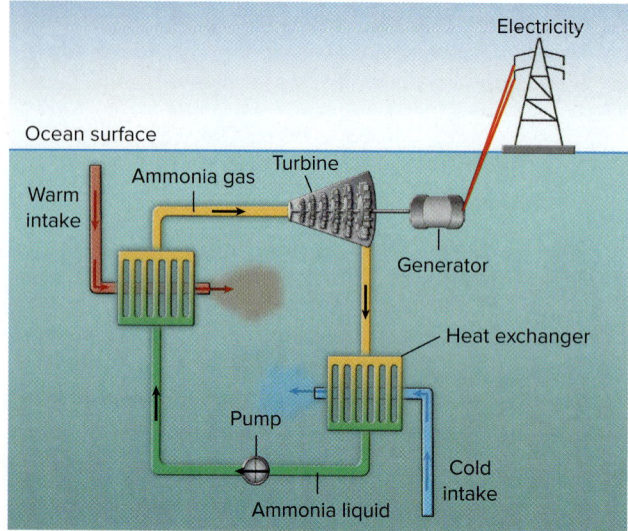

conditioning systems. Despite the increased efficiency, builders and homeowners tend to avoid installing geothermal heat pumps due to the relatively high installation costs. However, the savings in electrical consumption usually allow these costs to be recovered in 2–10 years.

Ocean Thermal Energy Conversion

Every day the oceans absorb about 70% of the solar radiation that reaches the Earth, converting sunlight into vast amounts of thermal energy. In 1881, a French physicist proposed tapping this energy, which eventually led to the development of a technique called **ocean thermal energy conversion (OTEC)** that makes use of a simple heat engine driven by thermal energy stored within the oceans. The first successful OTEC plant was developed in 1930 by France at a site in Cuba. Then in 1974 the United States established a laboratory in Hawaii dedicated to developing OTEC technology.

An OTEC plant draws in warm water from the ocean surface along with cold water from great depths, then uses the temperature difference of the two waters to drive a turbine and electrical generator. Modern OTEC plants built by Japan and the United States have shown the technology to be a cost-effective means of generating electricity. One OTEC design, shown in Figure 14.27, incorporates a closed-loop system that uses an ammonia-based fluid. When warm seawater is pumped through a heat exchanger, the ammonia begins to boil and turns into a gas. The ammonia vapor is then used to spin a turbine and electrical generator. From there the gas is piped through a second heat exchanger that is being flushed with cold seawater, causing the ammonia to condense back into a liquid. The other basic OTEC design puts the incoming warm seawater under a vacuum, which causes it to boil. The resulting steam, rather than ammonia, is then used to drive the electrical generator. Engineers included the additional step of vaporizing seawater so that valuable freshwater could be produced as a by-product, thereby making the overall system more cost-effective.

Although OTEC plants produce a net gain of electricity (generate more power than they consume), one of the problems that has limited their commercial development is that conventional land-based power plants are more cost-effective. Another problem is that for an OTEC plant to operate efficiently, it must be located where the temperature difference between the warm and cold seawater is at least 36°F (20°C). This restricts OTEC to tropical waters. In addition, to access water that is sufficiently cold, the plant's lower intake pipe must reach a depth of around 3,300 feet (1,000 m). This means that to locate an OTEC plant along a coastline, the slope of the seafloor must be quite steep. Otherwise, to reach sufficiently deep water would require building and maintaining an excessively long intake pipe, thereby making the entire operation uneconomical. This presents yet another problem in that most of the suitable tropical sites are located far from major population centers with large electrical demands. One proposed solution (Figure 14.28A) is to build a floating OTEC plant that would send power

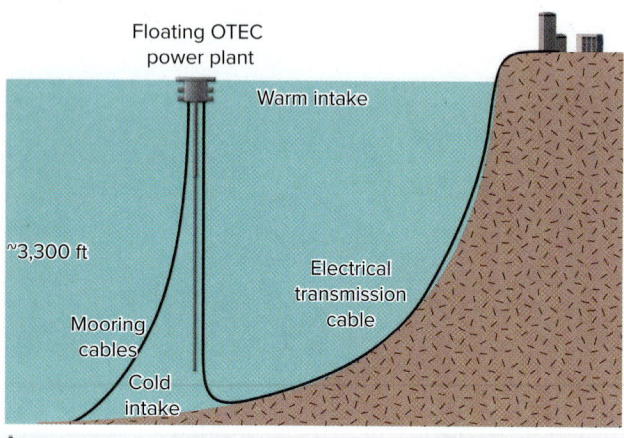

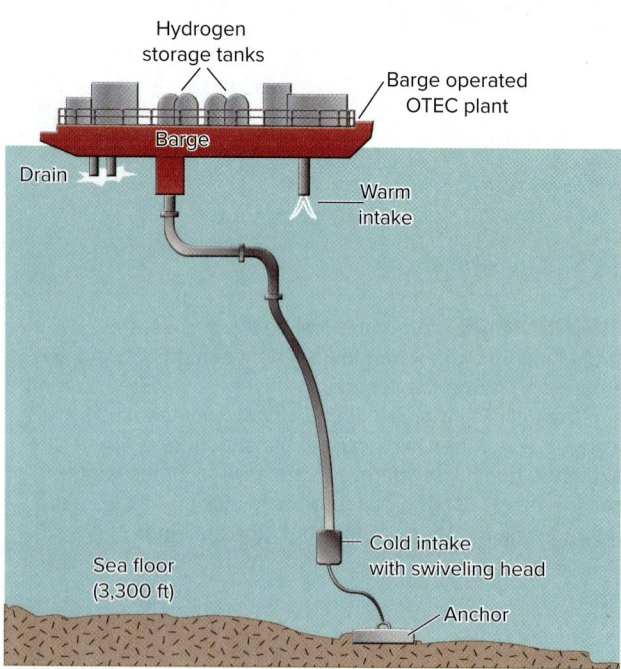

FIGURE 14.28   Because of the need to access cold water, land-based OTEC sites are restricted to tropical coastlines with a steep seafloor. A floating sea-based system (A) would easily access deep water, then send electricity to the mainland by a transmission cable. Another option is a floating barge system (B), where the electricity is used to produce hydrogen, which is shipped to the mainland via tankers.

to a city on the mainland via a transmission cable. Another solution is to have a floating plant use the electricity it produces to separate hydrogen from seawater (Figure 14.28B). The hydrogen would then be stored on the platform and eventually loaded onto tankers for transport to major population centers. There the hydrogen could be used as a transportation fuel.

In 2013 the international firms of Lockheed Martin and Reignwood Group signed a contract to design and install a 10-megawatt OTEC plant along the southern coast of China. The plant will be built on an offshore platform and send electrical power to the mainland via an undersea cable. If the prototype plant is successful, the companies plan to develop commercial plants ranging in capacity from 10 to 100 megawatts. Thus, OTEC could soon begin making a significant contribution to the world's supply of clean and renewable energy.

Tidal Power

Tidal power is distinctive in that it is not ultimately derived from sunlight or Earth's internal heat, but from the cyclic changes in sea level caused by the gravitational interaction between the Earth, Moon, and Sun (Chapter 9). On a nearly continuous 12-hour cycle, coastal estuaries and inlets around the world fill with seawater when an incoming tide moves up onto Earth's landmasses. This is followed by an outgoing low tide where the water drains back out to sea. Because Earth's oceans are so vast, this physical movement of water represents an enormous amount of kinetic energy. Similar to other forms of kinetic energy (e.g., wind and hydro), tidal power can be used to spin an electrical generator. However, tidal power is presently cost-effective only in places where the shoreline is shaped such that the tides are amplified.

There are two basic ways tidal power can be used to generate electricity. The first involves constructing a low-lying barrier called a *tidal barrage* across an enclosed inlet or estuary, as shown in Figure 14.29. The barrage is basically a dam with large gates (not shown in the figure) that remain open while the incoming tide fills the estuary. At high tide the gates close, trapping the water. After the reservoir is filled, water is allowed to fall through turbines and generate electricity, similar to a hydroelectric dam. The amount of electrical production naturally depends on the kinetic energy of the water as it falls through the turbines, which depends on the difference in reservoir height between high and low tide. It turns out that a tidal range of at least 16 feet (5 m) is required for a barrage system to function economically. Unfortunately, there are relatively few places in the world with this type of tidal range.

Currently there are only two commercial barrage plants in operation, one built in France in 1967 and the other in Canada in 1984. Although these tidal plants have been profitable and other suitable sites exist around the world, no other major plants have been built. The basic problem is that engineers did not take into consideration the potential environmental impact that barrage systems might have on coastal ecosystems. Recall from Chapter 9 that coastal estuaries and associated wetlands serve as the nursery grounds for nearly 70% of all commercial and sport fisheries. A tidal barrage, unfortunately, acts as a barrier for fish and other marine life attempting to enter an estuary to spawn. For those species that are able to get past a barrage, many end up being killed on the way back as they pass through the turbine blades. Another problem is that a barrage alters the timing of the tides, thereby impacting the entire food web in the estuary, which is highly sensitive to the natural ebb and flow of the tides.

Engineers have recently developed a new tidal technique that has less of an environmental impact than impounding water behind a barrage. This approach involves placing propeller-driven turbines in areas with naturally strong currents that result from tidal waters being funneled through narrow channels, similar to how mountain passes funnel the wind. In 2008 a British company installed the first set of commercial tidal turbines at its SeaGen project, located at the entrance

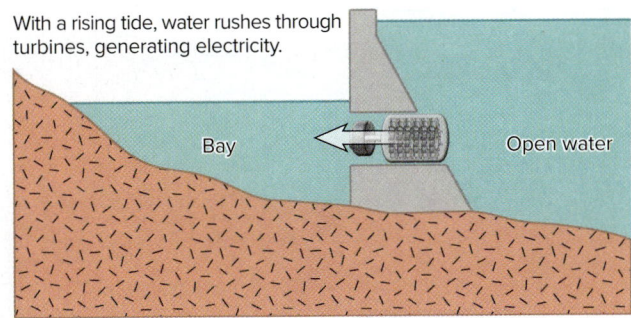

With a rising tide, water rushes through turbines, generating electricity.

Bay   Open water

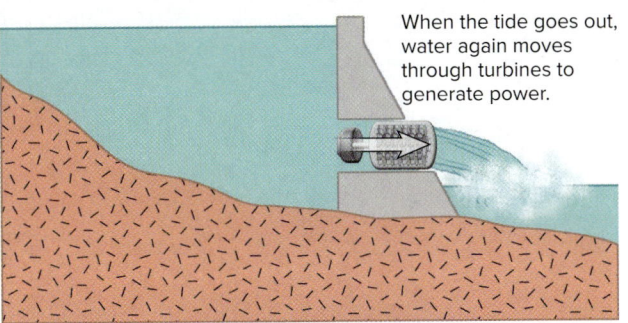

When the tide goes out, water again moves through turbines to generate power.

FIGURE 14.29 A tidal barrage is a barrier constructed across an inlet and is designed to trap the incoming high tide. Gates in the structure (not shown) are closed at high tide, and then opened to allow water to fall through turbines as the tide goes out. Systems can also be set up to generate electricity with incoming tides as well.

A

B

C

FIGURE 14.30   The first commercial tidal turbines were installed in 2008 at the entrance of Strangford Lough (A) in Northern Ireland. Here the incoming and outgoing tides are naturally forced to move through the restricted entrance to the bay, creating powerful currents that help make electrical production more cost-effective. An artist's illustration (B) of how the turbines are installed; an actual photo (C) showing the strength of the tidal current flowing past the turbine tower.
(b–c) © Marine Current Turbines LTD

to a restricted bay in Northern Ireland (Figure 14.30). These turbines swivel back and forth 180 degrees so that electricity can be produced during both the incoming and outgoing tides (the turbines can also be raised for maintenance). The two turbines are capable of producing 1.2 megawatts of power, which is enough to meet the electrical needs of over 1,000 homes. Now that the SeaGen project has proven to be successful, the company is planning to install multiple sets of turbines in various inlets around Great Britain and in the Bay of Fundy in Canada.

In addition the SeaGen project, a Welsh company has designed a set of tidal turbines called DeltaStream that rest on the sea floor and are connected in a triangular arrangement. One of the advantages of this system is that it stays in place on the sea floor due to its own weight, which eliminates the need to construct and maintain a central tower. After a successful pilot test in 2015, the company plans to install multiple sets of turbines in an inlet along the Wales coast as part of a long-term commercial-scale demonstration project.

Proponents of the British tidal power projects claim that the threat to marine life will be minimal because the propellers rotate at a relatively slow rate—10 to 15 revolutions per minute. Moreover, marine animals capable of swimming in areas with strong currents tend to be very agile and have great perceptive powers, which should enable them to avoid the rotating propellers. Although the risk to marine life is thought to be small, the SeaGen project included an extensive environmental impact study led by university researchers. The final report, issued in 2012, found no major impacts on marine mammals during the three-year monitoring program. Seals and porpoises regularly moved past the SeaGen turbines, demonstrating that the turbines did not act as a barrier. Likewise, no measurable changes were found in the ecology of organisms living on the seabed. Finally, the turbines themselves did not significantly change the velocity or direction of water flowing within the bay.

Because there are many sites around the world with sufficiently strong tidal currents, tidal-generated electricity has the potential for making a substantial contribution to the world's energy supplies. One drawback, however, is that electrical production is uneven. Since the tide comes in and goes out twice each day, tidal currents reach a maximum four times in a given day. This causes the rate of electrical production to reach a peak four times a day, with production falling to zero between each peak as the tide switches from incoming to outgoing. Despite the uneven nature of tidal power, it has an advantage over wind and solar in that the tides are both consistent and highly predictable. Finally, it should be noted that researchers in Florida are testing the feasibility of underwater turbines located within the steady currents of the Gulf Stream. Although the installation of these turbines poses certain technical and logistical difficulties, the use of open-ocean currents is an attractive option because the production of electricity would be more continuous.

FIGURE 14.31   A thermal camera reveals how heat preferentially escapes from a warm house on a winter night. Notice how the walls are losing much more heat than the roof. This type of analysis can help homeowners develop better strategies for insulating their homes, thereby conserving energy and saving money.
© Chris Moll, Radio 101 Thermography

Conservation

An often overlooked source of energy is conservation. **Energy conservation** involves reducing energy consumption through decreased human activity and more efficient use of what we do consume. An example of conservation through decreased activity would be if you cut back on the number of trips you made driving to the grocery store each week. An example of more efficient use would be if you bought a more fuel-efficient car or upgraded the insulation in your home (Figure 14.31). Conservation is important because when a society embarks on a coordinated effort to reduce energy consumption, the result can be a dramatic increase in supply. A classic example is the widespread energy conservation efforts that took place in developed nations in response to the oil crises of the 1970s (Chapter 13). During this period people reduced their personal consumption by changing their driving habits and by keeping their homes at a lower temperature during the winter. In the United States, Congress passed the Energy Policy Act of 1975 that gave the Department of Energy the authority to set energy conservation standards for residential, commercial, and industrial appliances and equipment. During this period Congress also gave the EPA the authority to phase in fuel efficiency standards for the nation's fleet of cars and light trucks. This resulted in the production of more fuel-efficient vehicles, as well as appliances that consumed less electricity. Buildings also became better insulated and engineers developed more efficient heating and cooling systems. All these efforts led to a sharp reduction in energy usage, particularly oil, which helped alleviate the supply problem. Also important was the fact that by using less energy, individual homeowners and businesses saved money.

Although major gains in energy efficiency were realized after the oil crises of the 1970s, progress slowed until energy prices rose again after 2000. For example, after the EPA set average fuel efficiency standards in 1975 for the nation's vehicles, efficiency went from 16 miles per gallon (MPG) to 25 MPG by 1990, representing a 56% improvement. However, it took 33 years before Congress would allow the EPA in 2008 to raise the standard to 35 MPG. Fuel standards have since been raised again, with the fleet average expected to reach 47 MPG by 2025.

In addition to more efficient vehicles, higher standards for homes, business, and industry have resulted in major improvements over the years. For example, American consumers are now saving an estimated $55 billion on their home utility bills each year. Perhaps the best example of improved home efficiency is the new standard for light bulbs. With the older incandescent bulbs, only 5% of

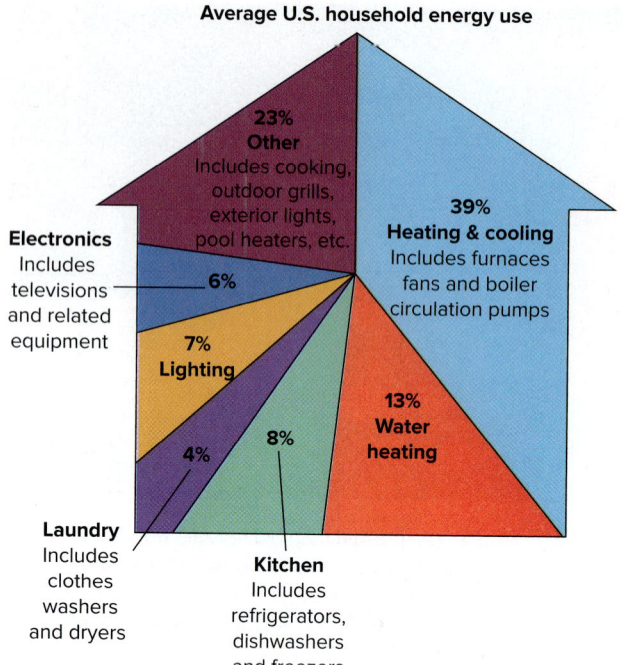

Average U.S. household energy use

23%
Other
Includes cooking, outdoor grills, exterior lights, pool heaters, etc.

39%
Heating & cooling
Includes furnaces fans and boiler circulation pumps

Electronics
Includes televisions and related equipment

6%

7%
Lighting

4%   8%

13%
Water heating

Laundry
Includes clothes washers and dryers

Kitchen
Includes refrigerators, dishwashers and freezers

FIGURE 14.32   Breakdown of energy used in the average American household. Since heating and cooling represent the single largest use of energy, they also have the largest potential savings in terms of energy conservation.

the electricity being consumed was converted into visible light; the other 95% was wasted as heat. In contrast, compact fluorescent bulbs convert 20% of the electricity into light and LED lights convert 80%, both of which represent a vast improvement over conventional bulbs. While improved lighting efficiency conserves energy and saves homeowners money, even greater savings can be obtained from other areas of the home. For example, note in Figure 14.32 that lighting represents just 7% of the energy consumed in the average American household, whereas heating and cooling account for 39%. Therefore, the largest potential source of savings for homeowners are more efficient heating and cooling systems, such as geothermal heat pump systems (described earlier in this chapter).

Despite increases in efficiency, per capita electrical consumption in the United States has actually risen over the years. One reason for this is that the average size of U.S. homes increased 73% between 1973 and 2014, going from 1,500 square feet (140 m^2) to 2,600 square feet (240 m^2). Another factor is the increased use of air conditioning. In 1973, 52% of American households had air conditioning compared to nearly 90% today. Electrical consumption has also risen because of the increased number and size of electrical appliances (e.g., TVs, computers, refrigerators). In addition, many electronic devices automatically go into a power saving or standby mode whenever they are not being used. Since many home electronics spend much of the day on standby, still consuming some electricity, about 75% of the power they use is wasted. Today, homeowners can purchase *energy monitoring systems* that allow them to track the power consumption of various appliances throughout the day. They can then identify where most of their electricity is being consumed and target their conservation efforts so as to maximize their savings. Also, utility companies can install *smart meters* that send homeowners information on their power consumption patterns and electrical rates. Homeowners can then save money by shifting some of their electrical consumption to off-peak periods when utility rates are lower.

Post-Petroleum World

In Chapter 13 you learned that the looming energy crisis is due to the natural depletion of conventional light oil supplies combined with population growth and increasing per capita consumption of oil. Moreover, the future oil shortage has the potential to cripple the global economy, with our transportation systems being the most vulnerable. For example, the transportation system in the United States is powered almost exclusively by gasoline, diesel, and jet fuels, all of which are refined from crude oil. While replacements for conventional oil do exist (oil sands, oil shale, coal), the amount of oil we consume is so large that a rapid scaling up of production of alternative sources is necessary. Moreover, the burning of massive amounts of coal, natural gas, and oil for the past 150 years has greatly increased the carbon dioxide concentration in the atmosphere, which is now affecting Earth's climate system (Chapter 16).

Because of the critical nature of the oil and climate crises, societies around the world are moving away from the traditional use of fossil fuels and toward a new era of alternative and renewable energy resources. To successfully make it through this transition, we will need to tackle the following challenges: (1) develop transportation systems that emit a minimal amount of carbon dioxide; (2) generate electricity with non-carbon-based energy sources; and (3) scale up the production of alternative sources to meet the current demand for energy. In the following sections we will briefly examine how these goals might be achieved.

Transportation Systems

The transportation system in most developed nations generally consists of cars, trucks, trains, ships, and airplanes. There are also heavy pieces of equipment, such as earth-movers, bulldozers, and backhoes. With the exception of jet aircraft, these machines are powered almost exclusively by internal combustion engines that burn either gasoline or diesel fuel. To transform our transportation system into one that produces a minimal amount of carbon dioxide will require limited use of internal combustion engines. At the present time the only feasible replacement for most applications is the electric motor—steam engines are not a viable option because they require a heat source for making steam. Although much of our transportation needs can be met with electric motors, certain vehicles, such as cargo ships and airplanes, will need to be powered by conventional diesel and jet engines for the foreseeable future.

Electric motors are already being used in a wide variety of applications. They range in size from ones that run the small fans for cooling our computers to those that supply massive amounts of power to construction and mining equipment. The problem with using electric motors to power nonstationary machines (e.g., trucks, trains) is that they must be attached to an electrical source via a transmission cable or run off a bank of batteries. The only other option is to have some means of producing electricity on the vehicle itself—examples include nuclear power plants on submarines and solar panels on satellites. Trains are perhaps the segment of our transportation network that could most easily be converted to electric motors. This would entail replacing diesel with electric-powered locomotives and installing transmission lines alongside existing rail beds. Because trains are one of the most efficient modes of transportation, they will likely play an increasingly important role in the future.

Although trains will be critical in helping us meet our transportation needs, we will still need automobiles since people in developed nations (particularly in the United States) tend to live in suburbs and must commute to work. By making use of both a conventional engine and an electric motor, today's hybrid vehicles represent a significant improvement in fuel efficiency. However, hybrids are not a long-term solution because gasoline is still used as either the primary source of power or the means of recharging the batteries while driving. Automobiles therefore will eventually need to be powered solely by electric motors. Getting the electricity from power lines is not going to work with cars as it would for trains, so our only other option is to use batteries or some means of generating electricity onboard the vehicle itself. The problem with batteries has always been storing enough electricity to meet the power demands of a vehicle for any useful length of time. Since conventional lead batteries are large and heavy, the only practical way of increasing the operating range of electric vehicles in the past was to reduce their power demand by lowering their weight. Electric cars therefore never became popular because they had to be small and have a relatively short driving range. Today's lithium-ion batteries are much lighter, yet provide significantly more power. The higher energy density of modern batteries has led to all-electric cars with greater driving ranges and with sizes comparable to conventional cars. Electric car manufacturers have also designed new battery packs that use individual lithium-ion batteries similar to those found in laptop computers, thereby significantly lowering the cost. These cheaper, lighter, more powerful batteries will likely make all-electric cars a more popular choice with consumers in the future. However, it remains to be seen whether improved battery technology will ever meet the power demands of all-electric semi-tractor trucks and heavy equipment.

Another option would be for electric cars to be powered by electricity generated onboard using a *fuel cell,* thereby eliminating the battery storage issue altogether. The most promising such device is a **hydrogen fuel cell,** which takes the chemical energy contained in hydrogen and oxygen atoms and converts it into electricity. As illustrated in Figure 14.33, hydrogen molecules

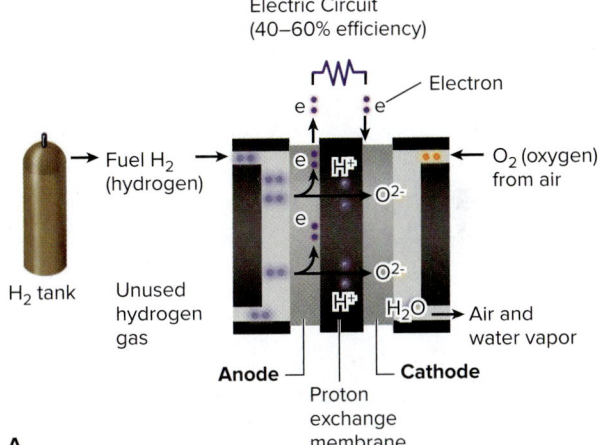

A

B

FIGURE 14.33   A hydrogen fuel cell (A) breaks down hydrogen (H_2) and oxygen (O_2) molecules into their respective ions (H^+ and $O2^-$), which are then combined chemically to produce electricity and water. The electricity can then be used to power a vehicle with an electric motor, thereby eliminating the need for banks of batteries. Hydrogen is stored onboard in tanks, whereas oxygen is drawn in from the outside air. Hydrogen fuel cell bus (B) used in a California demonstration project.

© Filmsight Productions, LLC

(H_2) are stored as a gas in a tank and then broken down into hydrogen ions (H^+) along a plate called a *cathode* within the fuel cell. On the opposite side of the cell is an *anode* plate where oxygen (O_2) is drawn in from the atmosphere and broken down into oxygen ions (O^{2-}). The hydrogen ions are then allowed to pass through a membrane where they chemically combine with the oxygen ions, producing an electric current. Note that the only by-products of this reaction are electricity and water (H_2O)—fuel cells do not generate any carbon dioxide.

Nearly every major automaker in the world has tested prototype electric vehicles powered by hydrogen fuel cells. Although the technology works, there are several technical and cost issues that must be overcome before these vehicles will ever be mass produced. One issue is that since hydrogen molecules are so small, they tend to escape through the walls of just about any type of container. Moreover, hydrogen gas reacts with metals to produce hydrogen ions, which will penetrate the metal itself and make it brittle. An even bigger problem is that because of its low density, hydrogen gas contains far less energy than conventional gasoline. For a vehicle to have a reasonable driving range, the hydrogen gas must be either highly compressed or cooled to a liquid. For example, it takes 26 gallons of hydrogen, compressed to 2,200 pounds per square inch (150 atmospheres), to generate the same amount of energy that is obtained from a single gallon of gasoline. The result is a hydrogen-powered car with a rather limited driving range. Cooling hydrogen into a liquid increases the driving range, but this process itself requires energy and therefore makes the use of hydrogen as a fuel less cost-effective. Similarly, the low density of hydrogen means that tanker trucks could only transport relatively small volumes to future filling stations, thereby increasing the cost of the fuel.

In addition to cost and storage problems, hydrogen gas (H_2) is a secondary source of energy, which means it must be produced using energy from some other source. Most hydrogen used today is generated by using steam to extract the hydrogen from natural gas (CH_4). Unfortunately, natural gas supplies are finite, plus this process generates carbon dioxide gas. The other method of producing hydrogen is through a process called *electrolysis,* in which an electric current is used to split the water molecule—basically a fuel cell in reverse. This process requires that electricity first be generated to produce hydrogen, only to have the hydrogen converted back into electricity again, a grossly inefficient method that results in a net loss of energy.

For society to develop electricity-based transportation systems, vast amounts of electricity will need to be produced using renewable and non-carbon-based energy sources. In the next and final section we will take a brief look at how we may be able meet the future demand for electricity in all sectors of an economy.

Generating Electricity in the Future

One of the major difficulties in using electricity instead of fossil fossils is that electricity itself is mostly produced using coal and natural gas. For example, 61% of all electricity in the world is generated by power plants that burn coal and natural gas. Consequently, society will not only need to generate greater amounts of electrical power in the future, but also produce progressively less with fossil fuels. Considering the vast amount of energy we consume, accomplishing this goal will be difficult. The solution is to make use of existing technology to ramp up electrical production on a massive scale, and from as many non-carbon-based sources as possible. At the same time we should reduce our electrical consumption by expanding the use of solar heating and geothermal heat pumps. Equally important will be increasing our conservation and efficiency efforts. Note that carbon sequestration (Chapter 16) may make it feasible to continue using coal and natural gas, but it is still unclear whether this new technology will be cost-effective.

TABLE 14.1 Comparison of the different energy sources commonly used to produce electricity.

	Electrical Generation Method	Percentage of World Power Demand	Leveled Cost (cents per kilowatt-hour)	Advantages and Disadvantages
Carbon-Based and Nonrenewable	Coal	39%	9.5	Coal reserves are plentiful, but coal produces acid rain, heavy metal fallout, and more carbon dioxide than other fossil fuels.
	Natural gas	22%	7.5	Gas-fired plants are less expensive to build than coal or nuclear plants, and lower gas prices have decreased production costs. Gas burns cleaner than coal, but still produces carbon dioxide.
Carbon-Free and Limited Fuel	Nuclear	11%	9.5	Risk of radioactive contamination and long-term waste disposal issues make nuclear power subject to public opposition. A nuclear plant produces large amounts of continuous power with no CO_2 emissions, and breeder reactors can create long-term fuel supplies.
Carbon-Free and Renewable	Hydro	17%	8.4	Hydro power currently generates more of the world's electricity than all other renewable sources combined. Future potential is limited due to the limited number of dam sites and concerns over damage to aquatic ecosystems.
	Wind	3%	7.4	Wind power is a highly cost-effective and renewable source that is readily available. However, wind is geographically restricted and intermittent, and wind farms face some public resistance over noise and aesthetic changes to the landscape.
	Geothermal	<1%	4.4	Geothermal energy is clean and renewable, but is geographically limited to areas of high heat flow through the Earth's crust. Some geothermal reservoirs become depleted if heat is removed at too fast a rate.
	Solar thermal	<1%	13	Concentrated solar power plants with heat storage are cost-competitive and largely eliminate the problem of sunlight being intermittent. Production is limited to regions with fewer clouds.
	Solar voltaic	<1%	11	High cost of photovoltaic panels had limited their use, but costs have been dropping as production increases. Although sunlight is an intermittent source, photovoltaics have the potential to revolutionize electrical generation and make distributed energy production possible.
	Tidal	<1%		New tidal turbines are cost-effective, and the environmental impact appears to be low. Tidal power is geographically limited and cyclic in nature, but is highly predictable.

Source: Data from International Energy Agency and the U.S. Dept. of Energy.

One of the common objections regarding the use of carbon-free and renewable sources for generating electricity is that costs will be higher, and this will harm the economy. From Table 14.1 one can see that coal and natural gas produce the vast majority of the world's electricity, but are not necessarily the least expensive. Hydro, wind, and geothermal sources are highly cost competitive with fossil fuels; solar is also quickly becoming more competitive. In addition, wind power can easily be scaled up to provide vast amounts of cheap electricity. Keep in mind that the costs in the table do not include all of the environmental costs associated with fossil fuels (e.g., pollution and global warming). Proposed carbon taxes and cap-and-trade systems (Chapter 16) are designed to address the impacts of fossil fuels and to make alternative sources even more cost-competitive.

Another important aspect of electrical power is that it must be produced on demand (i.e., as it is needed). Because electricity is not being stored in large quantities, utility companies are forced to constantly generate enough electricity to meet demand, twenty-four hours a day, seven days a week. As illustrated in Figure 14.34, electrical demand fluctuates, reaching a peak in the afternoon

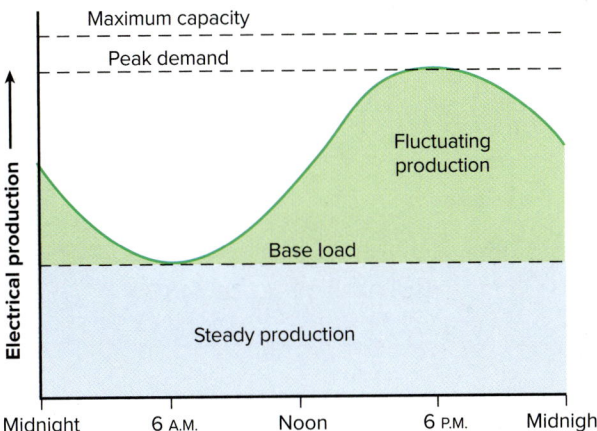

FIGURE 14.34 Electrical demand and production fluctuate on a daily cycle based on human and industrial activity. Demand goes up during the day as activity increases, then falls to a minimum or base level at night. Utility companies must provide a steady amount of power to meet base load requirements, yet maintain enough excess capacity to ensure that peak demand can be met throughout the year.

and then dropping to a minimum called *base load*. This base load is reached at night when human activity is at its lowest. Power plants therefore must continuously produce enough electricity to meet base-load demand, yet still have enough additional capacity to meet peak demand, each and every day. Electric utility companies generally prefer to use stored forms of energy (fossil fuels, nuclear, hydro) because this makes it easy to adjust production to meet the constant change in demand, simply by using more fuel. In contrast, intermittent sources of energy, such as solar and wind, do not provide the same flexibility for adjusting electrical production. Tidal power is highly reliable, but it is not continuous, due to the cyclic nature of tides. Hydro and geothermal energy are ideal because they are stored forms of energy and can produce power on demand, but suitable sites are geographically limited. Finally, there is ocean thermal energy conversion (OTEC), which also uses stored energy and can generate electricity on demand. The problem is that OTEC is restricted to tropical waters, and large-scale commercial production has yet to occur.

Ultimately there are two basic issues that need to be resolved for the world to replace fossil fuels as its primary means of generating electricity. The first is that, in order to help overcome the public fear of nuclear power, we have to develop a permanent means to safely dispose of the high-level radioactive waste being generated at nuclear power plants (see Case Study 14.1 and Chapter 15). The waste disposal and perceived safety issues must be addressed before nuclear energy, which provides large amounts of power on demand, can be expanded significantly. The other issue is that intermittent energy sources need to be integrated into the nation's electrical grid (collection of transmission lines) in such a way that utility companies are still able to meet daily peak demand. One solution would be to feed electrical power from solar, wind, and tidal sources into the grid intermittently, then use the flexibility of nuclear power plants to ensure that peak demand would always be met. Another option would be for power companies to produce excess electricity at night from wind and tidal, and then store the energy, for example, in a pumped storage facility as described earlier in this chapter (see Figure 14.9). Excess electricity could also be used to produce hydrogen. The hydrogen (stored energy) could later be used in fuel cells to generate electricity and help meet peak demand, or to fill intermittent gaps in production due to cloudy skies or decreased winds. The hydrogen could also be sold to filling stations to power cars that run on hydrogen fuel cells.

Some energy experts believe that developed nations, like the United States, will move away from the current centralized system where power is fed into the electrical grid from a relatively small number of large power plants. Instead, we will have a distributed system in which electricity is fed into the grid from numerous small producers as well as large power plants. As illustrated in Figure 14.35, a distributed system will likely include nuclear power plants to meet base-load demand and to ensure that demand can be met whenever intermittent sources are unable to provide sufficient power. Although more complicated, this system is feasible because solar, wind, and tidal resources are usually intermittent at different times and in different locations. For example, wind and tidal can be generating power at night when solar is not available, whereas solar can provide energy during the day to help fill gaps in wind and tidal production. Notice in Figure 14.35 that one of the key elements in a distributed power system is how individual homes and businesses would generate electricity during the day using photovoltaic panels. Individuals therefore could generate power to meet their own electrical needs, and transfer any excess power to the electrical grid or store it in home battery systems for use at night. The additional electricity could also be used to generate hydrogen gas from water, and then the gas could be stored in tanks for use in a hydrogen car. The hydrogen could also run a household fuel cell to provide electricity at night.

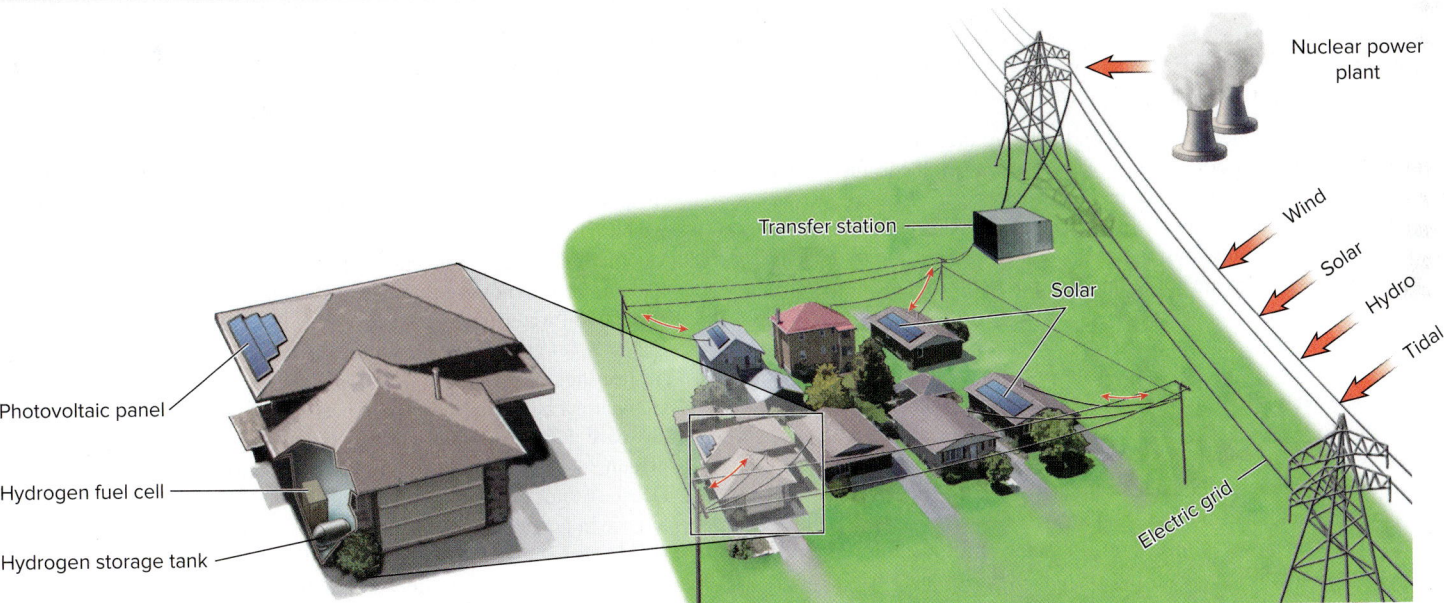

FIGURE 14.35 Homes and businesses receive power from an electrical grid system consisting of interconnected transmission lines. Most grids today receive power from a few centralized power plants, but as electrical production from alternative sources increases, grids will become more decentralized as shown here. Note how rooftop photovoltaic panels allow individual homes to feed excess electricity back into the grid, creating a more distributed system. Excess electricity could also be stored in home battery systems or used to produce hydrogen for filling cars that run on hydrogen fuel cells.

© Pioneer Valley PhotoVoltaics Cooperative

A distributed or decentralized power system will create challenges for engineers who will need to manage the fluctuating power contributions of all the various sources. Moreover, this type of system is becoming more and more economical as the cost of alternatives sources continue to fall. Consider that the cost of wind-generated electricity fell 40% since 2008 and that of electricity from solar photovoltaic panels fell 55% over the same period. Ultimately though, the need to make serious reductions in carbon dioxide emissions will likely force developed nations to meet more of their needs with carbon-free and renewable fuels.

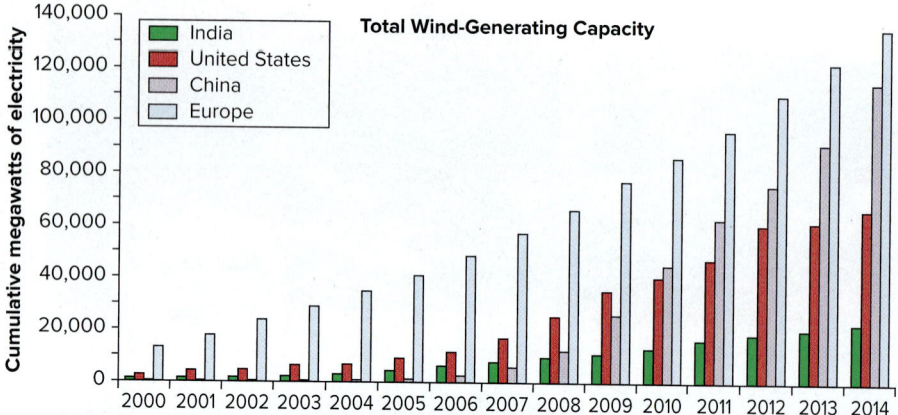

FIGURE 14.36 Graph compares the total wind-generating capacity of the world's top energy consumers. Although the United States and Europe produce about the same amount of electricity, Europe has been more aggressive in developing wind energy. Likewise, China overtook the United States in 2010 and is now the second-largest producer of wind power. Note that 1 megawatt of electricity will meet the needs of 400–900 American homes, depending on climate.

From the graph in Figure 14.36 one can see that European nations have already invested heavily and have taken the lead in the development of renewable energy, whereas the United States lags far behind. Also impressive is the rapid rate at which China is developing its wind resources, surpassing the United States in 2010. The basic difference has been that the European and Chinese governments have instituted policies and provided investment incentives that favor renewable energy. In contrast, U.S. government policies have tended to favor the use of conventional fossil fuels.

In the end, it will be necessary for all of the industrialized and developing nations to invest in alternative energy sources. By committing ourselves to the use of carbon-free and renewable fuels, we could help solve both the energy and global warming crises. In addition, such an effort would help the world develop a green economy in which workers would be needed to build and maintain new infrastructure, ranging from solar panels and wind turbines to expanded electrical grids and rail systems. Even more important, the move toward renewable fuels would be a critical step in creating a more sustainable global society.

SUMMARY POINTS

1. Tighter crude oil supplies and higher oil prices are making some nonconventional fossil fuels profitable to produce, such as synthetic fuels from coal, heavy oils, and oil sands. The problem is scaling up production to the point where these fuels can make a significant contribution to fuel supplies.

2. Oil shale and gas hydrates exist in large potential reserves, but these nonconventional fossil fuels are not yet produced in significant quantities due to economic and technical reasons.

3. Ethanol is considered a renewable substitute for gasoline, but provides only 72% of the energy of gasoline. Also, while ethanol made from sugarcane has been shown to be cost-effective, corn-based ethanol represents a net energy loss.

4. Hydroelectric power represents a clean and renewable source of energy, but further development will likely be limited due to the restricted number of suitable sites and the environmental impact of dams.

5. Nuclear fission is a proven means of generating large quantities of electricity and is carbon-free, but construction of new plants has slowed because of economic and safety issues and lack of long-term storage options for radioactive waste.

6. Passive solar heating involves having sunlight enter a building where it is converted into thermal energy. Active solar heating uses outdoor panels to collect sunlight and then convert the solar energy to thermal energy.

7. Photovoltaic or solar cells consist of sheets of silicon alloys that convert sunlight into electricity. The use of photovoltaic panels has been limited because they are somewhat expensive and have a long payback period. This technology is also less cost-effective at higher latitudes and in areas with frequent clouds.

8. Electricity can be generated by wind using a rotor-driven generator. Wind power has an advantage over solar in that it can generate electricity at night and on cloudy days, plus it is more cost-effective and has a shorter payback period.

9. Geothermal energy can cause shallow crustal rocks to become hot enough for groundwater to reach its boiling point. When brought to the surface the water turns to steam and is used to drive an electrical generator. If not at or above the boiling point, groundwater can still be used for heating.

10. Geothermal heat pumps exchange heat energy between the ground and the air space within a building. During the summer the constant ground temperature is used to cool a building, and in the winter it provides warmth.

11. Ocean thermal energy conversion (OTEC) is a technique that uses the difference in temperatures of ocean water from different depths to operate a simple heat engine for producing electricity. Land-based OTEC sites are restricted to tropical coastlines where the seafloor drops off rapidly.

12. Tidal power can be used to generate electricity when the incoming water is trapped behind a damlike structure called a barrage. Barrage systems have fallen out of favor because they are highly disruptive to marine ecosystems. Tidal turbines have less of an environmental impact and are now being used to produce electricity on a commercial basis.

13. To address global warming and the oil crisis, society needs to: (a) develop new transportation systems that emit minimal amounts of carbon dioxide; (b) generate electricity with non-carbon-based sources; and (c) scale up the production of alternative sources to meet the current demand for energy.

KEY WORDS

alternative energy sources   442
biofuels   449
energy conservation   471
gas hydrates   448
geothermal energy   465
geothermal heat pump   466
heavy crude oil   444

hydrogen fuel cell   473
hydroelectric power (hydro power)   451
light crude oil   444
nonconventional fossil fuels   443
nuclear fission   452
nuclear fusion   452
ocean thermal energy conversion (OTEC)   468

oil sands   445
oil shale   446
photovoltaic cell   459
solar heating   457
synthetic fuels   443
tidal power   469
wind farms   462

APPLICATIONS

Student Activity

Go through your house, apartment, or dorm and make a list of your activities that consume electricity—use of the refrigerator, dryer, hot water, lights, computer, TV, heating and air conditioning, and so on. Now make a list of things you could do to reduce your electrical consumption. Which would make the biggest difference in electrical usage? Do you think you could make a substantial reduction without too much inconvenience? Would paying a lower utility bill be worth the change in your personal habits? How would society benefit if almost everybody reduced his or her electrical consumption in a similar manner?

Critical Thinking Questions

1. Oil shale deposits in the Rocky Mountains region of the United States hold the equivalent of about twice the world reserves of conventional crude oil. Why then does the United States not develop this resource and become energy-independent?
2. Solar radiation can be transformed into both thermal (heat) and electrical energy. How is it that sunlight can do both?
3. Geothermal heat pumps transfer heat between a building and the ground, whereas conventional heat pumps transfer heat between a building and the outside air. Why does this difference make geothermal heat pumps cheaper to operate?

Your Environment: YOU Decide

© George Hammerstein/Corbis/Glow Images

NIMBY stands for "Not In My Back Yard." If a nuclear power plant were to be built near your home, would you object? Take into consideration that the plant would produce electricity without any greenhouse gas emissions and provide good jobs. The additional tax revenue would also boost the local school district and public services, and might even reduce local taxes. Would the risk of an accident outweigh the benefits to you?

Pollution and Waste Disposal

LEARNING OUTCOMES

After reading this chapter, you should be able to:

▶ Explain the difference between point source pollution and nonpoint source pollution.
▶ Describe why municipal waste is hazardous and how landfills can threaten our water supplies.
▶ Discuss why most communities no longer use local landfills, but instead ship waste to distant sites.
▶ Characterize how hazardous waste is disposed of.
▶ Describe the basic way sewage is broken down in septic systems and wastewater treatment plants.
▶ Explain how humans can cause excessive algae growth and oxygen depletion in rivers and lakes.
▶ Discuss why radioactive waste is a more difficult problem compared to other hazardous wastes.
▶ Describe what gases cause acid rain and the types of human activity that generate these gases.
▶ Explain how humans are increasing mercury in the atmosphere and how it threatens human health.

Humans create enormous amounts of waste that must ultimately be disposed of in the natural environment. Our waste material often ends up polluting the very water and air that we depend on for survival. Shown here is the daily routine of trucks unloading solid waste at a large landfill located near Hong Kong, China.

481

Also notice how some contaminants, such as fluoride, can be found at relatively high concentrations before posing a health risk, whereas dioxin is potentially deadly at exceedingly low concentrations.

Many of the contaminants regulated by the EPA are particularly worrisome because they impart no noticeable taste to drinking water and are potentially quite harmful at relatively low concentrations. Water that appears quite clean can actually be contaminated to the point where it is detrimental to human health. Moreover, scientists continue to be concerned because of the wide range of chemicals being introduced into the environment for which there is little health data. In the following sections we will examine the types of human activity that generate different contaminants and see how these contaminants enter our surface water and groundwater supplies.

Movement of Pollutants in Water

As illustrated in Figure 15.6, pollutants generally enter our water supplies in one of two ways. One is through a **point source,** which is a physically discrete point, such as a pipe discharging wastewater from a factory. Because point

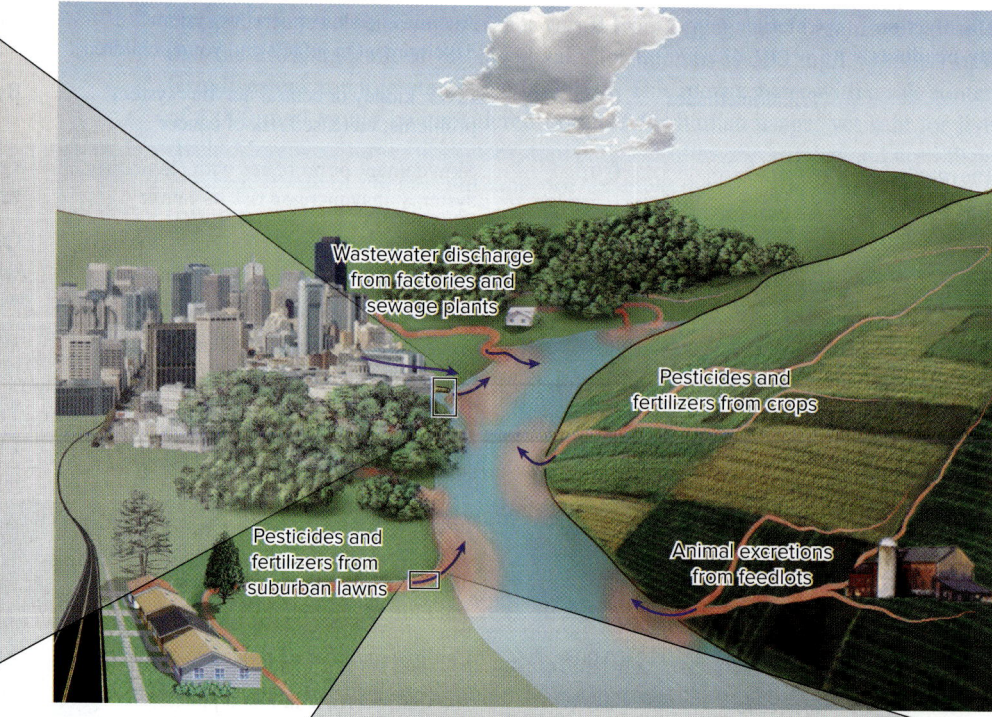

FIGURE 15.6   Pollutants entering our water supplies come from two types of sources. Point sources are discrete locations such as pipes discharging wastewater into streams from factories and sewage plants. Nonpoint sources are those from which contaminants get picked up by water as it moves across the landscape and then enter stream tributaries. Examples of contaminants from nonpoint sources include chemicals and animal wastes from agricultural fields and residential lawns and oil from parking lots and highways.

(left): © Sinisha Karich/123RF; (right): © Jim Reichard

sources are easy to identify, the EPA was able to quickly reduce water pollution by requiring wastewater be treated prior to being discharged into the environment. **Nonpoint sources,** on the other hand, are those which release contaminants over a broad area and commonly consist of multiple sites. Examples include agricultural fields, parking lots, golf courses, and lawns. Here overland flow (Chapter 8) carries both natural and human-made contaminants off the landscape and into nearby stream tributaries. At the same time, infiltrating water transports the contaminants into groundwater systems. Note that nonpoint source pollution has been more difficult for the EPA to control because the various sources are dispersed across the landscape. Consequently, nonpoint sources are now the leading cause of water pollution.

When a pollutant enters a body of water or an aquifer, it naturally begins to mix with the water so that it disperses and becomes less concentrated, forming what is called a *contaminant plume* (Figure 15.7). In the case of a surface body of water, differences in concentration will cause a plume to grow by simple *molecular diffusion.* If the water is also flowing, individual water molecules will cause contaminants to spread by what is called *advection.* Diffusion and advection also take place within aquifers, but here dispersion is enhanced due to the way water must flow around the individual particle grains. Surface and groundwater plumes are also affected by the way the pollutants are released. For example, a leaking underground storage tank may release pollutants continuously over a long period of time, whereas a tanker spill will release a large volume almost instantaneously.

When water becomes contaminated, its natural flow will eventually flush the contaminants from the system. In rivers and lakes this natural cleansing action can take place over a period ranging from days to years, whereas the sediment can remain contaminated for far longer. Because groundwater moves much more slowly than surface water (Chapter 11), it can take anywhere from decades to thousands of years for an aquifer to flush itself of contaminants. This means that once an aquifer is polluted, it is essentially no longer suitable as a drinking-water source. Although there are ways of remediating or reducing the concentrations of pollutants in groundwater, the low flow rate and granular nature of most aquifers make this exceedingly difficult—analogous to getting soap out of a sponge. It is far cheaper, therefore, to prevent groundwater contamination than to clean it up after the fact.

FIGURE 15.7   Contaminant plumes form when polluted water enters surface bodies of water like lakes and streams (A) or subsurface aquifers (B). Contaminants become diluted due to mixing as the plume grows in size, both laterally and in the direction of flow. Once a contaminant source has been eliminated, surface waters are generally able to flush themselves of pollutants relatively quickly. Groundwater moves much more slowly and can remain polluted for decades to thousands of years.
(a) US EPA

A Calumet River, near Chicago, Illinois

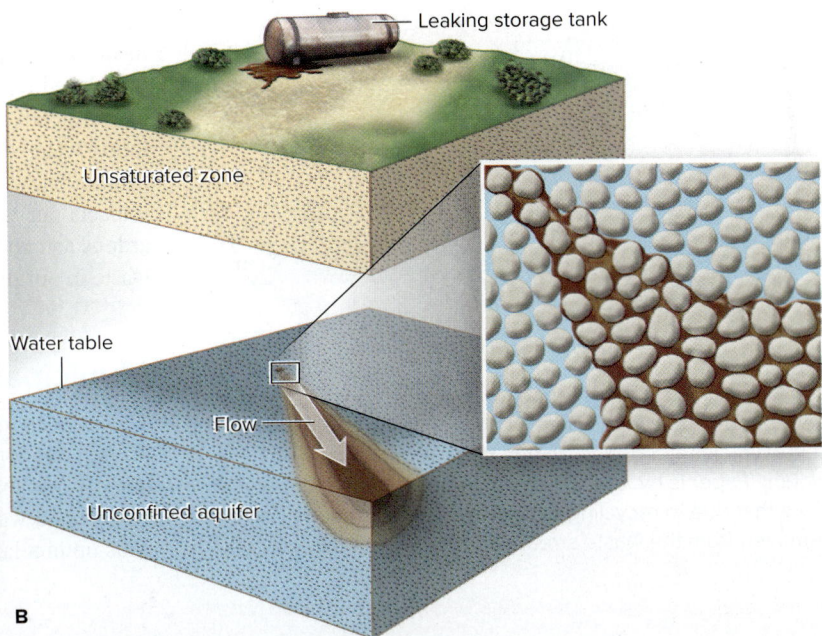

B

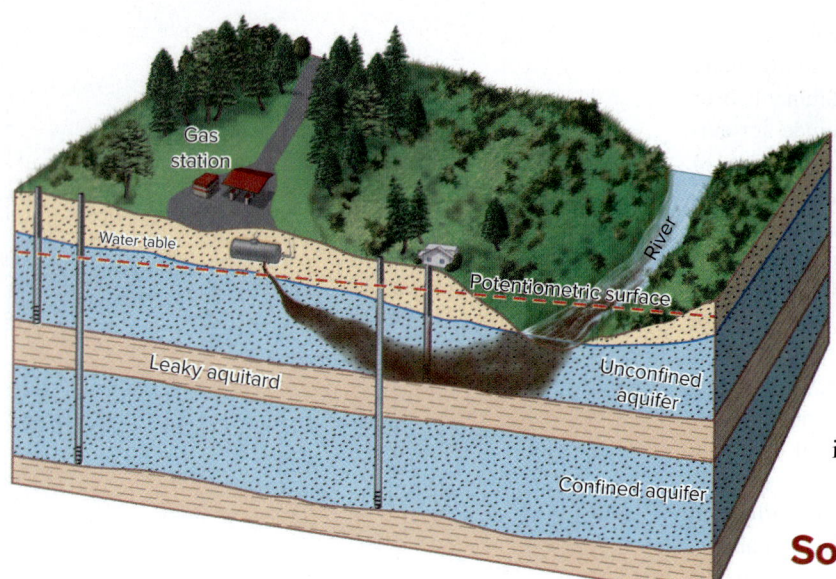

From Figure 15.8 one can see that an unconfined aquifer is highly susceptible to contamination from surface activities, whereas an aquitard provides a critical layer of protection for a confined aquifer. Notice that in areas where vertical leakage across an aquitard is upward, it is virtually impossible for an unconfined aquifer to contaminate the underlying confined system. Another key point is that a contaminant plume not only makes water-supply wells unsafe, but will eventually discharge into a stream or lake, polluting it as well. Finally, recall that streams receive overland flow only during storm events but receive groundwater on a continuous basis. Polluted groundwater then can cause serious water quality problems in streams, particularly during periods when stream flow is low and dilution is minimal.

FIGURE 15.8   Unconfined aquifers are highly susceptible to contamination from surface activities, whereas an aquitard provides a layer of protection for the underlying confined system. In areas where vertical flow is upward, confined aquifers are virtually immune from surface contamination. Note how human activity can easily pollute shallow aquifers, which eventually discharge into streams, polluting them as well.

Solid Waste Disposal

If you took a survey asking people in the United States what kind of solid waste they generate, common responses would include cardboard and plastic containers, beverage cans and bottles, and, of course, paper. Some might even remember to include discarded electronic items, tires, batteries, furniture, and so on. However, few people would think to mention all the industrial solid waste that is associated with the consumer goods they purchase. Take your car or truck, for example, which represents large volumes of rock removed from the earth and discarded as waste in order to extract the necessary metals. Each of us is also responsible for the waste generated to produce the food we eat, particularly meat. Farmers grow crops to feed cows, chickens, and pigs, which then turn the feed into large volumes of manure. Although we generally do not see industrial and agricultural wastes, we still are responsible for creating them.

The more familiar household garbage or trash typically collected by local governments is referred to as **municipal solid waste**—this includes trash from local businesses. The volume of municipal waste is surprisingly small, making up only 4% of all solid waste, whereas agricultural and mining wastes represent an astonishing 56% and 34%, respectively, and industrial wastes around 6%. Interestingly, only municipal and industrial wastes are collected and buried in landfills designed to keep hazardous substances from seeping into the environment. This means the vast majority of the waste from mining and agricultural activity is simply placed on the land surface. The problem is that much of this waste contains heavy metals or nutrients that can make their way into streams and aquifers, polluting water supplies and damaging ecosystems. In the following section we will examine how society handles different forms of solid waste and how this waste impacts our water supplies. Because solid industrial waste is generally nonhazardous and ends up in regular landfills, it will be lumped together with municipal waste in our discussion.

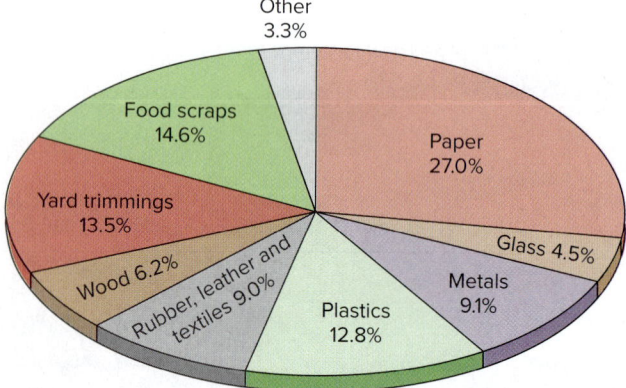

FIGURE 15.9   The stream of municipal solid waste generated in the United States can be broken down in several major categories, shown here as percentages by weight. Paper is by far the largest part of this waste stream. Note that due to recycling efforts much of this material is removed from the waste stream and is not buried in a landfill.

Municipal and Industrial Solid Waste

The term *solid waste stream* is often used to describe the steady flow of waste leaving our homes and businesses that society has to dispose of. From Figure 15.9 one can see that in terms of weight, paper products make up the largest portion of the solid waste stream in the United States. Of course, not all of this material ends up in a landfill, as much of it is either recycled or composted. Many people

FIGURE 15.10 Prior to 1976 municipal waste in the United States was routinely disposed of in open dumps. Health concerns eventually led to regulations requiring the use of sanitary landfills. Shown here is an open dump in Alaska, date unknown.
U.S. Fish and Wildlife Service/Photo by Luther Goldman

today are unaware that the disposal of solid waste in the United States has dramatically changed since the start of the environmental movement and onset of EPA regulations. Prior to those regulations municipal governments commonly placed local garbage in open dumps (Figure 15.10), which were often located in nearby ravines, wetlands, or abandoned gravel pits. However, the open nature of these sites attracted birds, rodents, and a variety of other pests. The dumps would also periodically catch fire, sending noxious fumes into the air. As early as the 1930s people recognized that open dumps posed a health threat to nearby residents, so engineers began developing so-called **sanitary landfills,** where the trash is compacted by heavy equipment and then covered each day with a layer of dirt (Figure 15.11). Open dumps were finally prohibited in 1983 when the EPA began requiring that all municipal waste be placed in sanitary landfills. These landfills effectively isolate the trash from atmospheric oxygen (O_2), greatly reducing decomposition rates due to the lack of oxygen-dependent (aerobic) bacteria. Landfills then became more like big preservation pits as opposed to compost piles.

Sanitary landfills were a big improvement over open dumps because the trash could no longer attract animals, catch on fire, or foul the air. As shown in Figure 15.11, the problem with sanitary landfills is that water can infiltrate through the trash, leaving shallow aquifers highly susceptible to contamination. Water percolating through a landfill can pick up bacteria, viruses, and various chemical compounds from the trash, forming a liquid referred to as **leachate**— similar to what would drain from a garbage can full of trash that had been soaking in water. The contamination of shallow aquifers by leachate is made worse by the fact that municipal trash commonly contains *hazardous household wastes,* including items such as batteries, paint, solvents, and insecticides. In the presence of water, hazardous compounds are leached from these materials and carried into the groundwater system.

In 1991 the EPA developed new standards for what are called **containment landfills,** which are designed to minimize the formation of leachate and prevent whatever does form from escaping into the environment. As shown in

FIGURE 15.11 Sanitary landfills are excavated pits where heavy equipment is used to compact and cover trash with dirt on a daily basis. Although an improvement over open dumps, these landfills do not prevent the escape of water that infiltrates and interacts with the trash to form leachate. The escape of leachate from landfills has led to widespread groundwater pollution.
© Doug Sherman/Geofile

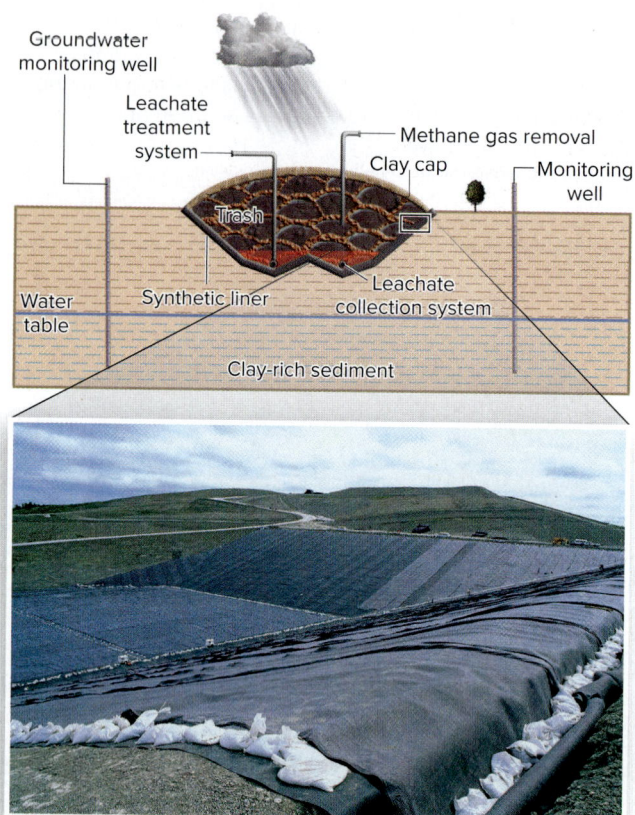

FIGURE 15.12 To prevent leachate from escaping, modern containment landfills use a compacted clay base overlain by an impermeable synthetic liner. For additional protection, landfills are typically located in areas of clay-rich sediment and surrounded by monitoring wells for detecting potential leaks. Clay caps are used to minimize infiltration, whereas drainage pipes allow for the collection and treatment of leachate. Methane gas is also recovered and can be used for heating or generating electricity.
© Doug Sherman/Geofile

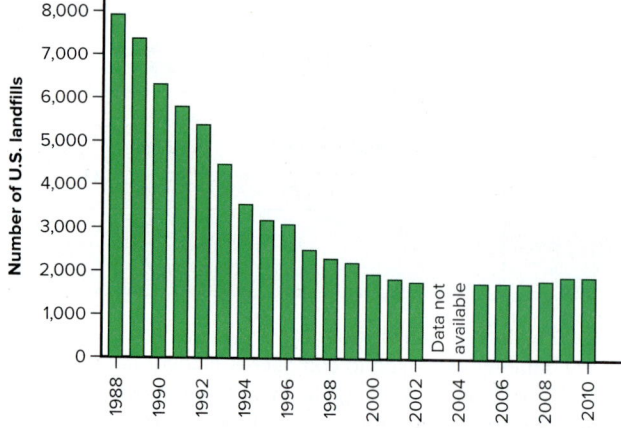

FIGURE 15.13 The number of landfills operating in the United States declined 75% after 1988 as new EPA standards caused many landfills to shut down. The shortage of landfills was compounded by public opposition to new landfills. This greatly increased the cost of solid waste disposal, as the remaining landfills were able to charge higher prices for their limited space.

Figure 15.12, modern containment landfills are typically excavated out of clay-rich materials. The bottom of the excavation is then covered with layers of compacted clay and synthetic liner material to prevent the escape of leachate; clay-rich sites are chosen because their low permeability will limit the movement of leachate should the containment system fail. Next, a system of pipes is installed above the liner to collect the leachate that forms after the landfill begins accepting trash. Once the landfill reaches its maximum capacity, it is covered with a clay cap and a final layer of topsoil. The cap reduces the infiltration of rainwater, thereby minimizing the formation of leachate—the vegetation helps reduce erosion and keeps the cap in place. Any leachate that does form is removed via the system of buried pipes and is either treated and discharged into the environment or disposed of as hazardous waste. The EPA also requires that a series of groundwater monitoring wells be installed around the landfill so that leaks can be detected. Because methane (natural) gas forms as the trash slowly decomposes, a containment landfill is also required to have a gas collection system. This prevents a dangerous buildup of the highly flammable gas, and provides the option of using the methane as an energy source for heating or generating electricity.

Although modern containment landfills have dramatically reduced the potential for groundwater contamination, thousands of older landfills still exist in the United States that were built prior to the newer EPA standards. Most of these older landfills were closed because they could not be upgraded to meet the new standards, which resulted in a 75% reduction in the number of active landfills after 1988 (Figure 15.13). Compounding the problem is the public opposition that invariably arises when new landfills are proposed. Understandably, people object over fears of groundwater contamination and lower property values. Although everyone wants to get rid of their trash, almost no one wants a landfill in their backyard—hence the expression NIMBY for "not in my backyard." This has resulted in a relatively small number of landfills servicing much broader areas than in the past. Consequently, most municipalities have built *waste-transfer stations,* in which local trash is pressed into bales and then loaded onto semi-tractor trailers for transport to distant landfills.

Another consequence of the dramatic decline in the number of U.S. landfills is that waste disposal is now more expensive. Companies that own landfills have something that is in short supply, namely, available space; hence they can command a higher price. The solid waste business today is quite lucrative, which helps explain the seemingly strange practice of trucking garbage hundreds of miles just to bury it in a landfill. Because of the increased cost, many municipalities have taken steps to reduce the amount of trash leaving their waste-transfer stations. The basic approach utilizes the concept of a **waste management pyramid,** which is a strategy for minimizing the amount of solid waste being sent to a landfill. As illustrated in Figure 15.14, the most desirable element of this strategy is *source reduction,* which involves efforts designed to keep material from entering the waste stream in the first place. For example, many municipalities no longer allow yard wastes to be mixed in with regular household trash, but instead have a separate curbside pickup for those wastes. The next most desirable element in the waste management pyramid is *recycling* and *reuse.* Recycling involves removing things from the waste stream that can be used again to make new products, such as paper, aluminum and glass containers, and various types of plastics and metals. Reuse entails using items like glass cups as opposed to disposable foam cups, as well as repairing items or donating them to a charity rather than simply throwing them away.

The last option in the waste management pyramid before landfilling is *incineration,* where combustible materials are removed from the waste stream and then burned at a high temperature. In the past, incineration was a common means of waste disposal for many hospitals, schools, and individual homeowners. However, there are materials in the solid waste stream that create dangerous by-products when combusted. The combustion of bleached (white) paper, commonly found in trash, is a particular problem as it releases the extremely hazardous compound known as dioxin (Table 15.1). Because of the air pollution issue, the EPA now requires that incinerator operators install air pollution controls to meet certain emission standards. Consequently, the incineration of municipal waste today is primarily handled by commercial operations where recyclables are recovered before the remaining waste is shredded for incineration. Some operations are also equipped to capture the heat from incinerated waste to generate electricity, thereby turning trash into valuable energy. Because burning trash generates carbon dioxide (CO_2), concern over global warming makes recycling and reuse a far more desirable option than incineration for reducing the volume of solid waste.

To appreciate the success of waste management programs, consider that the recycling rate in the United States went from 6.5% in 1960 to 34% in 2010. As shown in Figure 15.15, the combined effects of recycling, composting yard waste, and incineration have led to a reduction in the amount of solid waste being sent to landfills. Note that this reduction occurred despite the fact that the total amount of waste being generated increased, due to population growth and higher consumption rates. Although these waste reduction efforts have been successful, there is still room for improvement, as the recycling rates in Figure 15.16 indicate that considerable amounts of recyclable materials are still making their way into landfills.

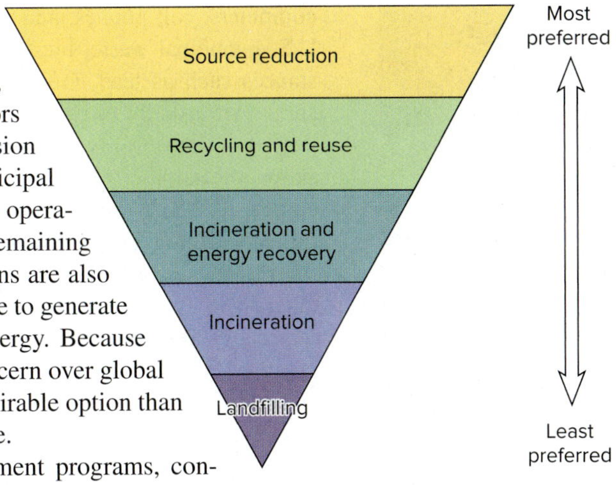

Solid Hazardous Waste

Earlier you learned that the EPA regulates a wide range of hazardous materials in a "cradle to grave" manner via RCRA. Industrial operations are required to keep records on the handling of all hazardous materials, including documentation on how and where the material was disposed. However, regular household trash is exempt from RCRA, which means citizens are allowed to place small amounts of hazardous household items in the municipal waste stream. Hazardous household material therefore ends up in a typical landfill. Industrial hazardous wastes, however, can only be disposed of in **secured landfills** that have special systems for collecting leachate and detecting leaks. Waste shipments require tracking permits and must be placed in separate containers and in specific locations within the landfill. Keep in mind that despite the greater safeguards, even secure landfills will someday begin leaking and allow leachate to escape into the environment.

In addition to secured landfills, solid hazardous waste can be disposed of in special high-temperature incinerators, commonly referred to as *hazardous-waste incinerators*. The waste is first shredded and sent into a primary combustion chamber, where the resulting ash and water collect at the bottom. Gases and unburned particulate matter then enter a secondary chamber and are burned at an even higher temperature. The remaining gases pass through a *scrubber system,* which basically removes everything but carbon dioxide (CO_2) gas and water vapor. However, scrubber systems are not completely efficient, so some hazardous material escapes into the atmosphere. In addition, the ash that remains is a hazardous solid and must be disposed of in a secured landfill. Opponents of high-temperature incineration point out that incineration does not actually destroy most of the hazardous waste, but converts it into fine ash and hazardous compounds that are released into the atmosphere. Nevertheless, high-temperature incineration is effective at destroying certain types of hazardous materials, such as pathogens in medical wastes and various types of petroleum-based solvents and oils.

A relatively new problem in solid waste management is discarded electronic products, called **electronic waste** or *e-waste*. The EPA reports that TVs,

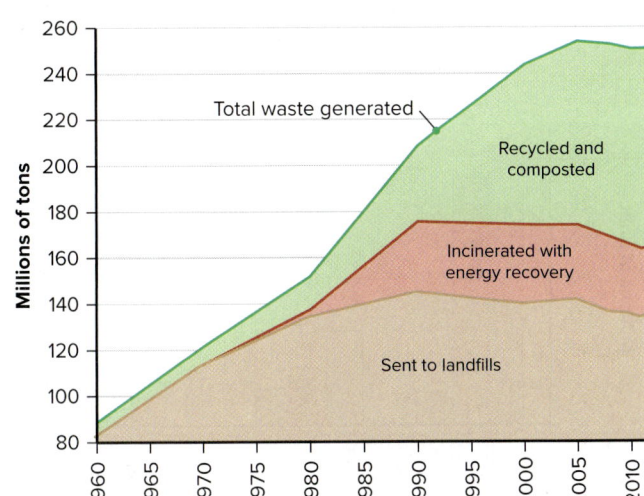

FIGURE 15.14   A waste management pyramid describes the strategy many communities use to minimize the amount of solid waste they must send off to a landfill. The most desirable option is to keep unnecessary waste from entering the waste stream in the first place. For the waste that is received, recycling, reuse, and incineration are ways of reducing the volume sent to a landfill.

FIGURE 15.15   Despite an overall increase in the amount of solid waste being generated, the combined efforts of recycling, composting, and incineration have actually reduced the amount of waste being sent to landfills.

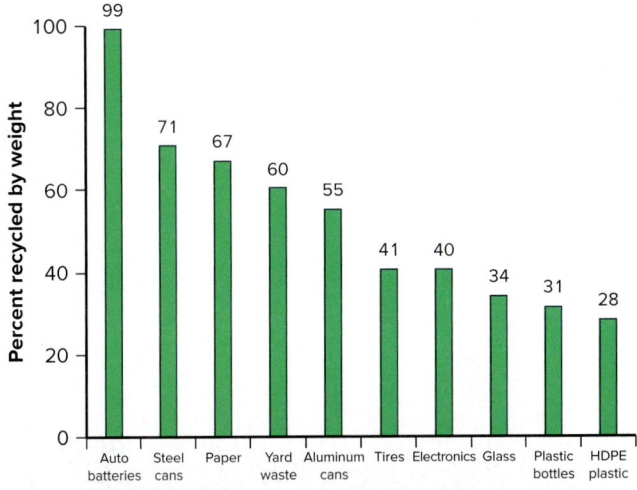

2013 U.S. Recycling Rates of Selected Materials

FIGURE 15.16   Percent recycling rates of selected materials in the United States in 2013. The recycling of these materials not only saves energy and valuable landfill space, but with the exception of yard wastes, also serves as a source of raw materials that can be used to make new products.

A

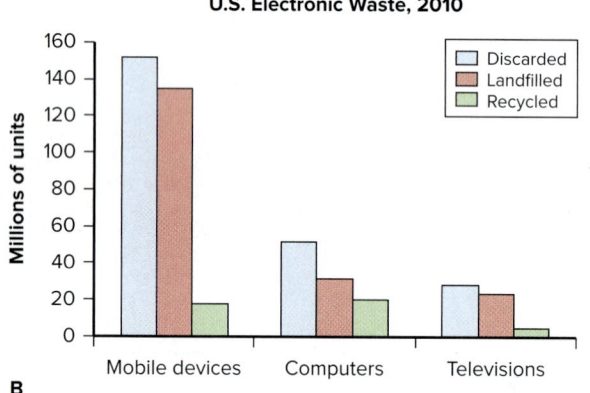

B

FIGURE 15.17   The volume of electronic waste has grown considerably in recent years (A), creating a disposal problem as many items contain toxic metals and plastics. To reduce the amount of electronic waste going to landfills, the EPA and state and local agencies have set up reuse and recycling programs. As indicated in the plot (B), only relatively small percentages of mobile devices, computers, and televisions are currently being recycled.
(a) © Johner Royalty-Free/Getty Images

computers, cell phones, and assorted other devices make up less than 2% of all U.S. municipal waste, but the electronic components contain highly toxic substances such as lead, mercury, and cadmium. Moreover, the plastic cases and cables are usually coated with bromine-based flame retardants, which are suspected of causing a variety of health problems, including cancer. Figure 15.17 shows the number of obsolete mobile devices, computers, and televisions discarded in 2010 in the United States. Of the 152 million mobile devices, 89% went into a landfill and only 11% were recycled. Computers and televisions had higher recycling rates, 40% and 17%, respectively, but the majority still ended up in a landfill. Recycling of electronics is important because it helps keep hazardous materials out of landfills, and it is a valuable source of metals, such as gold and copper. Unfortunately, many of the electronic components are being recycled in countries with weak environmental laws, creating serious health risks to workers. Because of the threat to people and the environment, the EPA and different state agencies have instituted electronic waste management programs. Although the programs vary considerably, the basic purpose is to increase the recycling rate of electronics and help ensure that recycling is done by certified companies that follow standard safe practices.

The proliferation of electronics in recent years has also caused a sharp rise in the use of *dry-cell batteries,* which contain toxic heavy metals. The EPA estimates that Americans purchase nearly 3 billion dry-cell batteries per year (Figure 15.18). In contrast, the batteries used in automobiles and boats are referred to as *wet-cell batteries* because they contain lead plates surrounded by acid. Retailers now exchange used wet-cell batteries with the purchase of a new one, which has resulted in a 99% recycling rate in the United States. Until recently, consumers have had few options for dry-cell battery disposal other than tossing them into the trash. Similar to electronic waste, special battery recycling programs are being implemented across the United States at the state and local levels.

FIGURE 15.18   This pile of depleted dry-cell batteries, collected in just one year by the author, illustrates the size of the battery disposal problem. Because batteries contain heavy metals that are highly toxic, many state and local governments are instituting battery recycling programs to help remove batteries from the municipal waste stream.
© Jim Reichard

Scrap Tires

There are literally hundreds of millions of cars around the world, all of which go through several sets of tires in their lifetime. Back in 2003 the EPA estimated that the United States alone generates nearly 290 million scrap tires each year, taking up landfill space and creating hazards in open dumps on the land surface (Figure 15.19). Due to the way tires hold rainwater, tire dumps make excellent breeding grounds for mosquito-borne diseases. These dumps also pose serious environmental hazards when the tires catch fire. The burning tires not only produce toxic smoke, but the intense heat releases liquid petrochemicals that end up contaminating shallow aquifers and nearby streams. All but five U.S. states currently have laws and recycling programs that deal specifically with the handling of scrap tires. Most states now require that tires be shredded prior to being placed in landfills—some states ban even shredded tires in landfills. A key part of most of these programs has been developing markets for the shredded tires themselves. Today, scrap tires are used as a combustion fuel and for making rubberized asphalt for running tracks, dock bumpers for boats, highway crash barriers, floor mats, soaker hoses, and more. According to the U.S. Rubber Manufacturers Association, in 1990 there were approximately 1 billion scrap tires being stockpiled in the United States, but due to the market for used rubber, the stockpile was reduced to around just 75 million tires by 2013. Approximately 53% of the discarded tires are now used as combustion fuel, 24% are processed into ground rubber, and only 8% are placed in landfills. It should be noted that researchers are currently investigating reports that recycled tire material in athletic and playground surfaces poses a serious human health risk. As the recycled material becomes weathered, it breaks down into small granules, at which point toxic compounds in the rubber can be released into the environment.

Liquid Waste Disposal

Recall from Chapter 11 that in terms of per capita water consumption, each of us shares responsibility for the water being used to manufacture products, irrigate crops, and generate electricity. What is important here is that nearly all of our nondrinking applications cause the water to become contaminated to one degree or another. Scientists and engineers use the term **wastewater** to refer to water that has become contaminated during some human-related process. Perhaps the best example is how the use of flush toilets turns high-quality drinking water into wastewater, or *sewage*. Another example is the wastewater generated when drinking water is used for processing animals into the meat we purchase in grocery stores.

In addition to wastewater, modern societies generate a variety of hazardous liquids, including acids, bases, and numerous petroleum-based products, ranging from motor oil and gasoline to agricultural pesticides. Prior to the Clean Water Act, industries commonly disposed of both liquid wastes and wastewater by piping it directly into nearby streams. Because most petroleum-based solvents are *immiscible* and do not mix with water, those molecules that are denser than water will sink and the lighter ones will float. Dense hydrocarbon compounds therefore tend to accumulate in the sediment of streams and lakes, whereas the lighter compounds float on the water's surface. Today, the EPA regulates a wide variety of toxic substances, setting limits on the amounts a company can discharge along with its wastewater. In some cases the liquid waste is simply too toxic, or the volume too large, for it to be included in the wastewater. When this is the case, the waste must either undergo a treatment process or be disposed of as a hazardous material and permanently isolated from the environment. Compared to solid wastes, liquid wastes much more difficult to isolate because their fluid nature enables them to migrate more easily.

FIGURE 15.19   In the past, scrap tires have been either placed in landfills or stockpiled on the surface in large open dumps, such as the one shown here in Saskatchewan, Canada. Such dumps created problems with mosquito-borne diseases and produced toxic fumes when the tires caught fire. In the United States the recent market for used tires has greatly reduced the stockpile of discarded tires.
Government of Saskatchewan

FIGURE 15.20   Considerable amounts of liquid hazardous waste stored on-site have been left abandoned by bankrupt companies, leaving costly cleanup efforts to government agencies, as shown in this photo taken in Illinois.
Illinois EPA

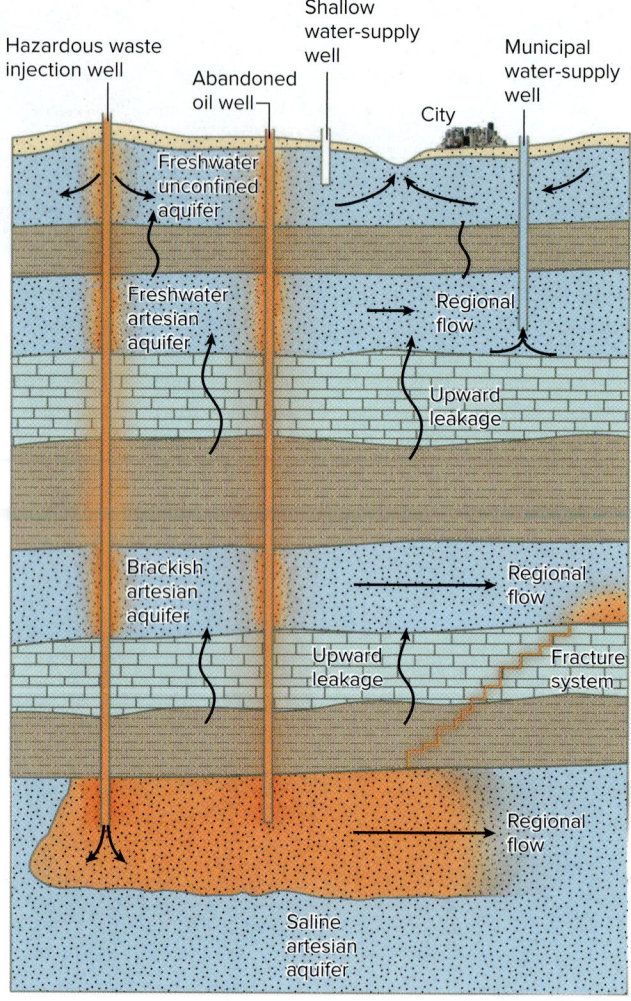

FIGURE 15.21   Deep-well injection involves disposing of liquid hazardous waste by pumping the material into a deep, saline aquifer that is not used as a water supply. Shown here is a worst-case scenario where groundwater naturally leaks upward, allowing the waste to contaminate freshwater aquifers via fractures and abandoned oil wells that have not been properly sealed. Contamination can also occur due to leakage along the casing of the injection well.

Liquid Hazardous Waste

The EPA estimates that of all the industrial waste generated in the United States, 97% is in the form of wastewater. With passage of the Clean Water Act and RCRA regulations, pollutants could no longer be freely dumped into rivers and lakes. Industries were then forced to develop better management strategies for handling their liquid wastes. As with solid waste management, businesses learned that the most cost-effective approach is to reduce liquid waste and recycle as much of it as possible, thereby minimizing more expensive disposal options, particularly landfilling. The result has been more efficient manufacturing processes that generate less hazardous waste. For the waste that is created, companies employ different wastewater treatment processes in which harmful substances are chemically broken down or neutralized before being released into the environment. Unfortunately, wastewater treatment is not effective or practical for removing all hazardous compounds, leaving EPA-approved disposal methods as the last option. Note that companies can temporarily store wastes on-site while waiting for treatment or permanent disposal. In the past, however, this has led to leaks and the escape of hazardous liquids. Also, bankrupt companies have often simply abandoned the waste, leaving costly cleanup efforts to government agencies (Figure 15.20).

One common disposal technique is to take containers filled with liquid waste to a *secured landfill.* Another option is known as *deep-well injection* or *deep-well disposal,* whereby hazardous liquids are pumped into a deep aquifer that is far removed from shallower aquifers being used as water supplies (Figure 15.21). This practice originated in Texas during the 1930s when oil companies began injecting wastewater from production wells back into petroleum reservoirs in order to maintain fluid pressure and increase oil production. Waste operators eventually began using injection wells and old petroleum reservoirs to dispose of hazardous liquids from manufacturing industries. Because of its long history of oil and gas production, Texas today leads the United States in deep-well disposal.

Deep-well injection is rather controversial despite the fact abandoned petroleum reservoirs typically contain only saline groundwater, and are separated from water-supply aquifers by a thick sequence of sedimentary rocks. The problem is that freshwater aquifers can become contaminated if the injected waste flows toward the surface along fractures or around the steel casing of the injection well, as shown in Figure 15.21. Of particular concern are abandoned oil and gas wells that have not been properly plugged and sealed. These wells serve as open conduits and allow the waste to move quite rapidly. Proponents of deep-well injection, however, claim that such contamination is actually quite rare and that the practice is still far safer than placing hazardous liquids in a landfill at the surface. In addition to concerns over groundwater contamination, the U.S. boom in tight oil and gas production (Chapter 13) has resulted in large volumes

FIGURE 15.22 In order to reduce the amount of hazardous household materials entering the municipal waste stream, some state and local governments now hold hazardous waste roundups where citizens can drop off household wastes for proper disposal by trained personnel. Photo from the State of Illinois.
Illinois EPA

of wastewater from hydraulic fracturing being disposed of by deep-well injection. The injection of this additional wastewater from hydraulic fracturing has coincided with a rise in earthquake activity around injection centers in parts of Oklahoma, Texas, Colorado, New Mexico, and Ohio. The injected wastewater increases the pore pressure in deep rocks, which reduces the frictional resistance along faults such that the rocks rupture and release stored seismic energy, causing an earthquake (Chapter 5). Although the majority of the earthquakes have been small, some have been large enough to cause damage.

The last option for disposing of hazardous liquid wastes is high-temperature incineration. Incineration is becoming a popular alternative to a secured landfill or deep-well disposal since the waste material is "destroyed," thereby eliminating the potential risk of leaks associated with long-term storage. However, high-temperature incineration is controversial because it creates hazardous ash, plus the scrubber systems for removing hazardous gas emissions are not 100% efficient. In the case of hazardous organic liquids derived from petroleum products, emissions from incineration are less of an issue because if done properly, the only combustion by-products are carbon dioxide (CO_2) gas and water vapor.

As with solid waste, typical household waste contains hazardous liquids, but these are exempt from RCRA regulations and therefore end up in the municipal waste stream. Examples of hazardous household liquids include used motor oil, insecticides, pesticides, oil-based paints, pool chemicals, and a wide variety of solvents and cleaning fluids. While the contribution from a single household is generally small, landfills are laced with hazardous liquids because nearly everyone at some point discards unused or partially used liquid products. To help reduce the amount of hazardous liquids entering the municipal waste stream, the EPA and state agencies have developed public awareness campaigns that encourage people to buy hazardous liquids more sparingly. Also, some state and local governments hold hazardous waste roundups where citizens are encouraged to bring unwanted household hazardous wastes to a collection site on specific days (Figure 15.22). Trained personnel can then safely place the materials in drums for proper disposal. Note that some states and municipalities now have regulations that are more stringent than RCRA regarding hazardous household wastes. Therefore, depending on where you live, it may be illegal to place certain household chemicals, batteries, electronics, and so forth, in the trash. Check your state or local government's website to find out where to dispose of such things in your area.

Human Waste

As late as the 1930s it was common for towns and cities in the United States to collect raw sewage (human waste mixed with water) from homes and businesses, and then discharge it directly into rivers and lakes. In rural areas people would dig a pit and place a wooden structure called an *outhouse* or *pit toilet* over the hole.

FIGURE 15.23 Photo showing a section of the Canoochee River in coastal Georgia that is completely covered with algae and experiencing severe oxygen depletion. The algae are fed by excessive nutrients, then die and fall to the bottom of the river where aerobic bacteria break down the material and remove oxygen from the water.
© Sylvia Smith-Reynolds

In the pit human waste is partially broken down by *anaerobic bacteria,* which do not require oxygen, and converted into compounds such as ammonia (NH_3). The wastewater within the pit then seeps into the ground and the remaining solids, called *sludge,* continue to accumulate, eventually forcing the owner to dig a new pit and move the outhouse. Because of the threat that serious diseases can spread into water supplies, pit toilets and the discharging of raw sewage are no longer acceptable practices in many countries.

Due to the things people eat and drink, human waste largely consists of organic compounds rich in carbon, nitrogen, and phosphorus, which are all natural fertilizers. However, our waste also contains dangerous bacteria and viruses that threaten our water supplies and can lead to serious human illness and disease. Moreover, when *aerobic bacteria* break down organic matter in an oxygen-rich environment, the bacteria consume free oxygen (O_2) and produce carbon dioxide (CO_2). The problem is that the bacteria will keep consuming oxygen until all of the organic matter is decomposed, or the available oxygen is used up. Scientists use the term *oxygen depletion* to refer to the undesirable situation where aerobic bacteria remove enough of the dissolved oxygen from a body of water that fish and other aquatic life begin to die. To gauge the risk of oxygen depletion, biologists measure what is called *biochemical oxygen demand (BOD),* which is the amount of oxygen microorganisms would need to break down whatever organic matter is present within the water. When the BOD is unusually high, it is a sign that raw sewage may be entering a water body, creating the risk for disease and oxygen depletion. Oxygen depletion, however, can also occur due to *excess nutrient loading,* which typically takes place in areas where agricultural fertilizers are used and then rain washes nitrogen and phosphorus off the land into streams and lakes. These nutrients then promote excessive growth of algae (Figure 15.23), which eventually die and are broken down by aerobic bacteria. Large amounts of aerobic bacteria will then lower the oxygen content of the water to the point where aquatic organisms begin to die. Next, we will examine how pathogens and oxygen-demanding wastes are removed from wastewater before it is discharged into the environment.

Septic Systems

A septic system is a means of treating wastewater from homes and small businesses that are not connected to municipal sewage systems. The EPA estimates that septic systems are used in 25% of all American homes.

As illustrated in Figure 15.24, the system consists of two components: a *septic tank* and a *drain field.* Here wastewater is piped into a sealed concrete tank, which is designed to prevent the waste from escaping. Most of the suspended organic matter settles out within the tank, forming a layer of sludge; the remainder either stays suspended or floats as a layer of scum. Because the tank is sealed, it contains very little oxygen, which means the organic wastes are only partially broken down by anaerobic bacteria—similar to what occurs in a pit toilet. When additional wastewater enters the tank, some of the partially clarified wastewater will flow into the drain field, where perforated pipes allow it to percolate through a bed of gravel. Because the gravel bed and underlying sediment are in the unsaturated zone, most of the remaining organic matter and harmful pathogens are either destroyed by aerobic bacteria or filtered out as the wastewater percolates through the sediment.

Although a properly constructed and maintained septic system can last almost indefinitely, the EPA estimates that 10–20% of systems fail to operate properly. Perhaps the most common cause is that the drainage field is located in an area that is poorly drained. The gravel bed and underlying sediment

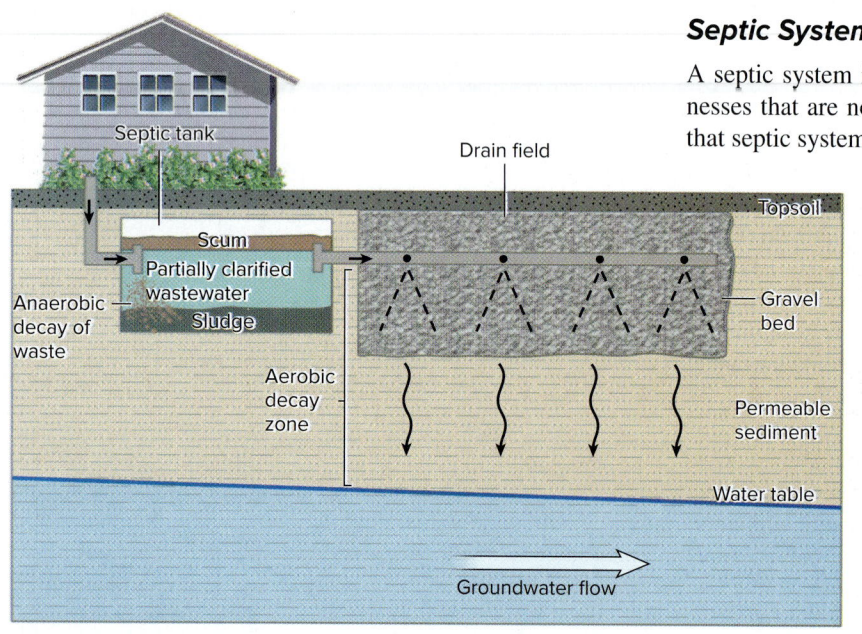

FIGURE 15.24 Septic systems are used for household wastewater in areas that are not connected to municipal sewage systems. Wastewater first goes into a holding tank where solids settle out and anaerobic bacteria partially break down the organic matter. Water and suspended waste eventually leave the tank and flow into a drain field and then percolate through an unsaturated gravel bed. Here the remaining organic matter and pathogens are removed by aerobic bacteria.

then become saturated, preventing the final aerobic breakdown of the waste. Poor drainage is usually the result of the system being built too close to the water table, or the underlying sediment not being sufficiently permeable. A septic system can also fail when a homeowner neglects to have a service company periodically pump the sludge from the tank. Over time, the sludge will build up to the point where solids begin moving into the pipe that leads to the drain field (Figure 15.24). This eventually causes the drain field to become plugged with particulate matter, thereby limiting the amount of oxygen and inhibiting the aerobic breakdown of the waste.

While a properly functioning septic system is an effective means of treating wastewater, environmental problems can still occur. Because septic systems remove very little of the nitrogen and phosphorus from the wastewater, these nutrients commonly end up in nearby water bodies, causing excessive algae growth and oxygen depletion. Excessive nutrient loading can be a real problem around recreational lakes, where large numbers of homes and cottages use septic systems. Another problem is that hazardous household chemicals are often incorporated in the wastewater, but are not broken down by either anaerobic or aerobic bacteria. The hazardous chemicals end up either settling out in the septic tank or percolating through the drain field, contaminating shallow aquifers and nearby streams.

Municipal Wastewater Treatment

Wastewater from homes and businesses in U.S. cities is typically collected by sanitary sewer lines and then piped to a centralized facility called a *wastewater treatment plant*—commonly known as a *sewage plant*. Here the wastewater undergoes a series of treatment processes before being discharged into a body of surface water. As shown in Figure 15.25, *primary treatment* involves using a large screen to first remove any coarse debris. The wastewater then flows into a series of chambers where solid particles are allowed to settle out, forming a layer of sludge. After this, the wastewater undergoes *secondary treatment* in aeration tanks, where fine bubbles of air move upward through the waste. With a nearly unlimited supply of oxygen, aerobic bacteria are able to flourish and convert

FIGURE 15.25   Modern wastewater treatment uses a series of processes to separate out solids and break down oxygen-demanding organic matter. Primary treatment involves the settling out of solids, and is followed by secondary treatment where aeration allows aerobic bacteria to break down most of the remaining waste. A final disinfection step is often used to eliminate any harmful bacteria that may remain. Tertiary treatment is sometimes included to reduce the amounts of nutrients and certain hazardous compounds.
Photo by Lynn Betts, USDA Natural Resources Conservation Service

Las Vegas, Nevada

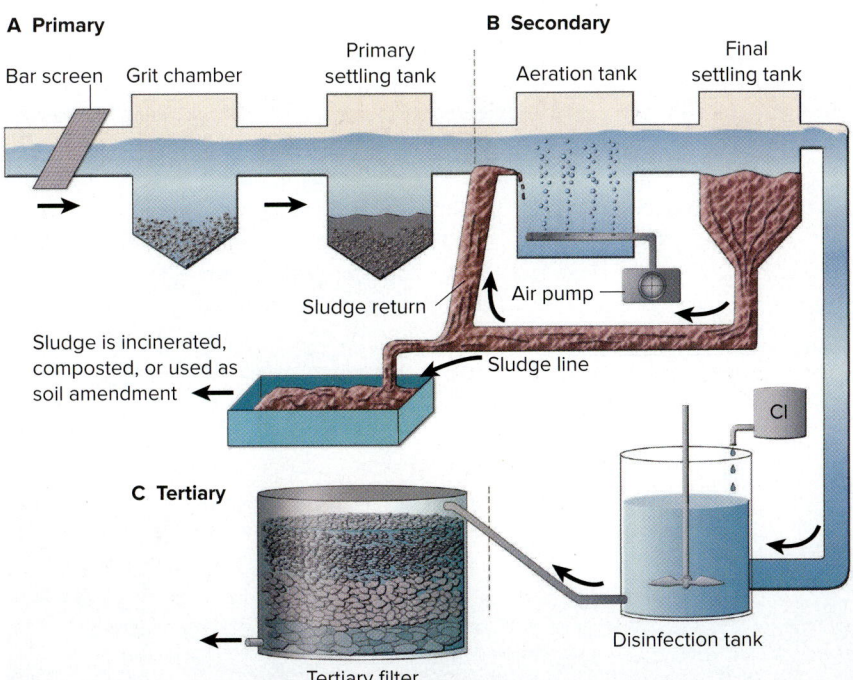

much of the remaining organic matter into harmless solids and carbon dioxide gas. Some plants include a final step where the wastewater is disinfected with either chlorine or ozone in order to kill any remaining pathogens.

The combination of primary and secondary treatment results in approximately 99% of all harmful bacteria being removed from the wastewater along with 90% of the oxygen-demanding organic wastes. However, as with septic tanks, the treatment process removes only small amounts of nitrogen and phosphorus. The discharge from a wastewater treatment plant therefore can lead to excessive nutrient loading and oxygen depletion. Note that primary and secondary treatment is also not very effective in removing heavy metals and many types of hazardous compounds. Consequently, some municipalities use *tertiary* or *advanced treatment* techniques that can remove nutrients and certain types of hazardous compounds.

Although wastewater treatment systems in the United States have been effective in eliminating waterborne disease, they also produce large amounts of sludge that must be disposed of. Sewage sludge has historically been applied to agricultural fields, as it contains many valuable nutrients necessary for plant growth. However, heavy metals and other toxic substances tend to accumulate in sludge during wastewater treatment. When sewage sludge is applied to the land surface, the heavy metals become immobile and accumulate in the soil, potentially making soil toxic over time. Because of the heavy metal issue and the fact that sludge has a foul smell, people living in the area commonly oppose the spreading of sewage sludge. As a result, sludge disposal is now moving away from land applications and toward the more expensive options of landfilling and incineration.

Another common problem involving wastewater treatment dates to when cities used to discharge raw sewage into rivers, lakes, and oceans. Municipalities built **combined sewer systems** consisting of a single set of underground pipes for collecting both sewage and the storm water runoff from streets and parking lots (Figure 15.26). When the time came to build sewage treatment plants, municipalities were faced with the prospect of having to lay separate collection lines. Rather than laying the additional lines, many cities found it more cost-effective to use their existing collection systems and run both storm water and sewage through the new treatment plants. The problem is that during large storms, the combined flow

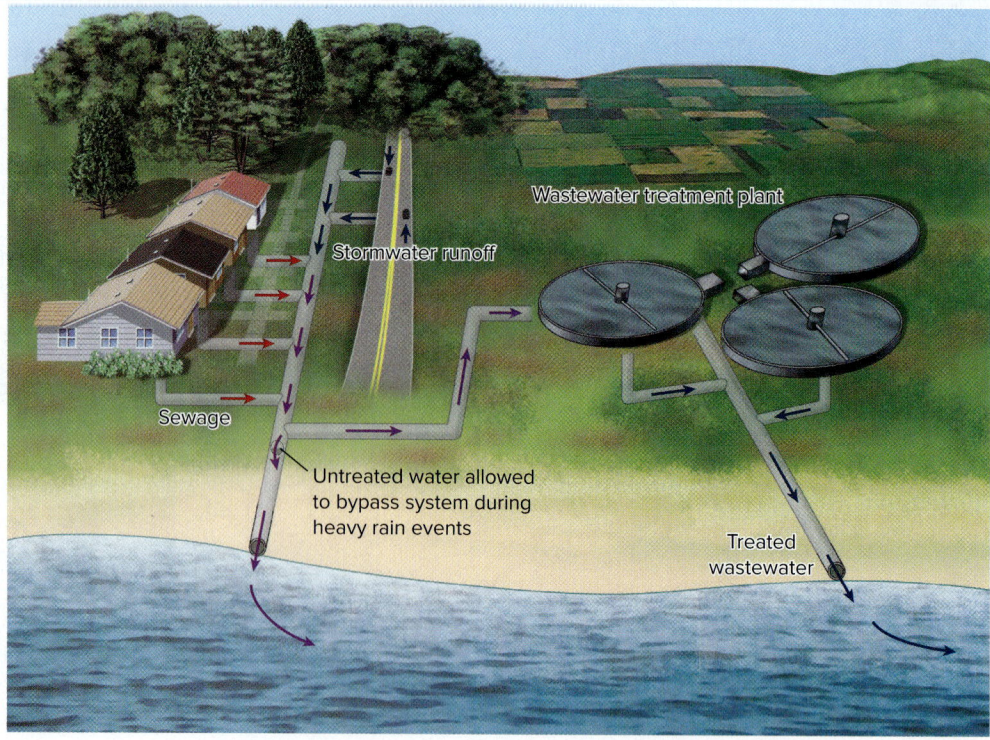

FIGURE 15.26   Many older cities have combined sewer systems rather than separate collection systems for sewage and storm runoff. During heavy rains or rapid snowmelt, the runoff is mixed with sewage and allowed to discharge directly into streams so as to not overwhelm the treatment plant. The raw sewage can cause a host of problems, particularly for cities located downstream that use the river for their water supply.

can become greater than what a treatment plant is designed to handle. When this occurs, the excess flow is allowed to bypass the plant and discharge directly into a body of surface water. This periodic release of raw sewage not only has a negative impact on aquatic ecosystems, it creates a potential human health hazard. Municipalities that use rivers as their source of drinking water must take extra care to ensure that their water-supply plants can eliminate the harmful bacteria associated with the periodic release of raw sewage from cities located upstream.

Agricultural and Urban Activity

There are many land use activities in rural and urban settings that end up releasing toxic substances or excessive sediment and nutrients into our water supplies. The result is widespread nonpoint source pollution. For example, consider the chemicals that farmers and suburban homeowners apply to different types of plants. A significant portion of what they apply either infiltrates or is washed off the landscape by overland flow, polluting rivers, lakes, and aquifers. In this section we will explore some of the more important land use activities related to nonpoint source pollution.

Agricultural Chemicals

Over the past 50 years, food production in the United States and other developed nations has increased dramatically due in part to modern machinery and irrigation techniques. Equally important has been the widespread use of inorganic fertilizers for promoting plant growth and petroleum-based synthetic pesticides for killing insects and herbicides for controlling weeds (Figure 15.27). These same chemicals are also used in urban areas to maintain landscaped areas around homes and businesses. For an example of how much we rely on these chemicals, consider that for pesticides alone, there are over 20,000 products registered with the EPA. While the benefits of synthetic chemicals for increasing food production and beautifying urban areas is quite clear, they also cause water pollution.

Given the widespread use of synthetic chemicals, it is not surprising that they are being detected in our rivers and shallow aquifers. Pesticides are of particular concern with respect to human health because they are specifically designed to kill living organisms. For two decades, from 1992 to 2011, the U.S. Geological Survey (USGS) tested streams for pesticides in both rural and urban areas and compared the results to human health and aquatic life benchmarks. These studies detected pesticides in more than 90% of the streams that were tested. However, only one stream had concentrations above benchmark standards for human health— keep in mind that just because trace amounts of pesticides can be detected does not necessarily mean that the concentrations are harmful to people. With respect to aquatic organisms, the studies found that approximately half the streams sampled had pesticide levels likely to have an adverse impact on aquatic life. Also of concern was that a large number of the streams had contaminated fish, whose pesticide concentrations were high enough to pose a risk to people who ate the fish.

FIGURE 15.27   Pesticides and herbicides are widely used in modern agriculture to control insects and weeds. These toxic substances end up in our streams and aquifers where they are found in relatively low concentrations. Studies have shown that these concentrations seldom exceed standards that ensure human health, but often are above levels that are safe for aquatic organisms.
© Digital Vision/Punchstock

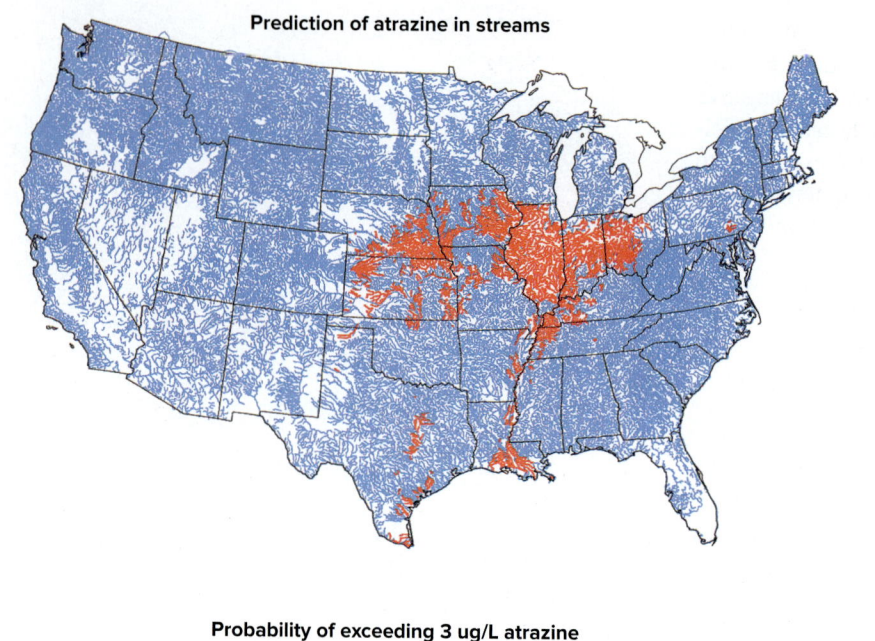

Prediction of atrazine in streams

Probability of exceeding 3 ug/L atrazine

Less than 5% 5% or greater

A

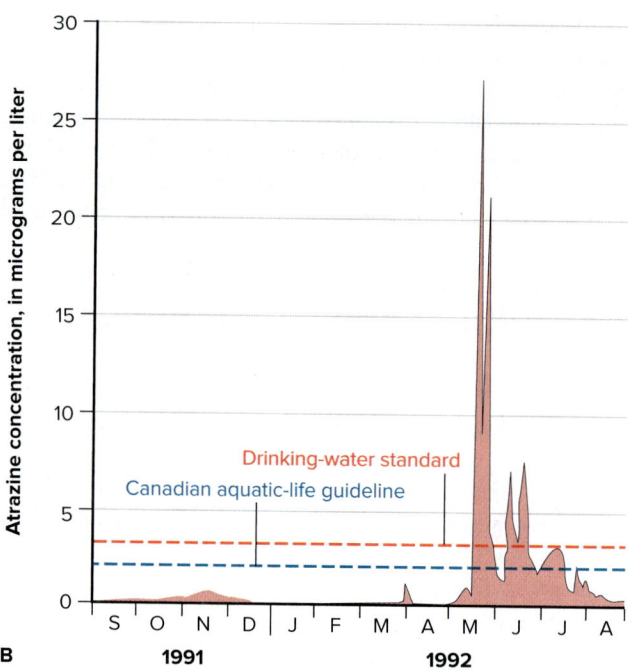

B

FIGURE 15.28 Map (A) showing areas of the U.S. Corn Belt where streams can be expected to contain elevated levels of the pesticide atrazine. Graph (B) showing how the atrazine concentration in the Platte River in Nebraska typically spikes in the spring and early summer after the pesticide is applied to farm fields.
(a) National Water Quality Assessment Program, USGS

Although the concentration of agricultural chemicals in U.S. streams and aquifers is seldom at levels likely to affect humans, the risk varies seasonally and is much higher in areas of intensive agriculture. For example, based on sampling data, USGS predictions show that the highest concentrations of the pesticide called atrazine should be found throughout the Corn Belt of the Midwest as shown in Figure 15.28. Note in the graph for the Platte River in Nebraska how atrazine concentration typically spikes in the month of May, reaching levels far above drinking-water and aquatic standards. Spring rains wash some of the recently applied atrazine into nearby streams and shallow aquifers, causing concentrations to spike.

Some of the fertilizer that is applied to agricultural fields and urban landscapes also finds its way into streams and shallow aquifers. Both manure and synthetic fertilizers typically have high concentrations of nitrogen, phosphorus, and potassium, which are essential plant nutrients. When rainwater carries excess fertilizer off the landscape and into bodies of water, these nutrients promote excessive algae growth, which leads to oxygen depletion as described earlier in this chapter.

Animal Wastes

Farmers have historically collected livestock excretions called *manure*, and then applied it to their fields as a natural means of fertilizing their crops—*organic farming* typically uses manure instead of synthetic fertilizers. Today there are farms that specialize in raising large numbers of livestock; some are considered industrial operations where thousands of animals are confined to relatively small areas called *feedlots* (Figure 15.29). Here the volume of manure is so large that it presents a serious disposal problem. In order to reduce the volume, managers commonly put manure and urine into large pits or lagoons, where it then undergoes partial decomposition by anaerobic bacteria. From there the consolidated waste is normally spread out onto nearby fields so it can be taken up by plants or broken down by aerobic bacteria as it infiltrates the unsaturated zone.

Feedlot operators commonly find it difficult to acquire an adequate amount of land on which to spread the concentrated waste, creating the incentive to apply more waste than the land can handle. This overapplication allows unbroken-down waste, containing dangerous pathogens and substantial amounts of nitrogen and phosphorus, to make its way into nearby streams and shallow aquifers. In addition to creating a biological hazard, the additional nutrients can lead to excessive algae growth and oxygen depletion. With respect to groundwater, the most common problem is that a plume of nitrate (NO_3^-) develops beneath the areas where the waste is being applied, thereby contaminating shallow drinking-water supplies.

The use of animal feedlots and synthetic fertilizers in the United States has also led to a growing problem of oxygen depletion in the Gulf of Mexico. As shown in Figure 15.30, the Mississippi River drains a large portion of the agricultural land in the United States. During this process the river transports the excess nutrients from the agricultural fields to the Gulf of Mexico. Here the nutrients fuel the familiar cycle of algae growth, followed by decay and oxygen depletion. In the Gulf of Mexico this has created what scientists refer to as a *hypoxic zone* or *dead zone,* where oxygen levels are too low for many types of marine organisms to survive. Although hypoxic zones are not uncommon, the dead zone near the mouth of the Mississippi is of concern because it has been growing in size since the 1980s, presenting a threat to the marine ecosystem in the Gulf of Mexico.

Sediment Pollution

Recall from Chapters 8 and 10 that *sediment pollution* occurs when agricultural and construction activity causes excessive amounts of sediment to move off the landscape and into drainage systems. The filling of stream channels with sediment not only results in an increased risk of flooding, it is also highly destructive to aquatic ecosystems. The fine sediment can cover up coarse sands and gravels on the streambed that serve as critical habitat for many aquatic species. Fine sediment also reduces visibility within the water since it stays suspended for relatively long periods of time. Reduced visibility is highly detrimental to fish that require fairly clear water to see their prey. Because sediment pollution results from agricultural and construction activity that leaves soils exposed to erosion, the problem can be

FIGURE 15.29   Animal feedlots generate large amounts of manure and urine. To reduce the volume, the waste is put in holding lagoons where it is partially decomposed by anaerobic bacteria. The remaining waste is normally applied to fields. However, overapplication commonly leads to pollution and nutrient-loading problems in shallow aquifers and nearby streams. USDA

FIGURE 15.30   The runoff of agricultural nutrients within the Mississippi River drainage basin has led to a zone of oxygen-depleted water in the Gulf of Mexico. This hypoxic zone or "dead zone" has been growing in recent years and is a threat to the marine ecosystem. © Dr. Nancy Rabalais, Louisiana Universities Marine Consortium

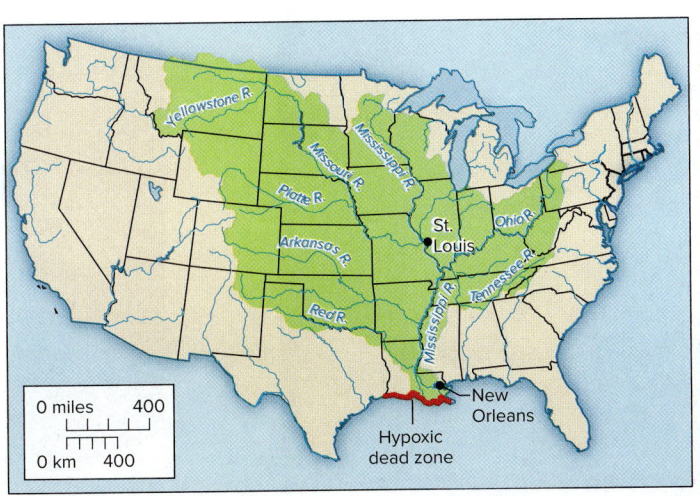

mitigated by using various techniques to minimize soil erosion. Examples include no-till farming and the use of silt fences around construction sites. See Chapters 8 and 10 for details.

Radioactive Waste Disposal

Radioactive waste is different from chemical waste in that it emits dangerous forms of energy, collectively referred to as *radiation*. Recall from Chapters 2 and 3 that radiation originates from nuclear reactions, and that the number of protons in the nuclei of atoms of a particular element is fixed, whereas the number of neutrons can vary. Scientists use the term *isotopes* when referring to the various combinations of protons and neutrons that can exist for a given element. What is important to our discussion here is that the nuclei of some isotopes are unstable. Moreover, such an unstable isotope, called a *radioactive isotope,* will eventually reach a stable configuration through the process of **radioactive decay,** in which the nucleus undergoes a spontaneous nuclear reaction and releases energy (radiation).

Ever since the nuclear age began over 70 years ago, the number of applications involving radioactive isotopes has expanded from making atomic bombs to include electricity generation, industrial and scientific research, and medical treatments for various forms of cancer. All of these applications create radioactive waste. As the number of applications has grown, so too has the amount of waste that is generated. The waste problem is compounded by the fact that anything that comes into contact with radioactive materials becomes contaminated, and therefore must be disposed of as radioactive waste. Unlike typical hazardous waste, radioactive waste cannot be treated or broken down by chemical processes. This is because nuclear decay is not affected by chemical reactions, which only involve the electrons orbiting the nucleus.

Radiation Hazard

Radioactive isotopes pose a human health risk because they emit radiation consisting of energetic particles and waves that can damage human tissue, which leads to development of cancer cells. As shown in Figure 15.31, *alpha particles* are composed of two neutrons and two protons and are relatively large and slow; hence they rapidly lose energy as they encounter other forms of matter. Interestingly,

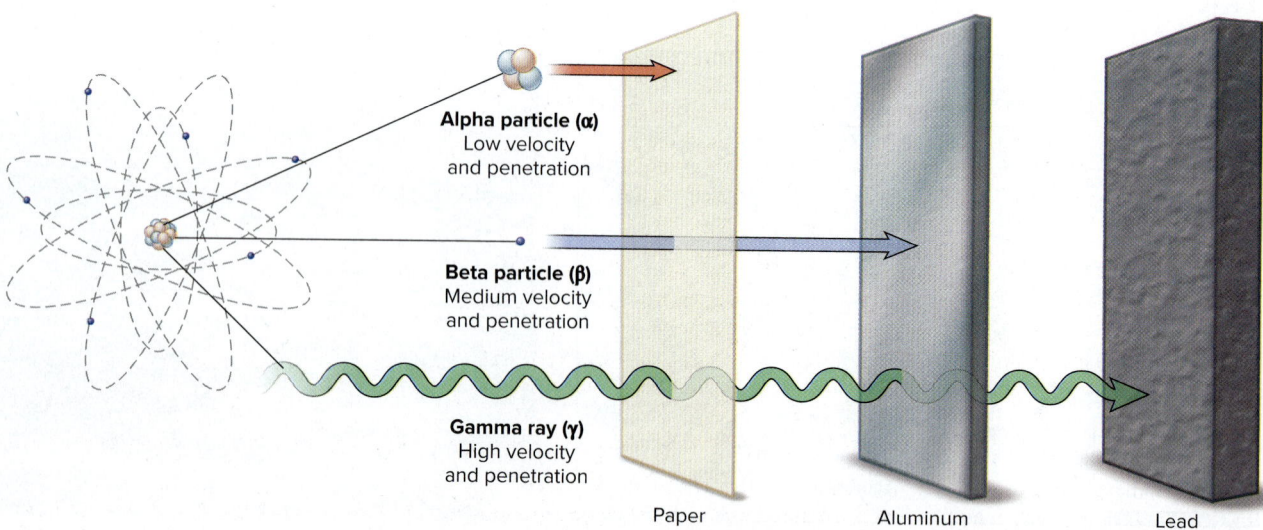

FIGURE 15.31
When the nucleus of an unstable isotope undergoes nuclear decay, it emits radiation, consisting of alpha and beta particles and energy waves called gamma rays. Because gamma rays possess the most energy and are not particles, they have the greatest ability to penetrate objects and threaten human health.

Alpha particle (α)
Low velocity and penetration

Beta particle (β)
Medium velocity and penetration

Gamma ray (γ)
High velocity and penetration

Paper   Aluminum   Lead

alpha particles generally travel only a few inches through air, and do not penetrate even thin paper or the outer layer of human skin. In contrast, *beta particles* are electrons that are ejected from the nucleus at high velocities. Although beta particles are much smaller than alpha particles, their higher velocity gives them enough energy to easily go through human skin and damage internal cells, but not enough to penetrate solid objects such as a door or wall. Another form of radiation is *gamma rays*, which are not particles but high-energy waves that are part of the electromagnetic spectrum (Chapter 2). Note that gamma rays are 10,000 times more powerful than visible light rays, making them the most energetic of all electromagnetic waves. Because gamma rays are very powerful and do not consist of particles, they can easily penetrate solid objects, thereby posing the greatest risk to humans. People can be protected from gamma rays by thick layers of concrete or lead shields, similar to those used by dentists.

The radiation risk to humans largely depends on the type and intensity of radiation emitted by a particular isotope. Relatively few isotopes emit all three types of radiation (alpha and beta particles and gamma rays); most tend to emit two of the three types, and a few emit just one. Any isotope emitting high-energy gamma rays is a serious risk, whereas those that emit only alpha particles generally pose a threat only if the material is ingested or inhaled. Radioactive isotopes also have different decay rates, which means that the intensity and duration of the radiation varies among the different isotopes. Scientists use the term *half-life* to describe the exponential decay rate of a particular isotope, which is simply the time required for half of the radioactive atoms in a sample to decay into stable isotopes. Interestingly, the half-lives of natural and human-made isotopes range from fractions of seconds to hundreds of millions of years. Although short-lived isotopes decay more rapidly, they tend to emit more intense radiation, and thus pose more of a direct health risk. In contrast, long-lived isotopes remain radioactive for longer periods of time, so they present greater problems in terms of waste disposal.

The Disposal Problem

The U.S. Nuclear Regulatory Commission classifies radioactive waste into two categories. **High-level waste** consists of intensely radioactive by-products from nuclear power and weapons reactors, as well as the wastes generated from the reprocessing of spent reactor fuel. During nuclear fission reactions, uranium is converted into heat and highly radioactive by-products that consist of new isotopes and elements such as plutonium (Chapter 14). Once the original uranium is consumed, the now highly radioactive fuel rods must either be discarded as waste or reprocessed into new fuel. During reprocessing, elements that can be used for fuel, mostly plutonium, are selectively removed and the remaining material is regarded as high-level waste. Until a national repository is operational (Case Study 15.1), high-level wastes in the United States are being stored on-site at commercial power plants and U.S. Department of Energy facilities. The spent nuclear fuel is either placed in special water-filled pools or put in dry storage casks. At federally operated plants, high-level liquid wastes are also being stored in large underground storage tanks made of stainless steel and concrete.

All radioactive waste in the United States that is not spent nuclear fuel is classified as **low-level waste**—tailings from uranium mines fall under mining regulations. This rather broad category includes contaminated materials that have come into contact with radioactive substances. Examples include contaminated reactor water residues, equipment and tools, protective clothing, wiping rags and mops, filters, medical needles and syringes, and remains of laboratory animals. Consequently the amount of radiation from low-level wastes can vary considerably, depending on the quantity of radioactive particles that have adhered to these materials.

Since 1962 commercial generators of low-level waste have been able to place the waste in sealed drums and ship the material to a federally licensed disposal site. Several disposal sites were shut down due to subsidence and surface and groundwater

Long-Term Storage of Nuclear Waste in the United States

One of the reasons for the lack of public support in the United States for building additional nuclear power plants is that the nation does not have a permanent means of storing high-level waste. The opening of a centralized repository has been delayed for many years in part because of the need to ensure that the chosen site can "permanently" isolate radioactive waste from the biosphere. This isolation is necessary because the isotopes in high-level waste are long-lived and require thousands to tens of thousands of years before decaying to nonthreatening levels. Consequently, as early as the 1950s scientists recommended that high-level waste be buried in deep geologic formations. The ideal location is one that is geologically stable and where severe earthquakes or volcanic eruptions are highly unlikely. Another important criterion is that the repository be located in a thick unsaturated zone, providing a buffer of dry material between the surface and underlying groundwater system.

Although scientists found three suitable geologic sites in different U.S. states, the final selection stalled because of public opposition. Citizens in these states were generally opposed to having theirs be the one to receive nuclear waste from across the country. The opposition focused on the potential risks and the fact that the waste would need to be safely contained for approximately 10,000 years. Despite all of the scientific and engineering efforts that have gone into finding a suitable site and in developing physical containers to hold the waste, no one can guarantee the integrity of the repository for that length of time. In addition to potential changes in climate and geologic conditions, people naturally asked what type of society, if any, would be monitoring the site for 10,000 years. Keep in mind that the Egyptian civilization ended 3,000 years ago, and the U.S. government has been around for less than 250 years.

In 1982 Congress passed the Nuclear Waste Policy Act and designated that of the three initial sites, Yucca Mountain, Nevada, would be the sole candidate for the nation's centralized repository. Although the geology at all three sites was suitable, Yucca Mountain was chosen partly because it is located on a remote and highly restricted piece of federal property that has been used for conducting nuclear weapons tests since 1945 (Figure B15.1). The repository was to be built within a 1,200-foot-high, flat-topped ridge composed of compacted volcanic ash (tuff) that formed more than 13 million years ago. Because the site is in a desert climate that receives about 6 to 7 inches of precipitation a year, the unsaturated zone is extremely thick. The repository itself would be located in the middle of the unsaturated zone and consist of tunnels mined within the volcanic rocks. Carefully packed containers of high-level waste would be placed in the tunnels and stored approximately 1,000 feet below the surface and 1,000 feet above the groundwater system.

Although extensive scientific and engineering studies have determined Yucca Mountain to be a suitable site, concerns remain. Much of the concern centers on the fact that the repository lies in an active earthquake zone. Here geologists have mapped numerous faults as well as a relatively recent volcanic cinder cone. Geologists have also found evidence of hydrothermal activity (Chapter 12) and numerous interconnecting fractures within the volcanic rock making up the repository. Should climate change bring increased precipitation and a higher water table, the fractures could allow infiltrating water to carry radioactive waste away from the site much more rapidly. Nevertheless, in 2002 President G. W. Bush approved the Department of Energy's recommendation that Yucca Mountain be the national repository for high-level nuclear waste. Then in 2010 the Department of Energy under the Obama administration withdrew its application to open the facility. Several states and other interested parties challenged the decision in federal court. In 2013 the court ruled that earlier studies be completed before the application could be withdrawn. Since then, supporters have been trying to gather political support in Congress to resume the project. It remains to be seen, therefore, if Yucca Mountain will be used to store nuclear waste, or if a new strategy must be developed to meet the nation's needs.

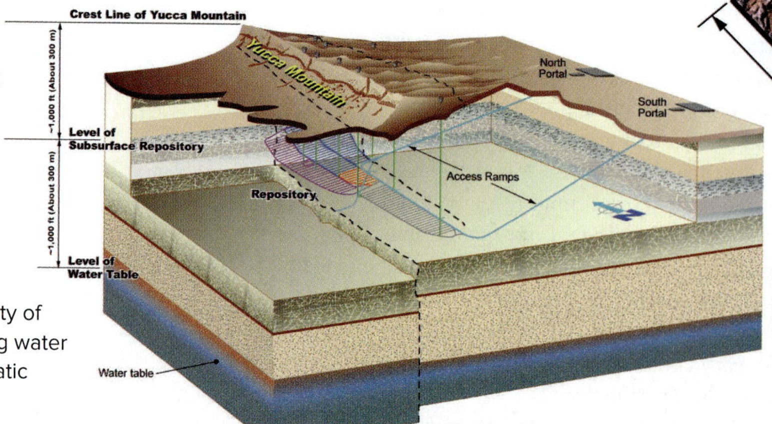

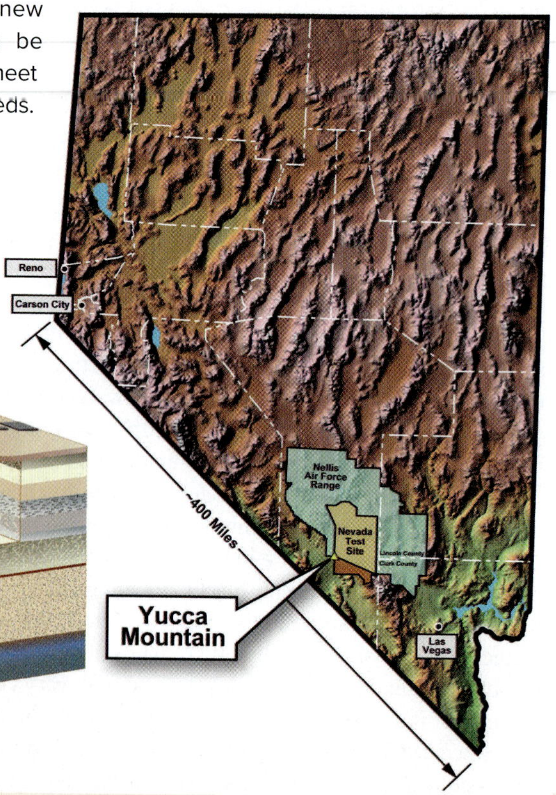

FIGURE B15.1 Yucca Mountain in Nevada was selected by the U.S. federal government to be a central repository for high-level nuclear wastes. The waste would be stored in specially designed containers and placed in mined-out shafts within volcanic rocks approximately 1,000 feet above the water table. Although official scientific studies have found the site suitable, concerns remain over the possibility of seismic activity and rising water levels due to future climatic changes.
(both): U.S. Dept. of Energy

contamination, which until recently left only three facilities in operation—one each in Washington, Utah, and South Carolina. In 1985 a federal law made each state responsible for properly disposing of its own low-level waste. The law allows states to form compacts with other states in order to share construction and operating costs of new disposal facilities. These new facilities must meet federal guidelines requiring that low-level wastes be contained for as much as 500 years, depending on the type of waste. The first new disposal site opened in Texas and began receiving waste in 2012, bringing the total to four facilities nationwide. Note that these facilities have compacts with other states for handling their waste.

Air Pollutants and Fallout

Another major form of waste discarded by society consists of the gases and particulate matter generated during the combustion of fossil fuels by our cars, factories, and power plants. These combustion gases and particulates are simply released into the atmosphere (Figure 15.32), and are considered pollutants because they have an

FIGURE 15.32   Burning fossil fuels releases large volumes of combustion gases into the atmosphere. Carbon dioxide is affecting the global climate system, whereas sulfur dioxide and nitrogen oxides are causing acid rain. Another serious problem is the fallout of mercury associated with burning coal. Photo shows a tall smokestack releasing combustion gases from a coal-burning power plant.
© Design Pics/Chris Knorr

adverse impact on human health and natural ecosystems. Scientists use the term *anthropogenic air pollution* to differentiate pollution that occurs due to human activity from natural forms of air pollution, such as volcanic eruptions and wildfires.

There are many different types of anthropogenic air pollutants, but the most common are particulate (solid) matter, ozone (O_3), carbon monoxide (CO), nitrogen oxides (NO_x), and sulfur dioxide (SO_2)—the subscript x is used to represent the numbers of oxygen atoms in the different types of nitrogen oxides. Like all combustion by-products, these pollutants have a limited residence time in the atmosphere. Eventually the pollutants react chemically with other gases in the atmosphere, fall back to the surface with rain or snow, or fall directly as particulate matter. We will focus our attention on the fallout of acid rain and heavy metals and their impact on the geologic environment. Note that anthropogenic carbon dioxide (CO_2) can also be considered a pollutant because it affects the Earth's climate system, which itself can have an adverse effect on human health. This topic will be covered extensively in Chapter 16.

Acid Rain

The fact that natural precipitation is slightly acidic has had a profound impact on the evolution of Earth's biosphere and weathering of rocks (Chapters 1 and 3). Recall that the pH of water is a measure of its hydrogen ion (H^+) concentration. The pH scale itself is logarithmic, where a pH of 1.0 (10^{-1} H^+ ions) is strongly *acidic,* a pH of 14.0 (10^{-14}) is strongly *alkaline,* and a pH of 7 is *neutral.* Because gaseous carbon dioxide (CO_2), sulfur dioxide (SO_2), and nitrogen oxides (NO_x) readily dissolve in atmospheric water to form acids, the pH of natural precipitation averages around 5.6, which is slightly acidic. The term **acid rain** is commonly used to describe precipitation with abnormally high levels of sulfuric and nitric acid resulting from the anthropogenic emission of SO_2 and NO_x. The EPA estimates that approximately 65% of SO_2 and 25% of NO_x emissions in the United States come from coal and natural gas power plants. Other important sources of SO_2 include the refining of crude oil and smelting of sulfide minerals (Chapters 12 and 13).

FIGURE 15.33   Map (A) showing the locations of U.S. coal-fired power plants in 2007—those equipped with pollution controls for removing sulfur dioxide are shown in green. Map (B) shows the average pH of precipitation in 2007. Due to the prevailing winds, acid rain was most severe in the northeast region, which lies downwind of the major concentration of coal-fired power plants. Acid rain has lessened significantly in recent years due to emissions reductions.

(b) National Atmospheric Deposition Program (NRSP-3). 2009. NADP Program Office, Illinois State Water Survey

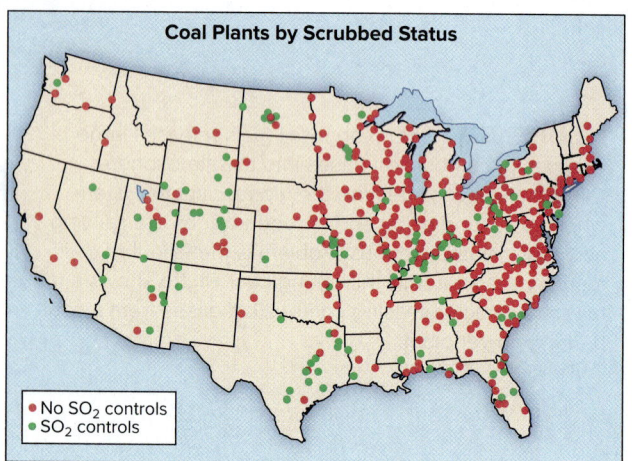

A

During combustion, sulfur compounds in fossil fuels are transformed into SO_2, and atmospheric nitrogen is converted into NO_x. Table 15.2 shows that burning coal produces the highest SO_2 and NO_x emissions, whereas the emission from natural gas is relatively minor. The reason for this is that many coal deposits contain significant amounts of sulfide minerals, whereas oil and gas contain sulfur compounds in much smaller proportions. Since coal is used to meet 39% of the U.S. electrical demand compared to 27% for natural gas, burning sulfur-rich coal has been considered to be the leading cause of acid rain—refining of petroleum makes a significant, but smaller contribution. Figure 15.33 shows the strong correlation between the location of coal-burning power plants and the areas of acid rain

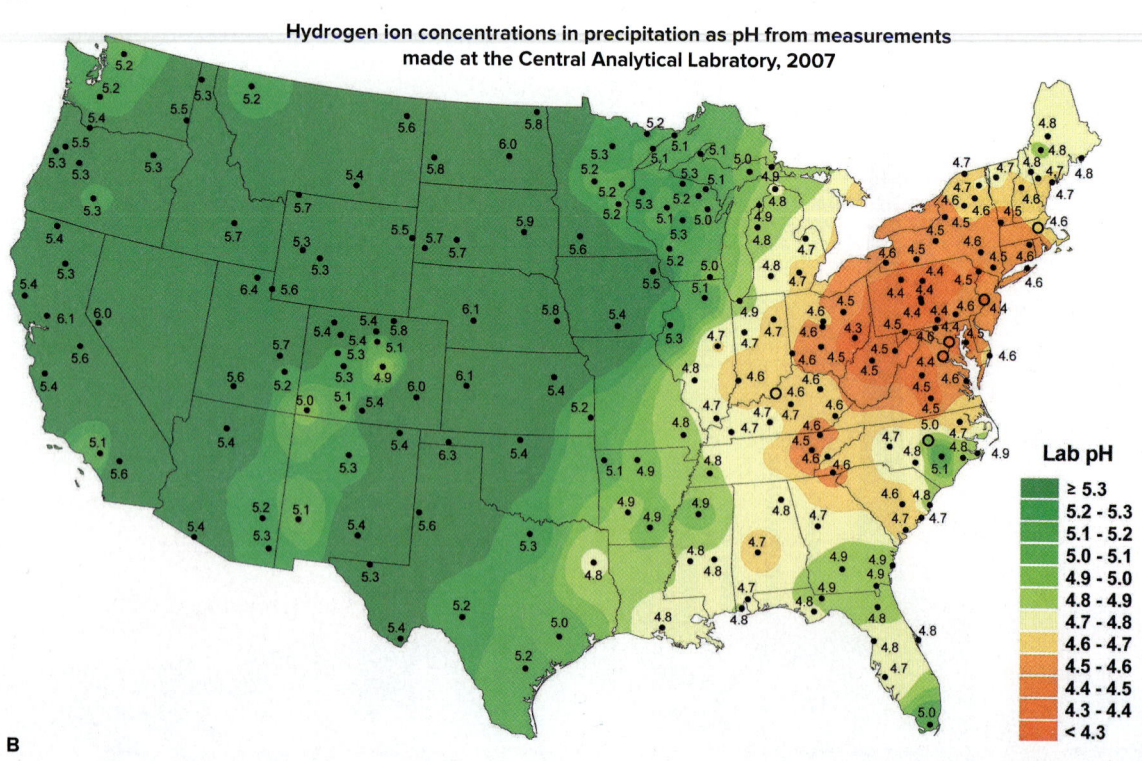

B

FIGURE 15.34 Photo from the Appalachian Mountains in North Carolina showing trees that were killed by acid rain, the source of which could have been hundreds of miles away upwind.
© Dr. Parvinder Sethi

in the United States. Because the smokestacks at these plants are quite tall (see Figure 15.32), the emission gases can be carried downwind far to the east before falling back to the surface in the form of acid rain.

One of the reasons acid rain is such a serious environmental problem is that Earth's biosphere has evolved in the presence of rainfall with a pH around 5.6. However, by the 1990s the average pH of rainfall in the eastern United States was as low as 4.0. Because the pH scale is logarithmic, this means that the precipitation was as much as 40 times more acidic than what fell prior to the Industrial Revolution. This is a dramatic increase, particularly for the many aquatic organisms that are highly sensitive to pH changes. Keep in mind that acid rain impacts both aquatic organisms and land plants, such as forests and agricultural crops (Figure 15.34). Acid rain also accelerates the chemical weathering of concrete, limestone, and marble building materials. These materials all contain the mineral calcite, which readily dissolves in acidic solutions. The result has been significant damage to concrete highways and bridges and monuments made of limestone and marble (Figure 15.35).

TABLE 15.2 Data showing the pounds of pollutants that are emitted for every billion BTUs of energy obtained from burning fossil fuels. On an equivalent energy basis, coal generates by far the most pollutants.

Pollutant	Natural Gas	Oil	Coal
Carbon dioxide (CO_2)	117,000	164,000	208,000
Carbon monoxide (CO)	40	33	208
Nitrogen oxides (NO_x)	92	448	457
Sulfur dioxide (SO_2)	1	1,112	2,591
Particulates	7	84	2,744
Mercury	0.0	0.007	0.016

Source: Energy Information Administration, Office of Oil and Gas; and Environmental Protection Agency.

FIGURE 15.35 Photo showing how acid rain has damaged the stonework and gargoyles of Winchester Cathedral in England. This church was built in 1079 and endured nearly 800 years of weathering by natural rainfall. Most of the damage shown here has occurred since the Industrial Revolution and the large-scale burning of fossil fuels.
© Sally A. Morgan/Corbis

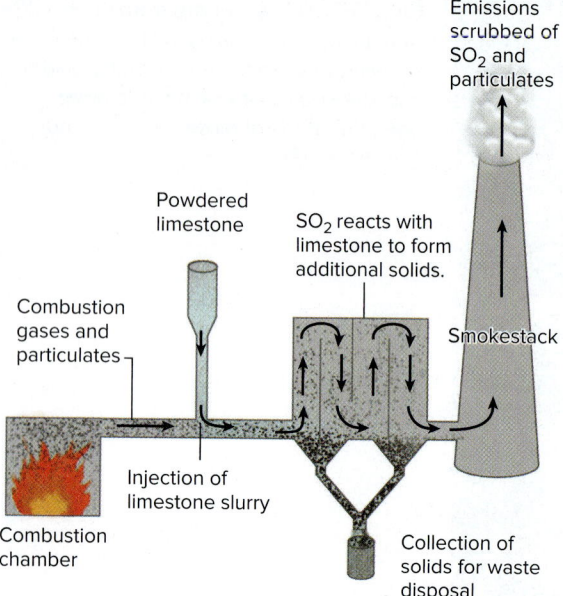

Emissions scrubbed of SO₂ and particulates

Powdered limestone

SO₂ reacts with limestone to form additional solids.

Smokestack

Combustion gases and particulates

Injection of limestone slurry

Combustion chamber

Collection of solids for waste disposal

FIGURE 15.36 Scrubbers are a means of chemically removing pollutants from emission gases. In general, some form of powdered limestone is injected into the emission gases as they leave the combustion chamber of a coal-burning plant. The limestone then bonds with the SO_2 gas to form solids, which fall to the bottom of a collection chamber, where they are later removed and disposed of in a landfill.

Mitigating the Effects of Acid Rain

In addition to reducing nitrogen oxide emissions by cars and trucks through the use of oxygenated fuels (Chapter 14), efforts to mitigate acid rain in the United States have focused on lowering sulfur dioxide emissions from electrical power and industrial plants. Most methods employ a system called a **scrubber** that chemically removes SO_2 and other pollutants before the emission gases enter the smokestack (Figure 15.36). In a scrubber finely ground limestone is injected into the emission gases as either a dry powder or mixed with water to form a slurry. Limestone particles then react with the SO_2 gas, forming a solid residue that falls to the bottom of a collection chamber. Modern scrubbers are capable of removing over 95% of the sulfur dioxide from emission gases. There are also scrubbing systems for nitrogen oxides, the most familiar being the catalytic converters required on the exhaust systems of U.S. autos and trucks since 1975.

Initial efforts at reducing acid rain began in 1970 when Congress passed the Clean Air Act, requiring power and industrial plants to start reducing their emission of SO_2 and other pollutants. Congress recognized that requiring industries to retrofit all the existing plants with scrubbing systems would be costly and an undue burden. Consequently, emission controls were mandated only on newly constructed facilities. The idea was that as old plants reached the end of their life expectancy, they would be replaced by new plants with scrubbers, allowing the additional costs to be phased in over time. By 1977 it was clear that new air-quality standards were not being met, in part because companies were upgrading old plants rather than building new ones with pollution controls. Congress then added a provision to the Clean Air Act stipulating that if a plant was modified beyond a certain dollar amount, the upgrade could no longer be considered "routine maintenance," but instead would trigger the mandatory installation of a scrubber system.

With only modest improvements being made with acid rain, Congress amended the Clean Air Act again in 1990 and included a cap on the amount of sulfur dioxide that could be emitted by the nation's electric generating plants. This market-based approach, commonly referred to as *cap and trade,* divides the total SO_2 limit for the year (i.e., the cap) by the number of power plants, then allocates limits for individual power plants based on their past history. As the overall emission cap becomes progressively lower, owners of power plants can either find ways of reducing their SO_2 emissions or purchase allowances from other plants that come in under their emission targets. This approach provides a clear financial incentive for companies to reduce the overall emissions of their plants. For example, they could choose to switch over to low-sulfur coal (Chapter 13), install scrubbing systems, or replace aging plants with new ones that use cleaner-burning natural gas.

The EPA's sulfur dioxide cap-and-trade program has proven to be quite successful in reducing acid rain in the United States. For example, the EPA reports that SO_2 emissions went from 17.3 million tons in 1980 to just 2.7 million tons in 2013, an incredible 85% reduction. As illustrated in Figure 15.37, this effort has led to lower amounts of sulfate (the acid form of SO_2) being deposited with precipitation, thereby dramatically improving the acid rain problem over large parts of the United States.

Mercury Fallout

Similar to carbon and nitrogen, the element mercury (Hg) moves in a cyclic manner between the hydrosphere, biosphere, atmosphere, and geosphere. Mercury and other heavy metals tend to accumulate in the same oxygen-poor environments in which fossil fuels originate (Chapter 13). Heavy metals in fossil fuels are usually found in trace amounts bonded with sulfur atoms. In addition to being present in fossil fuels, heavy metals are found in various manufactured products, many of which eventually get placed in our municipal and medical waste streams. The problem is that when fossil fuels and waste materials undergo combustion, some of the metals get carried into the atmosphere along with emission gases. Our focus here will be on mercury due to its extreme toxicity and the ease with which it moves through the food chain, posing a serious threat to human health. Notice in Table 15.2 that the combustion of coal releases more than twice as much mercury per BTU of energy than does oil—natural gas emits virtually none.

In studies of layered sediments, scientists have found mercury levels in modern layers to be approximately two to five times higher than in layers deposited prior to the Industrial Revolution. It is generally accepted that this increase is due to anthropogenic inputs of mercury into the atmosphere, which were estimated to total 2,160 tons per year worldwide in 2010. Of this the United States accounted for less than 3% (56 tons), whereas China was estimated to contribute nearly 30% (642 tons). Of considerable interest to scientists and policymakers is how and where this anthropogenic mercury returns to the land surface. As could be expected, studies have found mercury concentrations in soils and aquatic systems to be significantly higher downwind from emission sources. However, scientists have also found elevated levels in polar regions, far removed from any emission source, indicating that mercury deposition is more complex than originally thought.

Recent research indicates that the key to mercury deposition lies in the element's physical and chemical properties, especially its low melting point, which allows it to be a liquid at room temperature. As illustrated in Figure 15.38, when mercury compounds undergo combustion, the high temperature liberates elemental mercury atoms (Hg^0), which form a gas. During this process some of the gaseous mercury becomes oxidized and forms ions (Hg^{2+}) that are chemically reactive and able to bond to particulate matter and also dissolve in water. This oxidized mercury tends to fall out of the atmosphere downwind of emission sources during precipitation events—dry particulates can fall out on their own without precipitation.

Compared to its oxidized form, elemental mercury is nonreactive and may stay aloft in the atmosphere and circle the globe. Scientists have recently learned that the presence of certain oxidizing gases, such as ozone (O_3) and those

FIGURE 15.37   Maps showing significant improvements in acid rain as reflected by the lower amounts of atmospheric sulfate (SO_4^{2-}) deposited via precipitation over the eastern United States. The changes are largely the result of new emission standards for electric power plants.
(a–b) National Atmospheric Deposition Program/National Trends Network

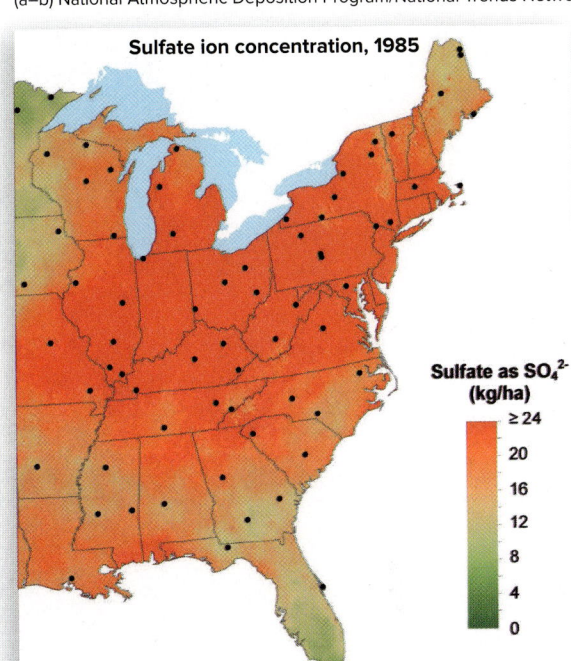

A

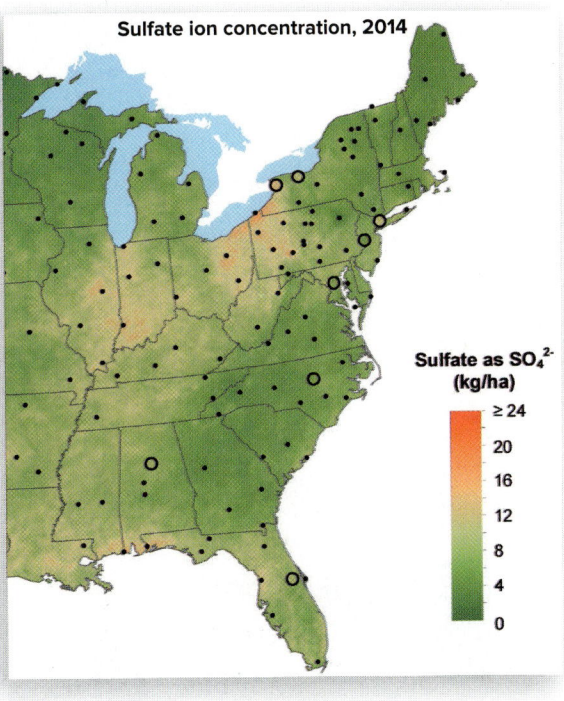

B

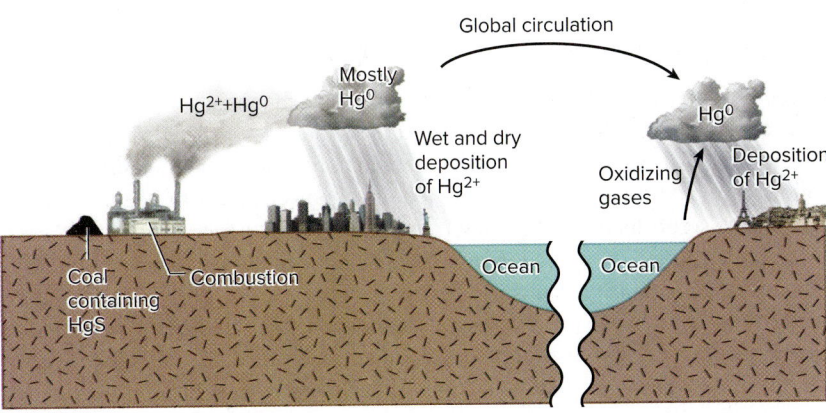

Hg^0 = Elemental mercury Hg^{2+} = Oxidized or reactive mercury

FIGURE 15.38   Anthropogenic mercury enters the atmosphere when compounds containing mercury undergo combustion. During combustion two different forms of gaseous mercury are liberated: elemental (Hg^0) and oxidized (Hg^{2+}). Oxidized mercury is highly reactive and tends to fall out of the atmosphere relatively close to the emission source as dry particles and in precipitation. Elemental mercury is nonreactive and stays in the atmosphere, circling the globe until encountering oxidizing gases that convert it into oxidized mercury.

containing chlorine and bromine, act as triggers that cause the fallout of elemental mercury (Figure 15.38). The unusually high concentrations of mercury found in polar regions are now believed to be related to the accumulation of these oxidizing gases on snow and ice crystals during the long winter period of complete darkness. When the Sun finally rises in the spring, these compounds are released and quickly convert the elemental mercury into oxidized mercury, which then falls from the sky. Researchers now understand that mercury fallout occurs regionally, downwind of emission sources, and also globally in a more complex manner.

Mercury and Human Health

When mercury falls onto the landscape, it takes a rather complicated path because of the way it bonds with different organic and inorganic substances. Of particular concern is the compound called **methylmercury,** in which mercury bonds with carbon and hydrogen atoms. Methylmercury is easily ingested by aquatic life-forms and then stored in their tissues. Consequently, larger and older aquatic species tend to contain more mercury. This also means that as smaller organisms are eaten by larger ones, the concentration of methylmercury increases or magnifies going up the food chain. For example, the concentration of methylmercury in fish ranges from 1 million to 10 million times greater than the original concentration in the water. A recent study of modern polar bears found that they have mercury levels 11 to 14 times greater than bears that lived prior to global industrialization. This demonstrates that human activity has caused mercury levels in the food chain to increase.

Because humans are at the top of the food chain, anthropogenic mercury represents a serious health risk to people who eat mercury-contaminated fish, shellfish, or mammals that eat seafood. During digestion, methylmercury is almost completely absorbed by the blood and then distributed to tissues throughout the body. The primary problem is the mercury that accumulates in the brains of children whose nervous system is still developing. Here it acts as a powerful neurotoxin that is known to cause a reduction in cognitive ability (IQ). A pregnant woman who eats contaminated fish easily passes methylmercury to the developing fetus, where it is known to cause serious birth defects and irreversible loss of cognitive ability. Because mercury concentrations greatly magnify in fish, the EPA and the U.S. Food and Drug Administration (FDA) recommend that young children and women of child-bearing age restrict their fish consumption, particularly certain types of large fish.

Reducing Anthropogenic Mercury Inputs

Because of the serious health risks, Congress included mercury in the list of regulated air pollutants when the Clean Air Act was amended in 1990. The EPA then developed new rules governing the incineration of medical and municipal wastes, followed by additional rules in 2000 requiring power plants to begin installing systems to scrub mercury from their emissions. As illustrated in Figure 15.39, this resulted in a 77% reduction in U.S. mercury emissions from 1990 to 2011, Notice that the majority of this decrease was due to a dramatic 99% reduction for medical and municipal waste incineration along with a nearly 60% reduction in emissions from both industrial processes and coal-burning power plants.

It should be noted that reductions in mercury emissions from coal-fired plants were delayed somewhat because before the EPA could implement its updated regulations in 2000, the new Bush administration issued a revised rule exempting electric utility companies. It was argued that emission controls were an expensive and unnecessary burden, and mercury is a global problem to which U.S. industries contribute only 3% of worldwide emissions. Independent scientists pointed out that while this is true with respect to elemental mercury, oxidized mercury is known to be deposited regionally. Nevertheless, the EPA later wrote a new rule in 2006 aimed at reducing mercury emissions using a cap-and-trade approach similar to that described earlier in this section for sulfur

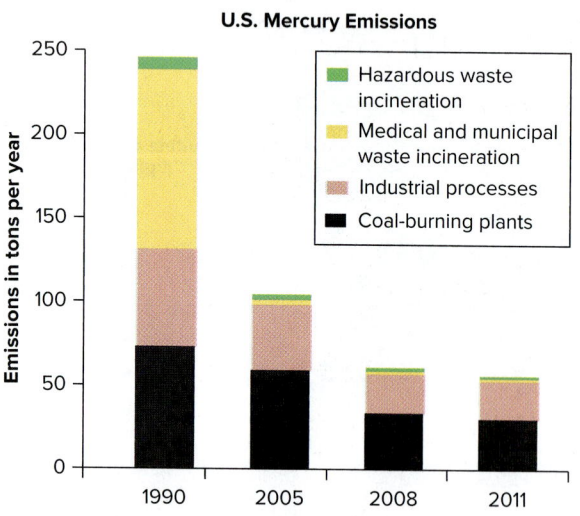

FIGURE 15.39 Anthropogenic mercury emissions in the United States have decreased significantly due to EPA regulations requiring scrubbing systems on medical and municipal waste incinerators. Today the single largest source of mercury released into the atmosphere is old coal-fired power plants.

dioxide. Environmentalists opposed the cap-and-trade system because the reduction targets were higher than those proposed under the original Clean Air Act amendment, and because old coal-fired plants could keep discharging untreated emissions, thereby exacerbating the existing problem of local and regional hot spots of mercury fallout. The issue then went to federal court, which ruled that the Bush-era exemption to the Clean Air Act was not valid. Finally, in 2011 the EPA adopted new mercury standards for coal-fired plants that are expected to reduce emissions another 30% by 2016. Perhaps even more significant is that China adopted mercury emission standards for its coal-fired power plants in 2012. Although China's standards are not as stringent as those of the United States, they represent a step forward. Their adoption is particularly important given the fact China accounts for nearly 30% of all mercury emissions worldwide.

Radon Gas

One of the most dangerous natural air pollutants is **radon gas,** which is an odorless radioactive gas that can cause lung cancer when it is allowed to accumulate in homes. According to the EPA, radon gas is responsible for 21,000 lung cancer deaths per year in the United States, making it the second leading cause of lung cancer—for comparison, 160,000 Americans die each year from cancer-related smoking, and another 3,000 from secondhand smoke. Radon (Rn) is one of several different elements that form from the radioactive decay of small amounts of uranium-bearing minerals in igneous rocks. When igneous rocks undergo weathering, the uranium ends up accumulating in the resulting sediment. Consequently, radon gas is found in a variety of geologic materials and also in the groundwater that comes into contact with these materials.

As illustrated in Figure 15.40, radon becomes hazardous when it moves up through the soil zone and seeps into a building. Radon usually enters a dwelling through cracks and pipe joints and in well water brought within the house. Once radon gas enters a building, it becomes trapped and can accumulate to dangerous levels. When the radon undergoes radioactive decay, it emits alpha particles and decays into the radioactive element polonium (Po). Radon gas is dangerous because it is easily inhaled, and then produces solid particles of polonium that become lodged in the lungs. Here the polonium emits alpha particles (radiation) that cause cell damage within the lungs, which can eventually lead to cancer. Polonium particles can also form in the airspace of a house and become attached to dust particles that are then inhaled. Note that the most common radon isotope (Rn-222) has a half-life of 3.8 days, whereas the half-life of its polonium decay product ranges from fractions of a second to several minutes. This means that once radon gas is eliminated from the home, the radiation hazard will quickly diminish.

The EPA estimates that 1 out of every 15 homes in the United States has elevated levels of radon gas. People at highest risk are those living in well-insulated homes and in areas with relatively high concentrations of uranium minerals in the underlying rock or sediment. Fortunately, homeowners can determine their radon levels through simple and inexpensive tests. If high concentrations are found, the radiation hazard can be minimized by (1) sealing cracks in the foundation and other possible entry points of radon; and (2) installing a simple ventilation system that prevents the gas from building up inside the home. For more information on radon in your area and how to have your home tested, see the list of state radon offices at the EPA's radon website—simply type "EPA and radon" into any Internet search engine.

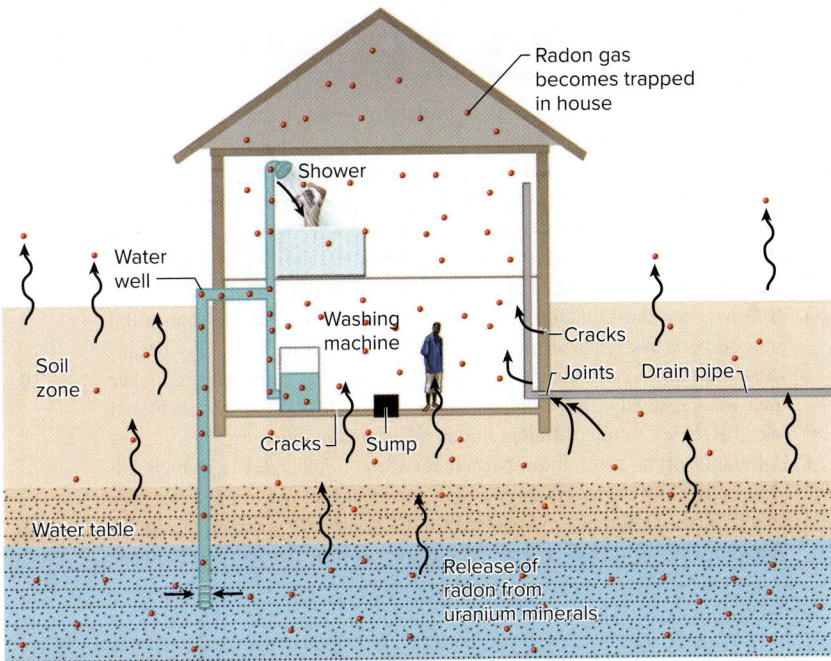

FIGURE 15.40 Uranium minerals release radioactive radon gas that can enter a house through cracks and joints and in well water. Once in the house, radon can become trapped and build up to dangerous levels. Radon is easily inhaled into the lungs, where it decays into radioactive particles of polonium that become trapped and damage lung tissue. The health risk due to radon can be reduced by eliminating entry points into the home and installing a ventilation system that prevents the gas from building up.

SUMMARY POINTS

1. Modern societies generate a wide variety of solid, liquid, and gaseous wastes. Because many wastes are harmful to humans and other living organisms, they cause pollution if not disposed of properly.

2. Accelerated industrialization and population growth in the United States after World War II led to pollution levels that were unacceptable to the public. Congress and the president responded by creating the EPA and passing the Clean Air Act and Clean Water Act in the 1970s.

3. *Polluted* describes situations in which a substance's concentration in the environment has reached the point where it is harmful to living organisms. *Contaminated* means the concentration of a substance is higher than natural levels, but is not necessarily harmful. What is considered harmful varies among different organisms and contaminants.

4. Pollutants often enter the environment from point sources, which are physically discrete locations such as smokestacks and wastewater discharge pipes. Pollutants also enter through nonpoint sources that release contaminants over a broad area, as in the runoff from agricultural fields or parking lots.

5. Municipal and industrial solid wastes typically contain hazardous materials that react with water to form leachate, which can easily migrate and contaminate water supplies. Today the EPA requires that municipal and industrial solid wastes be buried in containment landfills.

6. Tighter EPA standards have led to a dramatic decline in the number of U.S. landfills, resulting in local communities having to ship their waste to distant landfills. To lower costs most municipalities have implemented waste management strategies that reduce the amount of waste being shipped. Elements of such strategies include recycling and rules that exclude yard wastes from municipal waste.

7. Hazardous waste can be buried in specially designed landfills or burned in hazardous waste incinerators; liquid hazardous wastes can also be disposed of via deep-well injection. Some states still permit household hazardous wastes to be placed in the municipal waste stream, but others do not.

8. Human waste in the United States is required to undergo treatment before being released into the environment. Treatment options include individual septic systems or municipal sewage plants servicing an entire town or city. Both options utilize bacterial action to break down human waste into harmless compounds, but also generate sludge that is usually considered hazardous.

9. Animal wastes, fertilizers, overflow from combined sewer systems, and improperly operating septic systems can lead to algae blooms and oxygen depletion in lakes and streams.

10. Large amounts of fertilizers, pesticides, and insecticides are applied each year to agricultural fields, lawns, and gardens in order to increase food production and beautify urban landscapes. A significant portion of these chemicals either infiltrates or washes off the landscape, contaminating rivers, lakes, and aquifers.

11. Radioactive waste presents unique disposal issues due to the long half-life of some isotopes and the fact that radiation levels cannot be reduced using chemical methods. Low-level U.S. radioactive waste is currently being disposed of in one of four facilities, whereas high-level wastes are being stored on-site until a national repository becomes available.

12. Emission of combustion gases, primarily from fossil fuels, causes air pollution and other environmental problems. Carbon dioxide contributes to global warming, whereas sulfur dioxide and nitrogen oxides result in acid rain. The fallout of mercury is a serious problem because this element is a powerful neurotoxin that moves easily through the food chain and poses a threat to human health.

13. Both acid rain and mercury fallout can be greatly reduced by the installation of scrubbing systems that strip SO_2, NO_x, and mercury from the emission gases of electrical power and industrial plants.

14. Radon gas is a natural form of air pollution that develops from the radioactive decay of uranium-bearing minerals. Radon is a serious hazard that can build up to dangerous levels in homes and cause lung cancer.

KEY WORDS

acid rain   508
anthropogenic pollution   482
Clean Air Act   484
Clean Water Act   484
combined sewer system   500
containment landfill   491
electronic waste   493
high-level waste   505
leachate   491

low-level waste   505
maximum contaminant levels (MCLs)   487
methylmercury   512
municipal solid waste   490
National Environmental Policy Act (NEPA)   483
nonpoint source   489
point source   488
polluted   486
radioactive decay   504

radon gas   513
sanitary landfill   491
scrubber   510
secured landfill   493
superfund   484
waste management pyramid   492
wastewater   495

APPLICATIONS

Student Activity   Few people wake up each day with the thought that they are adding pollution to our environment. Here you will do some research to find out how you might be contributing to pollution in your area.

1. Determine how much of the electrical power being supplied to your community is produced by coal, natural gas, nuclear power, or renewable sources—suggest using "power profiler" and "mapping U.S. electricity" to search the Internet. Describe the environmental problems that might be associated with each of these methods.

2. Do an Internet search to find out how close the nearest landfill is to your house—suggest typing "waste management find a facility" into a search engine. Likewise, find the nearest recycling center—this

time enter the name of your city and "recycling" in the search engine. Do you recycle at home or do you throw everything in the trash? Explain why.

3. Contact your landlord or local government and find out whether the wastewater in your house or apartment goes through a private septic system or a municipal sewage treatment plant. After it is treated, does your wastewater seep into the ground, or is it discharged into a nearby stream or lake? Describe the potential problems your treated wastewater can cause. Can any of the wastewater by recycled? Explain.

Critical Thinking Questions

1. You read in the local news that someone in a neighboring county was caught illegally dumping hazardous waste near a stream. Describe how the contaminants from this illegal dumping could move through the hydrologic cycle and eventually affect you personally.

2. U.S. government policy has played a major role in shaping the country's air and water quality and the ways in which we use and dispose of hazardous materials. Describe some examples of successful regulations that have improved the quality of our environment. What are some areas where the policy has had unintended consequences or remains ineffective?

3. The text states that consumers are responsible for the environmental impacts related to industrial wastes and agricultural practices. Do you agree with this? List and describe some products you personally use that are connected to environmental impacts that take place far from your home.

Your Environment: YOU Decide

© Jason Hawkes/The Image Bank/Getty Images

Much of the municipal and hazardous waste in the United States is sent to landfills located far from the populated areas where most of the waste is being generated. Even well-designed landfills cannot be expected to safely contain the waste forever, and pollution of aquifers and nearby streams is almost inevitable. Moreover, the landfills are generally located in rural areas where people have little political power to prevent such a facility from being built. Is it fair that society forces these people to accept the waste and potential pollution problems, despite the fact they create very little of the waste themselves? What actions would improve the situation?

Chapter 16

Global Climate Change

Emission gases leaving the smokestack of a coal-burning power plant. Our modern society was built on the energy from fossil fuels, but one of the by-products of this activity has been the release of greenhouse gases that have disrupted the Earth's heat balance. Ironically, these emission gases made our modern society possible, but now threaten to destabilize the climate system and make life more difficult for people.

CHAPTER OUTLINE

Introduction

Earth's Climate System
- Solar Energy
- The Greenhouse Effect
- Variations in Earth's Orbit
- Feedback Mechanisms

Earth's Past and Future Climate
- Climate Models
- Ways of Studying Earth's Past Climate
- Lessons from the Past
- Future of Our Climate System

Consequences of Global Warming
- Changing Weather Patterns and Biomes
- Melting Ice and Permafrost
- Acidification of Oceans

Mitigation of Climate Change
- Strategy for Reducing Emissions
- The Future

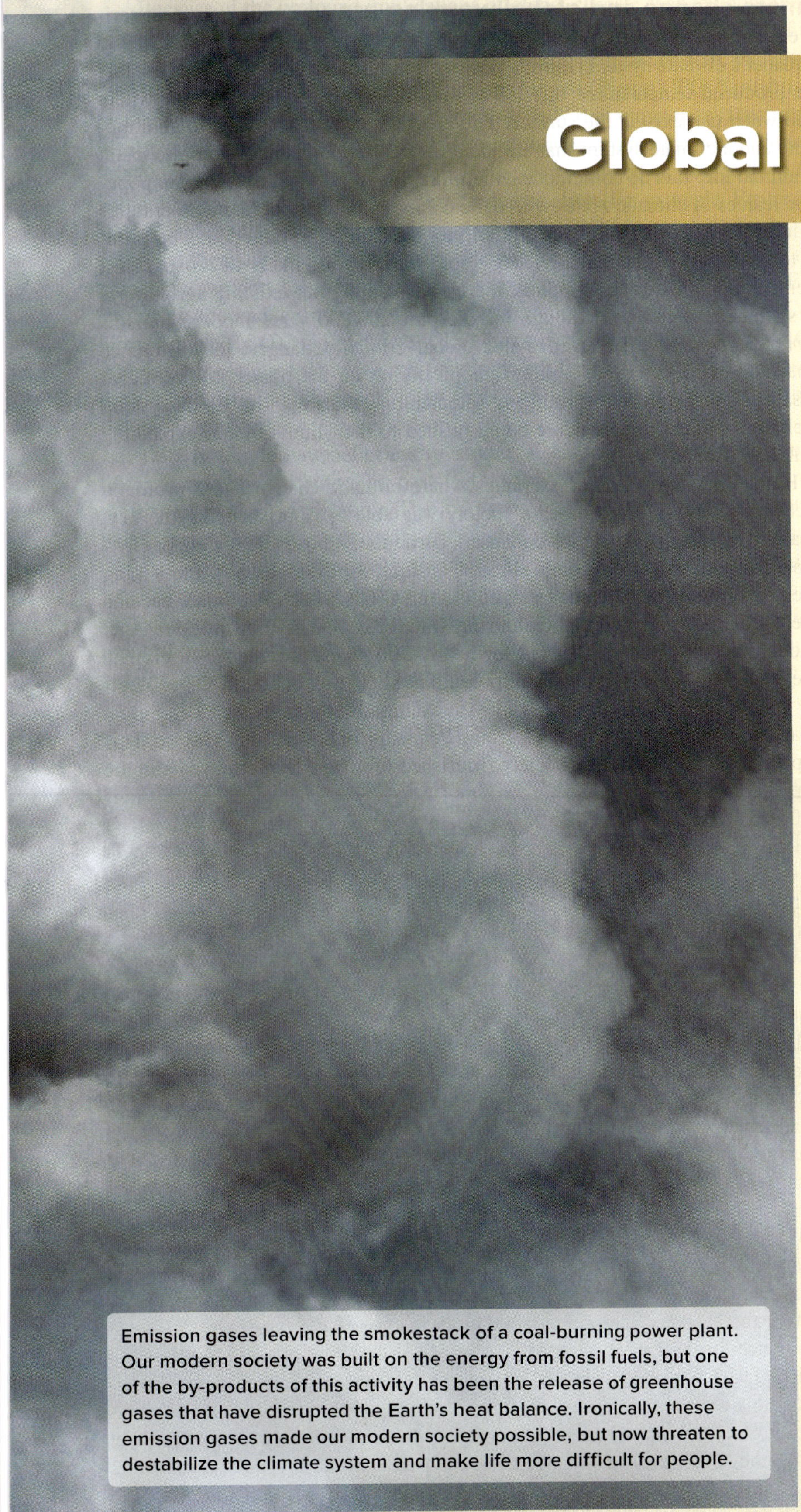

LEARNING OUTCOMES

After reading this chapter, you should be able to:

- Explain how the greenhouse effect operates.
- Describe how humans have disrupted the carbon cycle and how this affects Earth's climate system.
- Discuss how climate change can be triggered by orbital variations and then amplified by positive feedbacks.
- Describe the role the oceans play in shaping Earth's climate and weather patterns.
- Explain how the study of glacial ice has led to a better understanding of the climate system.
- Discuss what a threshold is and how it relates to abrupt and dramatic changes in climate.
- Describe the basic mitigation strategies for reducing carbon dioxide emissions and why it is imperative to begin making immediate and significant reductions.

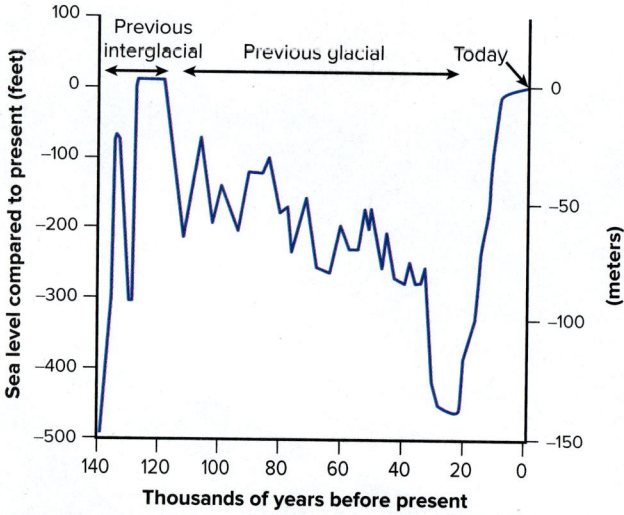

FIGURE 16.33 Graph showing dramatic changes in sea level during the previous glacial and interglacial periods. About 20,000 years ago enough freshwater was stored in glacial ice to lower sea level more than 400 feet compared to today. Most of this water has since been returned to the oceans, but continued melting is slowly raising sea level even further.

make up the largest store of freshwater on the planet after Antarctica and Greenland. Meltwater from these glaciers contributes as much as 70% of the summer flow in the Ganges, Indus, and Brahmaputra Rivers, which provide water for nearly 500 million people living in Pakistan, India, and Bangladesh. Researchers are increasingly worried that the rapid retreat of the Himalayan glaciers could soon limit the amount of water available for irrigating crops and generating hydroelectric power. Governments in the region, like those in the Andes, have very limited resources for constructing water-supply projects to ease future shortages.

Rising Sea Level

During glacial periods sea level naturally falls as additional water is removed from the oceans and stored on land as glacial ice. Sea level then rises when the climate warms and the ice begins to melt. As shown in Figure 16.33, around 20,000 years ago, during the depths of the last ice age, sea level was approximately 460 feet (140 m) lower than today. Because Earth has been moving between glacial and interglacial periods for the past 3 million years, such sea level changes are not unusual. Thus there is nothing special about the current coastlines that modern society has grown accustomed to. Although the climate had been quite stable for the past 10,000 years, sea level continued to rise as the Earth was slowly reaching equilibrium with respect to its heat balance. This has resulted in a 0.7-foot (0.2-m) rise in sea level since 1880. However, because the heat balance continues to be disrupted by greenhouse gas emissions, the remaining ice is now melting at an accelerated pace. For example, the rate of sea-level rise went from about 1 mm/year in 1900 to over 3 mm/year today. This is significant because 2.2% of Earth's water is still being stored on land as glacial ice. If all of that ice were to melt, sea level would rise an additional 260 feet (80 m). While this would likely take place over centuries, or even millennia, the current melting rate is still a major concern.

Based on temperature projections from climate models run under the low and high emissions scenarios described earlier (see Figure 16.19), the IPCC reported in 2013 that sea level will likely rise between 0.9 and 2.7 feet (0.26 and 0.82 m) by 2100. Note that the lower estimate is comparable to the rise since 1880. Keep in mind that even at today's comparatively low rate of 3 mm/year, people around the world are beginning to feel the effects of sea-level rise. For example, low-lying coastal areas are experiencing more frequent flooding during storms and unusually high tides. In fact, many of the Pacific island nations will likely have to be abandoned within the next 50 years. Also consider Bangladesh, one of the poorest nations on Earth, where many of the 150 million residents live on a low-lying river delta. According to an IPCC assessment report, nearly 17 million citizens in Bangladesh are expected to be displaced by 2100. The question is, where will these people go and who will feed them?

Although the IPCC's upper limit of a 2.7-foot (0.82-m) rise in sea level over the next century is based on good science, there is considerable evidence that the rise will be much greater. Here it is important to understand some of the assumptions and limitations of the modeling studies of the IPCC. Because the massive ice sheets covering Antarctica and Greenland are very complex, and good data are limited, the models had to assume that the ice sheets will melt somewhat similarly to blocks of solid ice. However, glacial ice sheets are actually slow-moving bodies with numerous fractures called *crevasses*. As shown in Figure 16.34A, lakes can form on the surface of ice sheets during the summer, which increases solar heating and accelerates the melting. This meltwater also moves downward through crevasses and large vertical conduits called *moulins*, causing the ice to melt from within (Figure 16.34B). Some of this water will build up and increase the pore pressure (Chapter 7) at the base of the ice, thereby reducing the friction between the glacier and underlying bedrock. The reduced friction allows the ice to flow more rapidly, and in some cases, triggers a sudden and rapid increase in flow called a **glacial surge.** The combined effects of internal melting and glacial surging act as a positive feedback, which can cause large ice sheets to break up much more rapidly than predicted by previous model calculations.

A

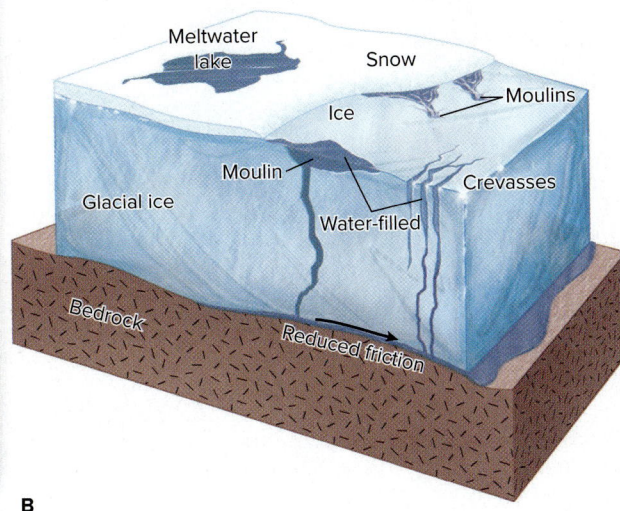

B

FIGURE 16.34 Aerial photo (A) showing a lake and numerous crevasses on the surface of the Greenland ice sheet. The presence of lakes increases solar heating and accelerates melting. Meltwater also infiltrates through crevasses and moulins (B), causing the ice to melt from within. Water that reaches the base reduces friction with the underlying bedrock and allows the ice to flow more rapidly.
(a) © James Balog/Aurora Photos

Glaciologists are now concerned that positive feedbacks will cause the Greenland and Antarctic ice sheets to reach a critical threshold such that large sections of the sheets will begin to disintegrate. Were this to occur, sea level would rise much higher and faster than predicted by the IPCC. In Greenland much of the ice is forced to flow through narrow glacial valleys, called *fjords,* before entering the sea (see Figure 16.1). Warmer ocean temperatures and the buildup of water along the base of the ice are believed to be responsible for the recent increase in glacial flow through Greenland's fjords. Since these narrow passageways act as buttresses that help support the overlying ice sheet, glaciologists believe that if the flow becomes too great, parts of the ice sheet could destabilize, sending much larger volumes of ice into the sea.

At the other end of the globe in Antarctica, there is growing evidence that the West Antarctic ice sheet is also being destabilized and will send even more ice into the sea than Greenland. As illustrated in Figure 16.35, ice in Antarctica flows away from the continental interior and is channeled into rapidly moving zones called *ice streams.* Upon reaching the coast, some ice streams extend great distances over the sea and form what is known as an *ice shelf.* Today there is mounting evidence that several of the major ice streams in West Antarctica are beginning to accelerate. It is suspected that rising sea level and warmer ocean temperatures are playing a role in destabilizing the ice streams where they extend offshore. The concern is that further destabilization could allow large volumes of ice from the main ice sheet to surge into the sea, resulting in rapid sea-level rise. Note that while the elevated parts of Antarctica are still gaining ice, even greater amounts are now being lost to the sea. Gravity measurements taken since 2002 show that the continent is steadily losing ice, similar to what is happening in Greenland.

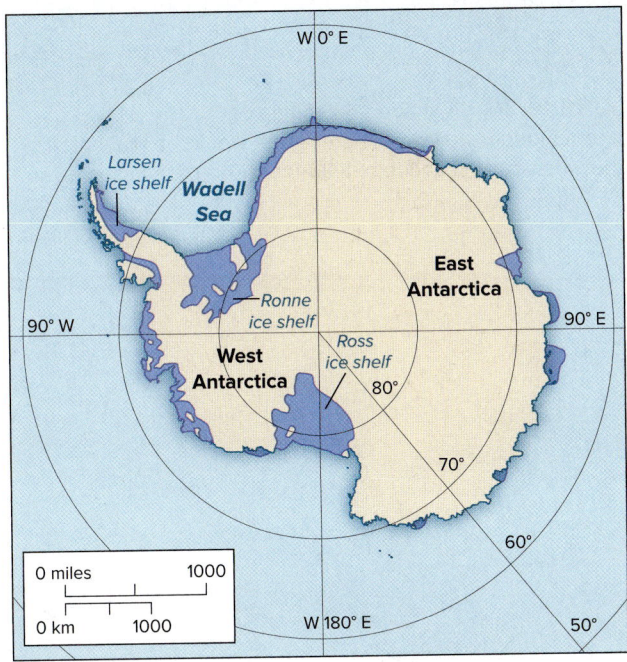

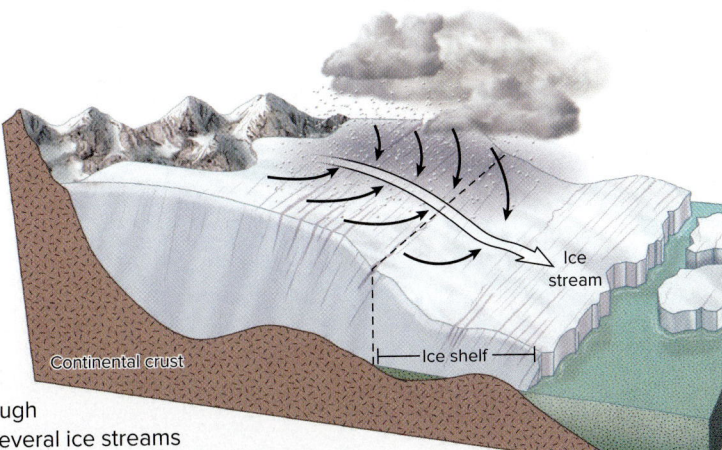

FIGURE 16.35 Glacial ice moving from the interior of West Antarctica funnels through areas called ice streams and moves out into the sea, forming extensive ice shelves. Several ice streams are now beginning to accelerate, creating the possibility that large sections of the West Antarctic ice sheet will destabilize and surge into the sea, causing a rapid sea-level rise.

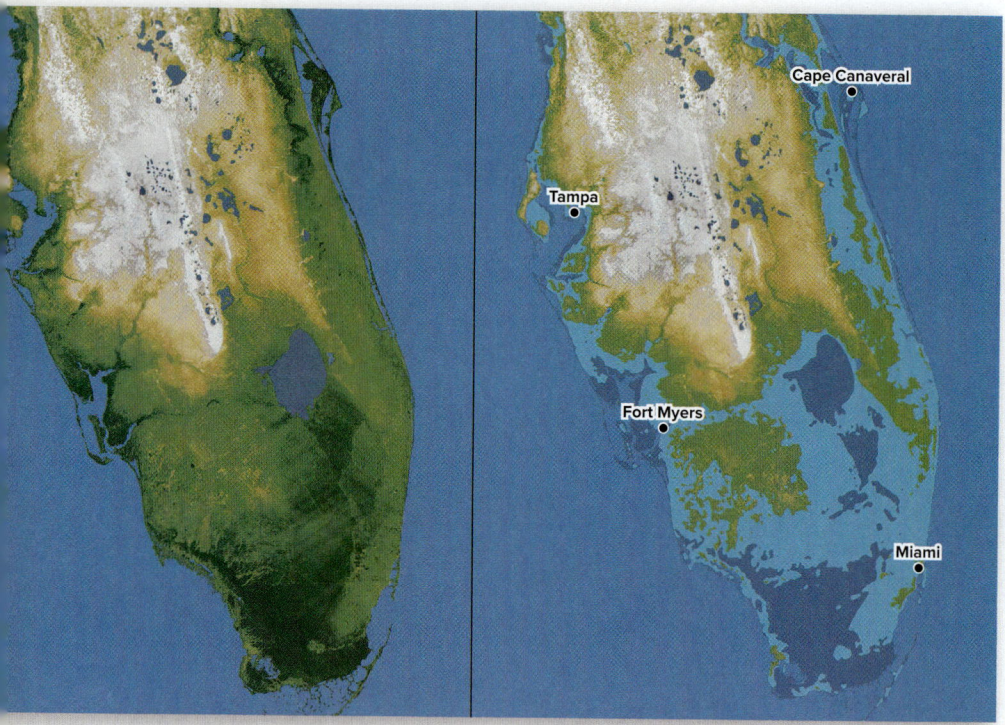

FIGURE 16.36 A shaded relief map of southern Florida showing the areas that would be inundated by the worst-case scenario of a 33-foot (10-m) rise in sea level. NASA/JPL/NGA

Scientists estimate that if both the Greenland and West Antarctic ice sheets were to destabilize, sea level would rise 33 feet (10 m) over the course of a few centuries, not millennia as projected by the IPCC. This worst-case scenario would have catastrophic consequences for nations around the globe. To help appreciate the severity of a 33-foot (10-m) rise, consider the inundation map of southern Florida in Figure 16.36. The consequences of such a rise would be severe considering that nearly 25% of the current U.S. population lives within the inundation zone. Flooding would progressively destroy much of the nation's coastal infrastructure and displace tens of millions of people, leading to serious economic and social upheaval. Keep in mind that regardless of how fast sea-level rise accelerates, we are already starting to feel the negative effects of higher ocean levels. Since what we have set in motion is largely irreversible, the only question now is how we will respond (see Case Study 16.1).

Melting Sea Ice and Release of Stored Methane

In addition to the retreat of glaciers, important changes are taking place in areas of the planet covered by sea ice and permafrost. Unlike the South Pole, there is no landmass situated over the North Pole. Here sea ice covers much of the Arctic Ocean, expanding in size during the winter, only to shrink again each summer. By using satellites to monitor the amount of ice that survives at least two summer melting seasons, referred to as *multi-year ice,* scientists have found a dramatic reduction in sea ice since the early 1980s (Figure 16.37). After nearly 35 years of sea ice decline, it appears that the Arctic Ocean is in an accelerated warming mode. Note that because sea ice floats on the ocean, this melting will not lead to sea-level rise (similar to how melting ice cubes do not cause a cup of water to overflow).

While there has been considerable media attention on the possible extinction of polar bears as Arctic sea ice is lost, a more pressing problem is how the loss of ice may trigger a positive feedback that amplifies global warming. As the ice continues to shrink, the amount of reflected sunlight (albedo) decreases, which increases solar heating and causes the Arctic Ocean to become warmer. This sets up a positive feedback, where warmer ocean temperatures increase the rate of ice loss. The concern is that the loss of sea ice may soon pass a critical threshold, beyond which the ice would not return for the foreseeable future. Moreover, an ice-free Arctic Ocean would represent a significant change in Earth's overall albedo that would help amplify global warming. Climatologists also express concern that a newly exposed Arctic Ocean would alter existing global weather patterns and ocean currents in ways not yet fully understood. Despite the possible disruptions in weather patterns, there are two positive benefits of an ice-free Arctic. The first is it would open up the famed Northwest Passage (Figure 16.37), greatly reducing the distance ships would have to travel from North America and Europe to Asian countries. Another benefit is that it would open up new areas of continental shelf for oil and gas exploration—burning additional fossil fuels, however, will only exacerbate global warming.

A warmer Arctic may trigger yet another powerful and dangerous feedback, namely the release of methane gas. Recall from Chapter 14 that *gas (methane) hydrates* are frozen substances containing methane (CH_4) gas. The Arctic currently contains enormous quantities of gas hydrates sequestered in marine sediment and permafrost. Another source of methane is from the bogs

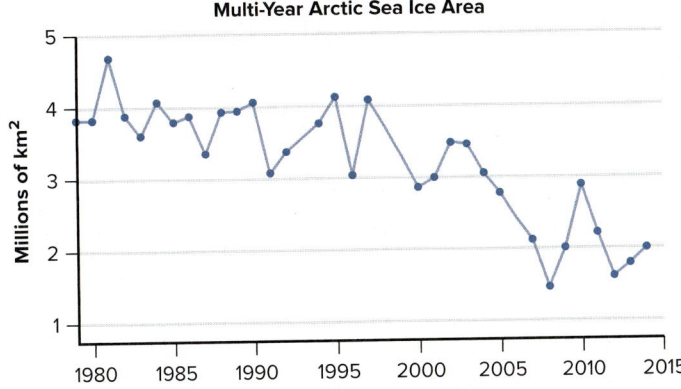

Multi-Year Arctic Sea Ice Area

FIGURE 16.37   Satellite data showing the amount of Arctic sea ice that survives several summer melting seasons (grey disk near the North Pole represents a lack of satellite coverage). As less ice survives the summer melting, more of the Arctic Ocean is exposed to solar radiation. This accelerates the warming of the ocean, which leads to an even greater loss of ice. Note how the Northwest Passage is now nearly ice-free during the summer.
(both): NASA Scientific Visualization Studio

1980

2015

found throughout the Arctic. Here thick layers of saturated organic material, known as *peat,* have been sequestering carbon in the cold climate. When bogs begin to warm, the carbon breaks down and is converted into CH_4 (similar to how food stored in a freezer will start to rot after it thaws). As the Arctic continues to warm, scientists expect that increasing quantities of methane gas will be released from both gas hydrates and bogs. Because methane has over 20 times the warming effect of carbon dioxide, a large release of methane from the Arctic should cause a sharp spike in atmospheric temperatures. In addition, methane has a residence time in the atmosphere of about a decade. Once in the atmosphere it converts to carbon dioxide, whose residence time is around 100 years, thereby extending the warming effect.

Acidification of Oceans

As part of the carbon cycle, the oceans have been absorbing roughly 30% of the carbon dioxide humans have released into the atmosphere. While this has been critical in reducing the rate of global warming, it also has caused the oceans to become more acidic, a process called **ocean acidification.** Recall from Chapter 15 that the pH scale is a measure of the number of hydrogen ions (H^+) in a solution, which is referred to as *acidity.* On this logarithmic scale a pH of 1.0 (10^{-1} H^+ ions) is strongly *acidic,* and, on the other end, a pH of 14.0 (10^{-14}) is strongly *alkaline,* or basic. In historical times the average pH of seawater had been 8.16, meaning the oceans were slightly alkaline (pH of 7.0 is neutral). However, because the oceans are absorbing some of the additional CO_2 released by humans, the dissolved CO_2 is reacting with water molecules and freeing more hydrogen ions (H^+), making the oceans more acidic. Notice in Figure 16.38 that after about 1950

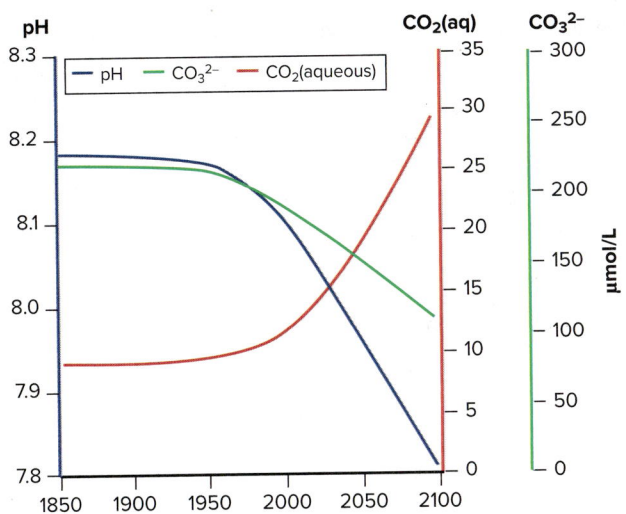

FIGURE 16.38   Graph showing how the increase in dissolved CO_2 is causing the pH of the oceans to lower, or become more acidic. With increased acidity there are fewer carbonate ions (CO_3^{2-}) available for marine organisms to make protective shells. If present CO_2 emission trends continue, by 2100 ocean acidity would increase 125% over preindustrial conditions.

Miami and South Florida: On a Collision Course with Sea-Level Rise

Due to our burning of fossil fuels and disruption of Earth's heat balance, glacial ice is going to continue to melt and sea level will keep rising, putting coastal areas at an ever-increasing risk of flooding. Since an estimated 200 million people live within just 6 feet of average high tide, periodic coastal flooding and eventual inundation will be one the most serious consequences of global warming. Of course, sea level has risen before, forcing early humans to slowly abandon their coastal settlements and agricultural fields. The situation today is far different, not just because of the hundreds of millions of people that will be affected, but because of the vast amount of infrastructure (e.g., ports, bridges, power plants, utility lines, highways, railroads) that is lying in harm's way would be extremely expensive and difficult to relocate.

In terms of population and property, the urban area of Miami in southern Florida ranks as one of the areas in the world at greatest risk from sea-level rise (Figure B16.1). For example, consider that the average elevation of Miami is only 6 feet above sea level and that about half its population of 5.5 million lives less than 4 feet above sea level. Keep in mind that the IPCC projects sea level will rise as much as 3 feet by 2100, but due to the accelerated melting of Greenland and Antarctic ice sheets, many scientists fear the rise will be far greater, perhaps 10 feet or more. Miami therefore, along with most of South Florida, can expect more frequent and severe flooding in the years to come, and eventually, to be completely inundated (see Figure 16.36).

The warning signs of this impending disaster have been evident for some time, and are now becoming more obvious as the sea slowly rises. Flooding has always been a problem in South Florida due to the flat topography and heavy rainfall events. To minimize flooding, cities like Miami have relied on a system of canals and storm sewers that drain excess rainwater from the landscape and carry it to the sea. The problem is that higher water levels are reducing the ability of the system to carry rainwater off the landscape. In addition, unusually high tides combined with higher sea levels have allowed seawater to move up through the stormwater system and cause street flooding, even on dry and sunny days. To make matters worse, South Florida is underlain by highly permeable layers of limestone rock, which also serve as the regional water supply. During unusually high tides, seawater moves through the aquifer system and floods the city from below. This process is also slowly causing saltwater contamination of the drinking-water aquifer on which millions of people depend. In the future, Miami and other cities in the region will have to rely on costly desalination plants to obtain freshwater.

To try and combat the problem of more frequent flooding, Miami has developed plans for installing additional pumps, strengthening seawalls, building more desalination plants, and moving critical utilities to higher ground. For example, the city recently embarked on a $300 million project designed to improve its stormwater drainage system. The new system will eventually have a total of 80 pumps, which officials hope will be enough to minimize flooding for several more decades. To help reduce street flooding, the city is also planning to raise roadways.

Unfortunately though, Miami's improvements will eventually be overwhelmed by the rising sea. While seawalls and pumps have been effective in holding back the sea in places like the Netherlands, this approach will not work in Miami due to the highly permeable limestone that underlies South Florida. No matter how high the seawalls are built, the rising seas will simply flow under the barriers and flood the city from below. Since it is not possible to keep the floodwaters out, the best long-term solution will be to move people and infrastructure to higher ground. Because of South Florida's flat topography, however, high ground is relatively far away (see Figure 16.36).

Despite all the warnings, many people in South Florida, particularly political leaders, have tended to ignore or minimize the threat of sea-level rise. This reaction is understandable given the fact that the worst impacts are perhaps decades away. Relatively few residents therefore have decided to relocate. Ironically, just the opposite is occurring: Miami has become one of the fastest-growing urban areas in the country. The recent real estate boom has been one of the main economic drivers of the region, which is why many politicians are reluctant to acknowledge the problem of sea-level rise.

the dissolved CO_2 content of the oceans began to increase, causing the average pH to decline from 8.16 to its current level of 8.06. A decline of 0.10 pH units may not sound like much, but since the pH scale is logarithmic, this means that the acidity of the oceans has increased by a factor of 1.25, or 25%.

Ocean acidification is a serious problem because many marine organisms, including corals, are sensitive to pH changes. Of particular concern are tiny marine organisms, such as plankton, that form the base of the marine food web. Similar to many other marine organisms, these tiny creatures make protective shells from carbonate minerals such as calcite ($CaCO_3$). As the oceans

FIGURE B16.1 Aerial view of Miami Beach in southern Florida. Much of the larger urban area of Miami lies just a few feet above sea level and will be at increasing risk of coastal flooding and inundation as sea level continues to rise. Defending the city is made more difficult because it is underlain by highly permeable limestone, which allows floodwaters to move up from below.
© Getty Images/Tetra images RF

Since sea-level rise will continue, the real estate market is bound to eventually collapse once the flooding becomes so great that insurers stop selling insurance policies. With flood insurance unavailable, banks will stop issuing mortgages, which means buying and selling homes will become quite difficult. Faced with foreclosure and a weak economy, many people will likely pack up and leave, similar to what happened during the Dust Bowl in the 1930s (Chapter 10). The ideal solution would be to plan ahead and begin relocating to higher ground in a controlled manner, as opposed to waiting and moving in a sudden and chaotic manner. If done properly, it might even be possible to extend the life of Miami as a scaled-down and elevated version of itself, similar perhaps to Venice, Italy.

become progressively more acidic, there are fewer carbonate ions (CO_3^{2-}) available to make the protective shells. Consequently, the release of anthropogenic CO_2 into the atmosphere will ultimately affect the entire marine food web. If carbon dioxide emissions continue to increase at their present rate, scientists expect that the pH of seawater will fall below 7.8 by 2100 (Figure 16.38), representing a 125% increase in acidity over preindustrial, conditions. Should this occur, it would represent a level of ocean acidity not seen for at least the past 20 million years and would likely cause a major collapse within the marine ecosystem.

Mitigation of Climate Change

Based on what you have learned in this chapter, there should be little question that human activity is altering the carbon cycle and impacting the climate system. Moreover, if carbon dioxide emissions keep increasing (Figure 16.39A), this could possibly push the system beyond critical thresholds. The result could not only be higher temperatures, but more rapid climate change and sea-level rise. Although the human species has proven to be extremely adaptable, the speed and degree of these changes might make life very difficult for people in many parts of the world that are currently considered habitable. Simply put, the changes could be catastrophic for civilization as we know it.

The fact that global warming is largely due to human activity also means that we have the ability to reduce its impact. Fortunately, the technology currently exists to begin making substantial reductions in greenhouse gas emissions. The basic problem lies in generating the political will necessary to make significant emission reductions, and to do so fast enough to avert more serious consequences. Many policymakers are working under the assumption that the warming will be gradual, as in the IPCC model projections. This has made some hesitant to take steps toward reducing emissions that they perceive will impact economic growth. However, the costs of inaction are likely to be far greater than those associated with reducing emissions.

The first serious effort to reduce greenhouse gas emissions was the 1997 international treaty known as the *Kyoto Protocol,* which was signed by 170 countries. Under the Kyoto agreement, developing nations were exempt from having to meet specific emission reductions. The rationale behind the exemption was that most of the excess CO_2 currently in the atmosphere was released over many decades by industrialized nations. With this exemption, developing nations would be able to first grow their economies, and then reduce emissions later. However, in 1998 the U.S. Senate failed to ratify the Kyoto Protocol largely because China was exempt from emission reductions due to its developing nation status. As can be seen in Figure 16.39B China and the United States are by far the largest producers of CO_2. Although China's per capita emissions are still relatively low, continued economic growth

FIGURE 16.39   Graph (A) illustrates the growth in worldwide CO_2 emissions since 1965. Plot (B) shows the world's largest carbon dioxide producers, as a percentage of total world emissions, along with their per capita emission rates in metric tons of CO_2.

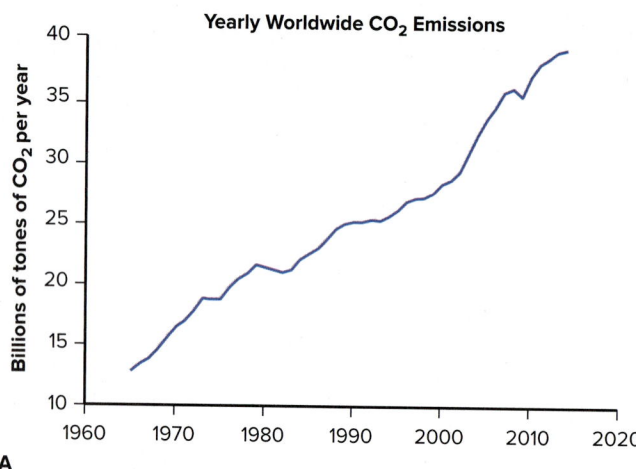

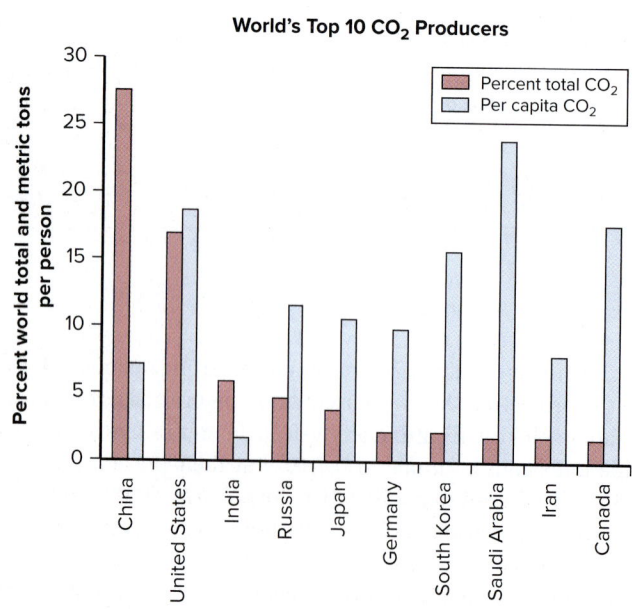

combined with a large population means that its total CO_2 emissions could be extremely high in the future.

While the U.S. objection to the Kyoto Protocol seemed reasonable to some, others felt the Senate should have ratified the agreement, setting an example that would have pressured China to follow suit. Because of the impasse, the world's two largest producers of greenhouse gases continued to increase their emissions after 1998. During this period other industrialized nations began meeting their reduction targets under Kyoto. Then in 2010, representatives from nations around the world met in Copenhagen, Denmark, and agreed to reduce their CO_2 emissions by 2020. However, the reductions in the so-called *Copenhagen Accord* were nonbinding. Moreover, the agreement did not extend the binding caps previously agreed to under the 1997 Kyoto Protocol.

Finally, in 2015, 193 nations signed the so-called *Paris Climate Agreement,* which is designed to reduce greenhouse gas emissions to a level scientists believe will keep temperatures from rising beyond the IPCC's 2.0°C (3.6°F) target over preindustrial levels. Unlike previous agreements, there were no exemptions for developing countries. Under the Paris Agreement, all of the nations are required to submit individual plans for reducing carbon emissions through 2030. While the national plans vary in scope, they all represent real reductions. For example, China has committed to reductions that will cause it emissions to peak by 2030. The United States has pledged to reduce emissions by 26–28% below its 2005 levels, whereas the European Union plan calls for a 40% reduction from 1990 levels. Because the agreement requires even developing countries to reduce emissions, a $100 billion per year commitment was included to help poorer nations mitigate the effects of climate change—note that this financial commitment is nonbinding. Finally, since many climatologists believe that the climate system will pass critical thresholds earlier than previously expected, the Paris Agreement set a long-term goal of keeping the temperature rise below 1.5°C (2.7°F).

It is important to realize that although the individual national emission plans in the Paris Agreement are all voluntary, each nation is legally required to monitor and publicly report on its progress every five years. Nations are also legally required to meet every five years and submit updated plans that include even more stringent reductions. It should be noted that because of political opposition to emission reductions in the United States, the Paris Agreement was specifically designed not to be a formal treaty. Despite the nonbinding nature of the emission reductions, the hope was that the legal reporting requirements will create public pressure that helps gives nations the incentive to honor their commitments. Participants also hoped that the agreement will encourage the global energy and financial markets to begin making larger investments in carbon-free energy sources such as wind, solar, and nuclear. Although it is not ideal, the Paris Agreement represents humanity's best, and perhaps last, effort at keeping climate change from spiraling beyond our control.

For industrialized nations whose economies rely on cheap fossil fuels, making rapid and substantial reductions in greenhouse gas emissions will not be an easy task. Switching to alternative, non-carbon-based energy sources will naturally take time, primarily because of the sheer scale of our energy needs (Chapters 13 and 14). This switch will be particularly difficult for China given its heavy dependence on coal. Chinese citizens though have been putting enormous pressure on their government to reduce the nation's serious air pollution, which is caused largely by burning coal (Chapter 15). Note that this pressure played a key role in the Chinese government's decision to cut greenhouse gas emissions under the Paris Agreement. In the United States emissions have already declined 8% since 2005, mostly due to conservation and the expanded use of natural gas and wind for generating electricity. Nevertheless, under the Paris Agreement the United States pledged to reduce greenhouse gas emissions by 28% from 2005 levels.

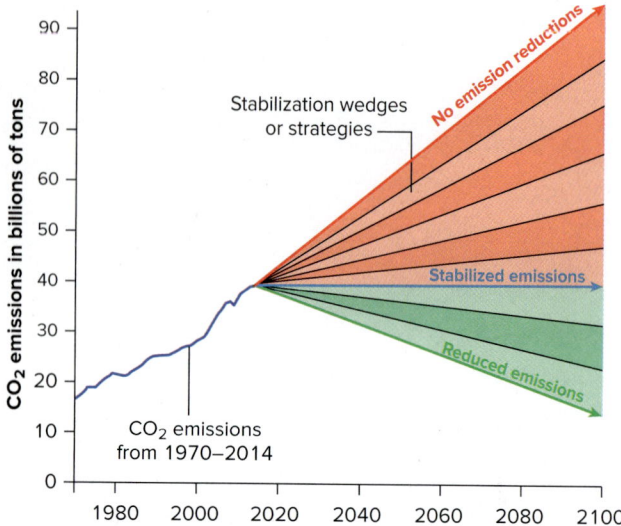

FIGURE 16.40   A stabilization triangle represents the amount that CO_2 emissions need to be reduced to keep future emissions at some stable level. This reduction can be accomplished through the collective contribution of individual wedges, or CO_2 reduction strategies. To prevent continued warming, climate experts recommend additional wedges to produce an overall reduction in emissions. Note that the emission levels shown for 2100 are for illustrative purposes and do not represent actual reduction targets.

Since no single alternative energy resource or conservation measure will lead to an overall reduction in CO_2 emissions, the collective contribution of multiple reduction strategies is needed to avoid accelerated global warming. In the United States researchers at Princeton University developed a useful concept called a *stabilization triangle*. As depicted in Figure 16.40, a **stabilization triangle** represents the amount that CO_2 emissions need to be reduced to keep future emissions at a steady or declining rate. Individual *stabilization wedges* represent the contribution of different reduction strategies, such as the use of wind and solar to replace coal for generating electricity, improved energy efficiency, reforestation, and conservation. When taken collectively, the individual wedges or strategies will make it possible for us to stabilize or reduce our carbon emissions.

Strategy for Reducing Emissions

To prevent the climate system from reaching the point where positive feedbacks cause warming to accelerate beyond our control, most climatologists believe that emissions must decrease to the extent that the temperature does not rise above the Paris Agreement's long-term goal of 1.5°C (2.7°F). To accomplish this, it will be necessary to concentrate our efforts on stabilization wedges that produce the greatest reduction in emissions in the shortest period of time, and at the least cost. For example, conservation is an approach that has an immediate impact and involves little or no cost to society. Building nuclear power plants, on the other hand, is expensive, and they can take 10 years or more to complete. It is also important that we focus our initial efforts on reducing methane emissions, since the warming effect of methane is 21 times greater than that of carbon dioxide. After concentrating on those wedges that are easy to implement and have the greatest impact, we can turn our attention to more costly and difficult measures. Keep in mind that as technology improves, additional wedges will become feasible and more cost-effective.

Because it is beyond the scope of this text to provide a thorough examination of possible stabilization wedges or strategies, the following sections are intended to give a general idea of what can be done. Ultimately, the best solution is to minimize the use of fossil fuels and begin switching over to carbon-free fuels. For details on energy usage, efficiency, and conservation, refer to Chapters 13 and 14.

Conservation

Without question, conservation is the most effective means of making immediate reductions in greenhouse gas emissions. Businesses and individual citizens can cut back on energy use wherever possible. Individual efforts can include such things as eliminating unnecessary trips, car-pooling, raising or lowering the thermostat, and turning off the lights when leaving a room. Businesses can also find ways of reducing energy consumption that do not add to their overall operating costs. In fact, one of the advantages of conservation efforts is that users end up saving money because of reduced energy consumption.

Increased Efficiency

In contrast to conservation, increasing efficiency typically requires some physical change to mechanical systems that use energy. This usually involves making an investment in new equipment or making modifications to existing equipment. Similar to conservation, increasing efficiency can be purely voluntary or government mandated. A good example is the higher fuel efficiency standards passed by Congress in 1975, resulting in a 56% increase in fuel efficiency by 1990. This heightened efficiency translated into car owners saving 56% on their annual fuel costs. In addition to automobiles, manufacturers began producing a wide range of more energy-efficient appliances, and builders began constructing well-insulated homes and businesses, all of which led to a significant reduction in energy consumption.

Although great progress was made after the 1970s, there are still many areas in which efficiency gains can be realized. One key area is transportation. Not only can automakers develop even more efficient vehicles, but policymakers can encourage the development and use of mass-transit systems. Getting people out of their cars and into buses and light-rail systems will reduce emissions and ease traffic congestion. Another big efficiency gain with respect to transportation could come from a national effort to rebuild the vast network of railroads that once existed in the United States. Although trucks have proven to be highly flexible and convenient in terms of delivering goods, trains are two to four times more fuel-efficient on a ton-per-distance-traveled basis. Trains would be far better suited for long-distance transport, leaving trucks for short-distance deliveries. Likewise, passenger rail service could shuttle people between regional destinations far more efficiently than airplanes or automobiles.

Incentives

Because fossil fuels have historically been a convenient and inexpensive source of energy, most policymakers agree that some type of incentive is needed for businesses and industries to switch to carbon-free energy sources. One option is a **carbon tax** levied on fossil fuels and phased in over a period of time to soften the impact. In this approach, the tax rate is based on the amount of CO_2 released by a particular type of fuel, including what is released during the fuel's extraction and processing. For example, the rate for coal would be higher than for natural gas because more CO_2 is released during the mining and burning of coal, per unit of energy supplied. The total amount of tax levied would then depend on total usage and the type of fossil fuel being used. This would produce a strong economic incentive to use fuels that produce the least amount of CO_2. Moreover, the tax revenues could be used to provide tax credits to encourage the development of carbon-free energy sources such as wind and solar power.

The other type of incentive is a market-based approach called a **cap-and-trade system,** where government regulators set a limit (cap) on the amount of emissions for a particular type of user, such as power plants. Similar to the successful cap-and-trade system that reduced acid rain (Chapter 15), a cap-and-trade system for carbon emissions would set a total that is lower than what all the plants combined had previously been releasing. The cap would then be divided up into small permits and allocated to individual plants based on their previous emission totals. After an initial phase-in period, companies would no longer be allowed to exceed their total number of permits. Because the cap-and-trade system allows plants to buy and sell individual permits, a financial incentive exists for operators to develop more-efficient systems or use a cleaner fuel. In this way, more-efficient plants can sell unused permits to other plants that are unable to meet their emission limits. Some policy experts prefer a carbon tax over a cap-and-trade system because it would lead to more immediate reductions in CO_2 emissions.

Due to political opposition in the United States to both a carbon tax and a cap-and-trade system, in 2015 the Obama administration used the EPA's authority under the Clean Air Act (Chapter 15) to institute the *Clean Power Program* for reducing carbon emissions as part of the Paris Agreement described earlier. Under this new program, the EPA set emission standards for electric generating plants that burn fossil fuels. By 2030 the EPA projects that the Clean

Answers to End-of-Chapter Questions

Chapter 1

Critical Thinking

1. Because healthy ecosystems are what provides us with the clean air and water that humans depend on. We also depend on a clean environment to enjoy a wide variety of outdoor activities.
2. Humans tend to view time on a much finer scale than the scale over which geologic processes operate. To an individual, decades or even hundreds of years may seem like a long time, but in terms of geology, it commonly takes millions of years for significant changes to take place. This is important because many environmental problems result from our failure to understand natural processes that operate over longer time scales.
3. Answers should include the need to maintain the Earth system so it can continue to support future generations. Students should recognize the need to limit population growth and to limit their personal consumption of resources.

Your Environment: YOU Decide

Answers will vary, but most students should see the parallel to Earth's limited resources and the need for humans to live in a sustainable manner. Supporting evidence includes global warming, widespread loss of habitat, and how humanity is reaching the limits of freshwater supplies and soil resources in many areas.

Chapter 2

Critical Thinking

1. Evidence supporting the hypothesis includes the fact that all the planets and moons revolve around the Sun in the same counter-clockwise direction, and whose orbits lie in the same plane. Scientists believe that planetary systems are a common occurrence since the same physical processes that operate in our solar system should operate throughout the universe. This is now being proven as astronomers keep discovering new planetary systems within our galaxy.
2. Perhaps the most compelling evidence is that all of the galaxies continue to be moving away from one another, which means the universe is still expanding.
3. Geologic evidence shows that all life on Earth has evolved from extremophile bacteria. It is believed that life on other planets would probably begin in a similar manner, which is why scientists are looking for signs of bacterial life in extraterrestrial rocks.

Your Environment: YOU Decide

Answers will vary. Some students may feel that due to budget restrictions, money would be better spent on problems here on Earth. Others will believe that space exploration is a small part of the overall budget, and is worthwhile since it pays dividends in terms of improved technology.

Chapter 3

Critical Thinking

1. At the end of their life cycle, many stars will explode in a supernova, sending most of their mass out into space. This material, which is composed of almost every element in the periodic table, becomes part of interstellar clouds called nebulas. These clouds can later collapse and form new planetary systems composed of both light and heavy elements.
2. If they have the same chemical formula, then their internal structure must be different, which also means their physical properties have to be different. A good example is diamond and graphite, both of which are composed of only carbon atoms.
3. There are many different paths that rocks can take through the rock cycle. A mountain composed of sedimentary rock does not have to undergo metamorphism. It can just as easily be exposed to weathering and erosion, generating sediment that forms new sedimentary rock.

Your Environment: YOU Decide

Most students should recognize that observing these features as they form in modern environments makes them a very reliable means of interpreting Earth history. Since the laws of physics and chemistry are universal, it is quite reasonable to use sedimentary features to study the history of other planets.

Chapter 4

Critical Thinking

1. There are many solid materials that will go from being brittle to ductile simply by raising their temperature. Good examples include plastic, rubber, and most metals.
2. Convection occurs in the atmosphere when sunlight warms the land surface, which causes the overlying air mass to become warm. This reduces the air's density, forcing it to rise, similar to a hot air balloon. In the oceans, convection begins when cold, saline water sinks because of its greater density.
3. A volcanic eruption in Los Angeles is highly unlikely because the subduction process does not occur along transform plate boundaries, similar to the one that passes through southern California.

Your Environment: YOU Decide

Answers will vary, but students will perhaps find that the most compelling evidence is how the age of the seafloor and magnetic reversals are symmetrical on opposite sides of spreading centers.

Chapter 5

Critical Thinking

1. Both seismic and sound waves represent energy that travels through matter—unlike electromagnetic wave energy, which can travel through the vacuum of space. Seismic waves differ from sound waves in that they travel through solid material as opposed to air.
2. Some faults experience repetitive earthquakes because the forces that are causing the strain to build remain active.
3. Yes, a large magnitude earthquake could have a low Mercalli ranking if it occurred in an unpopulated area. The opposite could occur if a moderate earthquake were to strike a populated area where the buildings are not designed to withstand lateral ground shaking.

Your Environment: YOU Decide

Answers will vary, but students will hopefully see the value in doing both. Since large numbers of people die in earthquakes due to structural failure, spending most of the money on building improvements in earthquake-prone areas would save the most lives. However, we should continue funding research on short-term earthquake predictions because, if we succeed, this too could save many lives.

Chapter 6

Critical Thinking

1. During the subduction process, water-rich sediment from the seafloor is pulled into the subduction zone along with the descending tectonic plate. As the water and sediment become involved in the melting, it ultimately produces a gas and SiO_2-rich magma that is highly explosive.
2. The relatively low viscosity of basaltic magmas allows them to travel farther from a volcano than andesitic magmas. Basaltic magmas have lower viscosity due to their higher temperature and lower SiO_2 content.
3. Volcanic mudflows can occur even when a volcano does not erupt. Early warning systems for mudflows consist of seismic sensors placed in stream valleys on the flanks of volcanoes where mudflows originate.

Your Environment: YOU Decide

Answers will vary depending on what an individual student perceives as an acceptable risk. With respect to volcano monitoring, most students should see the societal value in saving lives and limiting economic losses.

Chapter 7

Critical Thinking

1. The delay is due to the amount of time it takes for water to infiltrate deep into the slope. The deeper it can infiltrate, the more it can destabilize a slope.
2. Wet sand has adhesive forces between the water and solid grains, which are lacking in dry sand. Hence, wet sand can form a steeper slope. If the sand takes on too much water and becomes saturated, then the increased pore pressure reduces friction between the grains, thereby destabilizing the slope.
3. By definition, a fall occurs when earth materials fall through the air. Mud is a fluid that flows on the land surface, so it does not free-fall through air—the only exception would be if the mud flowed over a cliff, similar to a waterfall.

Your Environment: YOU Decide

Answers will vary, but should include a discussion about the money that logging companies would have to spend to reduce mass wasting, versus the societal costs associated with the environmental consequences, such as sediment pollution and increased flooding.

Chapter 8

Critical Thinking

1. Infiltration on bare ground is less compared to areas with natural ground cover because the land surface almost always has a slope. Natural ground cover helps keep rain and melting snow from moving downslope, giving water more time to infiltrate.
2. You would find gold with coarse gravel because the first types of material to be deposited when water velocity decreases are the largest and most dense grains. Although flecks of gold may be small, their high density causes them to be deposited with larger sediment grains.
3. In a natural system, excess water during a flood is stored in a river's floodplain. Although levees are effective in keeping a river from overflowing its banks, they disconnect the river from its floodplain, thereby forcing floodwaters to be stored elsewhere. This causes floods to occur more frequently in areas upstream of levees.

Your Environment: YOU Decide

Answers will vary, but students should recognize the trade-off between free choice and managing the risk so that people can obtain affordable flood insurance.

Chapter 9

Critical Thinking

1. Wave energy causes water molecules to move in a circular or elliptical manner within a water body. As the wave passes, a boat will rise and fall, whereas the circular motion causes it to roll in the direction of the passing wave.
2. Increased wave energy expands the size of the zone where the sand grains are in motion. This results in a greater volume of sand being transported, thereby increasing beach drift.
3. You should look for zones where the incoming waves are not breaking into surf because of the outgoing rip currents.

Your Environment: YOU Decide

Answers regarding a general versus a user tax will vary depending on whether one thinks the economic value of coastal tourism benefits only the immediate region or society as a whole. For those who prefer a purely natural coastline, hopefully they will recognize that this is feasible only for select areas specifically set aside for this purpose.

Chapter 10

Critical Thinking

1. The soil in the humid climate will have more organic matter due to denser vegetation growth. Also, higher infiltration would cause greater weathering and allow for more clay-sized particles to be transported deeper into the soil profile.

2. As aluminum-rich silicate rich minerals undergo chemical weathering, aluminum remains within the crystalline structure of the minerals. Therefore, aluminum stays behind and becomes more concentrated in the soil.
3. Nutrients such as calcium and magnesium are produced in soils by the chemical weathering of rock and mineral fragments. As nutrients are carried through the soil by infiltrating water, they become attached to clay, mineral, and organic particles, which store the nutrients for later use by plants.

Your Environment: YOU Decide

Answers will vary depending on whether one believes that businesses should bear the cost of environmental problems they create, or whether costs should be passed on to the consumer who creates the demand. Hopefully, students will recognize that it is less expensive to prevent a problem than paying for the consequences.

Chapter 11

Critical Thinking

1. Dams impact downstream ecosystems by lowering the water temperature, reducing the sediment load, and eliminating the occurrence of extreme low and high flows.
2. Contaminants can easily enter a limestone aquifer system through sinkholes, which are also places where people have historically placed trash. There is also very little attenuation of contaminants in open karst systems.
3. Conservation efforts involve very little costs, if any, whereas building new reservoirs and installing new wells requires considerable capital investment.

Your Environment: YOU Decide

Students can be expected to decide between the use of forced conservation for the collective benefit of society versus the self-interest of individuals.

Chapter 12

Critical Thinking

1. Certain elements remain dispersed throughout the crust because they rarely become concentrated by geologic processes.
2. Slate shingles are extremely durable and will not need to be replaced, provided that the underlying support structure remains intact.
3. Many water resources, particularly rivers and lakes, are being replenished within the hydrologic cycle on human time scales. In contrast, mineral resources form on a geologic time scale, which makes them finite and irreplaceable resources.

Your Environment: YOU Decide

Answers will vary, but students should see the value in purchasing only conflict-free, registered diamonds. Some students should question the registration process as a marketing tool. Others should argue the value of a bargain.

Chapter 13

Critical Thinking

1. There are many important uses of petroleum for which electricity cannot be used as a substitute. Examples include fuel for jet aircraft and heavy machinery and the making of plastics, chemical fertilizers, pesticides, and insecticides.

2. Its liquid nature and high energy density made it ideal for use in internal combustion engines. Coal on the other hand was bulky and had to be used with steam engines.
3. Natural gas consists of short chains of hydrocarbon molecules, which cannot easily be transformed into the long hydrocarbon chains found in crude oil. This means that we cannot use natural gas to make gasoline and many of the petrochemicals that we depend on.

Your Environment: YOU Decide

Students should see the value in adding to the global supply of crude oil, but may also feel it is not worth putting environmentally sensitive areas at risk only to make a modest and temporary improvement in supplies.

Chapter 14

Critical Thinking

1. Producing synthetic crude from oil shale is very energy intensive and has not been shown to be commercially viable under present crude oil prices. There is also the problem of limited water supplies in areas where oil shale deposits are found.
2. The answer lies in the fact that solar radiation has both wave and particle properties. When wave energy strikes matter, such as the land surface, it is transformed into heat energy. When light particles (photons) strike a semiconductor composed of layers of silicon, it causes electrons within the layers to flow, creating an electrical circuit.
3. There can be a large difference in temperature between the outside air and the inside of a building, whereas the difference between the ground and the building is rather small. Geothermal heat pumps therefore use less electricity when exchanging heat between the ground and a building.

Your Environment: YOU Decide

Answers will vary, but students should weigh the risks against the benefits of nuclear power. There is a small risk of a meltdown and radioactive contamination of water supplies, but there is also the need for society to generate large amounts of base load power without producing greenhouse gas emissions. Some might argue that the risk of global warming is greater than that of a major nuclear accident.

Chapter 15

Critical Thinking

1. The illegal waste can pollute both groundwater and surface water bodies, which, in turn, may be your source of drinking water. The pollution can also severely impact the ecology of streams and lakes, and thereby have a negative effect on fishing and other types of recreational activity.
2. Students can cite numerous successful examples of improved air and water quality that are directly related to the Clean Air and Clean Water Acts. There are also examples of unintended consequences, but these will likely be harder for students to think of—an example from the text is the removal of asbestos without regard to the actual risk.
3. Answers will vary, but students should recognize that nearly everybody in a consumer society contributes to environmental degradation. There are countless products that can be tied to environmental problems; examples include most commercial foods and the use of energy and mineral resources.

Your Environment: YOU Decide

Answers will vary, but students should understand that everyone in society generates waste, and that the waste must be stored where it can cause the least amount of harm. Student solutions to the social inequities of situating landfills are likely to include ways of compensating those who must live near landfills, and finding ways of minimizing the potential pollution problems.

Chapter 16

Critical Thinking

1. The main reason is that the variation in solar output is too small to account for the amount of warming that is taking place. The other reason is that solar output has been at a minimum at the same time the rate of warming has been accelerating.

2. Only a very small percentage of the gases in the atmosphere are able to trap some of the planet's heat that radiates into space. Next to water vapor, carbon dioxide is the most abundant greenhouse gas. Largely through the burning of fossil fuels, humans have increased the carbon dioxide concentration by nearly 40%, which is enough to disrupt Earth's heat balance and cause the planet to warm.

3. Scientists are concerned because the concentrations of these gases are currently far outside of Earth's normal operating range for the past 800,000 years, therefore temperatures will likely climb far above their normal range.

Your Environment: YOU Decide

Answers should vary, but students should recognize that binding action must be taken quickly before positive climate feedbacks result in accelerated warming. Students should also recognize that since China and the United States are by far the largest greenhouse gas producers, both nations need to make substantial emission reductions in order to avert accelerated warming. While some students may think it is more important to have cheap electricity as a means of supporting economic growth, others will see that in the long term, the costs of global warming will far out-weigh having higher electrical rates.

Glossary

A

absolute age The actual age of rocks in terms of years. This age is normally determined by *radiometric dating,* a technique that is based on the known decay rates of different radioactive elements.

accretion Process whereby planets eventually form when gravitational attraction causes individual particles to clump together into larger masses.

acid mine drainage Acidic leachate, normally rich in heavy metals, that drains from tailings and underground mines and that forms from the chemical reactions between water and sulfide minerals commonly found in ore deposits and coal.

acid rain Precipitation containing abnormally high levels of sulfuric and nitric acid, which results from emissions of sulfur dioxide and nitrogen oxides by human activities.

albedo The fraction of solar radiation that is reflected off a solid body. Light-colored objects, such as clouds and land covered with snow or ice, reflect more light and therefore have a high albedo.

alternative energy sources Energy sources other than the conventional coal, natural gas, and light crude oil that currently make up the bulk of the world's energy supply.

anthropogenic pollution The pollution resulting from various human activities and society's consumption of natural resources, in contrast to pollution that occurs solely due to natural processes.

aquifer A permeable geologic material (rock or sediment) that readily transmits water. Often used as a supply source for groundwater.

aquifer storage and recovery (ASR) A water management technique that involves storing surplus surface water in an aquifer, which is then removed at a later time when surface water supplies are more limited. Sometimes referred to as *water banking.*

aquitards A geologic material (rock or sediment) with low permeability that inhibits the flow of groundwater. Also act as confining layers for *artesian* (confined) aquifer systems.

artificial levees Earthen mounds or concrete panels constructed along river banks by humans for the purpose of keeping a river from overflowing its banks and inundating its floodplain. Those made of concrete are often called *floodwalls.*

asteroids Small bodies orbiting the Sun that are composed primarily of rocky and metallic materials. Most asteroids lie in what is known as the main asteroid belt between Mars and Jupiter.

asthenosphere The weak zone within the Earth that lies near the top of the mantle where silicate minerals are near their melting points. Tectonic plates ride over this weak layer.

B

back swamps Those areas of the floodplain that are poorly drained and can remain wet for extended periods of time. Drainage is inhibited by natural levees, high water tables, and fine-grained sediment beneath the swamps.

barrier islands Elongate sediment deposits that parallel a shoreline and are separated from the mainland by open water, lagoons, tidal mudflats, or salt-water marshes.

basalt A common type of extrusive igneous rock found in oceanic crust that is rich in plagioclase feldspar and ferromagnesian minerals.

base level The lowest level to which a stream can erode its channel. Sea level is often referred to as *ultimate base level* because the oceans represent the end or low point of most rivers.

bauxite A sedimentary deposit of highly enriched, aluminum minerals that form from the chemical weathering of silicate rocks over a long period of time in climatic zones with considerable rainfall.

beach nourishment The process of artificially adding sand to a beach in order to reduce erosion, and to enhance its recreational value.

big bang theory The scientific explanation for the origin of the universe in which all matter at one time existed at a single point, but then about 14 billion years ago began expanding outward in all directions.

biofuels Combustible materials derived from modern (nonfossilized) organic matter. Examples include ethanol and biodiesel fuels.

body waves Seismic waves that travel through Earth's interior, which includes *primary waves (P-waves)* and *secondary waves (S-waves).*

breakwater An engineering structure consisting of large rocks placed parallel to shore and designed to keep waves from breaking onto land, thereby creating a protected area.

C

calcite A calcium carbonate ($CaCO_3$) mineral that is a major constituent in the group of rocks called *limestone,* and used as the raw material for making cement and concrete.

caldera A circular depression that forms after a volcanic eruption when rocks begin collapsing or subsiding into the now empty magma chamber. Some geologists also refer to large *craters* that form during explosive eruptions as calderas.

cap rock A low-permeability rock layer, typically composed of clay or evaporite minerals, that overlies a petroleum reservoir and greatly limits the ability of oil and gas to escape from the reservoir.

cap-and-trade system A strategy for reducing carbon emissions where government regulators provide permits and set emissions limits for a particular type of user. Because users are allowed to buy and sell individual permits, there is a financial incentive to develop more efficient systems or switch to a cleaner fuel.

carbon capture and storage A technique under development for removing carbon dioxide from the emission gases of industries burning fossil fuels, then compressing the CO_2 and injecting it into permeable rock formations for permanent storage. Also see *carbon sequestration.*

carbon cycle The cyclical and natural movement of carbon atoms through Earth's solid interior and surface environment (biosphere, hydrosphere, and atmosphere).

carbon sequestration The process of removing carbon from the surface environment and placing it in storage for long periods of time. Can occur naturally (formation of limestone rock and fossil fuels) or artificially by humans (subsurface injection of carbon dioxide). Also see *carbon capture and storage.*

carbon tax A strategy for reducing carbon emissions by levying a tax on fossil fuels, thereby providing an incentive for businesses and industries to switch to non-carbon-based energy sources.

channelization A flood control technique that involves straightening and deepening a stream channel so that its *discharge capacity* is increased, thereby reducing the probability that water will overflow the banks.

chemical weathering Any chemical process that causes minerals within rocks to decompose into simpler compounds and individual ions.

cinder cones Relatively small volcanic features that form when lava is ejected into the air and cools into *cinders,* which then fall and accumulate around the vent.

clay minerals A group of aluminum-rich, silicate minerals that typically form by the chemical weathering of other silicate minerals.

Clean Air Act Act passed by the U.S. Congress in 1970 requiring the EPA to set minimum levels of air quality for cities and emission standards for sources that emit hazardous substances, including automobiles.

Clean Water Act Act passed by the U.S. Congress in 1972 giving the EPA the authority to regulate water quality and the discharge of pollutants into surface waters—officially called the *Water Pollution Control Act of 1972*.

climate The long-term average weather and its statistical variation for a given region; contrasts with *weather*, which refers to the state of Earth's atmosphere at any given time and place.

climate feedbacks Processes within the Earth system that respond to a disruption in the planet's heat balance and act to further increase or decrease the energy imbalance.

climate threshold The concept in which some component of Earth's climate system is pushed beyond a critical point, causing the entire system to suddenly change. Positive feedback mechanisms would then amplify the change.

coal A combustible sedimentary rock that forms when accumulated plant material becomes deeply buried under new sediment. The corresponding higher levels of heat and pressure drive off water and other volatile compounds, leaving behind a solid that is highly concentrated in carbon.

combined sewer system A wastewater collection system consisting of a single set of underground pipes for collecting both sewage and storm water runoff from city streets and parking lots.

comets Relatively small bodies, 0.6 to 6 miles in diameter, composed of small rocky fragments embedded in a mass of ice and frozen gases. Most have highly elliptical orbits around the Sun.

composite cone A cone-shaped volcano with steep slopes that consists of alternating layers of pyroclastic material and lava flows; also called a *stratovolcano*.

compressibility A property that describes the ability of a material to compact and reduce its volume when placed under a force or load.

compression A force that pushes on a rock body from opposite directions, causing it to become shorter.

cone of depression A cone-shaped depression in the water table or potentiometric surface of an aquifer, which forms when the water level (*hydraulic head*) within a well is lowered by pumping.

confined aquifer An aquifer that is overlain by an aquitard, which seals the aquifer off from the atmosphere and surface environment, creating pressurized conditions within the aquifer. Also known as an *artesian* aquifer.

consumptive water use Refers to those human activities in which water is removed from its source and is lost or consumed. Examples include irrigation of crops and production of concrete.

containment landfill A type of sanitary landfill designed to minimize the formation of leachate and to collect what does form. Such a landfill is lined with a synthetic fabric in order to prevent the escape of any leachate.

continental arc A volcanic mountain range associated with a subduction zone in which magma rises up through a continental tectonic plate to form a string of volcanoes.

convection cells The circular motion of heat and matter within the Earth that is driven by temperature-induced changes in the density of material.

convergent boundary A tectonic plate boundary that is dominated by compressive forces such that the two plates move toward one another.

crater A circular depression around a volcanic vent that forms during an eruption as pyroclastic material is ejected into the air.

creep An exceptionally slow type of mass wasting process where repeated expansion and contraction causes unconsolidated materials to move downslope.

crude oil The liquid phase of petroleum consisting of a mixture of heavier hydrocarbon molecules, which are separated during a distillation process (refining) to produce a variety of products, including gasoline, lubricating oil, and agricultural chemicals.

crust Earth's outermost layer consisting primarily of silicate-rich rocks whose density is lower than those in the underlying mantle.

D

deposition The process in which transported earth materials begin to accumulate. Solid sediment will accumulate in low-lying areas of the terrain, whereas dissolved ions accumulate in either a body of surface water or a groundwater system.

desalination The process of removing the dissolved ions from saline water, by either distillation or reverse osmosis, in order to generate freshwater that is suitable for human consumption.

disseminated deposits Low-grade hydrothermal deposits where the ore minerals are widely dispersed in a zone surrounding an igneous intrusion and within the intrusion itself. Also see *hydrothermal deposits*.

dissolution A type of chemical weathering reaction in which minerals completely dissolve or disassociate in water, leaving only individual ions in the solution.

divergent boundary A tectonic plate boundary that is dominated by tension forces such that the two plates move away from one another.

downstream flood A large-volume flood that tends to rise slowly and stay above flood stage for extended periods of time. Commonly occurs in the lower portion of a drainage system where a river has a wide channel and natural floodplains.

drainage basin The land area that collects water for an individual stream or river; also referred to as a *watershed*.

drainage divide A topographic line that follows the crests in the landscape and marks the point where surface water is forced to flow into different drainage systems.

E

earth resources Any natural resource that comes from the solid earth. Examples include water, soil, mineral, and energy resources.

Earth systems science The field of study that views the Earth as a dynamic, constantly changing system composed of four major components: the *atmosphere, hydrosphere, biosphere,* and *geosphere*.

earthquake Ground shaking associated with the vibrational wave energy that results when a rock body suddenly fails and releases its accumulated strain.

earthquake precursors Various physical phenomena that can occur just prior to the release of energy associated with a main earthquake shock.

ecological footprint Refers to the amount of *biologically productive* land/sea area needed to extract the resources needed by humans, and to absorb the waste they generate.

El Niño The periodic change in water and air currents in the equatorial parts of the Pacific Ocean, which brings dry conditions to the western Pacific and heavy rains to parts of coastal North and South America.

elastic limit The maximum amount of strain that a rock body can accumulate before either fracturing or undergoing plastic deformation.

elastic rebound theory Explains how earthquakes originate when a rock body deforms and accumulates strain such that it reaches its elastic limit, at which point the rock suddenly fails and releases its stored energy.

electromagnetic radiation A type of energy that travels in a continuous series of waves in which individual waves vary in terms of their wavelength and amount of energy they contain.

electronic waste Discarded electronic products, including TVs, computers, cell phones, and assorted audio and video equipment. Typically contains highly toxic substances such as lead, mercury, and cadmium.

energy The capacity to perform work or transfer heat. See *work*.

energy conservation The process of reducing energy consumption through decreased human activity and more efficient use of the energy itself.

energy efficiency A measure used to describe the amount of unwanted heat that is lost during energy transformations, such as during the conversion of electrical energy to radiant light.

enrichment factor The degree to which geologic processes have concentrated a mineral resource above its average concentration in the crust. A deposit may be deemed economical to mine only if it reaches a certain concentration factor.

environmental geology A branch of geology that examines the interaction between humans and the geologic environment. Common issues include resources, hazards, and pollution.

environmental risk The chance that some natural process or event will produce negative consequences for an individual, or society as a whole. Risk is characterized in terms of probability and consequences.

epicenter The point on the surface that lies directly above the focus, which is the place where an earthquake originates in the subsurface.

erosion The process whereby rock or sediment is removed from a given area through chemical reactions or by being physically picked up or worn down by abrasion.

essential nutrients The most critical elements necessary for plant growth, which include nitrogen, phosphorus, potassium, calcium, magnesium, and sulfur.

evaporites An economically important class of sedimentary minerals that form when water bodies undergo evaporation, causing salts to precipitate. Deposits of evaporites can be either marine or freshwater in origin.

expanding clays Those clay minerals that are capable of incorporating large numbers of water molecules within their structure, thereby producing significant volume changes. Synonym: *swelling clays.*

exponential growth When the amount added over successive time increments keeps increasing so that the growing total plots as a curve, as opposed to a straight line.

extremophile bacteria Bacteria that thrive under extreme conditions, such as ancient Antarctic ice, superhot vents on the seafloor, and rocks located deep underground. Earth's complex plant and animal life is generally believed to have evolved from extremophile bacteria.

F

fall A type of mass wasting that involves the rapid movement of earth materials falling through air.

fault A fracture plane along which slippage or movement has occurred along opposite sides of the fracture.

feldspars A group of rock-forming silicate minerals that are rich in aluminum (Al), and are commonly transformed into clay minerals by chemical weathering.

ferromagnesian minerals Rock-forming minerals that contain relatively high proportions of iron (Fe) and magnesium (Mg).

fertilizers Natural or synthetic materials containing essential nutrients that are applied to soil in order to increase plant growth and agricultural productivity.

flash floods A flood in which a stream rises and falls rapidly. Also known as *upstream floods* since small streams in the upper reaches of a drainage system tend to quickly overflow their banks during heavy rain events.

flood stage The height at which a river begins to overflow its banks. Stage height changes along a river because the channel continually decreases in elevation as it moves downstream.

flow A type of mass wasting involving loose material that accumulates enough water so that internal friction is reduced, allowing it to behave like a fluid and start flowing downslope.

focus The point within a rock body where accumulated strain is suddenly released, causing an earthquake. Also called the hypocenter.

foliated texture The parallel realignment of minerals within a rock caused by the increased pressure associated with regional metamorphism.

fossil fuels A collective term used to describe a combustible form of energy that has been derived from the remains of ancient plants and animals. Major types include coal, oil, and natural gas.

freshwater Low-salinity water (contains few dissolved ions) that originates as precipitation within the hydrologic cycle.

G

galaxies Groupings or clusters of stars within the universe, some of which form a planar, rotating disk of stars.

gamma-ray burst A short-lived burst of very high-energy waves that can destroy ozone molecules in the upper atmosphere, which shield the biosphere from dangerous ultraviolet radiation.

gas giants The four outermost planets that are largely composed of hydrogen and helium gas and have surfaces marked by clouds of swirling gases.

gas hydrates Ice-like substances composed of methane and water molecules and found in thick accumulations in shallow polar waters, deeper parts of continental shelves at lower latitudes, and in permafrost throughout the Arctic. Represents a potentially large source of new energy. Also called *methane hydrates.*

gas window The depth range where subsurface temperatures are high enough to transform organic compounds into natural gas. Beyond this window higher temperatures cause the hydrocarbons to be transformed into the mineral graphite.

general circulation models (GCMs) Sophisticated mathematical models that use a three-dimensional grid to simulate atmospheric processes on a global scale, including the many feedbacks involving the atmosphere, oceans, and land.

General Mining Act A law passed by the U.S. Congress in 1872 that still governs the mining of precious metals on public lands. The act allows prospectors to purchase public land for no more than $5 per acre, provided that they extract the existing mineral resource.

geologic hazard A geologic condition, natural or artificial, that creates a potential risk to human life or property. Examples include earthquakes, volcanic eruptions, floods, and pollution.

geologic time A term used to imply extremely long periods of time over which geologic processes take place, typically measured in intervals of millions or billions of years. In contrast, human time is often measured in intervals ranging from seconds to decades.

geologic time scale The time scale that classifies all the Earth's rocks according to their relative or chronological ages; uses various names to subdivide Earth's rock record into progressively smaller time intervals.

geology The study of the solid earth, which includes the materials it is composed of and the various processes that shape the planet.

geothermal energy The heat energy contained within the Earth that can be used to drive a steam turbine for producing electricity or serve as a heat source and sink for a geothermal heat pump.

geothermal gradient The rate of temperature increase in the Earth with increasing depth.

geothermal heat pump A mechanical system that supplies a building with warm or cool air by making use of the relative difference in temperature between the inside air and the ground.

glacial surge A sudden and rapid increase in the flow of glacial ice, believed to be triggered by the buildup of meltwater at the base of glacier, which reduces the friction between the glacier and underlying bedrock.

global warming A rise in the average global temperature of Earth's atmosphere. The term is commonly used to refer to the current warming trend.

granite A common type of intrusive igneous rock found in continental crust that contains relatively few ferromagnesian minerals, but is rich in quartz and potassium feldspar.

greenhouse effect The natural warming of Earth's atmosphere due to the presence of certain gases, which absorb outgoing radiation and release the corresponding heat energy into the atmosphere.

groin An engineering structure built perpendicular to the beach consisting of rocks too large to be moved by wave action. The purpose is to reduce shoreline retreat by widening the beach through the trapping of sand moving with the longshore current.

ground amplification A phenomenon that occurs when seismic waves encounter weaker materials and begin to slow down, causing an increase in wave amplitude and ground shaking.

ground fissures Large open cracks that form over a wide area of the landscape during an earthquake.

groundwater Fresh or saline water that resides within the voids or pore spaces of subsurface materials.

groundwater baseflow Groundwater that discharges into the surface environment, such as a stream, lake, wetland, or ocean.

groundwater mining The undesirable situation where the human withdrawal of groundwater is greater than the rate of natural recharge, causing water levels in an aquifer to get progressively lower over time.

H

habitable zone That relatively narrow zone around a star where the surface temperature of orbiting planets would be such that liquid water could exist, creating a greater potential for the development of life.

hardpan A soil layer whose physical characteristics limit the ability of either roots or water to penetrate the soil. Commonly consists of dense accumulations of clay minerals or soil particles that have been cemented together by minerals.

headwaters The upper portion of a drainage network where there are numerous small channels, which eventually merge to form progressively larger streams.

heat balance The dynamic relationship in which Earth naturally attempts to balance the amount of heat energy it radiates into space with the amount of energy it receives from the Sun. When the heat balance is disrupted, Earth either gains or loses heat until a new equilibrium can be established.

heavy crude oil Crude oil with a higher proportion of heavier hydrocarbon molecules and more sulfur atoms.

high-level waste Intensely radioactive by-products from nuclear power and weapons reactors, as well as the wastes generated from the reprocessing of spent reactor fuel.

historical geology A branch of geology that examines Earth's past by unraveling the information held in rocks.

hot spots Rising plumes of mantle material that cause partial melting in the overlying lithospheric plate, creating magma that moves upward through weak zones within the plate.

Hubbert's peak The projected peak in world oil production as predicted by the peak oil theory; named in honor of M. King Hubbert who developed the theory.

hurricane A large, rotating low-pressure storm system that originates in tropical oceans, and has sustained winds of over 74 miles per hour. Also called a *typhoon* or *cyclone* depending on where it forms in the tropics.

hydraulic conductivity A measure of the ability of a material to transmit a fluid, which is dependent on both a property of the material (permeability) and properties of the fluid itself (density and viscosity).

hydraulic gradient The slope or steepness of a water table or potentiometric surface, which is calculated from the difference in hydraulic head and distance between any two points in an aquifer system.

hydraulic head A measure of the potential energy within an aquifer system, as represented by the height of the water table or potentiometric surface.

hydraulic sorting The process whereby flowing water separates sediment grains based on their size, shape, and density.

hydrocarbons Organic molecules composed chiefly of hydrogen and carbon atoms that are the basis of petroleum (oil and gas), a combustible substance that forms when organic-rich shales become deeply buried and exposed to progressively higher levels of heat and pressure.

hydroelectric power (hydroelectricity) Mechanical energy obtained from falling water used for generating electricity.

hydrogen fuel cell A device that produces electricity by allowing hydrogen and oxygen atoms to chemically combine to form water molecules.

hydrologic cycle The cyclic movement of water within the Earth system, driven by solar radiation that causes water to evaporate from the oceans and land surface.

hydrolysis A type of chemical weathering reaction in which water molecules directly take part in the breakdown of minerals, releasing both dissolved ions and producing new minerals called *weathering products*.

hydrothermal deposits Mineral deposits associated with igneous intrusions, which result from the chemical interaction of hot, mineral-rich fluids with the surrounding rock. Commonly occur as sulfur-rich deposits in veins and disseminated grains in a zone around the intrusion.

hypothesis A scientific explanation of data or facts. Hypotheses must be testable such that it is possible to show them to be false or incorrect. Supernatural explanations are not considered scientific because they are not testable and cannot be shown to be false.

I

igneous rocks A major class of rocks that form when minerals crystallize from cooling magma.

inertia The physical tendency of objects at rest to stay at rest, which plays a key role in causing damage during earthquakes when structural foundations are suddenly forced to move.

infiltration capacity The ability of the land surface to absorb water. Water that is unable to infiltrate is generally forced to move downslope as overland flow.

inner core The innermost part of the Earth that consists of a solid metallic sphere.

intraplate earthquakes Earthquakes that occur far from a plate boundary or active mountain belt.

ion exchange The process where dissolved ions attach themselves to soil particles, and are then removed in a selective manner by growing plants and by water moving through the soil zone.

ions Individual atoms that have either gained or lost electrons, thereby acquiring either a positive or negative electrical charge.

island arc A landmass surrounded by ocean, which forms when magma rises up through a buckled tectonic plate along a subduction zone, producing a string of volcanic islands.

J

jetties Engineering structures built perpendicular to shore at the mouths of inlets, whose purpose is to keep inlets open for navigation by blocking the longshore movement of sand.

L

La Niña A pattern of water and air currents in the Pacific Ocean that sometimes follows a periodic El Niño event, suppressing rainfall along the western United States. Also see *El Niño*.

lag time The time difference between a rain event and the resulting peak discharge in a stream.

land subsidence The lowering of the land surface due to the closing of void spaces within subsurface materials; commonly triggered by the withdrawal of subsurface fluid (water or oil) or by the collapse of natural cavities or mining voids.

lava domes A steep-sided mound of cooling lava that is built from more viscous magma that does not flow very readily.

lava flow A body of lava that flows out onto the land surface and eventually cools and solidifies into igneous rock.

law A scientific term that describes some phenomena in which the relationship between different data occurs regularly and with little deviation. Such a relationship often can be described in terms of mathematics.

layered intrusions A type of igneous intrusion in which ore minerals had become separated from the magma during the crystallization process, forming a concentrated layer of valuable minerals.

leachate The liquid that forms when rainwater percolates into a landfill and interacts with the trash, incorporating bacteria, viruses, and various chemical compounds in the water. Another form of leachate forms during the chemical process where fluids are used for extracting metals from crushed ore. See *leaching*.

leaching A technique of processing metallic ores in which a solution is allowed to permeate through crushed ore, with the resulting chemical reactions then liberating the desired element(s). See *leachate*.

leakage A term used to describe the process where groundwater moves between two different aquifers by flowing across an aquitard (i.e., a confining layer).

light crude oil Crude oil with a higher proportion of lighter hydrocarbon molecules and fewer sulfur atoms.

limestone A type of clastic or chemical sedimentary rock that is composed chiefly of the mineral calcite ($CaCO_3$).

linear growth When the amount added over successive time increments remains the same so that the growing total plots as a straight line, as opposed to a curve.

liquefaction An earthquake phenomenon that occurs when the ground shaking causes sand-rich layers of sediment to behave as fluid.

lithosphere Rocks from both the crust and upper mantle, forming a brittle layer that is broken up into individual tectonic plates that move over the relatively weak asthenosphere.

longshore current The parallel movement of water and sand to land, which occurs in the surf zone when waves approach a shoreline at an angle.

low-level waste A broad category of radioactive waste, which incorporates any radioactive material that is not spent nuclear reactor fuel. Low-level waste includes contaminated materials that have come into contact with radioactive substances, such as protective clothing and equipment.

M

magma Molten rock material that forms within the Earth and can cool to form igneous rock. *Lava* refers to magma that cools on Earth's surface.

magma chamber A zone or reservoir of molten material that forms within the lithosphere.

magmatic earthquakes Earthquakes that result from the strain that accumulates as rising magma forces its way through crustal rocks.

mantle The rocky shell surrounding Earth's metallic center that is composed of iron-rich silicate minerals.

mass extinction When large numbers of Earth's species go extinct in a relatively short period of time.

mass wasting The general process of earth materials moving downslope due to gravity; the terms *landslide* and *avalanche* are often used synonymously, but technically refer to specific types of movement.

massive sulfide deposit A thick, hydrothermal deposit of sulfide minerals that is associated with mid-oceanic ridges, where active volcanism takes place along divergent plate boundaries. See *hydrothermal deposits*.

maximum contaminant levels (MCLs) Standards set by the EPA regarding the maximum allowable concentrations of specific contaminants in a water-supply system that provides drinking water to the public.

Mercalli intensity scale A qualitative means of ranking the intensity of earthquakes based on first-hand human observations, particularly the amount of structural damage.

metamorphic rocks A major class of rocks that form when preexisting rocks are altered by some combination of heat, pressure, and fluids.

methylmercury A highly toxic compound in which mercury bonds with carbon and hydrogen atoms. This form of mercury is easily ingested by aquatic life-forms and then gets passed on through the food chain, posing a human health risk.

mid-oceanic ridges The chain of submarine mountains, circling nearly the entire globe, where rising magma and seafloor spreading produce new oceanic crust.

Milankovitch cycles Refers to the cyclical changes in solar heating of the Earth caused by periodic changes in the way the planet orbits the Sun. When the orbital parameters (eccentricity, axial tilt, and precession) reinforce each other, the resulting change in heat balance can trigger an overall cooling or warming trend.

mine tailings Noneconomical rock and processed ore from mining operations, which is considered to be waste material and is placed on the land surface.

mineral A naturally occurring inorganic solid composed of one or more elements in fixed proportions, and where the individual atoms have an orderly arrangement called a *crystalline structure*.

mineral reserve A mineral deposit that is economical to extract using existing technology.

mineral resource Any rock, mineral, or element that has some physical or chemical property humans find useful. This term does not necessarily imply that any particular mineral deposit is economical to extract. See *mineral reserve*.

moment magnitude scale The modern earthquake magnitude scale that provides a more accurate measure of the amount of ground motion.

mouth The lowest point in a drainage system where a river discharges into an ocean, lake, or another river.

multiple working hypotheses Common in the early stages of an investigation where researchers develop more than one plausible hypothesis for a given set of data. The number of hypotheses normally decreases over time as new data show one or more hypotheses to be false.

municipal solid waste Trash or garbage from households and local businesses that is typically collected by local governments, and then disposed of in a landfill or by incineration.

N

National Environmental Policy Act Act passed by the U.S. Congress in 1969 stating the country's commitment to protecting the environment for the benefit of the people and the nation as a whole.

natural floodplain The flat portion of a river valley underlain by sediment, which has been deposited over time as the river periodically overflows its banks.

natural gas The gaseous phase of petroleum consisting of a mixture of lighter hydrocarbon molecules, including methane and propane.

natural levees A pair of sand ridges that run parallel to stream banks, and form over time as a river overflows its banks and experiences an immediate decrease in velocity, causing sediment to fall from suspension.

natural vibration frequency The frequency at which a building will naturally vibrate when the ground shakes during an earthquake.

nebular hypothesis Describes how all solar system objects originally formed from a rotating cloud of dust and gas called a *nebula*.

nonconventional fossil fuels All fossil fuels except for the traditionally used coal, natural gas, and light crude oils. Includes heavy crude and synthetic liquid fuels from coal and oil sands.

nonpoint source A pollution source that releases pollutants into the environment over a broad area, commonly consisting of multiple input sites. Examples include agricultural fields, parking lots, golf courses, and lawns.

nuclear fission The process of splitting the nucleus of an atom, resulting in the release of neutrons and conversion of mass into energy. Used in nuclear power plants to drive steam turbines for generating electricity.

nuclear fusion A natural process that takes place in the Sun in which the nuclei of hydrogen atoms combine or fuse, releasing large amounts of energy. Humans have used this process to create nuclear weapons, and may someday harness the energy to generate electricity.

O

ocean acidification The process where the acidity of the oceans increases over time due to higher carbon dioxide levels in the atmosphere. As the oceans absorb additional carbon dioxide, the dissolved gas is converted into carbonic acid, making the oceans more acidic.

ocean currents The physical movement of ocean water from one location to another. Localized *tidal currents* form in coastal areas when tidal forces funnel water through inlets and river channels. In the open ocean large-scale *surface currents* are driven mainly by winds, whereas *density currents* move throughout the ocean basins in response to differences in temperature and salinity.

ocean thermal energy conversion (OTEC) A technique for producing electricity that makes use of the thermal energy stored within the oceans to power a simple heat engine, which then drives a turbine and electrical generator.

ocean tides The periodic rise and fall of sea level along coastlines, caused by the combination of gravitational forces between the Earth, Moon, and Sun and the spinning motion of the Earth on its axis.

ocean trenches Narrow, steep-sided depressions associated with subduction zones that run parallel to adjoining landmasses.

off-stream use Refers to those human activities in which water is removed from one supply source, but then returned to a different source after being used. An example is the withdrawal of groundwater that is later discharged into a stream.

oil sands A near-surface sand deposit that contains a highly viscous hydrocarbon substance called bitumen, which can be extracted to produce synthetic crude oil. Also called *tar sands.*

oil shale A fine-grained sedimentary rock containing abundant organic matter, which can be converted to synthetic crude oil by simply heating the rock.

oil window The depth range where subsurface temperatures are high enough to transform organic compounds into oil and natural gas. Beyond this window higher temperatures cause all oil molecules to be transformed into natural gas.

open-pit mine A type of surface mining operation that commonly involves excavating large volumes of a low-grade ore, often in a disseminated deposit.

ore deposit A body of rock or sediment containing some mineral that has value to society and whose concentration is sufficiently high that the deposit is economical to extract. The term is often used to describe metallic deposits.

outer core The liquid metallic shell within the Earth that surrounds the solid metallic inner core.

overburden Undesirable rock or sediment that must be removed during a mining operation in order to gain access to a valuable mineral deposit.

overland flow The process where water moves downslope in thin sheets over the land surface.

oxidation/reduction A type of chemical weathering reaction in which electrons that are gained or lost take part in the breakdown of minerals, releasing both dissolved ions and producing new minerals called *weathering products.*

ozone depletion The thinning of the atmosphere's outermost layer, which has a relatively high concentration of ozone (O_3) molecules that shields the biosphere from dangerous ultraviolet radiation.

P

paleoclimatology The field of study that examines Earth's history of climate change using a variety of tools, including sedimentary deposits, hydrogen and oxygen isotopes, tree rings, and glacial ice.

paleosol A soil sequence that has been buried during some geologic event, representing a distinct time marker that can be useful in scientific investigations.

parent material The original weathering product or organic material from which soil horizons develop.

peak oil theory A statistical analysis, based on historical exploration and production data, which predicts that oil production will reach a peak, then follow a permanent production decline.

permafrost A subsurface horizon that remains frozen throughout most or all of the year.

permeability A property that describes the ease with which a fluid is able to flow through a porous material.

petroleum A general term geologists use to describe the various types of organic compounds that make up both crude oil and natural gas.

petroleum reservoir Permeable rock in the subsurface where hydrocarbons (oil and gas) have accumulated and are being stored in the rock's pore spaces.

petroleum trap Any configuration of rocks in the subsurface that allows hydrocarbons (oil and gas) to accumulate over time. Common traps include dome structures and faults.

photovoltaic cell A thin slab of material composed of silicon alloys, which produces electricity when sunlight is allowed to strike the slab.

physical geology The study of the solid earth and the processes that shape and modify the planet.

physical weathering Any process that causes rocks to disintegrate into smaller pieces or particles by some mechanical means.

placer deposit A concentrated sedimentary deposit consisting of dense, chemically resistant minerals that have been hydraulically sorted from the rest of the sediment load.

planets Solar system bodies that are large enough that their gravity is able to dominate their orbital zones, sweeping it clear of debris.

plasticity A property that describes the ability of a material to deform without breaking when a force is applied.

point source A pollution source that releases pollutants into the environment at a physically discrete point, such as a pipe discharging wastewater from a factory.

polluted Used to describe those situations where a substance has been introduced into the air or water and has reached concentrations that are harmful to living organisms.

pore pressure The pressure from fluids (water, oil, or gas) within the void spaces of rocks that acts outward in all directions.

porosity A measure that describes the amount of void (air) space within rock or sediment, usually expressed as a percentage of the material's overall volume.

potentiometric surface Represents the height that water in a well will rise above a confined aquifer; it is a measure of the amount of potential energy within the aquifer.

primary waves (P-waves) Seismic waves that cause solid particles to vibrate in the same direction the wave is traveling such that the rocks alternately compress and decompress.

pyroclastic flow A dry avalanche consisting of hot rock fragments, ash, and superheated gas that rushes down the side of a volcano at great speed.

pyroclastic material Particles of pulverized rock and lava that are ejected into the surface environment during explosive volcanic eruptions.

Q

quartz A mineral composed entirely of silicate ions (SiO_4^{4-}) and commonly found in continental crustal rocks along with feldspars. Because it is resistant to chemical weathering, quartz is also abundant in soil and sediment.

R

radioactive decay The process where the nucleus of an unstable isotope undergoes a spontaneous nuclear reaction, releasing radiant energy (radiation) along with *betaparticles* (electrons) and *alphaparticles* (neutrons and protons).

radiometric dating A general term applied to absolute dating techniques involving any type of radioactive element and its decay product. Since different radioactive elements decay at different rates, scientists can obtain reliable dates for events ranging anywhere from thousands to billions of years old.

radon gas An odorless, radioactive gas that forms naturally from the decay of uranium-bearing minerals found in igneous rocks and sediment. Radon can cause lung cancer in humans when it is allowed to accumulate in poorly ventilated homes.

rare earth hypothesis The scientific idea that life is probably common throughout the universe, but complex animal life similar to Earth's is likely to be exceedingly rare.

recurrence interval A statistical calculation representing the frequency at which a particular value of stream discharge can be expected to repeat itself.

relative age The geologic age of a rock, fossil, or event in relative terms as compared to the age of another rock, fossil, or event. The relative age of sedimentary layers is based on the principle that the bottom layers were deposited first, and thus are the oldest.

resonance A phenomenon that occurs when the natural vibration frequency of a building matches that of the seismic waves, causing the building to shake more violently.

retaining wall An engineering structure made of steel, concrete, rock, or wood and designed to strengthen an oversteepened slope; commonly used when a flat surface is needed in sloping terrain for a roadway, building, or parking lot.

retention basins Engineering structures designed to reduce flooding by temporarily storing excess surface water before it can reach a stream channel.

Richter magnitude scale The original earthquake magnitude scale, developed by Charles Richter and Beno Gutenberg, used to quantify the amount of ground motion.

rift valley A linear valley, also called a *graben*, that forms when tension forces cause the land to down-drop in a stair-step fashion along parallel faults.

rip current A dangerous current that flows away from a beach, created when the backwash from breaking waves funnels through a breech in an underwater sand bar; sometimes inappropriately referred to as a *rip tide*.

rock An aggregate or assemblage of one or more types of minerals.

rock bolts An engineering technique that utilizes a steel rod and anchoring system to prevent fractured blocks of rock from falling onto highways and rail lines. Also used for stabilizing walls and ceilings in tunnels and underground mines.

rock cycle A geologic concept that describes the recycling of rocks from one rock type to another by various geologic processes.

rock-forming minerals The relatively small set of minerals that make up most of the rocks in Earth's outermost layer, called the *crust*.

S

salinization A process in which the salinity of soil water increases to the point that plant growth is reduced. Common in arid areas where agricultural irrigation causes mineral salts in the soil to dissolve.

saltwater intrusion The undesirable situation where the human withdrawal of groundwater allows saline water to migrate into the freshwater portion of an aquifer that is being used as a water supply.

sandstone A type of clastic sedimentary rock that is dominated by sand-sized rock and mineral fragments.

sanitary landfill An excavation used for disposing of municipal solid waste, in which the trash is compacted and covered with dirt on a daily basis.

scientific method The process by which the physical world is examined in a logical manner. Data or facts are gathered via observations or experiments, and are then explained through hypotheses, theories, and laws.

scrubber A system that is installed at industrial boilers and power plants to chemically remove sulfur dioxide and other pollutants from emission gases prior to their release into the atmosphere.

seafloor spreading A hypothesis that describes how new oceanic crust forms as mid-oceanic ridges spread or open up over time, and are then filled by erupting magma.

seawall An engineering structure constructed of concrete, steel, or large rocks that is placed parallel to shore at the top of a beach. The purpose is to prevent shoreline retreat by creating a barrier that absorbs the impact of breaking waves.

secondary waves (S-waves) Seismic waves that cause solid particles to vibrate perpendicular to the wave path, which creates a shearing (side-to-side) motion.

secured landfill A highly regulated type of landfill designed to handle industrial hazardous wastes. Uses special systems for collecting leachate and detecting leaks and requires tracking permits for individual waste shipments. Waste must also be placed in separate containers within the landfill.

sediment Fragments of rock and mineral grains that are produced when rocks are broken down by undergoing physical weathering.

sediment pollution The movement of excessive sediment off the landscape and into drainage systems. Channels can become filled with sediment, destroying the ecology of streams and increasing the frequency and severity of flooding.

sedimentary rocks A major class of rocks that form when weathered rock fragments, or mineral grains that chemically precipitate from dissolved ions, are reassembled to form a layered rock sequence.

seismic gap A tool for predicting the likelihood of an earthquake, based on sections of an active fault where the strain has not been released for an extended period of time.

seismic waves Vibrational waves that travel through solid earth materials that are caused by earthquakes, sudden impacts, or explosions.

seismographs Instruments that are used to measure (quantify) the amount of ground motion during an earthquake.

sensitivity An engineering term used to describe how easily a soil will lose its strength when it is disturbed.

shale A type of clastic sedimentary rock that is dominated by fine, clay-sized particles.

shear A force that pushes on a rock body in an uneven manner, causing it to become skewed such that different sides of the body move in opposite directions.

shield volcanoes Large volcanic landforms that are composed primarily of a series of basaltic lava flows.

shoreline retreat The landward migration of a shoreline caused by the erosion that results from the interaction between waves and a landmass.

silt fences A technique for reducing sediment pollution and flooding. Consists of a barrier, placed downslope of construction sites, made of a synthetic fabric that is fine enough to trap sediment, but yet allows some water to pass.

sinkholes Circular depressions created by the collapse of solution cavities in limestone rock. In areas with large numbers of sinkholes, the landscape takes on a pitted or cratered appearance referred to as *karst* terrain.

slate A fine-grained and highly foliated type of metamorphic rock that forms when clay minerals in a shale are transformed into platy minerals of the mica family.

slide A type of mass wasting involving a mass of rock, earth, or debris (mixture of rock and earth) that moves in a sliding manner along a zone of weakness (bedding plane, fault, fracture, or foliation plane).

slump A complex form of mass wasting involving unconsolidated material where sliding takes places near the top of the slump, transitioning to a flowing mechanism toward the bottom.

smelting A technique of processing metallic ores in which the minerals are heated to the point where their chemical bonds are broken, producing liquid metal that is then allowed to cool.

snow avalanche The term used to describe a mass wasting event involving snow as opposed to rock or sediment.

soil A natural mixture of weathered mineral and/or organic material that is capable of supporting plant life.

soil erosion The movement or transport of soil particles away from their place of origin.

soil fertility The ability of a soil to supply the elements necessary for plant growth.

soil horizons Horizontal layers within the soil zone that form as the result of chemical and biological processes and the physical transport of material vertically within the soil.

soil loss The net loss of soil that occurs when soil erosion exceeds the rate of natural soil formation.

solar heating The process that takes place when solar radiation strikes a solid object and is transformed into thermal or heat energy.

spring A place where groundwater discharges into the surface environment in a concentrated manner.

stabilization triangle A concept illustrating the amount that carbon dioxide emissions would need to be reduced in order to keep future emissions at a steady level. Emission reductions are achieved through *stabilization wedges*, which represent individual techniques, such as increased use of wind power and higher fuel mileage.

storm surge The rapid rise in sea level during a hurricane that inundates areas above the normal high-tide line. Results from a dome of water that follows a hurricane ashore, which itself is created by the storm's high winds and low air pressure.

strategic minerals Minerals that a particular country considers critical to its civilian and defense industries, but which it must import in significant quantities.

stream discharge The volume of water moving through a channel over a given time interval, commonly measured in units such as cubic feet per second (ft^3/s).

stream gradient The rate of elevation change or steepness of a stream channel. Water velocity increases in areas where the gradient is higher.

strength A property that describes the ability of a material to resist being deformed.

strip mine A type of surface-mining operation in which the excavation follows a mineral deposit that lies parallel to the surface; commonly associated with sedimentary deposits.

subduction The process of one lithospheric plate descending beneath another, where it then undergoes melting and becomes incorporated into the mantle.

subduction zone earthquakes Earthquakes that occur when an oceanic plate is overridden by another plate, generating some of the more powerful earthquakes on record.

Superfund A trust fund created by the U.S. Congress in 1982, whose purpose was to pay the cleanup costs for toxic waste sites that had been abandoned, or for which the responsible parties did not have sufficient resources. The fund's primary source of revenue was an excise tax on the petroleum and chemical industries.

surface waves Seismic waves that travel along Earth's surface; examples include Rayleigh and Love waves.

sustainability The ability to maintain a system or process for an indefinite period of time. A *sustainable society* is one that lives within the Earth system's capacity to provide resources such that they remain available for future generations.

swelling clays Those clay minerals that are capable of incorporating large numbers of water molecules within their structure, thereby producing significant volume changes. Synonym: *expanding clays*.

synthetic fuels Gas or liquid fuels produced from coal or heavy oil; also called *synfuels*.

T

talus pile A cone-shaped deposit of rocks that accumulates at the base of an exposed rock body due to mass wasting processes.

tar sands A near-surface sand deposit that contains a highly viscous hydrocarbon substance called bitumen, which can be extracted to produce synthetic crude oil. Also called *oil sands*.

tectonic plates Individual slabs of the brittle lithosphere (crust and upper mantle) that move over the relatively weak asthenosphere.

tension A force that pulls on a rock body from opposite directions, causing it to become stretched or lengthened.

terracing A technique where a series of flat surfaces are cut into a hillside, with retaining walls commonly used to support oversteepened portions of the slope. The flat areas (terraces or benches) can be used for growing food, constructing buildings, or reducing the chance of rocks tumbling onto highways.

terrestrial planets The four planets closest to the Sun that have outer shells composed of rocky, earth-like materials.

theory A scientific term used to describe the relationship between several different and well-accepted hypotheses, providing a more comprehensive or unified explanation of how the world operates.

theory of plate tectonics A major theory in geology that describes how Earth's lithosphere is broken up into rigid slabs that are in motion due to forces associated with the planet's interior heat.

thermohaline circulation The collection of large-scale density- and wind-driven currents that move in a convective manner through the ocean basins. This circulation plays a key role in Earth's climate by transferring heat energy around the globe.

tidal power Energy that is obtained from the cyclical lowering and raising of sea level due to the tides and is used to turn a generator to produce electricity.

tight oil and gas Petroleum deposits extracted from very low permeability rock layers by hydraulic fracturing.

tragedy of the commons An environmental concept in which the self-interest of individuals results in the destruction of a common or shared resource.

transform boundary A tectonic plate boundary that is dominated by shear forces such that the two plates slide past one another.

transportation The natural process of moving earth materials from one location to another through some combination of gravity, running water, glacial ice, and wind.

triggering mechanisms Processes or events that lead to a mass wasting event by reducing the frictional forces on a slope and/or increasing the effect of gravity.

tsunami A series of ocean waves that form when energy is suddenly transferred to the water by an earthquake, volcanic eruption, landslide, or asteroid impact.

U

unconfined aquifer A type of aquifer that is open to the surface environment and includes a water table that marks the boundary between the saturated and unsaturated zones.

V

viscosity The ability of a fluid to resist flow that is caused by internal friction within the fluid. In magma, greater SiO_2 (silica) content and cooler temperatures result in greater internal friction, hence increased viscosity.

volcanic ash Fine pyroclastic fragments that are ejected from a volcano, then fall from the sky downwind of the volcano, sometimes traveling hundreds or even thousands of miles.

volcanic landslide The rapid downslope movement of rocks, snow, and ice that can occur when the steep flanks of a volcano become unstable and then fail; also called a *debris avalanche*.

volcanic mudflow A mixture of ash, rock, and considerable amounts of water that tends to rush down the stream valleys that lead away from a volcano; also called a *lahar* or *debris flow*.

volcano An accumulation of extrusive materials around a vent through which lava, gas, or pyroclastics are ejected into the surface environment.

W

waste management pyramid A strategy designed to minimize the amount of solid waste being sent to a landfill. Common elements of such a strategy include source reduction, recycling, reuse, and incineration.

wastewater Water that has become contaminated during some human-related process. Examples include water used in industrial processing and for flushing toilets.

water table The depth that marks the boundary between the unsaturated and saturated zones, below which all the pore spaces are filled with water.

water waves The rhythmic rise and fall in the surface of a water body that normally results from the transfer of energy from wind to water. As the energy moves through a water body, water molecules move in a circular manner such that the surface rises and falls.

wave attenuation The steady decrease in seismic wave energy that occurs as the waves travel away from their point of origin at the focus.

wave base The water depth at which water molecules are no longer affected by passing waves. The depth of wave base is directly proportional to the amount of wave energy.

wave refraction The bending of wave fronts caused by a progressive decrease in velocity that occurs when wave energy encounters the seafloor, thereby generating greater frictional resistance.

weather The state of Earth's atmosphere at any given time and place; contrasts with *climate,* which represents the long-term average weather and its statistical variation for a given region.

wind farms A collection of individual wind turbines used to generate electricity from the wind.

work Involves moving an object (mass) some distance against a given force (work = mass × distance).

X

xeriscaping A water conservation technique in which landscaping is done with native plants that require little to no irrigation.

Index

A

abandoned mine hazards, 400, 400f
abrasion, 231, 232f
absolute age, 14
accretion
    in element formation, 68
    in origin of solar system, 43–45, 43f, 68
acidification, of oceans, 547–548, 547f
acid mine drainage, 396–397, 397f
    coal and, 414
    constructed wetlands for, 397
    definition of, 396
acid rain, 508–510
    burning fossil fuels and, 507f, 508, 508f
    deterioration of buildings/monuments and, 72, 375
    mitigating effects of, 510, 510f
    pH levels of, 508, 508f
    sulfide minerals and, 399
active solar heating, 458, 458f
adhesive forces, 200
advection, 489
aerobic bacteria, 304
Africa
    arid belt, 541, 542f
    Congo River, 230, 230t
    Great Rift Valley, 110, 111f
    mid-Atlantic ridge and, 110, 111f
African plate, 106f, 109f
aftershocks, 126, 129, 130f
Agassiz, Louis, 523
age
    of Earth, 16, 99
    relative, 14, 14f, 15f
aggregates, 301, 376
agricultural waste
    animal, 502–503
    chemical, 501–502, 501f, 502f
agriculture. *See also* food production; soil loss
    groundwater withdrawal and, 348
    irrigation and, 24, 320, 334, 335f, 350–351, 361–362, 361f
    Ogallala Aquifer and, 350–351, 350f, 351f
A horizon, 300, 300f
air pollution, 12, 421, 507–513. *See also* acid rain; greenhouse gases; pollution
Alamosa River (Colorado), 398, 398f
Alaska
    Exxon Valdez oil spill, 422, 423f
    Muir Glacier, 527f
    Prudhoe Bay, 434
    Redoubt, 180–181
albedo, 526, 526f, 535
Aleutian Islands, 111, 112f
alfisols, 306t, 307f
alkali rocks, 384
Alley, Richard, 538
alluvial fans, 235, 236f
alluvium, 230, 302
alpha particles, 504

Alps, 112, 523
alteration zone, 375, 378f
alternative energy sources, 404–405, 441–478. *See also* specific types
    carbon-free and renewable fuels, 449–471
    conservation, 471–472, 471f
    defined, 442
    distributed power system, 477–478
    geothermal power, 465–468, 465f
    nonconventional fossil fuels, 443–449
    ocean thermal energy conversion, 468–469, 468f, 476
    in a post-petroleum world, 472–478
    solar power, 457–462
    tidal power, 469–471, 469f, 471f, 476
    wind power, 462–465, 462f, 464f, 478f
aluminosilicates, 298, 316, 379
aluminum, 316
Alvarez, Luis, 58–59
Alvarez, Walter, 58–59
Amazon River, 230, 231t
amorphous solids, 69
amphibole, 71t, 72
Amu Darya river, 338
anaerobic bacteria, 498
andesite, 74, 111
andesitic magmas, 111, 162, 164, 165t
Andes Mountains, 112, 113f, 169, 543
andisols, 306t, 307f
Andrew, Hurricane (1992), 274
angle of repose, 203
animal testing, 486, 486f
Animas River (Colorado), 397, 397f
Antarctica, 286
    climate record of, 535, 538–539, 538f
    coal in, 7, 7f, 100
    ozone hole over, 54, 54f
    sea level rising and, 264, 286
Antarctic plate, 106f
anthophyllite, 377f
anthracite, 411
anthropogenic climate change, 523
anthropogenic greenhouse gases, 523
anthropogenic pollution, 482, 508
anticline, 416, 416f
Appalachian Mountains
    acid rain and, 509f
    formation of, 112
    mountaintop removal, 412, 413f
    rainfall and, 276
aquatic ecosystems, 250–251
aqueducts, 336, 337, 337f, 339, 355
aquifers
    contamination of, 489
    defined, 341
    leakage of, 343–344, 343f
    Ogallala, 322, 350–351, 350f, 351f
    selecting a water supply source, 354–355, 354f
    types of, 341–342, 341f
aquifer storage and recovery (ASR), 358

aquitards, 341, 343–344
Arabian plate, 106f
Aral Sea (former Soviet Union), 338, 338f
Arctic Ocean, 538, 546
Argonne National Laboratory, 454
arid belt (Africa and Middle East), 541, 542f
aridisols, 306t, 307, 307f, 326
Arizona, Tucson, 349, 349f
Armero (Colombia), 160, 161f, 172, 183, 195, 205
Arrhenius, Svante, 523
artesian aquifers, 354
artesian wells, 342
artificial earthquakes, 126
artificial levees, 251–253, 251f, 252f, 280
asbestos, 375, 377–378, 377f, 378f
ash, volcanic, 180–182, 181f, 182f
Asia
    pollution and, 485, 485f
    water supply problems, 543–544
aspect, of slope, 304
assimilation, 111, 162
asteroids, 41–42f
    earth-crossing, 60–61, 61f, 62f
    impact risk, 17, 55–62, 57f
    in rock record, 85, 89
asthenosphere, 97f, 98
Atlantic Ocean
    mid-oceanic ridges, 101
    seafloor spreading, 103, 103f
atmosphere, 40
    in Earth systems science, 20, 32
    gases and their percent volume in, 522–523, 523f
atoms, 67–68, 67f
Australian plate, 106f
automobiles. *See* vehicles
avalanches
    loose-snow avalanche, 210
    slab avalanche, 210
    snow, 208, 210–212, 211f
aviation fuels, 420f, 421, 438
axial tilt, 524–525, 524f, 525

B

back swamps, 237
bacteria
    aerobic, 304
    extremophile, 48, 49
banded iron formation, 380, 381f
barrier islands, 269, 270f, 274–275
Barringer Crater, 56–57
bars (stream deposit), 234
basalt, 73f, 74
basaltic magmas, 107, 109, 110–111, 162–164, 163f–165f, 170f, 171
baseflow conditions, 228
base isolation, 152
base level, 233, 233f
base load (electrical demand), 476
bathymetric map, 101